Primers in Biology

The Cell Cycle

Principles of Control

Primers in Biology: Published titles

Protein Structure and Function

Gregory A Petsko & Dagmar Ringe

Forthcoming titles:

Immunity

Anthony DeFranco Richard Locksley & Miranda Robertson

Molecular Biology

Nancy L Craig Orna Cohen-Fix Rachel Green Carol W Greider Gisela Storz & Cynthia Wolberger

Genetics

Philip Ingham & Tanya Whitfield

Cell Signaling

Wendell Lim Bruce Mayer & Anthony Pawson

Primers in Biology

The Cell Cycle

Principles of Control

David O Morgan

Editor: Eleanor Lawrence
Managing Editor: Karen Freeland
Editorial Assistants: Kerry Gardiner and Joanna Miles
Design and Illustration: Matthew McClements, Blink Studio Ltd
Structure Graphics: Lore Leighton
Copy Editor: Bruce Goatly
Indexer: Liza Furnival
Production Director: Adrienne Hanratty

Distributors:

Inside North America:

Sinauer Associates, Inc., Publishers,
23 Plumtree Road, PO Box 407, Sunderland, MA 01375, USA
orders@sinauer.com
www.sinauer.com

Outside North America:

Oxford University Press
Saxon Way West
Corby, Northants
NN18 9ES
UK

Customers in the UK may use the OUP freepost address:
Oxford University Press
FREEPOST NH 4051
Corby, Northants NN18 9BR
bookorders.uk@oup.com
www.oup.co.uk

ISBN-13: 978-0-9539181-2-6 (paperback) New Science Press Ltd
ISBN-10: 0-9539181-2-2

ISBN-13: 978-0-19-920610-0 (paperback) Oxford University Press
ISBN-10: 0-19-920610-4

ISBN-13: 978-0-87893-508-6 (paperback) Sinauer Associates, Inc.
ISBN-10: 0-87893-508-8

British Library Cataloguing-in-Publication Data

A catalogue record for this book is available from the British Library

Published by New Science Press Ltd
Middlesex House
34-42 Cleveland Street
London W1P 6LB
UK
www.new-science-press.com

in association with
Oxford University Press
and
Sinauer Associates, Inc., Publishers

Printed by Stamford Press PTE Singapore

15 14 13 12 11 10 9 8 7 6 5 4 3 2 1

The Author

David O Morgan graduated in animal physiology from the University of Calgary in 1980 and then did his doctoral and postdoctoral work in endocrinology with Richard A Roth and William J Rutter, and in virology with Harold Varmus at the University of California, San Francisco, where he is now a Professor in the Departments of Physiology and Biochemistry & Biophysics.

Primers in Biology: a note from the publisher

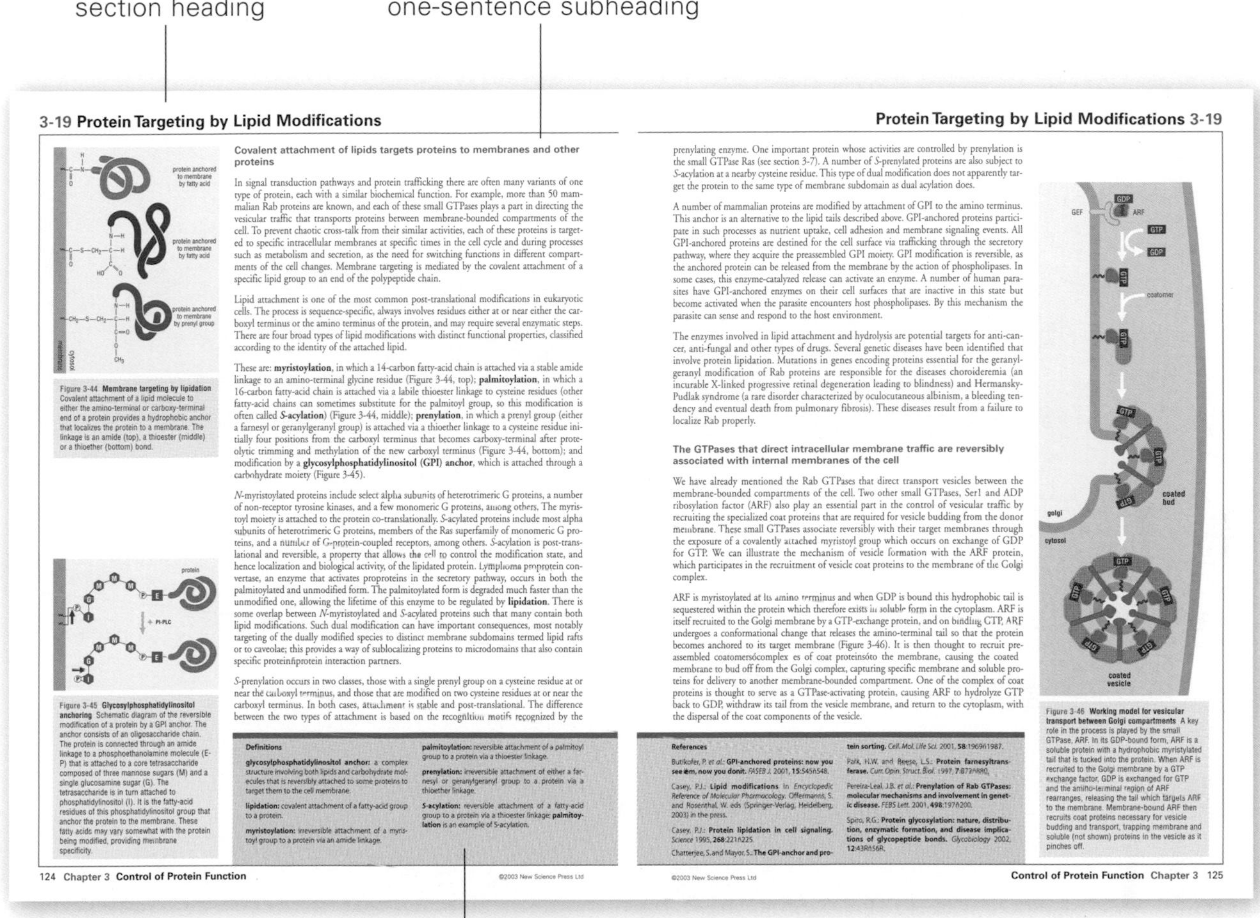

bottom margin:
definitions and references

The Cell Cycle: Principles of Control is part of a series of books constructed on a modular principle that is intended to make them easy to teach from, to learn from, and to use for reference, without sacrificing the synthesis that is essential for any text that is to be truly instructive. The diagram above illustrates the modular structure and special features of these books. Each chapter is broken down into two-page sections each covering a defined topic and containing all the text, illustrations, definitions and references relevant to that topic. Within each section, the text is divided into subsections under one-sentence headings that reflect the sequence of ideas and the global logic of the chapter.

The modular structure of the text, and the transparency of its organization, make it easy for instructors to choose their own path through the material and for students to revise; or for working scientists using the book as an up-to-date reference to find the topics they want, and as much of the conceptual context of any individual topic as they may need.

All of the definitions and references are collected together at the end of the book, with the section or sections in which they occur indicated in each case. Glossary definitions may sometimes contain helpful elaboration of the definition in the text, and references contain a full list of authors instead of the abbreviated list in the text.

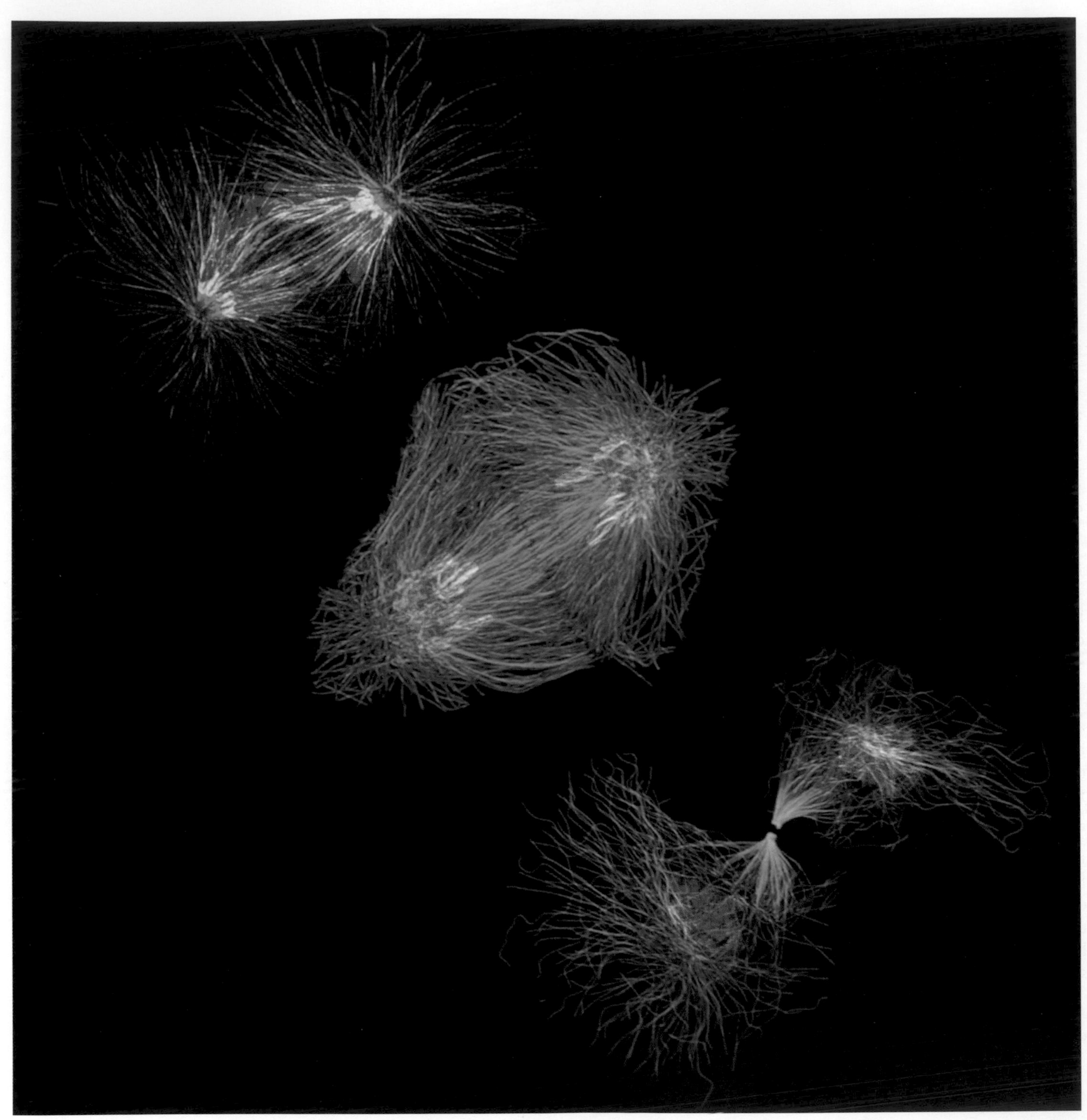

In the final stages of cell division, the duplicated chromosomes (pink) are pulled to opposite ends of the mother cell along tracks made of protein polymers called microtubules (green). Images kindly provided by Julie Canman.

Preface

The first century of cell biology belonged to cytologists, whose painstaking observations with microscopes revealed that all living things are composed of fundamental units called cells, that all cells arise by the division of preexisting cells, and that each daughter cell contains a set of chromosomes like that of the mother cell. At the turn of the 20th century, the collision of cytology and the newly minted field of genetics led to the discovery that chromosomes are the physical determinants of heredity. Later in that century came the next great convergence, when the disparate fields of cytology, genetics and biochemistry together gave rise to the realization that all eukaryotic cells use similar molecular machines and regulatory mechanisms to perform and guide the events of chromosome duplication and cell division. We can now look back on twenty years of astonishing expansion in our knowledge of these mechanisms. But a new problem has arisen: the flood of information has not been accompanied by a clear understanding of how the pieces fit together into a coherent whole.

This book is an attempt to help solve this problem. My goal is to provide a clear and concise guidebook that organizes our vast knowledge on a coherent framework that emphasizes the key problems in cell division and the molecular mechanisms that have evolved to solve those problems. Although organized around key principles, the book does not avoid the so-called details: on the contrary, it includes a glimpse of every layer in our knowledge of cell division, ranging from the cytologist's descriptions of major events to the biochemist's atomic-level analysis of the protein structures and chemical reactions underlying those events. All of these layers are important—and fascinating. The architect Le Corbusier, writing in 1935 of the stunning confluence of form and function in modern aircraft, was more eloquent: "There are no 'details'. Everything is an essential part of a whole. In nature microcosm and macrocosm are one."

I owe a major debt of gratitude to the many colleagues who provided thoughtful and constructive suggestions during the writing of this book (see Acknowledgements). The information contained herein remains the full responsibility of the author, however. It is well known that the teaching of principles requires exaggeration of some facts and omission of others. My apologies to those scientists whose discoveries I have over- or under-emphasized.

The writing of this book began, in a state of wilful ignorance about the effort involved, during an eight-month sabbatical leave at the University of Uppsala in Sweden. In the six years since then, I have returned to Uppsala every summer to do much of the writing in the calm confines of the Biomedical Center library. I am grateful to my hosts on those wonderful visits: Carl-Hendrik Heldin and his colleagues at the Uppsala branch of the Ludwig Institute for Cancer Research, with whom I enjoyed countless discussions—and lunches—between paragraphs.

Every step in the synthesis of this book was catalyzed by the inspirational guidance of Miranda Robertson and her team at New Science Press. I am immensely grateful for the brilliant editing of Eleanor Lawrence and the beautiful illustrations of Matthew McClements. And thanks also to Karen Freeland, who skillfully and enthusiastically piloted the project with the help of Kerry Gardiner and Joanna Miles. Lore Leighton provided valuable assistance with protein structural figures and Bruce Goatly did a masterful job of copy-editing.

And finally, my apologies and heartfelt thanks to my family, who stood by me with patient understanding and unwavering support.

David O Morgan

Online resources for The Cell Cycle

For everybody

All of the 273 colour illustrations in this book are freely available in the Oxford University Press Online Resource Centre and on the New Science Press website and can be downloaded for use in teaching.

Visit http://www.oxfordtextbooks.co.uk/orc/morgan/ or
http://www.new-science-press.com/browse/cellcycle/resources

For instructors

For instructors adopting the book for courses with enrolments of fifteen or more students:

Free access to

- the full text online for a year, for personal use only
- updates – revised, expanded, or new sections and updated references available online only
- PowerPoint functionality allowing instructors to compile any selection of illustrations into a slide show

Visit http://www.oxfordtextbooks.co.uk/orc/morgan/ to register for access to the instructor resources.

Access the resources, once registered, by visiting
http://www.new-science-press.com/browse/cellcycle/resources

Acknowledgements

The following individuals provided expert advice on entire chapters or parts of chapters:

Chapter 1 John Gerhart, University of California, Berkeley; Rebecca Heald, University of California, Berkeley; Andrew Murray, Harvard University; Kim Nasmyth, University of Oxford

Chapter 2 Bruce A. Edgar, Fred Hutchinson Cancer Research Center

Chapter 3 William G. Dunphy, California Institute of Technology; Bruce Futcher, Stony Brook University; J. Wade Harper, Harvard Medical School; Douglas Kellogg, Sinsheimer Laboratories, University of California, Santa Cruz; Charles Sherr, St. Jude Children's Research Hospital; Michael D. Tyers, Samuel Lunenfeld Research Institute, Mount Sinai Hospital

Chapter 4 John Diffley, Cancer Research UK; Paul Kaufman, University of Massachusetts Medical School; Matthew Michael, Harvard University; Johannes Walter, Harvard Medical School

Chapter 5 William G. Dunphy, California Institute of Technology; Tatsuya Hirano, Cold Spring Harbor Laboratory; Douglas Koshland, Carnegie Institution of Washington; Jonathon Pines, University of Cambridge

Chapter 6 Arshad Desai, University of California, San Diego; Rebecca Heald, University of California, Berkeley; Tarun Kapoor, Rockefeller University;

Chapter 7 Angelika Amon, Massachusetts Institute of Technology; Sue Biggins, University of Washington; Jonathon Pines, University of Cambridge; Frank Uhlmann, Cancer Research UK

Chapter 8 Bruce Bowerman, University of Oregon; Christine Field, Harvard Medical School; Michael Glotzer, The University of Chicago; Rong Li, Stowers Institute for Medical Research

Chapter 9 Angelika Amon, Massachusetts Institute of Technology; R. Scott Hawley, Stowers Institute for Medical Research; Neil Hunter, University of California, Davis; Nancy Kleckner, Harvard University

Chapter 10 Nicholas Dyson, Massachusetts General Hospital Cancer Center; Bruce A. Edgar, Fred Hutchinson Cancer Research Center; Martin Raff, University College London

Chapter 11 Karlene Cimprich, Stanford University School of Medicine; Stephen J. Elledge, Harvard Medical School; David Toczyski, University of California, San Francisco; Rodney Rothstein, Columbia University

Chapter 12 J. Michael Bishop, The George Williams Hooper Foundation, University of California, San Francisco; Paul Edwards, Hutchison/MRC Research Centre, University of Cambridge; Gerard Evan, University of California, San Francisco

We are grateful to the following for providing or permitting the use of illustrations:

Figure 1-4 Polytene chromosomes packed in the nucleus of a cell from the *Drosophila* salivary gland. Courtesy of John Sedat.

Figure 2-1 The budding yeast *Saccharomyces cerevisiae* and the fission yeast *Schizosaccharomyces pombe*. Panel (a) courtesy of Greg Tully; panel (b) courtesy of Kathleen Gould.

Figure 2-2 Early divisions in the frog *Xenopus laevis*. Courtesy of James C. Smith and Huw Williams.

Figure 2-3 Patterns of cell division in the early embryo of the fly *Drosophila melanogaster*. Courtesy of Tony Shermoen and Patrick O'Farrell.

Figure 2-4 Mammalian cells growing in culture. Courtesy of Susanne Steggerda.

Figure 2-7 Cell-cycle arrests in budding yeast *cdc* mutants. Courtesy of Greg Tully.

Figure 2-17 Analysis of cellular DNA content by flow cytometry. Courtesy of Liam Holt.

Figure 2-18 Synchronous progression through the cell cycle. Courtesy of Pei Jin.

Figure 4-6 Identification of a replicon cluster by radioactive labeling. Adapted from Huberman, J.A. and Tsai, A.: **Direction of DNA replication in mammalian cells.** *J. Mol. Biol.* 1973, 75:5–12.

Figure 4-7 Replication foci in nuclei of S-phase cells. Courtesy of Brian Kennedy.

Figure 4-8 Association of the ORC with replication origins in *Drosophila* follicle cells. Photographs kindly provided by Stephen P. Bell. From Austin, R.J., Orr-Weaver, T.L. and Bell, S.P.: ***Drosophila* ORC specifically binds to ACE3, an origin of DNA replication control element.** *Genes Dev.* 1999, 13:2639–2649. ©1999 Cold Spring Harbor Laboratory Press.

Figure 4-12 Rereplication in yeast cells with deregulated ORC, Mcm2–7 and Cdc6. Courtesy of Van Nguyen and Joachim Li.

Figure 4-25b Basic units of chromatin structure. From Bednar, J., Horowitz, R.A., Grigoryev, S.A., Carruthers, L.M., Hansen, J.C., Koster, A.J. and Woodcock, C.L.: **Nucleosomes, linker DNA, and linker histone form a**

unique structural motif that directs the higher-order folding and compaction of chromatin. *Proc. Natl Acad. Sci. USA* 1998, **95**:14173–14178. Copyright 1998 National Academy of Sciences, U.S.A.

Figure 5-16 Localization of Plk in mitotic cells. Kindly provided by Francis Barr and Ulrike Grunewald. From Barr, F.A., Sillje, H.H. and Nigg, E.A.: **Polo-like kinases and the orchestration of cell division.** *Nat. Rev. Mol. Cell Biol.* 2004, **5**:429–440.

Figure 5-18 Localization of aurora kinases in mitotic cells. Courtesy of Toru Hirota.

Figure 5-21 Models of SMC and cohesin structure. Panels (a) and (b) reproduced from **The Journal of Cell Biology, 2002, 156, 419–424** by copyright permission of The Rockefeller University Press.

Figure 5-22 Condensation and resolution of human sister chromatids in early mitosis. Kindly provided by Adrian T. Sumner. From Sumner, A.T.: **Scanning electron microscopy of mammalian chromosomes from prophase to telophase.** *Chromosoma* 1991, **100**:410–418. With kind permission of Springer Science and Business Media.

Figure 5-25 Structure of condensin. Panels (a) and (b) reproduced from **The Journal of Cell Biology, 2002, 156, 419–424** by copyright permission of The Rockefeller University Press.

Figure 5-26 Plk and aurora B are required for the removal of cohesin from chromosome arms in early mitosis. From Losada, A., Hirano, M. and Hirano, T.: **Cohesin release is required for sister chromatid resolution, but not for condensin-mediated compaction, at the onset of mitosis.** *Genes Dev.* 2002, **16**:3004–3016. ©2002 Cold Spring Harbor Laboratory Press.

Figure 6-1 Anatomy of the mitotic spindle. Panels (b) and (d) courtesy of Andrew Bajer.

Figure 6-6 Control of microtubule dynamics by associated proteins. Courtesy of Kazuhisa Kinoshita.

Figure 6-8 The mammalian centrosome Micrograph kindly provided by William R. Brinkley. Reprinted from *Ultrastruct. Res.*, Volume **57**, McGill, M., Highfield, D.P., Monahan, T.M. and Brinkley, B.R.: **Effects of nucleic acid specific dyes on centrioles of mammalian cells**, Pages 43–53, ©1976, with permission from Elsevier.

Figure 6-10 The spindle pole body of budding yeast. Panel (a) kindly provided by Thomas H. Giddings and Mark Winey. Reprinted from *Curr. Opin. Cell Biol.*, Volume **14**, Fisk, H.A., Mattison, C.P. and Winey, M.: **Centrosomes and tumour suppressors**, Pages 700–705, ©2002, with permission from Elsevier. Panel (c) kindly provided by Ian R. Adams. Reprinted from *Trends Cell. Biol.*, Volume **10**, Adams, I.R. and Kilmartin, J.V.: **Spindle pole body duplication: a model for centrosome duplication?**, Pages 329–335, ©2000, with permission from Elsevier.

Figure 6-12 Reduplication of centrosomes in prolonged S-phase arrest. Courtesy of Edward H. Hinchcliffe.

Figure 6-13a Kinetochore structure. Courtesy of Jeremy Pickett-Heaps.

Figure 6-14a A possible mechanism for dynamic kinetochore–microtubule attachment. Photograph kindly provided by Stefan Westermann and Georjana Barnes. Reprinted from *Mol. Cell*, Volume **17**, Westermann, S., Avila-Sakar, A., Wang, H.W., Niederstrasser, H., Wong, J., Drubin, D.G., Nogales, E. and Barnes, G.: **Formation of a dynamic kinetochore–microtubule interface through assembly of the Dam1 ring complex**, Pages 277–290, ©2005, with permission from Elsevier.

Figure 6-16 Recruitment of γ-tubulin to mitotic centrosomes. Images kindly provided by Alexey Khodjakov. Reproduced from **The Journal of Cell Biology, 1999, 146, 585–596** by copyright permission of The Rockefeller University Press.

Figure 6-18 Nuclear envelope breakdown in mitosis. Photographs kindly provided by Jan Ellenberg and Brian Burke. From Burke, B. and Ellenberg, J.: **Remodelling the walls of the nucleus.** *Nat. Rev. Mol. Cell Biol.* 2002, **3**:487–497. Reprinted with copyright permission from Nature.

Figure 6-19 Fragmentation of the Golgi apparatus in mitosis. Photographs kindly provided by Joachim Seemann. Reprinted, with permission, from the *Annual Review of Cell and Developmental Biology*, Volume **18** ©2002 by Annual Reviews www.annualreviews.org

Figure 6-21b Stabilization of microtubules around chromosomes by Ran–GTP. Kindly provided by Rebecca Heald. Reprinted with permission from Kalab, P., Weis, K. and Heald, R.: **Visualization of a Ran-GTP gradient in interphase and mitotic *Xenopus* egg extracts.** *Science* 2002, **295**:2452–2456. Copyright 2002 AAAS.

Figure 6-23b Kinetochore-derived microtubule formation. Kindly provided by Helder Maiato and Alexey Khodjakov. Reproduced from **The Journal of Cell Biology, 2004, 167, 831–840** by copyright permission of The Rockefeller University Press.

Figure 6-25 Accumulation of syntelic attachments in the absence of aurora B kinase activity. Kindly provided by Michael A. Lampson and Tarun M. Kapoor. From Lampson, M.A., Renduchitala, K., Khodjakov, A. and Kapoor, T.M.: **Correcting improper chromosome-spindle attachments during cell division.** *Nat. Cell Biol.* 2004, **6**:232–237. Reprinted with copyright permission from Nature.

Figure 6-27b Poleward force generation by the kinetochore. Kindly provided by Stefan Westermann and Georjana Barnes. Reprinted from *Mol. Cell*, Volume **17**, Westermann, S., Avila-Sakar, A., Wang, H.W., Niederstrasser, H., Wong, J., Drubin, D.G., Nogales, E. and Barnes, G.: **Formation of a dynamic kinetochore–microtubule interface through assembly of the Dam1 ring complex**, Pages 277–290, ©2005, with permission from Elsevier.

Figure 7-6 Spindle checkpoint component Mad2 at unattached kinetochores. Kindly provided by Jennifer C. Waters. Reproduced from **The Journal of Cell Biology, 1998, 141, 1181–1191** by copyright permission of The Rockefeller University Press.

Figure 7-7 Alternative conformations of the Mad2 protein. Adapted from Curr. Biol., Volume **15**, De Antoni, A., Pearson, C.G., Cimini, D., Canman, J.C., Sala, V., Nezi, L., Mapelli, M., Sironi, L., Faretta, M., Salmon, E.D. and Musacchio, A.: **The Mad1/Mad2 complex as a template for Mad2 activation in the spindle assembly checkpoint**, Pages 214–225, ©2005, with permission from Elsevier. Original structure graphics kindly provided by Andrea Musacchio.

Figure 7-15 Anaphase defects in the presence of nondegradable cyclin mutants. Photographs kindly provided by Devin Parry and Patrick O'Farrell. Reprinted from *Curr. Biol.*, Volume **11**, Parry, D.H. and O'Farrell, P.H.: **The schedule of destruction of three mitotic cyclins can dictate the timing of events during exit from mitosis**, Pages 671–683, ©2001, with permission from Elsevier.

Figure 7-16 The APC helps promote spindle disassembly in budding yeast. Photographs kindly provided by David Pellman. Reprinted, with permission, from Juang, Y.-L., Huang, J., Peters, J.-M., McLaughlin, M.E., Tai, C.-Y. and Pellman, D.: **APC-mediated proteolysis of Ase1 and the morphogenesis of the mitotic spindle.** *Science* 1997, **275**:1311–1314. Copyright 1997 AAAS.

Figure 7-17 Nuclear envelope assembly in *Xenopus* embryo extracts. Photographs kindly provided by Martin Hetzer and Iain Mattaj. From Hetzer, M., Meyer, H.H., Walther, T.C., Bilbao-Cortes, D., Warren, G. and Mattaj, I.W.: **Distinct AAA-ATPase p97 complexes function in discrete steps of nuclear assembly.** *Nat. Cell Biol.* 2001, **3**:1086–1091. Reprinted with copyright permission from Nature.

Figure 8-7 Control of cytokinesis by the RhoGEF Pebble in the *Drosophila* embryo. Kindly provided by Sergei N. Prokopenko and Hugo J. Bellen. From Prokopenko, S.N., Brumby, A., O'Keefe, L., Prior, L., He, Y., Saint, R. and Bellen, H.J.: **A putative exchange factor for Rho1 GTPase is required for initiation of cytokinesis in *Drosophila*.** *Genes Dev.* 1999, **13**:2301–2314. ©1999 Cold Spring Harbor Laboratory Press.

Figure 8-9a Microtubule behavior in the cleaving *Xenopus* embryo. Photograph kindly provided by Michael Danilchik and Kay Larkin. Reprinted from *Dev. Biol.*, Volume **194**, Danilchik, M.V., Funk, W.C., Brown, E.E. and Larkin, K.: **Requirement for microtubules in new membrane formation during cytokinesis of *Xenopus* embryos**, Pages 47–60, ©1998, with permission from Elsevier.

Figure 8-13 Positioning the contractile ring in *S. pombe*. Photographs kindly provided by Rafael R. Daga and Fred Chang. From Daga, R.R. and Chang, F.: **Dynamic positioning of the fission yeast cell division plane.** *Proc. Natl Acad. Sci. USA* 2005, **102**:8228–8232. Copyright 2005 National Academy of Sciences, U.S.A.

Figure 8-14 Positioning of cytokinesis by the mitotic spindle of embryonic cells. Kindly provided by Charles B. Shuster and David R. Burgess. Reproduced from **The Journal of Cell Biology, 1999, 146, 981–992** by copyright permission of The Rockefeller University Press.

Figure 8-16 Mitosis without cytokinesis in the *Drosophila* embryo. Courtesy of Barbara Fasulo and William Sullivan.

Figure 8-18 Membrane transport during cellularization. Kindly provided by John C. Sisson. From Papoulas, O., Hays, T.S. and Sisson, J.C.: **The golgin Lava lamp mediates dynein-based Golgi movements during *Drosophila* cellularization.** *Nat. Cell Biol.* 2005, 7:612–618. Reprinted with copyright permission from Nature.

Figure 8-20 Asymmetric division in a *Drosophila* neuroblast. Kindly provided by Silvia Bonaccorsi. From Giansanti, M.G., Gatti, M. and Bonaccorsi, S.: **The role of centrosomes and astral microtubules during asymmetric division of *Drosophila* neuroblasts.** *Development* 2001, **128**:1137–1145. Reprinted with permission from The Company of Biologists Ltd.

Figure 9-6 Early steps in homolog pairing. Photographs kindly provided by Denise Zickler. From Tessé, S., Storlazzi, A., Kleckner, N., Gargano, S. and Zickler, D.: **Localization and roles of Ski8p protein in *Sordaria* meiosis and delineation of three mechanistically distinct steps of meiotic homolog juxtaposition.** *Proc. Natl Acad. Sci. USA* 2003, **100**:12865–12870. Copyright 2005 National Academy of Sciences, U.S.A.

Figure 9-7 Homolog pairing defects in a *spo11* mutant. Photographs kindly provided by Denise Zickler. From Storlazzi, A., Tessé, S., Gargano, S., James, F., Kleckner, N. and Zickler, D.: **Meiotic double-strand breaks at the interface of chromosome movement, chromosome remodeling, and reductional division.** *Genes Dev.* 2003, 17:2675–2687. ©1999 Cold Spring Harbor Laboratory Press.

Figure 9-8 Electron microscopic analysis of chromosome structure in leptotene and early zygotene. Photographs kindly provided by Jim Henle and Nancy Kleckner. Panel (b) from Stack, S.M. and Anderson, L.K.: **Two-dimensional spreads of synaptonemal complexes from solanaceous plants. II. Synapsis in**

Lycopersicon esculentum (tomato). *Am. J. Bot.* 1986, 73:264–281. Panels (c) and (d) from Albini, S.M. and Jones, G.H.: **Synaptonemal complex spreading in *Allium cepa* and *A. fistulosum*. I. The initiation and sequence of pairing.** *Chromosoma* 1987, 95:324–338.

Figure 9-9 The synaptonemal complex. Photographs in panel (b) kindly provided by Karin Schmekel. Top photograph from Schmekel, K. and Daneholt, B.: **Evidence for close contact between recombination nodules and the central element of the synaptonemal complex.** *Chromosome Res.* 1998, 6:155–159; with kind permission from Springer Science and Business Media. Photographs in panel (c) kindly provided by Carole Rogers and Shirleen Roeder. Reproduced from **The Journal of Cell Biology, 2000, 148, 417–426** by copyright permission of The Rockefeller University Press.

Figure 9-10 Chiasmata. Reprinted from *Cell*, Volume 111, Blat, Y., Protacio, R., Hunter, N. and Kleckner, N.: **Physical and functional interactions among basic chromosome organizational features govern early steps of meiotic chiasma formation**, Pages 791–802, ©2002, with permission from Elsevier. Photograph taken from John, B.: *Meiosis* (Cambridge University Press, New York, 1990).

Figure 9-12 Microtubules of the first meiotic spindle in budding yeast. Courtesy of Mark Winey.

Figure 9-16 Securin destruction is required for meiotic anaphase I in mouse oocytes. Photographs kindly provided by Mary Herbert. From Herbert, M., Levasseur, M., Homer, H., Yallop, K., Murdoch, A. and McDougall, A.: **Homologue disjunction in mouse oocytes requires proteolysis of securin and cyclin B1.** *Nat. Cell Biol.* 2003, 5: 1023–1025. Reprinted with copyright permission from Nature.

Figure 9-18 Inhibition of Cdk1 triggers DNA synthesis after meiosis I. Photographs kindly provided by Keita Ohsumi. Reprinted by permission from Macmillan Publishers Ltd: *The EMBO Journal*, Iwabuchi, M., Ohsumi, K., Yamamoto, T.M., Sawada, W. and Kishimoto, T.: **Residual Cdc2 activity remaining at meiosis I exit is essential for meiotic M-M transition in *Xenopus* oocyte extracts.** *EMBO J.* 2000, 19:4513–4523, copyright 2000.

Figure 10-9 Antagonistic functions of the two E2F homologs in *Drosophila*. Photographs kindly provided by Maxim Frolov. From Frolov, M.V., Huen, D.S., Stevaux, O., Dimova, D., Balczarek-Strang, K., Elsdon, M. and Dyson, N.J.: **Functional antagonism between E2F family members.** *Genes Dev.* 2001, 15:2146–2160. ©2001 Cold Spring Harbor Laboratory Press.

Figure 10-20 Patterns of *cdc25 (string)* expression in the fly embryo. Photographs kindly provided by Bruce Edgar. From Edgar, B.A., Lehman, D.A. and O'Farrell, P.H.: **Transcriptional regulation of string (cdc25): a link between developmental programming and the cell cycle.** *Development* 1994, 120:3131–3143. Reprinted with permission from The Company of Biologists Ltd.

Figure 10-24 Analysis of growth control in the *Drosophila* eye. Kindly provided by Duojia Pan. Reprinted by permission from Macmillan Publishers Ltd: *Nature Cell Biology*, Gao, X., Zhang, Y., Arrazola, P., Hino, O., Kobayashi, T., Yeung, R.S., Ru, B. and Pan, D.: **Tsc tumour suppressor proteins antagonize amino-acid–TOR signalling.** *Nat. Cell Biol.* 2002, 4:699–704, copyright 2002.

Figure 11-7 RPA-dependent recruitment of ATR to sites of DNA damage. Kindly provided by Stephen J. Elledge. Reprinted with permission from Zou, L. and Elledge, S.J.: **Sensing DNA damage through ATRIP recognition of RPA-ssDNA complexes.** *Science* 2003, 300:1542–1548. Copyright 1997 AAAS.

Figure 11-8 Recruitment of the 9-1-1 complex to sites of DNA damage. Courtesy of Justine Melo and David Toczyski.

Figure 11-14 Abnormal DNA structures at stalled replication forks in yeast *chk2* mutants. Courtesy of Massimo Lopes and Marco Foiani.

Figure 11-18 Generation of a DNA damage response in senescent human cells. Photographs kindly provided by Fabrizio d'Adda di Fagagna. Reprinted by permission from Macmillan Publishers Ltd: *Nature*, d'Adda di Fagagna, F., Reaper, P.M., Clay-Farrace, L., Fiegler, H., Carr, P., Von Zglinicki, T., Saretzki, G., Carter, N.P. and Jackson, S.P.: **A DNA damage checkpoint response in telomere-initiated senescence.** *Nature* 2003, 426:194–198, copyright 2003.

Figure 12-11 Chromosomal abnormalities in cancer cells. Courtesy of Kylie Gorringe, Mira Grigorova and Paul Edwards.

Figure 12-13 Telomere degeneration in the formation of carcinomas. Photograph kindly provided by Ronald A. DePinho. From Artandi, S.E., Chang, S., Lee, S.L., Alson, S., Gottlieb, G.J., Chin, L. and DePinho, R.A.: **Telomere dysfunction promotes non-reciprocal translocations and epithelial cancers in mice.** *Nature* 2000, 406:641–645. Reprinted with copyright permission from Nature.

Figure 12-15 Mitotic spindle defects arising from abnormal centrosome number. From Pihan, G.A., Wallace, J., Zhou, Y. and Doxsey, S.J.: **Centrosome abnormalities and chromosome instability occur together in pre-invasive carcinomas.** *Cancer Res.* 2003, 63:1398–1404.

Figure 12-18 Inhibition of the protein kinase Abl by imatinib. Reprinted from *Cancer Cell*, Volume 2, Shah, N. P., Nicoll, J.M., Nagar, B., Gorre, M.E., Paquette, R.L., Kuriyan, J. and Sawyers, C.L.: **Multiple BCR-ABL kinase domain mutations confer polyclonal resistance to the tyrosine kinase inhibitor imatinib (STI571) in chronic phase and blast crisis chronic myeloid leukemia**, Pages 117–125, ©2002, with permission from Elsevier.

Contents summary

Contents in full

CHAPTER 9 Meiosis

1

The Cell Cycle

Cell reproduction occurs by an elaborate series of events called the cell cycle, whereby chromosomes and other components are duplicated and then distributed into two daughter cells. A complex network of regulatory proteins governs progression through the steps of the cell cycle.

1-0 Overview: Cell Reproduction

Cell reproduction is a fundamental feature of all living things

All cells arise by division of existing cells, and every cell living today is thought to be descended from a single ancestral cell that lived 3 or 4 billion years ago. Throughout this vast period of time, the evolution of cells and organisms—and thus the continued success of life on Earth—has depended on the transmission of genetic information by cell division.

Cell reproduction is fundamental to the development and function of all life. In single-celled organisms, cell division generates an entire new organism. In the development of multicellular organisms, countless cell divisions transform a single founder cell into the diverse communities of cells that make up the tissues and organs. In the adult, cell division provides the cells that replace those that die from natural causes or are lost to environmental damage.

Cells reproduce in discrete steps

How can a machine as complex as a cell reproduce itself, and do so with such remarkable precision? Part of the answer is that the problem of cell reproduction is simplified—both for the cell and for those of us who hope to understand it—by dividing the process into a series of distinct and more easily managed events: first, duplication of the cell's contents, and second, equal distribution of those contents, by division, into a pair of daughter cells.

The highly regulated series of events that leads to eukaryotic cell reproduction is called the **cell cycle** (Figure 1-1). The first part of the cycle, and generally by far the longest, is devoted to duplication of the cell's components. Most of a cell's components—cytoplasmic organelles, membrane, structural proteins and RNAs—are replicated continuously throughout the cell cycle, resulting in the gradual doubling of cell size by the end of the cycle. The chromosomes, however, are present in only single copies and must therefore be duplicated only once per cycle. This occurs during a discrete stage called the synthetic or *S phase*. The distribution of duplicated components into individual daughter cells occurs in a brief but spectacular final stage called the mitotic or *M phase*.

The ordering of cell-cycle events is governed by an independent control system

The duplication and division of cellular components must be achieved with extreme precision and reliability over countless generations. This is especially true of the genetic information stored in the DNA of the chromosomes, whose near-perfect transmission is so important for the perpetuation and evolution of species. One important solution to this problem of precision is that the machinery that carries out the events of the cell cycle—such as the enzymes of DNA synthesis and the apparatus that segregates the duplicated chromosomes—have evolved an astonishing level of speed and accuracy.

The fidelity of cell reproduction also depends on regulatory mechanisms that ensure that the events of the cell cycle occur in the correct order. Chromosome duplication, for example, should begin and end before an attempt is made to distribute the chromosomes into daughter cells. To ensure the correct order of events, the eukaryotic cell contains a complex regulatory network—called the **cell-cycle control system**—that controls their timing and coordination. This control system is essentially a robust and reliable biochemical timer that is activated at the beginning of a new cell cycle and is programmed to switch on cell-cycle events at the correct time and in the correct order.

Definitions

cell cycle: sequence of events that leads to the reproduction of the cell.

cell-cycle control system: network of regulatory proteins that controls the timing and coordination of cell-cycle events.

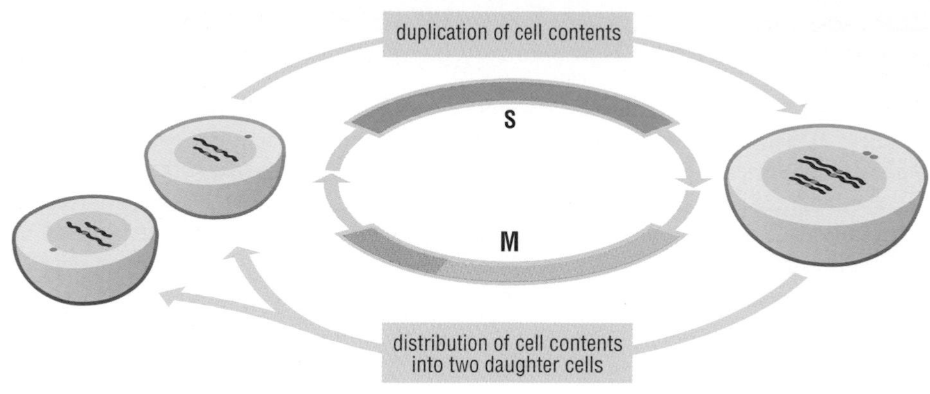

Figure 1-1 The cell cycle Cell reproduction begins with duplication of the cell's components, including the exact duplication of each chromosome in S phase. These components are then divided equally between two daughter cells in M phase.

In its simplest form, the programming of this control system is independent of the events it controls. In most cells, however, the order and alternation of cell-cycle events are reinforced by the dependence of one event on another and by feedback from the cell-cycle machinery to the control system. Entry into M phase is dependent on the completion of DNA synthesis, for example, ensuring that M phase occurs after S phase and does not overlap with it. The cell has mechanisms for monitoring the progress of cell-cycle events and transmitting this information to the cell-cycle control system. If the control system detects problems in the completion of an event, it will delay the initiation of later events until those problems are solved.

A daunting array of problems must be overcome to achieve the accurate duplication and distribution of a cell's contents. Chromosomes must be duplicated only once per cell cycle and no more. Duplicated chromosomes must be separated from each other and distributed accurately into daughter cells so that each cell gets a complete copy of the genome, and nothing more or less. The cell cycle must be coordinated with cell growth to maintain cell size. Cell growth and proliferation in multicellular organisms must be regulated in such a way that new cells are produced only when needed. In the past two or three decades, the solutions to these and many other problems in cell division have begun to emerge, resulting in a deluge of information that can be as confusing to those outside the field as it is enlightening to the experts within. The purpose of this book is to synthesize this information into a concise and current overview of the general principles underlying the events of the eukaryotic cell cycle, with an emphasis on how these events are governed by the cell-cycle control system. We begin in this chapter with a simple overview of cell-cycle events and key concepts in cell-cycle control. In Chapters 2 and 3 we continue with a discussion of the organisms in which the cell cycle is studied, and the components and design of the cell-cycle control system. With these important foundations in place, we then progress in the remainder of the book to a more detailed discussion of the stages of cell division, beginning with chromosome duplication in Chapter 4 and cell division in Chapters 5 to 8. Chapters 9 to 11 address a variety of related topics, until finally, in Chapter 12, we discuss cancer, a complete understanding of which is not possible without an understanding of all the chapters that come before.

References

Baserga, R.: *The Biology of Cell Reproduction* (Harvard University Press, Cambridge, MA, 1985).

Harris, H.: *The Birth of the Cell* (Yale University Press, New Haven, 2000).

Mitchison, J.M.: *The Biology of the Cell Cycle* (Cambridge University Press, Cambridge, 1971).

Murray, A.W. and Hunt, T.: *The Cell Cycle: An Introduction* (Freeman, New York, 1993).

Prescott, D.M.: *Reproduction of Eukaryotic Cells* (Academic Press, New York, 1976).

Wilson, E.B.: *The Cell in Development and Heredity* 3rd ed. (Macmillan, New York, 1925).

1-1 Events of the Eukaryotic Cell Cycle

Chromosome duplication and segregation occur in distinct cell-cycle phases that are usually separated by gap phases

The stages of the eukaryotic cell cycle are typically defined on the basis of chromosomal events (Figure 1-2). Early in the cell cycle, the DNA is replicated and chromosomes are duplicated in **S phase**. This process begins at specific DNA sites called *replication origins*, which are scattered in large numbers along the chromosomes. At these sites, proteins open the DNA double helix, exposing it to the enzymes that carry out DNA synthesis, which move outward in both directions from the origins to copy the two DNA strands. Chromosome duplication also requires increased synthesis of the proteins, such as histones, that package the DNA into chromosomes. Additional proteins are deposited along the duplicated chromosomes during S phase, resulting in a tight linkage, or *cohesion*, between them. The duplicated chromosomes are called *sister chromatids*.

The second major phase of the cell cycle is **M phase**, which is typically composed of two major events: nuclear division (**mitosis**) and cell division (**cytokinesis**). The period between the end of one M phase and the beginning of the next is called **interphase**.

Mitosis is the complex and beautiful process that distributes the duplicated chromosomes equally into a pair of daughter nuclei. The pairs of sister chromatids are attached in early mitosis to the *mitotic spindle*, a bipolar array of protein polymers called *microtubules*. By the midpoint of mitosis (*metaphase*), sister chromatids in each pair are attached to microtubules coming from opposite poles of the spindle. At the next stage (*anaphase*), sister-chromatid cohesion is destroyed, resulting in *sister-chromatid separation*. The microtubules of the spindle pull the separated sisters to opposite ends of the cell (*sister-chromatid segregation*) and the two sets of chromosomes are each packaged into new daughter nuclei.

Following mitosis, the cell itself divides by cytokinesis. Although the appearance of this process varies greatly among different organisms, the underlying requirement in all cells is the deposition of new plasma membrane, and new cell wall in some cases, at a position that bisects the long axis of the mitotic spindle, ensuring that the newly separated chromosome sets are distributed into individual cells. In many organisms, deposition of new membrane components is guided by a contractile ring of actin filaments and myosin motor proteins that forms beneath the cell membrane at the site of cell division. The cell cycle is completed when contraction of this ring pinches the cell in two.

Most cell cycles contain additional phases, known as gap phases, between S and M phases. The first gap phase, **G1**, occurs before S phase, whereas **G2** occurs before M phase. Gap phases provide additional time for cell growth, which generally requires much more time than is needed to duplicate and segregate the chromosomes. Gap phases also serve as important regulatory transitions, in which progression to the next cell-cycle stage can be controlled by a variety of intracellular and extracellular signals.

G1 is a particularly important regulatory period because it is here that most cells become committed to either continued division or exit from the cell cycle. In the presence of unfavorable growth conditions or inhibitory signals from other cells, cells may pause for extended periods in G1 or even enter a prolonged nondividing state, sometimes called **G0** (G zero). Many of the cells in the human body, for example, are in a nondividing, terminally differentiated state from which it is difficult or impossible to reenter the cell cycle. Such cells originate in populations of cells called stem cells, which retain the capacity to divide.

Definitions

cytokinesis: cell division, the process in late **M phase** by which the duplicated nuclei and cytoplasmic components are distributed into daughter cells by division of the mother cell.

G0: a prolonged nondividing state that is reached from **G1** when cells are exposed to extracellular conditions that arrest cell proliferation.

G1: the cell-cycle gap phase between **M phase** and **S phase**.

G2: the cell-cycle gap phase between **S phase** and **M phase**.

interphase: the period between the end of one **M phase** and the beginning of the next.

mitosis: nuclear division, the process in early **M phase** by which the duplicated chromosomes are segregated by the mitotic spindle and packaged into daughter nuclei.

M phase: the cell-cycle phase during which the duplicated chromosomes are segregated and packaged into daughter nuclei (**mitosis**) and distributed into daughter cells (**cytokinesis**).

S phase: the cell-cycle phase during which DNA replication and chromosome duplication occurs.

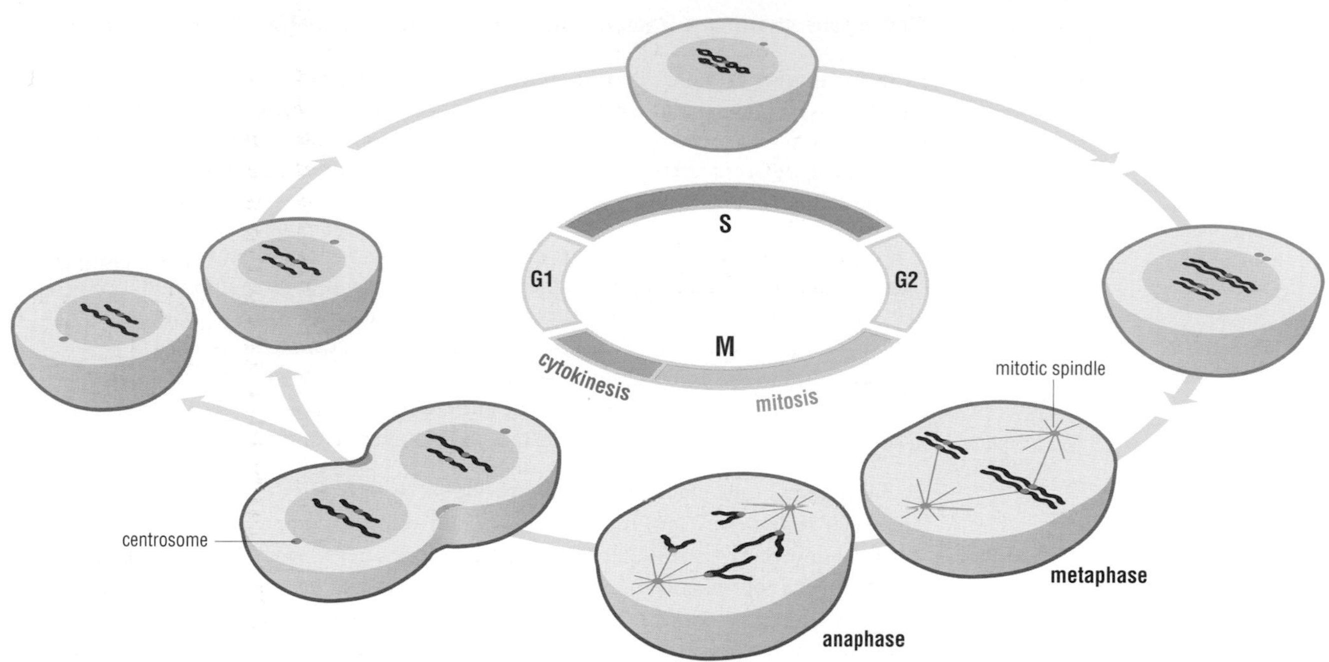

Figure 1-2 **The events of the eukaryotic cell cycle** The central events of cell reproduction are chromosome duplication, which takes place in S phase, followed by chromosome segregation and nuclear division (mitosis) and cell division (cytokinesis), which are collectively called M phase. G1 is the gap phase between M and S phases; G2 is the gap phase between S and M phases. Metaphase is the stage in mitosis when sister chromatids are aligned on the mitotic spindle and anaphase is the stage when sister chromatids are separated and pulled to opposite spindle poles. In most cells, the discrete events of the cell cycle occur against a background of continuous cell growth.

Cytoplasmic components are duplicated throughout the cell cycle

Most proteins, RNAs and other cellular macromolecules are synthesized continuously during the cell cycle. Because these components are present in many copies throughout the cytoplasm, their equal partitioning into daughter cells is readily achieved when the cell is pinched in two by cytokinesis.

Membrane-bounded cytoplasmic organelles reproduce by continuous growth and division of preexisting organelles. The smaller organelles, such as mitochondria and lysosomes, are present in large numbers and are distributed evenly between daughter cells when the cytoplasm is divided at cytokinesis, as are macromolecular complexes such as ribosomes. Larger organelles, such as the Golgi apparatus, are thought to be fragmented into smaller vesicles during mitosis and then distributed evenly along with the other cytoplasmic components. In some cells the large tubular network that forms the endoplasmic reticulum also breaks down into smaller vesicles, whereas in others it remains intact and is cut in two when the cell divides.

Not all protein components are duplicated by continuous synthesis during the cell cycle. The *centrosome* of animal cells (equivalent to the *spindle pole body* in yeast) is a large protein assembly whose function is to organize the long microtubules that radiate throughout the cytoplasm during interphase and form the spindle during mitosis. The centrosome, like the chromosomes, is present in only one copy per cell, and is therefore duplicated strictly once per cell cycle, typically in S phase. During mitosis, the duplicated centrosomes move to opposite poles of the cell to form the spindle, thereby allowing their distribution into daughter cells at cytokinesis (see Figure 1-2).

Cell growth is usually coordinated with cell division

In most cells, the discrete chromosomal events defining the phases of the cell cycle are coupled to the synthetic processes that duplicate the cell's mass of proteins and organelles. As a result, the mass of the cell doubles in each cell cycle and average cell size therefore remains constant. The mechanisms that couple cell growth and division are poorly understood.

In many cell types growth and division are not coordinated. Mammalian oocytes, neurons and muscle cells are all capable of considerable growth in the absence of division, whereas the early divisions of a fertilized animal egg occur in the absence of cell growth. These and many other examples demonstrate that cell growth and the chromosomal cell cycle are independently regulated processes that are coordinated in some, but not all, cell types.

Cell-cycle structure varies in different cells and organisms

Although most eukaryotic cells carry out the major chromosomal events of cell reproduction in discrete cell-cycle phases separated by gaps, the details of this basic scheme are modified in some species and cell types (Figure 1-3). A brief discussion of these variations illustrates some of the different strategies that are used to achieve the common goal of chromosome duplication and segregation.

The early cleavage divisions of many animal embryos, such as those of the South African clawed frog *Xenopus laevis*, rapidly subdivide the giant fertilized egg into thousands of individual cells, allowing the organism to quickly reach a free-living state in which it can fend for itself. Cell growth does not occur during these divisions, and gap phases tend to be minimal or absent. The cell proceeds directly from S phase to M phase and then on to the next S phase—indeed, DNA synthesis begins in many embryonic cells before M phase is complete. This early embryonic cell cycle is focused entirely on rapid duplication and distribution of the genome.

In most animal cells, the two stages of M phase—mitosis and cytokinesis—occur in rapid succession before entry into the following G1. In some cell types, however, cytokinesis is a relatively slow process that is not completed until long after mitosis, and can even occur after S phase of the following cell cycle. In the early embryos of sea urchins, for example, DNA synthesis begins in the last stage of mitosis and is completed shortly thereafter, while cytokinesis takes longer and is not finished until after the next S phase (Figure 1-3). Similarly, in rapidly dividing fission yeast, S phase occurs soon after nuclear division but the final separation of daughter cells occurs after S phase is complete (Figure 1-3). In such cell types, it is useful to focus on nuclear events and define the end of the cell cycle as the completion of mitosis, not that of cytokinesis.

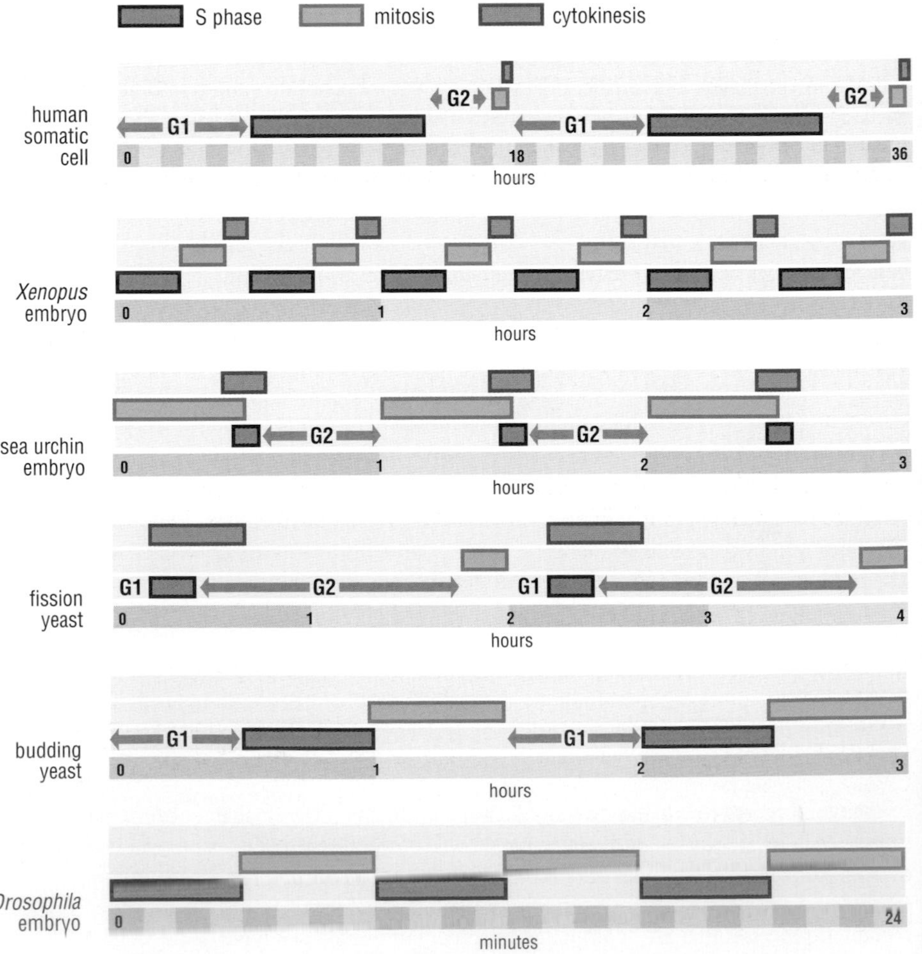

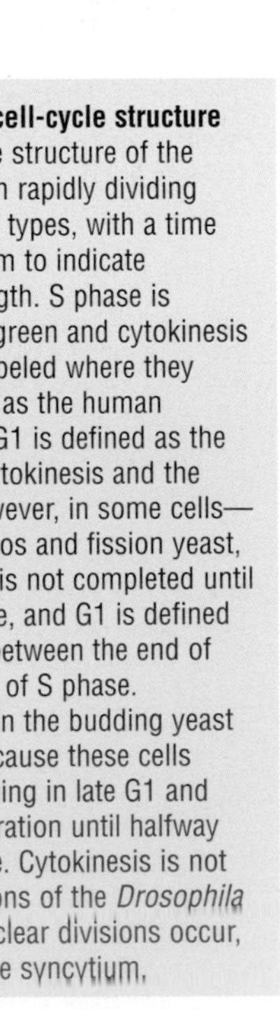

Figure 1-3 Variations in cell-cycle structure in different cell types The structure of the cell cycle is diagrammed in rapidly dividing populations of various cell types, with a time scale beneath each diagram to indicate approximate cell-cycle length. S phase is colored in red, mitosis in green and cytokinesis in blue. Gap phases are labeled where they exist. In animal cells such as the human somatic cell shown here, G1 is defined as the gap between the end of cytokinesis and the beginning of S phase; however, in some cells—those of sea urchin embryos and fission yeast, for example—cytokinesis is not completed until well into the next cell cycle, and G1 is defined as the gap, if one exists, between the end of mitosis and the beginning of S phase. Cytokinesis is not shown in the budding yeast diagram for simplicity, because these cells begin the process of budding in late G1 and do not complete cell separation until halfway through the next cell cycle. Cytokinesis is not present in the early divisions of the *Drosophila* embryo, in which only nuclear divisions occur, resulting in a multinucleate syncytium.

The early embryos of the fruit fly *Drosophila melanogaster* provide an extreme example of the dissociation of cell division from nuclear division (Figure 1-3). The remarkably rapid cell cycles of these embryos (about 8 minutes each) occur without cytokinesis, resulting in the formation of a multinucleate cell, or **syncytium**, with thousands of nuclei in a shared cytoplasm. Only after about 13 rounds of nuclear division does a plasma membrane form around each nucleus, resulting in the division of the syncytial cytoplasm into individual cells.

In the majority of cells, M phase does not begin until after the end of S phase. The budding yeast, *Saccharomyces cerevisiae*, is an exception to this general rule. In this organism the entry into mitosis is not as extensively regulated as in most higher eukaryotes, and preparations for mitosis, such as the assembly of the mitotic spindle, can sometimes begin before DNA synthesis is completed. These cells therefore lack a clearly defined G2 between S and M phases (Figure 1-3).

Multiple rounds of chromosome duplication or segregation can occur in the same cell cycle

Successful cell reproduction generally demands that S phase and M phase occur only once per cell cycle. In certain specialized cell types, however, multiple rounds of S phase or M phase can occur in a single cycle.

The *meiotic program* is a specialized form of nuclear division in which two rounds of chromosome segregation—called *meiosis I* and *II*—occur after a single round of DNA replication. As discussed in Chapter 9, this complex process is of fundamental importance in all sexually reproducing organisms. Most cells in these organisms are **diploid**: that is, they normally contain two nearly identical versions of each chromosome. These two versions of a chromosome are called **homologs**. The meiotic cell cycle reduces the number of chromosomes by half, resulting in **haploid** cells carrying one homolog of each chromosome. Fusion of two haploid cells generates a diploid **zygote**, completing the reproductive cycle.

Many cells in the larvae of *Drosophila*, as well as numerous cell types in other animals, undergo a specialized cell cycle called **endoreduplication**, in which multiple rounds of S phase occur without intervening M phases. In the *Drosophila* salivary gland, for example, ten rounds of S phase give rise to cells containing a thousand copies of the genome. The increase in gene copies that results from this process is thought to allow these cells to increase the rate of production of numerous gene products. In some larval cells the many copies of the DNA are bundled together in parallel, forming thick **polytene chromosomes** that are clearly visible under the light microscope (Figure 1-4).

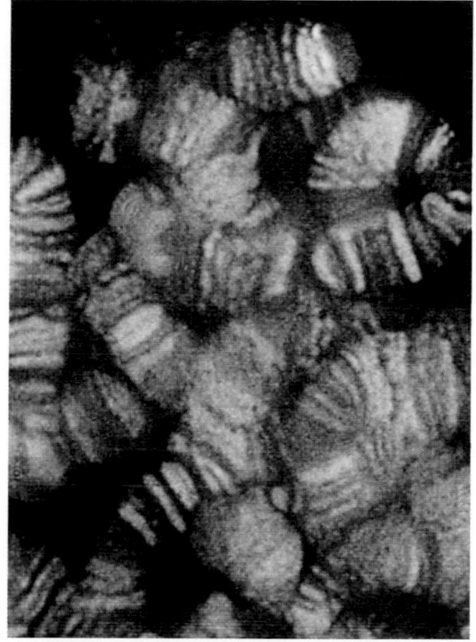

Figure 1-4 Polytene chromosomes packed in the nucleus of a cell from the *Drosophila* salivary gland A thousand identical strands of DNA are aligned in parallel in these chromosomes. After staining with fluorescent dyes, dark bands reveal regions where the DNA is more tightly compacted by associated protein components. The banding patterns therefore provide useful information about differences in chromatin structure along the chromosomes. Courtesy of John Sedat.

The symmetry of cell division varies in different cell types

Cell division usually gives rise to daughter cells with identical chromosomal DNA, but the distribution of other cellular components is not always symmetrical. In budding yeast, for example, the cytoplasmic components of the cell are divided unequally during cytokinesis, resulting in daughter cells of unequal size.

During the development of multicellular organisms, asymmetric cell division is often used to produce daughter cells with different functions or developmental fates. Various mechanisms are employed in these cells to localize regulatory proteins or specific organelles at one pole of the mother cell, resulting in their unequal distribution at cytokinesis.

Definitions

diploid: (of a cell) possessing two copies, or **homologs**, of each chromosome. The somatic cells of most multicellular organisms are diploid.

endoreduplication: the repeated replication of chromosomes without accompanying mitosis or cell division. This can result in large **polytene chromosomes** consisting of many copies in parallel.

haploid: (of a cell) possessing one copy, or **homolog**, of each chromosome. The egg and sperm cells of animals are haploid.

homolog: in sexually reproducing organisms, either of the two copies of each chromosome normally present in the **diploid** somatic cells. For each chromosome, one homolog is inherited from one parent and the other homolog from the other parent.

polytene chromosome: giant chromosome arising from repeated rounds of DNA replication in nondividing cells.

syncytium: multinucleate cell.

zygote: in sexually reproducing organisms, the **diploid** cell produced by the fusion of two **haploid** cells, such as egg and sperm, from the parents.

1-3 The Cell-Cycle Control System

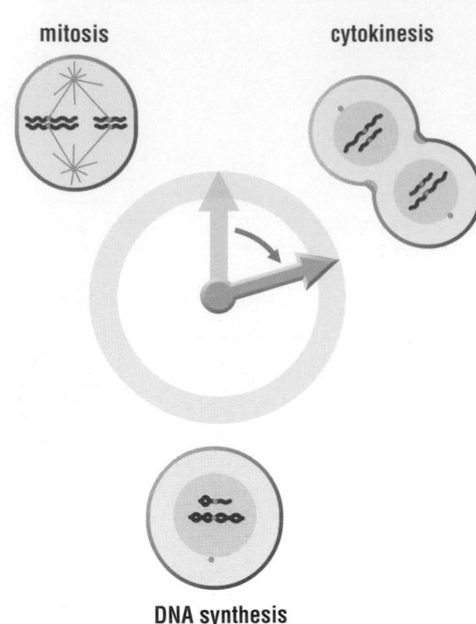

Figure 1-5 A simple cell-cycle control system In some early embryonic cell cycles, the cell-cycle control system behaves like an autonomous one-handed clock whose rotation is programmed to sequentially trigger cell-cycle events with appropriate timing. The clock continues to operate normally even when cell-cycle events fail.

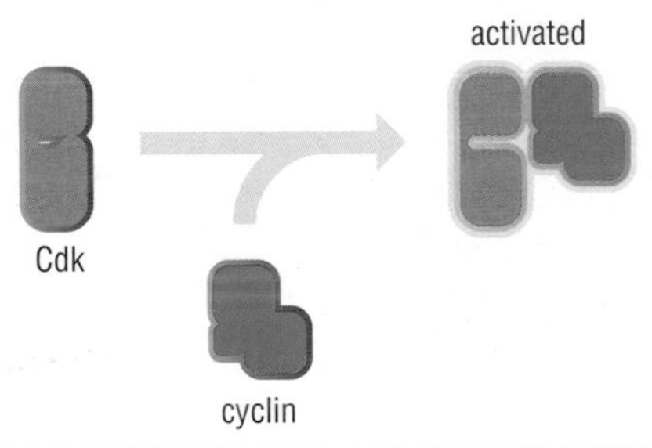

Figure 1-6 Cyclin-dependent kinase activation The cell-cycle control system is based on cyclin-dependent kinases (Cdks) that are activated at specific cell-cycle stages by regulatory subunits called cyclins. For simplicity, this diagram does not include changes in Cdk phosphorylation that also contribute to Cdk activation. These are discussed in Chapter 3. The activated complex is indicated by a gold outline.

Cell-cycle events are governed by an independent control system

The army of protein machines that execute the events of the cell cycle is under the strict control of a regulatory network called the cell-cycle control system. The basic features of this system are readily apparent in the simple cell cycles of early animal embryos. There the control system behaves like an autonomous biochemical timer that is precisely programmed to initiate cell-cycle events in the correct order and at specific intervals, such that each event is allowed just enough time to be completed before the next event is triggered. In these cells, the control system is independent of cell-cycle events and continues to operate even if those events fail (Figure 1-5).

The cell-cycle control system is based on oscillations in the activities of cyclin-dependent protein kinases

The central components of the cell-cycle control system are a family of enzymes called the **cyclin-dependent kinases (Cdks)**. Like other protein kinases, Cdks catalyze the covalent attachment of phosphate groups derived from ATP to protein substrates. This phosphorylation results in changes in the substrate's enzymatic activity or its interaction with other proteins.

Cdk activities rise and fall as the cell progresses through the cell cycle. These oscillations lead directly to cyclical changes in the phosphorylation of components of the cell-cycle machinery, resulting in the initiation of cell-cycle events. Thus, for example, an increase in Cdk activity at the beginning of S phase causes the phosphorylation of proteins that then initiate DNA synthesis.

Cdks are activated by binding to regulatory proteins called **cyclins** (Figure 1-6). Oscillations in Cdk activity during the cell cycle are due primarily to changes in the amounts of cyclins. Different types of cyclins are produced at different cell-cycle phases, resulting in the periodic formation of distinct cyclin–Cdk complexes that trigger different cell-cycle events. A wide range of mechanisms contributes to the control of cyclin levels and Cdk activity, resulting in a complex Cdk regulatory network that forms the core of the cell-cycle control system.

Cell-cycle events are initiated at three regulatory checkpoints

The cell-cycle control system drives progression through the cell cycle at regulatory transitions called **checkpoints** (Figure 1-7). The first is called **Start** or the G1/S checkpoint. When conditions are ideal for cell proliferation, G1/S- and S-phase cyclin–Cdk complexes are activated, resulting in the phosphorylation of proteins that initiate DNA replication, centrosome duplication and other early cell-cycle events. Eventually, G1/S- and S-phase Cdks also promote the activation of M-phase cyclin–Cdk complexes, which drive progression through the second major checkpoint at the entry into mitosis (**G2/M checkpoint**). M-phase cyclin–Cdks phosphorylate proteins that promote spindle assembly, bringing the cell to metaphase.

The third major checkpoint is the **metaphase-to-anaphase transition**, which leads to sister-chromatid segregation, completion of mitosis and cytokinesis. Progression through this checkpoint occurs when M-phase cyclin–Cdk complexes stimulate an enzyme called the *anaphase-promoting complex*, which causes the proteolytic destruction of cyclins and of proteins that hold the sister chromatids together. Activation of this enzyme therefore triggers sister-chromatid separation and segregation. Destruction of cyclins leads to inactivation of all Cdks in the cell, which allows phosphatases to dephosphorylate Cdk substrates. Dephosphorylation of these substrates is required for spindle disassembly and the completion of mitosis, and for cytokinesis.

Definitions

Cdk: see **cyclin-dependent kinase**.

checkpoint: regulated transition point in the cell cycle, where progression to the next phase can be blocked by negative signals. This term is sometimes defined to include the signaling mechanisms that monitor cell-cycle events and transmit the information to the control system; in this book the term is used to define the transition point in the cell cycle where these mechanisms act.

cyclin: positive regulatory subunit that binds and activates **cyclin-dependent kinases**, and whose levels oscillate in the cell cycle.

cyclin-dependent kinase (Cdk): protein kinase whose catalytic activity depends on an associated **cyclin** subunit. Cyclin-dependent kinases are key components of the cell-cycle control system.

G2/M checkpoint: important regulatory transition where entry into M phase can be controlled by various factors such as DNA damage or the completion of DNA replication.

metaphase-to-anaphase transition: cell-cycle transition where the initiation of sister-chromatid separation can be blocked if the spindle is not fully assembled. Also called the M/G1 checkpoint, but this is not an ideal term because it does not coincide with the boundary between M phase and G1.

Start: major regulatory transition at the entry into the cell cycle in mid to late G1, also called the G1/S checkpoint or the restriction point (in animal cells). Progression past this point is prevented if cell growth is insufficient, DNA is damaged or other preparations for

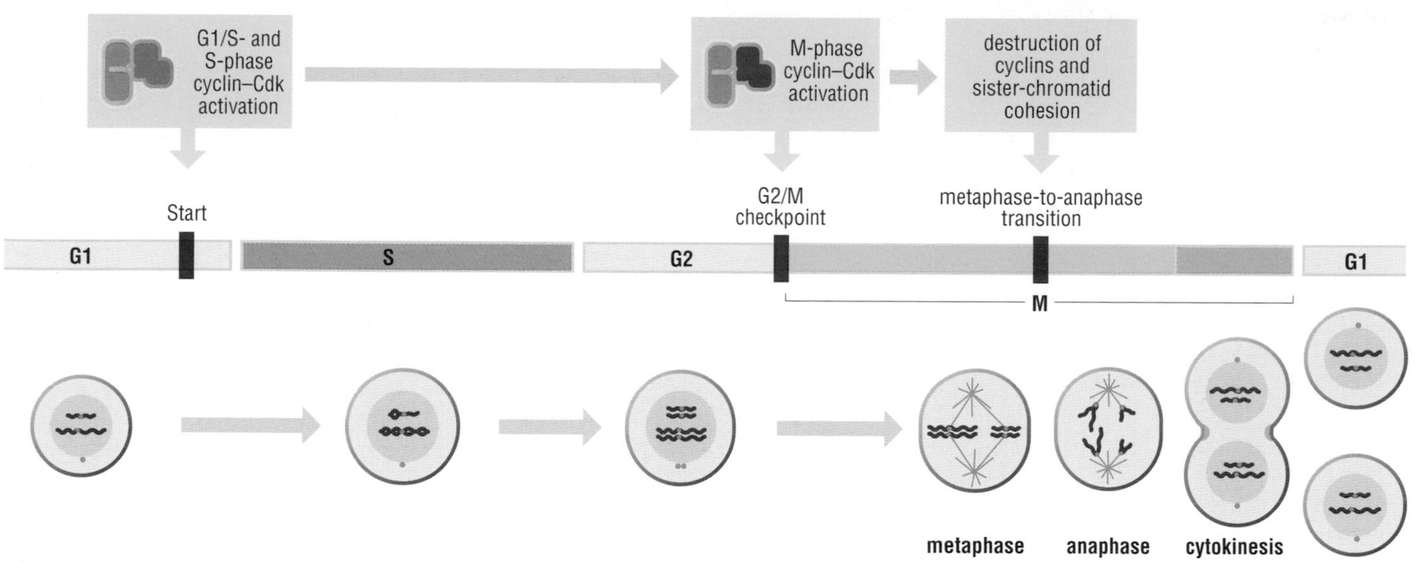

Figure 1-7 Overview of cell-cycle control Progression through the cell cycle is governed at three major checkpoints. In mid to late G1, Cdks are activated by G1/S- and S-phase cyclins, resulting in entry into the cell cycle at Start. Entry into mitosis (at the G2/M checkpoint) is triggered by activation of M-phase Cdk–cyclin complexes. Finally, the metaphase-to-anaphase transition is driven by a regulatory enzyme called the anaphase-promoting complex, which triggers the destruction of cyclins and other regulators.

Cell-cycle progression in most cells can be blocked at checkpoints

In the cells of the early animal embryo, the Cdk activities of the cell-cycle control system are linked together to form a rigidly programmed oscillator that is essentially autonomous—that is, it can generate appropriately timed waves of Cdk activity without external input. This system is ideal for cells that must divide as rapidly as possible and are not affected by external influences. The control system of most cell types, however, includes additional levels of regulation that allow cell-cycle progression to be adjusted by various intracellular and extracellular signals. Most cells, for example, initiate a new cell cycle only when stimulated by external signals, thus ensuring that new cells are made only when needed. Similarly, initiation of cell-cycle events in most cells is responsive to surveillance mechanisms that monitor the progress of previous events. If the cell fails to complete DNA replication, for example, a negative signal blocks the initiation of mitosis. Later events are thus dependent on the completion of earlier events.

To allow regulation of cell-cycle progression, the cell-cycle control system of most cells is supplemented by molecular braking mechanisms that can be used, if necessary, to inhibit the Cdks and other regulators that drive progression through the three major checkpoints. If environmental conditions are not appropriate for cell proliferation, inhibitory signals prevent activation of G1/S- and S-phase Cdks—thereby blocking progression through Start. Similarly, the failure to complete DNA replication blocks entry into mitosis by inhibiting M-phase cyclin–Cdk activation. Delays in spindle assembly inhibit the proteolytic machinery that drives the metaphase-to-anaphase transition, thereby preventing sister-chromatid segregation until the spindle is ready. By these and numerous other mechanisms, the cell arrests cell-cycle progression at an appropriate point when conditions are not ideal and continues it when they are.

The cell-cycle control system can thus be viewed as a linked series of tightly regulated molecular switches, each of which triggers the initiation of cell-cycle events at a specific regulatory checkpoint. We discuss the molecular components and design of this system in Chapter 3. First, in Chapter 2, we review the wide range of experimental organisms in which cell-cycle control is studied.

cell-cycle entry are not complete. Unlike cells arrested at the **G2/M checkpoint** or **metaphase-to-anaphase transition**, cells prevented from passing Start do not arrest at this point but typically exit the cell cycle into a prolonged nondividing state from which a return to the cycle is a lengthy process.

References

Hartwell, L.H. and Weinert, T.A.: **Checkpoints: controls that ensure the order of cell cycle events.** *Science* 1989, **246**:629–634.

Morgan, D.O.: **Cyclin-dependent kinases: engines, clocks, and microprocessors.** *Annu. Rev. Cell Dev. Biol.* 1997, **13**:261–291.

Murray, A.W. and Kirschner, M.W.: **Dominoes and clocks: the union of two views of the cell cycle.** *Science* 1989, **246**:614–621.

2

Model Organisms in Cell-Cycle Analysis

The fundamental features of the cell cycle have been conserved for a billion years of eukaryotic evolution, and can therefore be studied productively in a wide range of experimental organisms. The most commonly used model systems are unicellular yeasts, the embryos of frogs and flies, and mammalian cells growing in culture.

2-0 Overview: Cell-Cycle Analysis in Diverse Eukaryotes

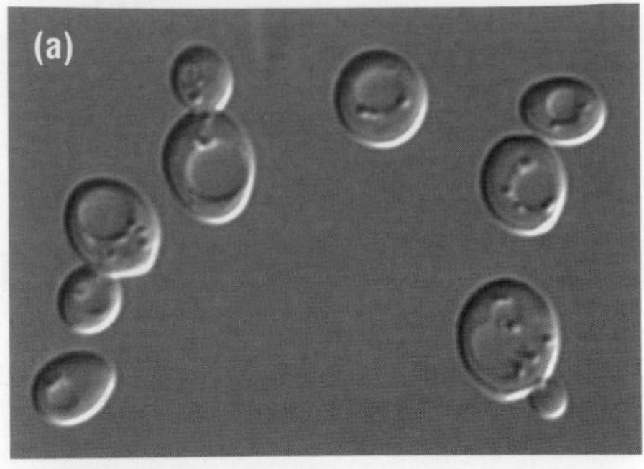

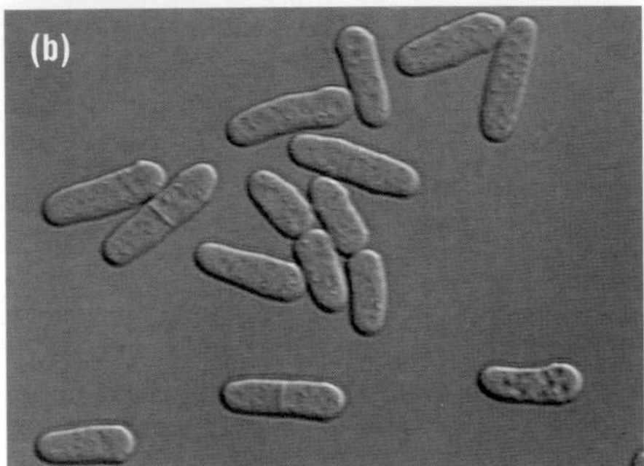

Figure 2-1 The budding yeast *Saccharomyces cerevisiae* and the fission yeast *Schizosaccharomyces pombe*
(a) In a typical population of *S. cerevisiae* cells, the size of the bud provides a rough indication of the cell's position in the cell cycle: unbudded cells are in early G1, small-budded cells in late G1 and larger-budded cells are in S phase or early M phase. **(b)** The cell-cycle stage in *S. pombe* can be estimated by cell length and by the presence of a septum, which indicates progression through cytokinesis. Panel (a) courtesy of Greg Tully; panel (b) courtesy of Kathleen Gould.

Mechanisms of cell-cycle control are similar in all eukaryotes

All eukaryotic cells employ similar machinery to duplicate and divide themselves, and all have a similar control system for the timing and coordination of cell-cycle events. It is therefore possible to obtain a comprehensive and unified view of cell-cycle control by studying cell division in widely different species, thereby exploiting the experimental advantages of each.

The most important systems for the study of basic cell-cycle control mechanisms are the single-celled yeasts and the early frog embryo. This chapter provides a brief overview of these model systems and of two others—the fruit fly and mammalian cells in culture—that are used extensively for the study of cell growth and division in multicellular organisms.

Budding and fission yeasts provide powerful systems for the genetic analysis of eukaryotic cell-cycle control

Yeast are small, single-celled fungi and are among the simplest eukaryotes (Figure 2-1). Two species in particular—the budding yeast *Saccharomyces cerevisiae* and the fission yeast *Schizosaccharomyces pombe*—have proved to be valuable model organisms for the study of cell-cycle control. They share a number of experimental advantages, but their greatest strength is the ease of genetic analysis. Both organisms are able to proliferate rapidly in simple culture conditions, and both have small, fully defined genomes. Most importantly, both organisms can proliferate in a haploid state, in which only a single copy, or homolog, of each chromosome is present in the cell. This makes it easy to generate mutations that inactivate a gene and to analyze them without the complications of a second gene copy. It is also relatively easy to delete specific genes, replace them with defined mutant versions, or express them under the control of promoters that are responsive to chemicals added to the medium.

Early animal embryos are useful for the biochemical characterization of simple cell cycles

The giant fertilized eggs of many animals carry large stockpiles of the proteins needed for cell division, and are capable of several rapid cell divisions in the absence of growth, gap phases and many of the control mechanisms that operate in most cells (Figure 2-2). These early embryonic cell cycles—particularly those of the frog *Xenopus laevis*—have been important in the discovery of the core components and behaviors of the cell-cycle control system. A key

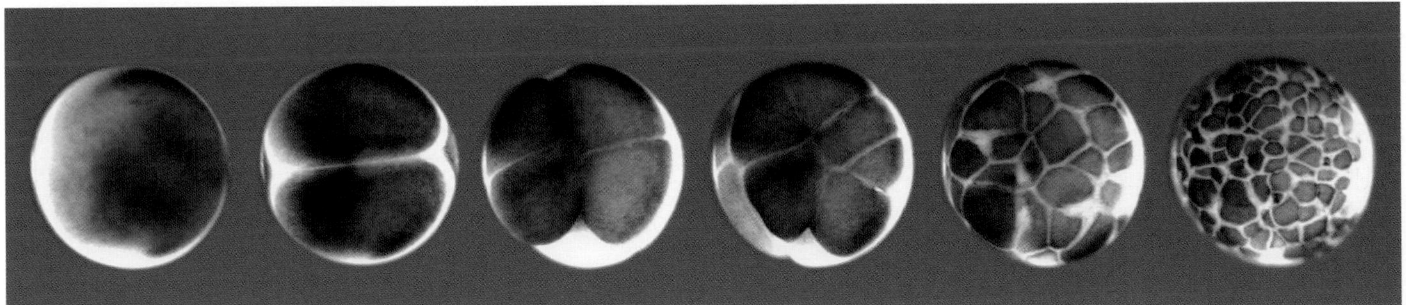

Figure 2-2 Early divisions in the frog *Xenopus laevis* Courtesy of James C. Smith and Huw Williams.

References

Baserga, R.: *The Biology of Cell Reproduction* (Harvard University Press, Cambridge, MA, 1985).

Mitchison, J.M.: *The Biology of the Cell Cycle* (Cambridge University Press, Cambridge, 1971).

Murray, A.W. and Hunt, T.: *The Cell Cycle: An Introduction* (Freeman, New York, 1993).

Figure 2-3 Patterns of cell division in the early embryo of the fly *Drosophila melanogaster* In this image of a fly embryo in the early stages of *gastrulation*, the DNA in all cells is stained light blue, while the chromosomes of cells in mitosis 14 are labeled red. Each stage in embryogenesis is characterized by a highly programmed pattern of cell division like that seen here. Courtesy of Tony Shermoen and Patrick O'Farrell.

advantage of eggs is their large size: the fertilized *Xenopus* egg, for example, is a millimeter in diameter and is easy to inject with test substances to determine their effect on cell-cycle progression. It is also possible to prepare almost pure cytoplasm from these eggs and reconstitute many events of the cell cycle in a test tube. These cell-cycle extracts provide a simple and highly controlled system in which cell-cycle events can be observed and manipulated.

Control of cell division in multicellular organisms can be dissected genetically in *Drosophila*

Although fundamental principles of cell-cycle control are most easily uncovered in yeast and frog embryos, some unique features of cell-cycle control in somatic animal cells are best studied in more complex systems. An important model organism for this purpose is the fruit fly *Drosophila melanogaster*, which has been particularly useful for studying mechanisms controlling cell growth and division in developing tissues (Figure 2-3). *Drosophila* has a relatively short generation time, a fully sequenced genome and well studied genetics, which can be deployed to generate mutants and analyze gene function *in vivo*.

The nematode worm *Caenorhabditis elegans* is another model invertebrate in which the fundamental features of metazoan biology can be revealed by genetic analysis. Considerable insights into the mechanics of mitosis and cytokinesis have come from studies of the early embryonic divisions of *C. elegans*. We will not discuss this organism in detail in this chapter, but it will be mentioned later in the book, particularly in our discussion of cytokinesis in Chapter 8.

Cultured cell lines provide a means of analyzing cell-cycle control in mammals

The ultimate goal of cell-cycle research is to understand the human cell cycle, which is more complex than that of any non-mammalian system. As experiments with intact humans are not possible, it is necessary to study mammalian systems that can be manipulated in the laboratory. Most commonly used are normal or tumor cells, typically mouse or human, that have been removed from the animal and grown in plastic dishes in the presence of essential nutrients and other factors (Figure 2-4). Cells taken from normal tissue (*primary* cells) cannot divide indefinitely in culture, whereas tumor cells and other immortalized cell lines carry mutations that allow unlimited proliferation in the culture environment.

Cell lines growing in culture often possess abnormal cell-cycle control mechanisms, in part because these cells carry genetic defects in these mechanisms, and in part because life in culture medium can never fully mimic that in the cell's normal environment in the animal. Nevertheless, cell lines are extremely useful for the biochemical and cytological characterization of features of cell-cycle control that are unique to mammals. Studies of tumor cells in culture can also provide important insights into mechanisms responsible for the inappropriate proliferative behavior of cancer.

Figure 2-4 Mammalian cells growing in culture In these photographs of a cultured mammalian cell line, the DNA of the chromosomes is labeled blue and the actin filaments that help form the cytoskeleton are labeled red. **(a)** A cell in interphase. The cells shown in **(b)** are completing M phase. Courtesy of Susanne Steggerda.

2-1 Life Cycles of Budding and Fission Yeasts

Budding yeast and fission yeast divide by different mechanisms

The budding yeast *Saccharomyces cerevisiae* and the fission yeast *Schizosaccharomyces pombe* are single-celled fungi with simple genomes, rapid growth rates and a long history of use in brewing, baking and molecular genetics. Despite being on the same branch of the phylogenetic tree, however, the two species are only distantly related. It is thought that they diverged in evolution several hundred million years ago and are no more related to each other than they are to us.

A glance at budding and fission yeasts under a microscope reveals clear differences in their external appearance and modes of division (see Figure 2-1). *S. cerevisiae* is an ovoid cell that is about 3–5 μm in diameter and encased in a tough cell wall. It divides by budding. The bud first appears at the end of G1 and grows continuously throughout S and M phases until it reaches a size slightly smaller than that of the mother cell. After distribution of one set of chromosomes into the bud at the completion of mitosis, the daughter cell pinches off the mother cell. Bud size is thus a useful marker of cell-cycle position. *S. pombe*, by contrast, is a rod-shaped cell (about 3 μm in diameter) that grows entirely by elongation at the ends. After mitosis, division occurs by the formation of a septum, or cell plate, that cleaves the cell at its midpoint. Separation of the daughter cells is a lengthy process that is usually not complete until the cell has entered S phase of the following cell cycle (see section 1-2). The position in the cell cycle can be assessed by cell length and the presence of a septum.

Differences in cell-cycle control are also apparent in the two yeast species. The budding yeast cell cycle has a long G1 but no clearly defined G2 between S phase and M phase, and entry into mitosis is not regulated as extensively as it is in other eukaryotic model systems (see section 1-2). Instead, the metaphase-to-anaphase transition is a more important regulatory checkpoint in this organism. Fission yeast, by contrast, governs entry into mitosis by mechanisms that are similar to those in multicellular animals, and studies of mitotic entry, and the control of mitotic Cdk activation in particular, have been particularly fruitful in this organism.

Although it can be argued that *S. pombe* is in some respects a better model of animal cells than *S. cerevisiae*, the latter tends to be more commonly used in laboratories studying cell-cycle control. Budding yeast has been used for longer by more laboratories, and studies in this organism are therefore supported by a stronger foundation of biological and methodological knowledge.

Yeast cells alternate between haploid and diploid states and undergo sporulation in response to starvation

The life cycles of *S. cerevisiae* and *S. pombe* are shown in Figures 2-5 and 2-6. Both species can proliferate in a haploid state, in which the cell carries only a single homolog of each

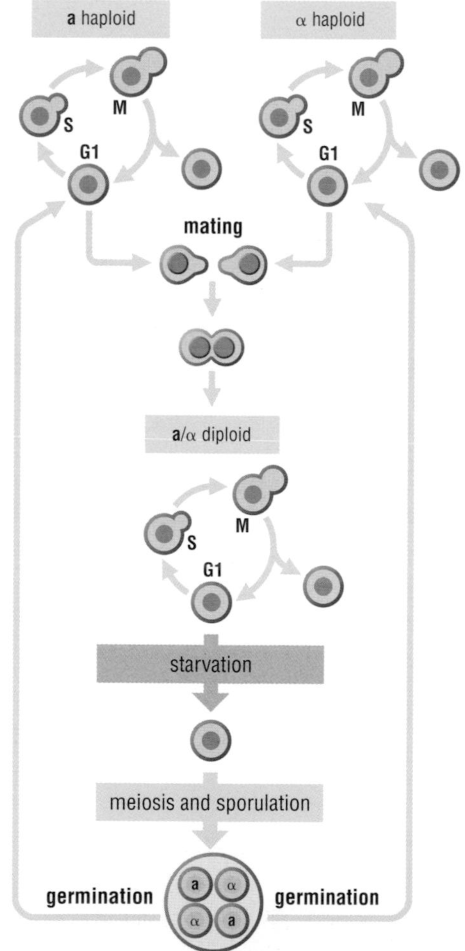

Figure 2-5 Life cycle of *Saccharomyces cerevisiae* Budding yeast has two mating types, **a** and α, that briefly proliferate in a haploid state (top) until they encounter mating factors secreted by cells of the opposite mating type. This triggers a G1 arrest and mating, resulting in the formation of an **a**/α diploid cell that proliferates as long as conditions are favorable. When diploid cells are starved, they undergo sporulation. The resulting tetrad of spores, two of each mating type, germinates when conditions improve, resulting in the formation of haploid cells. In nature these would immediately mate to restore the diploid proliferating state, but in the laboratory haploid cultures can be maintained.

Definitions

sporulation: the formation of spores. In yeast, refers to the formation of haploid spores from diploid cells by meiosis in conditions unfavorable for growth and proliferation.

tetrad: in yeast, the four haploid spores produced by meiosis from a single diploid cell.

chromosome. Haploid cells exist in two mating types (**a** and α in budding yeast; P and M in fission yeast), which secrete specific mating factors or pheromones (**a**-factor and α-factor in budding yeast, for example). When cells of opposite mating types encounter one another, these pheromones bind to cell-surface receptors on the opposite cell and trigger arrest in G1 and the synthesis of proteins required for mating. The two arrested cells then fuse to form a diploid cell carrying two homologs of each chromosome.

In response to adverse conditions such as starvation, diploid budding yeast cells undergo **sporulation**. This process begins with a specialized form of nuclear division, called *meiosis*, that reduces chromosome content to a haploid state (as discussed in Chapter 9). Following meiosis, the four haploid progeny are encased in individual cell walls, thereby forming a **tetrad** of dormant spores in a tough protective container called the ascus, which protects the spores from heat and desiccation. When conditions become favorable, the spores germinate and the cells return to the haploid proliferative state.

Despite the general similarities of budding and fission yeast life cycles, the relative importance of haploid and diploid states is very different in the two species (see Figures 2-5 and 2-6). Fission yeast cells normally proliferate in the haploid state and mate only in starvation conditions. The resulting diploid cells do not proliferate but immediately undergo sporulation. Budding yeast cells, by contrast, normally proliferate in the diploid state and undergo sporulation in response to starvation. After spore germination, haploid cells mate immediately to form new diploids.

In both yeast species, each spore tetrad contains two cells of each mating type, allowing immediate mating even if only a single ascus germinates. A mixture of mating types in a population is also promoted by the process of mating-type switching, by which the progeny of many haploid cells are converted genetically from one mating type to the other. As a result, even the progeny of a single spore can mate and form diploids.

Experimental analysis of gene function is greatly simplified in haploid cells (as discussed in section 2-2), and so yeast cells are generally studied in the haploid state. This is straightforward in fission yeast, whose cells normally proliferate as haploids and do not mate while growth conditions remain favorable. In budding yeast, however, stable proliferation of haploid cells requires some manipulation in the laboratory. First, spore tetrads must be physically separated with a microneedle to isolate single spores for propagation. Second, mating-type switching must be prevented to ensure that the progeny of a single spore will be the same mating type and unable to mate, so that they remain in a stable haploid state. Most laboratory strains therefore carry mutations that block mating-type switching.

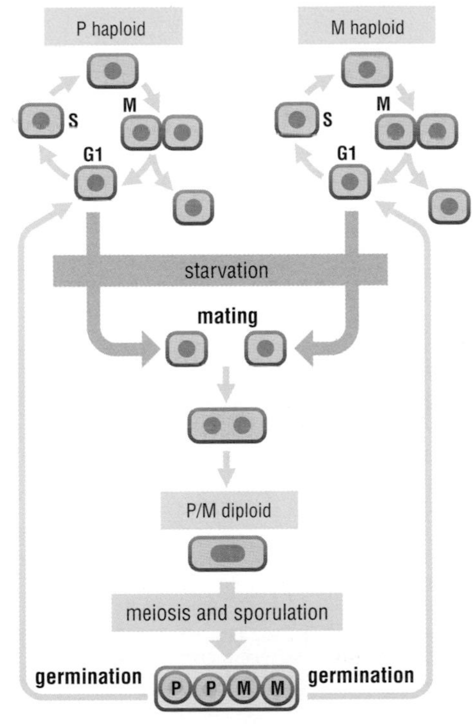

Figure 2-6 Life cycle of *Schizosaccharomyces pombe* Fission yeast normally proliferates in a haploid state. When starved, cells of opposite mating types (P and M) fuse to form a diploid zygote that immediately enters meiosis to generate four haploid spores. When conditions improve, these spores germinate to produce proliferating haploid cells. Note that the cell cycle is simplified here for clarity: G1 is normally very short or nonexistent, and cytokinesis is not normally completed until after S phase of the following cell cycle (see Figure 1-3).

References

Forsburg, S.L. and Nurse, P.: **Cell cycle regulation in the yeasts *Saccharomyces cerevisiae* and *Schizosaccharomyces pombe*.** *Annu. Rev. Cell Biol.* 1991, **7**:227–256.

Mitchison, J.M.: *The Biology of the Cell Cycle* (Cambridge University Press, Cambridge, 1971).

Murray, A.W. and Hunt, T.: *The Cell Cycle: An Introduction* (Freeman, New York, 1993).

Cell biological processes are readily dissected with yeast genetic methods

Budding and fission yeasts are remarkably powerful tools for the analysis of cell-cycle control and other fundamental cellular processes. Yeast cells divide rapidly in the laboratory, with a cell-cycle time of about 90 minutes to 2 hours, when cultured in a rich medium containing glucose, amino acids, nucleotides and other compounds. Because yeast can be propagated in a haploid state, it is easy to isolate and characterize mutations whose phenotypes might be difficult to analyze in the presence of a second, wild-type, copy of the gene. Identifying the mutant gene is straightforward in most cases, allowing the gene and its protein product to be characterized more fully. The power of yeast also lies in their relatively small, simple and fully sequenced genomes (about 15 million base pairs on 16 chromosomes in budding yeast; about 15 million base pairs on three chromosomes in fission yeast).

The expression of specific genes can be manipulated by a variety of methods in yeast. Most importantly, yeast have a high rate of *homologous recombination*, a complex process by which the gene sequence on a chromosome can be replaced with closely related sequences from another chromosome or from an artificial DNA plasmid. Thus, it is straightforward to use targeted recombination methods in yeast to disrupt specific genes, replace them with defined mutant forms or integrate genes at various defined locations in the genome. Genes can also be carried on plasmid DNA circles that can be stably propagated in the yeast cell without being integrated into a chromosome. Whether carried on a plasmid or inserted into a chromosome, specific genes can be expressed under the control of their natural promoter sequences or under the control of a conditional promoter that can be regulated by the addition of chemicals to the growth medium.

Conditional mutants are used to analyze essential cell-cycle processes

Many fundamental insights into cell-cycle control have been gained from studies of mutations in genes required for cell-cycle progression in yeast. Because these cell-division cycle (*cdc*) mutations inactivate functions that are essential for cell proliferation, mutant cells can be propagated only if the mutations are conditional: that is, the gene product is non-functional only in certain

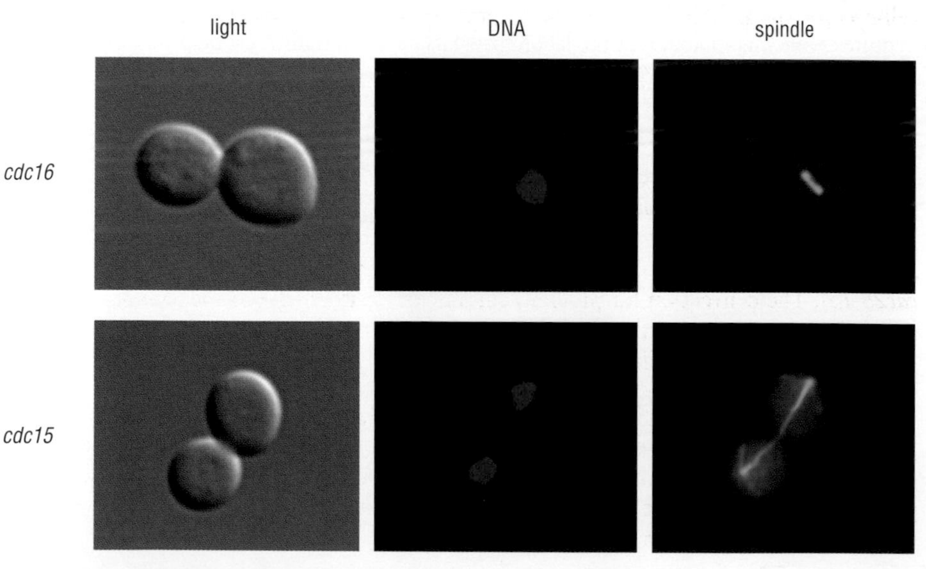

Figure 2-7 Cell-cycle arrests in budding yeast *cdc* mutants Two *cdc* mutants were grown at the restrictive temperature and analyzed by light microscopy (left panels). Cells were stained to reveal DNA (blue, middle) and microtubules of the mitotic spindle (green, right). A temperature-sensitive *cdc16* mutant (top) arrests with a short pre-anaphase spindle and chromosomes that have not segregated, because Cdc16 (a subunit of the APC) is required for sister-chromatid separation. Budding and bud growth, however, are not blocked by a *cdc16* mutation (see Figure 2-8), so a large bud is apparent in these mutants. A *cdc15* mutant (bottom) arrests in late mitosis with segregated chromosomes and an elongated anaphase spindle. Cdc15 is therefore required for the completion of mitosis and for cytokinesis (see Figure 2-8). Courtesy of Greg Tully.

References

Beach, D. *et al.*: **Functionally homologous cell cycle control genes in budding and fission yeast.** *Nature* 1982, **300**:706–709.

Forsburg, S.L. and Nurse, P.: **Cell cycle regulation in the yeasts *Saccharomyces cerevisiae* and *Schizosaccharomyces pombe*.** *Annu. Rev. Cell Biol.* 1991, **7**:227–256.

Hartwell, L.H.: **Twenty-five years of cell cycle genetics.** *Genetics* 1991, **129**:975–980.

Hartwell, L.H. *et al.*: **Genetic control of the cell division cycle in yeast.** *Science* 1974, **183**:46–51.

Hartwell, L.H. *et al.*: **Genetic control of the cell-division cycle in yeast, I. Detection of mutants.** *Proc. Natl Acad. Sci. USA* 1970, **66**:352–359.

Nurse, P.: **Genetic control of cell size at cell division in yeast.** *Nature* 1975, **256**:547–551.

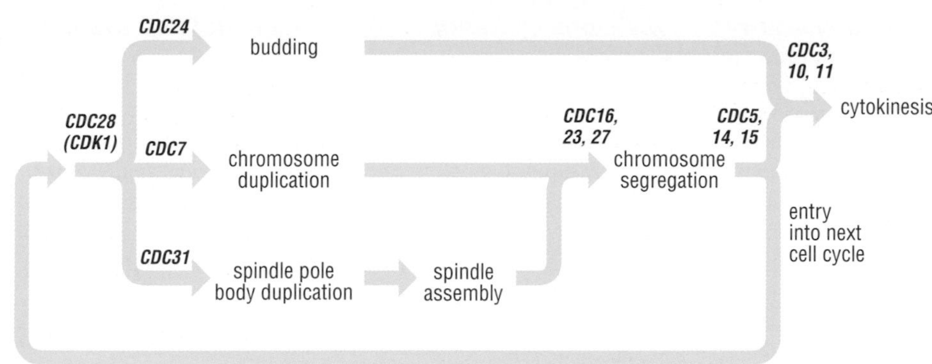

Figure 2-8 Functions of *CDC* genes in budding yeast Analysis of cell-cycle arrests in *cdc* mutants provides a general outline of the relationships between different cell-cycle events. The names of various *CDC* genes are shown above the points at which they are required for further progress in the cell cycle. All events flowing right from those points are blocked in cells with mutations in the indicated genes. Mutation of *CDC7*, for example, not only blocks DNA synthesis but also prevents the segregation of chromosomes; mitosis is thus dependent on completion of DNA synthesis. Spindle assembly and budding still occur in a *cdc7* mutant, however, because these events are not dependent on DNA synthesis. Similarly, inhibition of budding (in a *cdc24* mutant) does not prevent DNA synthesis or mitosis (although it does delay mitosis, which is not shown here). The events and gene products shown here are discussed in greater detail in later chapters.

specific conditions. Most *cdc* mutants are temperature-sensitive, so that the gene product is functional at low temperature (the permissive condition, typically room temperature) but can be inactivated by shifting the cells to high temperature (the restrictive condition, typically 37 °C). After the temperature is increased, *cdc* mutants continue to progress through the cell cycle until they arrest at the point at which the gene is required for further progress (Figure 2-7).

Some *cdc* genes encode enzymatic and structural machinery (DNA synthetic enzymes, for example) that carry out essential cell-cycle processes. Many *cdc* genes, however, encode components of the cell-cycle control system. One of the early searches for *cdc* mutants in fission yeast, for example, led to the discovery of a gene, named *cdc2*$^{+}$, that is required for entry into mitosis. A homologous gene, *CDC28*, was identified in mutant screens in budding yeast. The products of these genes are now known as Cdk1, the central component of the cell-cycle control system.

Early studies of *cdc* mutant phenotypes in budding yeast revealed that blocking early cell-cycle events, such as DNA replication or spindle pole body duplication, blocks the onset of later events, such as mitosis (Figure 2-8). These results provided the first clear evidence that later events in the cell cycle are dependent on the completion of earlier events. We now know that this dependence is due largely to surveillance mechanisms that monitor the completion of cell-cycle events and block the cycle at specific checkpoints when these events fail.

Homologous genes have different names in fission yeast and budding yeast

Fission yeast gene names are typically written in lower-case italics, with a plus sign to signify the wild-type gene (for example *cdc2*$^{+}$ for the wild type, *cdc2-1* for a mutant allele of that gene). This convention is different from that in budding yeast, for which wild-type genes are written in upper-case italics (for example *CDC28*) and mutant alleles in lower-case italics (for example *cdc28-13*). The convention for protein names is less established, but they are generally not italicized and are written with only the first letter capitalized (for example Cdc28); the letter p is sometimes added to signify protein (for example Cdc28p).

Because *cdc* genes were numbered simply in the order of their isolation in mutant screens, the name for a gene in fission yeast (for example *cdc2*$^{+}$) will not be the same as the name applied to the homolog of that gene in budding yeast (*CDC28*). Similarly, completely unrelated genes in the two species can have the same name: fission yeast *cdc25*$^{+}$, for example, bears no relationship to budding yeast *CDC25*.

2-3 The Early Embryo of *Xenopus laevis*

The early embryonic divisions of *Xenopus* provide a simplified system for cell-cycle analysis

Many animals begin life as giant fertilized eggs that are highly specialized for rapid cell division, so that the embryo can quickly reach a stage at which it can fend for itself. Because they contain large amounts of maternally supplied proteins, early embryos can divide rapidly without the need for synthesis of new RNA or protein. These simple early divisions also lack many of the checkpoint controls found in somatic cell cycles. As a result, the study of embryonic cell division, particularly in the frog *Xenopus laevis*, has led to many important insights into the fundamental logic and components of cell-cycle control.

Fertilization of the *Xenopus* egg triggers a remarkably rapid and synchronous series of 12 **cleavage** divisions (Figure 2-9; see also Figure 2-2). The first division, which lasts about 75–90 minutes, is followed by 11 synchronous divisions, each of which lasts only 20–30 minutes. These rapid divisions lack gap phases and quickly subdivide the egg into the **blastula**, a ball of 4,000 cells. The **midblastula transition** then occurs, at which transcription from the embryo's genes begins and cell divisions in different regions of the embryo slow down and become less synchronous. Gap phases appear in the cell cycle, cell growth occurs and the cell-cycle control system begins to assume its more complex adult form.

Studies of cell division in the pre-blastula *Xenopus* embryo were the first to suggest the existence of an autonomous cell-cycle oscillator, or clock, that continues to operate with normal timing even if cell-cycle events are severely crippled by removal of the nucleus (Figure 2-10). This oscillator is not readily apparent in most other cell types, in which checkpoint mechanisms arrest the clock when cell-cycle events fail to occur.

Unfertilized eggs develop from diploid oocytes by meiosis

The *Xenopus* egg is derived from a much smaller cell known as an **oocyte**. Soon after its birth in the ovary, the diploid oocyte enters the meiotic program and completes meiotic S phase (see Chapter 9 for a description of the stages of meiosis). It then arrests in meiotic prophase for several months, during which it grows to a diameter of about 1 mm. In response to hormonal cues from the pituitary gland, the follicle cells surrounding the oocyte then secrete the hormone progesterone, which interacts with the oocyte to initiate **oocyte maturation**. This process begins with the first meiotic division, meiosis I (Figure 2-11).

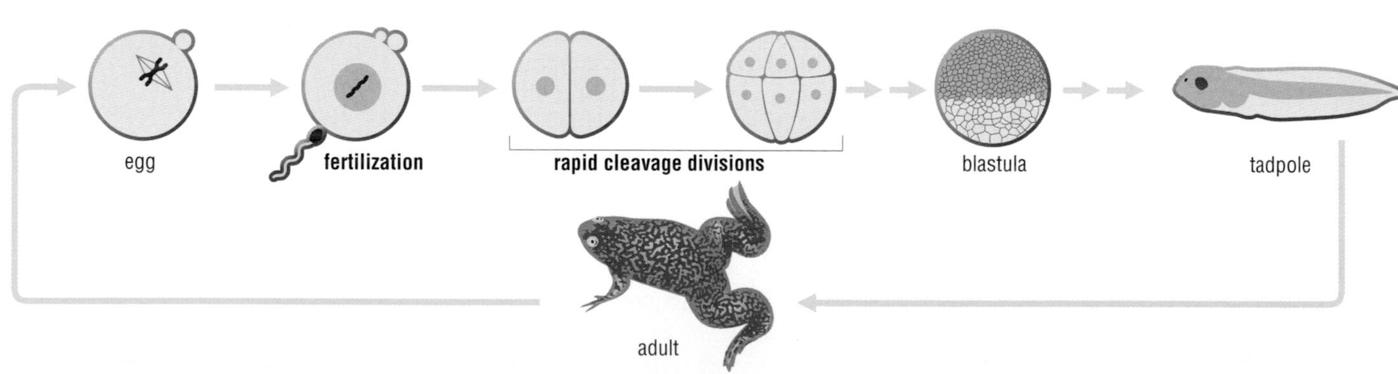

Figure 2-9 Early embryogenesis in *Xenopus* Fertilization of the mature egg initiates rapid cleavage divisions, which subdivide the cytoplasm into smaller and smaller cells until the embryo reaches the blastula stage (see Figure 2-2). The complex morphogenetic movements of gastrulation, or formation of the gut, then occur, and the embryo is quickly transformed into a free-swimming tadpole.

Definitions

blastula: hollow ball of cells that results from the early **cleavage** divisions in some animal embryos.

cleavage: the early cell divisions of animal embryos, which occur in the absence of growth and rapidly subdivide the large fertilized egg into thousands of smaller cells.

maturation-promoting factor: see **MPF**.

midblastula transition: in some animal embryos, transition from development based primarily on maternally supplied protein and RNA to development based on transcription of embryonic genes.

mitosis-promoting factor: see **MPF**.

MPF: maturation-promoting factor or **mitosis-promoting factor**: active complex of Cdk1 and cyclin B, which promotes the onset of meiotic maturation in immature **oocytes** and mitosis in somatic cells.

oocyte: precursor of the haploid egg cell in animals such as frogs, flies and mammals.

oocyte maturation: the process by which a frog **oocyte** arrested in meiotic prophase is induced by progesterone to undergo meiosis I and then arrest in metaphase of meiosis II.

polar body: small cell produced at each meiotic division in **oocytes**. At division, one chromosome set remains in the oocyte and the other is discarded in the polar body which is eventually resorbed.

After chromosome segregation in meiosis I, half of the chromosomes are expelled from the oocyte into a tiny separate cell called the first **polar body**. The oocyte then proceeds to the second meiotic division. After assembly of the spindle and attachment of the chromosomes, the oocyte arrests in metaphase of meiosis II. The oocyte has now been converted to an unfertilized egg and maturation is complete. Fertilization of the egg leads to the completion of meiosis II and the expulsion of a second polar body. This is followed by the fusion of the haploid egg nucleus with the haploid sperm nucleus, and the resulting diploid zygote begins the first mitotic cell division of the early embryo.

Because of their large size, it is possible to inject *Xenopus* oocytes, eggs and early embryos with various test substances. This technique has been of great importance in the study of the cell cycle. When a small amount of cytoplasm is removed from an unfertilized egg (in metaphase II) and injected into an immature, fully grown oocyte (in prophase), the oocyte is released from prophase arrest, completes meiosis I and enters meiosis II, even in the absence of progesterone. The cytoplasmic activity that stimulates meiotic division was originally called **maturation-promoting factor** (**MPF**) but was later found in mitotic cells, and so MPF can also signify **mitosis-promoting factor**. MPF is now known to be an active complex of the protein kinase Cdk1 and the mitotic cyclin, cyclin B, key components of the cell-cycle control system (see section 1-3).

The early embryonic cell cycle can be reconstituted in a test tube

A single *Xenopus* female produces several thousand unfertilized eggs that can be removed and placed in a dish. Mock fertilization, or activation, of these eggs can be achieved with electrical stimulation, which triggers a sudden influx of calcium ions into the eggs. Activated eggs complete meiosis II and begin a mitotic cell cycle despite the absence of a sperm nucleus.

Gentle centrifugation of activated frog eggs breaks them apart and stratifies their contents, allowing the isolation of essentially undiluted egg cytoplasm. When *Xenopus* sperm nuclei stripped of their membranes are added to this cytoplasm in a test tube, the sperm chromosomes decondense and are packaged in a nuclear envelope. Replication of the sperm DNA then occurs, after which the extracts proceed through mitosis and segregate the duplicated sperm chromosomes. This cycle of S and M phases can repeat itself in the test tube for several rounds, with cell-cycle lengths about twice that of a normal embryonic cell cycle.

The ability to reconstruct an early embryonic cell cycle in a test tube provides an unparalleled system for the biochemical dissection of basic cell-cycle control. Foreign proteins, chemicals or inhibitory antibodies can be added to these extracts to test their effects on cell-cycle progression. Specific proteins can be removed from extracts by using antibodies and then added back in pure form, allowing an assessment of the protein's normal function.

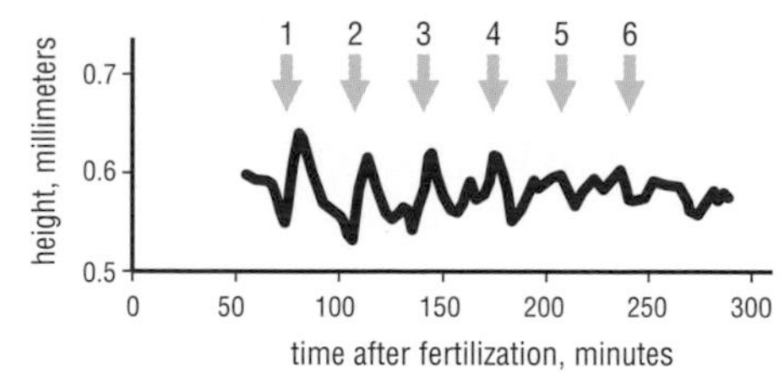

Figure 2-10 Cell-cycle oscillations in a denucleated *Xenopus* egg fragment Using a loop of fine human hair, a fertilized *Xenopus* egg can be divided into two unconnected halves, only one of which contains the nucleus. Cleavage divisions occur in the nucleated half with normal timing. In the non-nucleated half, divisions do not occur, but the surface of the egg exhibits periodic contractions, resulting in the indicated changes in the height of the non-nucleated fragment. These periodic contractions occur with about the same timing as the cleavage divisions occurring in the nucleated half of the egg (these divisions are indicated by the numbers 1–6). Thus, the cytoplasm of the non-nucleated half of the egg contains an autonomous oscillator that continues to cycle normally even when major cell-cycle events are prevented by removal of the nucleus. Adapted from Hara, K. *et al.*: *Proc. Natl Acad. Sci. USA* 1980, **77**:462–466.

Figure 2-11 Oocyte maturation and fertilization in *Xenopus* The oocyte arrests after meiotic S phase and grows to its final size. Progesterone then induces oocyte maturation: the oocyte completes the first meiotic division (resulting in the first polar body), enters the second meiotic division and arrests in metaphase II. Fertilization triggers the completion of meiosis II, resulting in a second polar body. Fusion of sperm and egg nuclei generates a diploid nucleus, and rapid cleavage divisions begin. The oocyte does not contain centrosomes, and the single centrosome of the fertilized egg is provided by the sperm.

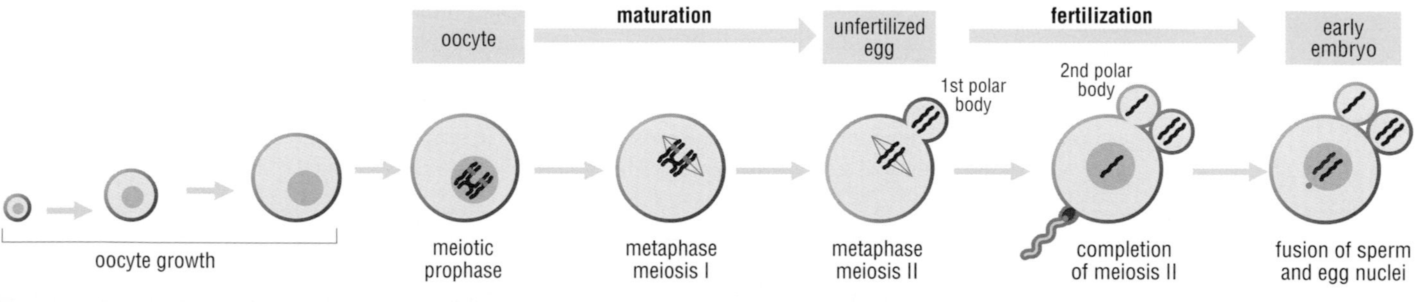

References

Hara, K. *et al.*: **A cytoplasmic clock with the same period as the division cycle in *Xenopus* eggs.** *Proc. Natl Acad. Sci. USA* 1980, **77**:462–466.

Lohka, M.J. *et al.*: **Purification of maturation-promoting factor, an intracellular regulator of early mitotic events.** *Proc. Natl Acad. Sci. USA* 1988, **85**:3009–3013.

Masui, Y. and Markert, C.L.: **Cytoplasmic control of nuclear behavior during meiotic maturation of frog oocytes.** *J. Exp. Zool.* 1971, **177**:129–146.

Murray, A.W. and Hunt, T.: *The Cell Cycle: An Introduction* (Freeman, New York, 1993).

Wolpert, L. *et al.*: *Principles of Development* 2nd ed. (Oxford University Press, Oxford, 2002).

2-4 The Fruit Fly *Drosophila melanogaster*

Drosophila allows genetic analysis of cell-cycle control in metazoans

Drosophila has several advantages as a model system for the study of cell-cycle control. It has a 2-week generation time and is grown easily in controlled laboratory conditions. The components and specialized features of its cell-cycle control system—particularly those unique to multicellular animals—are remarkably similar to those of humans. Most importantly, it is possible in *Drosophila* to identify and characterize genes that regulate cell-cycle progression.

The genome of *Drosophila* contains about 14,000 genes, about twice as many as yeast and half the number in human cells. Numerous techniques are available for the isolation of flies carrying defects in known genes. Genes can also be expressed ectopically in specific tissues or under the control of inducible promoters. For the purposes of classical genetics, flies are easily screened for mutant genes, although their diploid chromosome content makes the subsequent analysis and isolation of these genes more complex than in haploid yeast.

Cells of the early *Drosophila* embryo divide by a simplified cell cycle

Early development in *Drosophila* is understood in great detail and involves several unusual cell cycles. The fly therefore provides an excellent system in which to study changes in cell-cycle control during the stages of embryogenesis.

The *Drosophila* egg is a sausage-shaped cell about 400 μm long and 160 μm in diameter. As in many other animals, the fly egg is stocked by the mother with abundant copies of the components needed for cell division, so that the embryo can develop quickly—without the need for external nutrients—into a form that can move and feed.

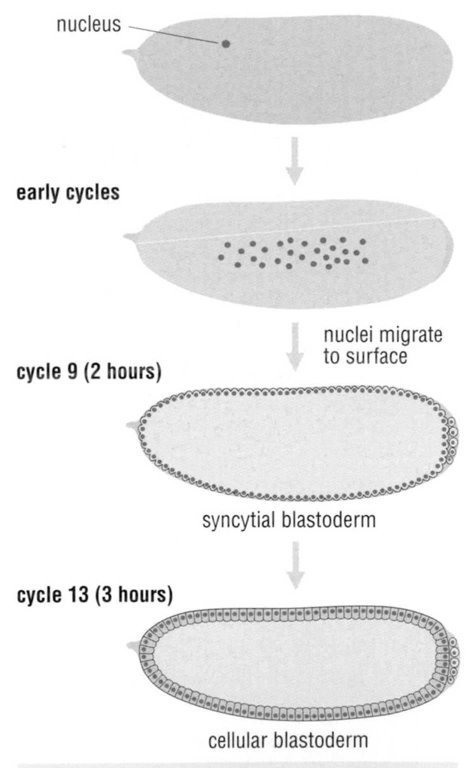

Figure 2-12 Early divisions in *Drosophila* After fertilization, rapid nuclear divisions occur without cytoplasmic division, resulting in the formation of a syncytium. After about nine nuclear divisions, nuclei begin to migrate to the surface of the embryo to form the syncytial blastoderm. Membranes grow inward from the cell surface and surround the nuclei at the end of the 13th division, resulting in the formation of the cellular blastoderm. About 15 pole cells form a separate group of cells in the posterior end of the embryo and will eventually give rise to the germ cells.

After fertilization, the fusion of sperm and egg nuclei results in a zygote nucleus that undergoes a series of rapid and synchronous divisions that do not contain gap phases and last less than 10 minutes each. Because these early nuclear divisions are not accompanied by cleavage of the cytoplasm, they rapidly convert the embryo to a syncytium, in which many nuclei share the same cytoplasm (Figure 2-12). After about nine divisions, the nuclei move to the surface of the embryo and continue to divide. At the end of the 13th division, membranes grow inward from the cell surface to surround the nuclei. This large-scale form of cytokinesis, called **cellularization**, results in the formation of a layer of about 6,000 cells beneath the embryo's surface. This is the **cellular blastoderm**.

Analysis of mutant phenotypes in early *Drosophila* embryos is complicated by the presence of stocks of the proteins needed for progression through the first 13 divisions. These are laid down by the mother when the egg is being made. Only after these 13 divisions does zygotic gene expression become significant; thus, mutations in a zygotic gene may not produce a discernible phenotype until the maternal gene product has been used up. Maternal proteins are depleted at varying rates: some proteins (for example, Cdc25) are depleted rapidly in cycle 14, others (for example, cyclins) slightly later in cycles 15 and 16, and still others (for example, the gene regulator dE2F1) much later in mid to late larval development. Thus, to study a gene's function in the early embryo, it is necessary to generate mutant mothers that lay eggs deficient in specific maternal gene products.

Gap phases are introduced in late embryogenesis

Blastoderm formation is followed by **gastrulation**, the process by which cell movement and differentiation transform the simple cellular blastoderm into a complex, multilayered embryo. The cell cycle lengthens and assumes a more complex form. Cycles 14–16 acquire a G2, whose length varies in different regions of the developing embryo. As a result, mitoses 14–16 are not synchronous, but instead occur in programmed spatial and temporal patterns that are coordinated with the complex morphogenetic movements that occur at this stage (see Figure 2-3).

The first G1 is introduced in cycle 17. Most cells in the embryo arrest at this point and eventually enter specialized endoreduplication cycles in which they undergo multiple rounds of DNA replication without intervening mitoses. Some G1-arrested epithelial cells in the embryo are programmed for a different fate: they become the **imaginal cells** that will give rise to the major structures of the adult fly, as we discuss next.

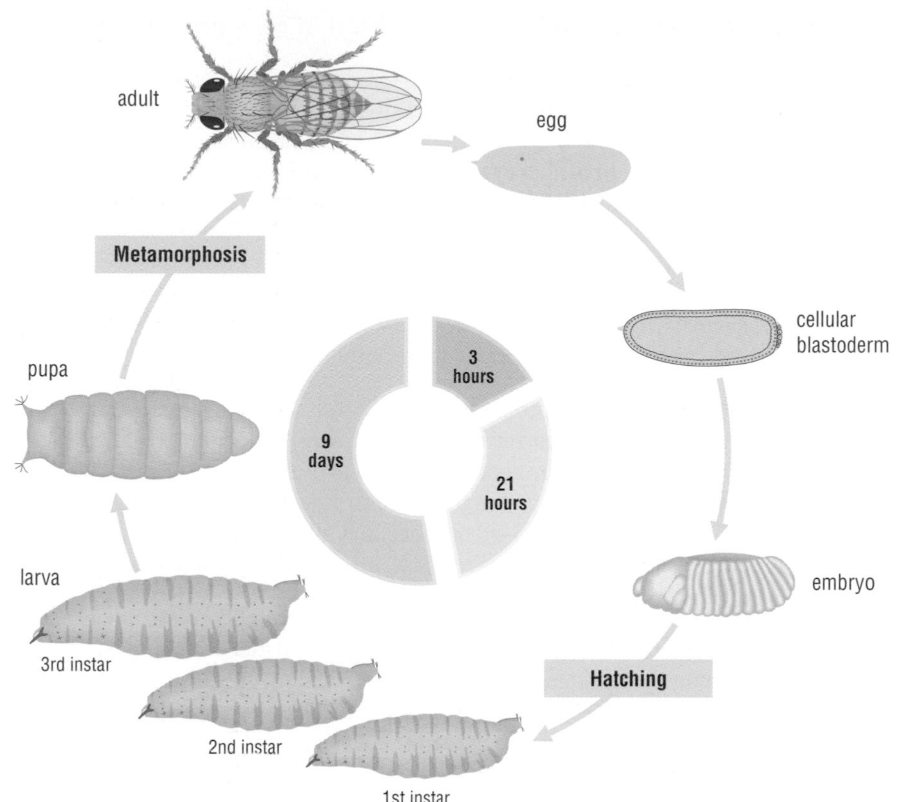

Figure 2-13 Life cycle of *Drosophila melanogaster* Early syncytial divisions are followed by cellularization, resulting in the single-layered cellular blastoderm, which is transformed by gastrulation and other processes into a segmented, multilayered embryo. Hatching of the embryo gives rise to the larva, which feeds, grows and goes through two molts (instars) before forming a pupa. Metamorphosis then leads to formation of the adult fly. An enormous amount of growth occurs during larval development, and these images are not drawn to scale.

Adult fly structures develop from imaginal cells

After embryogenesis, the embryo hatches as a feeding larva. This grows rapidly, not because of increased cell number but because endoreduplicating larval cells grow to an enormous size. The larva goes through two molts, or instars, before forming a pupa. After metamorphosis, the adult fly emerges (Figure 2-13).

Wings, eyes and several other adult structures are derived from small sheets of imaginal cells, called **imaginal discs**, in the larva (Figure 2-14). Each disc originates as a cluster of 10–50 G1-arrested imaginal cells in the embryo. When the larva begins to feed, these cells grow and divide until the disc contains several thousand cells, which then differentiate into an adult structure during metamorphosis. Some adult structures, such as the gut, are derived from imaginal cells that are not organized into discs.

The imaginal discs are useful for analyzing the role of regulatory proteins in the control of cell proliferation. The eye imaginal disc is particularly useful for this purpose and in screens for mutants with cell-cycle defects, because abnormal cell proliferation in this disc is not lethal but generates eye phenotypes that are easily observed in the adult. Another useful feature of the eye disc is that its cells divide and differentiate during the third larval stage in a synchronous wave, called the **morphogenetic furrow**, that passes across the disc—resulting in a linear array of cells in well defined stages of division and differentiation.

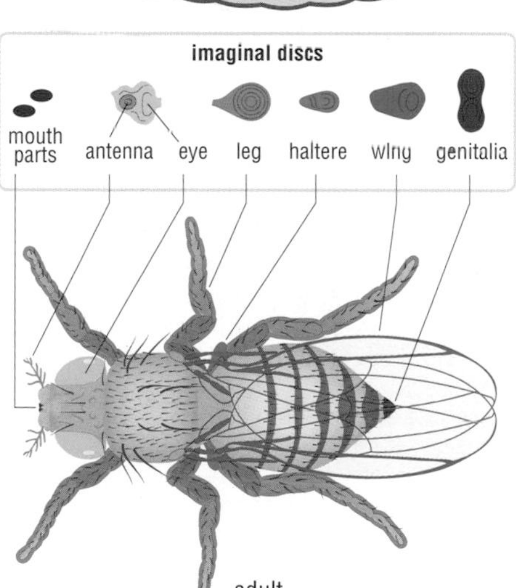

Figure 2-14 Imaginal discs in the *Drosophila* larva The imaginal discs are small sheets of epithelial cells in the larva. During metamorphosis, they give rise to a variety of adult structures.

Definitions

cellular blastoderm: early stage in *Drosophila* embryonic development, comprising a superficial epithelial layer of several thousand cells surrounding a yolky center.

cellularization: in insect development, the packaging of the nuclei of the syncytial embryo into individual cells, generating the **cellular blastoderm**.

gastrulation: cell movements that reorganize the blastula (**cellular blastoderm** in *Drosophila*) into an embryo with a gut surrounded by cell layers in place for the development of tissues and organs.

imaginal cells: cells of the *Drosophila* larva that are the precursors of adult structures.

imaginal discs: small sheets or pouches of **imaginal cells** in the *Drosophila* embryo, which will differentiate into major structures of the adult fly.

morphogenetic furrow: in the eye **imaginal disc** of *Drosophila*, a shallow indentation that results from a wave of cell differentiation that passes anteriorly across the disc during the third larval stage.

References

Edgar, B.A. and Lehner, C.F.: **Developmental control of cell cycle regulators: a fly's perspective.** *Science* 1996, **274**:1646–1652.

Lee, L.A. and Orr-Weaver, T.L.: **Regulation of cell cycles in *Drosophila* development: intrinsic and extrinsic cues.** *Annu. Rev. Genet.* 2003, **37**:545–578.

Wolpert, L. *et al.*: *Principles of Development* 2nd ed. (Oxford University Press, Oxford, 2002).

2-5 Mammalian Cell-Cycle Analysis

Mammalian cell-cycle control can be analyzed in cells growing in culture

The study of cell-cycle control in mammalian cells has one clear and unassailable advantage: direct relevance to human cells. Simple model systems are more effective for uncovering basic mechanisms of cell-cycle control, but the specialized features of the human cell cycle can be investigated only in cells from a human—or at least from a mouse or some other model mammal. Mammalian cells are particularly important for the study of the complex signaling networks that control the rate of cell division. Although the general features of these networks can often be revealed by genetic approaches in *Drosophila*, only in mammalian cells can we observe and dissect in detail the molecular components and signaling systems that are most relevant to humans. Understanding these systems is a key step toward the design of rational therapies to combat cancer and other diseases of uncontrolled cell proliferation.

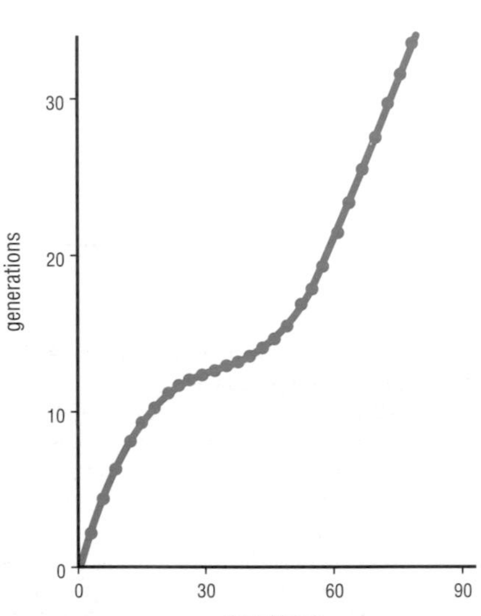

Figure 2-15 Proliferation of mouse embryonic fibroblasts in culture Cells from a mouse embryo were plated in a culture dish at low density (300,000 cells) in the presence of growth medium. Every 3 days, the cells were transferred to new dishes at the original density. This is the so-called 3T3 protocol of mouse cell culture, which has been used to generate numerous 3T3 cell lines. Cells were counted at each transfer to measure the number of population doublings occurring over a period of 90 days. The cells proliferated rapidly for the first ten generations but then slowed their rate of division after about 15 generations, a state known as senescence. Soon afterward, mutations led to the emergence from the population of an immortalized cell line that proliferated rapidly to the end of the experiment. Adapted from Todaro, G.J. and Green, H.: *J. Cell Biol.* 1963, **17**:299–313.

Mammalian cell-cycle analysis is best performed with cells that have been taken directly from the animal and grown in plastic dishes containing essential nutrients and serum, which provides the regulatory peptides required for cell division, growth and survival. These freshly removed cells—called **primary cells**—have a limited proliferative lifespan in culture: primary human cells, for example, stop dividing in culture after 25–50 divisions, and many rodent cell types stop dividing after even fewer divisions (Figure 2-15). This cell-cycle arrest is called **replicative senescence**.

Cell senescence results from at least two general mechanisms (discussed in detail in Chapter 11). First, many cell types (particularly rodent cells) are thought to gradually arrest the cell cycle in response to the nonphysiological conditions of the culture dish, in which cells lack normal cell–cell contacts and are usually bathed in abnormal levels of serum regulatory factors. There is evidence, for example, that many cell types can proliferate for longer periods, and sometimes indefinitely, if provided with a culture environment that more closely mimics that of the intact animal (Figure 2-16). Second, senescence in some cells, particularly human cells, is due to changes in the structure of the *telomere*, a complex DNA–protein structure that caps the end of each chromosome. Many human primary cells, for example, do not express *telomerase*, the enzyme that is primarily responsible for maintaining telomere structure. As a result, telomeres gradually degenerate during proliferation of these cells, until the DNA damage response system senses the damage and triggers cell-cycle arrest (see Figure 2-16). Telomere attrition is less important in cultured rodent cells, which often express higher levels of telomerase and have longer and more stable telomeres.

Mutations lead to immortalization and transformation of mammalian cells

When grown for long periods in culture, cells in a population can accumulate spontaneous mutations that prevent senescence, resulting in immortalized or established **cell lines** that grow indefinitely in culture. The nature of the mutations that cause immortalization varies from cell line to cell line. In many rodent cells, immortalization is a relatively rapid event (see Figure 2-15) that is achieved by mutations in components that arrest the cell cycle in response to nonphysiological culture conditions. In human cells, immortalization is more difficult because it requires multiple mutations that both inactivate these components and increase the activity of telomerase, allowing the formation of stable telomeres.

A wide range of immortal cell lines, derived from a variety of tissues, are used in cell-cycle analysis. Although immortalized cells possess defects in certain cell-cycle checkpoint mechanisms, they are immensely useful because, unlike primary cells, they provide an essentially unlimited supply of a cell type that is genetically homogeneous.

Definitions

cell line: a genetically homogeneous cell population that can proliferate indefinitely in culture. It is also called an immortalized or established cell line.

dominant negative: refers to a mutant gene product that inhibits the function of the wild-type gene product in a genetically dominant fashion, often by interfering with the ability of the wild-type protein to interact with other proteins.

knock out: to render a gene inactive by disrupting it in the animal, usually by replacing most of the gene with an inactivating insertion.

primary cells: cells taken directly from the tissue of an intact animal. They are generally susceptible to **replicative senescence** after several generations of proliferation in culture.

replicative senescence: general term for the eventual cessation of division by **primary cells** when grown in artificial culture conditions.

RNAi: see **RNA interference**.

RNA interference (RNAi): mechanism by which short fragments of double-stranded RNA lead to the degradation of homologous mRNAs.

transformed cell line: cell line that has acquired mutations that render it independent of normal proliferation controls, and typically capable of forming tumors when injected into mice.

Transformed cell lines have accumulated widespread genetic damage that not only results in immortalization but also disrupts the controls that normally limit the rate of cell proliferation. For example, transformed cells fail to stop dividing when cultured in the absence of serum or when deprived of attachment to a culture dish; they also form tumors when injected into mice. Such cells are not ideal for studying the normal control of cell proliferation, but are clearly useful for the direct analysis of molecular defects underlying the cancer phenotype.

Specific gene disruption is the ideal approach for assessing protein function in mammalian cells

A variety of methods are available for performing cytological and molecular analyses of cell-cycle control in cultured cell lines. The large size of mammalian cells makes them ideal for detailed microscopic analysis of cell-cycle events and for studying the subcellular location of regulatory molecules. Plasmid-borne foreign genes are readily expressed from natural or inducible promoters, and can be integrated into chromosomes to produce stable cell lines expressing various mutant versions of the desired protein.

The major weakness of mammalian cells as an experimental system is the lack of rapid and specific methods for blocking gene and protein function in the intact cell. Proteins are typically inhibited in the cell by injection of inhibitory antibodies or by the overexpression of a mutant form of the protein—called a **dominant negative**—that interferes with the function of the normal protein. These methods rarely offer the high specificity and rigorous interpretations possible with specific mutations of the gene encoding the protein.

A more specific approach to disrupting gene expression in mammalian cells is **RNA interference (RNAi)**, which triggers the destruction of a targeted mRNA in the intact cell. Cultured cells are treated with short double-stranded RNA fragments (called short interfering RNAs or siRNAs) carrying part of the sequence of the target RNA. With assistance from cellular proteins, the antisense strand of the siRNA targets the desired mRNA and cellular enzymes destroy it, thus preventing its translation and reducing the amount of the target protein in the cell.

Although more difficult than in yeast, specific gene disruption can be achieved in mammalian cells and in the whole animal. The standard approach is homologous recombination using specialized DNA vectors that target and disrupt, or **knock out**, the desired gene in cultured cells. This procedure is most frequently applied to cultured mouse embryonic stem cells (ES cells), enabling strains of mice with the target gene knock-out to be subsequently generated by transgenic techniques. It is then possible to assess the effect of the gene deletion in the context of the whole animal, which is particularly important in the study of cell-cycle regulatory proteins involved in the control of cell proliferation and cancer. The role of the disrupted gene in individual cells can also be analyzed in primary cell cultures, typically of embryonic fibroblasts, taken from the transgenic embryo.

However, gene disruption cannot be used to study gene products that are essential for cell-cycle progression. There are no simple methods for generating conditional mutant alleles of essential genes in mice or in mammalian cell lines, although RNAi can sometimes be a useful approach. In general, it is still difficult to apply rigorous genetic approaches to determining the normal function of essential genes in mammals.

(a)

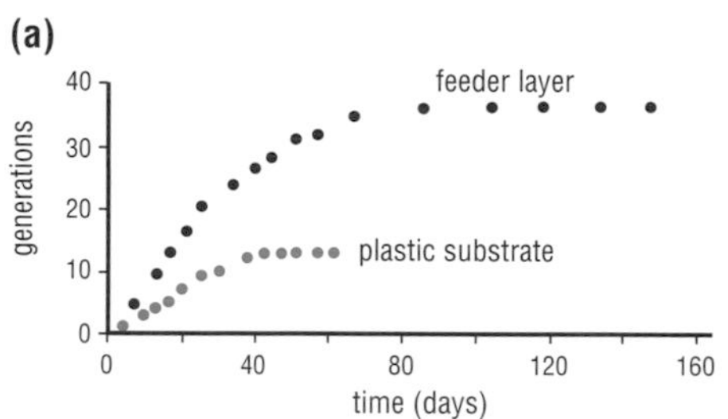

(b)

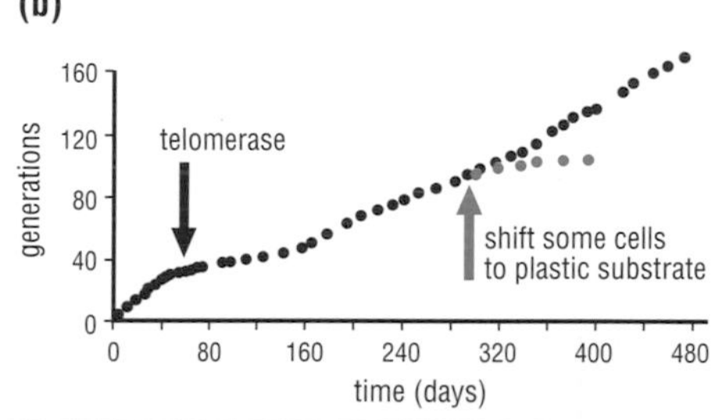

Figure 2-16 Effects of culture conditions and telomerase expression on human cell proliferation **(a)** Human keratinocytes (an epithelial cell type from skin) were grown in plastic culture dishes in the presence of a chemically defined medium (blue dots), leading to cell-cycle arrest after about 12–15 generations. In a parallel experiment, another batch of keratinocytes was cultured in dishes precoated with a layer of fibroblast cells, known as a feeder layer, in an attempt to better mimic the normal environment of the body (red dots). Keratinocytes growing on this layer did not stop dividing until about 35 generations, revealing the importance of culture conditions in the timing of senescence. **(b)** Human keratinocytes were cultured on a feeder layer of fibroblasts as in (a), until their rate of division began to slow (after about 35 generations). The cells were then engineered to produce human telomerase (red arrow). As a result, the population continued to proliferate indefinitely (red dots), indicating that telomere degeneration is the primary cause of senescence in human cells growing on feeder layers. After 280 days (blue arrow), some cells were transferred to uncoated plastic dishes instead of feeder layers, causing a cell-cycle arrest (blue dots). Thus, telomerase alone does not immortalize cells growing under suboptimal culture conditions. Adapted from Ramirez, R.D. *et al.*: *Genes Dev.* 2001, **15**:398–403.

References

Novina, C.D. and Sharp, P.A.: **The RNAi revolution.** *Nature* 2004, **430**:161–164.

Ramirez, R.D. *et al.*: **Putative telomere-independent mechanisms of replicative aging reflect inadequate growth conditions.** *Genes Dev.* 2001, **15**:398–403.

Todaro, G.J. and Green, H.: **Quantitative studies of the growth of mouse embryo cells in culture and their development into established lines.** *J. Cell Biol.* 1963, **17**:299–313.

2-6 Methods in Cell-Cycle Analysis

Cell-cycle position can be assessed by many approaches

Experimental analysis of the cell cycle generally requires a way of determining the cell-cycle stage of the cells being studied. The simplest approach is conventional light microscopy. As mentioned earlier (see section 2-1), the cell-cycle position of budding yeast and fission yeast can be estimated by the size of the bud and the length of the cell, respectively. Cultured mammalian cells tend to remain flat and attached to the dish during most of the cell cycle, but cells in M phase often reveal themselves by reducing their attachments and becoming round and refractile.

Microscopic analysis is made easier and more precise by fluorescent labeling of specific cellular components such as the chromosomes or mitotic spindle. Numerous fluorescent DNA dyes are available and can be added to cells that have been chemically fixed on a microscope slide. In a mammalian cell population, such dyes clearly reveal the condensed chromosomes of mitosis, and thus can be used to measure the **mitotic index**—the fraction of cells in a population that are in mitosis. Immunofluorescence methods are also very useful in cell-cycle analysis. Fixed cells can be incubated with an antibody that recognizes a specific cellular component, such as the mitotic spindle, and these antibodies can then be detected by a fluorescently tagged secondary antibody. Protein structures can also be fluorescently labeled in living cells. Typically, the gene encoding the desired protein is joined by molecular genetic methods to the gene encoding green fluorescent protein (GFP), a naturally fluorescent protein from jellyfish. The result is a fluorescent fusion protein that labels the desired intracellular structure. Such methods allow the continuous microscopic observation of the spindle or other structure in cells traversing the cell cycle.

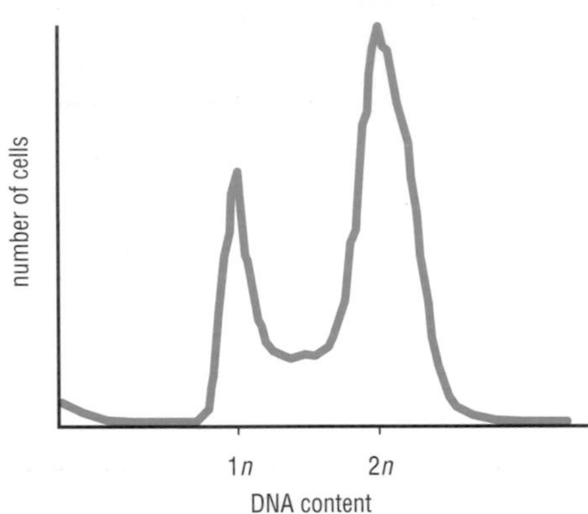

Figure 2-17 Analysis of cellular DNA content by flow cytometry A large population of haploid budding yeast cells was incubated with a fluorescent DNA dye and then passed through a flow cytometer, which measures the fluorescence content of each cell in the population. The number of cells with specific DNA contents is plotted as shown here. In an asynchronous, rapidly proliferating cell population, two peaks of cells are normally apparent. Cells in the left peak contain a single copy of each chromosome and are therefore in G1. Cells in the right peak contain twice as much DNA and are therefore in G2 or early M phase. Cells with DNA contents between the two peaks have partly duplicated their DNA and are therefore in S phase. The number of cells in the various positions reveals the fraction of cells in different cell-cycle phases. The actual chromosome content of cells will depend on whether the cells are haploid or diploid. The left peak in a haploid cell population, as shown here, represents a haploid state (called 1*n*), whereas the right peak represents the temporary diploid state (2*n*) in G2/M cells before they undergo mitosis. The chromosome content of a diploid population (see Figure 2-18) is doubled from diploid (2*n*) in G1 cells to tetraploid (4*n*) in G2/M cells. Courtesy of Liam Holt.

Another powerful method in cell-cycle analysis involves the precise measurement of cellular DNA content by **flow cytometry**. A large population of cells is treated with a fluorescent DNA dye and then injected into an instrument called a **flow cytometer**, which can rapidly measure the fluorescence intensity—and thus the DNA content—of every cell in the population (Figure 2-17). Cells in S phase can also be specifically labeled by treatment with bromodeoxyuridine (BrdU), an analog of the nucleotide thymidine. DNA that incorporates this nucleotide during S phase can be detected with fluorescent antibodies against BrdU, and a microscope or flow cytometer can reveal the fraction of cells that are actively synthesizing DNA during the treatment period.

Cell populations can be synchronized at specific cell-cycle stages

It is often necessary in cell-cycle research to synchronize a population of cells at a particular stage in the cell cycle, so that the cellular or biochemical features of that stage can be analyzed. There are various methods for doing this. In general, these involve treatment of the cells with a chemical or hormone, or a change in an environmental condition (such as temperature in the case of temperature-sensitive mutants) that blocks cell-cycle progression at some specific point. Ideally, these arrests are reversible, so that removal of the arresting condition allows the progression of cells into the next cell-cycle stage.

Many cultured mammalian cells can be arrested in a quiescent G1-like state by depleting the culture medium of serum factors that drive growth and division. Adding serum back to the medium results in the gradual progression of cells back into the cell cycle and S phase. Similarly, budding yeast cells can be arrested reversibly in a G1-like state by treatment with mating pheromone. The quiescent G1 state produced by these treatments is different from the relatively transient G1 state that occurs in cells traversing directly from mitosis to S phase.

Definitions

double thymidine block: a method for synchronously arresting a mammalian cell population at the beginning of S phase. Asynchronous cells are treated with thymidine, which causes arrest throughout S phase. These cells are released from arrest to allow progression out of S phase. A second thymidine treatment is then used to arrest all cells at the beginning of the subsequent S phase.

FACS: see **fluorescence-activated cell sorter**.

flow cytometer: instrument through which a stream of cells is passed and their fluorescence measured in the technique of **flow cytometry**.

flow cytometry: technique used to enumerate and analyze a sample of cells by incubating them with one or more fluorescently labeled antibodies and/or other molecules that can bind to cellular components and measuring the fluorescence intensity of each fluor for each cell. It is used to count the numbers of cells of different types, or at different stages in development or the cell cycle.

fluorescence-activated cell sorter (FACS): a modified **flow cytometer** that sorts individual cells into different containers according to their fluorescence.

mitotic index: fraction of cells in a population that are undergoing mitosis.

Chemical inhibitors of DNA synthesis and spindle assembly are also used to synchronize cell populations at specific points in the cycle (Figure 2-18). Numerous drugs, including thymidine and hydroxyurea, block DNA synthesis by inhibiting the synthesis of specific nucleotides, resulting in a reversible arrest in S phase with partly synthesized DNA. Drugs that inhibit microtubule function, such as nocodazole and benomyl, block the normal assembly of a spindle, which causes cells to arrest in early mitosis.

These methods suffer from two problems. First, cell-cycle progression never occurs at the same rate in all cells in the population, and so the synchrony of cell-cycle progression is lost soon after release from an arrest point. This is especially true of the prolonged process by which serum-starved mammalian cells return to the cycle. Second, artificially induced arrests, such as those produced by conditional mutations or inhibitors of DNA synthesis and spindle function, are never equivalent to the natural state of a cell passing through that stage of the cell cycle. Cells arrested by DNA synthesis inhibitors, for example, stop replicating their DNA but continue to grow and may also continue to make preparations for mitosis.

Non-invasive methods are therefore preferable in some cases. Useful information can often be gained simply from microscopic analysis of single cells in an untreated, asynchronous population. Alternatively, cells in specific stages can be purified from an asynchronous population by gentle methods. In the technique of centrifugal elutriation, for example, a specialized centrifuge is used to separate cells on the basis of their size: this allows large numbers of newly formed G1 cells to be obtained from an untreated cell population. Partly purified subpopulations of cells can also be obtained with a **fluorescence-activated cell sorter (FACS)**, a flow cytometer that sorts cells into different test tubes on the basis of their fluorescence content. If cells are labeled with a DNA dye, for example, then cells with G1 DNA content can be separated from cells with G2/M DNA content. Finally, as mentioned above, cultured mammalian cells typically lose their attachment to the plastic dish during mitosis, and so unperturbed mitotic cells can be released into the medium of these cells by tapping the culture dish; this is called mitotic shake-off.

Complete understanding of cell-cycle control mechanisms requires the analysis of protein structure and enzymatic behavior

The future of cell-cycle research lies not only in studies of cells and organisms but also in the analysis of protein structure and function. Many components of the cell-cycle control system are now being analyzed at the level of high-resolution three-dimensional structure, and these structural studies are accompanied by increasingly sophisticated studies of the chemical mechanisms underlying the function and regulation of the many enzymes that drive the cell through the steps of reproduction. In addition, considerable effort is being devoted to understanding how the many components of the cell-cycle control system interact in networks that generate complex behaviors—autonomous Cdk oscillations, for example. These important issues lie at the heart of cell-cycle control and form the central theme of the next chapter.

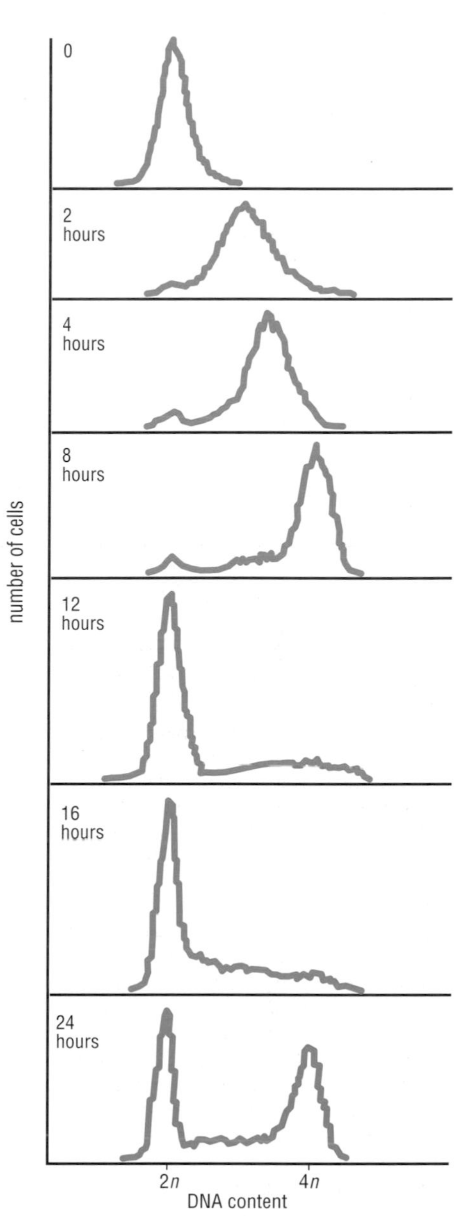

Figure 2-18 Synchronous progression through the cell cycle A human HeLa cell line was treated with a **double thymidine block** to arrest cells at the beginning of S phase. This method begins with treatment of cells with thymidine, which inhibits DNA synthesis and results in a cell population that is arrested throughout S phase. Cells are then released from this arrest for several hours. After all cells have completed S phase, they are treated with thymidine for a second time. All cells then arrest at the beginning of the next S phase. In the experiment shown here, the second thymidine treatment was washed out at time zero, and DNA content was then measured in the population at the indicated times. The cells start with unreplicated ($2n$) DNA content and then progress through S phase and achieve a $4n$ DNA content after about 8 hours. Progression through M phase results in an abrupt return to a $2n$ DNA content after 12 hours, and the cells then enter the next cell cycle. After 24 hours, the cell population is becoming asynchronous. Courtesy of Pei Jin.

References

Baserga, R.: *The Biology of Cell Reproduction* (Harvard University Press, Cambridge, MA, 1985).

Mitchison, J.M.: *The Biology of the Cell Cycle* (Cambridge University Press, Cambridge, 1971).

Murray, A.W. and Hunt, T.: *The Cell Cycle: An Introduction* (Freeman, New York, 1993).

3

The Cell-Cycle Control System

The cell-cycle control system is composed of a series of biochemical switches that trigger the events of the cycle in the correct order. The key components of this system are the cyclin-dependent kinases and their regulators, which are assembled into a robust and versatile regulatory network that is responsive to a variety of intracellular and extracellular information.

3-0 Overview: The Cell-Cycle Control System

The cell-cycle control system is a complex assembly of oscillating protein kinase activities

The cell-cycle control system is the regulatory network that controls the order and timing of cell-cycle events. A series of biochemical switches triggers progression through the three major regulatory checkpoints of the cell cycle: Start, which defines the entry into the cycle in late G1; the G2/M checkpoint, where entry into mitosis is controlled; and the metaphase-to-anaphase transition, where the final events of mitosis are initiated. We will see in this chapter how the biochemical switches that comprise the cell-cycle control system are put together and regulated.

The central components of the cell-cycle control system are the cyclin-dependent kinases (Cdks). As the cell progresses through the cycle, abrupt changes in the enzymatic activities of these kinases lead to changes in the phosphorylation state, and thus the state of activation, of proteins that control cell-cycle processes. Concentrations of Cdk proteins are constant throughout the cell cycle; oscillations in their activity depend primarily on corresponding oscillations in levels of the regulatory subunits known as cyclins, which bind tightly to Cdks and stimulate their catalytic activity. Different cyclin types are produced at different cell-cycle stages, resulting in the formation of a series of cyclin–Cdk complexes. These complexes govern distinct cell-cycle events and we will therefore call them G1–, G1/S–, S– and M–Cdks. In this chapter we will be concerned chiefly with the latter three complexes, which control passage through the three major checkpoints (Figure 3-1).

Multiple regulatory mechanisms govern Cdk activity during the cell cycle

Each cyclin–Cdk complex promotes the activation of the next in the sequence, thus ensuring that the cycle progresses in an ordered fashion. The precise timing of changes in Cdk activity is governed by multiple mechanisms. Cyclin concentrations are particularly important, and we will see how these are regulated by a combination of changes in cyclin gene expression and rates of cyclin degradation. The activity of cyclin–Cdk complexes is further modulated by the addition or removal of inhibitory phosphorylation, and by changes in the levels of Cdk inhibitor proteins.

The G1/S–, S– and M–Cdks are inactive in G1, ensuring that cell-cycle events are not triggered inappropriately before the cell commits itself to a new cell cycle. Three inhibitory mechanisms suppress the activity of these Cdks during G1. Two of these affect cyclins: expression of the major cyclin genes is suppressed by inhibitory gene regulatory proteins, and rates of cyclin degradation are greatly increased through the activation of an important protein complex called the *anaphase-promoting complex* or *APC*, which specifically targets the S and M cyclins (but not the G1/S cyclins) for degradation (see Figure 3-1). The third is the presence of high concentrations of Cdk inhibitors in G1.

Entry into a new cell cycle begins when signals from outside the cell (mitogens, for example) and inside (systems monitoring cell growth, for example) trigger a combination of events that unleash G1/S- and S-cyclin gene expression and activation of G1/S–Cdks. G1/S–Cdk activity rises immediately because the G1/S cyclins are not targeted by the APC and because the G1 Cdk inhibitor proteins either do not act on G1/S–Cdks (in yeast and flies) or are removed from G1/S–Cdks by other mechanisms (in mammals). The G1/S–Cdks directly initiate some early cell-cycle events, but their major function is to activate the S–Cdks—primarily by triggering the destruction of Cdk inhibitor proteins and the inactivation of the APC, both of which restrain S–Cdk activity in G1. S–Cdks then phosphorylate the proteins that initiate chromosome duplication, thereby launching S phase. As S phase proceeds, G1/S–Cdks promote their own inactivation by stimulating destruction of G1/S cyclins, and G1/S-cyclin gene expression is reduced.

Toward the end of S phase, M-cyclin gene expression is switched on and M-cyclin concentration rises, leading to the accumulation of M–Cdk complexes during G2. In most cell types, these complexes are initially held in an inactive state by inhibitory phosphorylation of the Cdk subunit. At the onset of mitosis, the abrupt removal of this phosphorylation leads to the activation of all M–Cdks. These then trigger progression through the G2/M checkpoint. Spindle assembly and other early mitotic events lead to the alignment of duplicated sister chromatids on the mitotic spindle in metaphase.

In addition to driving the cell to metaphase, M–Cdks eventually stimulate activation of the APC, which triggers the metaphase-to-anaphase transition. A central function of the APC at

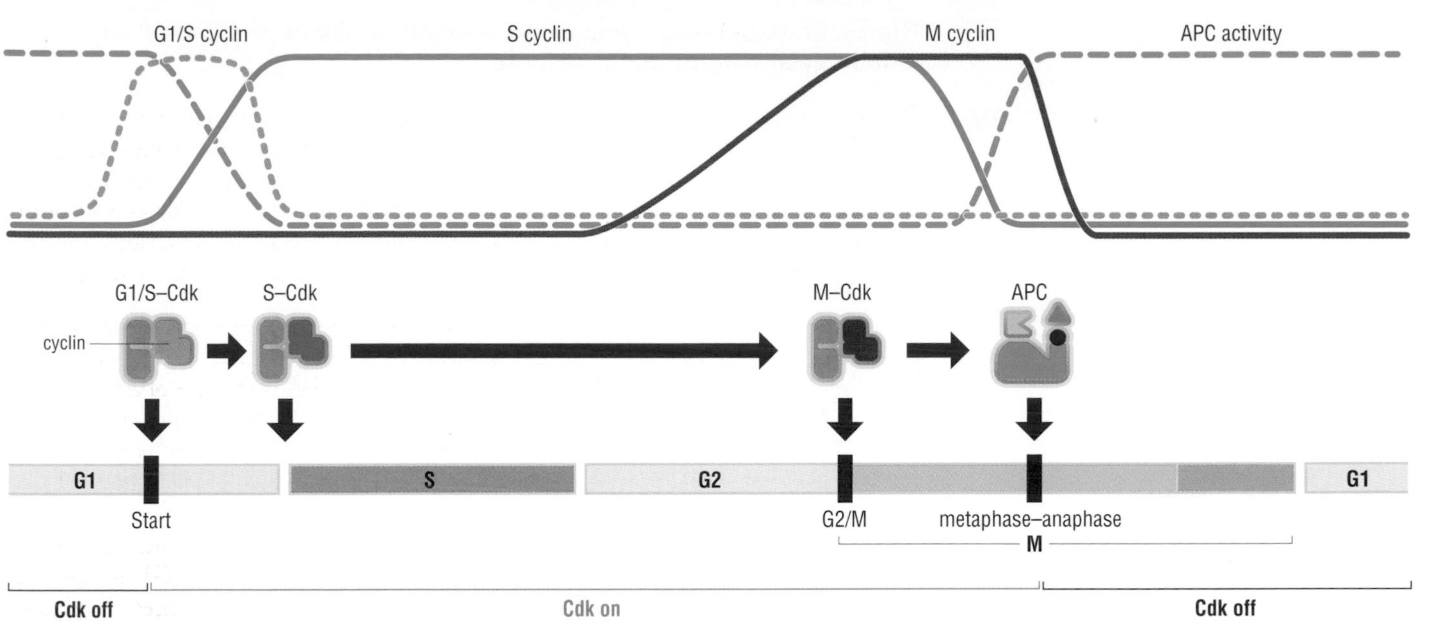

Figure 3-1 A simplified view of the cell-cycle control system Levels of the three major cyclin types oscillate during the cell cycle (top), providing the basis for oscillations in the cyclin–Cdk complexes that drive cell-cycle events (bottom). In general, Cdk levels are constant and in large excess over cyclin levels; thus, cyclin–Cdk complexes form in parallel with cyclin levels. The enzymatic activities of cyclin–Cdk complexes also tend to rise and fall in parallel with cyclin levels, although in some cases Cdk inhibitor proteins or phosphorylation introduce a delay between the formation and activation of cyclin–Cdk complexes. Formation of active G1/S–Cdk complexes commits the cell to a new division cycle at the Start checkpoint in late G1. G1/S–Cdks then activate the S–Cdk complexes that initiate DNA replication at the beginning of S phase. M–Cdk activation occurs after the completion of S phase, resulting in progression through the G2/M checkpoint and assembly of the mitotic spindle. APC activation then triggers sister-chromatid separation at the metaphase-to-anaphase transition. APC activity also causes the destruction of S and M cyclins and thus the inactivation of Cdks, which promotes the completion of mitosis and cytokinesis. APC activity is maintained in G1 until G1/S–Cdk activity rises again and commits the cell to the next cycle. This scheme serves only as a general guide and does not apply to all cell types.

this stage is to stimulate the destruction of proteins that hold the sister chromatids together. The APC also causes destruction of S and M cyclins, resulting in the inactivation of all major Cdk activities in late mitosis. Decreased S- and M-cyclin gene expression and increased production of Cdk inhibitor proteins also occurs in late mitosis. The resulting inactivation of Cdks allows dephosphorylation of their mitotic targets, which is required for spindle disassembly and the completion of M phase. Low levels of Cdk activity are maintained until late in the following G1, when rising G1/S–Cdk activities again commit the cell to a new cycle. The actions of the different cyclin–Cdk complexes and the APC in the course of the cell cycle are summarized in Figure 3-1.

The cell-cycle control system generates robust, switch-like and adaptable changes in Cdk activity

The cyclin–Cdk complexes and other regulators that drive the cell cycle are assembled into a highly interconnected regulatory system whose effectiveness is enhanced by a number of important features. First, the cell-cycle control system includes feedback loops and other regulatory interactions that lead to irreversible, switch-like activation and inactivation of most cyclin–Cdk complexes. Thus, cell-cycle events are generally triggered in an all-or-none fashion, allowing the cell to avoid the damage that might result if events were only partly initiated. Second, regulatory interactions between the different cyclin–Cdk switches ensure that they are properly ordered and coordinated with each other. Third, the cell-cycle control system is highly robust: the activation and inactivation of every cyclin–Cdk switch is governed by multiple mechanisms, allowing the system to operate well under a variety of conditions and even if some components fail. Finally, the system is adaptable, allowing the timing of the major regulatory switches to be adjusted by regulatory inputs from various intracellular and extracellular factors.

This chapter provides a general overview of the biochemical features of the major components of the cell-cycle control system, and describes the key principles underlying the assembly of these components into robust biochemical switches that oscillate with the appropriate timing. Detailed discussions of how the cell-cycle control system operates at specific cell-cycle stages and in specific organisms are found in later chapters.

3-1 Cyclin-Dependent Kinases

The cyclin-dependent kinases are a small family of enzymes that require cyclin subunits for activity

The cyclin-dependent kinases (Cdks) are a family of serine/threonine protein kinases whose members are small proteins (~34–40 kDa) composed of little more than the catalytic core shared by all protein kinases. By definition, all Cdks share the feature that their enzymatic activation requires the binding of a regulatory cyclin subunit. In most cases, full activation also requires phosphorylation of a threonine residue near the kinase active site.

Although originally identified as enzymes that control cell-cycle events, members of the Cdk family are involved in other cellular processes as well. Animal cells, for example, contain at least nine Cdks, only four of which (Cdk1, 2, 4 and 6) are involved directly in cell-cycle control (Figure 3-2). Another family member (Cdk7) contributes indirectly by acting as a *Cdk-activating kinase* (*CAK*) that phosphorylates other Cdks, as we discuss in section 3-3. Cdks are also components of the machinery that controls basal gene transcription by RNA polymerase II (Cdk7, 8 and 9) and are involved in controlling the differentiation of nerve cells (Cdk5).

We will focus on the small number of Cdks for which there is clear evidence of a direct role in cell-cycle control (see Figure 3-2). In the fission yeast *Schizosaccharomyces pombe* and the budding yeast *Saccharomyces cerevisiae* (see section 2-1), all cell-cycle events are controlled by a single essential Cdk called Cdk1. Cell-cycle events in multicellular eukaryotes are controlled by two Cdks, known as Cdk1 and Cdk2, which operate primarily in M phase and S phase, respectively. Animal cells also contain two Cdks (Cdk4 and Cdk6) that are important in regulating entry into the cell cycle in response to extracellular factors.

Cdk function has been remarkably well conserved during evolution. It is possible, for example, for yeast cells to proliferate normally when their gene for Cdk1 is replaced with the human one. This and other evidence clearly illustrates that Cdk function, and thus the function of the cell-cycle control system, has remained fundamentally unchanged over hundreds of millions of years of eukaryotic evolution.

Cdks exert their effects on cell-cycle events by phosphorylating a large number of proteins in the cell. During mitosis in particular, when many aspects of cellular architecture and metabolism are altered, Cdks phosphorylate hundreds of distinct proteins. These Cdk substrates are phosphorylated at serine or threonine residues in a specific sequence context that is recognized by the active site of the Cdk protein. In most cases, the target serine (S) or threonine (T) residue is followed by a proline (P); it is also highly favorable for the target residue to have a basic amino acid two positions after the target residue. The typical phosphorylation sequence for Cdks is [S/T*]PX[K/R], where S/T* indicates the phosphorylated serine or threonine, X represents any amino acid and K/R represents the basic amino acid lysine (K) or arginine (R).

The active site of cyclin-dependent kinases is blocked in the absence of cyclin

All protein kinases have a tertiary structure comprising a small amino-terminal lobe and a larger carboxy-terminal lobe. ATP fits snugly in the cleft between the lobes, in such a way that the phosphates are oriented outwards, toward the mouth of the cleft. The protein substrate

Cyclin-Dependent Kinases

Species	Name	Original name	Size (amino acids)	Function
S. cerevisiae	Cdk1	Cdc28	298	all cell-cycle stages
S. pombe	Cdk1	Cdc2	297	all cell-cycle stages
D. melanogaster	Cdk1	Cdc2	297	M
	Cdk2	Cdc2c	314	G1/S, S, possibly M
	Cdk4	Cdk4/6	317	G1 promotes growth
X. laevis	Cdk1	Cdc2	301	M
	Cdk2		297	S, possibly M
H. sapiens	Cdk1	Cdc2	297	M
	Cdk2		298	G1/S, S, possibly M
	Cdk4		303	G1
	Cdk6		326	G1

Figure 3-2 Table of cyclin-dependent kinases that control the cell cycle

Definitions

L12 helix: small alpha helix adjacent to the **T-loop** in the active site of Cdk2 (residues 147–151), which changes structure to a beta strand upon cyclin binding.

PSTAIRE helix: alpha helix in the amino-terminal lobe of Cdks (also known as the $\alpha1$ helix), which interacts with cyclin and is moved inward upon cyclin binding, resulting in reorientation of key active-site residues. The name of this helix comes from its amino-acid sequence, which is conserved among all major Cdks.

T-loop: flexible loop adjacent to the active site of Cdks, named for the threonine whose phosphorylation is required for maximal activity. Sometimes called the activation loop.

binds at the entrance of the cleft, interacting mainly with the surface of the carboxy-terminal lobe. Nearby residues catalyze the transfer of the terminal γ-phosphate of ATP to a hydroxyl oxygen in the protein substrate.

Cdks have the same two-lobed structure as other protein kinases (Figure 3-3), but with two modifications that make them inactive in the absence of cyclin. These modifications have been revealed by detailed crystallographic studies of the structure of human Cdk2. First, a large, flexible loop—the **T-loop** or activation loop—rises from the carboxy-terminal lobe to block the binding of protein substrate at the entrance of the active-site cleft. Second, in the inactive Cdk several important amino-acid side chains in the active site are incorrectly positioned, so that the phosphates of ATP are not ideally oriented for the kinase reaction. Cdk activation therefore requires extensive structural changes in the Cdk active site.

Two alpha helices make a particularly important contribution to the control of Cdk activity. The highly conserved **PSTAIRE helix** of the upper kinase lobe (also known as the α1 helix) interacts directly with cyclin and moves inward upon cyclin binding, causing the reorientation of residues that interact with the phosphates of ATP. The small **L12 helix**, just before the T-loop in the primary sequence, changes structure to become a beta strand upon cyclin binding, also contributing to reconfiguration of the active site and T-loop. We discuss the structural basis of Cdk activation in more detail in section 3-4. First, we will describe the cyclins and other regulators that influence activation.

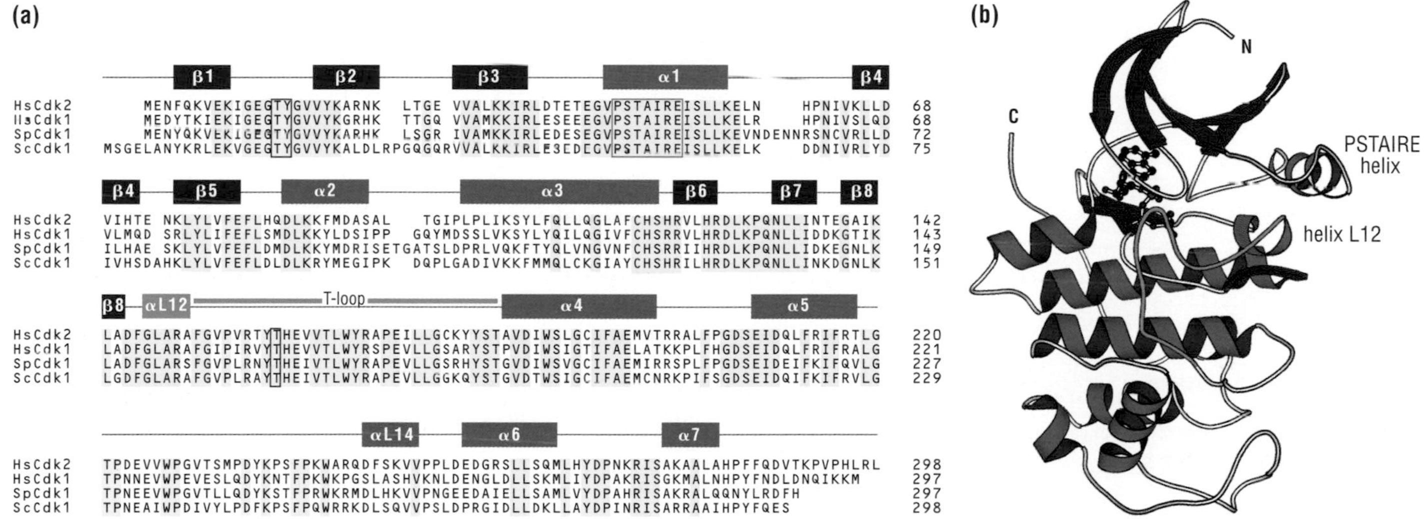

Figure 3-3 Cyclin-dependent kinase structure (a) Amino-acid sequences of major Cdks controlling the cell cycle in humans (*H. sapiens* (Hs) Cdk1 and Cdk2) and yeast (*S. pombe* (Sp) Cdk1 and *S. cerevisiae* (Sc) Cdk1). Yellow residues are identical in all four kinases. Above the alignment, secondary structure elements in human Cdk2 are shown for comparison with the tertiary structure in panel (b). Key landmarks are highlighted, including the PSTAIRE or α1 helix, the inhibitory phosphorylation sites Thr 14 and Tyr 15, the activating phosphorylation site (Thr 160 in human Cdk2) and the T-loop or activation loop where Thr 160 is found. **(b)** Tertiary structure of human Cdk2, determined by X-ray crystallography. Like other protein kinases, Cdk2 is composed of two lobes: a smaller amino-terminal lobe (top) that is composed primarily of beta sheet and the PSTAIRE helix, and a large carboxy-terminal lobe (bottom) that is primarily made up of alpha helices. The ATP substrate is shown as a ball-and-stick model, located deep within the active-site cleft between the two lobes. The phosphates are oriented outward, toward the mouth of the cleft, which is blocked in this structure by the T-loop (highlighted in green). (PDB 1hck)

References

De Bondt, H.L. *et al.*: **Crystal structure of cyclin-dependent kinase 2.** *Nature* 1993, **363**:595–602.

Morgan, D.O.: **Cyclin-dependent kinases: engines, clocks, and microprocessors.** *Annu. Rev. Cell Dev. Biol.* 1997, **13**:261–291.

Ubersax, J.A. *et al.*: **Targets of the cyclin-dependent kinase Cdk1.** *Nature* 2003, **425**:859–864.

Cyclins are the key determinants of Cdk activity and can be classified in four groups

Cyclins are a diverse family of proteins whose defining feature is that they bind and activate members of the Cdk family. Most cyclins display dramatic changes in concentration during the cell cycle, which help to generate the oscillations in Cdk activity that form the foundation of the cell-cycle control system. The regulation of cyclin concentration, primarily by changes in cyclin gene expression and destruction of cyclins by proteolysis, is therefore of fundamental importance in cell-cycle control.

Cyclins, like their Cdk partners, are involved in a number of processes other than cell-cycle control. We will focus only on those cyclins that directly regulate Cdks controlling cell-cycle progression (Figure 3-4). These can be divided into four classes, based primarily on the timing of their expression and their functions in the cell cycle. Three of these classes—the **G1/S cyclins**, **S cyclins** and **M cyclins**—are directly involved in the control of cell-cycle events, as discussed earlier (section 3-0, Figure 3-1). The fourth class, the **G1 cyclins**, contributes to the control of cell-cycle entry in response to extracellular factors.

The G1/S cyclins (Cln1 and Cln2 in the budding yeast *S. cerevisiae*, cyclin E in vertebrates) oscillate during the cell cycle, rising in late G1 and falling in early S phase (see Figure 3-1). The primary function of G1/S cyclin–Cdk complexes is to trigger progression through Start and initiate the processes leading to DNA replication, principally by shutting down the various braking systems that suppress S-phase Cdk activity in G1. G1/S cyclins also initiate other early cell-cycle events, such as duplication of the centrosome in vertebrates and its equivalent, the spindle pole body, in yeast.

The rise of G1/S cyclins is accompanied by the appearance of the S cyclins (Clb5 and Clb6 in budding yeast, cyclin A in vertebrates), which form S cyclin–Cdk complexes that are directly responsible for stimulating DNA replication. Levels of S cyclin remain high throughout S phase, G2 and early mitosis, when they help promote early mitotic events in some cell types.

M cyclins (Clb1, 2, 3 and 4 in budding yeast, cyclin B in vertebrates) appear last in the sequence, their concentration rising as the cell approaches mitosis and peaking at metaphase. M cyclin–Cdk complexes are responsible for the striking cellular changes that lead to assembly of the mitotic spindle and the alignment of sister-chromatid pairs on the spindle at metaphase. Their destruction in anaphase brings on mitotic exit and cytokinesis.

The G1 cyclins, typified by Cln3 in budding yeast and cyclin D in vertebrates, help coordinate cell growth with entry into a new cell cycle and are required in many cell types to stimulate entry into a new cell cycle at the Start checkpoint. The G1 cyclins are unusual among cyclins in that their levels do not oscillate in a set pattern during the cell cycle, but increase gradually throughout the cycle in response to cell growth and external growth-regulatory signals.

The division of cyclins into four classes is based on their behavior in the cell cycles of yeast and of vertebrate somatic cells. This is a useful simplification but is not universally applicable. The same cyclin can have different functions or timing of expression in different cell types,

Cell-Cycle Control Cyclins

	Cyclin class (with Cdk partner)			
Species	**G1**	**G1/S**	**S**	**M**
S. cerevisiae	Cln3 (Cdk1)	Cln1,2 (Cdk1)	Clb5,6 (Cdk1)	Clb1,2,3,4 (Cdk1)
S. pombe	Puc1? (Cdk1)	Puc1, Cig1? (Cdk1)	Cig2, Cig1? (Cdk1)	Cdc13 (Cdk1)
D. melanogaster	cyclin D (Cdk4)	cyclin E (Cdk2)	cyclin E, A (Cdk2,1)	cyclin A, B, B3 (Cdk1)
X. laevis		cyclin E (Cdk2)	cyclin E, A (Cdk2,1)	cyclin A, B (Cdk1)
H. sapiens	cyclin D1,2,3 (Cdk4,6)	cyclin E (Cdk2)	cyclin A (Cdk2,1)	cyclin B (Cdk1)

Figure 3-4 **Table of major cyclin classes involved in cell-cycle control**

Definitions

G1 cyclins: cyclins that bind and activate Cdks that stimulate entry into a new cell cycle at Start; their concentration depends on the rate of cell growth or on growth-promoting signals rather than on the phase of the cell cycle.

G1/S cyclins: cyclins that activate Cdks that stimulate progression through Start; their concentration peaks in late G1.

M cyclins: cyclins that activate Cdks necessary for entry into mitosis; their concentration rises at the approach to mitosis and peaks in metaphase.

S cyclins: cyclins that activate Cdks necessary for DNA synthesis; their concentrations rise and remain high during S phase, G2 and early mitosis.

References

Brown, N.R. *et al.*: **The crystal structure of cyclin A.** *Structure* 1995, **3**:1235–1247.

Evans, T. *et al.*: **Cyclin: a protein specified by maternal mRNA in sea urchin eggs that is destroyed at each cleavage division.** *Cell* 1983, **33**:389–396.

Kim, K.K. *et al.*: **Three dimensional structure of human cyclin H, a positive regulator of the CDK-activating kinase.** *Nat. Struct. Biol.* 1996, **3**:849–855.

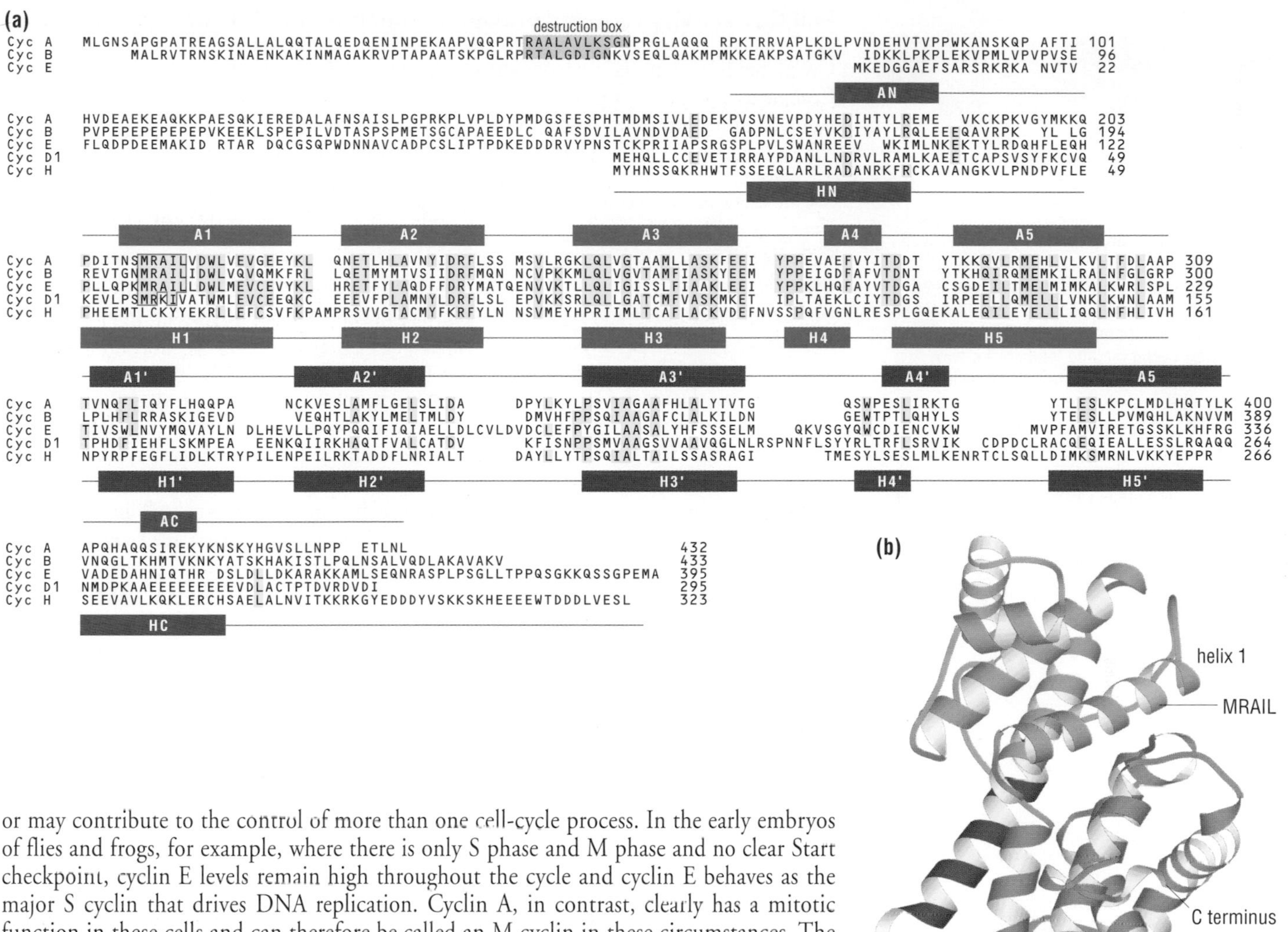

or may contribute to the control of more than one cell-cycle process. In the early embryos of flies and frogs, for example, where there is only S phase and M phase and no clear Start checkpoint, cyclin E levels remain high throughout the cycle and cyclin E behaves as the major S cyclin that drives DNA replication. Cyclin A, in contrast, clearly has a mitotic function in these cells and can therefore be called an M cyclin in these circumstances. The precise functions of cyclins acting at each phase of the cell cycle will be discussed in later chapters (S cyclins in Chapter 4, M cyclins in Chapter 5 and G1 and G1/S cyclins in Chapter 10).

Cyclins contain a conserved helical core

Members of the cyclin family are often quite distinct from each other in amino-acid sequence (Figure 3-5a). Sequence similarity among distantly related cyclins is concentrated in a region of about 100 amino acids known as the cyclin box, which is required for Cdk binding and activation.

Despite variations in their primary structure, all cyclins are thought to possess a similar tertiary structure known as the cyclin fold, which comprises a core of two compact domains each containing five alpha helices (Figure 3-5b). The first five-helix bundle corresponds to the conserved cyclin box. The second five-helix bundle displays the same arrangement of helices as the first, despite limited sequence similarity between the two subdomains.

Outside the cyclin fold, cyclin sequences are highly divergent (see Figure 3-5a). The length of the amino-terminal region is particularly variable, and contains regulatory and targeting domains that are specific for each cyclin class. For example, the amino-terminal regions of S and M cyclins (cyclins A and B in Figure 3-5a) contain short destruction-box motifs that target these proteins for proteolysis in mitosis.

The cyclin fold of twin five-helix bundles is also found in other proteins, including members of the pRB family, which regulate gene expression at the G1/S checkpoint. The cyclin fold is also found in the RNA polymerase II transcription factor, TFIIB. These structural relationships raise the intriguing possibility that cyclins and transcriptional regulators evolved from some common origin.

Figure 3-5 Cyclin structure (a) Amino-acid sequences and secondary structure of the major human cyclins. The alignment also includes cyclin H, which is not one of the key cell-cycle regulatory cyclins, to illustrate the conservation of secondary structure among distantly related cyclins. Residues conserved in three of the five cyclins are highlighted in yellow. Colored boxes above the alignment indicate the alpha helices in human cyclin A, for comparison with the structure in panel (b). Helices in cyclin H are indicated by boxes under the alignment. Sequence similarity is found primarily in the cyclin box, which contains the first five helices (A1–A5) of the protein core. Cyclins A and B also contain a destruction box (pink) that targets these proteins for degradation. **(b)** Tertiary structure of human cyclin A (lacking the amino-terminal 170 amino acids), showing the central core of two five-helix bundles, colored as in panel (a), with additional helices at the amino terminus (black) and carboxyl terminus (grey). The yellow region in helix 1 is the MRAIL sequence or *hydrophobic patch*, which contributes to the recognition of some substrates, as discussed in section 3-5. (PDB 1fin)

3-3 Control of Cdk Activity by Phosphorylation

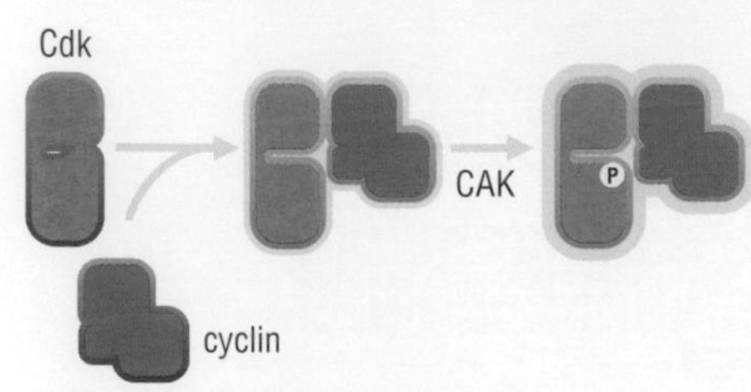

Figure 3-6 **Two steps in Cdk activation** Cyclin binding alone causes partial activation of Cdks, but complete activation also requires activating phosphorylation by CAK. In animal cells, CAK phosphorylates the Cdk subunit only after cyclin binding, and so the two steps in Cdk activation are usually ordered as shown here, with cyclin binding occurring first. Budding yeast contains a different version of CAK that can phosphorylate the Cdk even in the absence of cyclin, and so the two activation steps can occur in either order. In all cases, CAK tends to be in constant excess in the cell, so that cyclin binding is the rate-limiting step in Cdk activation.

Full Cdk activity requires phosphorylation by the Cdk-activating kinase

Cyclin binding alone is not enough to fully activate Cdks involved in cell-cycle control. Complete activation of a Cdk, and normal Cdk function in the cell, also requires phosphorylation of a threonine residue adjacent to the kinase active site (Figure 3-6). Phosphorylation at this site is catalyzed by enzymes called **Cdk-activating kinases (CAKs)**.

We are accustomed to thinking of phosphorylation as a reversible modification that is used to change enzyme activity under different conditions. Surprisingly, activating phosphorylation of Cdks does not seem to behave in this way. CAK activity is maintained at a constant high level throughout the cell cycle and is not regulated by any known cell-cycle control pathway. In addition, in mammalian cells, phosphorylation can occur only after cyclin is bound, whereas in budding yeast cells phosphorylation occurs before cyclin binding. In both cases, however, cyclin binding and not phosphorylation is the highly regulated, rate-limiting step in Cdk activation (see Figure 3-6). Activating phosphorylation can therefore be viewed as simply a post-translational modification that is required for enzyme activity. It is not clear why the requirement for Cdk phosphorylation has been so highly conserved during evolution if it is not exploited for regulatory purposes. The constant high level of activity of CAK may, however, be explained by the fact that it has a role in transcription as well as in cell-cycle regulation.

The identity of CAK varies dramatically in different species (Figure 3-7). In vertebrates and *Drosophila*, the major CAK is a trimeric complex containing a Cdk-related protein kinase known as Cdk7, along with its activating partner, cyclin H, and a third subunit, Mat1. In budding yeast, however, CAK is a small monomeric protein kinase known as Cak1, which bears only distant homology to the Cdks. The Cdk7 homolog in budding yeast (known as Kin28) does not possess CAK activity. Fission yeast seems to be intermediate between vertebrates and budding yeast, in that it uses two CAKs: one (a complex of Mcs6 and Mcs2) that is related to the vertebrate Cdk7–cyclin H complex, and another (Csk1) that more closely resembles budding yeast Cak1 (see Figure 3-7).

In addition to activating Cdks involved in cell-cycle control, CAK fulfills an entirely separate function in the control of basal gene transcription (Figure 3-8). Some of the vertebrate Cdk7–cyclin H–Mat1 complex is associated with TFIIH, a multimeric complex that is part of a giant protein assembly associated with RNA polymerase II at gene promoters and is involved in the control of polymerase II function or responses to DNA damage. In budding yeast, Cak1 indirectly influences basal transcription by phosphorylating and activating the Cdk7 homolog Kin28, which is also associated with yeast TFIIH. Fission yeast again seems to share features of both vertebrate and budding yeast systems (see Figure 3-8).

Cdk function is regulated by inhibitory phosphorylation by Wee1 and dephosphorylation by Cdc25

Whereas the activating phosphorylation of Cdks is not regulated, two inhibitory phosphorylations do have important functions in the regulation of Cdk activity. One is at a conserved tyrosine residue (Tyr 15 in human Cdks) that is found in all major Cdks. In animal cells,

Cdk-Activating Kinases

Species	Name	Alternative name	Comments
S. cerevisiae	Cak1	Civ1	monomer, no cyclin partner
	(Kin28)		(Cdk7-related, but no CAK activity)
S. pombe	Csk1		monomer, related to Cak1
	Mcs6		Cdk7-related, binds cyclin Mcs2
D. melanogaster	Cdk7		forms trimer with cyclin H and Mat1
X. laevis	Cdk7	M015	forms trimer with cyclin H and Mat1
H. sapiens	Cdk7		forms trimer with cyclin H and Mat1

Figure 3-7 **Table of Cdk-activating kinases**

Definitions

CAK: see **Cdk-activating kinase**.

Cdc25: protein phosphatase that activates cyclin-dependent kinases by removing phosphate from specific residues in the Cdk active site (Tyr 15 in most Cdks; also Thr 14 in animals).

Cdk-activating kinase (CAK): protein kinase that activates cyclin-dependent kinases by phosphorylating a threonine residue (Thr 160 in human Cdk2) in the T-loop.

Wee1: protein kinase that inhibits cyclin-dependent kinases by phosphorylating a tyrosine residue in the Cdk active site (Tyr 15 in most Cdks); a related protein kinase, Myt1, also phosphorylates this site and an adjacent threonine (Thr 14) in animals.

References

Dunphy, W.G.: **The decision to enter mitosis.** *Trends Cell Biol.* 1994, **4**:202–207.

Harper, J.W. and Elledge, S.J.: **The role of Cdk7 in CAK function, a retro-retrospective.** *Genes Dev.* 1998, **12**:285–289.

Morgan, D.O.: **Cyclin-dependent kinases: engines, clocks, and microprocessors.** *Annu. Rev. Cell Dev. Biol.* 1997, **13**:261–291.

Nurse, P.: **Universal control mechanism regulating onset of M-phase.** *Nature* 1990, **344**:503–508.

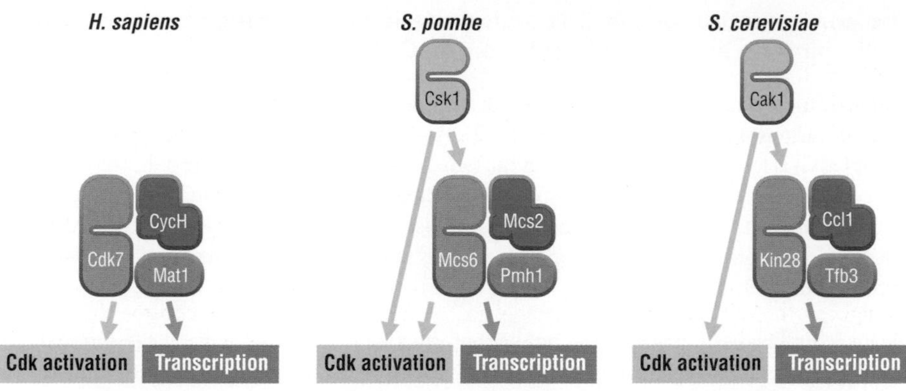

Figure 3-8 The functions of CAK in different species In animals (for example, *H. sapiens*, left), a trimeric CAK enzyme containing Cdk7 functions both in the activation of Cdks and in the regulation of transcription by RNA polymerase II. In the budding yeast *S. cerevisiae* (right) the homologous enzyme, Kin28, does not contribute to Cdk activation but is focused entirely on control of transcription. In this species, an unrelated protein kinase, Cak1, activates Cdks. The fission yeast *S. pombe* (center) occupies an intermediate position, in which Cdk activation can be achieved both by the Cdk7 homolog Mcs6 and by a Cak1 homolog, Csk1. Cdk7, Kin28 and Mcs6 are all Cdks whose activities are also enhanced by phosphorylation of residues in their T-loops. In budding and fission yeasts, this phosphorylation is carried out by Cak1 and Csk1, respectively. The kinase that phosphorylates Cdk7 in animals is not clear.

additional phosphorylation of an adjacent threonine residue (Thr 14) further blocks Cdk activity. Thr 14 and Tyr 15 are located in the roof of the kinase ATP-binding site and their phosphorylation probably inhibits activity by interfering with the orientation of ATP phosphates. Changes in the phosphorylation of these sites are particularly important in the activation of M–Cdks at the onset of mitosis, and they are also thought to influence the timing of G1/S- and S-phase Cdk activation.

The phosphorylation state of Tyr 15 and Thr 14 is controlled by the balance of opposing kinase and phosphatase activities acting at these sites. One enzyme responsible for Tyr 15 phosphorylation is **Wee1**, which is present (under various names) in all eukaryotes (Figure 3-9). Dephosphorylation of inhibitory sites is carried out by phosphatases of the **Cdc25** family, which has three members in vertebrates (Figure 3-10). The actions of these enzymes are shown in Figure 3-11. Fission yeast contains two kinases, Wee1 and Mik1, that both contribute to Tyr 15 phosphorylation. Vertebrates also contain a second protein kinase, Myt1, related to Wee1, which catalyzes the phosphorylation of both Thr 14 and Tyr 15.

Wee1 and Cdc25 provide the basis for the switch-like features of M–Cdk activation, which allows abrupt and irreversible entry into mitosis. Both enzymes are regulated by their mitotic substrate, the M-phase cyclin–Cdk complex: phosphorylation by M–Cdk inhibits Wee1 and activates Cdc25. Thus, M–Cdk activates its own activator and inhibits its inhibitor, and the resulting feedback loops are thought to generate switch-like Cdk activation at the beginning of mitosis, which is explained more fully later in this chapter. Wee1 and Cdc25 are also important targets for regulation of Cdk activity in response to factors such as DNA damage, as discussed in Chapter 11. We will next describe in more detail the structural basis of Cdk activation and the part played by cyclins in targeting activated Cdk molecules to their substrates.

Wee1 Enzymes

Species	Name	Comments
S. cerevisiae	Swe1	
S. pombe	Wee1	
	Mik1	
D. melanogaster	Dwee1	
	Dmyt1	
vertebrates	Wee1	phosphorylates Tyr 15
	Myt1	phosphorylates Thr 14 and Tyr 15

Figure 3-9 Table of inhibitory Cdk kinases in the Wee1 family

Cdc25 Enzymes

Species	Name	Comments
S. cerevisiae	Mih1	
S. pombe	Cdc25	
D. melanogaster	String	control of mitosis
	Twine	control of meiosis
vertebrates	Cdc25A	control of G1/S and G2/M
	Cdc25B	control of G2/M
	Cdc25C	control of G2/M

Figure 3-10 Table of stimulatory phosphatases in the Cdc25 family

Figure 3-11 Control of Cdk activity by inhibitory phosphorylation The fully active cyclin–Cdk complex (center) can be inhibited by further phosphorylation at one or two sites in the active site of the enzyme. Phosphorylation of Tyr 15 by Wee1, or phosphorylation of both Thr 14 and Tyr 15 by Myt1, inactivates the cyclin–Cdk complex. Dephosphorylation by the phosphatase Cdc25 leads to reactivation.

3-4 The Structural Basis of Cdk Activation

The conformation of the Cdk active site is dramatically rearranged by cyclin binding and phosphorylation by CAK

Cdk activation is understood in structural detail from X-ray crystallographic studies of human Cdk2 in various states of activity (Figure 3-12). As described earlier (see section 3-1), the active site of Cdk2 is located in a cleft between the two lobes of the kinase (Figure 3-12a). ATP binds deep within the cleft, with its phosphates oriented outward. The protein substrate would normally interact with the entrance of the active-site cleft, but this region is obscured in the inactive Cdk2 monomer by the T-loop. Key residues in the ATP-binding site are also misoriented in the Cdk2 monomer, further suppressing its activity.

Cyclin A binding has a major impact on the conformation of the Cdk2 active site (Figure 3-12b). Several helices in the cyclin box contact both lobes of Cdk2 in the region adjacent to the active-site cleft, resulting in extensive conformational changes in Cdk2. The most obvious change occurs in the T-loop, in which the L12 helix has been changed into a beta strand, and which no longer occludes the binding site for the protein substrate but lies almost flat at the entrance of the cleft. Major changes also occur in the ATP-binding site, leading to the correct positioning of the ATP phosphates for the phosphotransfer reaction. Cyclin A structure is unaffected by Cdk2 binding but provides a rigid framework against which the pliable Cdk2 subunit is molded.

The T-loop of Cdk2 contains Thr 160, the threonine residue whose phosphorylation by the Cdk-activating kinase (CAK) further increases the activity of the cyclin A–Cdk2 complex (see section 3-3). After phosphorylation, the phosphate on Thr 160 is inserted in a cationic pocket and acts as the central node for a network of hydrogen bonds spreading outward to stabilize neighboring interactions in both the Cdk and cyclin. The T-loop is flattened and moves closer to cyclin A (Figure 3-12c), and this region serves as a key part of the binding site for protein substrates containing the [S/T*]PX[K/R] phosphorylation site described in section 3-1 (Figure 3-12d).

Crystallographic studies of Cdk activation have so far focused primarily on human Cdk2 and its partner cyclin A. This complex probably serves as a good representative for the entire Cdk family, but the details of Cdk activation seem to be different in some complexes. There is evidence, for example, that the same Cdk, when bound by different cyclins, possesses different amounts of kinase activity toward the [S/T*]PX[K/R] sequence. It is therefore likely that different cyclins do not induce precisely the same conformational changes in the associated Cdk subunit.

Figure 3-12 The structural basis of Cdk activation These diagrams illustrate the structure of human Cdk2 in various states of activity. In each case, the complete structure is represented in the left column (PDB 1hck, 1fin, 1jst, 1gy3), while the right columns provide schematic views that emphasize key substructures, including the ATP in the active site, the T-loop (green) and the PSTAIRE helix (red). **(a)** In the inactive Cdk2 monomer, the small L12 helix next to the T-loop pushes out the large PSTAIRE helix, which contains glutamate 51 (E51), a residue important in positioning the ATP phosphates. The T-loop also blocks the active-site cleft. **(b)** When cyclin A binds, the PSTAIRE helix moves inward and the L12 helix changes structure to form a small beta strand; as a result, E51 moves inward to interact with lysine 33 (K33), while aspartate 145 (D145) also shifts position. These changes lead to the correct orientation of the ATP phosphates. The T-loop is also shifted out of the active-site entrance. **(c)** The phosphorylation of threonine 160 (T160) in the T-loop (yellow circle in left column) then causes the T-loop to flatten and interact more extensively with cyclin A. **(d)** Phosphorylation allows the T-loop to interact effectively with a protein substrate containing the SPXK consensus sequence (pink). The proline at the second position in this sequence interacts with the backbone of the T-loop, while the positively charged lysine residue at the fourth position (K^+) interacts, in part, with the negatively charged phosphate on T160. The hydroxyl oxygen of the serine residue (S) in the substrate is now positioned for nucleophilic attack on the γ-phosphate of ATP, which is catalyzed by several amino-acid side chains in the active-site region.

References

Brown, N.R. *et al.*: **The structural basis for specificity of substrate and recruitment peptides for cyclin-dependent kinases.** *Nat. Cell Biol.* 1999, **1**:438–443.

De Bondt, H.L. *et al.*: **Crystal structure of cyclin-dependent kinase 2.** *Nature* 1993, **363**:595–602.

Honda, R. *et al.*: **The structure of cyclin E1/CDK2: implications for CDK2 activation and CDK2-independent roles.** *EMBO J.* 2005, **24**:452–463.

Jeffrey, P.D. *et al.*: **Mechanism of CDK activation revealed by the structure of a cyclin A–CDK2 complex.** *Nature* 1995, **376**:313–320.

Pavletich, N.P.: **Mechanisms of cyclin-dependent kinase regulation: structures of Cdks, their cyclin activators, and CIP and Ink4 inhibitors.** *J. Mol. Biol.* 1999, **287**:821–828.

Russo, A.A. *et al.*: **Structural basis of cyclin-dependent kinase activation by phosphorylation.** *Nat. Struct. Biol.* 1996, [illegible]

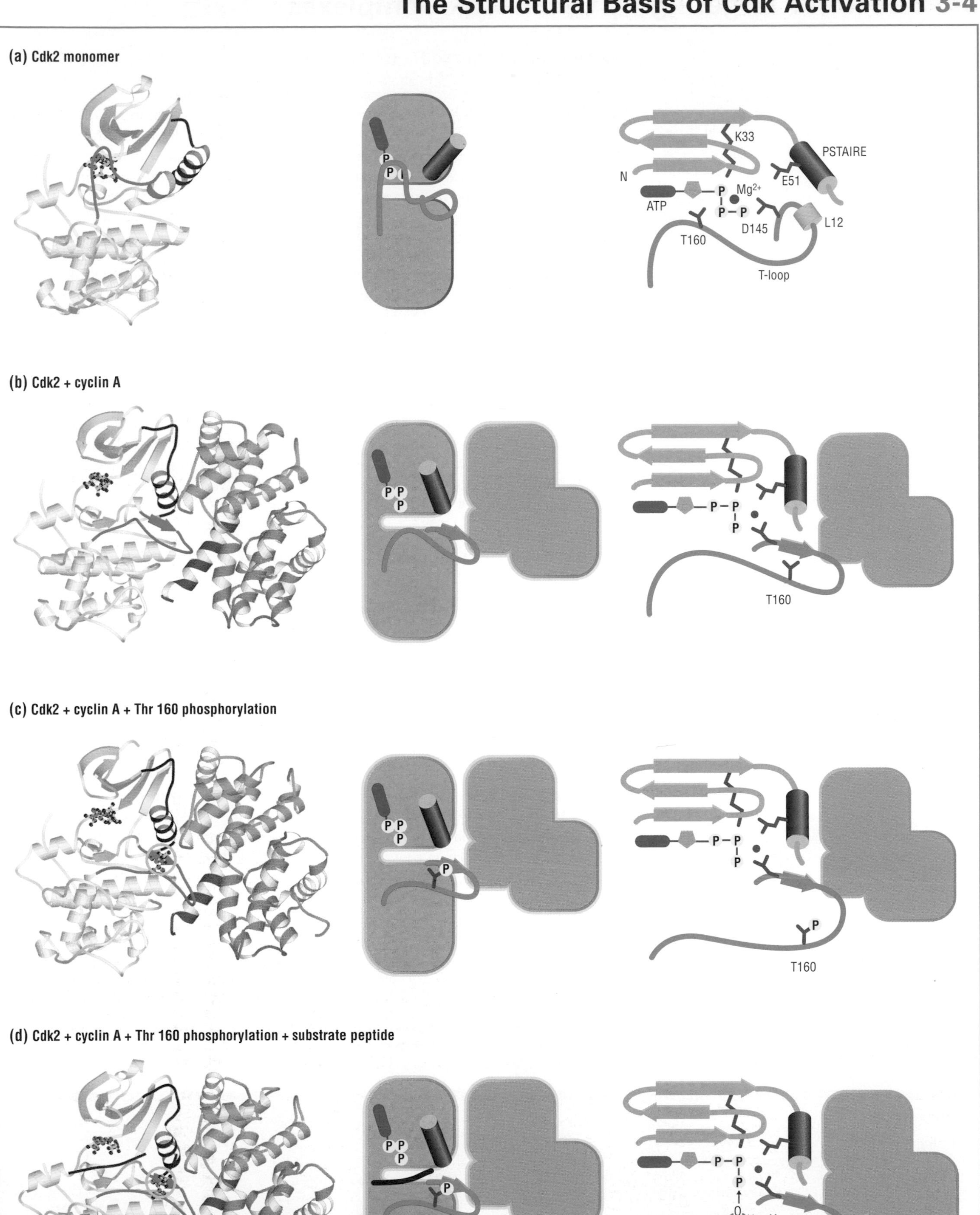
(a) Cdk2 monomer
P
P
K33
PSTAIRE
N
E51
ATP
P
Mg2+
P – P
D145
L12
T160
T-loop
(b) Cdk2 + cyclin A
P P
P
P – P
P
T160
(c) Cdk2 + cyclin A + Thr 160 phosphorylation
P P
P
P
P – P
P
P
T160
(d) Cdk2 + cyclin A + Thr 160 phosphorylation + substrate peptide
P P
P
P
P – P
P
S
O
H
X
K+
P
P

Cyclins are specialized for particular functions

If cell-cycle events are to occur in the correct order, then it is important that different cyclins stimulate different cell-cycle processes: S cyclins initiate DNA replication and M cyclins promote spindle assembly. What is the molecular basis of this cyclin specificity? One possibility, supported by considerable evidence in many species, is that cyclins are not simply activators of the associated Cdk subunit but also help direct that Cdk to specific substrates, either by directly binding the substrate or by taking the Cdk to a subcellular compartment where the substrate is found.

Functional specialization of cyclins helps ensure orderly and robust progression through the steps of the cell cycle, but it may not be absolutely essential in all species—particularly in yeast. Analysis of yeast cyclin mutants reveals considerable functional overlap between S and M cyclins: M cyclins can stimulate S phase to some extent, for example. In fission yeast, progression through S and M phase can be achieved in mutant cells lacking S cyclins and expressing only the M cyclin Cdc13. It is not clear how a single cyclin drives the correct sequence of S- and M-phase events in these cells. One possibility is that the specificity of the cyclin–Cdk complex is concentration-dependent, and a complex that promotes phosphorylation of S-phase substrates at one concentration promotes M-phase substrate phosphorylation when it accumulates to higher levels. Another possibility is that some Cdk substrates become available for phosphorylation only during a specific stage in the cell cycle. Cdks cannot initiate spindle assembly, for example, until the centrosome has been duplicated and the various spindle components have been produced as the cell nears mitosis.

Cyclins can interact directly with the substrates of the associated Cdk

In some cases, it is clear that the functional specialization of cyclins is due to a direct interaction between the cyclin and a specific subset of Cdk substrates. The S cyclins in particular—cyclin A in vertebrates and Clb5 in budding yeast—interact with numerous substrates involved in early cell-cycle events. In humans, for example, cyclin A–Cdk complexes, but not mitotic cyclin B–Cdk complexes, interact with and phosphorylate p107, which, as we will see in Chapter 10, is an important transcriptional regulator at the G1/S boundary. In budding yeast, Clb5–Cdk1 complexes, but not mitotic Clb2–Cdk1 complexes, bind and rapidly phosphorylate numerous proteins involved in DNA replication.

The substrate specificity of S cyclins depends on a region called the **hydrophobic patch**, which lies on the surface of the cyclin protein and is centered on the MRAIL amino-acid sequence in the first alpha helix of the cyclin box (Figure 3-13; see also Figure 3-5). This patch binds with moderate affinity to substrate proteins that contain a complementary hydrophobic sequence known as the **RXL** (or Cy) **motif**. The interaction increases kinase–substrate affinity and thereby enhances the rate of substrate phosphorylation. Mutation of the hydrophobic patch on the

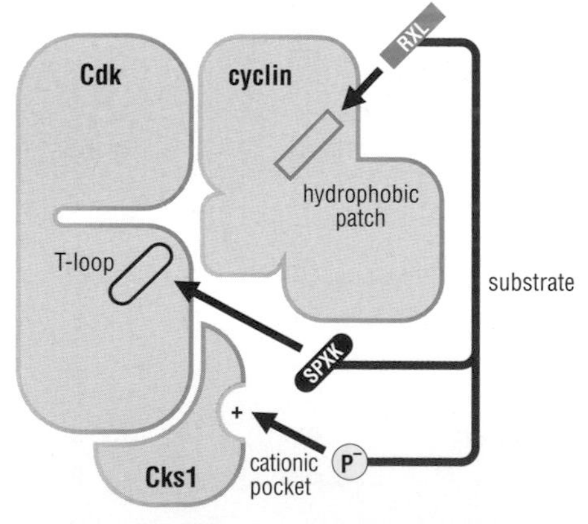

Figure 3-13 Substrate-targeting sites on cyclin–Cdk complexes The central substrate-recognition site on Cdks lies in the active-site T-loop, which interacts with the SPXK consensus sequence that contains the phosphorylation site (see Figure 3-12). An RXL motif in some substrates interacts with the hydrophobic patch on the cyclin, thereby enhancing the rate of phosphorylation. The presence of a phosphate-binding pocket on the accessory subunit Cks1 may facilitate interactions with targets that contain multiple phosphorylation sites.

Definitions

hydrophobic patch: small hydrophobic region on the surface of a protein. Many cyclins contain a hydrophobic patch that is based on the MRAIL sequence in the cyclin box. They interact with the **RXL motif** of Cdk substrates and inhibitors.

RXL motif: degenerate amino-acid sequence on Cdk substrates and inhibitors that interacts with the **hydrophobic patch** on the surface of cyclins. Also called a Cy motif.

cyclin, or the RXL motif on the substrate, greatly reduces the interaction between kinase and substrate, decreases the rate of substrate phosphorylation and reduces the biological action of the cyclin in the cell.

The hydrophobic patch not only binds substrates but also interacts with proteins that inhibit Cdk activity. As we discuss in section 3-6, several Cdk inhibitor proteins bind to the hydrophobic patch on cyclin, thus preventing binding and phosphorylation of Cdk substrates.

Cyclins can direct the associated Cdk to specific subcellular locations

Some cyclins contain sequence information that targets them and their Cdk partners to specific subcellular locations, providing another mechanism by which a cyclin can direct its catalytic partner to the right substrates. This mechanism is used to regulate the function of vertebrate cyclin B, for example. There are two forms of this cyclin, cyclins B1 and B2. One of the main targets of cyclin B1–Cdk1 is the nuclear lamina, the cytoskeletal network that lies under the nuclear envelope. Before mitosis, the cyclin B1–Cdk1 complex is held in the cytoplasm, preventing its access to targets inside the nucleus. In late prophase, however, cyclin B1–Cdk1 is rapidly translocated into the nucleus and immediately phosphorylates the proteins of the nuclear lamina, triggering the breakdown of the nuclear envelope. Accumulation of cyclin B1 in the nucleus depends on sequences in the amino-terminal part of the protein, outside the Cdk-binding domain. The mechanisms that control the localization of cyclin B1–Cdk1 are described in section 5-6.

The second type of vertebrate cyclin B, cyclin B2, associates with the Golgi apparatus and stimulates phosphorylation of proteins that cause fragmentation of this organelle during mitosis. As with cyclin B1, targeting of cyclin B2 depends on sequences outside the Cdk-binding domain.

The hydrophobic patch of some cyclins also influences their subcellular localization. In budding yeast, a subpopulation of the S cyclin Clb5 is localized throughout the cell cycle at origins of DNA replication, due to an interaction between the Clb5 hydrophobic patch and proteins bound to the origins.

Cks1 may serve as an adaptor protein that targets Cdks to phosphoproteins

Another Cdk-substrate targeting mechanism is thought to be important for substrates that contain clusters of multiple Cdk phosphorylation sites. The APC, for example, is activated in mitosis by M–Cdks, which phosphorylate it at multiple sites. Effective phosphorylation of such substrates is promoted by Cks1, a small (9–13 kDa) adaptor protein (Figure 3-14) that binds to the carboxy-terminal lobe of the Cdk, well away from the active site (see Figure 3-13). Cks1 binding has little effect on cyclin binding or on the conformation of the Cdk catalytic site, but instead seems to provide the Cdk with an accessory binding site that recognizes phosphorylated residues. Thus, after a cyclin–Cdk–Cks1 complex has phosphorylated one residue in a substrate, the ability of Cks1 to bind phosphate may increase the affinity of the Cdk for the substrate, facilitating the phosphorylation of other sites in the vicinity.

Cks1 Proteins

Species	Name	Size (amino acids)	Size (kDa)
S. cerevisiae	Cks1	150	17
S. pombe	Suc1	113	13
D. melanogaster	DmCks1	74	8
X. laevis	Xe-p9	79	9
H. sapiens	CksHs1	79	9
	CksHs2	79	9

Figure 3-14 Table of Cks1 proteins

References

Bourne, Y. *et al.*: **Crystal structure and mutational analysis of the human CDK2 kinase complex with cell cycle-regulatory protein CksHs1.** *Cell* 1996, **84**:863–874.

Brown, N.R. *et al.*: **The structural basis for specificity of substrate and recruitment peptides for cyclin-dependent kinases.** *Nat. Cell Biol.* 1999, **1**:438–443.

Loog, M. and Morgan, D.O.: **Cyclin specificity in the phosphorylation of cyclin-dependent kinase substrates.** *Nature* 2005, **434**:104–108.

Miller, M.E. and Cross, F.R.: **Cyclin specificity: how many wheels do you need on a unicycle?** *J. Cell Sci.* 2001, **114**:1811–1820.

Roberts, J.M.: **Evolving ideas about cyclins.** *Cell* 1999, **98**:129–132.

Cdk inhibitors help suppress Cdk activity in G1

In actively proliferating cells, most Cdk activity is suppressed during G1, resulting in a stable transition period during which cell growth and other regulatory influences, such as extracellular factors, can govern entry into the next cell cycle. As discussed earlier (section 3-0), a combination of three mechanisms suppresses Cdk activity during G1. Two of these mechanisms—increased cyclin destruction and decreased cyclin gene expression—are discussed later in this chapter. In this section we describe the third: the inhibition of Cdk activity by **Cdk inhibitor proteins (CKIs)** that bind and inactivate cyclin–Cdk complexes. These proteins are also important for promoting the arrest of the cell cycle in G1 in response to unfavorable environmental conditions or intracellular signals such as DNA damage.

Most, if not all, eukaryotic organisms possess a CKI that contributes to the establishment of a stable G1: these include Sic1 in budding yeast, Rum1 in fission yeast and Roughex (Rux) in *Drosophila* (Figure 3-15). Although these proteins display few if any similarities in amino-acid sequence, they share several important functional features. First, they are all potent inhibitors of the major S- and M-Cdk complexes, and all are expressed at high levels in G1 cells to help ensure that no S- or M-Cdk activity exists in those cells. Second, these proteins do not inhibit G1/S-Cdks; as a result, they do not block the activation of these kinases at the Start checkpoint. Finally, these inhibitors are all targeted for destruction when phosphorylated by Cdks. In late G1, rising G1/S-Cdk activity therefore leads to destruction of these inhibitors—allowing S-Cdk activation at the beginning of S phase.

Given the clear importance of Sic1, Rum1 and Rux in yeast and *Drosophila*, it is perhaps surprising that a clear functional homolog of these proteins has not been identified in mammalian cells. However, animal cells do possess another CKI protein—called Dacapo in *Drosophila* and p27 in mammals—that helps govern Cdk activity in G1, although by mechanisms that are somewhat distinct from those used by yeast Sic1. Most importantly, p27 inhibits G1/S-Cdks (cyclin E–Cdk2) as well as the S-Cdk cyclin A–Cdk2, but has relatively little effect on the M-Cdk cyclin B–Cdk1. Thus, in mammals the rise of G1/S-Cdks in late G1 requires the removal of p27, which is achieved by a combination of mechanisms. First, as we discuss below,

Cdk Inhibitors

Species	Name	Alternatives	Relatives	Targets, function
S. cerevisiae	Sic1		Rum1	inhibits S- and M-Cdks, suppresses Cdk activity in G1
	Far1		no relatives	inhibits G1/S-Cdk in response to mating pheromone
S. pombe	Rum1		Sic1	inhibits S- and M-Cdks, suppresses Cdk activity in G1
D. melanogaster	Roughex/Rux		no relatives	inhibits S- and M-Cdks, suppresses Cdk activity in G1
	Dacapo/Dap		Cip/Kip	inhibits G1/S-Cdks, suppresses Cdk activity in G1
X. laevis	Xic1	Kix1	Cip/Kip	inhibits G1/S- and S-Cdks
H. sapiens	p21	Cip1/Waf1	Cip/Kip	inhibits G1/S- and S-Cdks, activates cyclin D–Cdk4
	p27	Kip1	Cip/Kip	inhibits G1/S- and S-Cdks, activates cyclin D–Cdk4
	p57	Kip2	Cip/Kip	inhibits G1/S- and S-Cdks, activates cyclin D–Cdk4
	p15^{INK4b}		INK4	inhibits Cdk4, Cdk6
	p16^{INK4a}		INK4	inhibits Cdk4, Cdk6
	p18^{INK4c}		INK4	inhibits Cdk4, Cdk6
	p19^{INK4d}		INK4	inhibits Cdk4, Cdk6

Figure 3-15 Table of Cdk inhibitor proteins

Definitions

Cdk inhibitor protein (CKI): protein that interacts with Cdks or Cdk–cyclin complexes to block activity, usually during G1 or in response to inhibitory signals from the environment or damaged DNA.

Cip/Kip: small family of **CKIs** in animal cells, including mammalian p21 (Cip1) and p27 (Kip1), that inhibit Cdk activity by interaction with both subunits of the Cdk–cyclin complex.

CKI: see **Cdk inhibitor protein**.

INK4: small family of mammalian **CKIs**, including p15^{INK4b} and p16^{INK4a}, that bind the Cdk4 and Cdk6 proteins and reduce their binding affinity for cyclin D.

References

Brotherton, D.H. *et al.*: **Crystal structure of the complex of the cyclin D-dependent kinase Cdk6 bound to the cell-cycle inhibitor p19^{INK4d}.** *Nature* 1998, **395**:244–250.

Russo, A.A. *et al.*: **Crystal structure of the p27^{Kip1} cyclin-dependent-kinase inhibitor bound to the cyclin A–Cdk2 complex.** *Nature* 1996, **382**:325–331.

Russo, A.A. *et al.*: **Structural basis for inhibition of the cyclin-dependent kinase Cdk6 by the tumour suppressor p16^{INK4a}.** *Nature* 1998, **395**:237–243.

Sherr, C.J. and Roberts, J.M.: **CDK inhibitors: positive and negative regulators of G_1-phase progression.** *Genes Dev.* 1999, **13**:1501–1512.

the G1–Cdks (cyclin D–Cdk4) remove p27 from G1/S–Cdks. Second, p27 is destroyed in late G1 as a result of phosphorylation by multiple protein kinases, including the G1/S–Cdks themselves. We will discuss these regulatory mechanisms in Chapter 10.

Other CKIs help promote G1 arrest in response to specific inhibitory signals. Far1 in budding yeast and the INK4 proteins of mammals inhibit G1–Cdk activity when cells encounter anti-proliferative signals in the environment. The p21 protein of mammals blocks G1/S– and S–Cdks, and thus cell-cycle entry, in response to DNA damage, giving the cell time to repair the damage before starting to replicate its DNA.

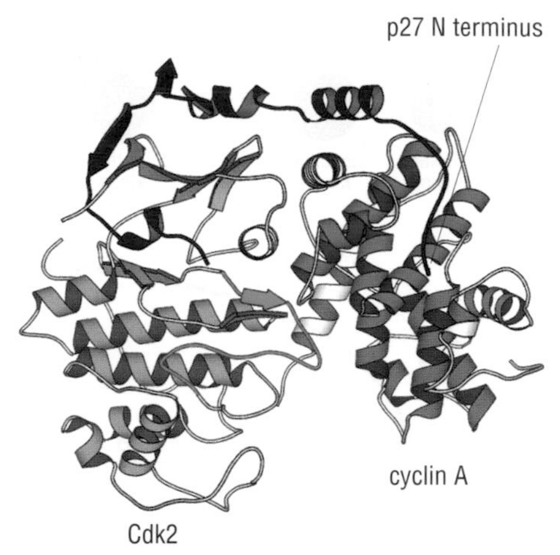

Figure 3-16 **Structure of Cdk2–cyclin A–p27** The structure of the Cdk2–cyclin A–p27 complex, as determined by X-ray crystallography, reveals that the inhibitor p27 (red) stretches across the top of the cyclin–Cdk complex. Only the amino-terminal region of p27 is shown in the structure. The amino-terminal end of this fragment contains an RXL motif that interacts with the hydrophobic patch of cyclin A. The carboxy-terminal end of the p27 fragment interacts extensively with the beta sheet of Cdk2, causing extensive disruptions to its structure; p27 also inserts into the ATP-binding site of Cdk2 and directly inhibits ATP binding. To appreciate the effects of p27, compare this structure with that of the active Cdk2–cyclin A complex in Figure 3-12. (PDB 1jsu)

Cip/Kip proteins bind both subunits of the cyclin–Cdk complex

The CKIs of animal cells are grouped into two major structural families, each with a distinct mechanism of Cdk inhibition. Members of the **Cip/Kip** family, including Dacapo and p27 (see Figure 3-15), control multiple cyclin–Cdk complexes by interacting with both the cyclin and Cdk. These proteins have complex biochemical actions: their primary function is to block cell-cycle progression by inhibiting G1/S– and S–Cdks, as just described, but they can also promote cell-cycle entry by activating G1–Cdks, as we discuss below. In contrast, members of the **INK4** family (see Figure 3-15) are strictly inhibitors that display a clear specificity for the monomeric forms of Cdk4 and Cdk6, and act in part by reducing cyclin binding affinity.

The amino-terminal half of the mammalian Cip/Kip proteins p21 and p27 is responsible for their Cdk inhibitory function and is composed of two key subregions: a short sequence containing an RXL motif that is required for cyclin binding, and a longer segment required for binding to the Cdk subunit. The structure of the Cdk2–cyclin A–p27 complex (Figure 3-16) reveals that the cyclin-binding portion of p27 interacts with the hydrophobic patch of cyclin A (see section 3-5). The Cdk-binding region of p27 interacts extensively with the kinase subunit. These interactions thoroughly distort and partly dismantle the structure of the amino-terminal lobe of the kinase above the active site, and also directly block the ATP-binding site, completely disrupting the catalytic function of the enzyme.

G1–Cdks are activated by Cip/Kip proteins and inhibited by INK4 proteins

Somewhat surprisingly, in view of their inhibition of cyclin–Cdk2 complexes, Cip/Kip proteins help activate the G1 kinases Cdk4 and Cdk6. Unlike most cyclin–Cdk pairs, cyclin D and Cdk4 or Cdk6 do not bind each other with high affinity in the absence of additional proteins. Assembly of cyclin D–Cdk4,6 complexes requires the assistance of Cip/Kip proteins, which enhance binding by interacting with both subunits. We do not know the structure of a Cip/Kip inhibitor bound to cyclin D–Cdk4,6, but it is clear that the Cdk-binding region of the inhibitor must interact with Cdk4 or 6 without creating the disruptive conformational changes seen in Cdk2.

In contrast to Cip/Kip proteins, members of the INK4 family are inhibitors of only Cdk4 and 6, binding preferentially to the Cdk monomer. Crystallographic structural studies indicate that these inhibitors bind both lobes of the Cdk on the side opposite the cyclin-binding site, disrupting the binding and orientation of ATP. The INK4 protein also twists the upper lobe of the kinase into an orientation that is incompatible with cyclin binding (Figure 3-17). INK4 proteins may also reduce cyclin D binding *in vivo* by blocking access of Cip/Kip proteins, and by blocking interactions with molecular chaperone proteins that are required for normal folding of Cdk4 or 6.

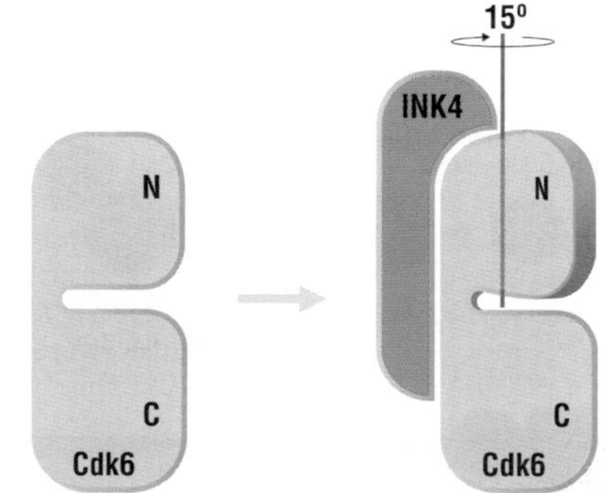

Figure 3-17 **Schematic of a Cdk6–INK4 complex** Crystal structures of Cdk6–p16^{INK4a} and Cdk6–p19^{INK4d} complexes reveal that INK4 proteins bind the face of the Cdk subunit that is opposite the active site and cyclin-binding site, resulting in the partial twisting of the amino-terminal lobe away from the correct orientation. This twist is predicted to reduce cyclin binding and distort the active site of the kinase.

The fact that Cip/Kip proteins activate cyclin D–Cdk4,6 but inhibit cyclin E,A–Cdk2 has interesting implications for the regulation of the different Cdk classes. As we will discuss in Chapter 10, extracellular mitogens often stimulate cell-cycle entry at Start by increasing the levels of cyclin D in the cell. This not only increases the activity of Cdk4 and 6 but also sequesters Cip/Kip proteins away from cyclin–Cdk2 complexes, thereby increasing their activity as well. Conversely, some extracellular anti-mitogens increase the levels of an INK4 protein in proliferating cells. This leads to disassembly of cyclin D–Cdk4,6 complexes and also induces the release from these complexes of Cip/Kip proteins, which can then inhibit cyclin–Cdk2 complexes—thereby preventing progression into S phase.

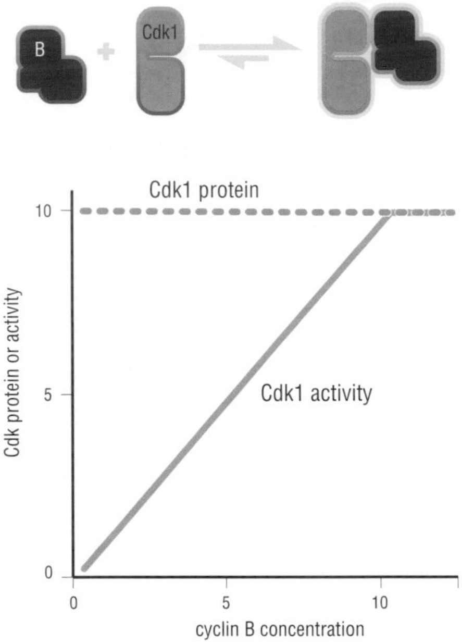

3-8 Switch-Like Activation of Cdk1

Cdk1 activation at mitosis is based on positive feedback

The activation of G1/S–, S– and M–Cdks all display switch-like behavior that is thought to be based on ultrasensitive and feedback mechanisms like those just described in section 3-7. In this section we describe the activation of cyclin B–Cdk1, a vertebrate M–Cdk that triggers spindle assembly and other early events at the onset of mitosis and is the best-understood example of a biochemical switch in cell-cycle control. In Chapter 10 we will discuss the mechanisms that lead to the switch-like activation of G1/S– and S–Cdks as the cell progresses through the Start checkpoint in late G1.

In vertebrate Cdk1 activation, cyclin B can be viewed as the stimulus and Cdk1 activity as the response. Let us first consider a simple hypothetical system containing only Cdk1 and cyclin B (Figure 3-21). We can ignore the Cdk-activating kinase because it is constitutively active. The state of Cdk activity therefore depends only on the binding of cyclin B, and so this system is similar to the simple ligand-activated kinase system described in the previous section (see Figure 3-18). The two systems behave differently, however, because in the cell the Cdk1 concentration is very high and the affinity of Cdk1 for cyclin B is also very high. Any cyclin present is therefore bound immediately by the kinase. Thus, a linear increase in cyclin B concentration results in a linear, not hyperbolic, increase in kinase activity. Now consider the response of this system to a gradual increase in cyclin B levels over time—like the linear increase that occurs in a frog embryonic cell cycle as it progresses through S phase (Figure 3-22a). As cyclin B levels rise, Cdk1 activity rises in parallel. Because the cyclin B concentration in the cell never reaches that of Cdk1, Cdk1 activity does not level off.

Our previous hypothetical system could be converted to a bistable, switch-like system by the addition of positive feedback (see Figure 3-20). Similarly, the cyclin B1–Cdk1 system just described can be modified to include positive feedback that generates switch-like increases in Cdk1 activity. The mechanism that generates positive feedback in Cdk1 activation is slightly different from that in our previous example, but the effect is the same: Cdk1 is able to activate itself. This is achieved in our cyclin B–Cdk1 system by adding the enzymes Wee1 and Cdc25, the kinase and phosphatase that act on the inhibitory phosphorylation sites in Cdk1 as described in section 3-3. The key to the effects of these enzymes on the dynamics of kinase activation is the ability of Cdk1 to activate its own activator (Cdc25) and inhibit its inhibitor (Wee1).

Figure 3-22b illustrates the effects of Wee1 and Cdc25 on our model system when we gradually increase cyclin B levels over time as before. Initially, when cyclin B is absent and Cdk1 activity is low, Wee1 activity is high and Cdc25 activity is low. As the concentration of cyclin B increases, Wee1 phosphorylates and inactivates the cyclin B–Cdk1 complexes as they accumulate, thereby keeping Cdk1 activity at a minimum. Eventually, the cell contains a large stockpile of inactive, phosphorylated cyclin B–Cdk1 complexes.

In our previous model system (see Figure 3-20), the positive feedback loop was initiated by the kinase itself when its activity rose above some threshold. This mechanism may not be used in the activation of Cdk1. Instead, Cdk1 activation is thought to involve a separate trigger mechanism that unleashes the positive feedback loop at the beginning of mitosis. The nature of this trigger mechanism is not well understood, but it is likely to involve multiple regulatory molecules. One possibility is that positive feedback is initiated in this system by cyclin A–Cdk2, which is active in G2 and can phosphorylate and partly activate the phosphatase Cdc25. According to this scheme, activation of some Cdc25 molecules by cyclin A–Cdk2 leads

Figure 3-21 Cdk1 activation by cyclin B
In a simple system comprising just Cdk1 and cyclin B, the activation of Cdk1 by cyclin B is similar to the ligand-dependent activation of the kinase shown in Figure 3-18. However, the stimulus–response relationships in the two systems are different because of differences in the concentration of the kinase. In the previous system, a hyperbolic response was generated because the concentration of the kinase was very low—far below the dissociation constant, or ligand concentration at which the kinase is 50% bound and activated. The cellular concentration of Cdk1 is, however, much higher than the very low dissociation constant for the cyclin B–Cdk1 interaction (cyclin B binds Cdk1 with very high affinity). As the cyclin B concentration increases, all cyclin B binds to Cdk1 to form active kinase molecules: there is essentially no free cyclin B in the cell. The result is a simple linear (not hyperbolic) stimulus–response relationship. When the cyclin B concentration reaches that of Cdk1, all Cdk1 is activated and activity does not rise further.

References

Ferrell, J.E. Jr: **Tripping the switch fantastic: how a protein kinase cascade can convert graded inputs into switch-like outputs.** *Trends Biochem. Sci.* 1996, **21**:460–466.

Ferrell, J.E. Jr: **Self-perpetuating states in signal transduction: positive feedback, double-negative feedback and bistability.** *Curr. Opin. Cell Biol.* 2002, **14**:140–148.

Pomerening, J.R. *et al.*: **[illegible] hysteresis and bistability in the activation of Cdc2.** *Nat. Cell Biol.* 2003, **5**:346–351.

Sha, W. *et al.*: **Hysteresis drives cell-cycle transitions in *Xenopus laevis* egg extracts.** *Proc. Natl Acad. Sci. USA* 2003, **100**:975–980.

Tyson, J.J. *et al.*: **Sniffers, buzzers, toggles and blinkers: dynamics of regulatory and signaling pathways in the cell.** *Curr. Opin. Cell Biol.* 2003, **15**:221–231.

to activation of some cyclin B–Cdk1 complexes. The active Cdk1 then phosphorylates and activates more Cdc25 molecules, while at the same time inactivating Wee1 molecules. More Cdk1 is activated, triggering the positive feedback loop. The system thus switches abruptly from a stable state of low Cdk1 activity to a stable state of high Cdk1 activity (see Figure 3-22b).

Once activated, Cdk1 in this system will remain active even if the trigger stimuli are removed. In other words, Cdk1 activation is essentially irreversible until some other regulatory component is introduced. In the cell this is the degradation of cyclin, which will be described in the next section. Irreversibility is a key requirement in a Cdk switch, as it helps ensure that cell-cycle events are completed in an all-or-none fashion.

Many features of the Cdk1 activation switch remain poorly understood. We have only a superficial understanding, for example, of the trigger mechanisms that initiate the feedback loop. Most importantly, we know little about the ultrasensitive mechanisms that presumably exist to prevent small fluctuations in Cdc25 or Wee1 activities from prematurely triggering the feedback loop. Numerous additional mechanisms and regulatory molecules are involved in Cdk1 activation. In Chapter 5 we will discuss these mechanisms and molecules in the context of mitotic control.

Cdk switches are robust as a result of multiple partly redundant mechanisms

Cdk activation is generally governed by multiple overlapping mechanisms, ensuring that Cdk switches are robust and reliable even if some components fail. In some cell types, for example, the rise in cyclin B–Cdk1 activity at mitosis promotes expression of the cyclin B gene, yielding another positive feedback loop that supplements the loop discussed above. Multiple mechanisms also contribute to the activation of Cdks at other cell-cycle stages. In every case, the loss of one regulatory mechanism seems to have only minor consequences for cell-cycle timing, because back-up mechanisms are present to ensure that normal Cdk regulation is maintained. Even the removal of both Wee1 and Cdc25, for example, has only minor effects on the cell cycle in fission yeast. It is, however, likely that every regulatory subsystem is critical for the long-term fidelity and reliability of cell-cycle control.

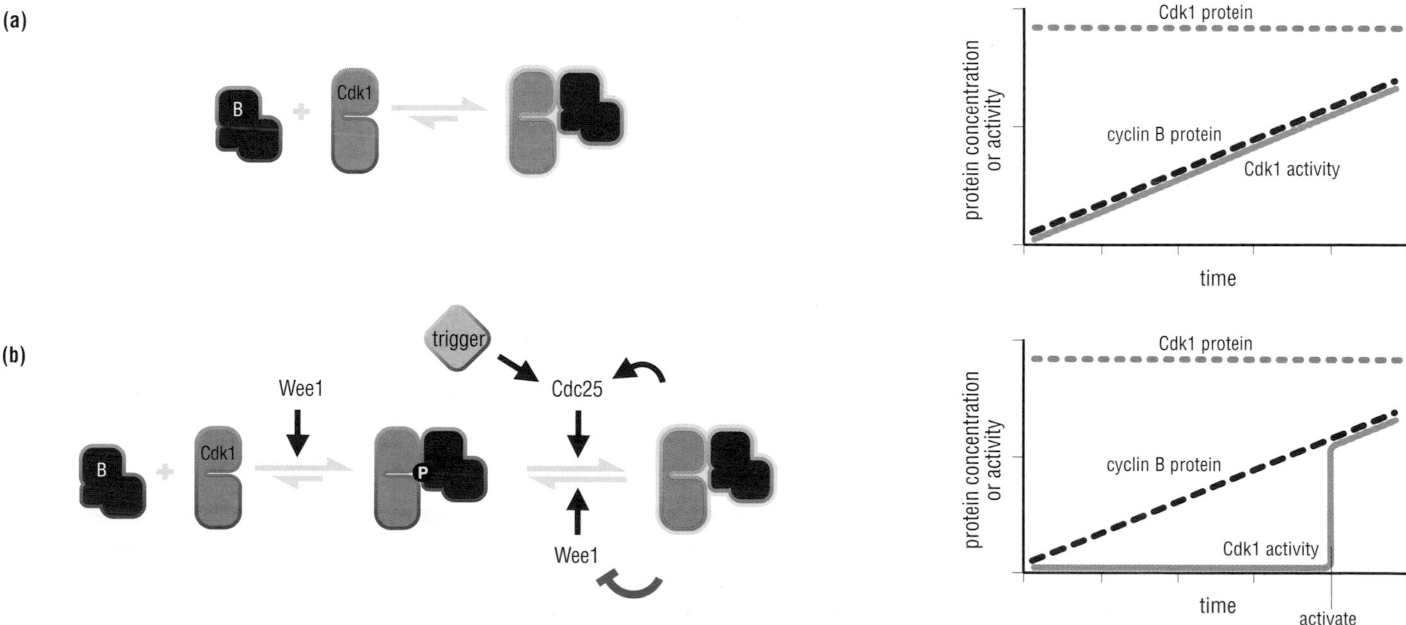

Figure 3-22 Assembling a Cdk1 switch at mitosis (a) Here we consider the behavior of the simple cyclin B–Cdk1 model system of Figure 3-21 as cyclin B levels are increased over time, as in the cell cycle. Note that the graph shown here is a plot of response against time, not a stimulus–response graph like that shown in Figure 3-21. The gradual accumulation of cyclin B protein (purple dashed line) leads to a parallel increase in Cdk1 activity (yellow line). In this example, as in the cell, Cdk1 activity does not level off because cyclin B levels never reach those of Cdk1. **(b)** The addition of Wee1 and Cdc25 to the system leads to the appearance of more switch-like Cdk1 activation, because Cdk1 activity inhibits Wee1 and stimulates Cdc25, resulting in the potential for positive feedback. This feedback loop is initiated by a trigger mechanism that is not well understood but is likely to act by stimulating Cdc25.

Negative feedback can generate a repeating oscillator

Oscillations in Cdk activity are fundamental to the cell-cycle control system. The mechanisms underlying these oscillations involve a combination of positive feedback, whereby Cdks promote their own activation (as discussed in section 3-8), with subsequent negative feedback, whereby they indirectly trigger their own inactivation. These mechanisms are best illustrated by the regulation of the ubiquitin-protein ligase APC^{Cdc20} by activated Cdks, and the consequent effect of the APC on Cdk activity, as described in section 3-10.

Regulatory networks that depend in some way on a negative feedback loop are often capable of generating repeated oscillations in the activity of one of their components. Such systems are called **negative feedback oscillators**. We can illustrate how these oscillators work by returning to the simple ligand–kinase model system containing a positive feedback loop, as described earlier in our discussion of switches (see Figure 3-20). Imagine that this system also includes a negative feedback loop: the activated kinase is able to phosphorylate and activate an inhibitor protein that then binds the kinase and blocks its activity (Figure 3-29). The activation of the kinase therefore leads indirectly to its own inactivation, and can thereby bring kinase activity back down to zero. A transient spike in kinase activity is the result.

If conditions are right, systems containing negative feedback have the potential to generate repeated oscillations. Imagine, for example, that our system also contains a phosphatase that slowly dephosphorylates both the inhibitor and the kinase, deactivating them (Figure 3-29a). When kinase activity is very high, this phosphatase is overwhelmed and extensive inhibitor phosphorylation occurs. When negative feedback inhibits the kinase, however, the reduced kinase activity allows the phosphatase to dephosphorylate both the kinase and the inhibitor. The system thereby returns to its basal state with unphosphorylated, inactive kinase and inhibitor. The presence of a small amount of activating ligand then triggers kinase activation, leading to another cycle of activation and inactivation. If the timing of the various reactions is optimized, this system will generate repeated spikes of kinase activity (Figure 3-29b).

Oscillatory behavior is, however, not the only possible outcome in a system with negative feedback. If, for example, the kinase activates its inhibitor too rapidly, then the kinase will be inhibited before it can achieve a high level of activity. The activities of the kinase and its inhibitor will not oscillate but will settle at some constant intermediate level. To generate oscillations, negative feedback systems must contain certain additional features. Two are particularly important: first, there must be a delay between activation of the kinase and activation of the inhibitor, and second, there must be a mechanism in the system that generates bistability in kinase activity. Both of these features are present in the model system illustrated in Figure 3-29 and can be summarized as follows.

There are numerous ways of introducing some sort of time delay between kinase activation and activation of its inhibitor. The simplest approach is to add further steps between the kinase and inhibitor. For example, the active kinase might phosphorylate some other protein kinase, which then phosphorylates the inhibitor. As a result, the signal generated by the first kinase takes some time to reach the inhibitor. This kinase would then be able to reach the fully activated state and remain there for a little time before delayed activation of the inhibitor brought kinase activity back down to zero.

Negative feedback systems can produce particularly robust oscillations if they also include mechanisms that generate bistability in the oscillating activity. In our model system, for example, the positive feedback loop allows the kinase to rapidly achieve maximal activity once initial activation by ligand has occurred (as discussed in section 3-7). If the strength of this positive feedback loop is optimal, it will generate a rapid, all-or-none kinase activation that will be

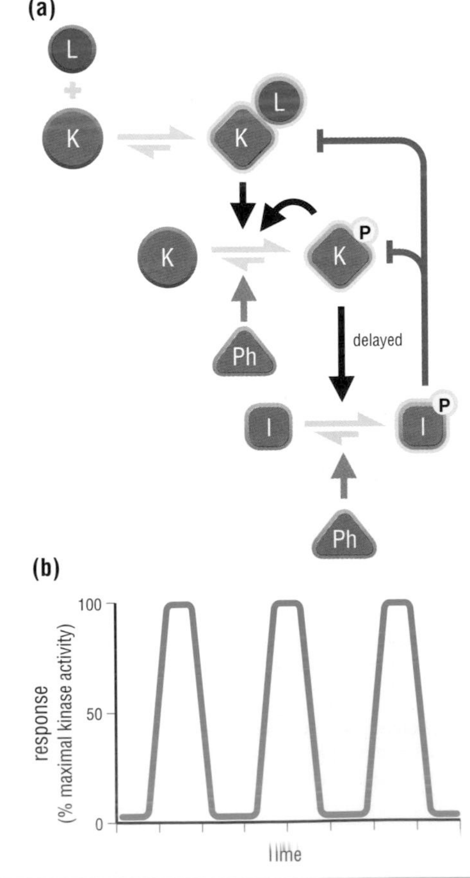

Figure 3-29 A biochemical oscillator (a) The positive feedback signaling system shown in Figure 3-20 is modified here to include a negative feedback loop: the activated kinase (K) is able to phosphorylate and activate an inhibitor (I), which inhibits the kinase. Activation of the kinase therefore leads, after a short delay, to its inactivation. A small amount of phosphatase (Ph) is also present, allowing dephosphorylation of the kinase and the inhibitor after the negative feedback loop is triggered. **(b)** This system has the potential, if the various parameters are optimal, to generate repeated oscillations in kinase activity.

sustained until sufficient inhibitor accumulates to flip the switch back to the kinase-off state. Negative feedback oscillators that flip between stable states are sometimes called **relaxation oscillators** because they relax from one state to the other.

These principles are clearly apparent in the oscillatory behavior of M–Cdk activity, particularly as seen in early embryonic cell cycles. Negative feedback is introduced into the M–Cdk activation system by the APC^{Cdc20} enzyme. As described in section 3-10, M–Cdks initiate the activation of APC^{Cdc20}, triggering the ubiquitination and destruction of cyclins—and thereby causing M–Cdk inactivation. Because APC^{Cdc20} activity is dependent on Cdk activity, its activity will also drop after Cdk inactivation. Cyclin can then begin to accumulate again, and the result is a system that has the potential to generate repeated oscillations in M–Cdk activity (Figure 3-30a). As in our model system, the M–Cdk oscillator includes two important features that help it generate robust oscillations: first, a delay, the mechanism of which is not well understood, occurs between M–Cdk activation and APC^{Cdc20} activation; and second, positive feedback (through Cdc25 activation and Wee1 inactivation—see section 3-8) generates bistability in M–Cdk activity.

Regulated braking mechanisms allow the Cdk oscillator to be paused in G1

Unlike embryonic animal cells, most cells do not begin to accumulate new cyclin immediately after mitosis, but instead pause in G1, allowing cell growth and other factors to regulate progression into the next cycle. As we have seen, this pause is due in part to the activation of the APC by Cdh1, which is prevented during early mitosis when Cdk activity is high but is triggered after the destruction of cyclins in late mitosis. Activation of APC^{Cdh1} ensures continued cyclin destruction and Cdk inactivity despite the absence of APC^{Cdc20} (Figure 3-30b).

Exit from this stable G1 state into a new cell cycle is regulated by growth or extracellular factors, which promote the accumulation of G1/S cyclins that are not targeted for destruction by APC^{Cdh1}. The resulting increase in G1/S–Cdk activity inactivates APC^{Cdh1}, allowing the accumulation of APC targets, such as S and M cyclins, to begin again.

A number of other factors stabilize the G1 state. These include Cdk inhibitors such as Sic1 of budding yeast (see section 3-6): like Cdh1, they are inhibited by Cdk activity, so their inhibitory effects are maximal in G1. In addition, the absence of Cdk activity in G1 prevents the expression of G1/S- and S-phase cyclin genes, because the gene regulatory proteins that promote their expression are activated by Cdks, as we will see in section 3-12.

The cell-cycle control system therefore has the potential to generate autonomous Cdk oscillations, particularly in simple cell cycles. Most cells, however, contain regulated braking systems that can block these oscillations and arrest the cell in G1 or at other checkpoints later in the cell cycle.

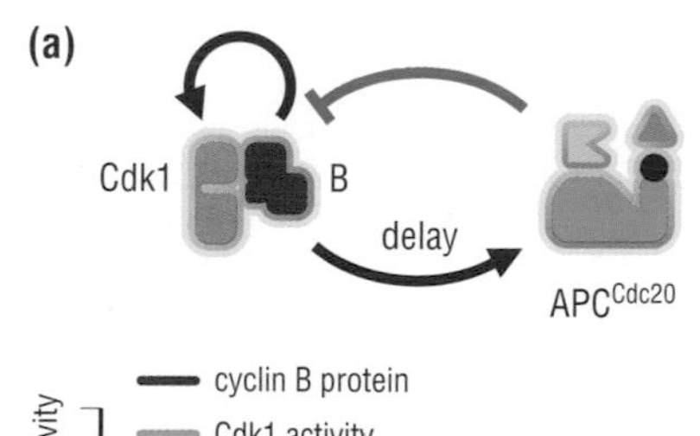

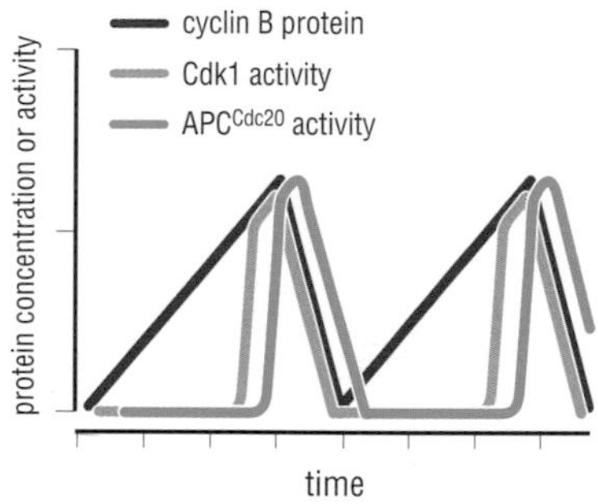

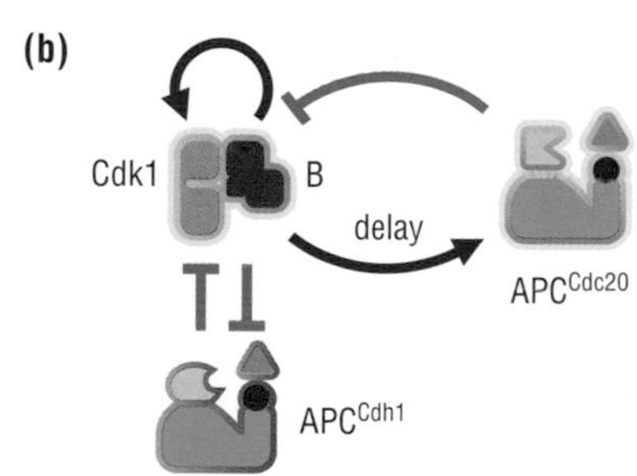

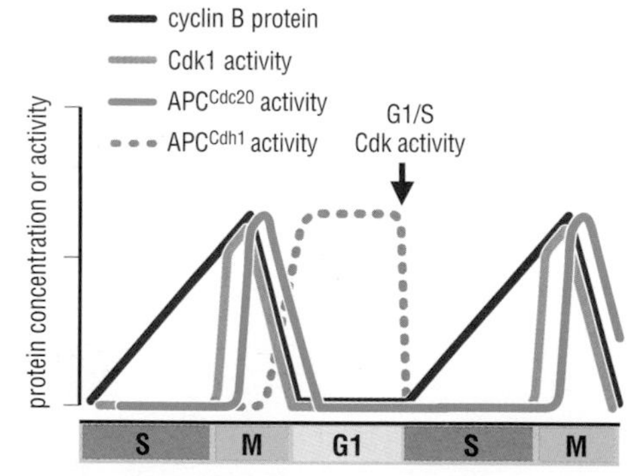

Figure 3-30 Assembling a Cdk oscillator **(a)** This simple model system resembles the cell-cycle control system of frog embryonic cells. Oscillations in the activity of cyclin B–Cdk1, an M–Cdk, occur when increased Cdk1 activity triggers the activation of APC^{Cdc20}, causing rapid cyclin B destruction and Cdk1 inactivation. After Cdk1 inactivation, APC^{Cdc20} activity also declines, allowing cyclin B accumulation—and a new cycle—to begin again. **(b)** The addition of APC^{Cdh1} to the system in (a) leads to the appearance of a prolonged state of Cdk inactivity after mitosis, because Cdh1 is activated by Cdk inactivation. Only when some externally regulated mechanism inactivates Cdh1 in late G1 (G1/S–Cdk activity, as shown here by the red arrow) can cyclin B accumulate and the cell cycle begin again.

Definitions

negative feedback oscillator: regulatory system in which a regulatory component activates its own inhibitor after a delay, resulting in oscillations in the activity of that component.

relaxation oscillator: oscillating system in which a regulatory component oscillates between two stable states, generally as a result of positive feedback.

References

Cross, F.R.: **Two redundant oscillatory mechanisms in the yeast cell cycle.** *Dev. Cell* 2003, **4**:741–752.

Ferrell, J.E. Jr: **Self-perpetuating states in signal transduction: positive feedback, double-negative feedback and bistability.** *Curr. Opin. Cell Biol.* 2002, **14**:140–148.

Pomerening, J.R. *et al.*: **Building a cell cycle oscillator: hysteresis and bistability in the activation of Cdc2.** *Nat. Cell Biol.* 2003, **5**:346–351.

Pomerening, J.R. *et al.*: **Systems-level dissection of the cell-cycle oscillator: bypassing positive feedback produces damped oscillations.** *Cell* 2005, **122**:565–578.

Tyson, J.J. *et al.*: **Sniffers, buzzers, toggles and blinkers: dynamics of regulatory and signaling pathways in the cell.** *Curr. Opin. Cell Biol.* 2003, **15**:221–231.

3-12 Transcriptional Control of Cell-Cycle Regulators

Cell-Cycle Transcription Factors		
Factor	**Subunits**	**Gene example**
Start		
SBF	Swi6 + Swi4	*CLN1,2*
MBF	Swi6 + Mbp1	*CLB5,6*
G2/M transition		
Mcm1–Fkh (SFF complex)	Mcm1 + Fkh1/2 + Ndd1	*CLB1,2*
Late mitosis		
Ace2	Ace2	*SIC1*
Swi5	Swi5	*SIC1*
Mcm1	Mcm1	*CLN3, SWI4*

Figure 3-31 **Table of major gene regulatory factors controlling Cdk activity in budding yeast**

A sequential program of gene expression contributes to cell-cycle control

Oscillations in Cdk activity during the cell cycle are driven not only by mechanisms involving protein phosphorylation, subunit binding and regulated proteolysis, but also by changes in the transcription of regulatory genes. Regulation of gene transcription is particularly important in controlling synthesis of the cyclins, and in the next section we shall see how the programmed sequential activation of the cyclin genes underlies the sequential activation of Cdks during the cell cycle. In this section we describe the known gene regulatory proteins and the general character of the genes they control. The main transcriptional control points are the Start and G2/M checkpoints, as well as the exit from mitosis. Each of these is controlled by a different set of gene regulatory proteins that are activated at the appropriate time by preceding cell-cycle events.

Expression of a large fraction of the genes in the yeast genome is regulated during the cell cycle

The importance of transcriptional regulation in cell-cycle control is reflected in the fact that in budding yeast about 800 genes, or about 15% of the protein-coding genes in the yeast genome, display significant changes in expression during the cell cycle. These genes can be roughly divided into groups according to the cell-cycle stage at which peak expression occurs. The largest group of genes, with about 300 members, comprises those expressed as the cell progresses through the Start checkpoint in late G1. These include the genes for the G1/S cyclins Cln1 and Cln2 and the S cyclins Clb5 and Clb6, as well as genes encoding the enzymes required for chromosome duplication and other S-phase events. Another large group of genes, numbering about 120, is expressed at the G2/M transition and during mitosis and includes the gene encoding the M cyclin Clb2. An additional group of roughly 110 genes, including the gene encoding the Cdk inhibitor Sic1, is expressed in late mitosis and G1.

Key gene regulatory proteins in yeast are activated at the major cell-cycle transitions

In budding yeast, gene expression at Start is controlled primarily by two gene regulatory complexes called SCB-binding factor (SBF) and MCB-binding factor (MBF), which bind to DNA sequence elements called SCBs and MCBs, respectively, in the promoter regions of their target genes. Both of these factors are heterodimers containing a DNA-binding protein (Swi4 and Mbp1, respectively) and a regulatory subunit (Swi6 in both factors) (Figure 3-31).

Before Start, these factors are suppressed by association with an inhibitor protein called Whi5. In late G1, the activity of the G1–Cdk, Cln3–Cdk1, promotes the inhibitory phosphorylation of Whi5, thereby unleashing active SBF and MBF. Activation of these factors results in the increased expression of a large group of G1/S genes, including the genes encoding G1/S and S cyclins, as mentioned above. Thus, the activation of SBF and MBF promotes G1/S- and S-phase Cdk activities and at the same time provides some of the enzymes and raw materials needed to begin S phase.

As the yeast cell approaches M phase, another gene regulatory protein, the Mcm1–Fkh1/2–Ndd1 complex, stimulates the expression of about 35 G2/M genes encoding mitotic regulatory proteins, including the M cyclin Clb2 and the APC activator Cdc20. In this way Mcm1–Fkh helps stimulate the M–Cdk activity that is required for mitotic entry, at the same time increasing production of regulatory components, such as Cdc20, that will eventually be needed for mitotic exit.

References

Breeden, L.L.: **Periodic transcription: a cycle within a cycle.** *Curr. Biol.* 2003, **13**:R31–R38.

Cho, R.J. *et al.*: **A genome-wide transcriptional analysis of the mitotic cell cycle.** *Mol. Cell* 1998, **2**:65–73.

Costanzo, M. *et al.*: **CDK activity antagonizes Whi5, an inhibitor of G1/S transcription in yeast.** *Cell* 2004, **117**:899–913.

de Bruin, R.A. *et al.*: **Cln3 activates G1-specific transcription via phosphorylation of the SBF bound repressor Whi5.** *Cell* 2004, **117**:887–898.

Dyson, N.: **The regulation of E2F by pRB-family proteins.** *Genes Dev.* 1998, **12**:2245–2262.

Futcher, B.: **Transcriptional regulatory networks and the yeast cell cycle.** *Curr. Opin. Cell Biol.* 2002, **14**:676–683.

Spellman, P.T. *et al.*: **Comprehensive identification of cell cycle-regulated genes of the yeast *Saccharomyces cerevisiae* by microarray hybridization.** *Mol. Biol. Cell* 1998, **9**:3273–3297.

Trimarchi, J.M. and Lees, J.A.: **Sibling rivalry in the E2F family.** *Nat. Rev. Mol. Cell Biol.* 2002, **3**:11–20.

Wittenberg, C. and Reed, S.I.: **Cell cycle-dependent transcription in yeast: promoters, transcription factors, and transcriptomes.** *Oncogene* 2005, **24**:2746–2755.

Zheng, N. *et al.*: **Structural basis of DNA recognition by the heterodimeric cell cycle transcription factor E2F–DP.** *Genes Dev.* 1999, **13**:666–674.

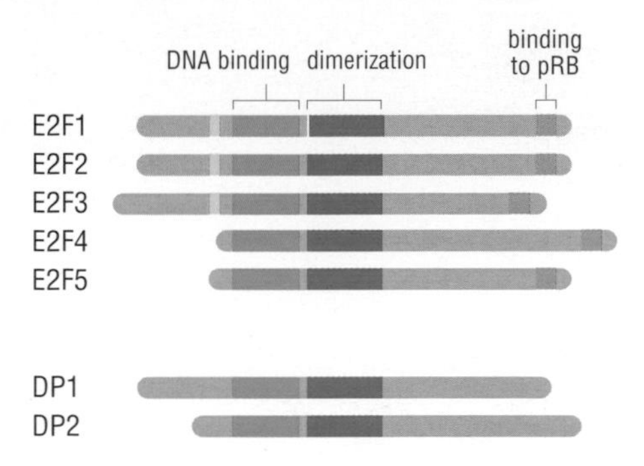

Figure 3-32 E2F family members and DP family members Schematic representations of the five major mammalian E2F proteins and two mammalian DP proteins are aligned to show their DNA-binding and dimerization domains, as well as domains that interact with members of the pRB family of regulatory proteins. Three of the E2F proteins contain short RXL motifs (yellow) that are thought to enhance their phosphorylation by cyclin–Cdk complexes (see section 3-5). Several additional E2F proteins exist, as discussed in Chapter 10.

In late mitosis, the activation of two gene regulatory factors, Swi5 and Ace2, leads to increased expression of about 30 genes—called the M/G1 genes—that encode various components involved in mitotic exit and the establishment of the next G1 phase. An important target of Swi5 and Ace2 is the gene encoding Sic1, the Cdk inhibitor protein that helps suppress Cdk activity in late mitosis and early G1.

Another group of M/G1 genes is regulated by Mcm1, the same factor that helps control G2/M gene expression. Mcm1 binds a DNA sequence called an early cell cycle box (ECB) in the promoters of several M/G1 target genes. From late G1 to early mitosis, expression of these genes is reduced because these promoters also contain binding sites for repressor proteins called Yox1 and Yhp1. In late mitosis, removal of these repressors allows Mcm1 to stimulate expression of its M/G1 target genes—including the genes encoding Cln3 and Swi4, whose slightly increased levels at the end of mitosis help prepare the cell for entry into the next cell cycle.

Although we have some knowledge of the regulatory proteins controlling the expression of many cell-cycle-regulated genes in yeast, several hundred genes seem to be controlled by unidentified mechanisms, emphasizing how much remains to be learned about this important aspect of cell-cycle control.

The E2F family controls cell-cycle-dependent changes in gene expression in metazoans

Cell-cycle-dependent gene regulation in animals is less well understood than in yeast and is focused primarily on the mechanisms governing cell-cycle entry at the Start checkpoint, where gene expression is controlled by members of a family of protein complexes collectively referred to as E2F. E2F complexes are heterodimers containing one subunit from the E2F family and one from the DP family (Figure 3-32). The three-dimensional structure of one E2F complex is shown in Figure 3-33. Each of these families contains multiple members: for example, there are at least eight E2F genes and two DP genes in mammals, and two E2F genes and one DP gene in *Drosophila*. In many respects, the function of E2F is roughly equivalent to that of SBF and MBF in yeast: E2F complexes regulate the expression of G1/S and S cyclins (cyclins E and A, respectively) as well as the expression of genes encoding various enzymes and other components required for the initiation of DNA synthesis.

The functions of E2F–DP complexes are regulated in part by interactions with members of the pRB family of proteins. These interact with E2F during G1 phase to inhibit the expression of G1/S genes that promote entry into the cell cycle. Lack of the protein pRB, for example, helps promote unregulated cell proliferation, leading to the retinal tumor retinoblastoma. The precise mechanisms by which pRB-family members inhibit gene expression vary with the different pRB and E2F proteins. In some cases, E2F complexes act as direct activators of G1/S gene expression, and the binding of the pRB protein simply inhibits this function. In other cases, the binding of a pRB protein to E2F creates a transcriptional repressor complex that prevents the expression of certain G1/S genes.

The pRB proteins thus serve as inhibitors of G1/S gene expression and entry into the cell cycle. In this respect they are functionally analogous to the Whi5 protein of budding yeast. Like Whi5, pRB proteins are inactivated at Start when they are phosphorylated by G1–Cdks—primarily cyclin D-Cdk4 complexes. Phosphorylation triggers dissociation of pRB–E2F complexes, thereby initiating G1/S gene expression and progression through Start and into S phase.

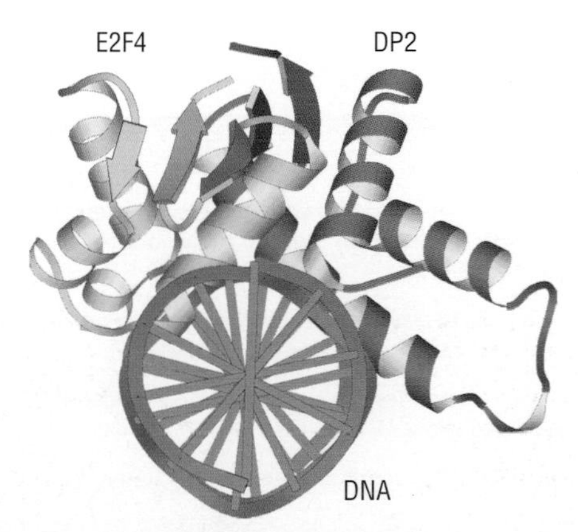

Figure 3-33 Structure of the E2F4–DP2–DNA complex The DNA-binding and dimerization domains of the E2F4–DP2 dimer were analyzed by X-ray crystallography, revealing similarities in the tertiary structure of the two proteins and showing the symmetrical basis of DNA binding. (PDB 1cf7)

The order of cell-cycle events is determined by regulatory interactions between multiple oscillators

Ancestral eukaryotic cell cycles may have operated with only a single cyclin–Cdk oscillator, but modern eukaryotes employ multiple different cyclin–Cdk complexes that are activated and inactivated in a fixed sequence. How is this order achieved? The answer lies in the intrinsic programming of the cell-cycle control system, which ensures that each Cdk promotes the activation of the next Cdk in the sequence. These mechanisms are best illustrated in budding yeast, in which the order of cyclin–Cdk activities is established by a complex regulatory network based on the gene regulatory factors discussed in section 3-12. The following brief description of this network serves as a foundation for more detailed discussions in later chapters.

In early G1, the activity of most Cdks is suppressed by three mechanisms: the low level of cyclin gene expression, the presence of the Cdk inhibitor Sic1 and cyclin ubiquitination by APC^{Cdh1} (see section 3-11). These inhibitory factors do not prevent the growth-dependent accumulation of the G1 cyclin Cln3, however, and thus they do not block the gradual increase in Cln3–Cdk1 activity in G1. In late G1, this activity reaches a threshold level that triggers activation of the gene regulatory factors SBF and MBF, which stimulate the expression of genes encoding G1/S cyclins (Cln1 and 2) and S cyclins (Clb5 and 6) (Figure 3-34).

A complex interplay between Cdks and their inhibitors at Start leads to the sequential activation of G1/S– and S–Cdks. Because G1/S cyclin–Cdk complexes are resistant to Sic1 and are not targeted by APC^{Cdh1}, G1/S–Cdk activity rises unchallenged in late G1 and begins to phosphorylate Cdh1, thereby reducing APC activity. Coupled with increased S-cyclin gene expression, this allows the accumulation of S–Cdk complexes. The kinase activity of these complexes is blocked initially by Sic1, resulting in the brief accumulation of a stockpile of inactive S–Cdks. Once G1/S–Cdk activity accumulates to high levels in late G1, however, G1/S–Cdks phosphorylate Sic1, resulting in its SCF-dependent ubiquitination and destruction (see section 3-9). Sic1 destruction unleashes the S–Cdks, which initiate chromosome duplication in S phase.

Also in late G1, activated S–Cdks collaborate with G1/S–Cdks to complete Cdh1 phosphorylation and thus APC inactivation. Together with Sic1 destruction this allows M cyclins to start accumulating despite low levels of M cyclin gene expression. The activation of G1/S gene expression therefore leads not only to the execution of S-phase events but also initiates a more gradual progression toward M phase. Rising M–Cdk activity activates the next gene regulatory factor in the sequence, Mcm1–Fkh, which stimulates the expression of M cyclin and other genes required for mitosis (see Figure 3-34). The resulting wave of M–Cdk activity initiates entry into mitosis.

The completion of mitosis, as discussed earlier, occurs when the mitotic Cdks activate APC^{Cdc20}, which occurs after a delay and triggers the metaphase-to-anaphase transition and cyclin destruction (see section 3-10). Cdk inactivation in late mitosis also activates APC^{Cdh1} and prevents destruction of the Cdk inhibitor Sic1. Reduced Cdk activity also leads to activation of the M/G1 gene regulators, Swi5 and Ace2, which are normally inhibited by phosphorylation by Cdks. Swi5 and Ace2 stimulate the expression of Sic1 and other proteins that cause Cdk inactivation (see Figure 3-34). After cell division, therefore, the system has returned to a stable G1 state of low Cdk activity, where it is poised to begin the cycle again.

An important feature of this system is that activation of each regulatory component, including gene regulatory factors and Cdks, is governed by ultrasensitive responses and positive feedback loops, resulting in switch-like behavior. We have already seen some of the mechanisms by which switch-like Cdk activation is achieved (see section 3-8), and others will be discussed in

References

Chen, K.C. *et al.*: **Integrative analysis of cell cycle control in budding yeast.** *Mol. Biol. Cell* 2004, **15**:3841–3862.

Cross, F.R.: **Two redundant oscillatory mechanisms in the yeast cell cycle.** *Dev. Cell* 2003, **4**:741–752.

Cross, F.R. *et al.*: **Testing a mathematical model of the yeast cell cycle.** *Mol. Biol. Cell* 2002, **13**:52–70.

Futcher, B.: **Transcriptional regulatory networks and the yeast cell cycle.** *Curr. Opin. Cell Biol.* 2002, **14**:676–683.

Li, F. *et al.*: **The yeast cell-cycle network is robustly designed.** *Proc. Natl Acad. Sci. USA* 2004, **101**:4781–4786.

Tyson, J.J. *et al.*: **Network dynamics and cell physiology.** *Nat. Rev. Mol. Cell Biol.* 2001, **2**:908–916.

Tyson, J.J. *et al.*: **Sniffers, buzzers, toggles and blinkers: dynamics of regulatory and signaling pathways in the cell.** *Curr. Opin. Cell Biol.* 2003, **15**:221–231.

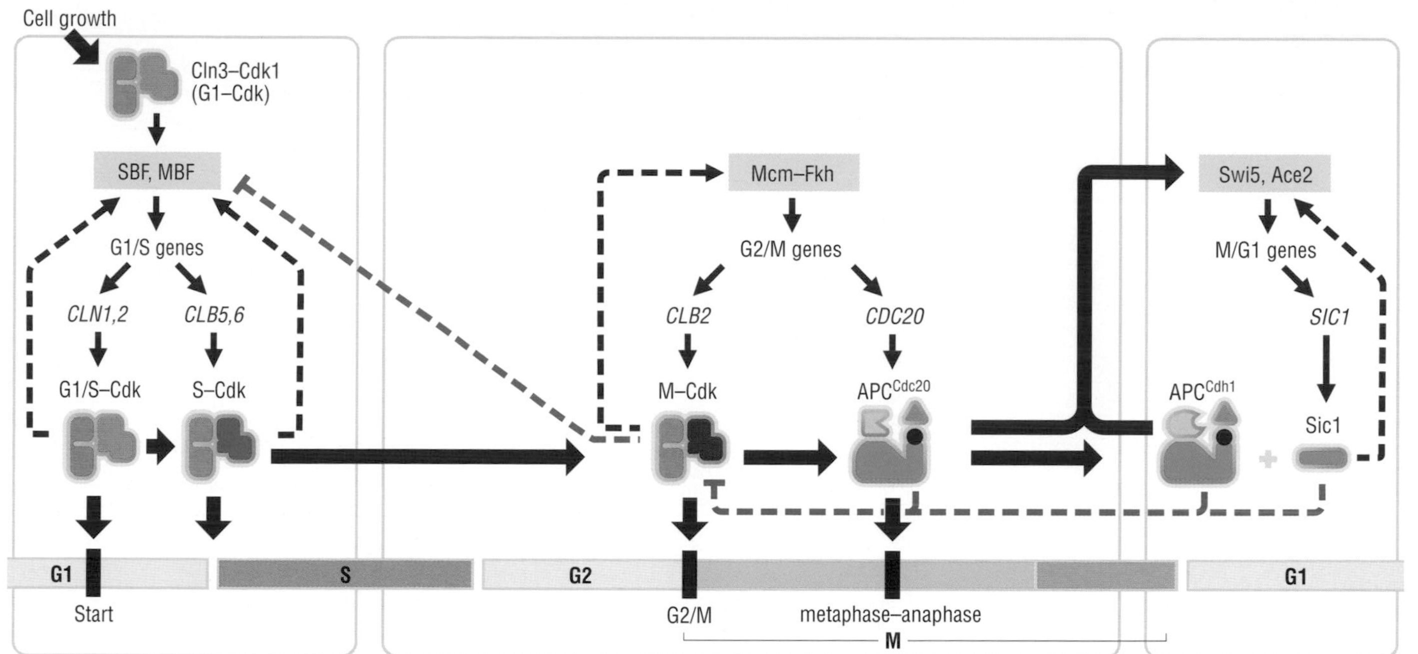

later chapters. Transcriptional regulators are also governed in part by positive feedback loops. SBF activation, for example, is promoted by G1/S–Cdks as well as G1–Cdks, whereas Mcm1–Fkh is activated by M–Cdks. The M/G1 gene regulatory protein Swi5 promotes its own activation, because the product of one of its target genes, Sic1, inhibits Cdk activity. These feedback relationships may not be absolutely essential for cell-cycle timing but they greatly enhance the reliability and robustness of the Cdk switches that drive cell-cycle events.

Negative feedback is also present at multiple points in this system, and helps to generate Cdk oscillations. The activation of SBF and MBF, for example, leads eventually to activation of M–Cdks, which then feed back to inhibit SBF and MBF. As a result, the expression of G1/S genes, including those encoding G1/S and S cyclins, declines as the cell proceeds into mitosis. Similarly, the inhibition of Cdk activity that occurs in late mitosis results in deactivation of Mcm1–Fkh, shutting down mitotic gene expression as the cell progresses into G1.

The elegant design of this network not only leads to the appropriate sequence of switch-like Cdk oscillations but also yields a system that is essentially autonomous: after SBF and MBF have been activated in G1, the system is committed to completing the entire sequence of events without the need for further external input.

The features of the cell-cycle control system can be reproduced and studied in some detail with mathematical models. These models are typically based on the integration of a large number of differential rate equations, each describing how the activity of a component of the system changes in response to changes in other components. Although the development of these models is still in its early stages, such approaches have the potential to provide important insights into the complex behaviors of the cell-cycle control system under various conditions.

Figure 3-34 Overview of the cell-cycle control system of budding yeast Three major sets of gene regulatory factors provide the underlying framework for an autonomous control system that triggers cell-cycle events in the correct order. The sequence of events begins with a stable resting state in G1. Rising Cln3–Cdk1 activity sets the system in motion by activating SBF/MBF, after which the flow of regulatory signals proceeds forward through the various Cdks and gene regulatory factors as shown by the solid red arrows, leading to ordered progression through the stages of the cell cycle and back to the stable G1 state. Positive feedback enhances the activation of each gene regulatory factor (dashed red arrows), and negative feedback allows some components to inhibit previous components in the sequence (dashed blue lines). Many of the regulatory interactions shown here are highly simplified and will be explored in more detail in later chapters.

The cell-cycle control system is responsive to many external inputs

Although the cell-cycle control system is programmed, in some cells at least, to drive essentially autonomous progression through the cell cycle, it is also equipped in most cells with a variety of reversible, regulated braking mechanisms that can be used to pause the cell-cycle control system when conditions are not ideal or when cell proliferation is not needed. Failure to complete mitotic spindle assembly, for example, leads to the generation of a negative signal that prevents APC^{Cdc20} activation, so that the onset of anaphase is blocked until the cell is fully prepared. Similar signaling pathways send negative signals to other components of the cell-cycle control system in the presence of DNA damage or when environmental factors restrict cell growth or cell proliferation. The control of progression through Start by extracellular factors is a particularly critical mechanism by which multicellular animals govern the production of new cells in the growth and maintenance of tissues.

4

Chromosome Duplication

During S phase, the long DNA molecule at the heart of each chromosome is copied with remarkable speed and accuracy. DNA synthesis begins at replication origins, where a complex assembly of proteins initiates replication, strictly once per cell cycle, in response to signals from the cell-cycle control system. The protein apparatus that packages the chromosomal DNA is also duplicated during S phase.

DNA synthesis begins at replication origins

A eukaryotic chromosome is an immense molecular assembly containing a long thread of DNA packaged by proteins into a durable and compact structure. Armies of enzymes move along the DNA, controlling gene expression or repairing DNA damage. Duplicating this vast and dynamic DNA–protein machine—like duplicating the cell itself—is not a simple process.

Chromosome duplication occurs in the S phase of the cell cycle. The central event in this process is the replication of the DNA (Figure 4-1). **DNA polymerases**, the enzymes that copy the template DNA strands into new complementary DNA strands, are assembled with a host of accessory components into a protein machine that travels along the DNA molecule, unwinding the helix and synthesizing complementary copies of each strand. Before DNA replication can start, however, the DNA helix must be opened and unwound. This occurs at specialized sites called **replication origins**, where a complex of *initiator proteins* binds and opens the DNA, making two Y-shaped DNA structures called **replication forks** (see Figure 4-1). Polymerases and other replication proteins are recruited to these forks, which then move outwards in both directions from the origin as the DNA is replicated. Duplication of chromosomal DNA therefore occurs in two distinct steps: initiation at replication origins and elongation away from them.

The giant DNA molecules of eukaryotic chromosomes need to be copied rapidly to ensure that replication is completed in a relatively short period of time. The DNA synthetic machinery replicates DNA at a rapid rate of about 5–100 nucleotides per second, depending on the amount of protein packaging present. In addition, eukaryotic chromosomes contain large numbers of replication origins, allowing many different regions in the same chromosome to be duplicated simultaneously. Replication forks traveling out from one origin meet forks from adjacent origins, or reach the end of a chromosome, until the entire chromosome is duplicated.

To ensure the accurate transmission of the genome, the DNA synthetic apparatus makes very few errors—about one in 10^9 nucleotides. The integrity of the DNA is also maintained by DNA damage response mechanisms, discussed in Chapter 11, which detect damaged DNA and delay its duplication until the damage is repaired.

The cell-cycle control system activates replication origins only once in each S phase

DNA replication is an all-or-none process. Once the duplication of a DNA molecule begins, it normally proceeds to completion. If this were not the case, then incompletely replicated chromosomes might be pulled apart and broken during mitosis. Once replication is complete, it does not occur again in the same cell cycle. This prevents the daughter cell from inheriting an excessive, and potentially unhealthy, amount of any DNA sequence.

The once-and-only-once nature of DNA replication is achieved by dividing the initiation process into two temporally distinct steps. First, in late mitosis and early G1, a large complex of initiator proteins, called the *prereplicative complex*, assembles at origins and prepares them for firing. This is sometimes called origin licensing. Second, in early S phase, the prereplicative complex is transformed into an active *preinitiation complex* that unwinds the origins and loads the DNA synthesis machinery (see Figure 4-1). Once an origin has been activated, the prereplicative complex disassembles, and its reassembly is prevented until the next G1. As a result, each origin is used once, and only once, per cell cycle.

Definitions

chromatin: the complex of DNA and protein that forms a chromosome.

DNA polymerase: enzyme that synthesizes new DNA by copying a single-stranded DNA template. The polymerase moves along the template and synthesizes a new strand of complementary DNA sequence by adding nucleotides, one at a time, onto the 3′ OH end of the new strand.

euchromatin: chromatin in which DNA is packaged in such a way as to be accessible to enzymes and gene regulatory proteins.

heterochromatin: chromatin in which DNA is packaged in such a way as to be poorly accessible to enzymes and gene regulatory proteins.

replication fork: the site at which DNA strands are separated and new DNA is synthesized. It is a Y-shaped structure and moves away from the site of replication initiation. Both strands of the DNA are copied at the replication fork.

replication origin: site or region in a chromosome which DNA synthesis is initiated by unwinding of the double helix and assembly of the DNA synthetic machinery.

References

Alberts, B. *et al.*: *Molecular Biology of the Cell* 4th ed. (Garland Science, New York, 2002).

Bell, S.P. and Dutta, A.: **DNA replication in eukaryotic cells.** *Annu. Rev. Biochem.* 2002, **71**:333–374.

Waga, S. and Stillman, B.: **The DNA replication fork in eukaryotic cells.** *Annu. Rev. Biochem.* 1998, **67**:721–751.

The cell-cycle control system governs the timing of both steps in the initiation of DNA synthesis. First, assembly of the prereplicative complex is inhibited by Cdks and stimulated, in some cells at least, by the ubiquitin-protein ligase APC. Thus, assembly occurs only in late mitosis and early G1, when Cdk activity is low and APC activity is high (see Figure 4-1). Second, initiation of DNA synthesis at licensed origins is triggered in S phase by S–Cdks. Because Cdk activities remain high (and APC activity is low) until late mitosis, as described in Chapter 3, new prereplicative complexes cannot be assembled at origins until the following G1. The mechanisms by which DNA replication is initiated and how it is controlled are described in the first part of this chapter (sections 4-2 to 4-8).

Chromosome duplication requires duplication of chromatin structure

The DNA in a chromosome is extensively packaged into an elaborate and poorly understood DNA–protein assembly called **chromatin**. The amount of protein packaging varies dramatically in different regions of the genome: some regions, called **heterochromatin**, are so tightly compacted that they are less accessible to the protein machines that govern gene expression or detect and repair DNA damage. For example, the ends of chromosomes—the *telomeres*—are extensively packaged by specialized proteins that block expression of genes in nearby regions. Other chromosome regions, called **euchromatin**, contain DNA that is more accessible to regulatory factors.

During S phase, it is not enough simply to duplicate the DNA molecule at the core of each chromosome. Chromatin structure must also be reproduced in each daughter chromosome. The cell synthesizes large amounts of the various proteins that package and maintain the DNA, and in most cases assembles these proteins on the new DNA as they were in the parent chromosome. Remarkably little is known about how this is achieved or regulated. The duplication of chromatin structure is discussed in the second part of this chapter (sections 4-9 to 4-13).

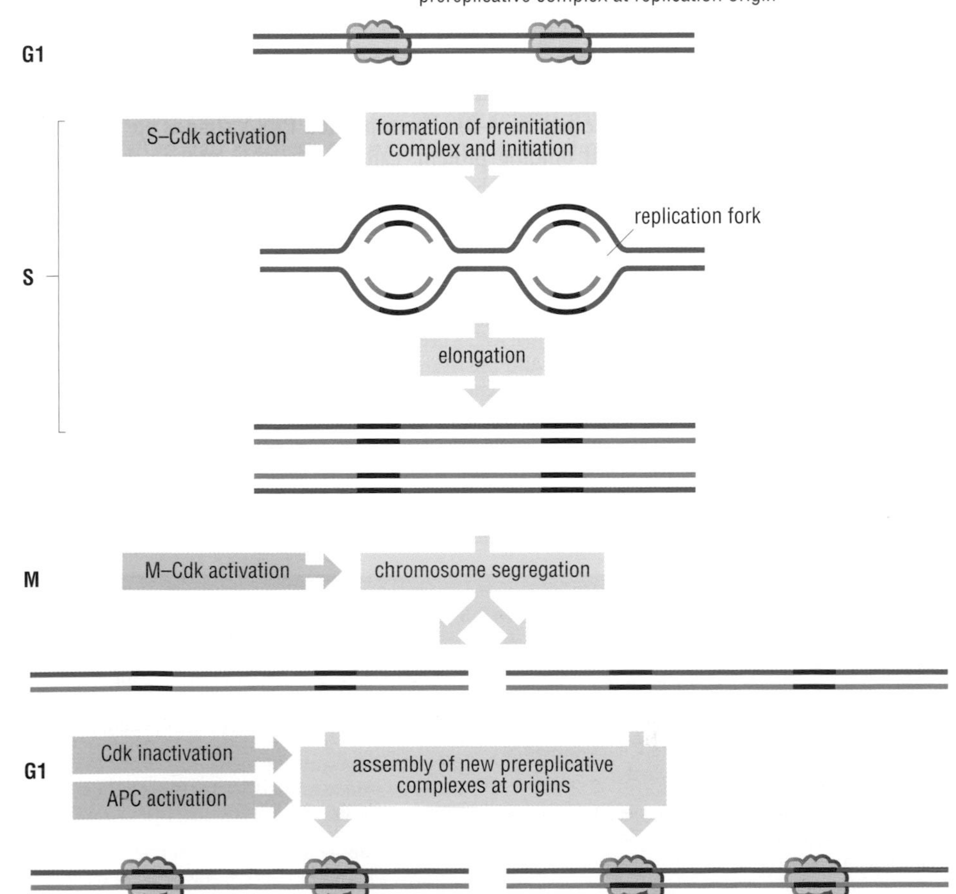

Figure 4-1 Overview of chromosome duplication in the cell cycle The chromosome is prepared for DNA duplication in G1, when prereplicative complexes are assembled at replication origins (red). Transformation of these complexes to preinitiation complexes and activation of the origin in S phase results in the unwinding of the DNA helix and initiation of DNA replication. Two replication forks move out from each origin until the entire chromosome is duplicated. Segregation of the duplicated chromosomes in M phase then results in two daughter cells with identical chromosomes. The activation of replication origins also causes disassembly of the prereplicative complex. Because new prereplicative complexes cannot be formed at origins until the following G1, each origin can be activated only once in each cell cycle.

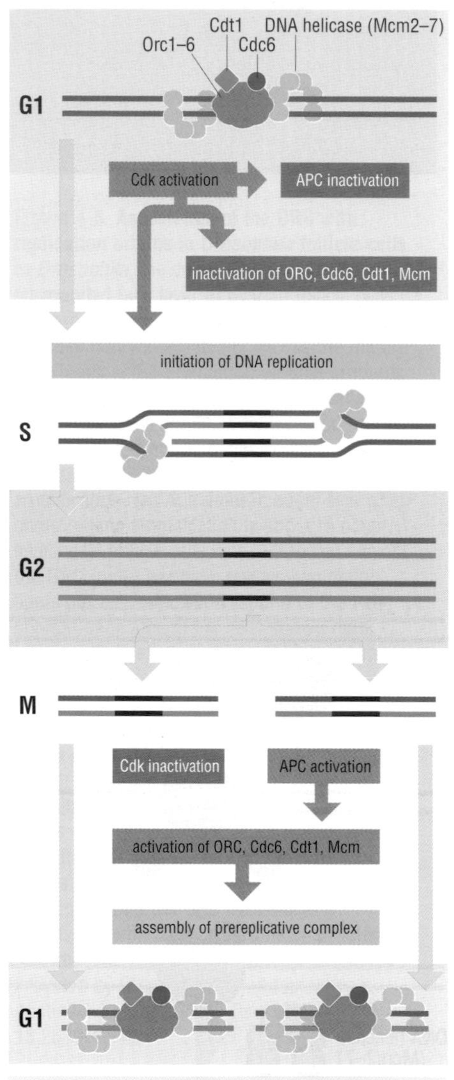

Figure 4-11 Mechanisms that limit prereplicative complex formation to G1 At the end of G1, the activation of Cdks triggers origin activation (see Figure 4-23) and also causes inactivation of the proteins that form the pre-RC. Inactivation is caused either by the direct action of Cdks on these proteins, or, in metazoans, by the indirect actions of Cdks in inactivating the APC and thereby stabilizing geminin, an inhibitor of Cdt1. The proteins that form the pre-RC remain inactive throughout S phase and into M phase, until Cdk activity declines in late mitosis and its inhibitory effects are lifted. APC activity also rises in late mitosis, promoting geminin destruction and thereby allowing Cdt1 to take part in pre-RC assembly. The ORC remains at the origin after G1 but is not shown here for simplicity.

Assembly of prereplicative complexes is restricted to G1 by multiple mechanisms

A simple and elegant mechanism ensures that DNA is replicated only once in each cell cycle. When DNA synthesis is initiated and the Mcm complex moves away from an origin along with the two replication forks, the pre-RC at that origin is dismantled and pre-RC components are destroyed or inhibited. A new Mcm complex cannot be loaded at a fired origin until pre-RC subunits are reactivated at the end of mitosis, thereby ensuring that one origin cannot be used twice in the same cell cycle.

In most cells, pre-RC assembly is limited to late mitosis and early G1 by the simple fact that assembly is inhibited by Cdk activity. When S–Cdks are activated in early S phase and trigger the initiation of DNA synthesis, they also promote the destruction or inhibition of individual pre-RC components, preventing immediate reassembly. S– and M–Cdks continue to block pre-RC assembly even after S phase is complete, thereby ensuring that assembly cannot occur again until all Cdk activities are reduced in late mitosis (Figure 4-11). In multicellular eukaryotes, pre-RC assembly is also limited to late mitosis and early G1 by the ubiquitin-protein ligase APC, which promotes pre-RC assembly by triggering the destruction of an inhibitor of assembly called geminin. Thus, APC inactivation in late G1 (see section 3-10) contributes to the inhibition of pre-RC assembly, and APC activation in late mitosis helps to promote it.

Prereplicative complex components are destroyed or inhibited in yeast as a result of Cdk activity

In budding yeast, in which regulation of pre-RC assembly is best understood, inhibition of assembly is due mainly to Cdk-dependent phosphorylation of various pre-RC components. This has a variety of consequences. Phosphorylation of Cdc6 by Cdk1, which is activated in late G1, results in binding of Cdc6 to the ubiquitin-protein ligase SCF^{Cdc4} (see section 3-9), leading to its proteolytic destruction. Cdc6 is therefore present in the cell only during G1, when Cdk1 is inactive, and is rapidly degraded at the onset of S phase and during the rest of the cell cycle. There is also evidence that phosphorylation of Cdc6 reduces its intrinsic ability to promote Mcm loading at the origin.

The assembly of Mcm proteins is prevented by their export out of the nucleus, which is promoted by Cdk-dependent phosphorylation. The pre-RC component, Cdt1, associates with the Mcm proteins and is thus also exported from the nucleus during S phase. By promoting the export of Mcm proteins rather than their inhibition, Cdks help prevent the loading of new Mcm complexes at origins during S phase but do not affect the Mcm complexes that are already in place on the DNA—where they are carrying out a vital function as DNA helicases in DNA replication (see section 4-3).

Pre-RC assembly is also inhibited by phosphorylation of ORC subunits, but the mechanism of inhibition is not clear in this case. Because the ORC is bound to replication origins throughout the cell cycle, even when Cdk activity is high, phosphorylation presumably acts by inhibiting its ability to bind other components of the pre-RC, rather than its binding to DNA. This inhibition of protein–protein interactions may also occur, at least in part, as a result of direct binding between the hydrophobic patch of the S-cyclin Clb5 and an RXL motif on ORC subunits (see section 3-5).

Multiple mechanisms thus ensure that the pre-RC is assembled in budding yeast only in G1. The system is an excellent example of a highly robust regulatory network that continues to function even if some mechanisms fail. If any single mechanism is blocked experimentally, pre-RC assembly is still limited to G1 and inappropriate reinitiation at origins does not occur. For example, budding yeast carrying a mutant Cdc6 protein that cannot be phosphorylated by Cdks still do not replicate their DNA more than once in each cell cycle, even though the mutant Cdc6 protein is stable throughout the cell cycle. Similarly, single mutations that prevent ORC phosphorylation or block Mcm nuclear export do not cause rereplication. Mutation of all three proteins in the same cell, however, does trigger reinitiation at many origins (Figure 4-12), indicating that the system eventually fails if several inhibitory mechanisms are inactivated.

Pre-RC assembly is controlled in animals by both Cdks and the APC

An important inhibitor of pre-RC assembly in animal cells is the protein geminin, which binds to Cdt1 and prevents it from binding to the ORC. Levels of geminin are kept low in G1, allowing pre-RC assembly, by the APC, which ubiquitinates geminin and thereby targets it for degradation. Destruction of geminin results in the release of Cdt1, which can then perform its function in pre-RC assembly. When the APC is inactivated at the end of G1, the accumulation of geminin takes Cdt1 out of action once again, binding Cdt1 from S phase through to late mitosis. Inhibition of geminin synthesis in fly and human cells by the technique of RNA interference (RNAi) (see section 2-5) leads to partial rereplication of DNA, arguing that geminin is a critical regulator of pre-RC assembly.

Cdks also contribute to pre-RC regulation in metazoan cells, but their effects are poorly understood and seem to vary in different organisms. In human cells, inhibition of Cdk activity causes some rereplication, suggesting that Cdks help inhibit pre-RC formation as they do in yeast. Cdks may act in part by phosphorylating Cdt1, thereby targeting it for ubiquitination by SCF.

The importance of Cdks as inhibitors of pre-RC formation remains unclear in some cases, however. In frog embryo cells, the Cdk that controls DNA replication (cyclin E–Cdk2) is only partly inhibited after mitosis, raising the question of how pre-RC formation can occur—and how premature origin activation is prevented—in the presence of Cdk activity. Inhibition of geminin in late mitosis by the APC is clearly one important mechanism. It cannot be the only one, however, as removal of geminin from frog egg extracts causes little rereplication. The key mechanism seems to depend on the destruction of Cdt1 during S phase. Phosphorylation of Cdt1 by Cdks is not required for this destruction; instead, it is triggered by a ubiquitin-protein ligase that recognizes Cdt1 when it binds to the sliding clamp at fired origins (see section 4-1). A similar mechanism operates in flies and worms. In human cells, this mechanism may act in parallel with Cdk-dependent Cdt1 destruction.

Paradoxically, Cdks stimulate pre-RC formation in mammalian cells under some conditions. In quiescent (G0) cells, Cdc6 is highly unstable and pre-RCs are absent. When these cells are stimulated to reenter the cell cycle, cyclin E–Cdk2 phosphorylates Cdc6. This stabilizes Cdc6, thereby allowing pre-RC assembly during the brief period in late G1 before geminin accumulates and inhibits Cdt1. It remains unclear how this mechanism is employed in cells progressing through G1 from mitosis. Indeed, much about pre-RC regulation in animals remains mysterious.

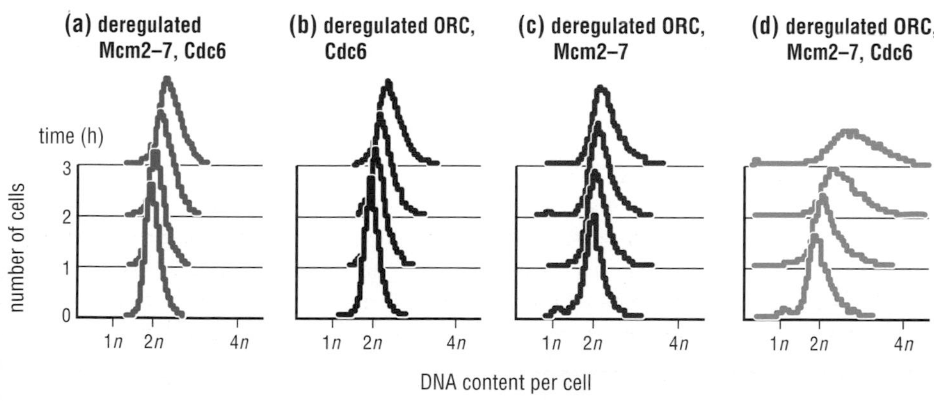

Figure 4-12 Rereplication in yeast cells with deregulated ORC, Mcm2–7 and Cdc6 DNA replication was analyzed in budding yeast strains containing mutations that block the ability of Cdks to inhibit the ORC, Mcm2–7 or Cdc6. Deregulated ORC cells contain Orc2 and Orc6 proteins in which Cdk phosphorylation sites are mutated, thereby preventing phosphorylation. Deregulated Mcm2–7 cells contain a version of Mcm7 carrying a short amino-acid sequence that forces the Mcm2–7 complex to stay in the nucleus even when phosphorylated by Cdks. Deregulated Cdc6 cells contain a truncated Cdc6 protein that lacks Cdk phosphorylation sites. All of these mutant proteins can perform their functions in pre-RC assembly but cannot be inhibited by Cdks. Yeast strains, each carrying two or three of these deregulated mutant proteins, were arrested in mitosis. Over a period of 3 hours, the occurrence of excess DNA replication was assessed using flow cytometry to measure cellular DNA content, as explained in section 2-6. **(a)** In cells containing deregulated Mcm2–7 and Cdc6, DNA content remains constant ($2n$) in mitotic cells, indicating that rereplication is not occurring (the slight rightward shift in apparent DNA content is an artifact caused by increasing cell size during the experiment). Similarly, deregulating the activity of **(b)** the ORC and Cdc6 or **(c)** the ORC and Mcm2–7 in the same cells does not cause excess DNA replication. Thus, mutation of any two of the three proteins does not cause rereplication. **(d)** In cells with mutant forms of all three proteins, excess DNA replication can be seen after 2–3 hours. Detailed analysis of replication origin firing in these cells (not shown) confirms that many (but not all) origins have reinitiated in these cells, indicating that additional mechanisms are still blocking reinitiation at some origins. Rereplication is lethal, and therefore cells carrying mutations in all three proteins do not survive; this experiment was performed by placing the mutant *CDC6* gene under the control of an inducible promoter that was turned on at the zero time point. Courtesy of Van Nguyen and Joachim Li.

References

Arias, E.E. and Walter, J.C.: **PCNA functions as a molecular platform to trigger Cdt1 destruction and prevent re-replication.** *Nat. Cell Biol.* 2006, **8**:84–90.

Bell, S.P. and Dutta, A.: **DNA replication in eukaryotic cells.** *Annu. Rev. Biochem.* 2002, **71**:333–374.

Diffley, J.F.: **Regulation of early events in chromosome replication.** *Curr. Biol.* 2004, **14**:R778–R786.

Li, A. and Blow, J.J.: **Non-proteolytic inactivation of geminin requires CDK-dependent ubiquitination.** *Nat. Cell Biol.* 2004, **6**:260–267.

Mailand, N. and Diffley, J.F.: **CDKs promote DNA replication origin licensing in human cells by protecting Cdc6 from APC/C-dependent proteolysis.** *Cell* 2005, **122**:915–926.

Nguyen, V.Q. *et al.*: **Cyclin-dependent kinases prevent DNA re-replication through multiple mechanisms.** *Nature* 2001, **411**:1068–1073.

Tanaka, S. and Diffley, J.F.: **Interdependent nuclear accumulation of budding yeast Cdt1 and Mcm2-7 during G1 phase.** *Nat. Cell Biol.* 2002, **4**:198–207.

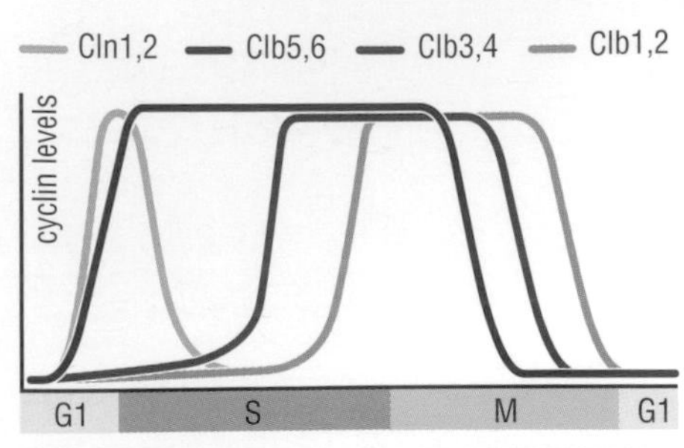

Figure 4-13 Cyclin levels during the budding yeast cell cycle Cln1,2 are G1/S cyclins, Clb5,6 are S cyclins and Clb1–4 are mitotic cyclins.

Cdks and Cdc7 trigger the initiation of DNA replication

Once the formation of prereplicative complexes (pre-RCs) has been completed by the loading of Mcm at origins, the DNA is primed for replication. The signal to start replication comes from the cell-cycle control system. Late in G1, if environmental and other conditions are right, G1 and G1/S cyclin–Cdk complexes are activated. This commits the cell to a new cell cycle and sets the stage for DNA replication by stimulating expression of genes that encode components of the DNA synthetic machinery. G1/S–Cdk activation also promotes the expression and activation of S–Cdk complexes. Replication origins are then activated directly by S–Cdks, G1/S–Cdks, or both, depending on the species and cell type. S-cyclin levels remain elevated throughout S phase, G2 and early mitosis. One important function of continued S–Cdk activity is to suppress the formation of new pre-RCs at spent origins, as discussed earlier (see section 4-4).

Origin activation also requires a second protein kinase called Cdc7, which is activated in late G1, probably as a result of Cdk activation. The function and regulation of Cdc7 are discussed later (see section 4-7). Cdks and Cdc7 collaborate in the firing of origins by promoting the formation of the preinitiation complex at the origin, as will be discussed in section 4-8.

In budding yeast, the cyclins Clb5 and Clb6 are key activators of replication origins

The regulation of DNA replication by Cdks is best understood in *S. cerevisiae*, where the S cyclins Clb5 and Clb6 are primarily responsible for initiating DNA replication. As described in Chapter 3 (sections 3-12 and 3-13), accumulation of the G1 cyclin Cln3 in late G1 triggers the expression of various G1/S genes, including those encoding DNA synthetic machinery, the G1/S cyclins Cln1 and Cln2 and the S cyclins Clb5 and Clb6 (Figure 4-13). The resulting activation of Cln1,2–Cdk1 complexes further stimulates G1/S gene expression and commits

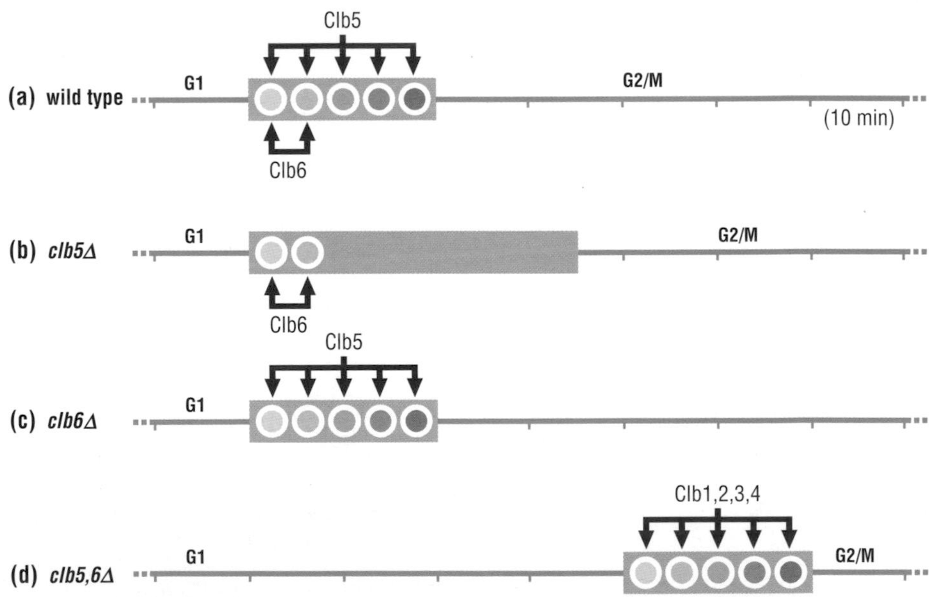

Figure 4-14 Effects of *clb5* and *clb6* mutations on timing of origin activation The S-phase functions of Clb5 and Clb6 in budding yeast are revealed by analysis of cells lacking *CLB5* or *CLB6* genes. In this diagram, progression through the cell cycle in various mutants is represented by the green lines, and the length and timing of S phase are indicated by the blue rectangles. Grey spots represent the firing of replication origins at various times during S phase, with early origins shaded in light grey and late origins shaded in dark grey. **(a)** In wild-type cells, Clb5 and Clb6 both activate early origins, but only Clb5 activates late origins. **(b)** Cells lacking Clb5 (*clb5Δ*) do not efficiently activate late origins, and S phase is prolonged because late-replicating regions are duplicated by the gradual spread of replication forks from early-replicating regions. **(c)** In cells lacking Clb6 (*clb6Δ*), Clb5 is able to activate all origins, and the length of S phase is not significantly affected. **(d)** In cells lacking both Clb5 and Clb6 (*clb5,6Δ*), the onset of S phase is greatly delayed but eventually occurs because one or more of the mitotic cyclins (Clb1,2,3,4) have some ability to activate origins. This delayed S phase is about the normal length, and origins are fired in the correct order. Adapted from Donaldson, A.D. *et al.*: *Mol. Cell* 1998, **2**:173–182.

References

Cross, F.R. *et al.*: **Specialization and targeting of B-type cyclins.** *Mol. Cell* 1999, **4**:11–19.

Donaldson, A.D.: **The yeast mitotic cyclin Clb2 cannot substitute for S phase cyclins in replication origin firing.** *EMBO Rep.* 2000, **1**:507–512.

Donaldson, A.D. *et al.*: ***CLB5*-dependent activation of late replication origins in *S. cerevisiae*.** *Mol. Cell* 1998, **2**:173–182.

Fisher, D.L. and Nurse, P.: **A single fission yeast mitotic cyclin B p34^{cdc2} kinase promotes both S-phase and mitosis in the absence of G_1 cyclins.** *EMBO J.* 1996, **15**:850–860.

Hu, F. and Aparicio, O.M.: **Swe1 regulation and transcriptional control restrict the activity of mitotic cyclins toward replication proteins in *Saccharomyces cerevisiae*.** *Proc. Natl Acad. Sci. USA* 2005, **102**:8910–8915.

Loog, M. and Morgan, D.O.: **Cyclin specificity in the phosphorylation of cyclin-dependent kinase substrates.** *Nature* 2005, **434**:104–108.

the cell irreversibly to passage through the Start checkpoint. Most importantly, Cln1,2–Cdk1 activity triggers destruction of the Cdk inhibitor Sic1 (see section 3-6), thereby allowing Clb5,6–Cdk1 activation. Clb5,6–Cdk1 complexes are then directly responsible for triggering the activation of replication origins. Cln1 and Cln2 therefore act as global promoters of cell-cycle commitment and S-phase entry, whereas Clb5 and Clb6 are the workhorses required throughout S phase to directly activate each origin.

There are subtle and poorly understood functional differences between Clb5 and Clb6. Clb6, for example, seems able to activate only those origins that normally fire early in S phase, whereas Clb5 activates both early and late origins (Figure 4-14). We do not understand the molecular basis for these differences, although they may arise from differences in the levels of expression of the two cyclins or from differences in their intrinsic activity at different origins.

Yeast cells lacking S cyclins can replicate their DNA

There is considerable overlap in the functional specificities of S and M cyclins in yeast. In budding yeast lacking both Clb5 and Clb6 but containing the mitotic cyclins Clb1 to Clb4, the initiation of DNA replication is greatly delayed but eventually occurs at both early and late origins (see Figure 4-14). Similarly, chromosome duplication occurs normally, but is delayed, in fission yeast cells carrying only the mitotic cyclin Cdc13 and lacking the S cyclin Cig2 and its close relative Cig1. DNA duplication in these cases is dependent on mitotic cyclins, which therefore seem able to promote replication origin firing, albeit with delayed timing.

Why is there a delay in the onset of S phase in cells expressing only mitotic cyclins? One simple possibility is that the timing of expression of different cyclins is critical: mitotic cyclins may be fully capable of activating origins but are not expressed until later in the cell cycle (see Figure 4-13). This seems not to be the answer, however. If mutant cells lacking Clb5 are engineered to contain active Clb2–Cdk1 complexes early in the cell cycle (when Clb5 would normally be expressed), S phase is still delayed (Figure 4-15). Clb5 therefore possesses a greater intrinsic ability to activate replication origins.

The ability of Clb5 to promote replication efficiently depends, at least in part, on its hydrophobic patch sequence (see section 3-5), which helps target the associated Cdk1 to specific S-phase protein substrates such as Sld2, a component of the preinitiation complex. The reduced ability of mitotic Clb–Cdk1 complexes to promote replication is due primarily to their reduced affinity for these substrates.

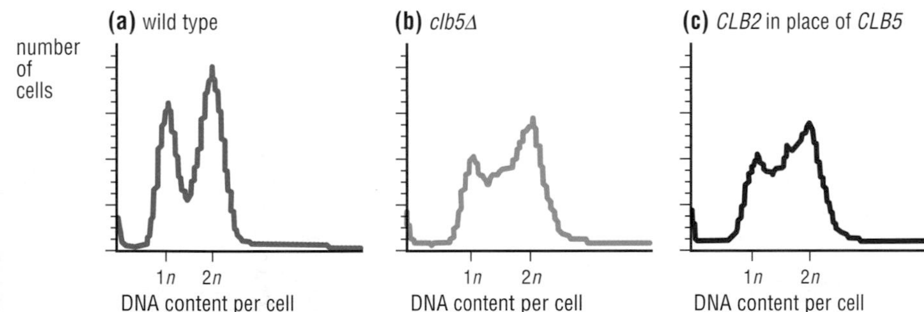

Figure 4-15 **Clb2 is intrinsically less effective than Clb5 in the activation of replication** This experiment was performed to compare the replication functions of the S cyclin Clb5 and the M cyclin Clb2. Progression through the cell cycle was assessed by analysis of cellular DNA content by flow cytometry (explained in section 2-6). **(a)** A wild-type population contains a mixture of cells in multiple cell-cycle stages, including a small number of cells in S phase (DNA content between 1*n* and 2*n*). **(b)** In cells lacking Clb5, the increased length of S phase results in a greater fraction of the cell population with a DNA content that is intermediate between 1*n* and 2*n*. **(c)** Clb2 is normally produced late in the cell cycle. In the cells shown here, however, the protein-coding region of the *CLB5* gene was replaced with that of the *CLB2* gene, so that Clb2 was produced early in the cell cycle when Clb5 would normally be present. The premature expression of Clb2 did not affect the DNA content of the cells in the population, however, indicating that S phase was still delayed. Clb2 is therefore intrinsically less active than Clb5 in the promotion of DNA replication. The higher activity of Clb5 results primarily from its higher affinity for specific S-phase Cdk substrates. In addition, Clb2–Cdk1, when forcibly expressed in S phase, has lower intrinsic kinase activity because it is more sensitive than Clb5–Cdk1 to inhibition by the protein kinase Wee1 (see section 3-3). Adapted from Cross, F.R. *et al.*: *Mol. Cell* 1999, **4**:11–19.

4-6 Cyclins Required for Activation of Replication Origins in Metazoans

Different cyclins control initiation of DNA replication in different stages of animal development

The control of replication initiation by Cdks is less well understood in animals than it is in yeast, and it also seems to be regulated differently in different stages of embryonic development. In the adult cells of vertebrates, as in yeast, a G1/S–Cdk (cyclin E–Cdk2) acts as the global regulator of cell-cycle entry at the Start checkpoint, where it stimulates the expression of cyclin genes and the activation of S–Cdks (cyclin A–Cdk2) that directly activate replication origins. In the early embryos of frogs and *Drosophila*, however, there is no G1, no Start checkpoint and no cyclin gene expression (see sections 2-3 and 2-4), and cyclin E and A have functions that are different from those in adult cells: cyclin E–Cdk2 is the major trigger of replication origin firing, whereas cyclin A–Cdk1 serves primarily in the control of mitosis. In the adult forms of these organisms, it seems that cyclin E and A assume more conventional G1/S- and S-cyclin functions, but these issues remain unclear.

Cyclin A is a major regulator of replication initiation in cultured mammalian cells

The control of replication origin activation in adult mammalian cells is poorly understood, in part because G1/S-regulatory mechanisms in these cells are so complex and in part because rigorous genetic or biochemical dissection of these mechanisms is difficult. Current evidence suggests a method of control similar to that described in the previous section for budding yeast. As we will discuss in more detail in Chapter 10, cell-cycle commitment in late G1 in mammalian cells is triggered by several regulatory mechanisms, including the accumulation of G1 cyclin (cyclin D, which binds and activates Cdk4 and Cdk6). This leads to an increase in the expression of proteins required for DNA synthesis, as well as an increase in the expression of G1/S cyclin (cyclin E, bound to Cdk2). Cyclin E acts as a global regulator of cell-cycle entry at the Start checkpoint and promotes further expression of S-phase proteins, including cyclin A, an S cyclin that binds Cdk2.

Cyclin A levels increase in S phase (Figure 4-16a), and cyclin A–Cdk2 complexes are concentrated at replication foci on the chromosomes. Injection of G1 cells with antibodies against cyclin A blocks the onset of DNA replication. It is therefore likely that, like the Clb5,6–Cdks of budding yeast, cyclin A–Cdk2 complexes are directly involved in firing replication origins in somatic mammalian cells.

DNA replication in frog embryos is triggered by cyclin E–Cdk2

Cyclin E, not cyclin A, is the major controller of replication in the cells of developing frogs and flies. The activation of replication origins in the early frog embryo, for example, is controlled primarily by cyclin E. In early embryonic divisions, the levels of cyclin E (bound to Cdk2)

(a) human somatic cell

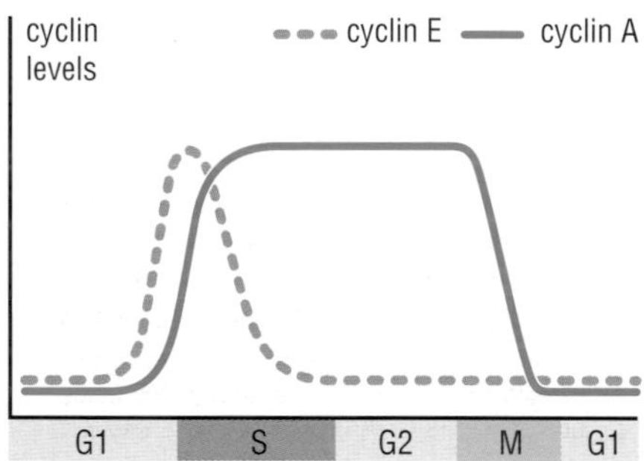

(b) frog early embryonic cell

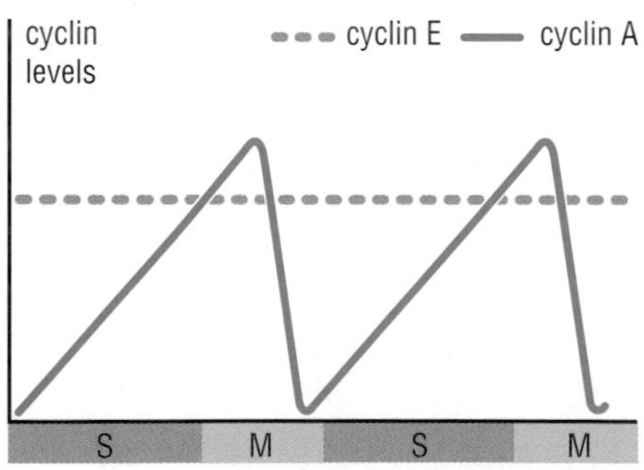

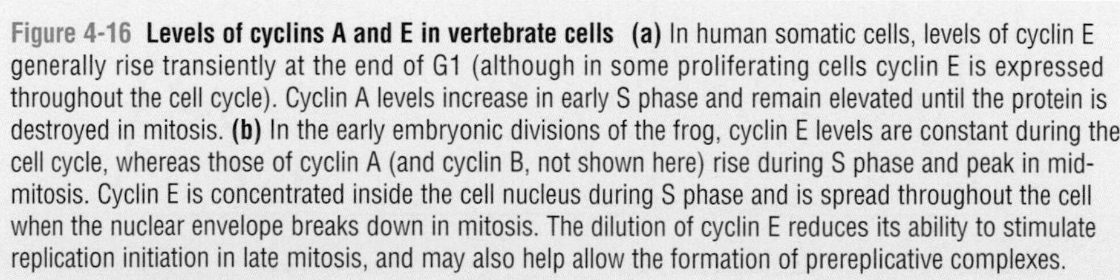

Figure 4-16 Levels of cyclins A and E in vertebrate cells (a) In human somatic cells, levels of cyclin E generally rise transiently at the end of G1 (although in some proliferating cells cyclin E is expressed throughout the cell cycle). Cyclin A levels increase in early S phase and remain elevated until the protein is destroyed in mitosis. **(b)** In the early embryonic divisions of the frog, cyclin E levels are constant during the cell cycle, whereas those of cyclin A (and cyclin B, not shown here) rise during S phase and peak in mid-mitosis. Cyclin E is concentrated inside the cell nucleus during S phase and is spread throughout the cell when the nuclear envelope breaks down in mitosis. The dilution of cyclin E reduces its ability to stimulate replication initiation in late mitosis, and may also help allow the formation of prereplicative complexes.

References

Follette, P.J. *et al.*: **Fluctuations in cyclin E levels are required for multiple rounds of endocycle S phase in *Drosophila*.** *Curr. Biol.* 1998, **8**:235–238.

Follette, P.J. and O'Farrell, P.H.: **Cdks and the *Drosophila* cell cycle.** *Curr. Opin. Genet. Dev.* 1997, **7**:17–22.

Girard, F. *et al.*: **Cyclin A is required for the onset of DNA replication in mammalian fibroblasts.** *Cell* 1991, **67**:1169–1179.

Jackson, P.K. *et al.*: **Early events in DNA replication require cyclin E and are blocked by p21^{CIP1}.** *J. Cell Biol.* 1995, **130**:755–769.

Lilly, M.A. and Spradling, A.C.: **The *Drosophila* endocycle is controlled by cyclin E and lacks a checkpoint ensuring S-phase completion.** *Genes Dev.* 1996, **10**:2514–2526.

remain relatively constant (Figure 4-16b), whereas those of cyclins A and B (bound to Cdk1) oscillate and peak in mitosis. If the production of cyclins A and B is blocked in frog embryo extracts by inhibitors of protein synthesis, DNA replication still occurs, but mitosis does not. If cyclin E is removed with antibodies, replication is blocked. Cyclin E is therefore necessary and sufficient for DNA duplication in the early embryonic cells of the frog.

It is likely, however, that cyclin A makes some contribution to origin activation in frog cells. If DNA replication is blocked in frog embryo extracts by the removal of Cdks, the addition of purified cyclin A–Cdk1 (but not cyclin B–Cdk1) stimulates DNA synthesis, indicating that frog cyclin A possesses S-phase-promoting activity. This activity may not be essential for embryonic DNA replication but could be important, and perhaps even essential, for a complete S phase in adult cell cycles.

Cyclin E–Cdk2 is a major regulator of DNA replication in *Drosophila*

In many tissues of the growing *Drosophila* larva, cells do not reproduce by typical four-phase cell cycles but instead undergo endoreduplication or endocycles (see sections 1-2 and 2-4), in which repeated rounds of DNA duplication occur in the absence of M phases. S phase in these cells is dependent on cyclin E, whose levels rise abruptly at the onset of S phase and decline as S phase proceeds (Figure 4-17a). Cyclin A is not produced in these cells and is not required for DNA duplication, clearly demonstrating that origin firing depends entirely on cyclin E–Cdk2 complexes.

An interesting feature of the *Drosophila* system provides further insight into the importance of cyclin E throughout S phase. In endoreduplicating cells, the chromosomes are not completely replicated as they are in normal mitotically cycling cells: the DNA in certain regions of the genome, primarily composed of heterochromatin, is not replicated. These regions can be duplicated, however, in experiments in which cyclin E expression is forced to continue for longer than normal (Figure 4-17b and c). In these cells, therefore, cyclin E is a limiting regulator of origin firing whose continuous presence is required for complete DNA synthesis.

The picture is less clear in *Drosophila* cell types that undergo the typical four-phase mitotic cycle (Figure 4-17d). Cyclin E expression does not continue throughout S phase in these cells, raising the question of how replication is completed. The most likely answer is that complete replication is promoted by cyclin A, which is expressed during S phase in these cells and has the capacity to stimulate DNA synthesis when forcibly overexpressed in certain cell types.

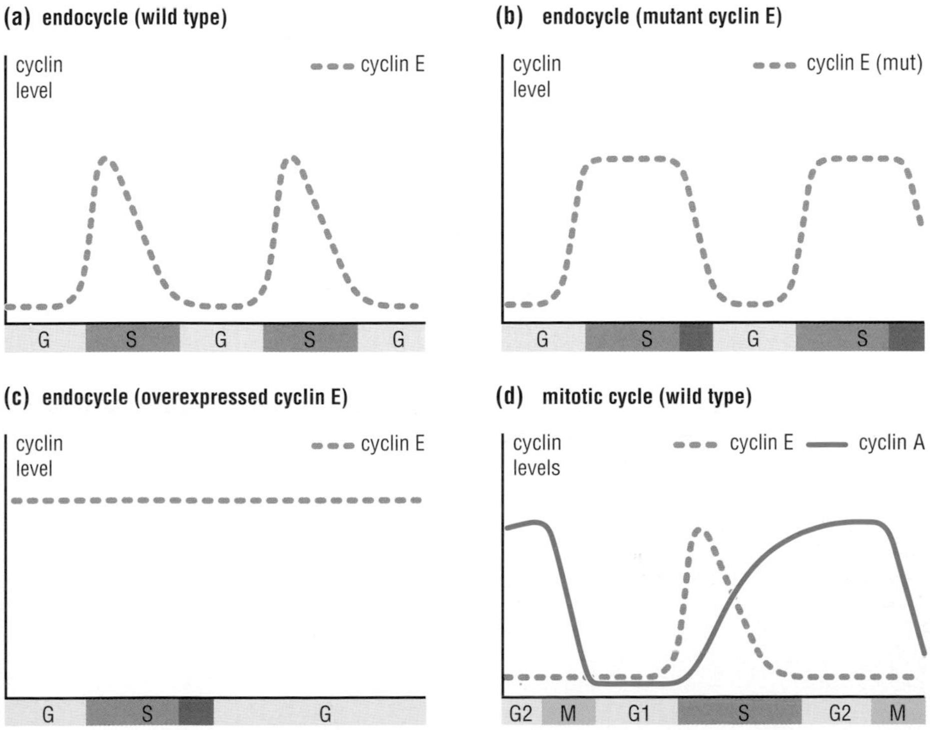

Figure 4-17 The functions of cyclins E and A in *Drosophila* DNA replication
(a) In endoreduplicating cells of the *Drosophila* larva, cyclin E levels rise at the onset of S phase and drop shortly thereafter. Cyclin A is not expressed in these cells. Deletion of the cyclin E gene prevents DNA replication (not shown). **(b)** Certain regions of heterochromatin are not duplicated during endocycle S phases. However, these regions can be duplicated in cells carrying a mutant form of cyclin E that is expressed for a longer time than wild-type cyclin E (indicated by dark pink extension of S phase). These results suggest that the levels of cyclin E normally limit the length of endocycle S phases, and that cyclin E can act throughout S phase to trigger origin activation. They also raise the question of how heterochromatic DNA duplication is prevented in wild-type cells: why, for example, are these regions not duplicated by replication forks from other regions? Finally, these results indicate that endocycling cells do not possess functional DNA damage response mechanisms, which block cell-cycle progression in the presence of unreplicated DNA (see Chapter 11). **(c)** In experiments where cyclin E is overexpressed at constant high levels in endocycling cells, a single S phase occurs but further replication is blocked. This effect is probably due to the ability of Cdk activity to block the assembly of pre-RCs at fired origins (see section 4-4). **(d)** In *Drosophila* cells with typical four-phase cell cycles, levels of cyclin E decrease before S phase is complete, raising the possibility that cyclin A contributes to replication control, at least in late S phase.

4-7 Control of Replication by the Protein Kinase Cdc7–Dbf4

Homologs of Cdc7 and Dbf4

Name	*S. cerevisiae*	*S. pombe*	Vertebrates
Cdc7	Cdc7	Hsk1	Cdc7
Dbf4	Dbf4	Dfp1	Dbf4/Ask Drf1

Figure 4-18 Table of alternative names for Cdc7 and Dbf4

Cdc7 triggers the activation of replication origins

The activation of replication origins is not carried out by Cdks alone: a second protein kinase, called Cdc7, is also required (Figure 4-18). The amino-acid sequence of Cdc7 is only distantly related to that of the Cdks, but there are many functional parallels between these kinases. As with the Cdks, Cdc7 activation depends on association with a specific regulatory protein, analogous to a cyclin, whose levels oscillate during the cell cycle. Like the yeast Clb5,6–Cdk1 complexes discussed in section 4-5, Cdc7 is not a global regulator of S-phase entry but is a direct activator of origin firing, as shown by the observation that Cdc7 function is required throughout S phase for the firing of late origins (Figure 4-19).

Cdc7 is a highly conserved and essential regulator of DNA replication. Mutations in the budding yeast gene *CDC7*, or the homologous gene in fission yeast, block the initiation of DNA replication. Cdc7-related proteins have also been identified in frogs and humans, and presumably exist in all eukaryotes. DNA replication is blocked when these proteins are inhibited with anti-Cdc7 antibodies—by the addition of antibodies to frog egg extracts or by the injection of antibodies into frog or human cells.

Cdc7 is activated during S phase by the regulatory subunit Dbf4

Like the Cdks, Cdc7 is not active throughout the cell cycle, and its activation is strictly timed to avoid premature initiation of DNA replication. In yeast and cultured mammalian cells, Cdc7 activity has been observed to rise and fall during the cell cycle, although the levels of the protein itself do not vary. Cdc7 activity is low during early G1, rises abruptly in late G1, and remains high until the exit from mitosis (Figure 4-20). Cdc7 is activated between late G1 and mitosis by association with a regulatory subunit, Dbf4, which binds directly to Cdc7 and stimulates its protein kinase activity—much as the binding of cyclins promotes Cdk activity. Indeed, Cdc7 is sometimes called the Dbf4-dependent kinase, or DDK. The timing of changes in Cdc7 activity is due primarily to changes in the level of Dbf4, which rises in late G1 and remains high until the exit from mitosis.

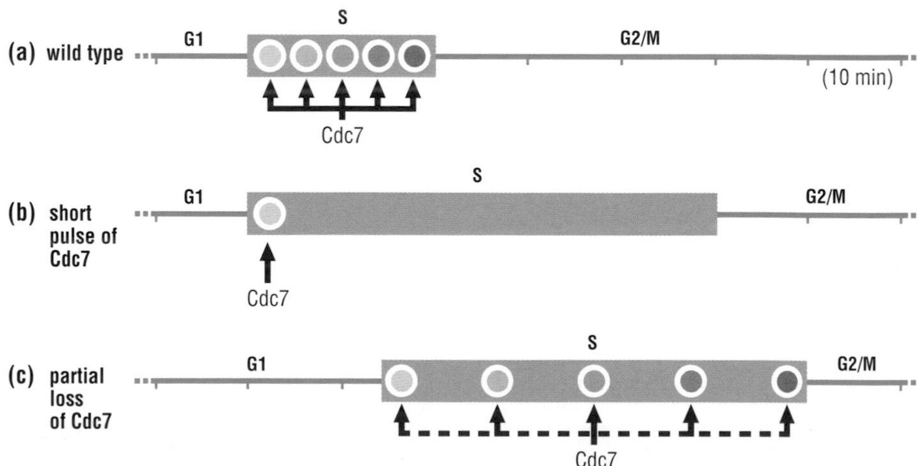

Figure 4-19 Function of Cdc7 in activation of replication origins The S-phase function of Cdc7 in budding yeast can be determined by analysis of origin activation in cells with a *cdc7* mutation. As in Figure 4-14, progression through the cell cycle is represented by green lines, and the length and timing of S phase is indicated by blue rectangles. Grey spots represent the firing of replication origins at various times during S phase, with early origins in light grey and late origins in dark grey. **(a)** In wild-type cells, Cdc7 is required for activation of early and late origins. When cells carrying a temperature-sensitive *cdc7* mutation are incubated at the restrictive temperature (37 °C), Cdc7 is not functional and the cells do not initiate DNA replication at any origin (not shown). **(b)** To test the importance of Cdc7 throughout S phase, *cdc7* mutant cells were arrested in late G1 by incubation at 37 °C. Cells were then incubated at a permissive temperature (23 °C) for a few minutes and then returned to 37 °C, resulting in a brief burst of Cdc7 activity. Early origins were activated, but not late ones, indicating that Cdc7 activity is required throughout S phase to activate late origins. S phase is prolonged because regions containing late origins are replicated passively by replication forks that have started from early origins further away. **(c)** To test the role of Cdc7 throughout S phase, the mutant cells were grown at a semi-permissive temperature (27 °C), resulting in partial loss of Cdc7 activity. Activation of all origins is less efficient, indicating that Cdc7 acts throughout S phase as a rate-limiting regulator of origin activity. Adapted from Donaldson, A.D. *et al.*: *Genes Dev.* 1998, **12**:491-501.

References

Bell, S.P. and Dutta, A.: **DNA replication in eukaryotic cells.** *Annu. Rev. Biochem.* 2002, **71**:333–374.

Bousset, K. and Diffley, J.F.: **The Cdc7 protein kinase is required for origin firing during S phase.** *Genes Dev.* 1998, **12**:480–490.

Donaldson, A.D. *et al.*: **Cdc7 is required throughout the yeast S phase to activate replication origins.** *Genes Dev.* 1998, **12**:491–501.

Takahashi, T.S. and Walter, J.C.: **Cdc7–Drf1 is a developmentally regulated protein kinase required for the initiation of vertebrate DNA replication.** *Genes Dev.* 2005, **19**:2295–2300.

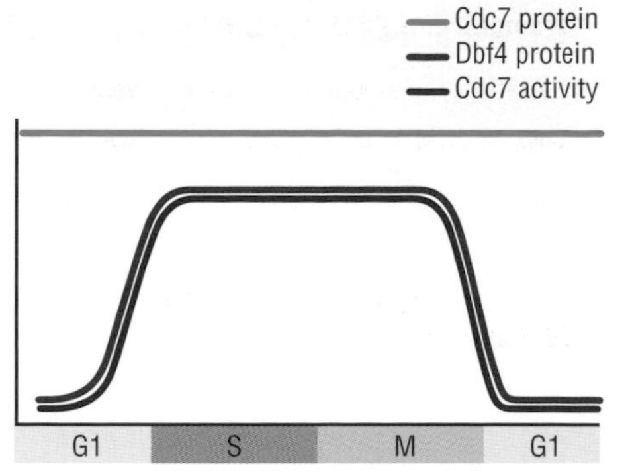

Figure 4-20 Changes in Cdc7 activity during the budding yeast cell cycle The amount of Cdc7 protein does not change during the cell cycle. The levels of its activator, Dbf4, increase in late G1 and decline in late M phase, resulting in parallel changes in Cdc7 activity.

Dbf4, like the cyclins, is probably more than just an activator of Cdc7 enzymatic activity: it may also serve as a targeting subunit that directs the kinase to the prereplicative complex (pre-RCs) at replication origins. There is evidence that Cdc7–Dbf4 phosphorylates Mcm subunits at activated replication origins, as discussed in section 4-8.

Dbf4 levels are regulated by multiple mechanisms

The levels of Dbf4 in the cell, like those of cyclins, are regulated by changes in rates of synthesis and degradation. The fall in Dbf4 levels in G1 is due both to decreased transcription of the gene *DBF4* and to increased degradation of the Dbf4 protein. Decreased *DBF4* expression in G1 is known to occur in fission yeast and humans and, to a lesser extent, in budding yeast, while a dramatic increase in Dbf4 degradation has been observed only in budding yeast. Dbf4 is thought to be targeted for degradation during G1 by the ubiquitin-protein ligase APC^{Cdh1} (see section 3-10), as Dbf4 degradation does not occur during G1 in yeast cells carrying APC mutations.

The evidence that Dbf4 degradation is dependent on the APC may provide an answer to the important question of how the activation of Cdc7 is triggered at the onset of a new cell cycle. In budding yeast, APC^{Cdh1} is inactivated late in G1 by the rise in G1/S– and S–Cdk activities (see section 3-11). It is therefore likely that Dbf4 stabilization, and thus the activation of Cdc7, is the indirect result of Cdk activation (Figure 4-21).

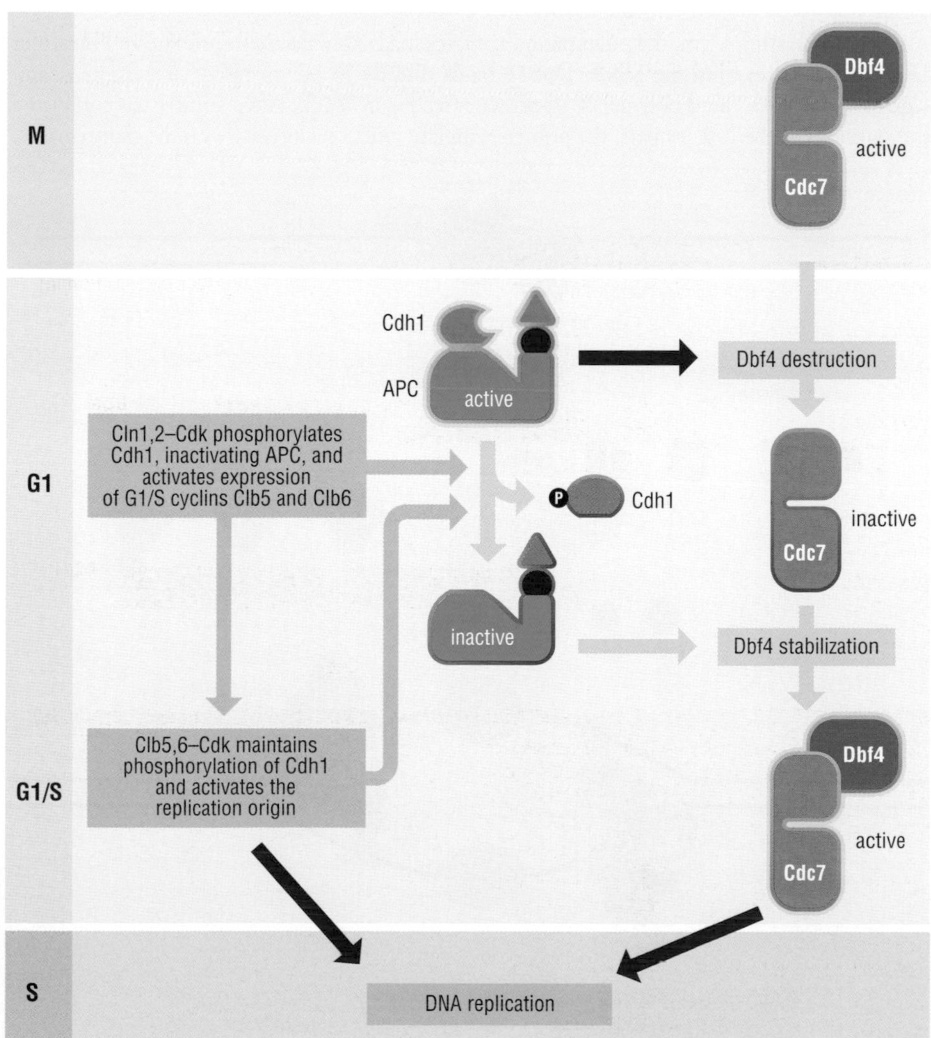

Figure 4-21 Control of Cdc7 activation by Cdk activity The activation of the ubiquitin-protein ligase APC in mitosis leads to destruction of Dbf4, thereby causing Cdc7 inactivation. In late G1, the activation of Cln1,2–Cdk1 and Clb5,6–Cdk1 complexes results in inhibitory phosphorylation of the APC activator Cdh1 (see section 3-13). The APC is thereby inactivated, allowing Dbf4 levels to rise and activate Cdc7. Cdc7–Dbf4 and Clb5,6–Cdk1 then collaborate to trigger the activation of replication origins. This scheme does not apply to the frog embryonic cell cycle, in which Dbf4 levels do not oscillate and Cdc7 regulation is not well understood.

4-8 Activation of the Replication Origin

Proteins in the Preinitiation Complex

Name	S. cerevisiae	S. pombe	D. melanogaster	Vertebrates
Mcm10	Mcm10/Dna43	Cdc23	Mcm10	Mcm10
Cdc45 complex				
Cdc45	Cdc45/Sld4	Sna41/Cdc45	Cdc45	Cdc45
Sld3	Sld3	Sld3	–	–
Dbp11 complex				
Dbp11	Dbp11	Cut5/Rad4	Mus101?	TopBP1?
Sld2	Sld2/Drc1	Drc1	–	–
GINS complex				
Sld5	Sld5	Sld5	Sld5	Sld5
Psf1	Psf1	Psf1	Psf1	Psf1
Psf2	Psf2	Psf2	Psf2	Psf2
Psf3	Psf3	Psf3	Psf3	Psf3

Figure 4-22 Table of components of the preinitiation complex Dashes indicate that a higher eukaryotic homolog has not been clearly identified.

Replication begins with DNA unwinding at the origin

By late G1, replication origins are loaded with prereplicative complexes (pre-RCs) in which the Mcm helicase is present but inactive (see section 4-4). For DNA replication to begin, the DNA helix must be separated, the helicase activated and DNA polymerases and the rest of the DNA synthetic machinery loaded onto the opened origin. This all depends on the assembly of a massive protein assembly called the **preinitiation complex**, which forms at the origin in response to the S–Cdk and Cdc7 activities discussed earlier (sections 4-5 to 4-7). The preinitiation complex activates the Mcm helicase and binds DNA polymerases, thereby recruiting them to the origin.

The components of the preinitiation complex are listed in Figure 4-22 and illustrated schematically in Figure 4-23. Most were first identified in budding yeast, where they are essential for the initiation of DNA replication, and homologs of almost all of these proteins have been identified in animals. Interaction of these proteins with the origin depends on Mcm2–7, suggesting that it is Mcm2–7, rather than other components of the pre-RC, that nucleates formation of the preinitiation complex. It is likely that most subunits of the complex are assembled simultaneously, as the association of several proteins with the complex is dependent on the presence of others. Binding of the four-subunit GINS complex, for example, depends on both Cdc45 and Dpb11, whereas Cdc45 and Dpb11 do not bind well in the absence of GINS.

Loading of the preinitiation complex onto the origin activates the Mcm helicase. We do not yet know how the preinitiation complex catalyzes activation of the Mcm helicase and its loading onto the DNA, or even the precise mechanism of action of Mcm2–7. As discussed earlier (section 4-3), a subcomplex of Mcm4, 6 and 7 possesses helicase activity *in vitro*. This activity is inhibited by Mcm2, and a complex containing all six Mcm proteins seems to have little helicase activity. It is likely that all six Mcm proteins are present in the pre-RC as an inactive hexamer, which is then activated in S phase by a rearrangement of inhibitory subunits, allowing an active helicase ring to form around the DNA.

As well as activating Mcm, the preinitiation complex loads polymerase α–primase and the other DNA polymerases onto the DNA. Dpb11 binds directly to polymerase ε, for example, and Cdc45 and Mcm10 bind polymerase α–primase (see Figure 4-23). After primase has completed the synthesis of the first primers, the primer–template junctions interact with the clamp loader,

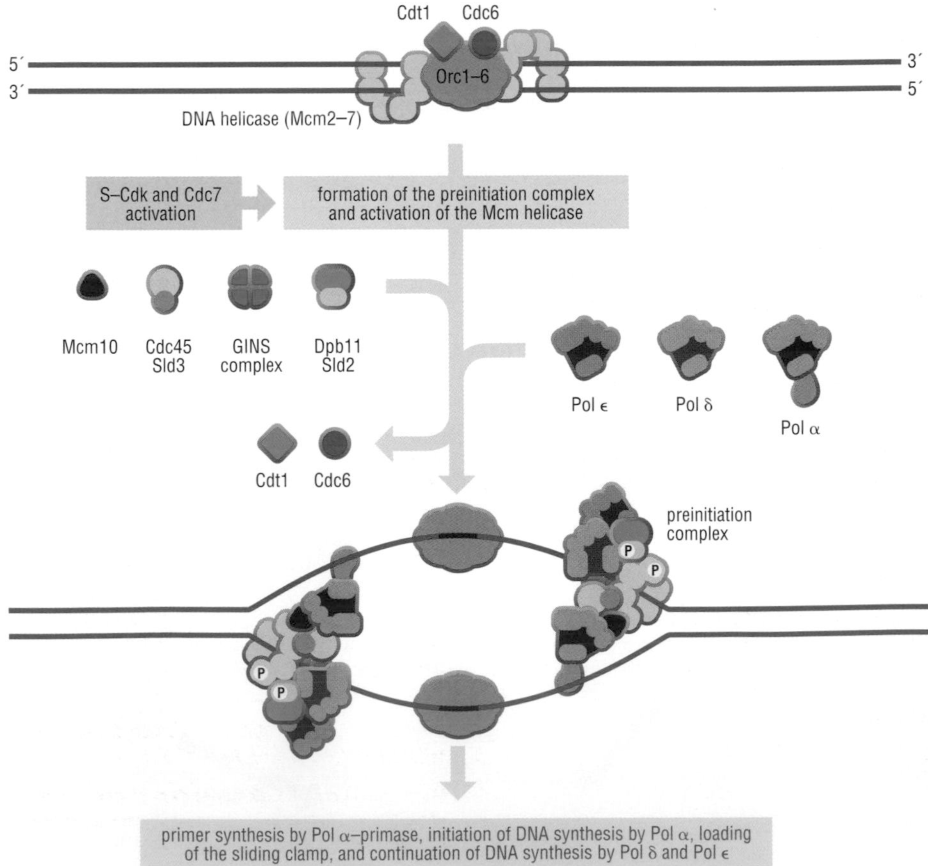

Figure 4-23 Initiation of DNA replication The assembly of the pre-RC during G1 readies the replication origin for firing (see Figures 4-10 and 4-11). The activation of S–Cdks and Cdc7 triggers origin activation by promoting the formation of the preinitiation complex, a large group of proteins, including the DNA polymerases, that assembles at the origin, probably through an interaction with the Mcm2–7 complex. Some of the components of the preinitiation complex bind as dimers (Dpb1–Sld2 and Cdc45–Sld3) and tetramers (the GINS complex) and it is likely that these core proteins of the complex, together with Mcm10, bind simultaneously. Formation of the preinitiation complex results in the displacement of Cdc6 and Cdt1 from the ORC, which remains at the origin. Preinitiation complex formation activates the Mcm helicase, leading to unwinding of the DNA helix. On activation, Mcm subunits may be lost (not illustrated, for simplicity) or rearranged. RPA proteins (not shown) bind the single-stranded DNA, preventing its reannealing, and primase is recruited to begin primer synthesis and DNA replication (see Figure 4-3). Synthesis of the leading strand begins first. The Mcm2–7 helicase and many components of the preinitiation complex travel out from the origin with the replication fork.

which loads the sliding clamp onto the DNA. DNA polymerases δ or ε then interact with the DNA and extend the primers to start synthesis of the two leading strands, after which synthesis of the lagging strands is initiated by a similar mechanism (see section 4-1). Most components of the preinitiation complex remain associated with the replication forks as they move out from the origin, and they seem to be important for the elongation phase of DNA synthesis.

All these events depend on the activities of S–Cdks and Cdc7, and a few potential substrates for these kinases have been identified. In budding yeast, phosphorylation of Sld2 by the S–Cdk Clb5–Cdk1 is required for DNA replication to occur. This phosphorylation enhances the affinity of Sld2 for its partner Dpb11 and may stimulate the recruitment of polymerase ε to the origin. If the phosphorylation sites on Sld2 are mutated, DNA replication is prevented. A likely target of Cdc7 is the Mcm2–7 complex. Phosphorylation of the inhibitory Mcm2 subunit, in particular, seems to be dependent on Cdc7, but it is not clear whether this phosphorylation promotes Mcm activation or other aspects of origin activation.

Late-firing origins are regulated independently

We know little about why origins fire at different times in S phase. In budding yeast, the order of origin firing is established in G1, indicating that late-firing origins are somehow marked as such before S phase begins. The marking of late origins is not simply a matter of local chromatin structure blocking all access to the DNA, because the Mcm complex is loaded at late origins, as it is at early origins, during G1. In contrast, the preinitiation complex in budding yeast does not assemble at late origins until late in S phase (Figure 4-24). This explains why S–Cdks and Cdc7 are required throughout S phase: they are needed to promote preinitiation complex formation and origin activation (see sections 4-6 and 4-7). We do not know what prevents preinitiation complex formation at late origins in early S phase even though S–Cdks and Cdc7 are active, or how this block is relieved in late S phase.

Replication must be completed before chromosome segregation occurs

When all origins have been activated, the resulting array of replication forks moves out along the DNA to complete the replication process. Normally, the timing of the cell-cycle control system is programmed to ensure that replication is completed before chromosome segregation is triggered. If, however, replication fails during S phase (as a result of depletion of nucleotides, for example), a regulatory system called the DNA damage response detects stalled replication forks and sends out a signal that blocks the firing of other replication origins and prevents entry into mitosis. The DNA damage response is discussed in Chapter 11.

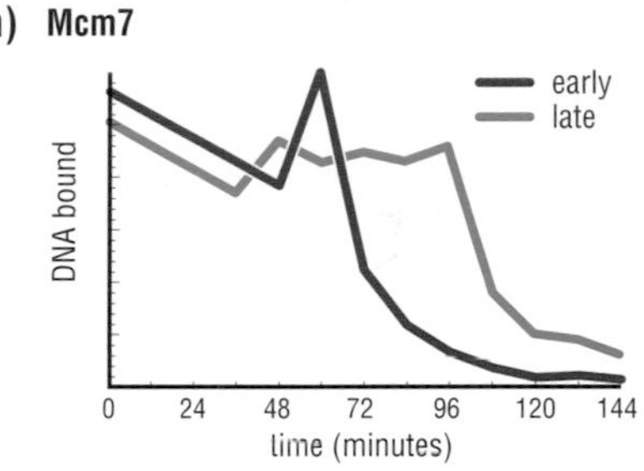

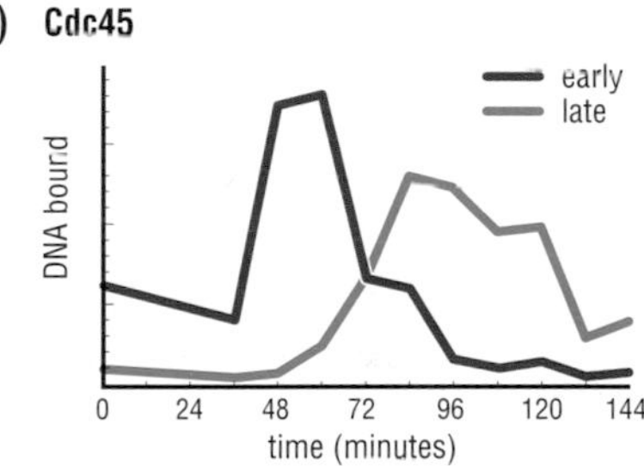

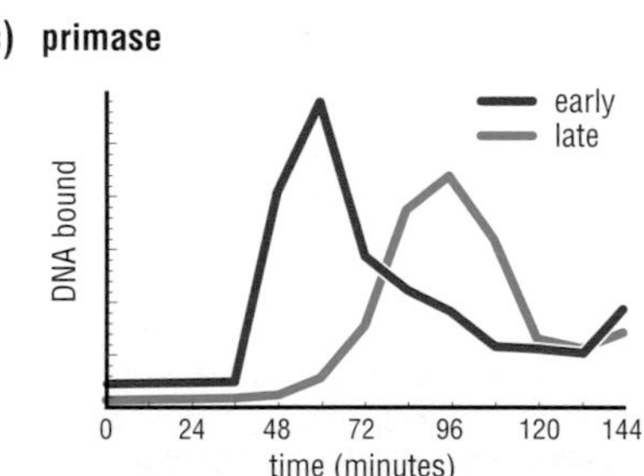

Figure 4-24 Binding of replication proteins to early and late origins of replication To measure their binding to origin DNA, proteins are purified from cells with specific antibodies and the amount of origin DNA associated with each protein is determined. In the experiment shown here, the binding of Mcm7, Cdc45 and primase was measured at two different origins: one that is normally fired in early S phase (purple) and one that is normally fired in late S phase (blue). Budding-yeast cells were arrested in G1 with mating pheromone and then released into the cell cycle at time zero; entry into S phase occurred after 36–48 minutes. **(a)** The amount of Mcm7 is initially high at all origins, reflecting the presence of the prereplicative complex in G1. Mcm binding drops as each origin is activated during S phase, presumably because the Mcm complex moves away with the replication forks. **(b)** Cdc45, a component of the preinitiation complex, is initially absent from origins in G1 but then binds to early origins in early S phase and late origins in late S phase. Like Mcm7, Cdc45 disappears from origins as it moves away with the replication forks. **(c)** Like Cdc45, primase associates with origins as they are activated. Adapted from Aparicio, O.M. *et al.*: *Proc. Natl Acad. Sci. USA* 1999, **96**:9130–9135.

Definitions

preinitiation complex: large complex of proteins that assembles at the replication origin when the origin is activated by S–Cdks and Cdc7. It includes DNA polymerases and other components that initiate DNA replication.

References

Aparicio, O.M. *et al.*: **Differential assembly of Cdc45p and DNA polymerases at early and late origins of DNA replication.** *Proc. Natl Acad. Sci. USA* 1999, **96**:9130–9135.

Bell, S.P. and Dutta, A.: **DNA replication in eukaryotic cells.** *Annu. Rev. Biochem.* 2002, **71**:333–374.

Masumoto, H. *et al.*: **S-Cdk-dependent phosphorylation of Sld2 essential for chromosomal DNA replication in budding yeast.** *Nature* 2002, **415**:651–655.

Ricke, R.M. and Bielinsky, A.K.: **Mcm10 regulates the stability and chromatin association of DNA polymerase-alpha.** *Mol. Cell* 2004, **16**:173–185.

Takayama, Y. *et al.*: **GINS, a novel multiprotein complex required for chromosomal DNA replication in budding yeast.** *Genes Dev.* 2003, **17**:1153–1165.

Walter, J. and Newport, J.: **Initiation of eukaryotic DNA replication: origin unwinding and sequential chromatin association of Cdc45, RPA, and DNA polymerase alpha.** *Mol. Cell* 2000, **5**:617–627.

4-9 Basic Chromatin Structure

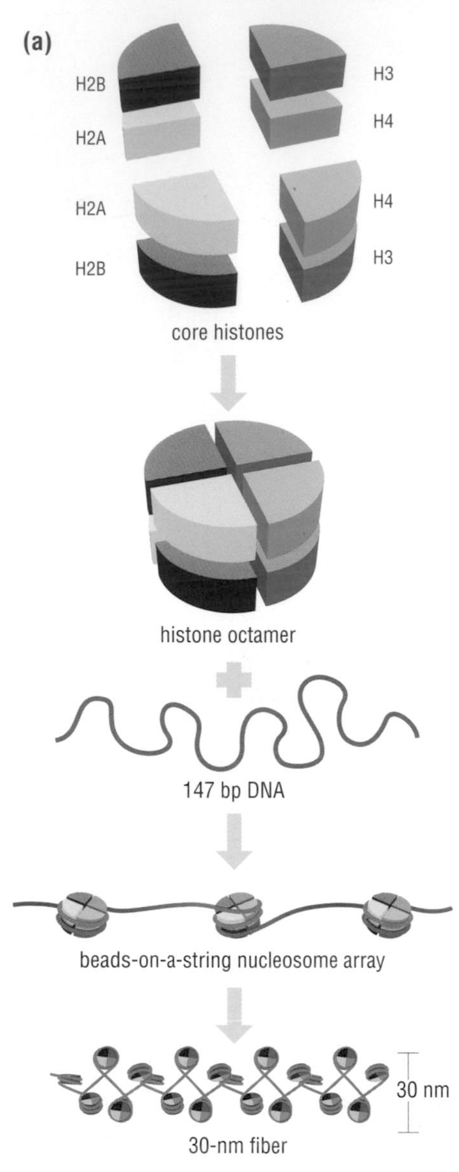

Chromatin is complex and dynamic

The thin thread of DNA in each chromosome is packaged into chromatin by a variety of proteins. This protein packaging not only folds the long DNA molecules into compact forms that can fit into a cell's nucleus but is also a dynamic structure that can be regulated and reshaped to change the degree of chromatin condensation or the accessibility of the DNA to proteins that control gene expression, DNA repair and DNA replication. The chromatin as a whole is duplicated in every cell cycle, which requires a doubling in the mass of proteins that package the DNA. The new proteins are generally loaded onto newly synthesized DNA in the same arrangement as in the original chromosome, thereby ensuring that chromatin structures that contribute to gene regulation are not lost when the cell divides.

The basic unit of chromatin structure is the nucleosome

The fundamental unit of DNA packaging is the **nucleosome**, in which about 147 bp of DNA makes 1.7 turns around a protein core—called the **histone octamer**—composed of eight histone proteins (two copies each of histones H2A, H2B, H3 and H4). Histone octamers are arranged along the DNA, like beads on a string, at variable intervals averaging about 200 bp, resulting in compaction of the DNA to about one-third its initial length (Figure 4-25).

Nucleosome structure can be adjusted to allow regulation of chromatin function. Histone–DNA and histone–histone interactions can be loosened, thereby altering the accessibility of the DNA to enzymes involved in transcription, repair and replication. Limited access is provided by spontaneous unwrapping of DNA from the histone octamer, which occurs at a significant rate in cells. In addition, members of a large and diverse class of enzymes, called **chromatin-remodeling complexes** or **ATP-dependent nucleosome-remodeling complexes**, use the energy of ATP hydrolysis to temporarily weaken the histone–DNA interaction. As well as providing direct access to the loosened DNA, remodeling can result in a change in the position of the nucleosome, and in changes in the histone content of the octamer. Thus, the regulation of these remodeling activities provides an important mechanism for controlling nucleosome position and DNA access in different regions of the genome.

Not all nucleosomes contain the conventional octamer of histones H2A, H2B, H3 and H4. In some parts of a chromosome, specific histones are replaced by **variant histones**—slightly different versions that contribute to local chromosome function. These include a variant of histone H3, called CenH3, which is found specifically in the nucleosomes of the *centromere*, the region of a chromosome that attaches to the mitotic spindle. Other variant histones are H2A.Z and H3.3, which tend to be found in highly transcribed chromatin; H2A.X, which is concentrated in regions of DNA damage; and macroH2A, which is found in the inactive X chromosome of mammals.

(b)

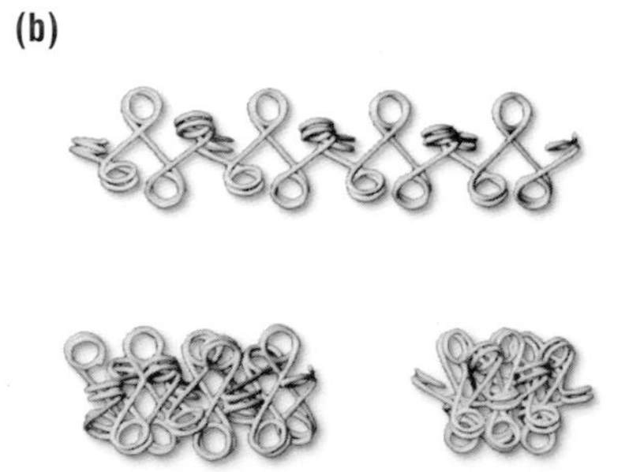

Figure 4-25 **Basic units of chromatin structure** **(a)** The histone octamer is formed from two subunits each of the four core histones. The four H3 and H4 subunits form a tightly packed tetramer that associates with two H2A/H2B dimers to form each octamer. About 147 bp of DNA is wrapped around the octamer to form a nucleosome. Nucleosomes can be arrayed in the loosely packed beads-on-a-string form of chromatin, but are generally more tightly packaged into the 30-nm fiber. Fiber formation requires histone tails and additional proteins, neither of which is shown here (but see Figure 4-26) **(b)** The 30-nm fiber can be packed at different densities by expansion or contraction of the fiber along its length. These changes in fiber structure probably require non-histone proteins, which are not shown here. From Bednar, J. *et al.*: *Proc. Natl Acad. Sci. USA* 1998, **95**:14173–14178.

Definitions

ATP-dependent nucleosome-remodeling complex: see **chromatin-remodeling complex**.

bromodomain: protein domain that binds to acetyl-lysine. Bromodomains are found in several non-histone chromatin proteins and interact with specific acetylated lysines on histone H3 or H4.

chromatin-remodeling complex: large protein complex that, through release of energy by ATP hydrolysis, causes changes in **nucleosome** structure that allow rearrangements in nucleosome position or access to proteins involved in DNA transcription, repair or replication.

chromodomain: protein domain that binds to methyl-lysine. Chromodomains are found in several non-histone chromatin proteins and interact with specific methylated lysines on histone H3 or H4.

histone octamer: protein core of the **nucleosome**, composed of eight histone subunits (two each of histones H2A, H2B, H3 and H4).

nucleosome: fundamental unit of eukaryotic chromatin structure, containing about 147 bp of DNA wrapped around a **histone octamer**.

variant histone: any histone other than the canonical five histones H2A, H2B, H3, H4 and H1 found in **nucleosomes** in specific chromosome regions.

References

Becker, P.B. and Hoerz, W.: **ATP-dependent nucleosome remodeling.** *Annu. Rev. Biochem* 2002, **71**:247–273.

Higher-order chromatin structure is also controlled by non-histone proteins, histone H1 and histone modifications

Few chromatin regions contain the simple beads-on-a-string structure. Typically, chromatin in the cell seems to be organized into more compact forms called 30-nm fibers. The structure of these fibers is not clear, but one possibility is that adjacent nucleosomes interact with each other in a zigzag arrangement (Figure 4-25a). According to this model, the density of packing in these fibers can be adjusted by their accordion-like expansion or contraction (Figure 4-25b).

The formation of 30-nm fibers depends on interactions between adjacent nucleosomes and between histones and other proteins. These interactions are mediated largely by the histone tails—flexible, positively charged stretches of 10–30 amino acids at the amino terminus of each histone, which extend from the surface of the histone octamer (Figure 4-26a). Formation of 30-nm fibers involves tails from one octamer interacting directly with the histones and DNA of neighboring nucleosomes, while other tails interact with a variety of non-histone proteins (Figure 4-26b). Chromatin structure also depends on an additional linker histone, called histone H1, which is bound to the DNA where it emerges from the nucleosome core (see Figure 4-26b). Histone H1 stabilizes the 30-nm fiber by neutralizing the negative charge on DNA and by interacting with the histone octamer and with non-histone proteins.

Much of the DNA is composed of loops of dynamic 30-nm fibers that are accessible to gene regulators and other DNA-binding proteins. Significant amounts of chromatin are, however, organized by non-histone chromosomal proteins into highly condensed and inactive heterochromatin. Examples of heterochromatic regions are telomeres and centromeres, whose structure and duplication are discussed at the end of this chapter (sections 4-12 and 4-13).

A major mechanism for controlling all levels of chromatin structure is the covalent modification of histone tails. These modifications both alter the properties of the tail itself and create binding sites for non-histone proteins that control chromatin structure and function. Histone tails undergo a variety of modifications, including serine phosphorylation and the acetylation, methylation and monoubiquitination of lysine residues. The best-characterized modification is acetylation, which can occur on several lysines in each histone, primarily in histones H3 and H4. Acetyl groups are added by enzymes called *histone acetyltransferases* and removed by *histone deacetylases*. Acetylation generally promotes a more open chromatin structure that enables genes to be expressed (see Figure 4-26b). It acts in part by neutralizing positive charge in the histone tail, thereby reducing interactions between the tail and the negatively charged DNA of other nucleosomes.

Chemical modifications of histone tails also influence chromatin structure and function by providing binding sites for non-histone proteins. Acetylation or methylation on specific lysines, for example, are recognized by non-histone proteins that promote more open or more condensed chromatin structure, respectively. Specialized domains in these proteins—**bromodomains** (which recognize acetylated lysines) and **chromodomains** (which recognize methylated lysines)—bind to modified residues and thus recruit the proteins to the modified histone. We will see in section 4-13, for example, that histone methylation promotes heterochromatin formation in part by recruiting the HP1 protein, which binds via its chromodomain to specific methylated lysines. The Sir protein complex of budding yeast, in contrast, helps induce heterochromatin formation by deacetylating histone tails, thereby preventing their interaction with the non-histone proteins associated with open chromatin. Before we discuss these mechanisms of heterochromatin structure and its replication, we will describe in the next two sections the duplication of the basic chromatin unit, the nucleosome.

(a)

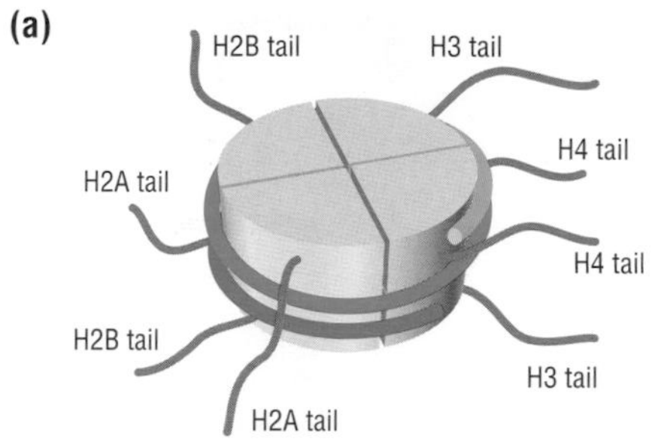

(b)

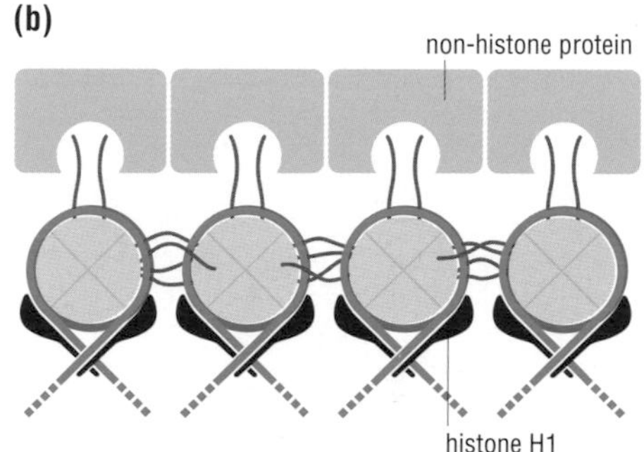

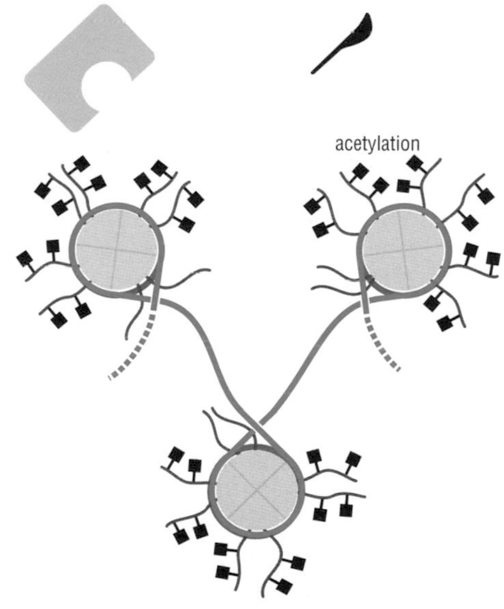

Figure 4-26 **Histone tails and their function in chromatin formation** **(a)** The flexible amino-terminal tail of each histone extends from the surface of the histone octamer. **(b)** In 30-nm fibers, the histone tails of one nucleosome interact with the histones and DNA of adjacent nucleosomes. In some chromatin, histone tails also interact with non-histone proteins (green) that help package the DNA. Linker histones (histone H1, red) also contribute to chromatin formation. Acetylation of histone tails alters their interaction with other nucleosomes and non-histone proteins, generally resulting in a more open chromatin structure.

Bednar, J. *et al.*: **Nucleosomes, linker DNA, and linker histone form a unique structural motif that directs the higher-order folding and compaction of chromatin.** *Proc. Natl Acad. Sci. USA* 1998, **95**:14173–14178.

Dorigo, B. *et al.* **Nucleosome arrays reveal the two-start organization of the chromatin fiber.** *Science* 2004, **306**:1571–1573.

Hayes, J.J. and Hansen, J.C.: **Nucleosomes and the chromatin fiber.** *Curr. Opin. Genet. Dev.* 2001, **11**:124–129.

Jenuwein, T. and Allis, C.D.: **Translating the histone code.** *Science* 2001, **293**:1074–1080.

Kamakaka, R.T. and Biggins, S.: **Histone variants: deviants?** *Genes Dev.* 2005, **19**:295–310.

Khorasanizadeh, S.: **The nucleosome: from genomic organization to genomic regulation.** *Cell* 2004, **116**:259–272.

Woodcock, C.L. and Dimitrov, S.: **Higher-order structure of chromatin and chromosomes.** *Curr. Opin. Genet. Dev.* 2001, **11**:130–135.

4-10 Histone Synthesis in S phase

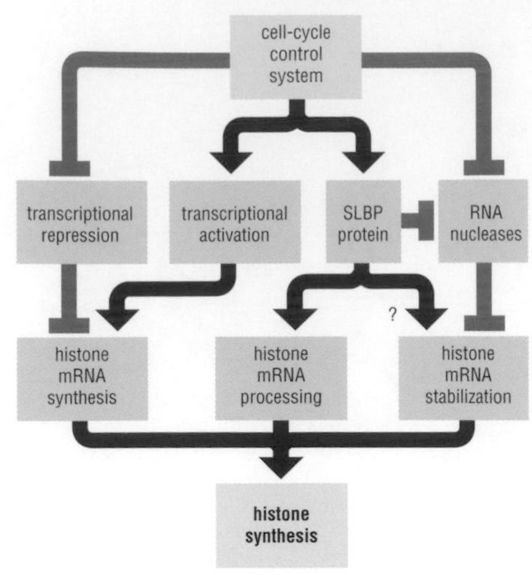

Figure 4-27 Mechanisms leading to increased histone synthesis in S phase At the end of G1, the cell-cycle control system promotes increased histone synthesis by multiple mechanisms. Transcription of histone genes is increased by activation of transcriptional activators and inhibition of transcriptional repressors. In metazoans, processing of histone mRNA also increases as a result of increased levels of an essential RNA-processing protein, SLBP. Histone mRNA stability also increases in S phase, probably as a result of reduced destruction of histone mRNA by RNA-specific nucleases.

Histone synthesis rises sharply during S phase

The first step in the duplication of chromatin structure is the assembly of new nucleosomes on the newly synthesized DNA. This produces a sudden increase in demand for the raw materials—the histones—needed to make nucleosomes. To meet this demand, the synthesis of the major histones (H2A, H2B, H3 and H4), as well as that of histone H1, increases dramatically during S phase. The rise in histone production is governed both by the cell-cycle control system and by regulatory mechanisms that couple histone production to DNA synthesis.

The rise in the level of histone proteins is due to a large increase in the amount of histone mRNA during S phase relative to the rest of the cell cycle. Several factors contribute to this increase. A higher rate of histone gene transcription occurs in all eukaryotes and is particularly important in yeast. In metazoans, increased histone mRNA levels are due primarily to an increase in the processing of histone mRNAs to their mature forms and a decrease in histone mRNA degradation (Figure 4-27).

Transcription of histone genes increases in S phase

Because of their extremely large genomes, metazoan cells require large amounts of histones to package the DNA. To allow the production of large quantities of histones in a limited amount of time, metazoans have multiple copies of the major histone genes, typically grouped together on the chromosome. Hundreds of copies are found in the genomes of some species, such as sea urchins and some amphibians, in which histone production must meet the demands of the extremely rapid cell cycles of early embryogenesis. The human genome contains a total of about 60 histone genes clustered on two chromosomes. In the nucleus, these gene clusters are localized in small structures called Cajal bodies or coiled bodies. Yeasts, in contrast, carry only one or two copies of each histone gene, scattered throughout the chromosomes.

Transcription of the major histone genes increases severalfold during S phase. These genes are sometimes called the replication-dependent histone genes, to distinguish them from the variant histone genes (see section 4-9), which are expressed throughout the cell cycle. Transcription of the replication-dependent histone genes is inhibited during the rest of the cell cycle, and their increased transcription in S phase is due to both increased activation and decreased repression.

Histone gene transcription is controlled by a combination of gene regulatory proteins that interact with specific sequences in histone gene promoter regions. A link between the cell-cycle control machinery and the activation of at least one of these regulatory proteins has been established in mammalian cells, where the transcription factor NPAT activates histone gene expression only after it has been phosphorylated by the G1/S–Cdk, cyclin E–Cdk2, in early S phase.

In budding yeast, the activators of histone gene expression are not clearly identified, but one candidate is SBF, a transcription factor whose activity is stimulated in late G1 by G1– and G1/S–Cdks (see section 3-12). The suppression of histone gene expression outside S phase in

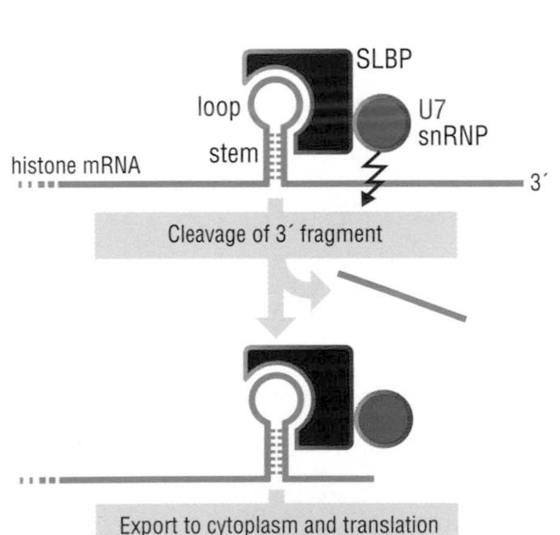

Figure 4-28 Processing of histone mRNA in metazoans The 3′ terminus of histone mRNAs contains a small stem–loop structure that interacts with SLBP (red) during S phase. In the nucleus, SLBP associates with an RNA–protein complex called U7 snRNP (blue), which contains a nuclease activity that removes a small fragment from the end of the mRNA. The mRNA–protein complex is then exported to the cytoplasm for translation.

References

Dominski, Z. *et al.*: **A novel zinc finger protein is associated with U7 snRNP and interacts with the stem-loop binding protein in the histone pre-mRNP to stimulate 3′-end processing.** *Genes Dev.* 2002, **16**:58–71.

Dominski, Z. *et al.*: **A 3′ exonuclease that specifically interacts with the 3′ end of histone mRNA.** *Mol. Cell* 2003, **12**:295–305.

Gunjan, A. and Verreault, A.: **A Rad53 kinase-dependent surveillance mechanism that regulates histone protein levels in *S. cerevisiae*.** *Cell* 2003, **115**:537–549.

Marzluff, W.F. and Duronio, R.J.: **Histone mRNA expression: multiple levels of cell cycle regulation and important developmental consequences.** *Curr. Opin. Cell Biol.* 2002, **14**:692–699.

Whitfield, M.L. *et al.*: **Stem-loop binding protein, the protein that binds the 3′ end of histone mRNA, is cell cycle regulated by both translational and posttranslational mechanisms.** *Mol. Cell Biol.* 2000, **20**:4188–4198.

Ye, X. *et al.*: **The cyclin E/Cdk2 substrate p220[NPAT] is required for S-phase entry, histone gene expression, and Cajal body maintenance in human somatic cells.** *Mol. Cell Biol.* 2003, **23**:8586–8600.

Zheng, L. *et al.*: **Phosphorylation of stem-loop binding protein (SLBP) on two threonines triggers degradation of SLBP, the sole cell cycle-regulated factor required for regulation of histone mRNA processing, at the end of S phase.** *Mol. Cell Biol.* 2003, **23**:1590–1601.

budding yeast is better understood, and involves changes in chromatin structure. Histone gene repression depends on a small group of proteins called the histone-regulatory (Hir) proteins. Together with other proteins involved in nucleosome assembly (discussed in section 4-11), the Hir proteins promote the formation of inactive chromatin structure in chromosomal regions containing histone genes, thereby blocking access to transcriptional activators. In S phase the inhibition is lifted by the removal of this repressive chromatin structure, probably by an ATP-dependent nucleosome-remodeling complex (see section 4-9) called the Swi/Snf complex. It is likely that similar mechanisms govern histone gene expression in metazoans.

Histone mRNA processing and stability increase in S phase

As with transcription, histone mRNA processing and degradation are both coupled to the cell-cycle control system. Newly transcribed metazoan RNAs generally have to be processed, by the addition of poly(A) tracts and removal of introns, before they can be exported from the nucleus to the cytoplasm for translation. Replication-dependent histone mRNAs, however, do not have introns or poly(A) tracts; instead, the only processing required for their maturation is the removal of a short piece at the 3′ end. This occurs at a more rapid rate during S phase, contributing to the increase in mature histone mRNA. This mechanism is not used in yeast, in which histone mRNAs end in the conventional poly(A) tract.

Histone mRNA processing in metazoans depends on the association of the RNA with a protein called the stem–loop binding protein (SLBP), whose levels are highest during S phase. SLBP binds to a structure called a stem–loop or hairpin at the 3′ end of the histone mRNA, and recruits ribonucleoproteins that process the RNA (Figure 4-28). The high levels of SLBP during S phase are the result of its increased synthesis at the end of G1 and increased degradation at the end of S phase; these processes are under the control of the cell-cycle control system but seem to be independent of DNA synthesis (Figure 4-29). SLBP degradation is triggered by phosphorylation, which seems to be catalyzed by S–Cdks, but we do not yet know why it occurs only at the end of S phase, long after S–Cdks are activated.

SLBP is also involved in the stabilization of histone mRNAs during S phase. In mammalian cells, histone mRNA half-life increases from about 10 minutes in G1 to about an hour in S phase. Stability then drops abruptly at the end of S phase, when the mRNA is destroyed by an RNA-specific exonuclease called 3′ hExo, which binds to the histone mRNA stem–loop. Because SLBP is bound to the stem-loop throughout S phase, this inhibits mRNA degradation by 3′ hExo until the level of SLBP drops at the end of S phase.

The level of free histones in the cell acts as a signal to link histone synthesis to DNA synthesis

As we have seen, histone mRNA levels are regulated in various ways by components of the cell-cycle control system. In addition, histone production is closely linked to DNA synthesis (Figure 4-29b and c). The level of free histones seems to be one of the signals that the cell monitors during S phase to ensure that replication-dependent histones are produced only when needed for packaging new DNA. If there are more histones than can be bound into nucleosomes, either because DNA synthesis is experimentally stopped or through the artificial addition of excess copies of histone genes to cells, histone gene transcription is rapidly repressed by the Hir-dependent change in chromatin structure described earlier. In metazoan cells, histone mRNA degradation is also triggered, probably by 3′ hExo. Degradation will occur even in the presence of SLBP, indicating that SLBP alone cannot stabilize mRNA when DNA synthesis is inhibited (Figure 4-29c). The nucleosome assembly machinery, discussed in section 4-11, may also contribute to the control of histone levels.

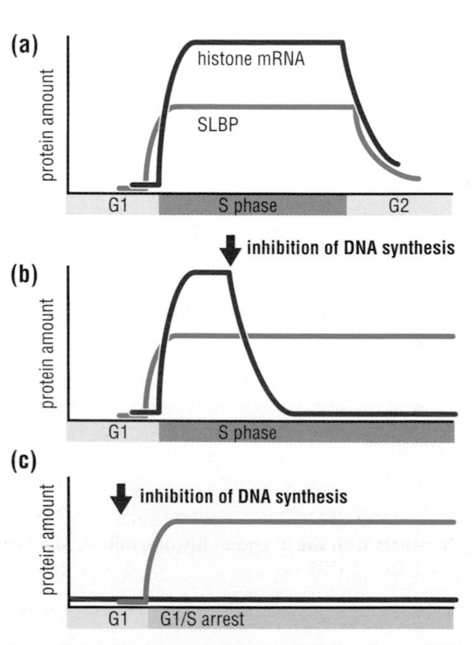

Figure 4-29 Independent regulation of SLBP and histone mRNA levels in mammalian cells **(a)** In cultured Chinese hamster ovary cells, the levels of SLBP rise at the end of G1 (owing to increased synthesis of the protein) and decrease after S phase (owing to increased degradation of the protein). Histone mRNA levels oscillate roughly in parallel with those of SLBP. **(b)** If chemical inhibitors of DNA synthesis are added before S phase is complete, histone mRNA levels drop rapidly owing to decreased gene transcription and increased mRNA degradation. SLBP levels remain elevated. **(c)** If DNA synthesis inhibitors are added before S phase begins, SLBP levels still rise on schedule but histone mRNA levels remain low. Adapted in part from Whitfield, M.L. *et al.*: *Mol. Cell Biol.* 2000, **20**:4188–4198.

4-11 Nucleosome Assembly on Nascent DNA

Nucleosomes are distributed to both new DNA strands behind the replication fork

The central event in chromatin assembly during S phase is the construction of nucleosomes on the two new DNA duplexes. As the replication machinery moves along the DNA, the old nucleosomes are transiently displaced, and new nucleosomes are then assembled on the duplicated DNA behind the replication fork (Figure 4-30). During this process, histones H3 and H4 from the old nucleosomes are somehow prevented from escaping into solution and are randomly distributed as intact tetramers onto either one of the new DNA duplexes. As a result, about half of the new nucleosomes on each daughter DNA contain histone H3 and H4 subunits from the nucleosomes of the parent DNA. In contrast, histones H2A and H2B from the old nucleosomes are released into solution during passage of the replication fork, and the new nucleosomes are composed primarily of newly synthesized H2A and H2B. As discussed later in this chapter, the random distribution of parental histone H3–H4 tetramers that have been covalently modified (see section 4-9) might be one of the mechanisms underlying the inheritance of chromatin structure.

Nucleosome assembly factors load histones on nascent DNA

The loading of new histones on DNA is catalyzed by specialized **nucleosome assembly factors** (also known as histone deposition complexes or histone chaperones). These proteins facilitate nucleosome assembly by binding histones and neutralizing their extensive positive charge, thereby preventing their nonspecific aggregation with DNA.

The first step in nucleosome assembly is the loading of a tetramer of histones H3 and H4 on the DNA. This process is catalyzed, at least in part, by an assembly factor called chromatin assembly factor-1 (CAF-1), which was originally identified as an activity from human cells that promotes the assembly of histone H3–H4 into nucleosomes during the replication of DNA *in vitro*. Human CAF-1 is a complex of three subunits (p150, p60 and p48) that binds tightly to newly synthesized histone H3–H4 complexes. The structure and function of CAF-1 have been conserved during evolution. Budding yeast, for example, contains a related trimer (Cac1, 2 and 3) that also contributes to nucleosome assembly. CAF-1 is located at the replication fork by an association with the sliding clamp and is thus ideally positioned to promote nucleosome assembly as the new DNA emerges from the DNA polymerase (Figure 4-31).

Several other proteins collaborate with CAF-1 to bring about nucleosome assembly. One important participant is a protein complex called replication coupling assembly factor (RCAF), which contains a subunit, Asf1, that binds to newly synthesized histones H3 and H4. In studies of DNA replication *in vitro*, Asf1 has been found to synergize with CAF-1 to promote nucleosome assembly on the nascent DNA.

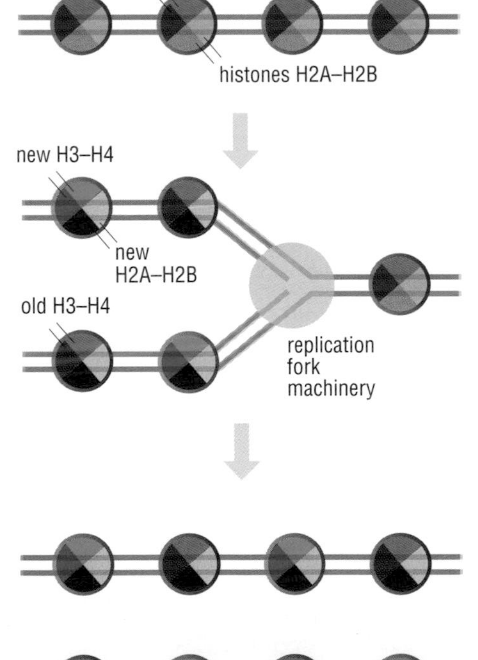

Figure 4-30 Conservative distribution of nucleosomes on nascent DNA As the replication fork moves along the DNA, nucleosomes are displaced from the DNA and then reassembled on the newly synthesized DNA behind the replication fork. Tetramers of histones H3 and H4 (green and blue) are not released during this process and are distributed randomly to the two daughter DNA molecules; newly synthesized H3 and H4 are shown in brighter green and blue. In contrast, old histones H2A and H2B (grey) are released into solution when the replication fork passes, and newly synthesized H2A and H2B (yellow and red) are used to construct the new nucleosomes. The end result is that the two new DNA molecules contain a random mixture of old and new histone H3–H4 tetramers.

Definitions

nucleosome assembly factor: protein that binds to histones and facilitates their assembly into nucleosomes.

References

Akey, C.W. and Luger, K.: **Histone chaperones and nucleosome assembly.** *Curr. Opin. Struct. Biol.* 2003, **13**:6–14.

Becker, P.B. and Hoerz, W.: **ATP-dependent nucleosome remodeling.** *Annu. Rev. Biochem.* 2002, **71**:247–273.

Haushalter, K.A. and Kadonaga, J.T.: **Chromatin assembly by DNA-translocating motors.** *Nat. Rev. Mol. Cell Biol.* 2003, **4**:613–620.

Henikoff, S. *et al.*: **Histone variants, nucleosome assembly and epigenetic inheritance.** *Trends Genet.* 2004, **20**:320–326.

Leffak, I.M. *et al.*: **Conservative assembly and segregation of nucleosomal histones.** *Cell* 1977, **12**:837–845.

Poot, R.A. *et al.*: **The Williams syndrome transcription factor interacts with PCNA to target chromatin remodelling by ISWI to replication foci.** *Nat. Cell Biol.* 2004, **6**:1236–1244.

Sogo, J.M. *et al.*: **Structure of replicating simian virus 40 minichromosomes. The replication fork, core histone segregation and terminal structures.** *J. Mol. Biol.* 1986, **189**:189–204.

Verreault, A.: **De novo nucleosome assembly: new pieces in an old puzzle.** *Genes Dev.* 2000, **14**:1430–1438.

Yamasu, K. and Senshu, T.: **Conservative segregation of tetrameric units of H3 and H4 histones during nucleosome replication.** *J. Biochem.* 1990, **107**:15–20.

Deposition of a histone H3–H4 tetramer on newly synthesized DNA is followed by the addition of two H2A–H2B dimers (see Figure 4-31). Neither CAF-1 nor Asf1, which seem to be specific for histones H3 and H4, are involved in this process. The best candidate for a histone H2A–H2B chaperone is a protein called nucleosome assembly protein-1 (NAP-1), which interacts preferentially with these histones and catalyzes their deposition on DNA *in vitro*.

Nucleosome assembly also depends on the acetylation of newly synthesized histones H3 and H4. This acetylation occurs at different lysines from those at which acetylation helps control the accessibility of mature chromatin (see section 4-9). Acetylation of new histones increases the efficiency of nucleosome assembly, but how it does this is not clear; acetylation does not, for example, improve binding of histones to nucleosome assembly factors. New histones are gradually deacetylated after nucleosome assembly and this is necessary for the maturation of chromatin. Inhibition of this deacetylation prevents the formation of functional heterochromatin at centromeres.

The formation of nucleosomes on newly synthesized DNA is called replication-coupled nucleosome assembly. Some histones, however, are deposited on DNA at other times by a process called replication-independent nucleosome assembly. In certain transcriptionally active chromosome regions, for example, replication-independent nucleosome assembly factors replace the histone H3 subunit of some nucleosomes with the variant histone H3.3, which may help promote the open chromatin structure in these regions.

In addition to nucleosome assembly factors, the loading and spacing of nucleosomes on newly synthesized DNA depend on ATP-dependent nucleosome-remodeling complexes (see section 4-9). Members of one class of these enzymes—the ISWI class—are found at the replication fork and are likely to be involved in nucleosome arrangement during S phase. Chromatin-remodeling complexes are also involved in the poorly understood processes by which the string of nucleosomes is assembled with additional non-histone proteins into higher-order chromatin structure (see section 4-9). We discuss the higher-order structure of heterochromatin in section 4-12, after which we describe mechanisms by which this structure may be duplicated in the cell cycle.

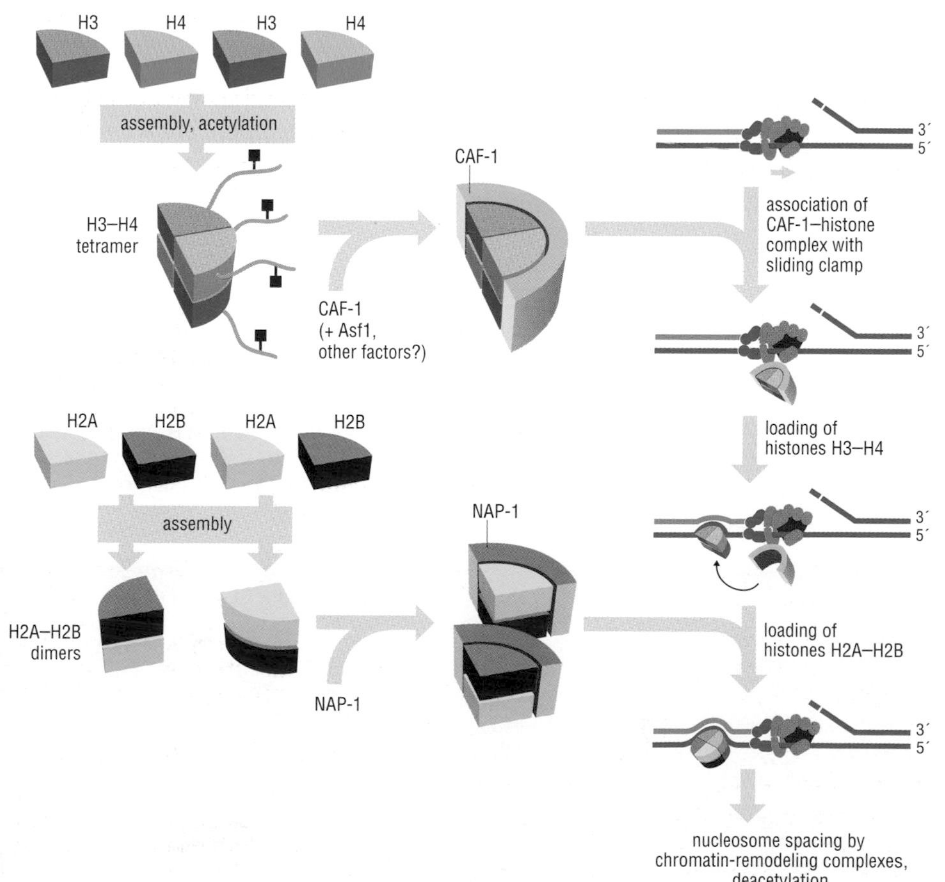

Figure 4-31 Steps in nucleosome assembly Nucleosome assembly begins with the binding of acetylated histone H3–H4 complexes to the assembly factor CAF-1. This complex is then recruited to the replication fork by an interaction with the sliding clamp, and a tetramer of histones H3 and H4 is loaded onto the nascent DNA to form half of a new nucleosome. Dimers of histones H2A and H2B are then loaded onto the DNA, probably with the assistance of other assembly factors such as NAP-1, to form the complete histone octamer. Chromatin-remodeling complexes contribute to nucleosome assembly and spacing. The number of histone H3 and H4 subunits that interact with CAF-1 is not clear. These subunits form highly stable tetramers in solution (unlike histones H2A and H2B, which form dimers) and it is therefore thought that CAF-1 loads a complete H3–H4 tetramer as shown here. There is some evidence, however, that CAF-1 can interact with an H3–H4 dimer, in which case the loading of a tetramer would require two CAF-1–histone complexes.

4-12 Heterochromatin at Telomeres and Centromeres

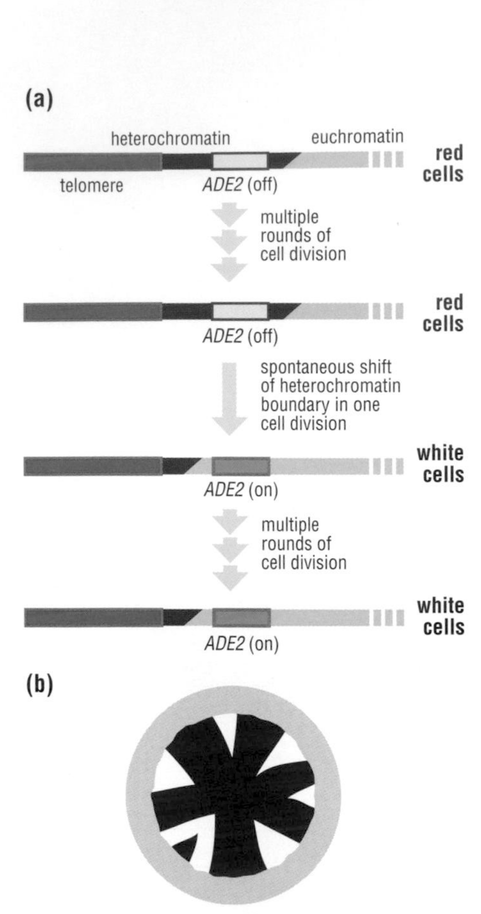

Figure 4-32 Epigenetic inheritance of heterochromatin structure at a yeast telomere **(a)** The *ADE2* gene of budding yeast encodes an enzyme involved in adenine biosynthesis, and the loss of *ADE2* expression causes the accumulation of a red pigment in the cells. If the *ADE2* gene is moved from its normal chromosomal location to a location next to the telomere, as shown here, its expression is silenced by the repressive structure of telomeric heterochromatin, and cells turn red. The heterochromatin structure at the telomere is stable over many cell divisions, so that most descendants of a red cell are also red. In an occasional cell division, however, the heterochromatin boundary spontaneously shifts toward the telomere, resulting in expression of the *ADE2* gene—and white cells. This new heterochromatin boundary is stably inherited—in an epigenetic fashion—over many cell divisions. **(b)** If a red yeast cell carrying *ADE2* next to the telomere is grown on solid agar medium, the resulting yeast colony, diagrammed here, is mostly red, reflecting the stable inheritance of the telomeric heterochromatin structure. As the colony grows, however, white sectors seem to fan out from the center. Each white sector contains the progeny of a cell in which the heterochromatin boundary shifted as in panel (a), allowing *ADE2* to be expressed. The heterochromatin boundary occasionally shifts back again, blocking *ADE2* expression and resulting in a red sector within the white, as shown at lower left of the colony.

Heterochromatin is inherited by epigenetic mechanisms

In some chromosome regions, nucleosome arrays are packed tightly with other proteins into a higher-order chromatin structure called heterochromatin (see section 4-9). These complex chromatin structures, like the DNA at their core, are accurately duplicated during the cell cycle. In some cases, the formation of new chromatin may be driven simply by the DNA sequence; for example, sequence-specific DNA-binding proteins such as transcriptional regulators will bind to newly synthesized DNA at precisely the same sequences they occupied on the parental DNA. In most cases, however, it is clear that chromatin is assembled by mechanisms that do not depend simply on DNA sequence. This form of chromatin inheritance is therefore called **epigenetic**, and evidence for it comes primarily from studies of heterochromatin formation at telomeres and centromeres. In this section we describe the basic features of heterochromatin structure and behavior in these regions. In section 4-13 we discuss the epigenetic mechanisms by which heterochromatin structure is duplicated during the cell cycle.

Telomeres are packaged in a heritable heterochromatin structure

As discussed in section 4-1, telomeres are long, repetitive DNA sequences at the ends of chromosomes. Telomeric DNA is packaged by specific proteins into a tightly condensed form of heterochromatin that also extends into adjacent regions of the chromosome. This packaging blocks potentially lethal fusions with other chromosome ends, and also helps prevent chromosome ends from triggering DNA damage responses, which are normally induced by double-stranded DNA ends caused by chromosome breakage, as described in Chapter 11.

The structure of telomeric heterochromatin is best understood in budding yeast. If a yeast gene that is normally expressed in euchromatin is placed in a region of heterochromatin near the telomere, its expression is repressed by the tight local chromatin structure; this is known as gene **silencing** (Figure 4-32). The boundary between silenced heterochromatin and adjacent euchromatin is usually inherited through multiple cell divisions. At an occasional division, however, the boundary moves to a new position, indicating that the heterochromatin has expanded or contracted along the chromosome. This new boundary position is also heritable and is preserved in most of the progeny of that cell. Heterochromatin structure is therefore dynamic, and its boundaries are not determined by DNA sequence but by epigenetic mechanisms.

A major component of telomeric heterochromatin is the Rap1 protein, which binds directly to telomeric DNA sequences in budding yeast and is thought to be arrayed along the length of the telomere. The other major components of yeast telomeric heterochromatin are the silent information regulator (Sir) proteins. Three of these proteins (Sir2, 3 and 4) form trimeric complexes that interact with Rap1 at telomeres and are also arrayed along adjacent regions of the chromosome.

The centromere nucleates a heritable and poorly understood form of heterochromatin

Every eukaryotic chromosome contains a **centromere**, a region where a specialized protein apparatus, the *kinetochore*, is assembled. The kinetochore, as we discuss in Chapter 6, mediates the attachment of duplicated chromosomes, or *sister chromatids*, to the mitotic spindle. The centromere therefore provides the essential foundation for the distribution of duplicated chromosomes to daughter nuclei at mitosis.

Definitions

alpha-satellite (alphoid) DNA: in human chromosomes, the repetitive DNA sequence found at the **centromere**.

centromere: region of the chromosome where kinetochores are assembled and attach to the mitotic spindle.

epigenetic: inherited through mechanisms that are not dependent on DNA sequence. Known epigenetic mechanisms often concern gene regulation and are dependent on modifications of the DNA or local chromatin structure.

neocentromere: chromosomal region that nucleates the formation of centromeric heterochromatin and kinetochore assembly at a position other than that of the normal centromere for that chromosome.

silencing: (of chromatin) establishment of a heritable state of chromatin, known as heterochromatin, characterized by repression of gene expression and recombination and delayed replication.

Although the function of the centromere is highly conserved, its DNA sequence varies widely in different species. Among the eukaryotes studied so far, only chromosomes in budding yeast contain a well defined centromeric DNA sequence, of about 125 bp, that provides all the information needed for chromosome transmission in mitosis (see section 6-5). In most other species, centromeric DNA is much larger, ranging in size from about 40–100 kb in fission yeast to hundreds and thousands of kilobases in flies and humans. Human centromeres, for example, contain a 171-bp sequence called **alpha-satellite**, or **alphoid**, DNA, that is tandemly arrayed in higher-order repeats that range in size from 200 to 9,000 kb. In some species, such as the nematode *C. elegans*, the centromere and its associated kinetochore–spindle attachments are scattered diffusely over the entire chromosome.

In most species, a centromeric DNA sequence does not seem to be an essential determinant of centromere function. In some human cells, for example, fragments of chromosomes are generated that do not contain conventional centromeric DNA but are still correctly transmitted in mitosis (Figure 4-33a). These chromosomes have apparently constructed a new centromere, or **neocentromere**, in a region where none existed before and where no alpha-satellite DNA is present. In contrast, other human cells contain a chromosome with two centromeric DNA regions as a result of abnormal chromosome fusion events. These abnormal chromosomes are also transmitted normally in mitosis, apparently because only one centromeric region per chromosome can mediate kinetochore assembly (Figure 4-33b). Thus, centromeric alpha-satellite DNA is neither necessary nor sufficient for centromere function in human cells.

How can we explain these findings? The answer is not clear, but the most likely explanation is that the protein composition of centromeric chromatin is the crucial determinant of centromere function and inheritance. Neocentromeres, for example, might form by the rare, spontaneous assembly of centromeric proteins at DNA that has some special feature, such as repetitive sequences. These neocentromeric chromatin structures are heritable: they are duplicated in every cell division, even when the underlying DNA sequence is not conventional centromeric DNA.

Many of the proteins that form centromeric chromatin are not yet identified. One known key protein is the variant histone CenH3 (CENP-A in mammals and Cse4 in budding yeast), which is found specifically in centromeric nucleosomes and is required for kinetochore assembly during mitosis (see section 6-5). Other centromeric proteins include HP1, which is present in many other regions of heterochromatin in fission yeast and animals. HP1 contains a chromodomain, which specifically binds methylated histone tails. In the next section we discuss how the binding of histones to non-histone proteins such as HP1 may contribute to the duplication of chromatin structure in S phase.

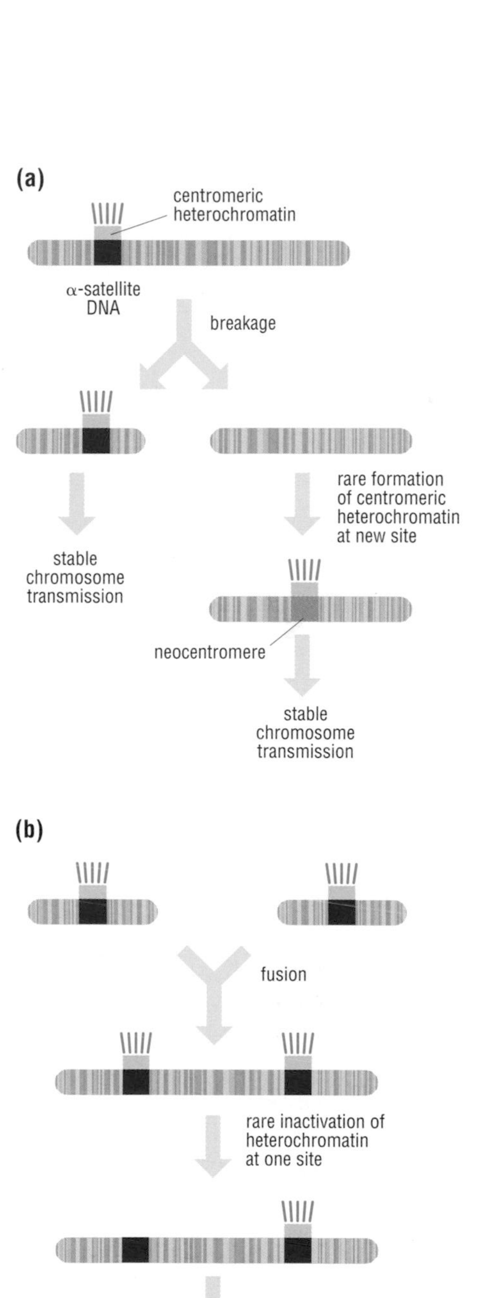

Figure 4-33 Epigenetic inheritance of centromeric chromatin structure (a) A normal chromosome (top) contains one region of centromeric DNA containing alpha-satellite repeats (red), which is packaged in heterochromatin (yellow). During mitosis, the centromere serves as the site of microtubule attachment (blue). If a chromosome breaks in two, the fragment carrying the centromere is propagated normally. The other fragment is usually lost, but in rare cases centromeric heterochromatin assembles spontaneously, despite the absence of alpha-satellite DNA, and the fragment is propagated normally. **(b)** A fusion can lead to a chromosome with two centromeres, but such chromosomes are usually not transmitted successfully at mitosis because they attach to both poles of the spindle, resulting in breakage. Occasionally, the heterochromatin at one centromere is spontaneously inactivated, resulting in a monocentric chromosome that is transmitted normally.

References

Cleveland, D.W. *et al.*: **Centromeres and kinetochores: from epigenetics to mitotic checkpoint signaling.** *Cell* 2003, **112**:407–421.

Gottschling, D.E. *et al.*: **Position effect at *S. cerevisiae* telomeres: reversible repression of Pol II transcription.** *Cell* 1990, **63**:751–762.

Maison, C. and Almouzni, G.: **HP1 and the dynamics of heterochromatin maintenance.** *Nat. Rev. Mol. Cell Biol.* 2004, **5**:296–304.

Moazed, D.: **Common themes in mechanisms of gene silencing.** *Mol. Cell* 2001, **8**:489–498.

Rusche, L.N. *et al.*: **The establishment, inheritance, and function of silenced chromatin in *Saccharomyces cerevisiae*.** *Annu. Rev. Biochem.* 2003, **72**:481–516.

Sullivan, K.F.: **A solid foundation: functional specialization of centromeric chromatin.** *Curr. Opin. Genet. Dev.* 2001, **11**:182–188.

4-13 Molecular Mechanisms of Heterochromatin Duplication

Duplication of heterochromatin structure involves proteins that recognize and promote localized histone modification

Epigenetic inheritance of heterochromatin structure is not well understood, in part because the structure of heterochromatin is itself so mysterious. Nevertheless, accumulating evidence allows some speculation about the molecular mechanisms underlying heterochromatin duplication.

The most popular model of chromatin inheritance is based on the presence of modified histones H3 and H4 in heterochromatic regions (Figure 4-34). These marked histones specifically bind proteins that promote the localized packaging of DNA into heterochromatin. During DNA duplication, nucleosomes containing marked histones H3 and H4 are distributed equally to both the new DNA helices (see Figure 4-32), and new, unmarked nucleosomes are assembled in the spaces between. Histone-binding proteins on old nucleosomes then recruit histone-modifying enzymes that mark the new histones, generating binding sites for more heterochromatin proteins. Heterochromatin is thereby inherited by mechanisms based primarily on protein interactions and not on DNA sequence. In this section we describe two systems in which this mechanism seems to operate.

The Sir proteins form a heritable polymer at telomeres in budding yeast

As mentioned earlier (see section 4-12), a complex of three Sir proteins forms a central component of telomeric heterochromatin in budding yeast. Several features of the Sir complex provide clues to the mechanism by which telomeric chromatin is duplicated during the cell cycle. Most importantly, the Sir complex interacts with specific underacetylated forms of histone tails that are concentrated in telomeric nucleosomes. In addition, one of the Sir subunits (Sir2) encodes a histone deacetylase. These observations support the appealing hypothesis that local histone modifications drive heterochromatin assembly. When newly synthesized DNA near the telomere is incorporated into new nucleosomes, the Sir complexes on old nucleosomes deacetylate the new nucleosomes, thereby generating more binding sites for Sir proteins. The ability of one Sir complex to bind another also contributes to the process by promoting cooperative polymerization of Sir complexes along the new DNA (Figure 4-35).

This mechanism provides an elegant explanation for the epigenetic inheritance of telomeric heterochromatin. Because the mechanism is based on reversible interactions between proteins, it may also explain how the heterochromatin boundary can sometimes shift to a new, stable

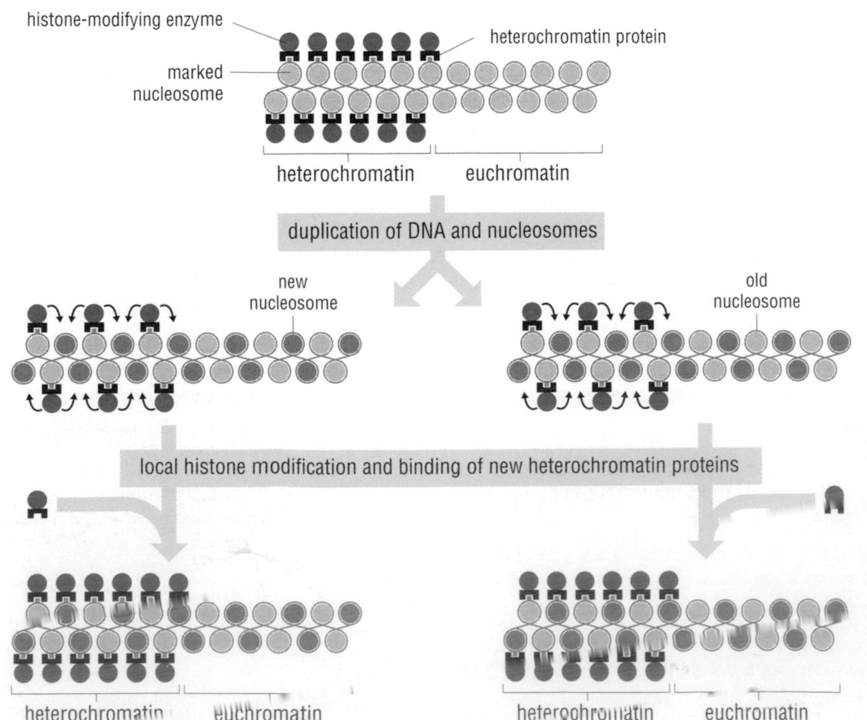

Figure 4-34 General model for duplication of heterochromatin during cell division Heterochromatin contains specialized proteins (red) that bind to histone H3 or H4 subunits that have been marked by a specific modification (green). The enzyme that performs this modification is also present in heterochromatin (blue), ensuring that the modification is maintained. When the chromosome is duplicated, the marked histone H3–H4 tetramers of the parent chromosome are distributed randomly to the two daughter strands, resulting in a mixture of old (light grey) and new (dark grey) nucleosomes. In heterochromatin, new nucleosomes are rapidly marked by the histone-modifying enzymes bound to old nucleosomes. This provides new binding sites for heterochromatin proteins. These proteins (such as the Sir complex or HP1) also have the ability to bind to each other, further promoting the assembly of a protein polymer along the chromosome. It can also be seen in this diagram that the boundary between heterochromatin and euchromatin is not rigidly fixed, because small local changes in the extent of histone modification could cause shifts in its position.

position. A small number of Sir complexes at the boundary might dissociate for a sufficiently long time to allow local histone acetylation, thereby shrinking the heterochromatic region in a heritable fashion. Similarly, excessive deacetylation in adjacent regions might allow Sir complexes to extend too far in some cells, resulting in expansion of heterochromatin.

Mechanisms exist to limit the distance that Sir complexes can reach along the chromosome. Specialized chromatin regions, called heterochromatin barriers or boundary elements, contain DNA that binds protein complexes that inhibit local nucleosome assembly. In budding yeast, highly transcribed regions next to silenced chromatin contain the histone variant H2A.Z, which blocks local spreading of Sir proteins by some unknown mechanism. Spreading is also prevented by the promotion of local histone acetylation.

Although the properties of Sir proteins alone suggest a model for the epigenetic inheritance of local chromatin structure, it is clear that telomeric heterochromatin is established in the first place by sequence-specific binding of binding of proteins to DNA. The Sir complex is localized at the telomere because it interacts with the protein Rap1, which binds specifically to the telomeric DNA sequence.

HP1 may nucleate heritable chromatin structure at the centromere and other regions

HP1 is a small protein found in regions of heterochromatin in many species, including fission yeast, flies and humans. In both yeast and flies, mutations in HP1 reduce gene silencing in heterochromatic regions, suggesting that HP1 is required for local chromatin packaging.

Various lines of evidence suggest that the function of HP1 in heterochromatin duplication is similar to that of Sir complexes at yeast telomeres. First, HP1 binds specifically, via its chromodomain (see section 4-9), to histone H3 tails that contain a methyl group at a specific lysine, lysine 9; this modification is found only in regions of heterochromatin. Second, HP1 is normally found in a complex with a histone methyltransferase, SUV39H, which adds the methyl group to lysine 9 on histone H3. Finally, HP1 is capable of self-association: one HP1 molecule can bind to another. Together, these features of HP1 suggest a mechanism by which heterochromatin can be self-perpetuating: the local methylation of histones provides binding sites for HP1–SUV39H complexes, which maintain the methylated histone state. When DNA is duplicated, the presence of these complexes on old nucleosomes ensures that new nucleosomes are quickly methylated as well, providing more binding sites that nucleate the local polymerization of HP1 arrays. Nascent DNA is thereby packaged in the same protein structure as that of the mother cell.

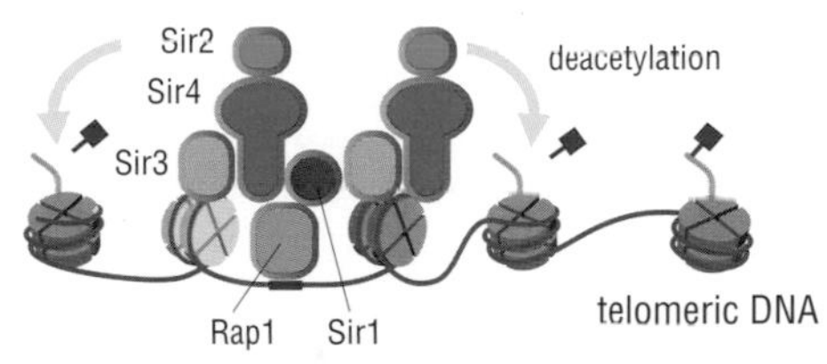

Figure 4-35 Formation of heterochromatin structure by the Sir complex at budding yeast telomeres Sir proteins are initially recruited to telomeres by the DNA-binding protein Rap1, which is bound at intervals along telomeric DNA. The Sir2, Sir3 and Sir4 proteins can then spread along the chromatin independently of Rap1 by binding to each other and to underacetylated histones. The Sir2 histone deacetylase maintains histones in the underacetylated state typical of heterochromatin. When a chromosome is duplicated, the presence of the Sir complex on old nucleosomes promotes the deacetylation of new nucleosomes, thereby providing binding sites for new Sir complexes and allowing the inheritance of chromatin structure.

Sister-chromatid cohesion in S phase prepares the cell for mitosis

S phase is not only the period when interphase chromatin is duplicated, but is also a time when specialized chromatin structures are constructed to prepare the chromosomes for segregation in mitosis. During and after DNA replication, the two new sister chromatids (see section 1-1) are linked together by *sister-chromatid cohesion*. Proteins called *cohesins*, which are concentrated at the centromere and are also present along the chromatid arms, attach the two sister chromatids along their length. The molecular basis of sister-chromatid cohesion is discussed in Chapter 5, where we will also describe the changes in chromatin structure that lead to chromosome condensation, another important step in the preparation of chromosomes for segregation.

References

Maison, C. and Almouzni, G.: **HP1 and the dynamics of heterochromatin maintenance.** *Nat. Rev. Mol. Cell Biol.* 2004, **5**:296–304.

Meneghini, M.D. *et al.*: **Conserved histone variant H2A.Z protects euchromatin from the ectopic spread of silent heterochromatin.** *Cell* 2003, **112**:725–736.

Moazed, D.: **Common themes in mechanisms of gene silencing.** *Mol. Cell* 2001, **8**:489–498.

Richards, E.J. and Elgin, S.C.: **Epigenetic codes for heterochromatin formation and silencing: rounding up the usual suspects.** *Cell* 2002, **108**:489–500.

Rusche, L.N. *et al.*: **The establishment, inheritance, and function of silenced chromatin in *Saccharomyces cerevisiae*.** *Annu. Rev. Biochem.* 2003, **72**:481–516.

5

Early Mitosis: Preparing the Chromosomes for Segregation

In early mitosis, the tightly linked pairs of duplicated chromosomes are prepared for separation and segregation in late mitosis. These events are governed by a complex regulatory system based on multiple mitotic cyclin–Cdk complexes.

5-0 Overview: The Events of Mitosis

Prophase

kinetochore
sister chromatids condensing
mitotic spindle forming
nuclear envelope
centrosome

Prometaphase

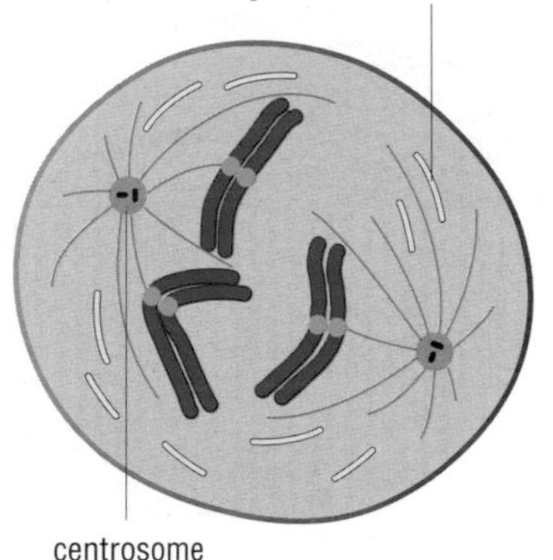

Metaphase

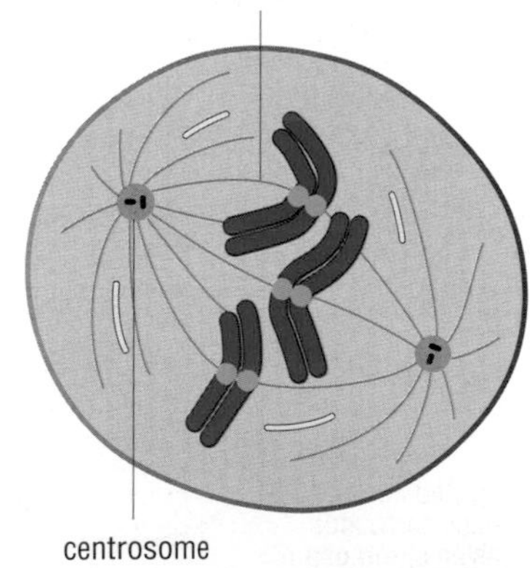

The central events of mitosis are sister-chromatid separation and segregation

At the end of S phase, the cell contains a duplicate set of chromosomes in tightly associated pairs called **sister chromatids**. During M phase, the paired sister chromatids are separated and one of each pair is distributed to each daughter cell. In all eukaryotes, the delivery of chromosomes to the daughter cell depends on the **mitotic spindle**, a bipolar array of *microtubules* that attach to the sister chromatids and pull them to opposite ends of the cell.

Before mitosis, sister chromatids are tightly interlinked by intertwining of their DNA (*DNA catenation*) and by specialized protein complexes called *cohesins*. This linkage, called **sister-chromatid cohesion**, is essential for successful mitosis because it allows each sister pair to be attached to the spindle with a bipolar orientation—that is, with each sister attached to the opposite spindle pole. Once all the sister pairs are attached to the spindle in this way, the disruption of cohesion then allows the pulling forces of the spindle to drag them apart. Two key processes therefore lie at the heart of mitosis: **sister-chromatid separation**, whereby cohesion is disrupted, and **sister-chromatid segregation**, the process by which the separated sisters are pulled to opposite poles of the cell.

The events of early mitosis set the stage for sister-chromatid segregation

Mitosis can be divided into two parts, with the onset of sister-chromatid segregation serving as the climactic midpoint. The first half of mitosis, sometimes called mitotic entry, encompasses two major processes that occur in parallel: the preparation of the sister chromatids for segregation, which we discuss in this chapter, and the assembly of the mitotic spindle, which we discuss in Chapter 6.

Sister chromatids are readied for segregation by several major structural changes. The chromosomes undergo *condensation*—packing into compact and flexible rods that are more easily moved by the mitotic spindle. In addition, sister-chromatid cohesion is loosened by removal of DNA catenation and partial loss of the cohesin proteins that hold sister chromatids together. This results in sister-chromatid *resolution*: the formation of distinct and easily separated sister chromatids.

Entry into mitosis also leads to the separation of the two *centrosomes* (or spindle pole bodies in yeast; see section 1-1). Each centrosome nucleates its own radial microtubule array, and as the two centrosomes separate and then migrate to opposite sides of the nucleus, the bipolar microtubule array of the mitotic spindle forms between them.

In vertebrate cells, the nuclear envelope breaks down in early mitosis, thereby dissolving the barrier between the growing mitotic spindle in the cytoplasm and the condensing sister chromatids in the nucleus. Spindle microtubules then attach to sister chromatids at a specialized chromatin structure, the **kinetochore**, which is built on the centromeric DNA. Initially, microtubules from one spindle pole attach to one kinetochore. Microtubules from the opposite pole then capture the kinetochore of the adjoining sister chromatid, resulting in the bipolar attachment of each sister-chromatid pair.

Figure 5-1 Stages of early mitosis in a vertebrate cell

Definitions

anaphase: the first stage of mitotic exit, when the **sister chromatids** are segregated by the **mitotic spindle**. In most species, anaphase is divided into anaphase A, the initial movement of chromosomes to the spindle poles, and anaphase B, the movement of spindle poles away from each other.

kinetochore: protein complex at the centromere of a chromosome, where the microtubules of the spindle are attached during mitosis.

metaphase: the last stage of mitotic entry, when the **sister chromatids** are fully attached to the spindle and await the signal to separate in **anaphase**.

mitotic spindle: bipolar array of microtubules, generally with a centrosome or spindle pole body at each pole, which segregates the **sister chromatids** during mitosis.

prometaphase: the second stage of mitosis in animal and plant cells, when the nuclear envelope breaks down and the **sister chromatids** become attached to the spindle.

prophase: the first stage of mitosis, when chromosome condensation, centrosome separation and spindle assembly begin.

sister chromatids: pair of chromosomes that is generated by chromosome duplication in S phase.

sister-chromatid cohesion: linkages that hold **sister chromatids** together between S phase and anaphase.

sister-chromatid segregation: the process by which separated **sister chromatids** are pulled to opposite poles of the cell for packaging in daughter nuclei.

Yeast do not dismantle their nuclear envelope in mitosis, and the mitotic spindle is a relatively primitive structure that forms inside the nucleus. There is evidence that kinetochores are linked to the spindle pole bodies throughout the cell cycle; bipolar attachment to the spindle may simply occur when the centromeres are duplicated in S phase.

In vertebrate cells, the events of early mitosis are traditionally divided into three phases (Figure 5-1). **Prophase** begins at the onset of chromosome condensation, the first mitotic event that is apparent through a microscope. Centrosome separation and the initiation of spindle assembly occur in mid to late prophase. **Prometaphase** begins with nuclear envelope breakdown and continues until the sister chromatids are completely attached to the spindle and have migrated to its central region. **Metaphase** is the period during which the sisters are aligned at the center of the spindle (the metaphase plate) awaiting the signal to separate.

These classical terms for the stages of mitotic entry are not useful in all eukaryotes. In particular, the distinction between prophase and prometaphase cannot be made in yeast, in which the nuclear envelope does not break down; similarly, nuclear envelope breakdown occurs only partly or later in mitosis in *Drosophila* and *C. elegans*. It is therefore appropriate in some cases to use the term prophase to describe all early mitotic events. In addition, sister-chromatid pairs do not align on a central metaphase plate in yeast, but instead are scattered throughout the spindle when segregation begins. The term metaphase can therefore be used more generally to describe the brief period just before sister-chromatid separation.

The completion of mitosis begins with sister-chromatid segregation

The most dramatic event in the cell cycle occurs at the onset of **anaphase**, when the cohesin links between sister chromatids are abruptly dissolved and the separated sisters are pulled to opposite poles of the spindle (Figure 5-2). This is known as anaphase A. In anaphase B, the spindle poles themselves move farther apart from each other, completing the segregation of the sister chromatids into the two halves of the dividing cell.

Mitosis is completed in **telophase**, when the chromosomes and other nuclear components are repackaged into identical daughter nuclei. The spindle is disassembled, leaving a single centrosome associated with each set of chromosomes. In vertebrates, a nuclear envelope reforms around the decondensing chromosomes; in yeast, the elongated anaphase nucleus is pinched in two during cytokinesis.

The events of mitosis, and the mechanisms that control them, are discussed in this chapter and in Chapters 6 and 7. In this chapter we begin with an overview of the key regulatory components—particularly the mitotic cyclin–Cdk complexes—that control mitotic entry, after which we discuss the processes that prepare the sister chromatids for segregation. In Chapter 6, we continue with the events of mitotic entry and review how a mitotic spindle is assembled and attached to the sister chromatids. Finally, in Chapter 7, we describe the segregation of sister chromatids in anaphase and the regulatory mechanisms that govern the completion of mitosis.

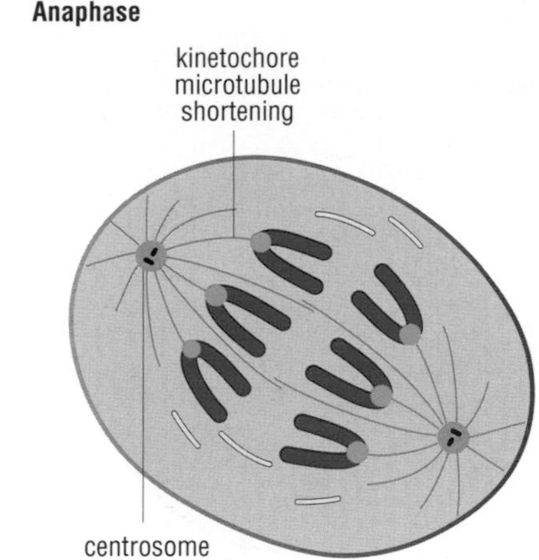

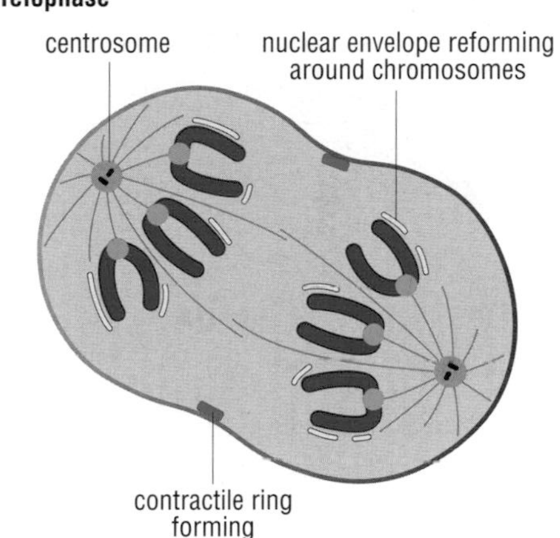

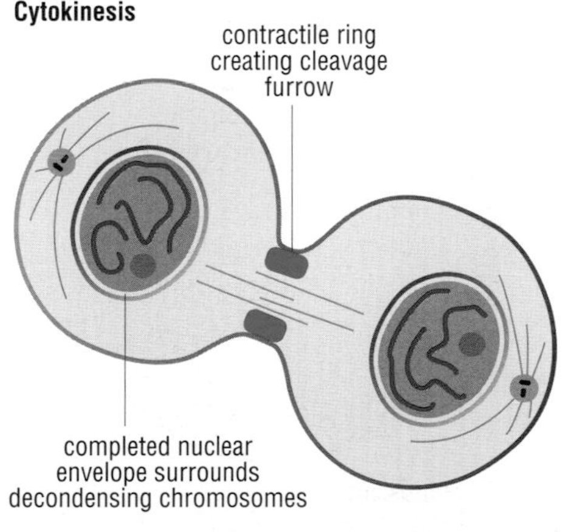

Figure 5-2 Stages of late M phase in a vertebrate cell

sister-chromatid separation: the process by which **sister-chromatid cohesion** is dissolved and **sister-chromatid** pairs dissociate at the metaphase-to-anaphase transition.

telophase: the final stage of mitosis, when the spindle is disassembled and, in multicellular eukaryotes, the chromosomes decondense and the nuclear envelope reforms.

References

Mazia, D.: **Mitosis and the physiology of cell division** in *The Cell Vol. III* Brachet, J. and Mirsky, A.E. eds (Academic Press, New York, 1961), 77–412.

Mitchison, T.J. and Salmon, E.D.: **Mitosis: a history of division.** *Nat. Cell Biol.* 2001, **3**:E17–E21.

Wilson, E.B.: *The Cell in Development and Heredity* 3rd ed. (Macmillan, New York, 1925).

5-1 Overview: Principles of Mitotic Regulation

Phosphorylation and proteolysis control progression through mitosis

Rigorous control of mitotic events is essential for the successful completion of sister-chromatid segregation and cell division. The timing and coordination of these events are governed primarily by the cell-cycle control system, whose basic operation can be outlined as follows (Figure 5-3). Entry into mitosis is triggered by mitotic cyclin–Cdk complexes and other protein kinases, which phosphorylate proteins that drive spindle assembly and other events leading up to alignment of sister chromatids on the metaphase spindle. Cdks also promote the activation of the APC^{Cdc20} ubiquitin-protein ligase (see section 3-10). This leads to the destruction of proteins that hold sisters together, allowing the spindle to move the sets of chromosomes to opposite poles of the cell. APC^{Cdc20} also brings on cyclin destruction and Cdk inactivation, which leads to dephosphorylation of Cdk targets. Cdk inactivation is required for spindle disassembly, the exit from mitosis, and cytokinesis.

Phosphorylation and proteolysis are therefore the two major regulatory mechanisms controlling mitotic events. Phosphorylation is a rapidly reversible protein modification and is thus ideally suited for the control of reversible mitotic processes such as spindle assembly, which is turned on by Cdks in early mitosis and then reversed when Cdks are inactivated in late mitosis. Proteolysis, by contrast, is an ideal mechanism for controlling events that must not be reversed. Because sister-chromatid separation is driven by proteolysis, for example, it is difficult for sisters to reattach after anaphase. Similarly, the proteolytic destruction of cyclins in late mitosis results in the essentially irreversible inactivation of Cdk1, thereby preventing the recurrence of early mitotic events.

As discussed in Chapter 3 (section 3-11), the ability of mitotic Cdks to promote APC^{Cdc20} activity, and thereby promote their own inactivation, may provide the basis for a regulatory system that generates autonomous oscillations in the activities of Cdks and APC^{Cdc20}. An oscillator of this sort may lie at the heart of the cell-cycle control system in simple embryonic cell cycles like those of the frog.

We also saw in Chapter 3 (sections 3-12 and 3-13) that the control of gene expression adds another layer of sophistication to M-phase regulation in most cells. In budding yeast, for example, mitotic entry is driven in part by the ability of Cdk1 to activate a gene-regulatory factor called Mcm1–Fkh, which increases expression of G2/M genes involved in early mitotic events. Similarly, Cdk1 inactivation in late mitosis results in the activation of the gene-regulatory factors Swi5 and Ace2, which help promote mitotic exit by stimulating expression of M/G1 genes.

Mitotic events must go to completion

The success of mitosis demands that processes such as chromosome condensation and sister-chromatid separation go to completion—or not at all. The molecular basis for the all-or-none behavior of mitotic events is not well understood. It is likely that this behavior arises in part from switch-like activation and inactivation of Cdks, APC^{Cdc20}, and other regulatory components. Indeed, as discussed in Chapter 3 (section 3-8), there is evidence that mitotic Cdk1 regulation involves feedback loops that generate irreversible, switch-like changes in Cdk1 activity. There is still much to be learned, however, about the behavior of these Cdk regulatory circuits. Even less is known about APC^{Cdc20} activation and the all-or-none behavior of anaphase events.

References

Morgan, D.O.: **Regulation of the APC and the exit from mitosis.** *Nat. Cell Biol.* 1999, **1**:E47–E53.

Murray, A.W. and Hunt, T.: *The Cell Cycle: An Introduction* (Freeman, New York, 1993).

Nurse, P.: **Universal control mechanism regulating onset of M-phase.** *Nature* 1990, **344**:503–508.

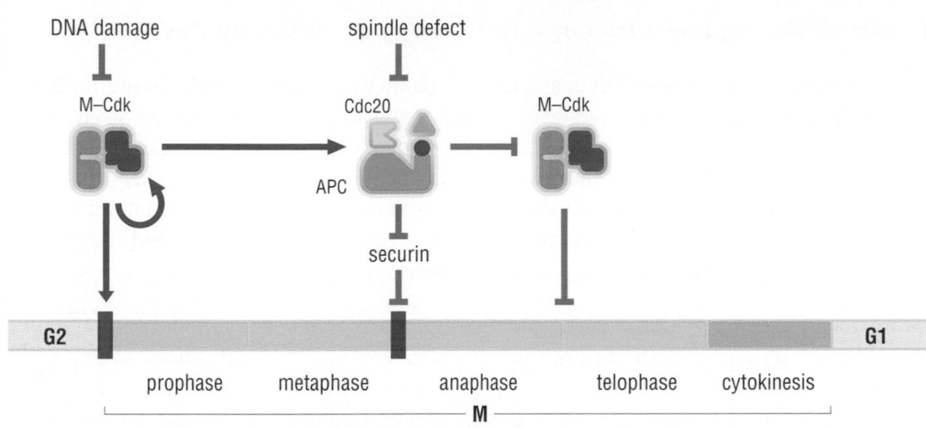

Figure 5-3 Overview of mitotic control mechanisms Early in mitosis, mitotic cyclin–Cdk complexes trigger the events leading up to metaphase. Mitotic Cdks promote their own activation, resulting in positive feedback. M–Cdks also activate the ubiquitin-protein ligase APC^{Cdc20}, resulting in the destruction of two proteins: securin, a protein that inhibits sister-chromatid separation; and mitotic cyclins. The destruction of securin results in sister-chromatid separation. Cyclin destruction causes Cdk inactivation, which leads to the completion of mitosis and cytokinesis. Progression through mitosis can be blocked at specific checkpoints. In fission yeast and metazoans, DNA damage inhibits the activation of mitotic Cdks, thereby blocking mitotic entry at the G2/M checkpoint. In all eukaryotes, defects in spindle assembly inhibit APC^{Cdc20}, thereby preventing progression through the metaphase-to-anaphase transition.

Mitotic entry and exit are major regulatory transitions with differing importance in different species

In most cells, the cell-cycle control system can arrest the cell cycle at specific checkpoints if conditions are not ideal (see section 1-3). Mitosis contains at least two major checkpoints (see Figure 5-3). The first is found at the G2/M boundary, controlling mitotic entry, where progression is normally triggered by the activation of mitotic cyclin–Cdk complexes. The second occurs at the metaphase-to-anaphase transition, where progression is driven by activation of APC^{Cdc20}.

To prevent the distribution of damaged or incomplete chromosomes, cell-cycle progression in most eukaryotes is blocked at the G2/M checkpoint if chromosomal DNA is damaged or not fully replicated. Damaged DNA or stalled replication forks send out inhibitory signals that block mitotic entry by preventing the activation of mitotic Cdks. If the damage is repaired or replication completed, the inhibitory signals are withdrawn, Cdks are activated, and mitosis begins. These mechanisms are discussed in Chapter 11.

Progression can also be blocked at the metaphase-to-anaphase transition under certain conditions. If sister chromatids are not attached properly to the spindle, for example, kinetochores send out inhibitory signals that block the activation of APC^{Cdc20}, thereby preventing anaphase and mitotic exit until correct spindle attachment has occurred.

Budding yeast, alone among the model systems used in cell-cycle studies, does not possess a well defined G2/M checkpoint. In this organism, S and M phases seem to overlap: spindle pole body separation and spindle assembly can begin before DNA replication is complete. If DNA damage occurs in budding yeast, the cell cycle is arrested not at G2/M but at the metaphase-to-anaphase transition. The end result is the same as in other species: the segregation of damaged or incompletely replicated chromosomes is prevented.

Cyclin–Cdk complexes trigger mitotic entry in all eukaryotes

The eukaryotic cell undergoes a dramatic reorganization as it enters M phase. In multicellular organisms in particular, almost every subcellular organelle and macromolecular structure is rebuilt or altered in some way as the cell assembles a spindle and prepares the sister chromatids for segregation. Remarkably, all of these processes depend on a single group of master regulators: the mitotic cyclin–Cdk complexes. At the onset of mitosis, these protein kinases phosphorylate a broad array of protein substrates—including structural components and regulatory enzymes—that bring about the events of early mitosis.

As we saw in Chapter 3, Cdk catalytic subunits are present in the cell at high concentrations that do not vary significantly during the cell cycle. The formation of mitotic cyclin–Cdk complexes is driven by cyclin concentration, which increases as the cell approaches mitosis. In most cell types, mitotic cyclin–Cdk complexes are initially held in an inactive state by inhibitory phosphorylation, and are then abruptly activated at the onset of mitosis. As the cell exits from mitosis, mitotic cyclins are destroyed by proteolysis and Cdks are thereby inactivated.

Mitotic cyclins are classified according to primary sequence relationships, particularly in the 100-amino-acid cyclin box, the Cdk-binding domain that lies at the heart of all cyclins (see section 3-2). Many mitotic cyclins are related to cyclin B, the major mitotic cyclin first identified in early studies of invertebrate embryos, and are therefore called B-type cyclins (Figure 5-4). All B-type cyclins contain short sequences that target these proteins to the APC in late mitosis, triggering their destruction.

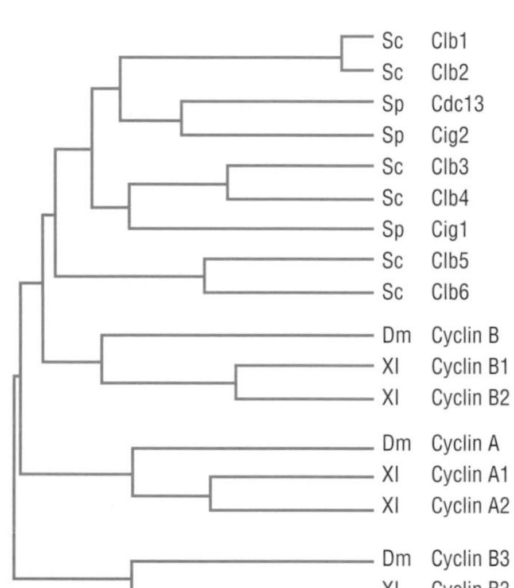

Figure 5-4 The B-type cyclin family This evolutionary tree summarizes the relationships among B-type cyclins from the budding yeast *S. cerevisiae* (Sc), the fission yeast *S. pombe* (Sp), the fruit fly *D. melanogaster* (Dm), and a vertebrate (the frog *X. laevis*, Xl). The amino-acid sequences of cyclin boxes from the indicated cyclins were compared. The total length of the horizontal lines separating two cyclins on this diagram provides an indication of how different they are. For example, closely related cyclins such as Clb1 and Clb2 have very short distances between them, whereas distantly related cyclins such as Clb1 and Clb5 are farther apart. Adapted from Jacobs, H.W. *et al.*: *Genes Dev.* 1998, **12**:3741–3751.

Fission yeast cells trigger mitosis with a single mitotic cyclin

Fission yeast cells possess a clear-cut G2/M transition that is controlled by a simplified version of the system found in metazoans. Unlike any other of the model organisms used in cell-cycle studies, fission yeast controls mitosis with just one cyclin–Cdk complex. Although this organism contains three proteins with B-type cyclin sequence (Cig1, Cig2 and Cdc13), only one—Cdc13—drives the periodic formation of cyclin–Cdk1 complexes that control mitotic entry. Temperature-sensitive mutations in the gene *cdc13*$^{+}$ prevent entry into mitosis—much like mutations in the gene encoding Cdk1.

Two pairs of mitotic cyclins control budding yeast mitosis

In contrast to fission yeast, budding yeast and metazoans control mitosis with multiple mitotic cyclins whose levels rise and fall at slightly different times. It is likely, but unproven, that differences in the timing of mitotic cyclin expression reflect differences in their function: mitotic cyclins expressed early in mitosis, for example, probably control early mitotic events, whereas later events are triggered by cyclin–Cdk complexes that are activated slightly later.

Budding yeast cells contain six B-type cyclins, Clb1 to Clb6, which all associate with a single Cdk, Cdk1. As discussed in Chapter 4 (section 4-5), Clb5 and Clb6 levels rise at the beginning of S phase, and these proteins function primarily in the control of DNA replication. The other four B-type cyclins, Clb1 to Clb4, are involved in the control of mitotic events. Of these, Clb3 and Clb4 are a closely related pair whose levels increase in mid S phase—at about the same time as separation of the spindle pole bodies occurs. Levels of the remaining pair, Clb1 and Clb2, rise shortly thereafter, as mitotic spindle assembly progresses (see Figure 4-13).

References

Fisher, D. and Nurse, P.: **Cyclins of the fission yeast *Schizosaccharomyces pombe*.** *Semin. Cell Biol.* 1995, **6**:73–78.

Fitch, I. *et al.*: **Characterization of four B-type cyclin genes of the budding yeast *Saccharomyces cerevisiae*.** *Mol. Biol. Cell* 1992, **3**:805–818.

Jacobs, H.W. *et al.*: ***Drosophila* Cyclin B3 is required for female fertility and is dispensable for mitosis like Cyclin B.** *Genes Dev.* 1998, **12**:3741–3751.

Richardson, H. *et al.*: **Cyclin-B homologs in *Saccharomyces cerevisiae* function in S phase and in G2.** *Genes Dev.* 1992, **6**:2021–2034.

Figure 5-5 **Viability of budding yeast cells lacking B-type cyclins** Diploid yeast cells were engineered to remove one of their two copies of a mitotic cyclin gene. These heterozygous mutant diploids were then induced to undergo meiosis and sporulation, resulting in two wild-type haploid cells and two haploid cells lacking the selected cyclin gene. The ability of mutant haploids to proliferate on culture plates was then used to determine whether individual cyclins, or various double and triple combinations, are essential for cell viability. Note that viability in this experiment does not necessarily imply normal cell division, as demonstrated by analysis of DNA content (see Figure 5-6) and spindle morphology. Adapted from Richardson, H. *et al.*: *Genes Dev.* 1992, **6**:2021–2034; and Fitch, I. *et al.*: *Mol. Biol. Cell* 1992, **3**:805–818.

Cyclin Function in Budding Yeast

Missing cyclin(s)	Viability
Clb1	+
Clb2	+
Clb3	+
Clb4	+
Clb1, 2	+/–
Clb1, 3	+
Clb1, 4	+
Clb2, 3	+/–
Clb2, 4	+
Clb3, 4	+
Clb1, 2, 3	–
Clb1, 2, 4	+/–
Clb1, 3, 4	+
Clb2, 3, 4	–
Clb1, 2, 3, 4	–

It is difficult to assign specific functions to each mitotic cyclin in budding yeast. Some clues can be gleaned from studies of mutant yeast cells lacking one or more of the four mitotic cyclins (Figure 5-5). Cells lacking any one cyclin are viable, indicating that no single cyclin is absolutely essential for mitosis. Analysis of various double and triple cyclin deletions, however, reveals that many mutant combinations are lethal, owing to a block in spindle pole body separation and mitotic entry. One cyclin in particular, Clb2, seems most important for progression into mitosis. Cells lacking Clb2 are viable but display delayed mitotic entry (Figure 5-6), and mitosis is blocked in cells lacking both Clb2 and either Clb1 or Clb3. The only triple mutant combination that gives rise to viable cells is the *clb1Δclb3Δclb4Δ* mutant (see Figures 5-5 and 5-6), indicating that Clb2 is the only mitotic cyclin that can promote entry into mitosis on its own.

These results should not be taken as evidence that Clb2 normally stimulates all mitotic events. They simply mean that Clb2 has the capacity to promote these functions well enough to allow survival if other cyclins are missing. It is likely, as mentioned above, that different cyclins have unique functions that are not essential when simply testing the ability of a mutant cell to proliferate in ideal laboratory conditions. A more complete understanding of the normal functions of each cyclin requires more detailed analyses of mutant phenotypes, and also requires more knowledge of the protein substrates phosphorylated by the various cyclin–Cdk1 complexes in the cell.

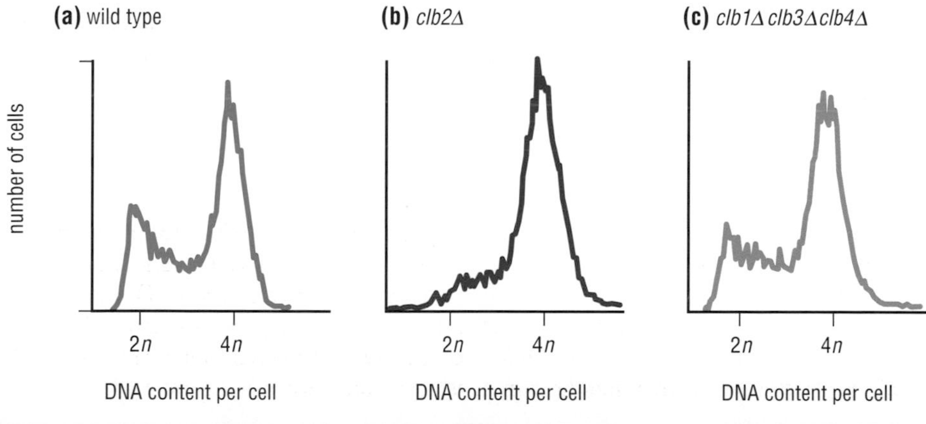

Figure 5-6 **Importance of Clb2 in mitotic entry** **(a)** A normal population of diploid, wild-type budding yeast cells was analyzed for DNA content by flow cytometry (as explained in section 2-6). **(b)** A diploid yeast strain with a homozygous deletion of *CLB2* is viable but has a higher content of $4n$ DNA than wild-type cells, indicating a delay between S phase and the completion of M phase. Analysis of spindles in these cells also indicates a defect in mitotic entry. **(c)** A diploid yeast strain lacking *CLB1*, *CLB3* and *CLB4* does not have a significant mitotic delay. Clb2 is the only mitotic cyclin in this strain, indicating that this cyclin alone has the ability to drive mitosis. Adapted from Richardson, H. *et al.*: *Genes Dev.* 1992, **6**:2021–2034.

5-3 Cyclins that Promote Mitotic Entry in Metazoans

Mitosis in metazoans is governed by cyclins A and B

In multicellular eukaryotes, mitosis is controlled primarily by two cyclins, cyclin A and cyclin B (see Figure 5-4). A third protein, cyclin B3, also contributes to mitotic control in some organisms but is less critical. In addition to containing related cyclin box sequences, all these proteins contain destruction sequences that target them to the APC in late mitosis (see section 3-10). As in budding yeast, the individual functions of each cyclin remain uncertain, and the relative importance of different types varies among species. In addition, one of these cyclins, cyclin A, has non-mitotic functions in the control of DNA replication in some cells (see section 4-6).

Drosophila possesses one version each of cyclins A, B, and B3. Deletion of one, two or all three of these proteins reveals that all are involved in progression through mitosis, but that their relative importance varies (Figure 5-7). Cyclin A is the most potent stimulator of mitosis; indeed, it is the only cyclin of the three that is essential for entry into mitosis. Cyclin B has an intermediate potency, and cyclin B3 seems to have a relatively minor function. A close examination of mutant embryos also suggests that different cyclins have some unique functions. Cyclin A, for example, seems to be especially important in chromosome condensation, whereas cyclin B is more critical for mitotic spindle assembly.

Cyclin Function in *Drosophila*

Missing cyclin(s)	Phenotype
cyclin A	cell-cycle arrest before mitosis 16
cyclin B	delayed mitosis and slight spindle defects, but flies viable
cyclin B3	no major mitotic defect, flies viable
cyclins A, B	cell-cycle arrest before mitosis 15
cyclins A, B3	defects in mitosis 15; cell-cycle arrest before mitosis 16. Defective chromosome condensation
cyclins B, B3	severe defects in mitosis 15, cell-cycle arrest before mitosis 16. Defective spindle assembly
cyclins A, B, B3	defects in early embryonic (syncytial) cycles; cell-cycle arrest before mitosis 15

Figure 5-7 Table of mitotic phenotypes of *Drosophila* cyclin mutants This list summarizes the key mitotic defects in *Drosophila* strains with mutations in the indicated cyclin genes. These studies focused on fly embryos, where the phenotypes caused by cyclin mutations are obscured by the presence in the mutant embryo of abundant cyclin mRNA deposited in the egg by the non-mutant mother (see section 2-4). Mutant phenotypes are therefore not clear until the maternal cyclin supply runs out after about 15 or 16 embryonic divisions, and the embryo begins to depend on its own cyclin supply. Moderate defects therefore cause a cell-cycle arrest before mitosis 16, whereas more severe defects cause an arrest before mitosis 15.

Vertebrate mitosis is driven by multiple forms of cyclins A and B

In frog embryonic cells, somatic mammalian cells and presumably all vertebrates, cyclins A and B are the two major, essential regulators of mitotic entry. The distantly related cyclin B3 protein has no clear mitotic function.

As discussed in Chapter 4 (section 4-6), cyclin A in vertebrates is an important S-phase regulator in some cells and is generally expressed earlier in the cell cycle than cyclin B: its levels (and associated Cdk activity) rise early in S phase and remain elevated until the protein is destroyed soon after nuclear envelope breakdown (Figure 5-8). In contrast, cyclin B levels rise as the cell approaches mitosis, and its associated Cdk activity increases abruptly in prophase. Cyclin B is destroyed during metaphase.

Vertebrate cyclin A exists in two forms, cyclin A1 and cyclin A2, each encoded by separate genes. Cyclin A1 is expressed in germ cells and early embryonic cells, and is the version of cyclin A that is studied in frog embryonic cells (where its major partner is Cdk1). Mice lacking cyclin A1 are viable, but males are sterile as a result of a defect in the first meiotic division of spermatogenesis. Cyclin A2 is expressed in early development and in adult tissues, and mice lacking this protein die as early embryos. Cyclin A2 is the version of cyclin A typically studied in cultured mammalian cells (where its major partner is Cdk2). Because there are no obvious differences in the cell-cycle functions of the two cyclin A subtypes, we will not generally distinguish between them here.

Vertebrate cyclin B also exists in two forms—cyclin B1 and cyclin B2. Both are present in frog embryonic cells and in cultured mammalian cells, both display the typical mitotic pattern of expression, and both bind only to Cdk1. Mice lacking cyclin B2 are viable, whereas deletion of cyclin B1 results in early embryonic lethality. This and other evidence indicates that cyclin B1 is the more important of the two, and for this reason it is the one most commonly studied. It will be the primary focus of our discussion.

References

Furuno, N. *et al.*: **Human cyclin A is required for mitosis until mid prophase.** *J. Cell Biol.* 1999, **147**:295–306.

Jacobs, H.W. *et al.*: ***Drosophila* Cyclin B3 is required for female fertility and is dispensable for mitosis like Cyclin B.** *Genes Dev.* 1998, **12**:3741–3751.

Lehner, C.F. and O'Farrell, P.H.: **The roles of *Drosophila* Cyclins A and B in mitotic control.** *Cell* 1990, **61**:535–547.

Minshull, J. *et al.*: **Translation of cyclin mRNA is necessary for extracts of activated *Xenopus* eggs to enter mitosis.** *Cell* 1989, **56**:947–956.

Murray, A.W. and Kirschner, M.W.: **Cyclin synthesis drives the early embryonic cell cycle.** *Nature* 1989, **339**:275–280.

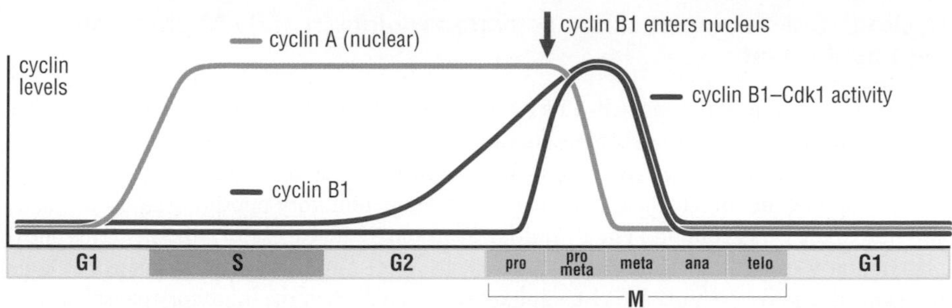

Figure 5-8 Mitotic cyclin levels in a somatic mammalian cell Cyclin A is primarily a nuclear protein whose levels and associated Cdk2 activity rise in early S phase and drop in prometaphase (pale purple line). Cyclin B1 (dark purple) begins to accumulate in G2 and reaches peak levels just before its destruction in metaphase. The protein kinase activity of cyclin B1–Cdk1 complexes (red) remains low in G2, owing to inhibitory phosphorylation of Cdk1. Removal of these inhibitory phosphorylations in prophase results in the sudden, all-or-none activation of Cdk1. In late prophase, cyclin B1–Cdk1 complexes are translocated from the cytoplasm to the nucleus. The nuclear envelope breaks down several minutes later, defining the onset of prometaphase (see section 5-0).

The active cyclin B1–Cdk1 complex moves from cytoplasm to nucleus in late prophase

The two forms of cyclin B are found at different locations in mammalian cells. Cyclin B2 is associated predominantly with membranes of the Golgi apparatus throughout G2 and mitosis. Cyclin B1 localization is more complex. As its levels rise in G2, cyclin B1 is found entirely in the cytoplasm. Early in prophase, when the cyclin B1–Cdk1 complex is first activated, it remains in the cytoplasm and is focused primarily at the duplicated centrosomes, just outside the nucleus, as they begin to separate. In late prophase, however, most of the active complex is abruptly translocated into the nucleus, where it is involved in promoting nuclear envelope breakdown. Soon thereafter, the nuclear envelope disintegrates and cyclin B1–Cdk1 is distributed throughout the cell.

Vertebrate cyclins A and B drive different mitotic events

The individual functions of cyclins A and B in vertebrate mitosis are not clearly defined. The earliest nuclear events of mitosis—chromosome condensation in particular—are probably initiated by cyclin A–Cdk, because these complexes, unlike cyclin B–Cdk1, are active and found entirely within the nucleus in early prophase. There is also evidence that chromosome condensation can be blocked—and even reversed—in human cells injected with a protein inhibitor of cyclin A–Cdk2. Cyclin A–Cdk complexes may also contribute to the activation of cyclin B–Cdk complexes, as we discuss in more detail later in this chapter.

Cyclin B–Cdk1 is thought to stimulate the major mitotic events that occur slightly later in prophase (centrosome separation) and thereafter (nuclear envelope breakdown and spindle assembly). It also promotes the completion of chromosome condensation. Unlike the effects of cyclin A–Cdk2, those of cyclin B–Cdk1 are irreversible once set in motion. This is due in part to the all-or-none, irreversible nature of cyclin B–Cdk1 activation, which we discuss in the next sections of this chapter.

Cyclin B–Cdk1 complexes are activated rapidly in early M phase by dephosphorylation

The activation of mitotic cyclin B–Cdk1 complexes is controlled by rapid changes in Cdk1 phosphorylation. Cyclin B levels typically rise as the cell approaches mitosis, resulting in the formation of cyclin B–Cdk1 complexes. As described in Chapter 3 (sections 3-3 and 3-8), these complexes are initially held in an inactive state by inhibitory phosphorylation at one or two sites in the Cdk1 subunit (Thr 14 and Tyr 15 in vertebrate Cdk1, for example). This phosphorylation is carried out by protein kinases of the Wee1 family. In early M phase, the large stockpile of inactive cyclin B–Cdk1 complexes is activated when the inhibitory phosphates on Cdk1 are removed by phosphatases of the Cdc25 family. In this and the next two sections of this chapter we describe some of the features of this complex system and discuss how it generates rapid and irreversible Cdk1 activation in mitosis.

Elegant studies in fission yeast provided the first glimpse of this regulatory system (Figure 5-9). In this organism, the protein kinase Wee1 is primarily responsible for inhibitory phosphorylation of Cdk1 at Tyr 15 (a closely related kinase, Mik1, also contributes but is less important). Mutations in the gene *wee1*$^{+}$ result in premature Cdk1 activation and mitosis (causing the cells to divide at a smaller size—hence the name of the mutant gene). Overproduction of Wee1 blocks entry into mitosis. The phosphatase that opposes Wee1 and catalyzes the dephosphorylation of Cdk1 is Cdc25, also identified first in fission yeast from a mutant that arrests in late G2 because Cdk1 cannot be activated. As expected if Wee1 and Cdc25 have opposing functions, the cell-cycle arrest caused by *cdc25* mutations is prevented by simultaneous mutation of *wee1*.

Budding yeast is the exception to the general rule that Cdk1 activation at mitosis is triggered by dephosphorylation. Inhibitory phosphorylation and dephosphorylation of Cdk1 in this organism are not required for cell-cycle progression. Cells display only minor cell-cycle defects, for example, when the inhibitory tyrosine in budding yeast Cdk1 (Tyr 19) is changed to a residue that cannot be phosphorylated, or when the *WEE1* homolog is deleted. In this species, mitotic Cdk1 activity seems to be determined primarily by the cellular concentrations of the mitotic cyclins.

Multiple Wee1-related kinases and Cdc25-related phosphatases govern Cdk1 activity in animal cells

In animal cells, including those of flies and vertebrates, two distinct Wee1-related kinases—called Myt1 and Wee1—collaborate in the inhibition of Cdk1 before mitosis. The key inhibitory kinase is Myt1, which is inserted in the membranes of the endoplasmic reticulum and Golgi apparatus. Myt1 phosphorylates both Thr 14 and Tyr 15 of Cdk1. Wee1 is a soluble and predominantly nuclear protein that phosphorylates only Tyr 15. The activities of both Myt1

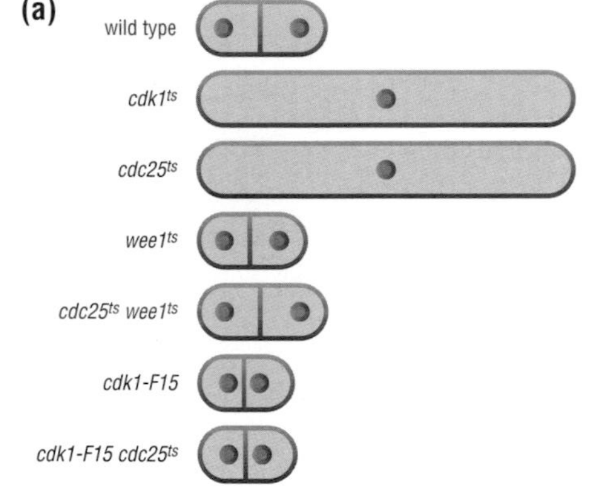

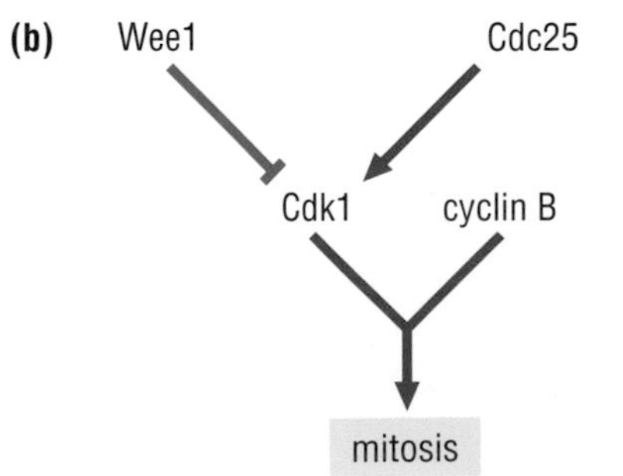

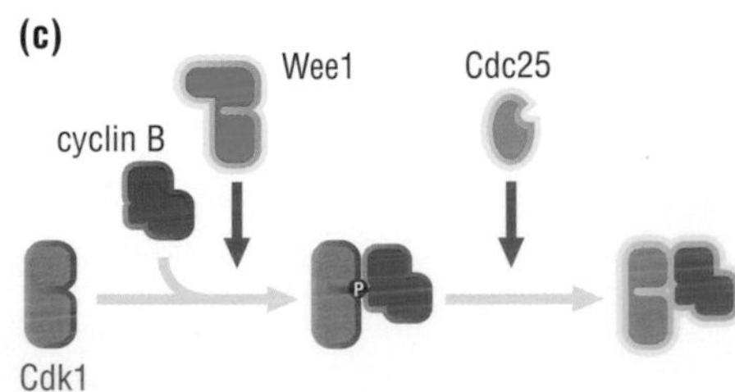

Figure 5-9 The logic of mitotic control in fission yeast In the fission yeast *S. pombe*, the timing of mitotic entry can be estimated by measuring cell length during septation, as diagrammed here. Cells that enter mitosis prematurely are smaller than wild-type cells at septation, whereas cells that are delayed in mitosis are larger than wild type. **(a)** When grown at the restrictive temperature, temperature-sensitive (ts) *cdk1* or *cdc25* mutants arrest before mitosis (but continue to grow), indicating that these genes encode positive regulators of mitosis. In contrast, a *wee1*^{ts} mutant enters mitosis early, resulting in cells that are smaller—suggesting that this gene encodes an inhibitor of mitosis. Cells carrying mutations in both *cdc25* and *wee1* have normal morphology, indicating that neither gene is essential for mitosis, and that the two genes act antagonistically to control mitotic entry. Cells also enter mitosis prematurely if the inhibitory phosphorylation site in Cdk1 (Tyr 15) is changed to a non-phosphorylatable amino acid (phenylalanine, to give the *cdk1-F15* mutant; note that Thr 14 is not phosphorylated in yeast as it is in metazoans). These cells often undergo a so-called mitotic catastrophe in which cells die from entering mitosis too early, presumably because the activity of Cdk1 cannot be inhibited before mitosis. The *cdk1-F15* phenotype is more severe than the *wee1*^{ts} phenotype because a small amount of Cdk1 phosphorylation occurs in the *wee1*^{ts} mutant (because of the presence of another inhibitory protein kinase, Mik1). **(b)** Analyses of mutants, like those in panel (a), are consistent with the simple model that Wee1 and Cdc25 have opposing actions on the mitosis-promoting function of Cdk1. **(c)** Mutant analyses in yeast, combined with biochemical studies in numerous organisms, indicate that Cdk1 activation proceeds as shown here. During G2, cyclin binding and phosphorylation by Wee1 results in formation of an inactive cyclin–Cdk complex, which is then activated in mitosis by Cdc25.

and Wee1 are high during most of the cell cycle but then decrease abruptly during mitosis, thereby allowing Cdk1 dephosphorylation (and thus activation) by phosphatases of the Cdc25 family (Figure 5-10).

Because cyclin B1–Cdk1 complexes are found almost entirely in the cytoplasm until late prophase (see section 5-3), Myt1 is best positioned to serve as the critical inhibitory kinase. The function of Wee1 may be to maintain the inhibition of the small amount of Cdk1 that is found in the nucleus before mitosis. This function does not seem to be essential, because deletion of the *Drosophila* gene for Wee1 is not lethal, presumably because Myt1 provides sufficient Cdk1 inhibition before mitosis. In addition, Wee1 is not present in *Xenopus* oocytes; inhibitory Cdk1 phosphorylation in these cells is controlled by Myt1 alone.

Animal cells contain multiple Cdc25-related phosphatases. In *Drosophila*, a single Cdc25, called String, is essential for Cdk1 activation in mitosis. A related enzyme called Twine is involved primarily in the control of Cdk activity in meiotic cells. Vertebrate cells contain three versions of Cdc25, called Cdc25A, Cdc25B and Cdc25C. All these enzymes contribute to the activation of cyclin B–Cdk1 in mitosis, but it is not clear whether any single one is essential. Deletion of both the Cdc25B and Cdc25C genes has little apparent effect in mouse cells, arguing that the activation of Cdc25A alone (coupled with Myt1 and Wee1 inhibition) is sufficient to allow normal mitotic entry. It is likely, however, that the robustness of Cdk1 regulation depends on the presence of all three Cdc25 isoforms.

The three vertebrate Cdc25 isoforms display different patterns of activity before and during mitosis, providing clues to their specific functions in Cdk1 activation. Cdc25B is activated early and is therefore thought to be involved in the initiation of Cdk1 activation. Cdc25B levels and activity increase in late S phase and G2, peak in prophase and decline in prometaphase (Figure 5-10). Part of the Cdc25B population is cytoplasmic and thus co-localizes with the cyclin B–Cdk1 complex when it first becomes activated in prophase cells. However, given that Cdc25B is active in G2 cells, in which Cdk1 is largely inactive, there is likely to be little activation of Cdk1 by Cdc25B before mitosis—perhaps because Myt1 and Wee1 counteract it.

Cdc25A and Cdc25C are relatively inactive in G2 and are activated abruptly in prophase. These phosphatases therefore seem to be important for generating the dramatic increase in Cdk1 activity that occurs in early mitosis (see Figure 5-10). Cdc25A activity rises in mitosis as a result of an increase in Cdc25A protein levels, caused by a decrease in the rate of its degradation. The level of Cdc25C does not change during the cell cycle but its catalytic activity increases in mitosis. Cdc25A and Cdc25C are both found at locations where they have access to cyclin B1–Cdk1: Cdc25A is located mainly in the nucleus and to a lesser extent in the cytoplasm, whereas Cdc25C—like cyclin B1–Cdk1—resides in the cytoplasm in early prophase and then moves into the nucleus in late prophase.

Rapid Cdk1 activation in prophase therefore results from sudden increases in the activities of Cdc25A and Cdc25C, combined with simultaneous decreases in the activities of Myt1 and Wee1 (see Figure 5-10). How are these abrupt changes generated? The major mechanism is phosphorylation of all these enzymes, which activates the Cdc25 proteins and inhibits the Myt1 and Wee1 kinases. As discussed in Chapter 3 (section 3-8), phosphorylation of these proteins is catalyzed, at least in part, by Cdk1—resulting in positive feedback loops by which Cdk1 activates its own activators and inhibits its inhibitors. We describe this complex regulatory system in section 5-5.

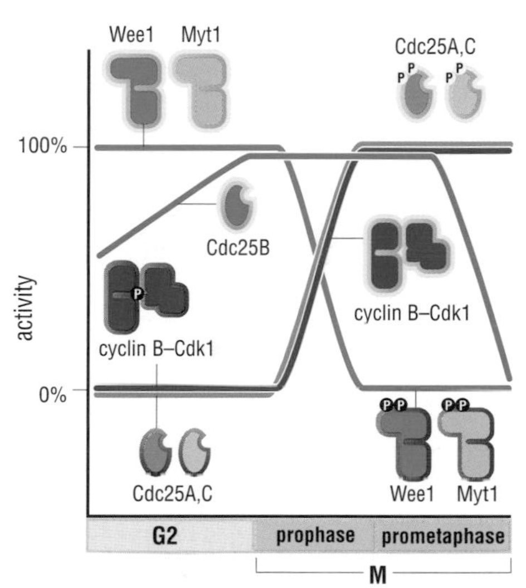

Figure 5-10 Activities of key regulators during entry into mitosis Before mitosis, Myt1 and Wee1 activities are high and Cdc25A and Cdc25C activities are low, resulting in inhibitory phosphorylation of cyclin B–Cdk1 complexes. Cdc25B activity is present in G2 but promotes only minor amounts of Cdk1 activation in the face of abundant Myt1 and Wee1 activity. In early mitosis, Wee1 and Myt1 are inhibited by phosphorylation and Cdc25A and Cdc25C are stimulated by it. As a result, Cdk1 is rapidly dephosphorylated and activated. The regulatory circuits underlying these changes are described in section 5-5. For simplicity, only one inhibitory phosphorylation site on Cdk1 is shown, and the activating phosphate (see section 3-3) is not shown.

References

Ferguson, A.M. *et al.*: **Normal cell cycle and checkpoint responses in mice and cells lacking Cdc25B and Cdc25C protein phosphatases.** *Mol. Cell Biol.* 2005, **25**:2853–2860.

Gould, K.L. and Nurse, P.: **Tyrosine phosphorylation of the fission yeast *cdc2*$^+$ protein kinase regulates entry into mitosis.** *Nature* 1989, **342**:39–45.

Lindqvist, A. *et al.*: **Characterisation of Cdc25B localisation and nuclear export during the cell cycle and in response to stress.** *J. Cell Sci.* 2004, **117**:4979–4990.

Lindqvist, A. *et al.*: **Cdc25B cooperates with Cdc25A to induce mitosis but has a unique role in activating cyclin B1–Cdk1 at the centrosome.** *J. Cell Biol.* 2005, **171**:35–45.

Mailand, N. *et al.*: **Regulation of G_2/M events by Cdc25A through phosphorylation-dependent modulation of its stability.** *EMBO J.* 2002, **21**:5911–5920.

Nurse, P.: **Universal control mechanism regulating onset of M-phase.** *Nature* 1990, **344**:503–508.

Russell, P. and Nurse, P.: **Negative regulation of mitosis by *wee1*$^+$, a gene encoding a protein kinase homolog.** *Cell* 1987, **49**:559–567.

Solomon, M. *et al.*: **Cyclin activation of p34^{cdc2}.** *Cell* 1990, **63**:1013–1024.

Mitotic Cdk1 activation involves multiple positive feedback loops

We have seen that Cdk1 activation in mitosis results from sudden increases in Cdc25 phosphatase activities and decreases in Myt1/Wee1 kinase activities (see Figure 5-10). In this section we describe the regulatory circuitry that triggers these changes and generates abrupt, switch-like Cdk1 activation—ensuring that mitotic events are initiated completely and irreversibly.

Positive feedback lies at the heart of mitotic Cdk1 activation: changes in Cdc25 and Myt1/Wee1 activities in mitosis are caused, at least in part, by Cdk1 itself (Figure 5-11). Cdk1 phosphorylates and thereby stabilizes the Cdc25A protein, causing an increase in its levels. Cdc25C phosphorylation stimulates its enzymatic activity. Phosphorylation of Myt1 and Wee1 inhibits their enzymatic activities. These feedback relationships are expected to generate a regulatory system that is bistable: that is, it switches from a stable state of Cdk1 inactivity to a stable state of Cdk1 activity (see section 3-7). The presence of so many feedback loops in this system also ensures that the Cdk1 switch functions effectively even if some components fail.

It is likely that additional positive feedback loops exist in the Cdk1 activation system. One of these may involve another mitotic serine/threonine kinase called polo-like kinase or Plk. In frog egg extracts Plk is activated in early mitosis by a poorly understood mechanism that depends on Cdk1 activity—suggesting that Cdk1 stimulates Plk activation. In addition, Plk phosphorylates Cdc25C and Myt1 at some of the sites that are phosphorylated in mitosis, arguing that Plk can indirectly stimulate Cdk1 activation. A reasonable interpretation of these results is that Plk and Cdk1 activate each other in a positive feedback circuit (see Figure 5-11).

Cdc25B and cyclin A–Cdk help trigger cyclin B–Cdk1 activation

Bistable systems based on strong positive feedback need additional components, called trigger or starter mechanisms, to initiate a change in one component that drives the system from one stable state to the other (see section 3-7). The nature of these trigger mechanisms in Cdk1 activation is still unclear. One simple possibility is that Myt1 and Wee1 do not completely suppress cyclin B–Cdk1 activity: a very small amount of activity might remain and gradually increase as cyclin B levels increase, until a threshold is reached where the amount of active Cdk1 is sufficient to promote enough Cdc25 activation (and Myt1/Wee1 inactivation) to fire the feedback loops and switch the system to the on state. Although appealing in its simplicity, this model is unlikely to be correct: there is considerable evidence that Cdk1 activation involves additional components and regulatory interactions.

The phosphatase Cdc25B is well qualified to serve as a trigger of Cdk1 activation. As described in section 5-4, Cdc25B activity rises in late S phase and peaks in prophase. It seems that Cdc25B alone is not sufficiently active to drive more than a small amount of Cdk1 activation. Even partial Cdk1 activation, however, could eventually raise Cdk1 activity to a level that would result in partial activation of Cdc25A and Cdc25C or inhibition of Myt1 and Wee1. This would then stimulate further Cdk1 dephosphorylation, eventually setting the positive feedback in motion and switching the system to a state of Cdk1 activity. Cdc25B cannot be the only trigger mechanism, however, because mouse cells divide normally without it.

Cyclin A–Cdk complexes may also act as a trigger of cyclin B–Cdk1 activation (see Figure 5-11). The activity of cyclin A–Cdk complexes rises in early S phase—roughly in parallel with the concentration of cyclin A in the cell—and remains high until cyclin A is destroyed in

References

Abrieu, A. *et al.*: **The Polo-like kinase Plx1 is a component of the MPF amplification loop at the G_2/M-phase transition of the cell cycle in *Xenopus* eggs.** *J. Cell Sci.* 1998, **111**:1751–1757.

Barr, F.A. *et al.*: **Polo-like kinases and the orchestration of cell division.** *Nat. Rev. Mol. Cell Biol.* 2004, **5**:429–440.

Ferrell, J.E. Jr: **Self-perpetuating states in signal transduction: positive feedback, double-negative feedback and bistability.** *Curr. Opin. Cell Biol.* 2002, **14**:140–148.

Kumagai, A. and Dunphy, W.G.: **Purification and molecular cloning of Plx1, a Cdc25-regulatory kinase from *Xenopus* egg extracts.** *Science* 1996, **273**:1377–1380.

Lindqvist, A. *et al.*: **Cdc25B cooperates with Cdc25A to induce mitosis but has a unique role in activating cyclin B1–Cdk1 at the centrosome.** *J. Cell Biol.* 2005, **171**:35–45.

Mailand, N. *et al.*: **Regulation of G_2/M events by Cdc25A through phosphorylation-dependent modulation of its stability.** *EMBO J.* 2002, **21**:5911–5920.

Pomerening, J.R. *et al.*: **Systems-level dissection of the cell-cycle oscillator: bypassing positive feedback produces damped oscillations.** *Cell* 2005, **122**:565–578.

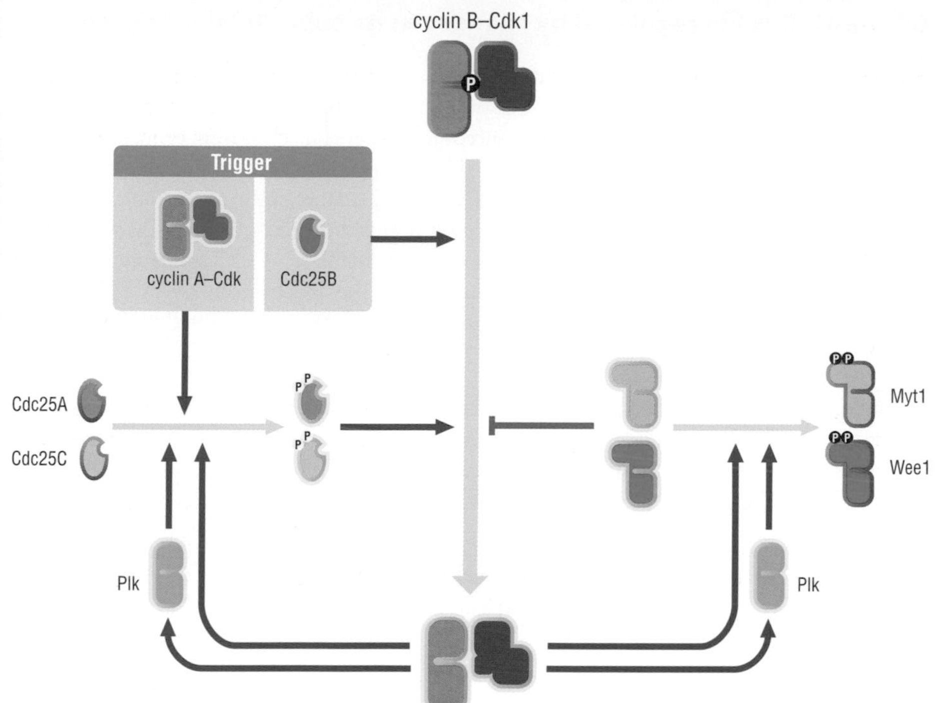

Figure 5-11 Multiple feedback loops governing Cdk1 activation in early mitosis Cdk1 enhances its own activity by a variety of feedback loops. Cdk1 phosphorylates and thus stimulates its activator Cdc25C and inhibits its inhibitors Myt1 and Wee1. Phosphorylation by Cdk1 also stabilizes the levels of another activator, Cdc25A. Cdk1 may also promote its own activation by stimulating Plk activity, which further stimulates Cdc25C and Myt1. Cdc25B and cyclin A–Cdk complexes may help trigger Cdk1 activation by promoting partial Cdk1 dephosphorylation in late G2. For simplicity, only one inhibitory phosphorylation is shown on Cdk1, and the activating phosphorylation is not shown (see section 3-3).

prometaphase. One might imagine, therefore, that cyclin A–Cdk activity in late G2 cells could help phosphorylate Cdc25A, Cdc25C, Myt1 or Wee1, thereby helping to trigger the feedback loops that activate cyclin B–Cdk1. Consistent with this possibility, there is evidence that inhibition of cyclin A–Cdk2 delays cyclin B–Cdk1 activation in human cells. In frog embryonic cells a similar function may be provided by cyclin E–Cdk2, which is active throughout S phase in these cells. Some contribution from cyclin A– or E–Cdks therefore seems likely, although it is not clear why these kinases do not activate cyclin B–Cdk1 when they are first turned on in S phase.

As discussed in section 3-7, regulatory systems containing strong positive feedback have the potential to be too sensitive: the feedback loop can be triggered accidentally by minor, random fluctuations in some input signal (such as Cdc25 activity, for example). It is therefore essential to incorporate mechanisms that reduce the effects of inappropriate signals and ensure that the feedback loops are fired only when appropriate. Little is known about the mechanisms that provide this filtering function in Cdk1 activation. One speculative possibility is that filtering is provided by employing multisite phosphorylation at several points in the system. Full activation of Cdc25C, for example, may occur only if the protein is fully phosphorylated at several sites by both Cdks and Plk. As a result, activation will not occur when small, random fluctuations in Cdk or Plk activity cause a low level of Cdc25C phosphorylation. It is also likely that these low levels of phosphorylation are also reduced by a basal level of phosphatase activity that removes this phosphorylation.

The switch-like features of cyclin B1–Cdk1 activation are also influenced by the cellular location of the complex and its regulators, as we discuss in the next section.

Cyclin B1–Cdk1 is regulated by changes in its subcellular localization

A key concept in cell-cycle control—and in cellular regulation in general—is that the function of regulatory proteins can be controlled not only by changing their intrinsic activity but also by changing their location in the cell. This concept is illustrated nicely during vertebrate mitosis, when the action of cyclin B1–Cdk1 is regulated in part by changes in its subcellular location. As mentioned in section 5-3, inactive cyclin B1–Cdk1 accumulates in the cytoplasm during G2 and early prophase. Cyclin B1–Cdk1 activation begins in the cytoplasm and is particularly prominent at the centrosomes. In late prophase, most cyclin B1–Cdk1 suddenly enters the nucleus, after which the nuclear envelope dissolves. By this mechanism, cyclin B1–Cdk1 is kept away from its nuclear targets until late prophase, thereby providing an additional layer of inhibition before prometaphase.

Not all cyclin B1–Cdk1 enters the nucleus in late prophase: some remains in the cytoplasm to promote other mitotic processes such as centrosome separation and reorganization of the Golgi apparatus. Golgi reorganization is also stimulated by cyclin B2–Cdk1 complexes, which are associated with the Golgi throughout early mitosis.

G2

cyclin B1–Cdk1

Cdc25C

nuclear accumulation promoted by phosphorylation of cyclin B1 and Cdc25C and dephosphorylation of Ser 216 in Cdc25C

late prophase

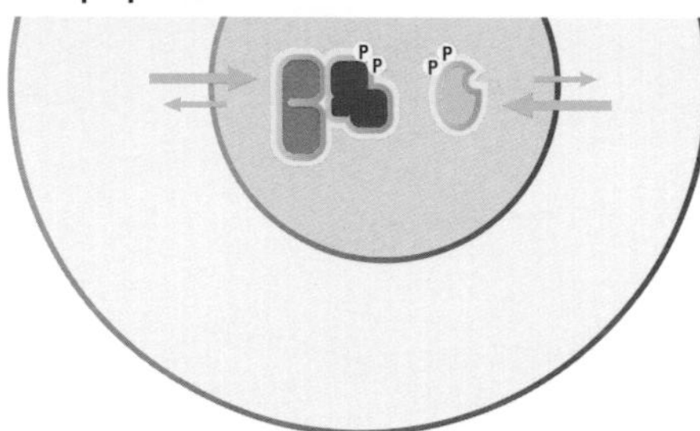

Figure 5-12 Nuclear accumulation of cyclin B1–Cdk1 and Cdc25C in late prophase
In G2, inactive cyclin B1–Cdk1 complexes and Cdc25C are located primarily in the cytoplasm because their rates of nuclear export are greater than the rates at which they are imported. When Cdc25C and Cdk1 are activated in late prophase, Cdk1 and Plk phosphorylate both cyclin B1 and Cdc25C. This obscures nuclear export signals on both proteins, reducing their nuclear export rate (see Figure 5-13). Phosphorylation of cyclin B1 also stimulates import of the protein by unknown mechanisms. Dephosphorylation of Cdc25C at a specific serine (Ser 216 in humans) activates a nuclear import signal on the protein, thereby enhancing the rate of its import into the nucleus. The overall effect is a rapid accumulation of cyclin B1–Cdk1 and Cdc25C inside the nucleus in late prophase.

Cyclin B1–Cdk1 location is controlled by phosphorylation of cyclin B1

The nuclear–cytoplasmic ratio of cyclin B1–Cdk1, like that of most proteins, is governed by the relative rates of the protein's nuclear import and export. Before mitosis, the rate of cyclin B1–Cdk1 import is low and its rate of export is high, resulting in net localization in the cytoplasm. In late prophase, however, the rate of import increases and the rate of export decreases, resulting in accumulation of cyclin B1–Cdk1 inside the nucleus (Figure 5-12).

Import and export of cyclin B1–Cdk1 complexes are directed by the cyclin subunit. Most proteins that are imported into the nucleus from their sites of synthesis in the cytoplasm contain a sequence called a nuclear localization signal, which is recognized by nuclear import receptors. No such sequence has been found in cyclin B1, and its mechanism of import remains unclear. The increased rate of import in early mitosis depends on phosphorylation at several sites in the cyclin amino-terminal region (Figure 5-13), including one site (Ser 113) that is also involved in the control of cyclin export, as we discuss next.

The export of cyclin B1–Cdk1 from the nucleus is better understood. Export is directed by a short sequence, called a nuclear export signal, that is located in the amino-terminal region of cyclin B1. This sequence binds to a transporter protein, Crm1, which carries cyclin B1–Cdk1 out of the nucleus. In early mitosis, Ser 113 within the nuclear export signal is phosphorylated, thereby blocking binding to Crm1 and reducing the rate of export (see Figure 5-13).

Interestingly, the kinases that phosphorylate the amino-terminal sites in cyclin B1 seem to be Cdk1 itself and Plk. Thus, there may be a striking parallel between the activation of Cdc25C, as discussed in section 5-5, and the nuclear accumulation of cyclin B1–Cdk1: both seem to be triggered in early mitosis by a synergistic combination of Cdk1 and Plk activities.

Cdc25C localization is regulated by phosphorylation

The subcellular localization of Cdc25C roughly parallels that of cyclin B1–Cdk1: it is cytoplasmic before mitosis and is translocated into the nucleus in prophase. Like cyclin B1, the localization of Cdc25C depends on the relative rates of its nuclear import and export (see Figure 5-12).

References

Ferrell, J.E. Jr: **How regulated protein translocation can produce switch-like responses.** *Trends Biochem. Sci.* 1998, **23**:461–465.

Hagting, A. *et al.*: **Translocation of cyclin B1 to the nucleus at prophase requires a phosphorylation-dependent nuclear import signal.** *Curr. Biol.* 1999, **9**:680–689.

Jackman, M. *et al.*: **Active cyclin B1–Cdk1 first appears on centrosomes in prophase.** *Nat. Cell Biol.* 2003, **5**:143–148.

Li, J. *et al.*: **Nuclear localization of cyclin B1 mediates its biological activity and is regulated by phosphorylation.** *Proc. Natl Acad. Sci. USA* 1997, **94**:502–507.

Takizawa, C.G. and Morgan, D.O.: **Control of mitosis by changes in the subcellular location of cyclin B1–Cdk1 and Cdc25C.** *Curr. Opin. Cell Biol.* 2000, **12**:658–665.

Toyoshima-Morimoto, F. *et al.*: **Plk1 promotes nuclear translocation of human Cdc25C during prophase.** *EMBO Rep.* 2002, **3**:341–348.

Yang, J. *et al.*: **Combinatorial control of cyclin B1 nuclear trafficking through phosphorylation at multiple sites.** *J. Biol. Chem.* 2001, **276**:3604–3609.

Cdc25C is imported and exported by separate mechanisms, but both are regulated by phosphorylation. Like many proteins that can shuttle between nucleus and cytoplasm, Cdc25C contains both a nuclear localization signal, which interacts with nuclear import transporters, and a nuclear export signal, which interacts with the Crm1 exporter. These signals are located in separate parts of the amino-terminal region of the protein. During interphase, Cdc25C is phosphorylated on Ser 216 (human) or Ser 287 (*Xenopus*), which is near the nuclear localization signal. Phosphorylation here creates a binding site for a small phosphoserine-binding protein called 14-3-3, which obscures the nuclear localization signal, thereby reducing nuclear import of Cdc25C (without affecting the nuclear export signal). Phosphorylation of Ser 216/287 probably inhibits the phosphatase activity of Cdc25C as well. This site is dephosphorylated at the onset of mitosis, which enhances Cdc25C import into the nucleus and may also contribute to its activation.

The kinases and phosphatases acting at Ser 216/287 are not well understood. Two protein kinases that phosphorylate this site, Chk1 and Chk2, are activated after DNA damage; their phosphorylation of Ser 216/287 is thought to provide one mechanism by which DNA damage inhibits mitotic entry, as discussed in Chapter 11. It is likely that these kinases are also key regulators of Ser 216/287 in normal cell cycles.

The nuclear export signal on Cdc25C is near the multiple sites at which phosphorylation triggers Cdc25C activation during mitosis. Phosphorylation in this region—probably by Plk and Cdk1—not only stimulates Cdc25C activity but also obscures the nuclear export signal, thereby decreasing the rate of nuclear export. Plk thus seems able to promote simultaneous nuclear accumulation and activation of both Cdc25C and its target, cyclin B1–Cdk1 (see Figure 5-12).

Wee1 and Myt1, the kinases that inhibit Cdk1, are also found in specific subcellular locations. Myt1 is associated throughout the cell cycle with membranes of the Golgi apparatus and endoplasmic reticulum. Wee1 is found in the nucleus during most of the cell cycle but seems to be exported to the cytoplasm during prophase, by an unknown mechanism.

Cyclin B1–Cdk1 activation and nuclear accumulation are partly interdependent

Although the activation and nuclear accumulation of cyclin B1–Cdk1 occur at roughly the same time, activation does not depend on import. In *Xenopus* embryos or egg extracts, for example, oscillations in cyclin B1–Cdk1 activity continue even in the absence of a nucleus, and careful measurements in numerous other cell types indicate that cyclin B1–Cdk1 activation begins before it is translocated into the nucleus.

It remains possible, however, that nuclear translocation enhances the rate and switch-like properties of cyclin B1–Cdk1 activation. The coincident nuclear import of cyclin B1–Cdk1 and its activator Cdc25C results in a sudden increase in their concentrations inside the nucleus, which alone is likely to enhance the rate of activation. Other features of this system are also expected to make the activation process more switch-like. First, nuclear import of cyclin B1–Cdk1 may saturate the Wee1 in the nucleus, reducing its effectiveness as an inhibitor; the export of Wee1 from the nucleus in prophase would enhance this effect. Second, nuclear import of Cdk1 also moves the kinase away from its cytoplasmic inhibitor, Myt1. This is particularly effective in *Xenopus* oocytes, which do not contain nuclear Wee1.

(a) *Xenopus* cyclin B1

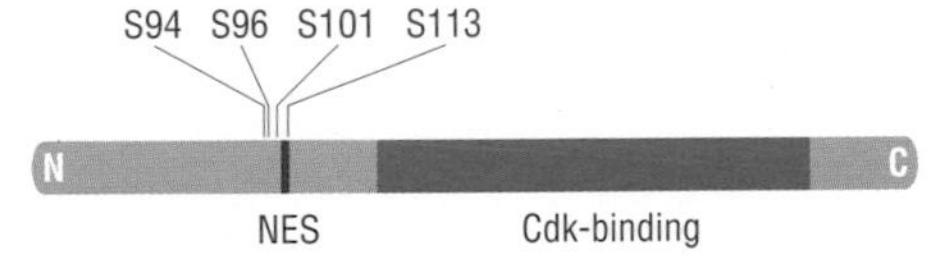

(b) Import defects in cyclin B1 mutants

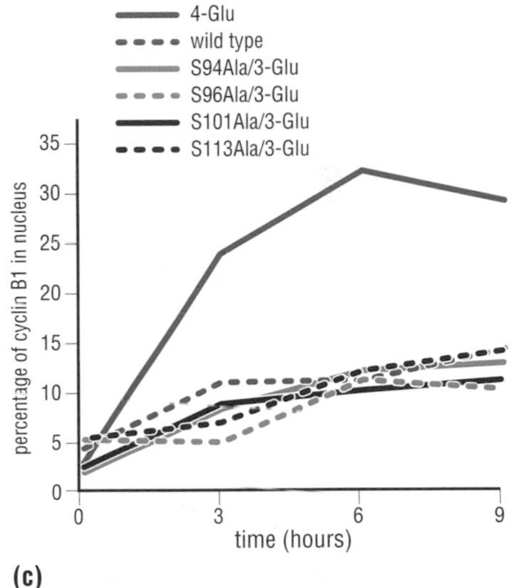

(c)

Export Defects in Cyclin B1 Mutants

Cyclin B1 mutant	Crm1 binding	Export rate
wild type	++	++
4-Ala	++	++
4-Glu	–	–
Ser113Glu	–	–
Ser113Ala + 3-Glu	++	++

Figure 5-13 Phosphorylation of cyclin B1 controls its localization The effects of cyclin B1 phosphorylation on its localization are studied with cyclin B1 mutants in which serine phosphorylation sites are changed either to alanine (to prevent phosphorylation) or to glutamate (to produce a negative charge that partly mimics phosphorylation). This figure provides an example of such studies with the *Xenopus* cyclin B1 protein **(a)**, which contains four serines near its amino terminus that are phosphorylated during mitosis (human cyclin B1 has one additional site in the same region). One of these serines (Ser 113 in *Xenopus*, Ser 147 in humans) is located within the nuclear export signal (NES). **(b)** Wild-type and mutant cyclin B1 proteins were injected into the cytoplasm of *Xenopus* oocytes treated with a drug that blocks nuclear export. The amount of each protein in the nucleus was then determined at various times after injection to assess the rate of nuclear import. When all four phosphorylation sites in cyclin B1 were changed to glutamate (the 4-Glu mutant), cyclin was imported more rapidly than wild-type cyclin B1, suggesting that phosphorylation increases the rate of import. Interestingly, changing any three serines to glutamate, as in the various 3-Glu mutants, does not stimulate import. This evidence suggests that all four serines must be phosphorylated to stimulate import. The nuclear import signal on cyclin B1 is not clearly identified. **(c)** Wild-type and mutant forms of the amino terminus of cyclin B1 were added to interphase extracts of *Xenopus* oocytes and tested for their ability to bind the nuclear export transporter Crm1. In parallel experiments, the various proteins were injected into oocyte nuclei and their rates of nuclear export were measured. No phosphorylation of cyclin B1 occurs in this extract and so the wild-type protein binds well to Crm1 and is exported efficiently. The same is true for the non-phosphorylatable mutant in which all serines have been changed to alanine (4-Ala). Changing all four serines to glutamate (4-Glu) inhibits Crm1 binding and export, suggesting that phosphorylation inhibits export. Closer analysis of mutants in individual sites reveals that just the mutation of Ser 113 to glutamate (Ser113Glu) blocks export, whereas mutation of the other three serines (Ser113Ala + 3-Glu) does not. Phosphorylation of Ser 113 alone thus seems sufficient to block cyclin B1 export. Adapted from Yang, J. *et al.*: *J. Biol. Chem.* 2001, **276**:3604–3609.

5-7 Protein Kinases of the Polo and Aurora Families

Polo-like kinases (Plks) help control spindle assembly and mitotic exit

Cdks are the master regulators of mitotic entry, but they do not act alone. Several other protein kinases are activated at the onset of mitosis and help control a subset of early mitotic events. The most important of these mitotic kinases are the **polo-like kinase, Plk** (see section 5-5), and two protein kinases called *aurora A* and *aurora B*. In this section we provide a brief overview of Plk and aurora regulation and function. Their functions will be described in more detail when we discuss specific mitotic processes later in this chapter and in Chapters 6 and 7.

A single Plk is employed in mitotic control in all eukaryotes (Figure 5-14) and these proteins all have a similar structure (Figure 5-15). The amino-terminal half of a Plk contains a protein kinase catalytic domain similar to that of other kinases, and the carboxy-terminal half contains a domain called a polo box domain, which targets the kinase to specific substrates and subcellular locations. The polo box domain has a high affinity for proteins that are phosphorylated at serine or threonine residues within specific sequence contexts. The interaction of Plk with these proteins therefore requires prior phosphorylation, or priming, by Plk itself or by some other kinase. Phosphorylation by Cdk1, for example, may promote an interaction between some proteins and Plk, providing one mechanism by which Cdk1 could influence the activity or location of Plk.

In most species, Plk is activated in early mitosis. Plk activation is thought to depend on prior Cdk1 activation, but the underlying mechanisms of activation remain unclear. Plk synthesis increases in early mitosis as a result of increased expression of the gene. The enzymatic activity of Plk also increases in mitosis, probably as a result of phosphorylation at specific activating sites. Several phosphorylation sites have been identified in Plk proteins of some species, but the identities of the protein kinases that act at these sites *in vivo* remain unclear. Plks also contain sequence motifs that target them to the ubiquitin-protein ligase APC^{Cdh1}, resulting in their proteolytic destruction in late mitosis and G1.

Plk has a wide range of functions in early and late M phase, in particular in spindle assembly and cytokinesis. In most eukaryotes, except budding yeast, Plk is required for centrosome separation and for the construction of a bipolar spindle. Mutational inactivation of Plk genes in *Drosophila* or fission yeast, as well as injection of anti-Plk antibodies into human cells or frog embryos, results in spindles that are monopolar or otherwise abnormal. Plk is located at the centrosomes in early mitosis, further supporting a function at that site (Figure 5-16).

Plk is also involved in late M-phase events. The inactivation of mitotic Cdk1 in late mitosis requires Plk function in several species, and Plk also helps to control cytokinesis. In fission yeast, for example, there is clear genetic evidence that Plk is a key promoter of the septation process that divides the cell in late M phase. In *Drosophila* and vertebrate cells, Plk is localized in late mitosis at the spindle midzone (see Figure 5-16), and cytokinesis fails in cells lacking Plk function.

Polo-Related Protein Kinases

Species	Name	Synonyms	Comments
S. cerevisiae	Cdc5		required for mitotic exit, not entry
S. pombe	Plo1		regulation of mitosis and cytokinesis
D. melanogaster	Polo		regulation of mitosis and cytokinesis
X. laevis	Plx1		control of mitotic entry and exit
mammals	Plk1		regulation of mitosis and cytokinesis
	Plk2	Snk	no clear function in mitosis
	Plk3	Fnk, Prk	no clear function in mitosis
	Plk4	Sak	control of centrosome duplication

Figure 5-14 Table of polo-related protein kinases

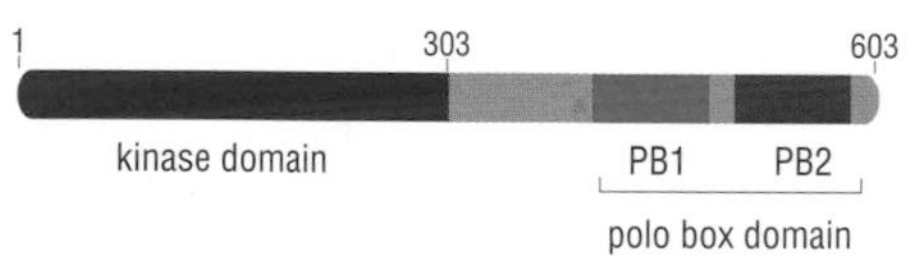

Figure 5-15 General structure of Plk The human Plk1 protein, like all members of the Plk family, contains an amino-terminal kinase domain and a carboxy-terminal polo box domain that interacts with other proteins. The polo box domain is composed of two distinct polo box sequences, PB1 and PB2.

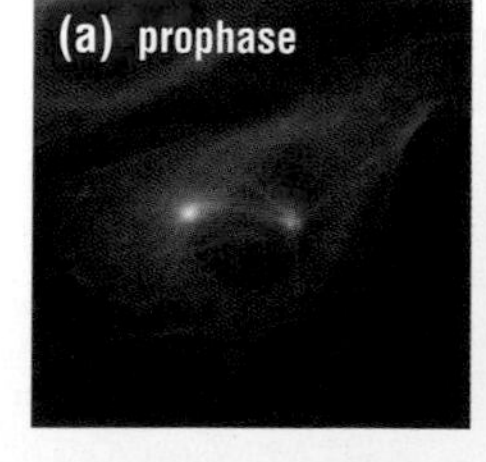

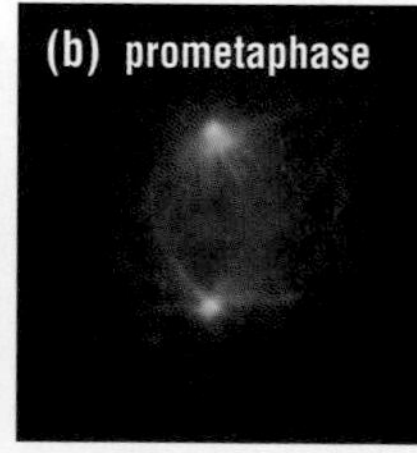

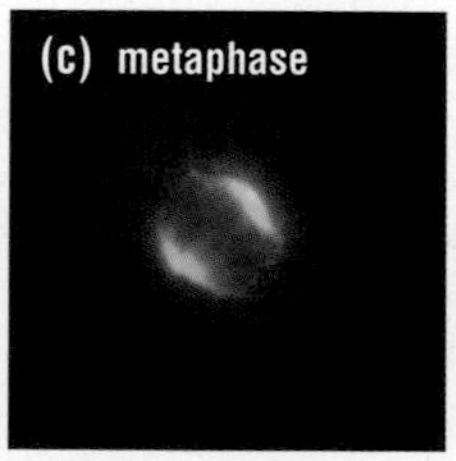

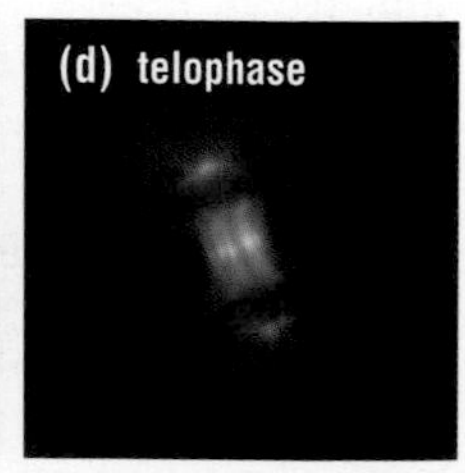

Figure 5-16 Localization of Plk in mitotic cells Human HeLa cells growing in culture were stained with an antibody against Plk1 (red). DNA is stained in blue and microtubules of the spindle are stained green. From prophase to metaphase, Plk co-localizes with the microtubules at the spindle poles, resulting in a yellow color. In prometaphase and metaphase, faint Plk staining is also seen at kinetochores (red patches). In telophase, Plk is localized at the spindle midzone. Kindly provided by Francis Barr and Ulrike Grunewald. From Barr, F.A. *et al.*: *Nat. Rev. Mol. Cell Biol.* 2004, **5**:429–440.

Definitions

aurora A: serine/threonine protein kinase that is activated at the beginning of M phase and is involved in centrosome function and spindle assembly.

aurora B: serine/threonine protein kinase that is activated at the beginning of M phase and is involved in chromosome condensation, spindle assembly, attachment of kinetochores, sister-chromatid segregation and cytokinesis.

Plk: see **polo-like kinase**.

polo-like kinase (Plk): serine/threonine protein kinase that is activated at the beginning of M phase and inactivated in late mitosis and G1, and is involved in a variety of mitotic processes including spindle assembly and kinetochore function, and in cytokinesis.

Spindle function and sister-chromatid segregation are controlled in part by aurora kinases

Another important group of mitotic kinases is the aurora family. Metazoans have two major members of this family, called **aurora A** and **aurora B**, whereas yeast have a single family member that most closely resembles metazoan aurora B (Figure 5-17). A third member, aurora C, is found in mammalian germ cells and its functions are not well understood. All aurora kinases contain a related protein kinase catalytic domain plus amino-terminal extensions of various sizes and sequences. As in Plks, the non-catalytic region of these proteins is thought to regulate their localization and activity.

Like Plks, aurora kinases are activated in mitosis. The levels and enzyme activities of both aurora A and B increase in mitosis, and both are phosphorylated at multiple sites, although little is known about the functions of these sites or the kinases responsible for phosphorylation. Aurora A and B also interact with numerous regulatory proteins that govern their activity. A particularly important activator of aurora A is a protein called TPX2, which targets the active kinase to spindle microtubules and thereby promotes its function in spindle assembly. Aurora B also interacts with activating proteins: throughout mitosis, it forms a complex with two other proteins, INCENP and survivin, that stimulate its activity and are required for its proper localization.

In metazoans, aurora A is found at the centrosome and on the spindle (Figure 5-18), and helps control bipolar spindle assembly and stability. Mutation of aurora A in *Drosophila* or *C. elegans*, or inhibition of aurora A function in vertebrate cells, results in spindles that are unstable and, in some cases, monopolar.

Aurora B helps control sister-chromatid structure and segregation. The aurora B protein (together with its partners INCENP and survivin) is found in early mitosis on condensing chromosome arms, and then becomes focused primarily at the centromeres and kinetochores during metaphase. At these locations, aurora B has at least two functions: first, it contributes to the stimulation of chromosome condensation and resolution, and second, it helps control kinetochore attachment to the spindle. Aurora B is probably involved in the regulation of cytokinesis as well. After sister-chromatid segregation, aurora B resides at the midzone of the spindle (see Figure 5-18) and then at the neck of the dividing cell during cytokinesis (much like Plk). Mutations or other defects in aurora B often lead to a failure of cytokinesis.

The single aurora kinase of budding and fission yeasts has functions that seem roughly equivalent to those of metazoan aurora B. Mutations in the yeast aurora kinases do not cause major spindle problems but result primarily in chromosome segregation defects.

Thus far in this chapter we have focused on the major regulators of mitosis and provided a brief overview of their regulation and functions. In the remainder of this chapter, and in Chapters 6 and 7, we turn to the mechanical and regulatory processes that prepare the cell for mitosis and then carry out the separation and segregation of sister chromatids.

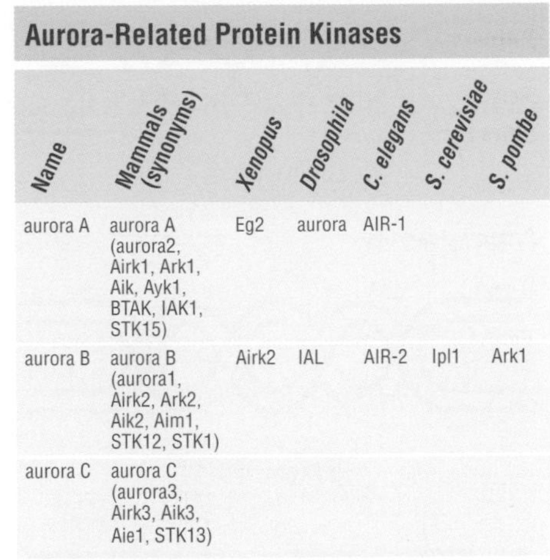

Aurora-Related Protein Kinases

Name	Mammals (synonyms)	Xenopus	Drosophila	C. elegans	S. cerevisiae	S. pombe
aurora A	aurora A (aurora2, Airk1, Ark1, Aik, Ayk1, BTAK, IAK1, STK15)	Eg2	aurora	AIR-1		
aurora B	aurora B (aurora1, Airk2, Ark2, Aik2, Aim1, STK12, STK1)	Airk2	IAL	AIR-2	Ipl1	Ark1
aurora C	aurora C (aurora3, Airk3, Aik3, Aie1, STK13)					

Figure 5-17 Table of aurora-related protein kinases Note that each yeast species contains only one aurora-like protein kinase. On the basis of their functions these proteins are classified with aurora B of metazoans.

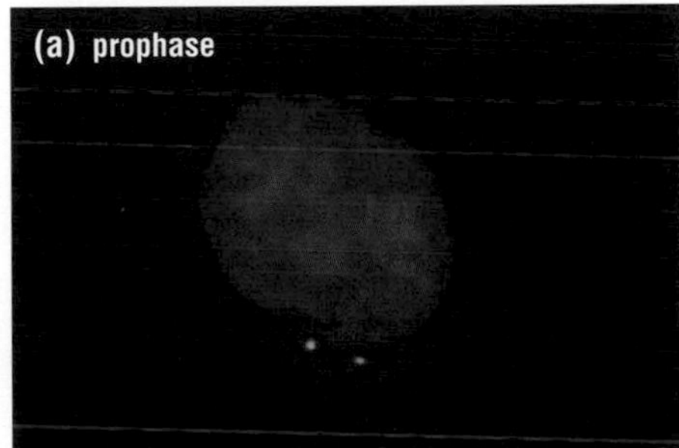

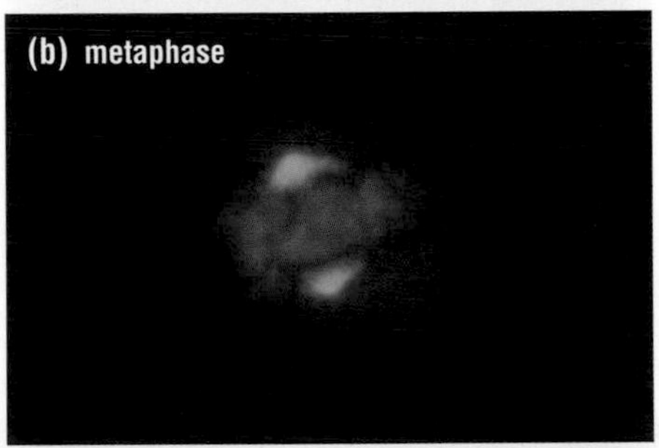

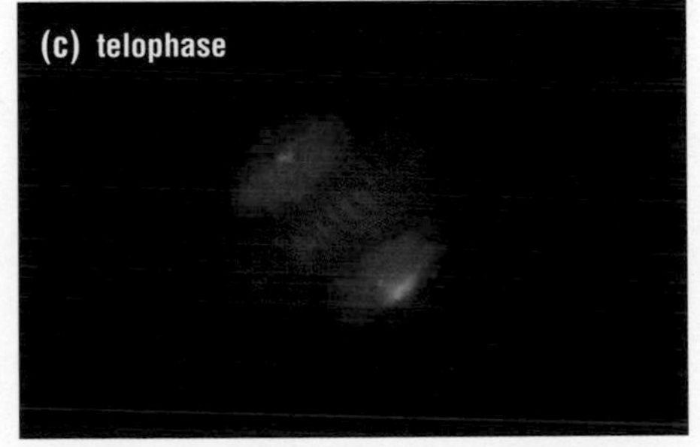

Figure 5-18 Localization of aurora kinases in mitotic cells Human HeLa cells growing in culture were stained with an antibody against aurora A (green) and aurora B (red). DNA was stained blue. Throughout mitosis, aurora A is found at the centrosomes. Aurora B stains the kinetochores in prophase and metaphase and is found at the spindle midzone in telophase. Courtesy of Toru Hirota.

References

Barr, F.A. *et al.*: **Polo-like kinases and the orchestration of cell division.** *Nat. Rev. Mol. Cell Biol.* 2004, **5**:429–440.

Bayliss, R. *et al.*: **Structural basis of Aurora-A activation by TPX2 at the mitotic spindle.** *Mol. Cell* 2003, **12**:851–862.

Carmena, M. and Earnshaw, W.C.: **The cellular geography of aurora kinases.** *Nat. Rev. Mol. Cell Biol.* 2003, **4**:842–854.

Elia, A.E. *et al.*: **Proteomic screen finds pSer/pThr-binding domain localizing Plk1 to mitotic substrates.** *Science* 2003, **299**:1228–1231.

Marumoto, T. *et al.*: **Aurora-A—a guardian of poles.** *Nat. Rev. Cancer* 2005, **5**:42–50.

Meraldi, P. *et al.*: **Aurora kinases link chromosome segregation and cell division to cancer susceptibility.** *Curr. Opin. Genet. Dev.* 2004, **14**:29–36.

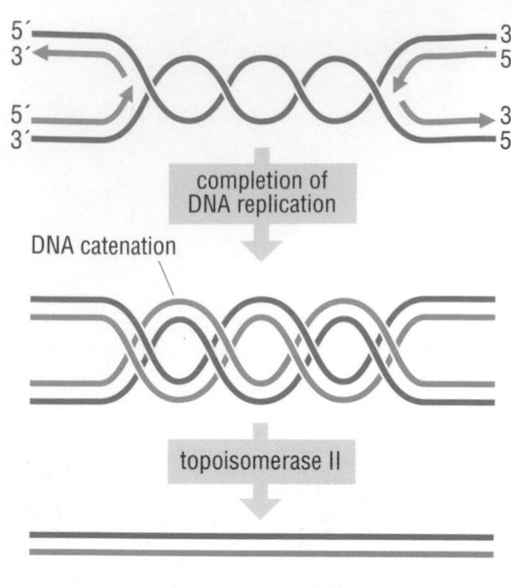

Figure 5-19 Formation of catenated DNA during DNA replication DNA catenation occurs wherever two replication forks collide, resulting in extensive DNA intertwining all along the chromosomes. This tangle is resolved by the enzyme topoisomerase II. When this enzyme encounters a pair of DNA strands that cross one another, it performs a three-step reaction: first, it creates a double-stranded break in one DNA strand; second, it passes the unbroken DNA strand through the break in the other; and third, it reseals the broken DNA strand and releases the two intact DNA molecules.

Sister chromatids are held together by two mechanisms

Long before M phase begins, the cell prepares for mitosis by initiating two processes: the establishment of sister-chromatid cohesion, which is discussed in this section, and the duplication of the centrosome or spindle pole body, which is discussed in Chapter 6.

To appreciate the importance of sister-chromatid cohesion, one need only imagine the chaos that would ensue if duplicated sister chromatids drifted apart after S phase. The reliable attachment of each of the two sisters to a different spindle pole—a prerequisite for accurate segregation—would be difficult to ensure under those conditions. Indeed, a variety of experimental evidence indicates that defects in sister-chromatid cohesion lead to errors in chromosome segregation.

At least two mechanisms are involved in sister-chromatid cohesion. The first, **DNA catenation**, is the extensive intertwining of duplicated DNA molecules that occurs when two adjacent replication forks meet during DNA synthesis (Figure 5-19). The enzyme topoisomerase II removes most of this catenation by the time of metaphase, so that it makes only a minor contribution to cohesion at this point. The second cohesive mechanism depends on a protein complex called **cohesin**, which links the duplicated DNA molecules together as they are synthesized. These complexes are almost solely responsible for holding sister chromatids together in metaphase, and their removal is the central event in sister separation at the metaphase-to-anaphase transition.

Cohesin is a key mediator of sister-chromatid cohesion

Cohesin is a complex of four subunits—Smc1, Smc3, Scc1 and Scc3—whose amino-acid sequences and cohesive function have been highly conserved in evolution (Figure 5-20). Studies of yeast mutants, as well as frog egg extracts depleted of these proteins, indicate that all four cohesin subunits are essential for sister-chromatid cohesion.

Two of the cohesin subunits, Smc1 and Smc3, are members of a structurally related family of proteins called **SMC proteins** (for structural maintenance of chromosomes). In prokaryotes as well as eukaryotes, SMC proteins contribute to numerous aspects of chromosome structure and dynamics (two other members of this family, Smc2 and Smc4, are involved in chromosome condensation, as we will discuss in section 5-9). All SMC proteins, including Smc1 and Smc3, are long thin proteins containing a coiled-coil region flanked by a globular domain with ATPase activity at one end and a dimerization domain at the other. The dimerization domain allows two SMC proteins to interact and form V-shaped dimers (Figure 5-21). When bound to ATP, the ATPase domains of the two proteins in a dimer interact, resulting in a giant ring structure (see Figure 5-21). ATP hydrolysis triggers a conformational change that causes separation of the ATPase domains. Cycles of ATP binding and hydrolysis might therefore drive opening and closing of the ring.

Subunits of the Cohesin Complex

Name	*S. cerevisiae*	*S. pombe*	*Drosophila*	Vertebrates
Smc1	Smc1	Psm1	DmSmc1	Smc1
Smc3	Smc3	Psm3	DmSmc3	Smc3
Scc1	Mcd1/Pds3	Rad21	DmRad21	Rad21
Scc3	Scc3/Irr1	Psc3	DmSA	SA1 and SA2

Figure 5-20 Table of cohesin complex subunits This table does not include related proteins that function in meiosis (see Chapter 9).

The ATPase domains of SMC proteins are regulated by interactions with non-SMC proteins. In cohesin, these are Scc1 and Scc3, which bind the ATPase domains of Smc1 and Smc3 (see Figure 5-21). Scc1 is a member of a family of related SMC-associated proteins called kleisins. Binding of the Scc1 subunit is particularly important and is thought to lock the ring structure in the closed position, perhaps by inhibiting ATP hydrolysis. Sister-chromatid separation in anaphase occurs when Scc1 is proteolytically cleaved and thereby released from cohesin.

The structural mechanism by which cohesin links sister chromatids is not understood in detail. One appealing possibility is that the cohesin ring encircles the two sister chromatids (see Figure 5-21). It is likely that cohesin function depends on direct interactions not only

Definitions

cohesin: a complex of four proteins that links sister chromatids together after S phase.

DNA catenation: the intertwining of sister DNA molecules that occurs when replication forks meet.

SMC (structural maintenance of chromosomes) proteins: family of large proteins composed of a long coiled-coil region with a terminal ATPase domain made up of the amino and carboxyl termini of the protein, and which play a part in chromosome segregation and DNA recombination and repair. SMC proteins typically form dimers with another SMC protein to form large ring structures that may encircle chromosomes and position them with respect to each other, as well as recruiting other proteins essential for their maintenance.

References

Anderson, D.E. *et al.*: **Condensin and cohesin display different arm conformations with characteristic hinge angles.** *J. Cell Biol.* 2002, **156**:419–424.

Hopfner, K.P.: **Chromosome cohesion: closing time.** *Curr. Biol.* 2003, **13**:R866–R868.

Hopfner, K.P. and Tainer, J.A.: **Rad50/SMC proteins and ABC transporters: unifying concepts from high-resolution structures.** *Curr. Opin. Struct. Biol.* 2003, **13**:249–255.

Ivanov, D. and Nasmyth, K.: **A topological interaction between cohesion rings and a circular minichromosome.** *Cell* 2005, **122**:849–860.

Nasmyth, K. and Haering, C.H.: **The structure and function of smc and kleisin complexes.** *Annu. Rev. Biochem.* 2005, **74**:[illegible]–648.

with DNA or nucleosomes but also with other chromatin proteins. There is evidence from yeast, for example, that kinetochore and heterochromatin proteins are required for cohesion in centromeric regions.

Cohesion is established during DNA replication

The cohesin linkage between sister chromatids is established during S phase and seems to be tied closely to DNA replication. Cohesin complexes associate with the chromosome before replication begins, but passage through S phase is required to transform these complexes into the cohesive structures that link the duplicated sister chromatids until metaphase. If yeast cells are engineered so that cohesin is not expressed until after S phase, sister-chromatid cohesion does not occur, indicating that cohesion can be established only if the cohesin complex is present during S phase.

We know little about how cohesin is initially loaded onto chromosomes in G1 or how it is rearranged to establish cohesion during S phase. The ATPase domains of the SMC subunits are known to bind DNA, and it is conceivable that the initial association between cohesin and chromosomes is mediated, at least in part, by this interaction. Given the importance of ATP binding and hydrolysis in the conformation of the ATPase domains, it has been proposed that regulation of the ATPase activity of these domains provides a mechanism by which the cohesin ring might be opened and then closed around the sisters during S phase.

Cohesin loading and sister linkage are likely to involve a complex series of steps that depend on assistance from other proteins. In budding yeast, a complex of two proteins, Scc2 and Scc4 (unrelated to Scc1 and Scc3), is required for cohesin loading on chromosomes, whereas the establishment of cohesion in S phase depends on a protein called Eco1. Normal cohesion also requires a protein complex called RFC–Ctf18, a modified form of the RFC clamp loader that promotes loading of the sliding clamp at the replication fork (see section 4-1). Cohesion in yeast also depends on chromatin-remodeling complexes (see section 4-9). We can only speculate about the molecular mechanisms used by these various proteins to govern cohesin function, and it remains unclear whether similar proteins contribute to cohesin regulation in organisms other than yeast.

When chromosome duplication is complete, cohesin complexes are arrayed at intervals of 10–15 kb along the arms of the sister chromatids, with a greater concentration of cohesin in centromeric regions—where it seems that the high levels of heterochromatin proteins contribute to the maintenance of cohesion. The high degree of cohesion at centromeres is presumably required to oppose the strong pulling forces exerted by the mitotic spindle on this region.

DNA decatenation prepares sister chromatids for separation

A central purpose of mitosis is to distribute the duplicated sister chromatids into two daughter nuclei. This clearly requires chromatid separation. Although the final, most important, step in separation occurs at the beginning of anaphase, the loosening of sister-chromatid cohesion actually begins long before, in S phase, when the enzyme topoisomerase II begins to disentangle the catenated sister DNA molecules (see Figure 5-19). DNA decatenation continues throughout G2 and is largely complete in the chromatid arms by early mitosis. Some catenation remains until the end of metaphase, particularly at the centromeres. Mutation or chemical inhibition of topoisomerase II therefore results in defective sister segregation in anaphase.

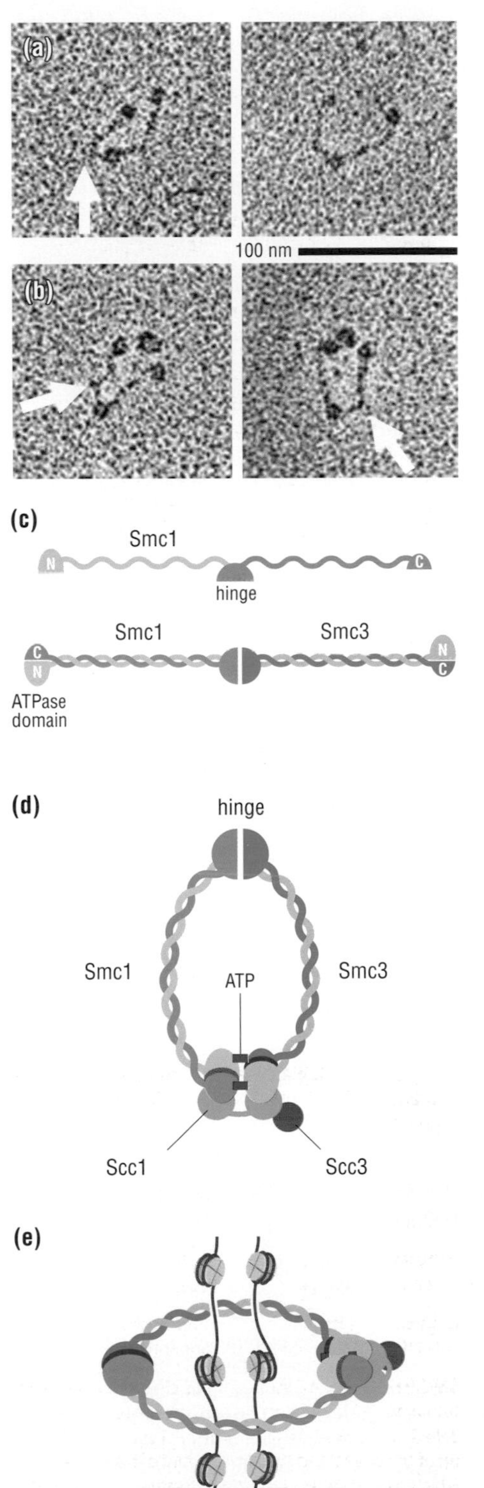

Figure 5-21 Models of SMC and cohesin structure (a) Electron microscopy of *Xenopus* Smc1–Smc3 dimers illustrates the V-shaped structure commonly seen with SMC proteins. The flexible 'hinge' region is at the bottom of the V and the two globular ATPase domains are at the top. The arrow indicates a kink that is sometimes seen in one arm. **(b)** Addition of the two non-SMC subunits (Scc1 and Scc3) of the cohesin complex results in the appearance of a globular structure next to the two heads of the Smc1–Smc3 dimer. **(c)** The linear structure of an SMC protein includes two globular domains at each terminus, linked by a long repetitive sequence and a central dimerization or hinge domain. When the SMC protein is folded, the two domains at the termini join to form a complete ATPase domain, while the arm regions form a helical coiled-coil. The hinge domain that forms at the other end of the arm interacts with the hinge domain of another SMC protein. In cohesin, this results in the formation of a Smc1–Smc3 heterodimer. **(d)** Binding of ATP (red) promotes binding of the two ATPase domains, resulting in closure of the SMC ring. The non-SMC protein Scc1 interacts with both ATPase domains and holds them together. Cleavage of Scc1 in anaphase therefore opens the ring. **(e)** The cohesin complex may form a 50-nm ring around two sister chromatids. Because of its small size, however, this ring could only link nucleosomal DNA and not more complex chromatin structures. Panels (a) and (b) from Anderson, D.E. *et al.*: *J. Cell Biol.* 2002, **156**:419–424.

5-9 Entry into Mitosis: Sister-Chromatid Condensation and Resolution

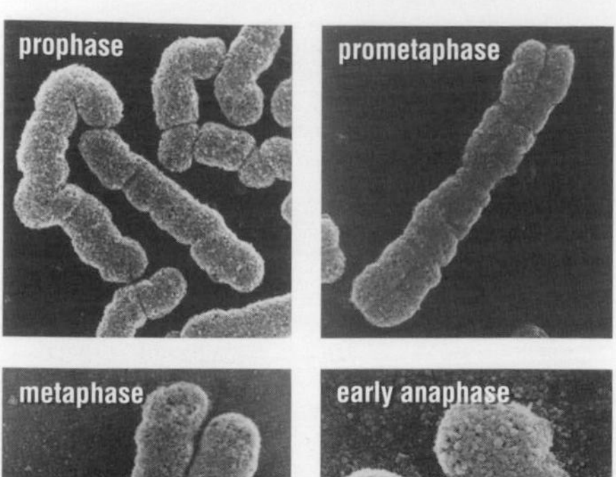

Figure 5-22 Condensation and resolution of human sister chromatids in early mitosis Scanning electron microscopy reveals that sister-chromatid pairs first condense into single rod-like structures during prophase. As mitosis proceeds, chromatid arms are gradually resolved and become almost completely distinct by the end of metaphase. Kindly provided by Adrian T. Sumner. From Sumner, A.T.: *Chromosoma* 1991, **100**:410–418.

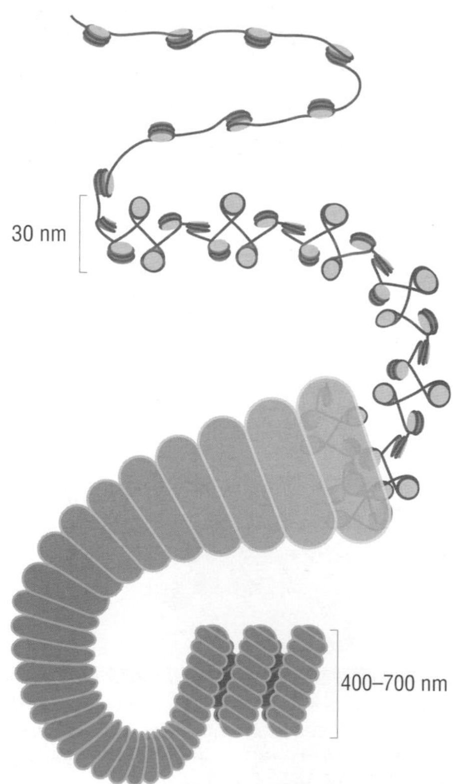

Figure 5-23 The hierarchical folding model of chromosome condensation This model of chromosome condensation in animal cells proposes that the 30-nm fiber of interphase chromatin (see section 4-9) is folded into larger and larger coils to form the thick, compact metaphase sister chromatid.

Chromosomes are dramatically reorganized in mitosis

Entry into mitosis initiates two major processes that occur in parallel. One of these is the assembly of the mitotic spindle, which we discuss in Chapter 6. Here we focus on the other major early mitotic event: the preparation of sister chromatids for segregation.

When S phase is complete, the cell contains several sister-chromatid pairs that form a fragile mass of tangled DNA and protein. Any attempt to segregate the intertwined chromatids in this state would almost certainly lead to DNA breakage. The immense length of interphase chromosomes would also cause problems: when the centromeres were being pulled apart by the spindle, the ends of the long chromatid arms would lie across the midline of the cell—and would be cut to pieces during cytokinesis.

To avoid these problems, entry into mitosis triggers dramatic structural changes in the chromosomes. The sister chromatids are compacted by **chromosome condensation** into durable, rod-like structures that are less likely to become entangled with each other and are short enough to ensure that chromatid arms are safely within the future daughter cells before cytokinesis. The intertwined sister chromatids are also reorganized by the process of **sister-chromatid resolution** into distinct units that can be pulled apart easily in anaphase. Resolution depends on decatenation of sister DNAs and the partial rearrangement or loss of cohesin complexes holding the sisters together. Condensation and resolution generally occur in parallel throughout early mitosis.

In animal cells, where chromosomes are particularly long, condensation results in a 10,000-fold reduction in chromosome length. In early prophase, condensation first results in the formation of thick, rod-like chromosomes in which the two sisters are not apparent (Figure 5-22). As condensation and resolution continue in parallel during prometaphase and metaphase, the chromosomes are gradually transformed into compact and distinct sister chromatids joined primarily at their centromeres. During these stages, most cohesin complexes are lost from sister-chromatid arms, which are almost completely distinct when the cell reaches metaphase. Cohesion remains strong at the centromeres, allowing bipolar attachment of sisters to the spindle and providing resistance to spindle pulling forces. Sister separation in anaphase therefore depends mainly on loss of cohesion at the centromeres. In both budding and fission yeasts, which have relatively small chromosomes, chromosome condensation is less extensive but is still important for facilitating the resolution and segregation of sister chromatids. In yeast, cohesin does not dissociate from chromatid arms, which remain linked along their entire length until anaphase.

The higher-order structure of condensed mitotic chromosomes is poorly understood. It has been proposed that the compaction of the mitotic chromosome involves the gradual folding of chromatin fibers into progressively more compact structures: 30-nanometer fibers (see section 4-9) are coiled to form thicker and shorter fibers, which are then coiled further to form even thicker fibers (Figure 5-23). This hierarchical folding may be guided by a structural protein scaffold that forms at the core of each sister chromatid and organizes the condensing chromatin around it. Several proteins, including the condensin protein discussed next, are concentrated along the midline of condensing chromatids and may be a part of some axial chromosome core.

Condensin complexes drive chromosome condensation and resolution

We know little about the structural basis of mitotic chromosome changes, but a central player in this process is **condensin**, a five-subunit protein complex that is related, both structurally and functionally, to the cohesin complex (Figures 5-24 and 5-25). The condensin subunits Smc2 and Smc4 are members of the SMC family and form Y-shaped heterodimers with globular ATPase domains at the ends of the arms. The structural integrity and function of condensin depend on three non-SMC subunits called CAP-D2, CAP-G, and CAP-H. CAP-H is a distantly related member of the kleisin family of proteins, which includes the Scc1 subunit of cohesin (see section 5-8); it is therefore possible that CAP-H, like Scc1, cross-links the ATPase domains of the SMC subunits (see Figure 5-25).

Animal cells possess two condensin complexes, called condensin I and II, that contain the same SMC heterodimer (Smc1–Smc3) but different non-SMC subunits. Condensin I contains the three non-SMC subunits named above, whereas condensin II contains related non-SMC subunits called CAP-D3, CAP-G2 and CAP-H2 (see Figure 5-24).

Figure 5-24 Table of subunits of the condensin complex The canonical condensin, found in all eukaryotes, is called condensin I in animal cells to distinguish it from a related complex called condensin II. Condensin II contains the same two SMC subunits as condensin I but different non-SMC subunits. Dashes indicate that a homolog of that subunit has not been identified in that species.

Subunits of the Condensin Complex

Name	*S. cerevisiae*	*S. pombe*	*Drosophila*	Vertebrates
Core subunits				
Smc2	Smc2	Cut14	DmSmc2	CAP-E
Smc4	Smc4	Cut3	DmSmc4	CAP-C
Non-SMC subunits (condensin I)				
CAP-D2	Ycs4	Cnd1	DmCAP-D2	CAP-D2
CAP-G	Ycs5/Ycg1	Cnd3	DmCAP-G	CAP-G
CAP-H	Brn1	Cnd2	Barren	CAP-H
Non-SMC subunits (condensin II)				
CAP-D3	–		DmCAP-D3	CAP-D3
CAP-G2	–	–	–	CAP-G2
CAP-H2	–	–	DmCAP-H2	CAP-H2

The condensin complex was first discovered on the basis of its ability to promote chromosome condensation in frog egg extracts—removal of condensin I from these extracts results in striking defects in chromosome condensation. Inhibition of condensin I or II in human cells (by RNA interference) also causes condensation defects. Similarly, mutations in any one of the five condensin subunits of fission yeast or budding yeast lead to defects in the relatively limited chromosome condensation that occurs in these species. It is therefore clear that both the structure and function of condensins have been well conserved during evolution.

Although discovered through their role in chromosome condensation, condensins are also required for sister-chromatid resolution. Mutations in condensin proteins are lethal in yeast and *Drosophila*, not simply because of defective chromosome condensation but also because of defects in chromatid segregation. These segregation defects are very similar to those seen in topoisomerase II mutants, suggesting that condensins are required, at least in part, to allow sister-chromatid decatenation, perhaps by governing the function of topoisomerase II.

Like cohesin, condensin is thought to form a ring structure in which the behavior of the ATPase domains is controlled by cycles of ATP binding and hydrolysis. One speculative possibility (see Figure 5-25) is that the condensin ring, like cohesin, encircles or cross-links DNA—except that condensin cross-links different parts of the same DNA molecule in a single sister chromatid, thereby compacting it, whereas cohesin cross-links DNA from different sister chromatids. Both condensin and cohesin should be viewed as dynamic protein machines that use energy provided by ATP hydrolysis to rearrange chromatin structure. Purified frog condensin I promotes the ATP-dependent supercoiling of DNA molecules *in vitro*, and it seems likely that this enzymatic activity is the basis for its ability to condense and resolve sister chromatids *in vivo*.

One might expect that compaction of chromosomes by a mechanism such as hierarchical folding (see Figure 5-23) would require removal of most sister-chromatid cohesion. Surprisingly, however, there is evidence from frog egg extracts that cohesin removal is not required for chromosome compaction but only for sister-chromatid resolution, arguing that the condensation machinery can somehow package a pair of tightly linked sisters into a single rod-like structure. In section 5-10, we discuss how this poorly understood interplay between condensation and cohesion may be regulated in mitosis.

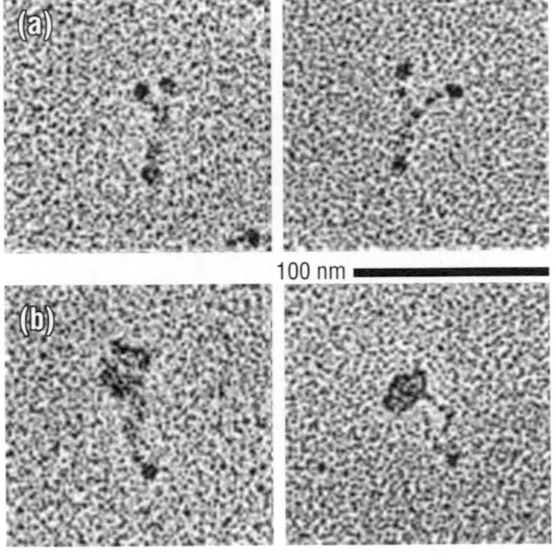

(c)

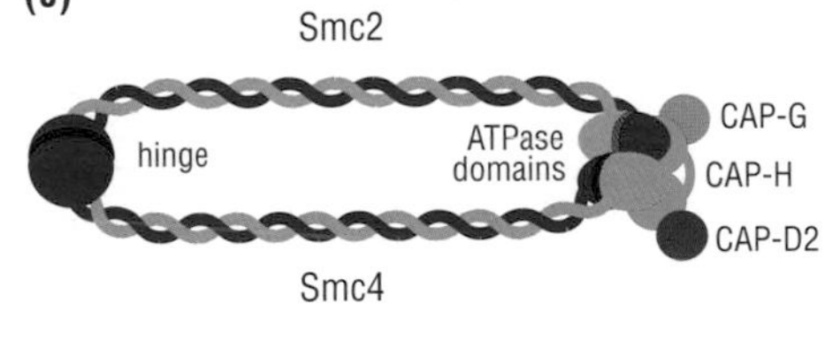

(d)

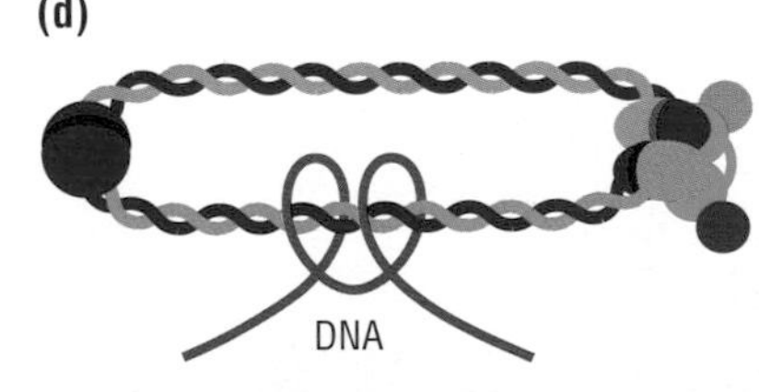

Figure 5-25 Structure of condensin (a) Electron microscopy of *Xenopus* Smc2–Smc4 dimers shows that the coiled-coil arms of condensin tend to interact with each other more than the arms of cohesin SMC subunits (see Figure 5-21), resulting in rods or Y-shaped structures. **(b)** Addition of the three non-SMC subunits (CAP-D2, CAP-G and CAP-H) of condensin I results in the appearance of a globular structure at the heads of the Smc2–Smc4 dimer. **(c)** Model of condensin structure, based on studies of cohesin and other SMC proteins (see Figure 5-21). **(d)** Experimental analysis of DNA rearrangements caused by purified condensin suggest that the complex coils the DNA twice around itself. One speculative model of condensin function proposes that cycles of ATP binding and hydrolysis cause conformational changes in condensin ATPase domains, resulting not only in ring opening and closure but also causing the twisting of two coils of DNA inside the ring as shown here. Panels (a) and (b) from Anderson, D.E. *et al.*: *J. Cell Biol.* 2002, **156**:419–424.

Definitions

chromosome condensation: in mitosis, the structural changes that result in compaction of the chromosomes into short, thick structures.

condensin: complex of five proteins that helps condense and resolve the sister chromatids in mitosis.

sister-chromatid resolution: the gradual disentangling of sister chromatids before separation in anaphase, which makes them visible as distinct structures under the microscope.

References

Anderson, D.E. *et al.*: **Condensin and cohesin display different arm conformations with characteristic hinge angles.** *J. Cell Biol.* 2002, **156**:419–424.

Hirano, T.: **Condensins: organizing and segregating the genome.** *Curr. Biol.* 2005, **15**:R265–R275.

Kireeva, N. *et al.*: **Visualization of early chromosome condensation: a hierarchical folding, axial glue model of chromosome structure.** *J. Cell Biol.* 2004, **166**:775–785.

Lavoie, B.D. *et al.*: ***In vivo* requirements for rDNA chromosome condensation reveal two cell-cycle-regulated pathways for mitotic chromosome folding.** *Genes Dev.* 2004, **18**:76–87.

Sumner, A.T.: **Scanning electron microscopy of mammalian chromosomes from prophase to telophase.** *Chromosoma* 1991, **100**:410–418.

Swedlow, J.R. and Hirano, T.: **The making of the mitotic chromosome: modern insights into classical questions.** *Mol. Cell* 2003, **11**:557–569.

Mitotic Cdks act on condensin to govern the timing of chromosome condensation

Mitotic changes in chromosome structure depend on coordinated changes in chromosome compaction, DNA decatenation, and partial loss of sister-chromatid cohesion. The control of these processes is achieved primarily through the regulation of condensin and cohesin by the mitotic protein kinases of the Cdk, Plk and aurora families (see section 5-7). These kinases act through multiple pathways to govern chromosome behavior in mitosis. The relative importance of each pathway seems to vary in different organisms.

The initiation of chromosome condensation in early mitosis is driven primarily by mitotic cyclin–Cdk complexes. In vertebrates, as discussed earlier in this chapter (see section 5-3), cyclin A–Cdk2 complexes are active in the nucleus at the beginning of mitosis and are thought to initiate condensation during prophase. Following their import into the nucleus and nuclear envelope breakdown, cyclin B1–Cdk1 complexes then seem to accelerate condensation.

Because the structural basis of chromosome condensation is so poorly understood, it is perhaps not surprising that we know little about how Cdks promote it. Nevertheless, the evidence so far indicates that condensin complexes are an important target of cell-cycle regulation. In vertebrates, two of the non-SMC subunits of condensin I (CAP-D2 and CAP-H) are phosphorylated by Cdk1 during mitosis, and this phosphorylation enhances the ability of condensin to supercoil DNA *in vitro*. It therefore seems likely that Cdk1 promotes condensation in part by acting directly on condensin.

In vertebrate cells, the steps in chromosome condensation may be controlled, at least in part, by condensin localization inside the cell. In prophase, condensin I is found in the cytoplasm and condensin II associates with the chromosomes inside the nucleus. Condensin II is therefore positioned to initiate chromosome condensation in prophase, perhaps in response to stimulation by cyclin A–Cdk. After breakdown of the nuclear envelope, condensin I gains access to the chromosomes and may be important for the cyclin B-dependent maturation of mitotic chromosome structure in prometaphase and metaphase.

Less is known about how Cdks regulate condensin function in yeast. In fission yeast, phosphorylation of the Smc4 subunit promotes the nuclear localization of the condensin complex during mitosis. In budding yeast, condensin is associated with chromatin throughout the cell cycle and therefore seems to be regulated, by unknown means, at some step beyond chromosome binding.

The regulation of chromosome condensation may depend in some species on sister-chromatid cohesion. In budding yeast, where cohesin remains bound to chromosome arms throughout early mitosis, mutation of cohesin results in defects in chromosome condensation. In this species, therefore, cohesin complexes are required for the normal function of condensin, perhaps because they stimulate condensin activation. This does not seem to be the case in vertebrates, however, where most of the cohesin is removed from chromosome arms as they condense, and where experimental removal of cohesins has only minor effects on chromosome compaction.

Species-specific differences in condensin regulation can also be seen in studies of the protein kinase aurora B. As discussed earlier (see section 5-7), this protein is found along chromosome arms as they condense in early mitosis. Loss of aurora B function in *Drosophila* cells reduces

References

Hauf, S. *et al.*: **Dissociation of cohesin from chromosome arms and loss of arm cohesion during early mitosis depends on phosphorylation of SA2.** *PLoS Biol.* 2005, **3**:e69.

Hirano, T.: **Condensins: organizing and segregating the genome.** *Curr. Biol.* 2005, **15**:R265–R275.

Kimura, K. *et al.*: **Phosphorylation and activation of 13S condensin by Cdc2 *in vitro*.** *Science* 1998, **282**:487–490.

Lavoie, B.D. *et al.*: ***In vivo* requirements for rDNA chromosome condensation reveal two cell-cycle-regulated pathways for mitotic chromosome folding.** *Genes Dev.* 2004, **18**:76–87.

Losada, A. *et al.*: **Cohesin release is required for sister chromatid resolution, but not for condensin-mediated compaction, at the onset of mitosis.** *Genes Dev.* 2002, **16**:3004–3016.

Sumara, I. *et al.*: **The dissociation of cohesin from chromosomes in prophase is regulated by Polo-like kinase.** *Mol. Cell* 2002, **9**:515–525.

Swedlow, J.R. and Hirano, T.: **The making of the mitotic chromosome: modern insights into classical questions.** *Mol. Cell* 2003, **11**:557–569.

Watanabe, Y.: **Shugoshin: guardian spirit at the centromere.** *Curr. Opin. Cell Biol.* 2005, **17**:590–595.

the binding of condensin to chromosomes and causes defects in chromosome structure. In contrast, depletion of aurora B from *Xenopus* egg extracts, or mutation of aurora B in budding yeast, has little impact on condensin loading or chromosome compaction in early mitosis. It seems likely that chromosome condensation is governed by multiple mechanisms whose importance varies in different organisms.

A potentially important target of aurora B is histone H3, which in all species tested is phosphorylated by the kinase at a conserved serine at position 10 in the histone tail. Phosphorylation at this site generally correlates with chromosome condensation. Mutation of this site has no significant effect on condensation in budding yeast, but it is possible that phosphorylation of H3 helps promote chromosome condensation in other species.

Sister-chromatid resolution is governed by Plk and aurora B in animal cells

Chromosome condensation occurs in parallel with sister-chromatid resolution. As might be expected, sister-chromatid resolution depends on the partial loss of sister-chromatid cohesion, which results from decatenation of DNA by topoisomerase II and, in vertebrate cells, by removal of most cohesin complexes along chromatid arms.

The partial removal of cohesin from chromatid arms is triggered by the protein kinases Plk and aurora B. If either of these kinases is removed from a mitotic frog egg extract, cohesin removal is somewhat reduced (Figure 5-26a). Inhibition of both kinases, however, completely blocks its removal. Inhibition of Plk and aurora B has little effect on condensin recruitment and chromosome compaction, but it does inhibit the normal resolution of sister-chromatid arms into distinct rods, suggesting that cohesin removal is required for resolution but not for condensation (Figure 5-26b).

It is likely that Plk and aurora B promote cohesin removal by different mechanisms. Aurora B may destabilize sister-chromatid cohesion by phosphorylating histone H3, as mentioned above. Plk, in contrast, seems to act by phosphorylating Scc3, one of the non-SMC subunits of cohesin. If mitotic Scc3 phosphorylation is prevented by mutation of its phosphorylation sites, cohesin remains partly associated with sister-chromatid arms in early mitosis and sister-chromatid resolution is defective. It therefore seems likely that the effects of Plk on sister cohesion and resolution are mediated through direct effects on cohesin.

Cohesin is not removed from centromeric regions in early mitosis, thereby ensuring that cohesion remains strong in the region where spindle forces are strongest. The loss of centromeric cohesin is prevented by a mechanism that depends on a protein called Sgo1, which localizes to centromeric chromatin and may act, at least in part, by blocking cohesin phosphorylation by Plk.

By metaphase, the sister chromatids are almost completely decatenated and held together only by cohesin complexes along their entire length (in yeast) or focused primarily at the centromeres (in animal cells). The sisters are now ready for the final step in separation, which is triggered at the metaphase-to-anaphase transition by proteolysis of the Scc1 subunit of cohesin. We describe the mechanism of sister-chromatid separation in Chapter 7. First, in Chapter 6, we discuss the other major process that occurs during early mitosis: the assembly of the mitotic spindle and its attachment to the sister-chromatid pairs.

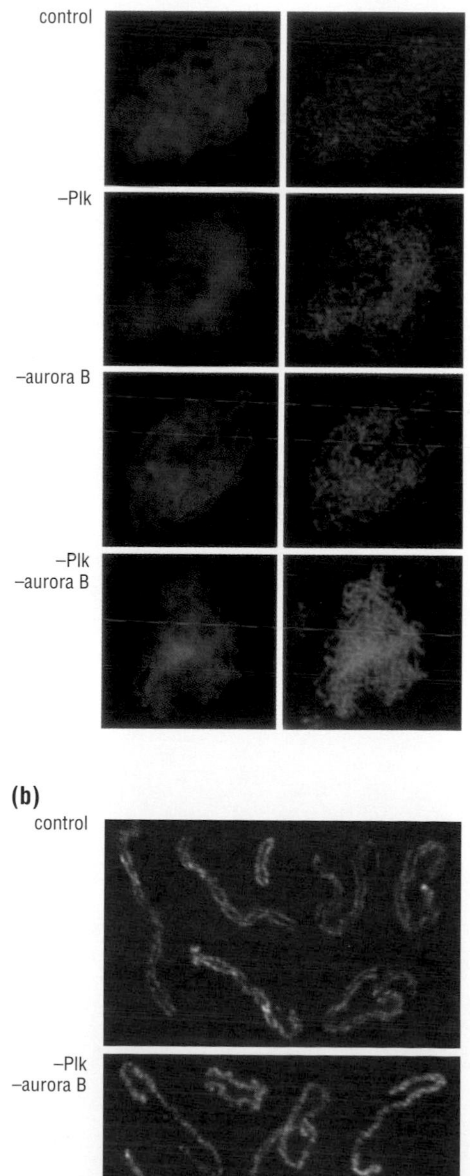

Figure 5-26 Plk and aurora B are required for the removal of cohesin from chromosome arms in early mitosis When interphase extracts from *Xenopus* eggs are incubated with sperm DNA it becomes packaged into intact nuclei. **(a)** Nuclei are stimulated to enter mitosis by the addition of mitotic cell extract. (Under these conditions, the many condensed chromosomes are seen as a tangled mass.) DNA in isolated nuclei is labeled with a red dye, and antibodies are used to label the Scc3 subunit of cohesin (green). In control experiments (top row), there is very little cohesin on chromosomes in mitosis—what little that remains is focused primarily at the centromeres, which are barely visible in these images. When either Plk or aurora B is depleted from the extracts with the use of antibodies, a small amount of cohesin remains bound to the chromatin in mitosis, whereas depletion of both kinases at the same time (bottom row) completely blocks cohesin removal. **(b)** Detailed analysis of chromosomes from the experiments in panel (a) reveals that depletion of both Plk and aurora B prevents the resolution of sister chromatids into distinct pairs. From Losada, A. *et al.*: *Genes Dev.* 2002, **16**:3004–3016.

6

Assembly of the Mitotic Spindle

Chromosome segregation is driven by the mitotic spindle, a bipolar array of microtubules and associated motors and other proteins. Spindle assembly begins early in mitosis and is completed when microtubules from both poles are attached to kinetochores on each sister-chromatid pair.

6-0 Overview: The Mitotic Spindle

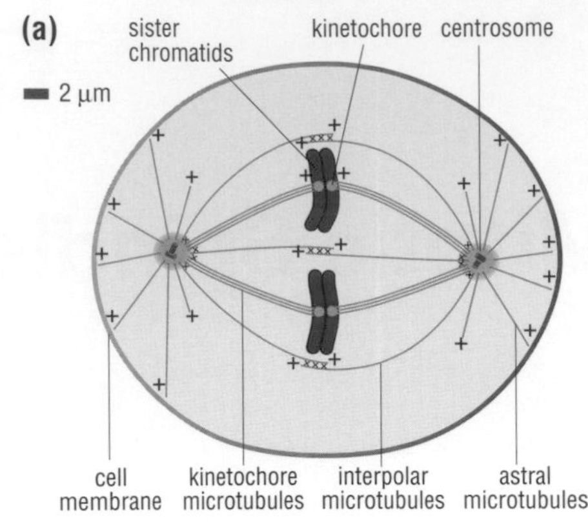

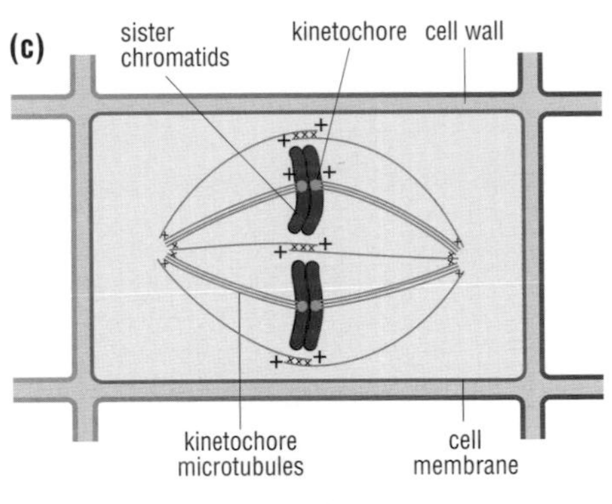

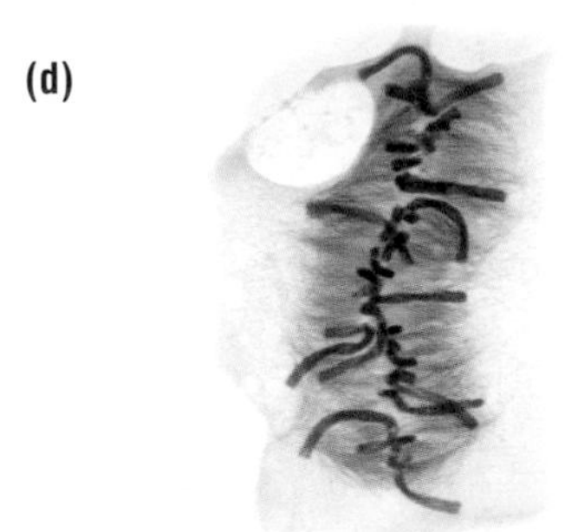

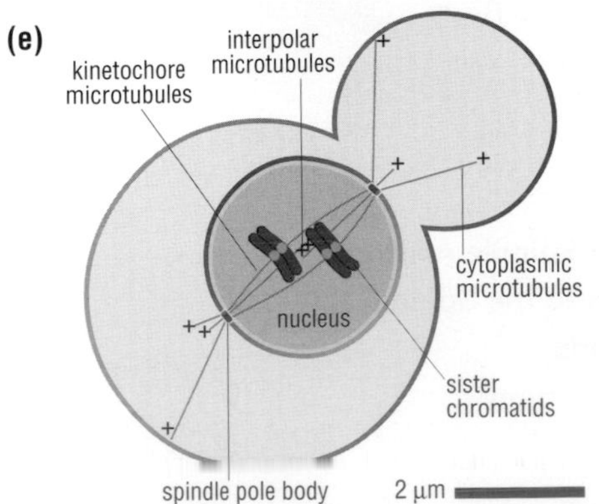

Chromosome segregation depends on the mitotic spindle

The central function of mitosis is to segregate the two sets of chromosomes that are present in the cell after S phase. Chromosome segregation is carried out by a complex and beautiful machine—the mitotic spindle—that pulls the sister chromatids apart and moves a complete set of chromosomes to each pole of the cell, where they are packaged into daughter nuclei.

The mitotic spindle is based on a bipolar array of *microtubules*, each of which is a polarized protein polymer with one end, the so-called minus end, embedded in a spindle pole and the other end, the plus end, pointing outward from the pole. Plus ends from one pole overlap with plus ends from the other, resulting in an antiparallel array in the spindle midzone (Figure 6-1). Microtubules are highly dynamic polymers that continuously grow and shrink, and in the spindle this behavior is regulated by many different proteins that bind to the sides or ends of microtubules. These include the *motor proteins*, which can travel along microtubules and have important roles in the assembly and stability of the microtubule array and the movement of chromosomes on the spindle.

The sister chromatids are active participants in spindle assembly and function. Each chromatid carries a kinetochore, a multiprotein complex that attaches the chromatid to microtubules connected to a spindle pole (see section 5-0). In addition, proteins in the kinetochore help generate forces that drive chromosome movement. Motors and microtubule-regulatory proteins in the chromatid arms also help govern microtubule growth and spindle assembly.

The mitotic spindle must be bipolar

Although some features of spindle structure vary in different organisms (see Figure 6-1), the underlying logic is always the same. The microtubule array must be bipolar, and the chromatid pairs must be oriented on the spindle such that each sister is connected to an opposite spindle pole. This is known as *bi-orientation*. The bilateral symmetry in spindle structure and chromosome attachment is critical to the ability of the spindle to pull apart chromatid pairs and transport a complete set of chromosomes to each end of the cell. Any defects in spindle bipolarity or chromosome bi-orientation lead to potentially lethal errors in chromosome segregation.

All spindles are bipolar, but the structure of the poles differs in different organisms. In most somatic animal cells, each spindle pole is focused in a large multiprotein organelle called the *centrosome*, which organizes spindle microtubules and also helps position the spindle within the cell (see Figure 6-1a and b). Some cell types, including those of higher plants (see Figure 6-1c and d) and the oocytes of many vertebrates, do not contain centrosomes and depend on the self-organizing properties of microtubules and microtubule-associated proteins to generate the two poles. In budding yeast, the spindle is constructed entirely within the nucleus, which remains intact throughout mitosis, and is organized by protein organelles called *spindle pole bodies* that are embedded in the nuclear envelope (see Figure 6-1e).

Figure 6-1 **Anatomy of the mitotic spindle** **(a)** Basic features of the mitotic spindle of somatic animal cells in metaphase. The microtubules (green) have one end (the minus end) embedded in the centrosome and the other (plus end) pointing outward. Numerous motors and other proteins (represented by Xs) cross-link the minus ends of microtubules at the spindle poles and the plus ends of interpolar microtubules in the spindle midzone. Additional motor and other proteins link astral microtubules to the cell cortex, kinetochore microtubules to the kinetochore, and interpolar microtubules to the chromatid arms (not shown). For simplicity, this diagram includes only two sister-chromatid pairs and a fraction of the thousands of microtubules that typically exist in the spindle. **(b)** Light micrograph of a centrosome-based spindle (stained red) from a salamander. **(c)** The mitotic spindle in plant cells is similar to that of animal cells except that there are no centrosomes at the spindle poles and no astral microtubules. **(d)** Light micrograph of an acentrosomal spindle from the African blood lily. **(e)** In budding yeast, the metaphase spindle is composed of microtubules that emanate from the nuclear face of a pair of spindle pole bodies embedded in the nuclear envelope. For simplicity, only two of the 16 sister-chromatid pairs are shown. The budding yeast spindle is composed of only about 18 intra-nuclear microtubules from each pole: one for each sister-chromatid pair and one or two interpolar microtubules. A small number of astral microtubules radiate out from the spindle poles and attach to the cell cortex to help position the nucleus in the bud neck. Panels (b) and (d) courtesy of Andrew Bajer.

The spindle contains three classes of microtubules. Kinetochore microtubules connect the spindle poles to kinetochores on the sister chromatids; in animal cells multiple kinetochore microtubules bundle together to form *kinetochore fibers*. Interpolar microtubules link the two spindle poles by interdigitating with each other in the midzone of the spindle. Astral microtubules extend from the poles away from the spindle and are typically involved in anchoring and positioning the spindle in the cell. Astral microtubules are generally found only in cells that use centrosomes or spindle pole bodies to form the spindle poles.

Multiple mechanisms drive spindle assembly

The mitotic spindle is assembled in early mitosis in parallel with the changes in chromosome structure that were discussed in Chapter 5. The two key problems in spindle assembly are how to construct a bipolar array of microtubules that surrounds the sister chromatids, and how to attach sister-chromatid pairs to the array with the correct bi-orientation.

In all eukaryotes, construction of a bipolar spindle depends in large part on the ability of the spindle components to self-organize. Motor and other proteins interact with microtubules to organize them into two antiparallel arrays in which the plus ends of each array overlap in the center (Figure 6-2a). The minus ends are cross-linked by other microtubule-associated proteins to form a pair of spindle poles. Spindle self-organization also depends on proteins associated with the sister chromatids, so that the microtubule array is built around them. As the microtubules grow, some plus ends become attached to kinetochores, thus connecting chromatids to the poles. Self-organization is the only mechanism of spindle assembly in cells lacking centrosomes, such as the plant cells shown in Figure 6-1c and d.

In many other cells, including those in humans, spindle microtubules grow out from the centrosomes, which act as prefabricated microtubule-organizing centers (Figure 6-2b). The centrosome is duplicated before mitosis, and upon mitotic entry the two centrosomes move apart to provide the poles of the spindle. As in an acentrosomal spindle, motor and other proteins cross-link the antiparallel microtubule array between the poles and also help focus microtubule minus ends in the centrosomes. Chromatids are attached to the spindle by a process known as search and capture, in which the plus ends of some of the microtubules radiating out from the centrosomes attach to kinetochores. These centrosome-dependent mechanisms are not essential, however, as animal cells can assemble spindles even when their centrosomes have been inactivated.

Errors in sister-chromatid attachment can sometimes occur during spindle assembly. Both sister kinetochores, for example, can become attached to the same spindle pole. How is the correct bi-orientation of sister chromatids achieved? By mechanisms that are discussed later in this chapter, the kinetochore monitors the orientation of microtubule attachment—and corrects any errors that occur.

This chapter provides an overview of the key principles in mitotic spindle assembly, with an emphasis on the centrosome-dependent mechanisms of animal cells, which are the best understood. The spindle machinery of microtubules, their associated regulators, and the centrosome and the kinetochore are introduced in sections 6-1 to 6-5. In the remainder of the chapter we will see how these components interact to assemble a bipolar spindle, how bi-orientation is achieved, and how the motor and other proteins generate forces that align chromatids in the center of the spindle to prepare the cell for anaphase.

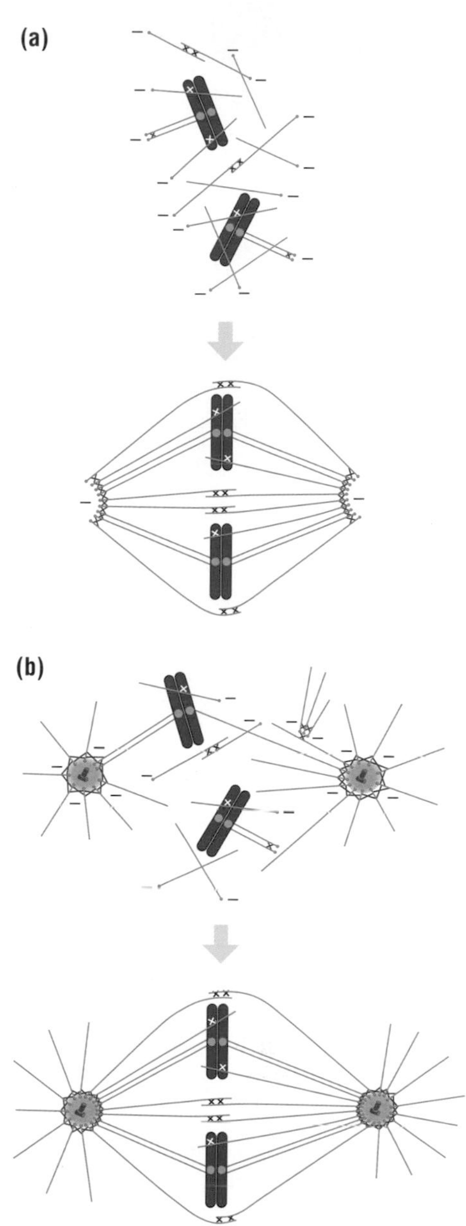

Figure 6-2 General mechanisms of spindle assembly (a) Spindle assembly depends on microtubule self-organization, whereby microtubules form in the vicinity of the sister chromatids and then become organized by motors and other proteins into a bipolar array. This process depends on numerous motor proteins (Xs), including motors that cross-link antiparallel plus ends (black), motors that focus minus ends at the poles (red), and chromatid-associated motors that help orient the array (white). **(b)** In most animal cells centrosomes facilitate the self-organization process shown in panel (a). Rapidly growing and shrinking microtubules that radiate from the centrosomes search the space between the poles and bind to kinetochores. Many sister-chromatid pairs are initially attached by this search and capture mechanism. Centrosomal microtubules can also capture preformed microtubule bundles and pull them into the poles. Bi-orientation eventually results when a pair of sister kinetochores is connected to both poles.

References

Gadde, S. and Heald, R.: **Mechanisms and molecules of the mitotic spindle.** *Curr. Biol.* 2004, **14**: R797–R805.

Kwon, M. and Scholey, J.M.: **Spindle mechanics and dynamics during mitosis in *Drosophila*.** *Trends Cell Biol.* 2004, **14**:194–205.

Mitchison, T.J. and Salmon, E.D.: **Mitosis: a history of division.** *Nat. Cell Biol.* 2001, **3**:E17–E21.

Segal, M. and Bloom, K.: **Control of spindle polarity and orientation in *Saccharomyces cerevisiae*.** *Trends. Cell Biol.* 2001, **11**:160–166.

Wadsworth, P. and Khodjakov, A.: ***E pluribus unum*: towards a universal mechanism for spindle assembly.** *Trends Cell Biol.* 2004, **14**:413–419.

Wittmann, T. *et al.*: **The spindle: a dynamic assembly of microtubules and motors.** *Nat. Cell Biol.* 2001, **3**:E28–E34.

6-1 Microtubule Structure and Behavior

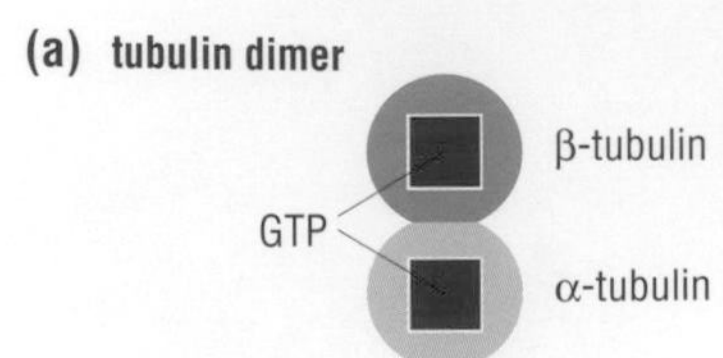

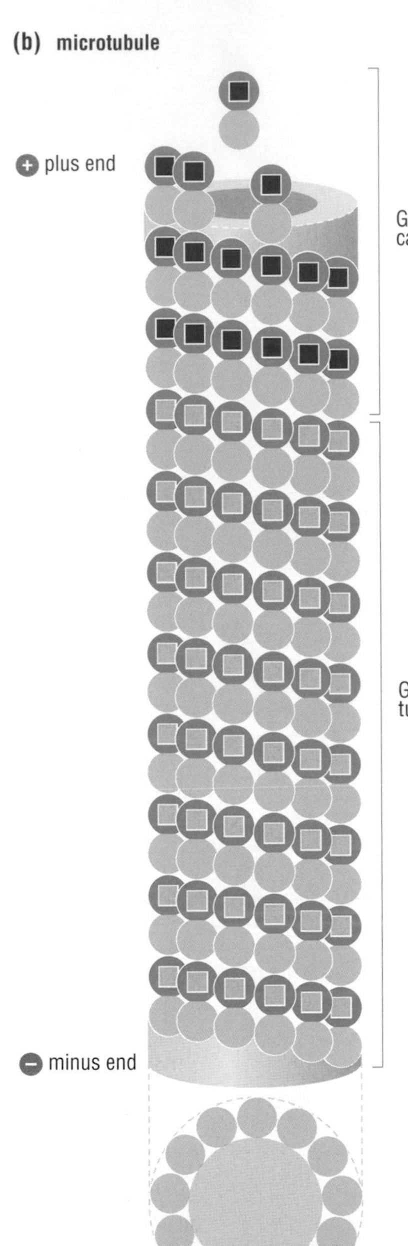

Microtubules are polymers of tubulin subunits

Microtubules are the main structural components of the mitotic spindle, and their structure, formation and behavior are key to spindle assembly and function. In this section we address the intrinsic properties of microtubules; section 6-2 will introduce some of the proteins that regulate microtubule behavior in the spindle.

The basic building block of a microtubule is a dimer of two closely related proteins, α-tubulin and β-tubulin (Figure 6-3a). Tubulin dimers are joined head to tail in the same orientation to form a linear protofilament, and a microtubule is made of 13 protofilaments aligned in parallel to form the wall of a hollow cylinder (Figure 6-3b). Although the interactions between tubulin subunits in a microtubule are noncovalent, the combined strength of the end-to-end and lateral interactions results in a strong, rigid filament. Because these interactions lock tubulin subunits tightly within the microtubule lattice, the addition and removal of subunits occur only at the microtubule ends.

Because all tubulin dimers in a filament are oriented in the same way, the two ends of a microtubule are structurally different. One end has β-tubulin exposed and is called the **plus end**, whereas the other ends in α-tubulin and is known as the **minus end** (see Figure 6-3). The two ends of a microtubule have different growth and shrinkage properties that result from differences in the rates at which tubulin subunits bind or dissociate. Most importantly, the rate of tubulin association at the plus end is higher than that at the minus end. As a result, the plus end grows more rapidly than the minus end when provided with an abundant supply of free tubulin subunits.

Microtubules exhibit dynamic instability

The behavior of microtubules is influenced greatly by the fact that tubulin subunits are not simply building blocks but are also GTPases: enzymes that bind the nucleoside triphosphate GTP and catalyze its hydrolysis to GDP and phosphate. Like many GTPases, tubulin uses the energy of GTP binding and hydrolysis to change its shape, or conformation, and this change in shape results in a change in the affinity of tubulin for the ends of microtubules, with GTP–tubulin binding more tightly than GDP–tubulin. Thus, the growth and shrinkage of microtubules can be controlled by altering the relative amounts of GTP- and GDP-bound tubulins at the microtubule tip.

Free tubulin dimers in solution have a low rate of GTP hydrolysis and therefore exist primarily in the GTP-bound form. Binding of GTP–tubulin to the microtubule end increases the GTPase activity of the β-tubulin, resulting in GTP hydrolysis and conversion of the β-tubulin subunit to GDP–tubulin. Because the GDP is trapped in the tubulin structure, it cannot

Figure 6-3 Microtubule structure (a) The basic microtubule building block is a dimer of α-tubulin and β-tubulin. Both subunits bind GTP (red squares). **(b)** A microtubule is a polymer of tubulin dimers, arrayed in 13 protofilaments to form a hollow tube. β-tubulin is oriented toward the plus end of the microtubule, whereas α-tubulin is oriented toward the minus end. Binding of a tubulin dimer to the microtubule end stimulates the GTPase activity of β-tubulin (but not that of α-tubulin), resulting in the hydrolysis of GTP (red squares) to GDP (grey squares). In a growing microtubule, the rate of new subunit addition is more rapid than the rate of GTP hydrolysis, and so the microtubule end contains a cap of GTP–tubulin. The α-tubulin subunit has little GTPase activity and remains GTP-bound throughout the polymer; GTP hydrolysis by this subunit is not a major factor in microtubule dynamics.

Definitions

catastrophe: (in **microtubules**) sudden shrinkage that occurs when GTP hydrolysis occurs at the microtubule tip.

dynamic instability: the tendency of **microtubules** to switch between states of rapid growth and rapid shrinkage.

microtubule: long hollow polymer of tubulin subunits with two distinct ends, a **plus end** and a **minus end**, that display different polymerization behaviors.

minus end: the end of a **microtubule** with α-tubulin exposed. Tubulin subunits are added more slowly at this end than the other.

plus end: the end of a **microtubule** with β-tubulin exposed. Tubulin subunits are added more rapidly at this end than the other.

rescue: (in **microtubules**) sudden shift from shrinkage to growth that occurs when a GTP cap forms at the microtubule tip.

treadmilling: (in **microtubules**) the addition of GTP–tubulin to the **plus end** while GDP–tubulin is dissociating from the **minus end**. It results in the net movement of tubulin subunits from the plus end to the minus end.

References

Alberts, B. *et al.*: *Molecular Biology of the Cell* 4th ed. (Garland Science New York, 2002)

Desai, A. and Mitchison, T.J.: **Microtubule polymerization dynamics.** *Annu. Rev. Cell Dev. Biol.* 1997, **13**:83–117.

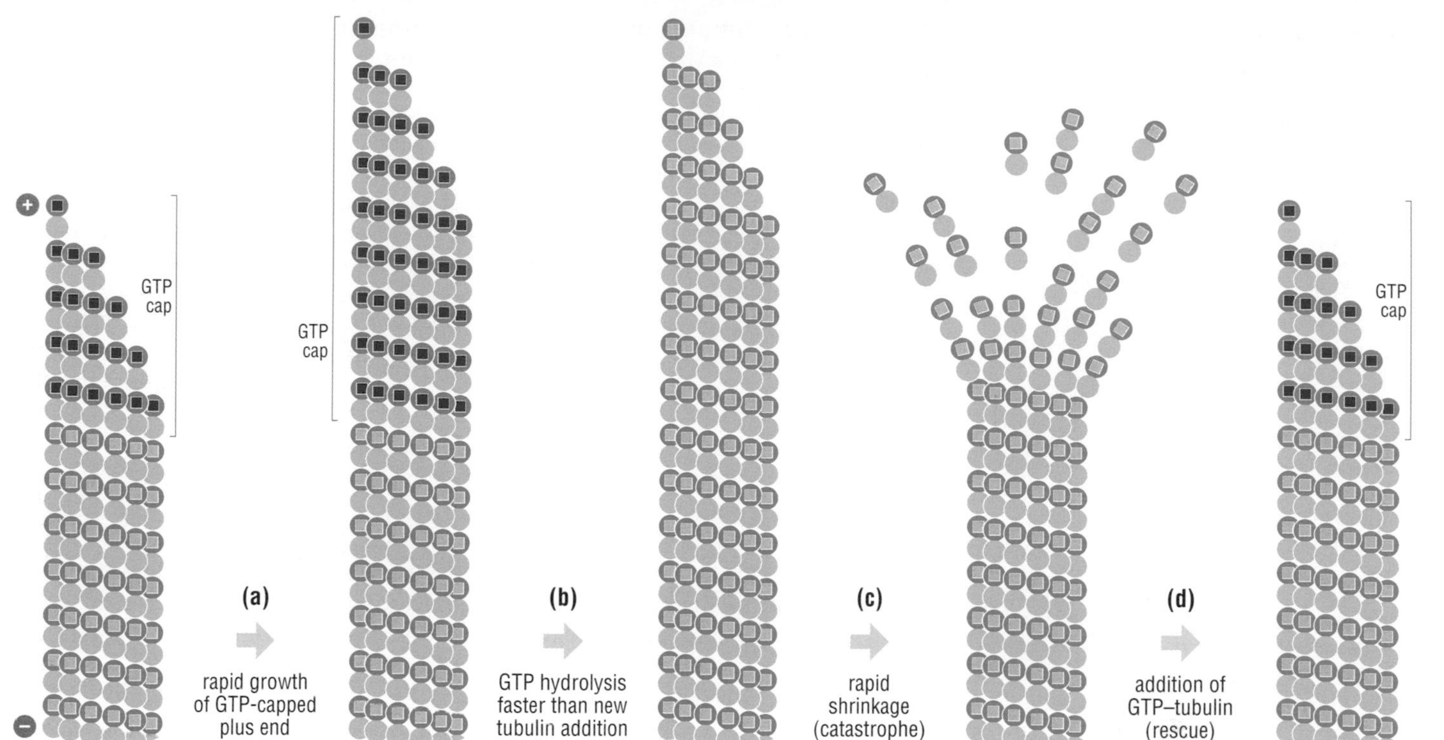

Figure 6-4 **Dynamic instability of microtubules** Microtubules switch between periods of rapid growth and rapid shrinkage, due to changes in the relative rates of new subunit addition and GTP hydrolysis. **(a)** At high tubulin concentrations, rapid growth occurs because new GTP–tubulin dimers are added at a rate that exceeds the rate at which GTP hydrolysis occurs at the microtubule end. **(b)** At intermediate tubulin concentrations, the rate of new GTP–tubulin addition can be similar to that of GTP hydrolysis. Hydrolysis can then occur in tubulin subunits at the plus end of the microtubule, resulting in loss of the GTP cap. **(c)** GDP–tubulin is less tightly bound to the microtubule and so rapidly dissociates, resulting in a switch from rapid growth to rapid shrinkage (catastrophe). GDP–tubulin protofilaments have a curved shape, so that they curl away from the microtubule as they dissociate. **(d)** Eventually, new GTP–tubulin subunits are added to the shrinking microtubule tip, allowing the reformation of a stable GTP cap and rapid growth (rescue). GTP is red; GDP is grey.

dissociate and be replaced by GTP. Most β-tubulin in microtubules therefore exists in the GDP-bound form, although the plus end of the microtubule is often composed of newly added GTP–tubulin molecules that have not yet hydrolyzed their GTP. This is called a GTP cap.

The different microtubule-binding affinities of GDP– and GTP–tubulin can lead to a behavior known as **dynamic instability**, which, as we will see later in the chapter, is important for the assembly and function of the mitotic spindle. Dynamic instability is the rapid interconversion between extended periods of growth and shrinkage at the microtubule end (Figure 6-4). When the rate of addition of GTP–tubulin is greater than the rate of GTP hydrolysis, the GTP cap is maintained and the microtubule will continue to grow rapidly. When the rate of GTP–tubulin association is similar to the rate of GTP hydrolysis, then some microtubule ends will be composed of relatively low-affinity GDP–tubulin. This will not only be less likely to bind a new tubulin dimer but will tend to dissociate from the microtubule, causing the latter to shrink. Even within a single microtubule, small random changes in the rates of tubulin binding and GTP hydrolysis will lead to seemingly random interconversions between a rapidly shrinking state, called **catastrophe**, and a rapidly growing state called **rescue**.

Another interesting microtubule behavior also results from the higher rate of tubulin association at the plus end than at the minus end. At certain concentrations of free tubulin, the plus end grows rapidly while the minus end shrinks, for the reasons outlined above. As a result, subunits added to the plus end make their way down the polymer and are released from the minus end. This is known as **treadmilling**. Treadmilling behavior is well established in actin filaments, the other major cytoskeletal polymer, but its importance in microtubule function in the cell remains uncertain.

Although superficially similar, treadmilling should not be confused with the dynamic process known as *microtubule flux*, which is seen in animal cell spindles. Kinetochore microtubules depolymerize at their minus ends at the spindle pole, while new tubulin is added at the plus ends attached to the kinetochores. The result is a net flux of tubulin subunits from the kinetochore to the pole while the microtubule as a whole remains intact. Unlike treadmilling, which results from the intrinsic properties of the microtubule, flux is driven by microtubule-associated motors and other proteins that move the microtubules poleward and govern tubulin binding at the minus and plus ends. We will discuss the function of microtubule flux in section 6-11.

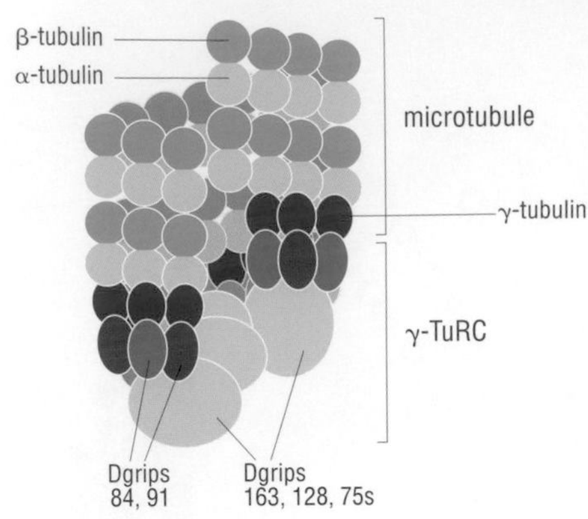

Figure 6-5 Structure of the γ-tubulin ring complex Proposed structure for the γ-tubulin ring complex (γ-TuRC) based on electron microscopic analysis of purified complexes from *Drosophila* embryos. A circular array of γ-tubulin subunits nucleates the assembly of the microtubule and caps the minus end. Other accessory subunits, called Dgrip proteins in *Drosophila*, stabilize the ring, help anchor it in the microtubule-organizing center, and are also likely to be involved in regulation of γ-TuRC activity. Adapted from Moritz, M. *et al.*: *Nat. Cell Biol.* 2000, **2**:365–370.

Cellular microtubules originate on preformed protein complexes that are usually concentrated in a microtubule-organizing center

Assembly of the mitotic spindle requires the formation of many new microtubules, and their efficient initiation is aided by accessory protein complexes that provide a foundation on which the microtubule can be built. The formation of a new microtubule from free tubulin dimers *in vitro* begins with the process of nucleation, the self-assembly of tubulin subunits into a stable aggregate, or nucleation center, upon which the new microtubule grows. Spontaneous nucleation of microtubules is very slow, however, and does not occur to any significant extent in cells. Instead, cellular microtubules usually originate from preexisting nucleation centers, the most important of which is called the **γ-tubulin ring complex (γ-TuRC)**. This large protein complex contains a specialized version of tubulin called γ-tubulin, which nucleates a microtubule at its minus end, allowing the plus end to grow outward (Figure 6-5).

In animal somatic cells and in budding yeast, γ-TuRCs are concentrated in subcellular organelles—the centrosome and the spindle pole body, respectively—which act as microtubule-organizing centers. These organelles nucleate an array of microtubules with their plus ends radiating outward. When cells enter the cell cycle, duplication of the centrosome or spindle pole body is an important step toward the construction of a bipolar mitotic spindle. Nucleating centers do not, however, have to be focused in specialized organelles to be able to organize microtubules into bipolar arrays. Centrosomes are not present in plant cells and frog oocytes (see section 6-0), and yet these cells construct functional spindles in which minus ends nucleated by γ-TuRCs are focused at the two poles. In such cells, microtubule formation is initiated by γ-TuRCs scattered around the sister chromatids, after which motor and other proteins organize the bipolar microtubule array and draw together the minus ends at the poles. In some acentrosomal spindles, such as those of some meiotic cells in *Drosophila*, γ-tubulin is not present at the spindle poles and the nucleating center is not clearly identified.

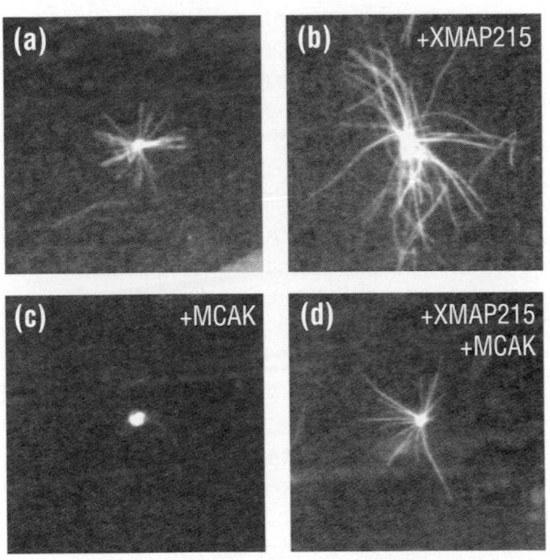

Figure 6-6 Control of microtubule dynamics by associated proteins This experiment reveals the opposing functions of XMAP215, a stabilizing factor, and MCAK, a catastrophe factor of the kinesin-13 family. **(a)** Purified centrosomes were incubated with purified tubulin dimers, resulting in the formation of microtubule asters: arrays of microtubules radiating outward from the centrosomes. **(b)** Addition of purified XMAP215 to these preparations increases the rate of microtubule growth, resulting in longer microtubules. **(c)** Addition of purified MCAK increases the frequency of catastrophe, resulting in decreased average microtubule length. **(d)** Addition of both proteins results in an intermediate microtubule length, revealing that the two regulatory proteins antagonize each other's function. This balance of activities provides the basis for the rapid growth and frequent catastrophes seen in early mitotic cells. Courtesy of Kazuhisa Kinoshita.

Microtubule dynamics are governed by a variety of stabilizing and destabilizing proteins

The assembly and function of the spindle also depend on a broad range of accessory proteins that modify the dynamic properties of microtubules—particularly their rates of growth, the frequency of microtubule catastrophe (see section 6-1), and their association with each other and with other proteins. The length of microtubules in the spindle can be restrained by proteins that increase the frequency of catastrophe. An important group of these so-called catastrophe factors is the kinesin-13 family (also called the KinI family). Although kinesin-13 proteins are related structurally to kinesin motor proteins, which will be discussed below, they are not motors but are proteins that bind the ends of microtubules and induce catastrophe by triggering a conformational change that disrupts lateral interactions between protofilaments. Another well known microtubule destabilizer is a small, highly conserved protein known as Op18 or stathmin, which can promote catastrophes when added to pure microtubules or frog egg extracts. Op18 may act in part by binding and inactivating tubulin dimers in solution, and may also act by binding to microtubule plus ends and enhancing tubulin dissociation.

The destabilizing effects of catastrophe factors are opposed by the stabilizing influences of other microtubule-associated proteins that bind along the sides or at the ends of microtubules. Particularly important in spindle function is the XMAP215 protein of frogs (called Dis1 in fission yeast and TOG in human cells). XMAP215 has complex effects on microtubules under different conditions and in different cell types, but one of its major functions is to promote microtubule growth. It acts in part by binding to microtubule plus ends and thus blocking the binding of destabilizers such as the kinesin-13 proteins (Figure 6-6).

Several important proteins bind to the plus ends of microtubules and control plus-end dynamics or link the plus end to other cellular structures. A prominent member of this family is the protein EB1, which binds to growing plus ends but not to shrinking ones, and interacts with a range of other proteins found at the kinetochore and the cell cortex. Another important group of microtubule-regulatory proteins stabilizes and cross-links microtubule minus ends to form focused spindle poles, often with the assistance of motor proteins that travel along microtubules in the direction of the minus ends.

Motor proteins move along microtubules

Perhaps the most fascinating of the microtubule-associated proteins are the **motor proteins**. These move along microtubules, transporting a molecular cargo or linking microtubules into force-generating arrays such as the spindle. There are two families of microtubule motors. The larger family, the kinesins or kinesin-related proteins, contains many members in all eukaryotes. Most kinesin-related motors move along microtubules toward the plus end, although there are some that move in the opposite direction. Members of the other major family of microtubule motors, the dyneins, all move toward the minus end. Microtubule motors are typically dimers of two identical subunits, each containing a globular head or motor domain that is also an ATPase. The two head domains associate to form a functional motor domain that is in direct contact with the microtubule and uses energy provided by the binding and hydrolysis of ATP to move along it. Motor proteins typically contain other domains or associated subunits that link the motor domain to a protein cargo or anchor it to a subcellular organelle, thus enabling the motor to move cargo along the microtubule or move the microtubule past an anchor point.

Four main classes of motor proteins are particularly important in spindle assembly (Figure 6-7). Kinesin-5 proteins (also called the bipolar or BimC kinesins) are plus-end-directed kinesins that contain two complete motor domains, each of which binds a separate microtubule. These multivalent motor proteins cross-link antiparallel microtubules in the spindle midzone and push the microtubules poleward to help separate the poles. Kinesin-14 proteins (also called C-terminal or Ncd kinesins) move toward the microtubule minus end. In addition to their motor domain, they contain other domains that can interact with another microtubule, enabling them to cross-link microtubules and slide them past each other or focus them at the spindle pole. Kinesin-4 and kinesin-10 proteins (also called chromokinesins) are plus-end-directed kinesins found in animal cells but not in yeast. They contain domains that associate with chromosome arms, linking them to the spindle and contributing to chromosome positioning and the separation of spindle poles. Cytoplasmic dynein is a minus-end-directed motor that links microtubules to other proteins at various cellular locations. At the cell cortex, for example, dynein associates with the plus ends of astral microtubules and connects them to proteins in the actin cytokeleton—thereby providing a way to position and separate the spindle poles. With assistance from other proteins, dynein also helps cross-link minus ends at the spindle pole.

Figure 6-7 Motor proteins in the spindle Four types of motor proteins are particularly important in spindle function. Kinesin-5 cross-links antiparallel interpolar microtubules and pushes the spindle poles apart. Kinesin-14 also cross-links interpolar microtubules but pulls the poles together, thereby balancing the actions of the kinesin-5 proteins. Cytoplasmic dynein anchors plus ends at the cortex and focuses minus ends at the spindle poles. Kinesin-4 and 10 family members link interpolar microtubules to chromosome arms and pull them toward the poles. The long green arrows indicate the direction of microtubule movement, the short dark-blue arrows indicate direction of movement of the motor along the microtubule. The motor domains are depicted by circles.

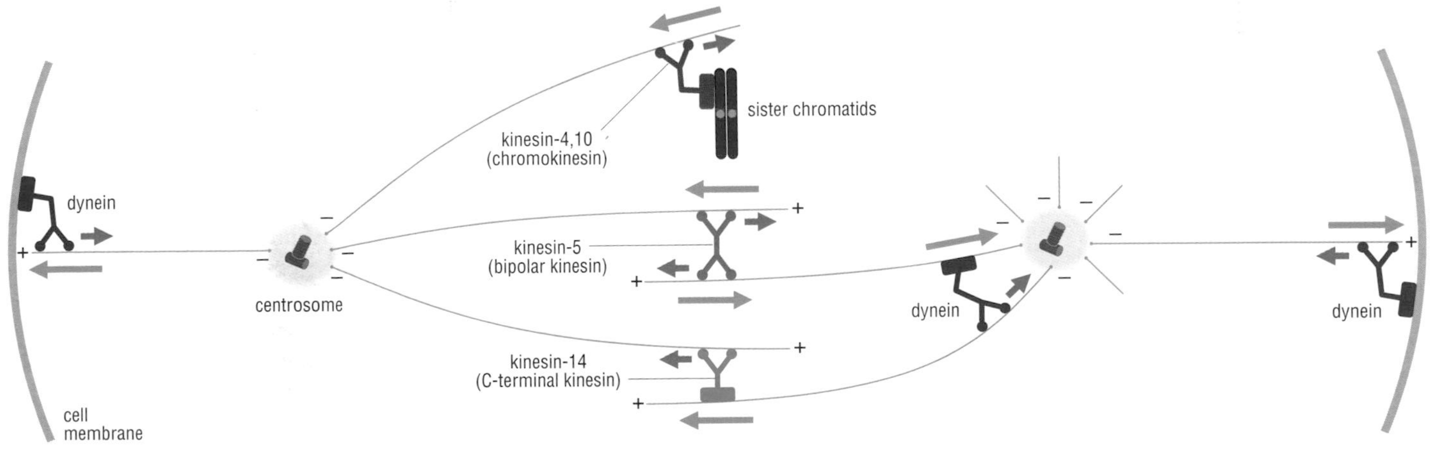

Definitions

γ-tubulin ring complex (γ-TuRC): large, multisubunit protein complex containing a ring of γ-tubulin subunits. It is thought to serve as a template for microtubule nucleation.

γ-TuRC: see **γ-tubulin ring complex**.

motor protein: any of a wide range of proteins with ATPase activity that can move along microtubules or actin filaments. Those connected with spindle assembly are the microtubule motors of the kinesin and dynein families.

References

Gadde, S. and Heald, R.: **Mechanisms and molecules of the mitotic spindle.** *Curr. Biol.* 2004, **14**:R797–R805.

Kinoshita, K. *et al.*: **Reconstitution of physiological microtubule dynamics using purified components.** *Science* 2001, **294**:1340–1343.

Kline-Smith, S.L. and Walczak, C.E.: **Mitotic spindle assembly and chromosome segregation: refocusing on microtubule dynamics.** *Mol. Cell* 2004, **15**:317–327.

Kwon, M. and Scholey, J.M.: **Spindle mechanics and dynamics during mitosis in *Drosophila*.** *Trends Cell Biol.* 2004, **14**:194–205.

Lawrence, C.J. *et al.*: **A standardized kinesin nomenclature.** *J. Cell Biol.* 2004, **167**:19–22.

Moritz, M. *et al.*: **Structure of the gamma-tubulin ring complex: a template for microtubule nucleation.** *Nat. Cell Biol.* 2000, **2**:365–370.

6-6 Early Steps in Spindle Assembly

Spindle assembly begins in prophase

Attachment of sister chromatids to spindle microtubules is an integral part of spindle assembly. In animal cells, however, the centrosomes and microtubules that form the spindle are located in the cytoplasm outside the nucleus. Assembly of a centrosome-dependent spindle therefore depends on breakdown of the nuclear envelope, which gives the microtubules access to the chromosomes. Many of the motors and other microtubule regulators that help organize the spindle are associated with the chromosomes inside the nucleus; removal of the nuclear envelope therefore allows these proteins to make their essential contributions to spindle assembly.

This section provides an overview of early mitotic changes in microtubule and centrosome behavior in animal cells, both before and just after nuclear envelope breakdown. In later sections of this chapter we describe nuclear envelope breakdown in more detail and discuss how chromosomal proteins collaborate with the centrosomes to create a bipolar microtubule array on which the sister chromatids are attached correctly.

Mitotic microtubules are highly dynamic

Interphase animal cells contain an extensive network of long, stable microtubules that originate in the centrosome and radiate throughout the cell. In mitosis this network is dismantled to allow the complete reorganization of the microtubules into a spindle. Entry into mitosis therefore triggers dramatic changes in microtubule dynamics. Microtubules tend to be shorter in mitosis, largely as a result of an increase in the rate of microtubule catastrophes—the rate at which microtubules change from the growing state to the shrinking state (see section 6-1). Mitotic microtubules tend to grow rapidly but then shrink back more frequently—the ideal behavior for the search and capture of kinetochores and other spindle components.

Mitotic changes in microtubule behavior begin in prophase but become more pronounced after nuclear envelope breakdown, when nuclear microtubule regulators gain access to the growing spindle. Numerous microtubule-stabilizing and cross-linking factors are associated with the chromosomes. Thus, despite the general instability of astral microtubules in the surrounding cytoplasm, microtubules around the chromosomes tend to be longer and relatively abundant. This results in the formation of a dense and dynamic interpolar microtubule array that is well suited for rapid attachment to sister chromatids.

Microtubule dynamics in mitosis are regulated by a large number of proteins. Some key players are MCAK and other catastrophe factors of the kinesin-13 family (see section 6-2; Figure 6-6), which are primarily responsible for the increased frequency of catastrophe in mitotic microtubules.

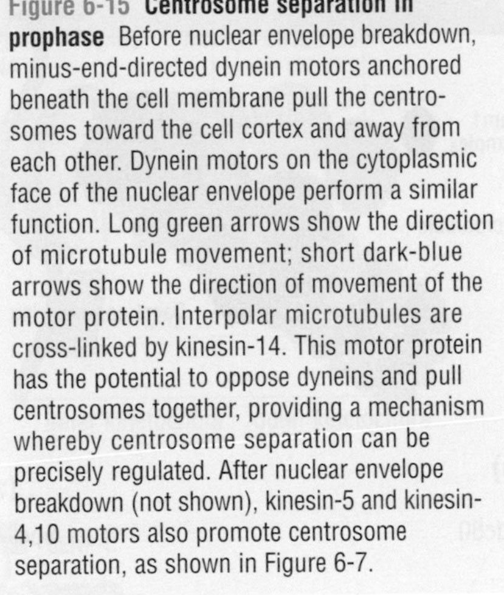

Figure 6-15 Centrosome separation in prophase Before nuclear envelope breakdown, minus-end-directed dynein motors anchored beneath the cell membrane pull the centrosomes toward the cell cortex and away from each other. Dynein motors on the cytoplasmic face of the nuclear envelope perform a similar function. Long green arrows show the direction of microtubule movement; short dark-blue arrows show the direction of movement of the motor protein. Interpolar microtubules are cross-linked by kinesin-14. This motor protein has the potential to oppose dyneins and pull centrosomes together, providing a mechanism whereby centrosome separation can be precisely regulated. After nuclear envelope breakdown (not shown), kinesin-5 and kinesin-4,10 motors also promote centrosome separation, as shown in Figure 6-7.

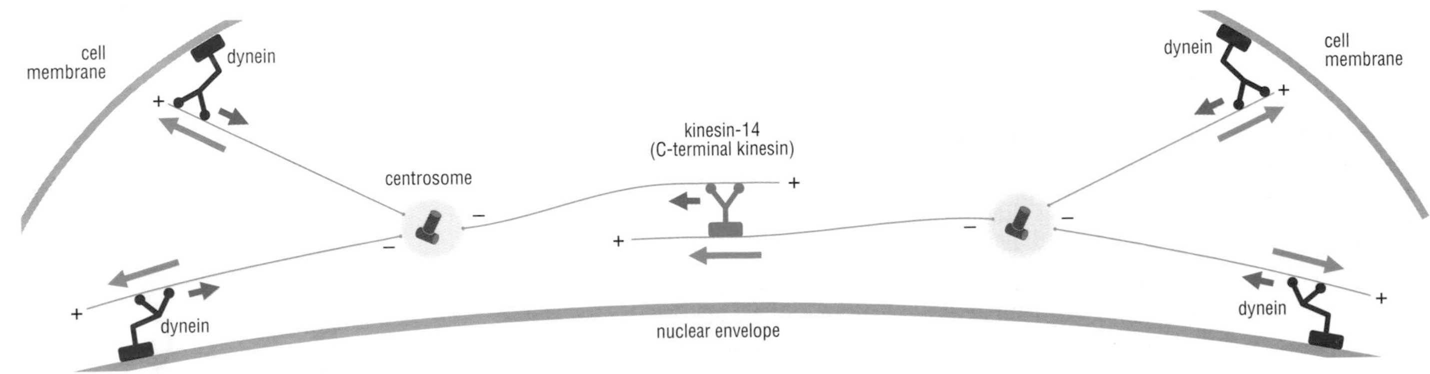

Definitions

centrosome disjunction: severing of the protein linkage between duplicated centrosomes in early mitosis.

centrosome maturation: improvement in the microtubule-nucleating ability of the centrosome that occurs during mitosis. It is due to an increase in the number of γ-tubulin ring complexes.

centrosome separation: the movement apart of duplicated centrosomes that follows **centrosome disjunction** in early mitosis.

References

Barr, F.A. *et al.*: **Polo-like kinases and the orchestration of cell division.** *Nat. Rev. Mol. Cell Biol.* 2004, **5**:429–440.

Khodjakov, A. and Rieder, C.L.: **The sudden recruitment of gamma-tubulin to the centrosome at the onset of mitosis and its dynamic exchange throughout the cell cycle do not require microtubules.** *J. Cell Biol.* 1999, **146**:585–596.

Kline-Smith, S.L. and Walczak, C.E.: **Mitotic spindle assembly and chromosome segregation: refocusing on microtubule dynamics.** *Mol. Cell* 2004, **15**:317–327.

Kwon, M. and Scholey, J.M.: **Spindle mechanics and dynamics during mitosis in *Drosophila*.** *Trends Cell Biol.* 2004, **14**:194–205.

Marumoto, T. *et al.*: **Aurora-A—a guardian of poles.** *Nat. Rev. Cancer* 2005, **5**:42–50.

Rosenblatt, J.: **Spindle assembly: asters part their separate ways.** *Nat. Cell Biol.* 2005, **3**:219–222.

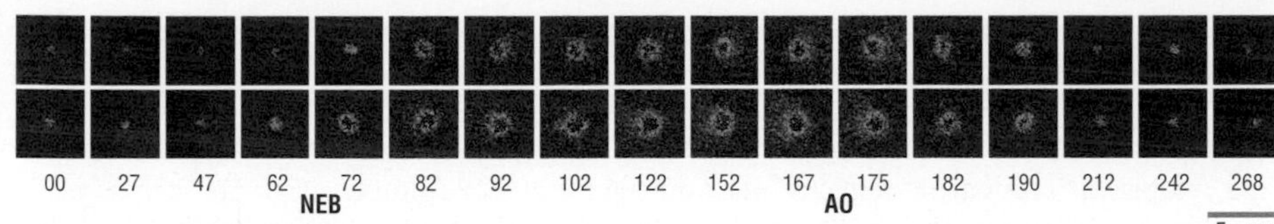

Figure 6-16 **Recruitment of γ-tubulin to mitotic centrosomes** Animal cells were engineered to express a version of γ-tubulin that is tagged with a fluorescent protein, so that the amount of γ-tubulin at the centrosomes could be measured by fluorescence microscopy. The centrosome pair in one cell was analyzed, starting in early prophase and ending soon after the completion of mitosis. At the indicated times (minutes), the intensity of fluorescent γ-tubulin in the two centrosomes (top and bottom rows) was measured. The data were then processed to produce pseudocolor images, in which a low fluorescent signal is shown as blue, a moderate signal as green, and an intense signal as red. These images show that the amount of γ-tubulin in the centrosomes increases dramatically around the time of nuclear envelope breakdown (NEB), and then declines rapidly soon after the onset of anaphase (AO). Images kindly provided by Alexey Khodjakov. From Khodjakov, A. and Rieder, C.L: *J. Cell Biol.* 1999, **146**:585–596.

The increased activity of these proteins during mitosis may be due to a decrease in the activity of stabilizing factors such as XMAP215 (see section 6-2). Plus-end-capping proteins, such as the CLASP proteins in *Drosophila*, also make important contributions to the control of microtubule dynamics.

Centrosome separation initiates spindle assembly

Most animal cells enter mitosis with a pair of tightly associated centrosomes that together form a single microtubule-organizing center. A key early step in spindle assembly is centrosome separation, which can be divided into two distinct steps. The first is **centrosome disjunction**, when the cohesive protein linkages between the duplicated centrosomes are dissolved by mechanisms that are still essentially unknown. The second step is **centrosome separation**, when the disjoined centrosomes move apart, primarily as a result of forces generated by motor proteins.

Centrosome separation is initiated by minus-end-directed dynein motors that are anchored to the cell cortex and nuclear envelope. The dynein motor domains attach near the plus ends of the astral microtubules that radiate from each centrosome (see Figure 6-7). Movement of these motors toward the minus end—that is, toward the centrosome—pulls each centrosome outward and away from the other (Figure 6-15).

Dynein-dependent separation of the centrosomes enables an array of overlapping microtubules to form gradually between the poles. In prophase, the plus ends of these antiparallel microtubules are cross-linked by motors of the kinesin-14 family. As described earlier in this chapter (see Figure 6-7), these proteins are minus-end-directed motors with a motor domain attached to one microtubule and another domain anchored to a microtubule from the other centrosome. These motors therefore tend to pull the centrosomes together and oppose the actions of dyneins (see Figure 6-15). This counterbalancing action presumably provides the system with greater control and precision, but little is known about how the opposing forces are adjusted to provide an optimal rate of centrosome separation.

Two other major classes of motor proteins—kinesin-5 (bipolar kinesins) and kinesin-4,10 (chromokinesins)—are located within the nucleus. After nuclear envelope breakdown, these plus-end-directed motor proteins associate with interpolar microtubules and help push the spindle poles apart by mechanisms discussed earlier in this chapter (see Figure 6-7).

Little is known about how centrosome separation is controlled by the cell-cycle control system. The initiation of centrosome separation depends on mitotic Cdk activity, but the Cdk targets that drive this separation are not clear. Mitotic Cdks and the protein kinase aurora A (see section 5-7) are both required for centrosome separation after nuclear envelope breakdown, and a likely target for both of these kinases is the kinesin-5 Eg5, whose phosphorylation is thought to stimulate its ability to push the centrosomes apart. A centrosomal protein known as C-Nap1 is phosphorylated in mitosis by a protein kinase called Nek2, and this modification may promote the loss of centrosome cohesion that initiates centrosome separation.

Centrosome maturation increases microtubule nucleation in mitosis

In late prophase, centrosomes display a dramatic increase in their ability to nucleate microtubules, which is an essential prerequisite for the assembly of the densely packed microtubule arrays of the spindle. This **centrosome maturation** process involves the recruitment of large numbers of new γ-TuRCs to the centrosomes (Figure 6-16). Maturation is probably controlled by the protein kinase Plk (see section 5-7), because it is blocked by microinjection of antibodies against this kinase. The targets of Plk are not clear, but there is evidence from *Drosophila* that Plk regulates a protein called Asp, which anchors γ-TuRCs in the centrosome. Centrosome maturation also requires the activity of aurora A, acting in part through phosphorylation of a centrosome protein called TACC, which helps stabilize microtubule minus ends at the spindle poles.

6-7 Nuclear Envelope Breakdown

The nuclear envelope is composed of two membranes on an underlying protein support

In all eukaryotes the nuclear envelope serves as a highly regulated barrier between the nucleus and cytoplasm. This barrier remains intact during mitosis in a simple eukaryote such as yeast, which constructs a mitotic spindle within the nucleus and undergoes what is known as a closed mitosis. In animal cells, however, the envelope must be removed to allow the cytoplasmic spindle to gain access to and segregate the sister chromatids; this is known as an open mitosis.

The nuclear envelope is made up of an inner membrane and an outer membrane enclosing a luminal space (Figure 6-17). The membranes and lumen of the nuclear envelope are continuous with those of the endoplasmic reticulum, which surrounds the nucleus. Scattered throughout the nuclear envelope are large pores, at which the inner and outer membranes are joined. These pores contain a giant protein assembly, the nuclear pore complex, that regulates the transport of proteins between the cytoplasm and nuclear interior. In animal cells, but not in yeast, a protein meshwork called the nuclear lamina provides a flexible structural framework on the nuclear side of the envelope. The lamina is composed of cross-linked intermediate filament proteins called nuclear lamins, as well as a variety of lamin-associated proteins that connect the lamina to the inner nuclear membrane and to chromatin located in the nuclear periphery.

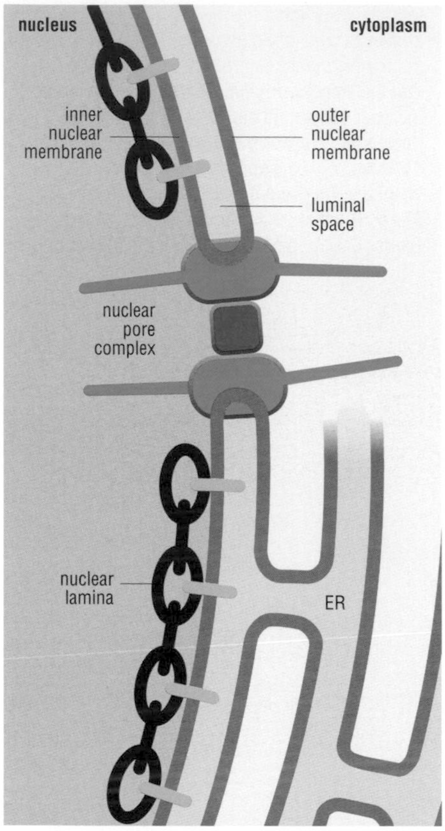

Figure 6-17 Structure of the nuclear envelope The nuclear envelope is a specialized extension of the endoplasmic reticulum. It is composed of inner and outer membranes surrounding a lumen that is continuous with that of the endoplasmic reticulum (ER). Nuclear pore complexes mediate the transport of proteins and RNA across the envelope. In animal cells, a network of proteins forms the nuclear lamina (red) on the internal face of the inner nuclear membrane. The nuclear lamina interacts with proteins in the inner membrane (yellow).

Nuclear envelope breakdown begins at nuclear pores

During mitosis, the nuclear envelope is dismantled into its various components. In vertebrates, the envelope and underlying lamina are completely dissolved in early mitosis (Figure 6-18), whereas in *Drosophila* embryos and *C. elegans* the process begins with localized breakdown near the spindle poles in early mitosis and is not complete until metaphase or anaphase.

Several mechanisms combine to bring about the breakdown of the nuclear envelope in vertebrate cells. A key early event seems to be the phosphorylation of several components of the nuclear pore complex, which triggers its disassembly into smaller subcomplexes that dissociate from the envelope membranes. Disassembly of nuclear pore complexes is an important early step in the loss of envelope integrity. It is likely that cyclin B–Cdk1 is directly responsible for the phosphorylation of pore components, but this possibility has not yet been investigated fully.

The growing mitotic spindle is directly involved in nuclear envelope breakdown. In cells entering mitosis, the separating centrosomes are anchored to the nuclear envelope by minus-end-directed dynein motors (see Figure 6-3). The movement of these motors toward the centrosomes not only helps pull the centrosomes apart but also has the effect of pulling the envelope toward the centrosomes. The envelope becomes bunched up and infolded next to the centrosomes, while becoming stretched, and eventually tearing, on the opposite side of the nucleus (see Figure 6-18). Chemicals that disrupt microtubule structure cause a delay, but not a block, in nuclear envelope breakdown, indicating that this mechanism is important but not essential in the process.

Breakdown of the nuclear envelope also requires breakdown of the nuclear lamina. After the import of cyclin B–Cdk1 into the nucleus in late prophase (see section 5-6), this complex directly phosphorylates nuclear lamins and thereby causes disassembly of the meshwork of lamin filaments. Phosphorylation of inner nuclear envelope proteins, possibly by cyclin

References

Barr, F.A.: **Golgi inheritance: shaken but not stirred.** *J. Cell Biol.* 2004, **164**:955–958.

Burke, B. and Ellenberg, J.: **Remodelling the walls of the nucleus.** *Nat. Rev. Mol. Cell Biol.* 2002, **3**:487–497.

Heald, R. and McKeon, F.: **Mutations of phosphorylation sites in lamin A that prevent nuclear lamina disassembly in mitosis.** *Cell* 1990, **61**:579–589.

Hetzer, M. *et al.*: **Pushing the envelope: structure, function, and dynamics of the nuclear periphery.** *Annu. Rev. Cell Dev. Biol.* 2005, **21**:347–380.

Lenart, P. and Ellenberg, J.: **Nuclear envelope dynamics in oocytes: from germinal vesicle breakdown to mitosis.** *Curr. Opin. Cell Biol.* 2003, **15**:88–95.

Lenart, P. *et al.*: **Nuclear envelope breakdown in starfish oocytes proceeds by partial NPC disassembly followed by a rapidly spreading fenestration of nuclear membranes.** *J. Cell Biol.* 2003, **160**:1055–1068.

Shorter, J. and Warren, G.: **Golgi architecture and inheritance.** *Annu. Rev. Cell Dev. Biol.* 2002, **18**:379–420.

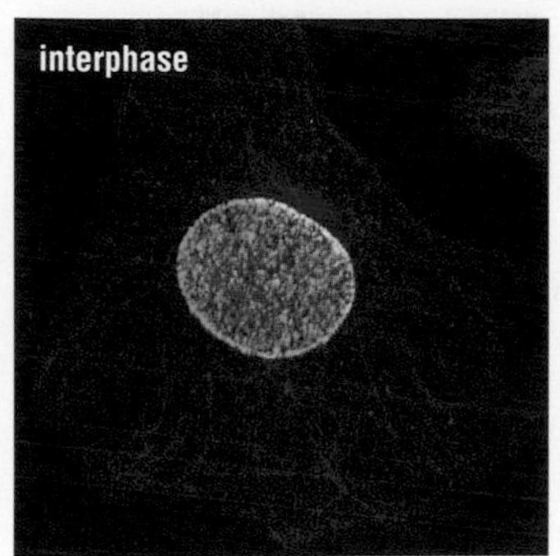

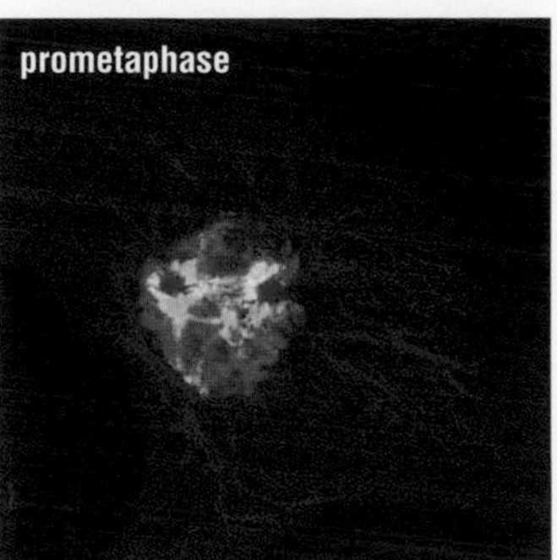

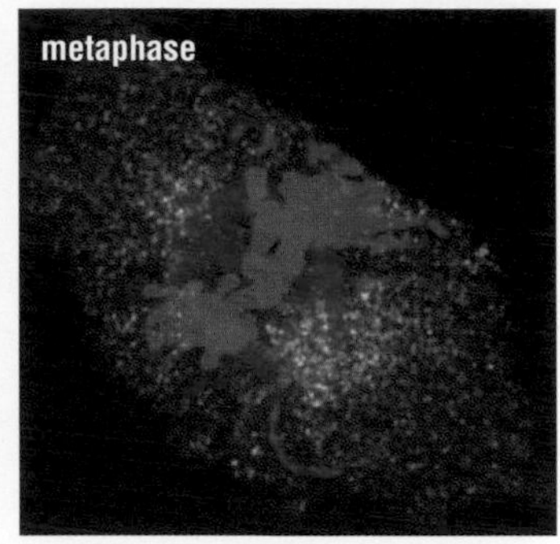

B–Cdk1 as well, causes the lamin meshwork to dissociate from the nuclear envelope. Disassembly of the lamina is essential for nuclear envelope breakdown, because breakdown does not occur in cells expressing mutant lamins that cannot be phosphorylated.

The combined effect of these processes is the dramatic loss of nuclear envelope structure. Eventually, the nuclear membrane disappears completely, probably being broken up into small vesicles or absorbed into the membranes of the endoplasmic reticulum. After mitosis, nuclear membrane proteins are again sorted into separate membranes to reform the nuclear envelope around the segregated chromosomes.

Figure 6-18 Nuclear envelope breakdown in mitosis Cultured vertebrate cells were fixed at various stages of mitosis and stained for DNA (blue), microtubules (red), and an inner nuclear membrane protein (green). In prophase, the normally smooth nuclear rim becomes infolded near the centrosomes, as a result of the actions of dynein motors anchored to the envelope and interacting with centrosome microtubules. This process, coupled with the disassembly of nuclear pore complexes and the nuclear lamina, causes the nuclear envelope to fragment at the onset of prometaphase. Photographs kindly provided by Jan Ellenberg and Brian Burke. From Burke, B. and Ellenberg, J.: *Nat. Rev. Mol. Cell Biol.* 2002, **3**:487–497, with permission.

The endoplasmic reticulum and Golgi apparatus are reorganized in mitosis

The endoplasmic reticulum is an extensive network of membrane tubules with multiple functions in the cell, including new membrane synthesis and the synthesis and transport of secreted and transmembrane proteins. Like all cytoplasmic organelles, the endoplasmic reticulum must be distributed equally during cell division. In most cells, it remains largely intact during mitosis, and is thought to be distributed into daughter cells by simply being pinched in two during cytokinesis. In some cell types, however, the tubular structure becomes broken up into smaller membrane vesicles that are equally distributed into daughter cells during cytokinesis, after which they reunite to form the interphase organelle.

The Golgi apparatus is a large membrane-bounded organelle to which secreted or transmembrane proteins are transported to undergo modification before delivery to the cell membrane or extracellular space. In animal cells, the Golgi apparatus is made up of stacks of disc-shaped membrane-bounded compartments. Several stacks are usually linked together to form a large, ribbon-like organelle that is typically located near the centrosome in animal cells. The Golgi apparatus is radically restructured during mitosis (Figure 6-19). The ribbon of stacks is dissociated into individual stacks, which then break down into tubules and vesicles. This dissociation depends on the activities of cyclin B–Cdk1 and the protein kinase Plk (see section 5-7), which phosphorylate proteins holding the stacks together, thereby disrupting their structure and promoting vesiculation.

Vesiculated Golgi membranes are supported by an underlying protein matrix and are thought to remain largely distinct from the membranes of the endoplasmic resticulum. Clusters of Golgi membranes remain close to each spindle pole throughout mitosis. In telophase they nucleate the reconstruction of a new Golgi apparatus in each daughter cell.

Figure 6-19 Fragmentation of the Golgi apparatus in mitosis Cultured vertebrate cells were fixed at various stages of mitosis and stained for DNA (blue), microtubules (red), and a Golgi protein (green). The Golgi apparatus is focused near the centrosome during interphase, and then becomes vesiculated and more widely dispersed during mitosis. Photographs kindly provided by Joachim Seemann. From Shorter, J. and Warren, G.: *Annu. Rev. Cell Dev. Biol.* 2002, **18**:379–420, with permission.

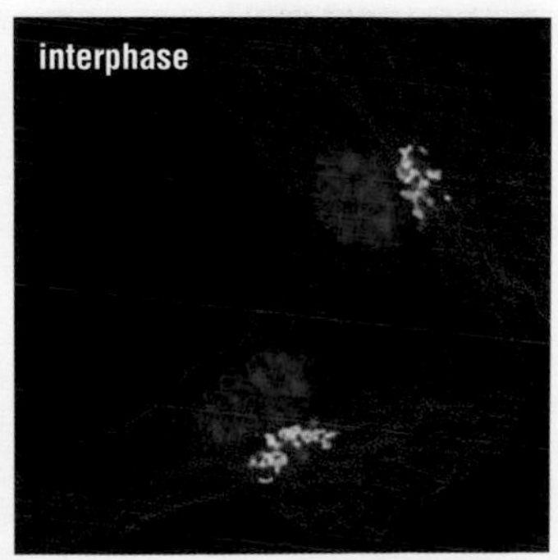

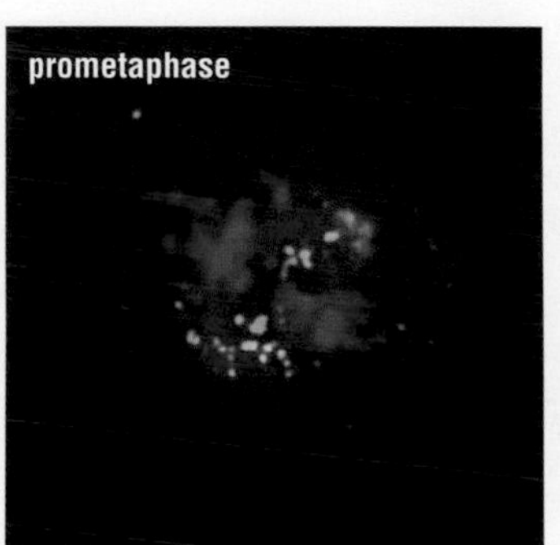

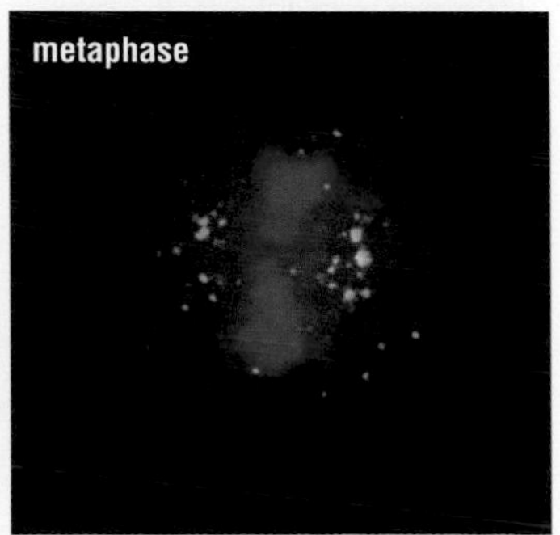

6-8 Mitotic Chromosome Function in Spindle Assembly

(a) Nucleation

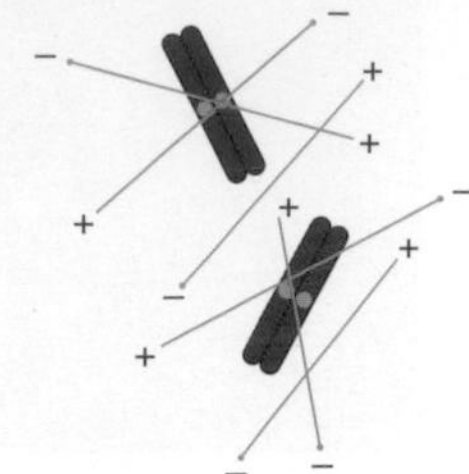

(b) Antiparallel cross-linking by kinesin-5

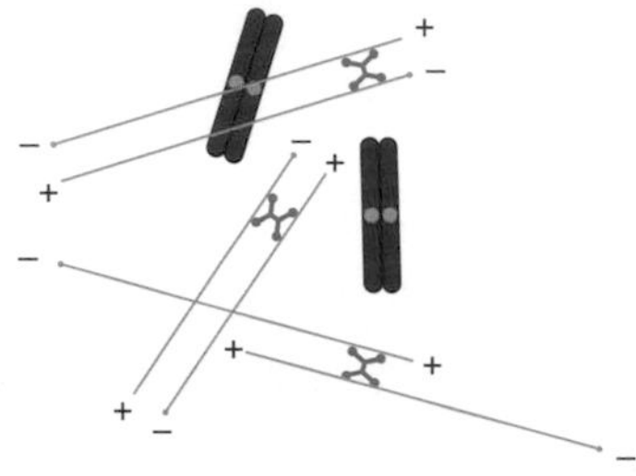

(c) Outward push by kinesin-4,10

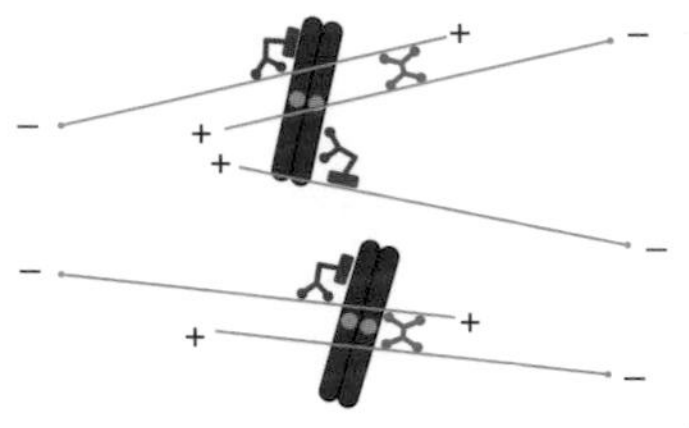

(d) Pole focusing by dynein and kinesin-14

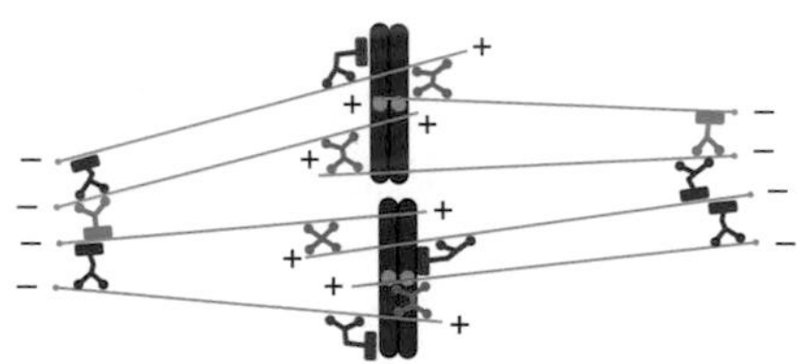

Figure 6-20 Self-organization of a bipolar microtubule array (a) Spindle self-assembly begins with the formation of stable microtubules in the vicinity of the DNA. These microtubules are nucleated by soluble γ-TuRCs in most cell types, but other nucleating factors may also contribute. **(b)** Kinesin-5 motors cross-link pairs of microtubules in an antiparallel orientation, resulting in the formation of microtubule bundles with minus ends pushing outward. Kinesin-14 motors (not shown here, but see Figure 6-7) also help cross-link antiparallel microtubules. **(c)** Kinesin-4 and 10 motors associate with chromosome arms and capture the microtubules. By moving toward the plus end of these microtubules, these kinesins push microtubule minus ends away from the chromosomes. **(d)** Microtubule minus ends are focused into spindle poles by dynein and its associated minus-end cross-linkers, as well as by minus-end-directed kinesin-14 motors.

Spindles self-organize around chromosomes

When the nuclear envelope breaks down, cytoplasmic microtubules gain access to the condensing mitotic chromosomes and other nuclear components, including regulatory proteins that stabilize microtubules and organize them into a bipolar array. This microtubule self-organization is thought to be a major driving force in spindle assembly—not only in cells that normally lack centrosomes but also in centrosome-containing animal cells.

Spindle self-organization depends on the four major groups of motor proteins discussed earlier, which work in collaboration with microtubule-nucleating, stabilizing and cross-linking factors (see Figure 6-7). We do not yet have a complete understanding of the process, but a likely sequence of events is shown in Figure 6-20. The first step is the nucleation of microtubules around chromosomes, which is promoted by soluble γ-TuRCs and possibly by other nucleating factors. Local stabilizing factors then promote the formation of a randomly oriented meshwork of long microtubules around the chromosomes.

Kinesin-5 motors begin to organize this meshwork by cross-linking antiparallel microtubules. These motors move toward microtubule plus ends, thereby pushing the minus ends outward, away from the chromosomes. Minus-end-directed kinesin-14 motors also cross-link antiparallel microtubules and oppose the actions of the kinesin-5 motors. A third group of motors, the kinesin-4 and 10 families of chromokinesins, are plus-end-directed motors attached to chromosome arms: by moving toward the plus end they cause the movement of minus ends away from the chromosomes.

Bipolarity is achieved in the microtubule array by the focusing of microtubule minus ends into two poles. Pole focusing probably depends on at least two motors, dynein and kinesin-14, both of which are thought to bind the minus end of one microtubule while moving toward the minus end of another (see Figure 6-20). Dynein carries out its pole-focusing function with assistance from numerous other nuclear proteins, including the microtubule-associated protein NuMA. Like dynein, NuMA is required for the establishment of focused spindle poles in many animal cell types and may form a localized protein matrix at the pole.

Microtubules can be stabilized by a gradient of Ran–GTP around chromosomes

Bipolar spindle assembly around the mitotic chromosomes depends in part on chromosome-associated proteins that promote highly localized microtubule stabilization and cross-linking. One major mechanism by which this occurs involves localized activation of a regulatory protein called Ran, which activates numerous spindle assembly-promoting factors around the chromosomes.

Ran is a member of a family of small GTPases that includes Ras and other proteins that act as regulatory switches in various cellular processes. Like other members of this family, the activity of Ran depends on the nucleotide that it binds: it is active when bound to GTP and becomes inactive when the bound GTP is hydrolyzed to GDP. Reactivation then requires the dissociation of GDP and its replacement with GTP. These reactions are stimulated by accessory proteins: GTP hydrolysis requires a GTPase-activating protein called RanGAP, whereas the exchange of GDP for GTP requires a guanine-nucleotide exchange factor called RCC1.

References

Blower, M.D. *et al.*: **A Rae1-containing ribonucleoprotein complex is required for mitotic spindle assembly.** *Cell* 2005, **121**:223–234.

Caudron, M. *et al.*: **Spatial coordination of spindle assembly by chromosome-mediated signaling gradients.** *Science* 2005, **309**:1373–1376.

Gadde, S. and Heald, R.: **Mechanisms and molecules of the mitotic spindle.** *Curr. Biol.* 2004, **14**:R797–R805.

Harel, A. and Forbes, D.J.: **Importin beta: conducting a much larger cellular symphony.** *Mol. Cell* 2004, **16**:319–330.

Kalab, P. *et al.*: **Visualization of a Ran-GTP gradient in interphase and mitotic *Xenopus* egg extracts.** *Science* 2002, **295**:2452–2456.

Sampath, S.C. *et al.*: **The chromosomal passenger complex is required for chromatin-induced microtubule stabilization and spindle assembly.** *Cell* 2004, **118**:187–202.

The amount of active Ran–GTP is therefore determined by the relative amounts of RanGAP and RCC1. In interphase cells, RanGAP is located in the cytoplasm and RCC1 is associated with chromatin inside the nucleus. Active Ran–GTP therefore tends to be concentrated inside the nucleus, whereas inactive Ran–GDP is in the cytoplasm.

Apart from its specific role in spindle assembly, Ran–GTP is involved in the general process of nuclear import. Proteins destined for transport into the nucleus bind in the cytoplasm to soluble transporter proteins, such as the dimeric complex called importin, and are transported across the nuclear pore. Inside the nucleus, Ran–GTP binds to the complex of importin and associated cargo, causing its dissociation and thereby releasing the cargo into the nuclear space.

During interphase and early mitosis, numerous microtubule-regulating proteins are imported by this mechanism into the nucleus, where they have no effect on cytoplasmic microtubules. Following nuclear envelope breakdown, these proteins spread throughout the cell but are bound and inhibited by cytoplasmic importin. However, RCC1 on the mitotic chromosomes generates a high concentration of Ran–GTP in the immediate vicinity. Ran–GTP triggers local dissociation of importin from the regulatory proteins, which then act to promote spindle assembly around the chromosomes (Figure 6-21).

Among the microtubule-regulating proteins thought to be liberated in this way by Ran–GTP are NuMA and a protein called TPX2, which stimulates microtubule bundling and spindle pole formation, in part by activating the protein kinase aurora A (see section 5-7). A kinesin-14 motor called XCTK2 also collaborates with dynein and NuMA in the cross-linking of microtubule minus ends at spindle poles. Another microtubule regulator liberated from importin by Ran–GTP is a protein called Rae1, which associates with several other proteins to form a large microtubule-stabilizing complex that also contains RNA. Experimental destruction of RNA inhibits spindle assembly, raising the possibility that RNA is somehow involved in spindle construction or maintenance.

Ran–GTP-dependent spindle assembly has so far been demonstrated primarily in extracts of *Xenopus* eggs, and it remains unclear whether it is the predominant mechanism by which chromosomes stimulate spindle assembly in all cells. Other mechanisms clearly exist. As mentioned earlier, kinesins on chromosome arms help promote spindle assembly. In addition, the protein kinase aurora B, together with its partners INCENP, survivin and other proteins (see section 5-7), associates with chromosomes in early mitosis and promotes the stability of local microtubules by inhibiting the catastrophe factor MCAK (see Figure 6-6). Through these and probably several other mechanisms, chromosome-associated proteins make a variety of important contributions to spindle assembly.

As microtubule self-organization around chromosomes can be so effective at constructing a bipolar spindle, what is the advantage to animal cells of having centrosomes? First, centrosomes provide a pair of preassembled spindle poles upon which self-organization mechanisms can more efficiently construct the bipolar microtubule array. A second advantage is that centrosomes, unlike acentrosomal spindle poles, nucleate astral microtubules that connect the spindle to the cell cortex, where dynein motors can pull on the microtubules. This provides a mechanism for pulling spindle poles apart and, perhaps more importantly, provides a means of positioning the spindle in the cell. As we will see in Chapter 8, the position of the spindle determines the plane of division in animal cell cytokinesis; during early embryonic development, for example, the regulation of spindle position provides an important way of changing the orientation of cell division and the relative sizes of the daughter cells.

Wadsworth, P. and Khodjakov, A.: **E pluribus unum: towards a universal mechanism for spindle assembly.** *Trends Cell Biol.* 2004, **14**:413–419.

Walczak, C.E. *et al.*: **A model for the proposed roles of different microtubule-based motor proteins in establishing spindle bipolarity.** *Curr. Biol.* 1998, **8**:903–913.

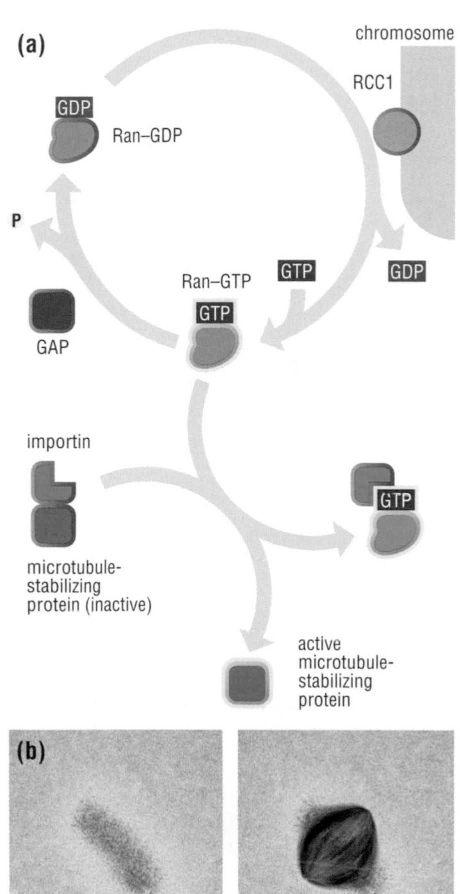

Figure 6-21 Stabilization of microtubules around chromosomes by Ran–GTP (a) The presence of the guanine-nucleotide exchange factor RCC1 on chromosomes results in local production of the active GTP-bound form of Ran. Ran–GTP causes local activation of microtubule-stabilizing proteins by releasing them from importin. **(b)** To measure Ran–GTP concentration around chromosomes, a protein was constructed that is fluorescent when dissociated from importin by Ran–GTP. In mitotic *Xenopus* egg extracts, the fluorescence of this protein is most intense around chromatin (left panel), reflecting the high local concentration of Ran–GTP. In the right panel, an image of the mitotic spindle is superimposed on the chromatin. Panel (b) kindly provided by Rebecca Heald. Reprinted, with permission, from Kalab, P. *et al.*: *Science* 2002, **295**:2452–2456.

Centrosomes search for and capture kinetochores in prometaphase

The construction of the mitotic spindle involves two key processes. The first is the construction of a bipolar microtubule array around the chromosomes, which is summarized in section 6-8. The second is the correct attachment of sister-chromatid pairs to the opposite poles of this array, which we describe in this section and in section 6-10.

The best-understood mechanism of chromosome attachment is that of centrosome-containing animal cells. In these cells, mitotic changes in microtubule dynamics, combined with centrosome maturation, result in a dense array of microtubules radiating from each centrosome. When the nuclear envelope dissolves at the end of prophase, these microtubules are stabilized by the sister chromatids, which are thereby surrounded by large numbers of microtubule plus ends from both poles. The growth and shrinkage of microtubules then provides a remarkably efficient mechanism for chromatid attachment that is known as search and capture.

Attachment to the spindle generally begins when one kinetochore in a sister-chromatid pair is captured by a microtubule from the nearest centrosome (Figure 6-22). Minus-end-directed dynein and kinesin-14 motors then transport the chromatid pair rapidly along the side of the microtubule toward the pole (Figure 6-22b). Near the pole, microtubule attachment is converted to the standard end-on orientation, with the plus end of the microtubule embedded in the kinetochore. Additional microtubules from the same pole are then attached to the kinetochore, resulting in the formation of a **kinetochore fiber** containing several microtubules (Figure 6-22c). This mono-oriented chromatid pair oscillates near the spindle pole and then moves toward the center of the spindle (a movement called **chromosome congression**). It is not clear how this movement occurs, but one simple possibility is that a microtubule from the opposite pole captures the unoccupied sister kinetochore and pulls it to the spindle center (Figure 6-22d and e). Alternatively, an unoccupied kinetochore can bind to the side of a kinetochore fiber of a preexisting bi-oriented pair. A plus-end-directed kinesin-7 motor called CENP-E then transports the mono-oriented pair along the fiber to the spindle center, where it is captured by a microtubule from the opposite pole. The correct bi-oriented attachment is stabilized by the addition of more microtubules to both kinetochore fibers, which eventually contain 10–40 microtubules each.

Some kinetochore microtubules originate at the kinetochore

Search and capture is not the only mechanism of chromosome attachment; indeed, this mechanism is unlikely to be effective in cells lacking centrosomes. An alternative attachment mechanism, which can be used in the presence or absence of centrosomes, depends on the

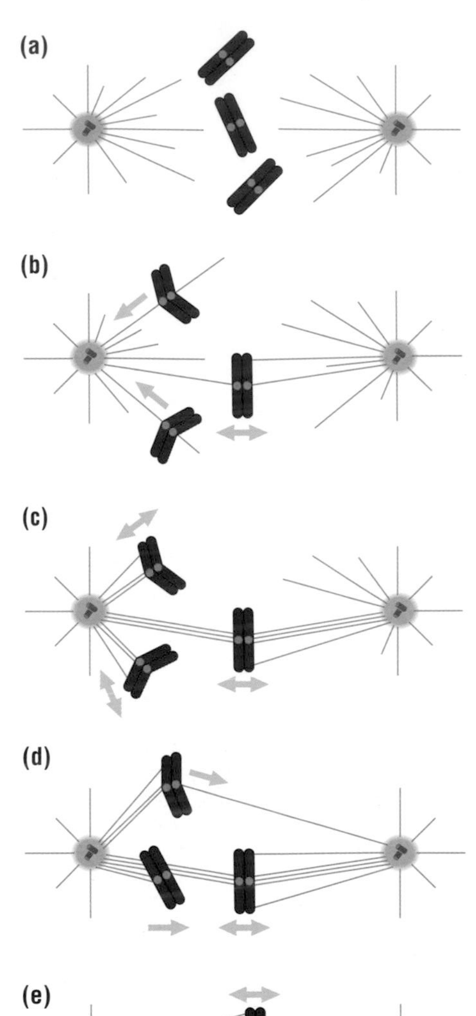

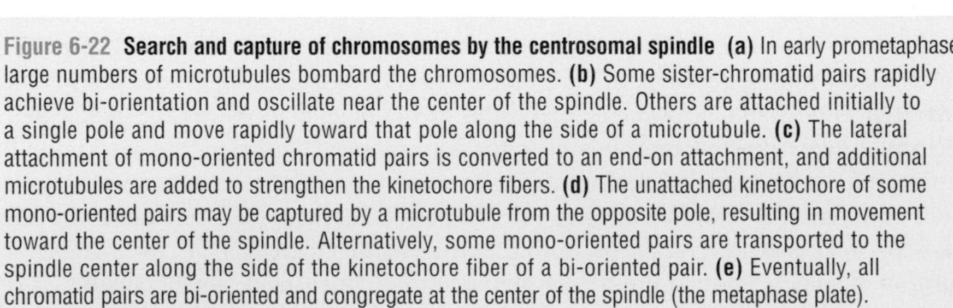

Figure 6-22 **Search and capture of chromosomes by the centrosomal spindle** **(a)** In early prometaphase, large numbers of microtubules bombard the chromosomes. **(b)** Some sister-chromatid pairs rapidly achieve bi-orientation and oscillate near the center of the spindle. Others are attached initially to a single pole and move rapidly toward that pole along the side of a microtubule. **(c)** The lateral attachment of mono-oriented chromatid pairs is converted to an end-on attachment, and additional microtubules are added to strengthen the kinetochore fibers. **(d)** The unattached kinetochore of some mono-oriented pairs may be captured by a microtubule from the opposite pole, resulting in movement toward the center of the spindle. Alternatively, some mono-oriented pairs are transported to the spindle center along the side of the kinetochore fiber of a bi-oriented pair. **(e)** Eventually, all chromatid pairs are bi-oriented and congregate at the center of the spindle (the metaphase plate).

Definitions

chromosome congression: (in animal cells) the alignment of sister-chromatid pairs at the center of the spindle in metaphase.

kinetochore fiber: bundle of microtubules that links a chromatid kinetochore to a spindle pole.

References

Gadde, S. and Heald, R.: **Mechanisms and molecules of the mitotic spindle.** *Curr. Biol.* 2004, **14**:R797–R805.

Kapoor, T.M. *et al.*: **Chromosomes can congress to the metaphase plate before biorientation.** *Science* 2006, **311**:388–391.

Maiato, H. *et al.*: **Kinetochore-driven formation of kinetochore fibers contributes to spindle assembly during animal mitosis.** *J. Cell Biol.* 2004, **167**:831–840.

Rieder, C.L. and Alexander, S.P.: **Kinetochores are transported poleward along a single astral microtubule during chromosome attachment to the spindle in newt lung cells.** *J. Cell Biol.* 1990, **110**:81–95.

Tulu, U.S. *et al.*: **Peripheral, non-centrosome-associated microtubules contribute to spindle formation in centrosome-containing cells.** *Curr. Biol.* 2003, **13**:1894–1899.

Wadsworth, P. and Khodjakov, A.: ***E pluribus unum*: towards a universal mechanism for spindle assembly.** *Trends Cell Biol.* 2004, **14**:413–419.

Waters, J.C. *et al.*: **Localization of Mad2 to kinetochores depends on microtubule attachment, not tension.** *J. Cell Biol.* 1998, **141**:1181–1191.

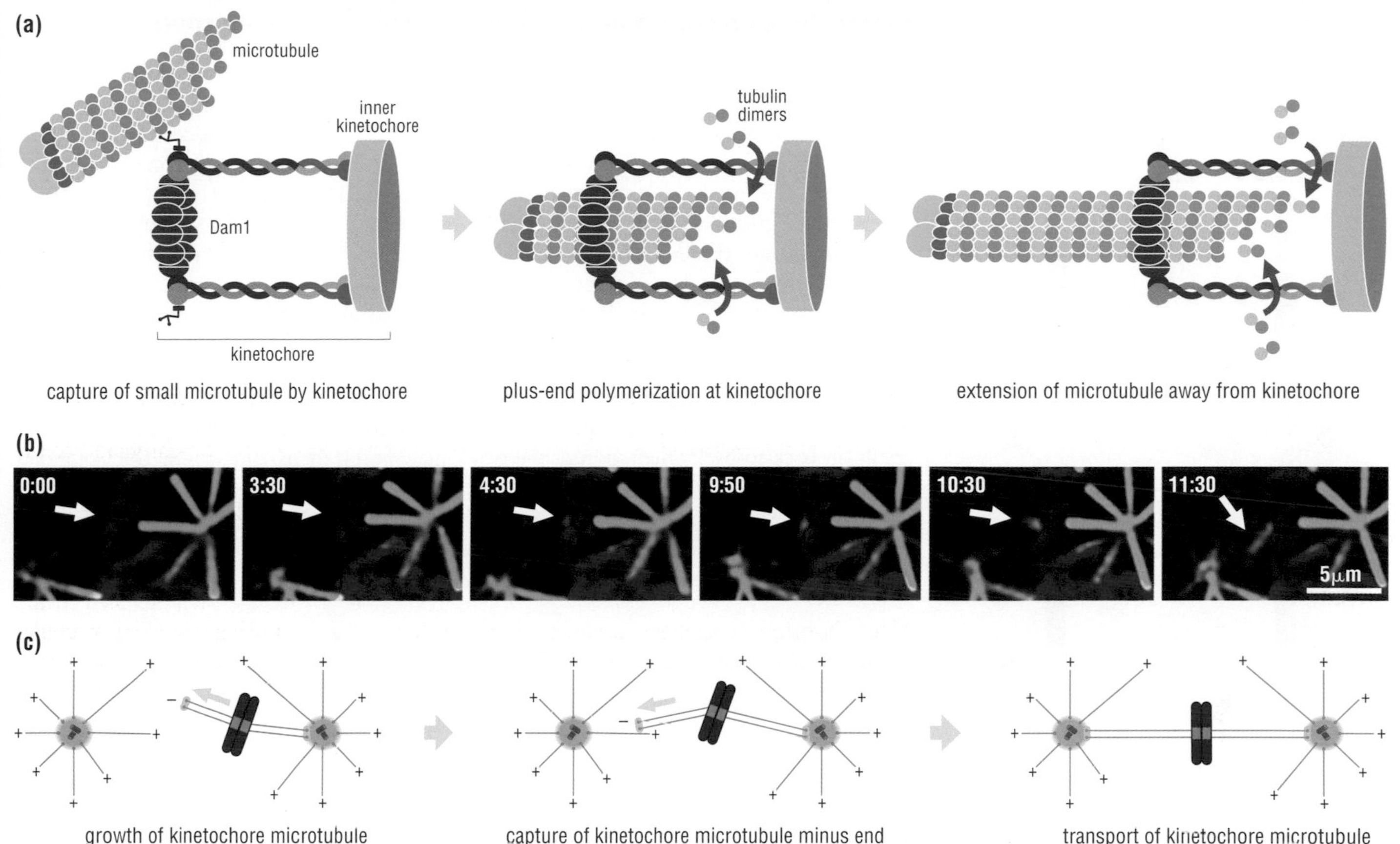

growth of microtubules at the kinetochore (Figure 6-23a and b). Short microtubules, probably nucleated by soluble γ-tubulin ring complexes, form near the chromosomes and are captured by dynein or other proteins at the unattached kinetochore. The plus ends are then embedded end-on in the kinetochore. Regulatory proteins associate with the microtubule plus end and stimulate polymerization, resulting in extension of the microtubule away from the kinetochore.

Several microtubule minus ends from the same kinetochore are then bundled together by minus-end-associated motors and cross-linkers like those involved in spindle pole focusing (Figure 6-23c). These minus-end proteins eventually associate with the side of a microtubule that originates in the spindle pole, and minus-end-directed motors then carry the kinetochore microtubule towards the pole, to which its minus ends become attached. Although centrosomes are not essential for attachment of kinetochore-derived microtubules to the pole, they are likely to facilitate it. Microtubules radiating from the centrosome can capture bundled kinetochore microtubule minus ends in much the same way as they capture kinetochores (see Figure 6-23c).

Figure 6-23 Kinetochore-derived microtubule formation (a) Microtubule formation at a kinetochore is thought to begin when a short microtubule is captured by proteins at the kinetochore. After assuming the correct end-on attachment, plus-end polymerization leads to growth of the microtubule away from the kinetochore. **(b)** Microtubule growth at the kinetochore can be seen in these video images of a centrosome-containing *Drosophila* cell in prometaphase; microtubules are green, chromosomal DNA red, and the numbers indicate time (minutes:seconds). In the first two images, a mono-oriented chromosome (arrow) is attached to the right-hand spindle pole but not to the other spindle pole at lower left. At time 4:30, a short microtubule appears at the unattached kinetochore. The microtubule grows toward the upper left until the last time point, when it suddenly becomes oriented toward the spindle pole at lower left, presumably because it has been captured by a microtubule from the pole. **(c)** These diagrams illustrate the behavior of the kinetochore fiber at the later time points in panel (b). Attachment of the kinetochore microtubule bundle to the spindle pole occurs when a polar microtubule captures the bundled minus ends of the kinetochore microtubules and reels them in to the pole, using a minus-end-directed motor protein such as dynein. This mechanism does not depend on the presence of centrosomes at the spindle poles and is likely to be the primary chromosome attachment mechanism in acentrosomal spindles. Panel (b) kindly provided by Helder Maiato and Alexey Khodjakov. From Maiato, H. *et al.*: *J. Cell Biol.* 2004, **167**:831–840.

Chromosome attachment results in tension between sister kinetochores

When a sister-chromatid pair is correctly attached to opposite spindle poles, poleward forces at the kinetochores pull the sisters in opposite directions, resulting in tension in the protein and DNA structures between them. This tension is revealed by microscopic analysis of the distance between sister kinetochores, which increases dramatically when bi-orientation is achieved. This local separation of the kinetochores is limited to a short region of the chromosome near the kinetochore and occurs despite the presence of abundant cohesin linkages at the centromere (see section 5-8). Centromeric chromatin may have unique elastic properties that allow kinetochore separation to occur without compromising sister-chromatid cohesion.

Sister-chromatid pairs sometimes become incorrectly attached to the spindle: both kinetochores can be attached to the same pole, for example, or one kinetochore can attach to both poles. In section 6-10 we describe the elegant mechanisms that are used to correct these errors.

6-10 Bi-Orientation of Sister Chromatids

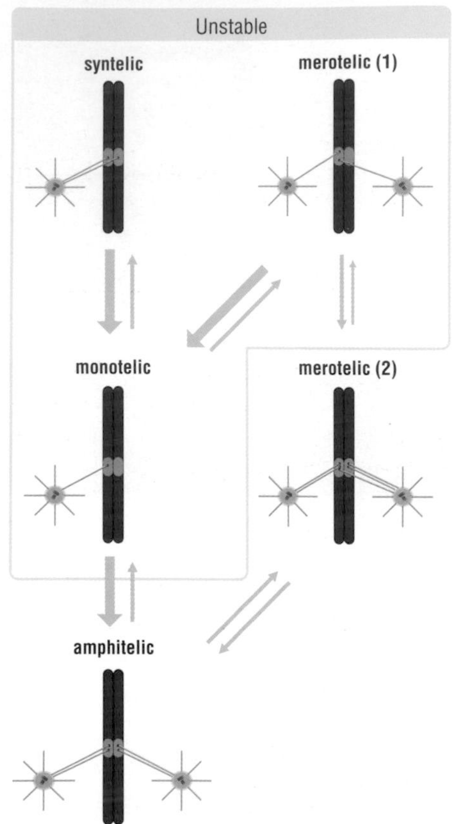

Figure 6-24 Stable and unstable chromosome attachments In the early stages of spindle assembly, unstable chromosome attachments like those shown here can occur. Monotelic attachment, although unstable on its own, is the first step to correct bipolar attachment. The strength of attachment is governed in part by kinetochore tension. Some incorrect attachments (syntelic and the simple merotelic attachment at top right) do not result in significant tension and are thus unstable, so that microtubule–kinetochore bonds dissociate rapidly. When amphitelic attachment is achieved, tension across the kinetochores stimulates microtubule attachment, thereby locking the correct orientation in place. Merotelic attachments involving both sister chromatids (center right) may also cause kinetochore tension, resulting in the stabilization of attachments. These merotelic attachments are corrected as shown in Figure 6-26.

Kinetochore–microtubule attachment is stabilized by tension

The success of chromosome segregation requires that each sister kinetochore in a pair is attached to the opposite spindle pole. Although incorrect attachments do sometimes occur, most are switched to the correct one before anaphase begins. When chromosome attachment begins in prometaphase, sister-chromatid pairs tend to start out, if only briefly, with one kinetochore attached to one spindle pole, a state known as **monotelic attachment** (Figure 6-24). Ideally, the second kinetochore then becomes attached to the opposite pole, resulting in **bi-orientation** or **amphitelic attachment**. The second attachment is not always the right one, however: both sister kinetochores can attach to the same spindle pole (**syntelic attachment**), or one kinetochore can attach to both spindle poles (**merotelic attachment**), which is possible only when kinetochores contain multiple microtubule-attachment sites.

Incorrect attachments are prevented to some degree by the geometry of sister kinetochores. The centromeric regions of the paired sisters are assembled into a heterochromatic structure that results in back-to-back orientation of the two kinetochores, thereby increasing the likelihood that the sisters will attach to opposite poles. Unattached kinetochores are, however, large crescent-shaped structures that extend slightly around the sides of the chromatid. It is therefore possible, if unlikely, for microtubules from opposite poles to contact binding sites in the same kinetochore, or for both kinetochores to attach to the same pole.

When incorrect attachments occur, they are corrected by a form of positive reinforcement. Incorrect attachments are thought to be weak and easily reversed, whereas proper amphitelic attachment triggers an increase in the strength of the kinetochore–microtubule bond. Bi-orientation is thereby locked in place once it occurs.

How is correct attachment detected by the kinetochore? The answer probably lies in the fact that only bi-orientation results in the generation of tension at kinetochores, as discussed in section 6-9. By mechanisms that remain unclear, it is thought that components of the kinetochore, possibly within each microtubule-attachment site, can detect the degree of tension. At low levels of tension, signals are generated to destabilize microtubule attachment; high tension shuts off these signals to stabilize attachment.

Aurora B is required for the correction of syntelic attachments

The mitotic protein kinase aurora B (see section 5-7) is an essential component of the regulatory system that governs chromosome attachments. In budding yeast and in human cells, inhibition of aurora B activity results in the accumulation of incorrect syntelic attachments (Figure 6-25), and reactivation of the kinase in these experiments leads to the rapid correction of these attachments to the amphitelic form.

How might aurora B promote the correction of improper attachments? The most likely possibility, based on numerous lines of evidence, is that aurora B destabilizes microtubule binding to kinetochores that are not under tension. Aurora B, together with its associated regulatory subunits, is located at the kinetochore and phosphorylates numerous kinetochore components, including subunits of the Dam1 and Ndc80 complexes (see Figure 6-14). In yeast, these phosphorylations decrease the binding of Dam1 complexes to Ndc80 without affecting Ndc80 localization to the kinetochore. An attractive possibility, therefore, is that in the absence of tension, aurora B reduces the affinity of microtubule binding by phosphorylating key components of the microtubule-attachment site. When bi-orientation results in tension across

Definitions

amphitelic attachment: see **bi-orientation**.

bi-orientation: correct bipolar attachment of a sister-chromatid pair to the spindle, with the two kinetochores attached to opposite poles.

merotelic attachment: incorrect attachment of a sister-chromatid pair to the spindle, with one kinetochore attached to both poles. Merotelic attachment is often seen in a form where one sister is attached to both poles and the other sister to just one, resulting in an incorrect form of **bi-orientation**.

monotelic attachment: attachment of a sister-chromatid pair to the spindle, with one kinetochore attached to one pole and the other kinetochore unattached.

syntelic attachment: incorrect attachment of a sister-chromatid pair to the spindle, with both kinetochores attached to the same pole.

References

Andrews, P.D. *et al.*: **Aurora B regulates MCAK at the mitotic centromere.** *Dev. Cell* 2004, **6**:253–268.

Cheeseman, I.M. *et al.*: **Phospho-regulation of kinetochore-microtubule attachments by the Aurora kinase Ipl1p.** *Cell* 2002, **111**:163–172.

Cimini, D. *et al.*: **Anaphase spindle mechanics prevent mis-segregation of merotelically oriented chromosomes.** *Curr. Biol.* 2004, **14**:2149–2155.

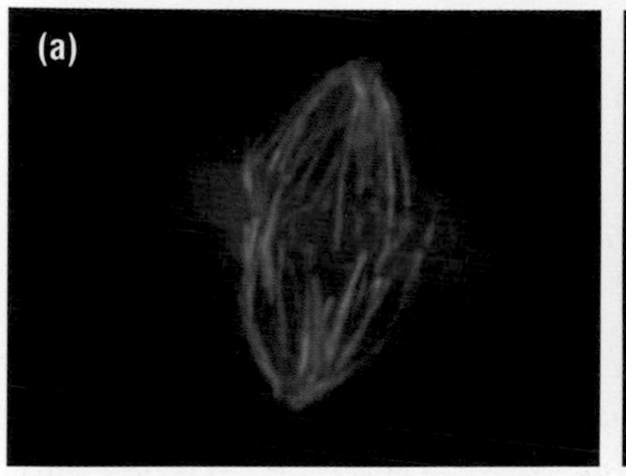

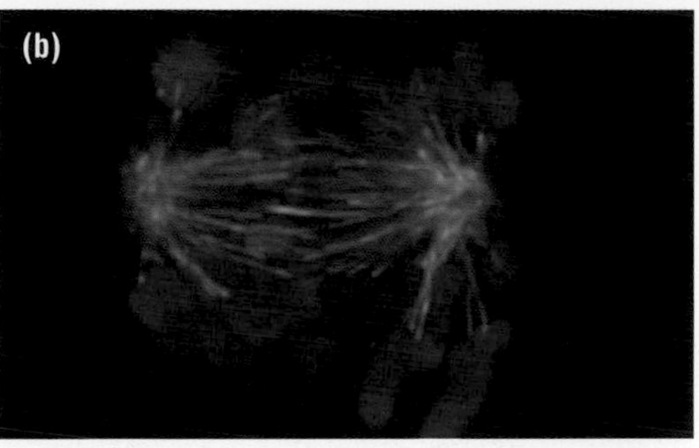

Figure 6-25 Accumulation of syntelic attachments in the absence of aurora B kinase activity (a) Vertebrate cells in culture were treated with a drug that inhibits centrosome separation, resulting in a monopolar spindle in which all sister-chromatid pairs are syntelically attached to the pole. The drug was then removed, leading to the rapid correction of attachments and the formation of the normal spindle shown here. Microtubules are green and chromosomes are blue. **(b)** If spindle assembly is initiated in the presence of a drug that inhibits aurora B, the result is a bipolar spindle in which most of the sister-chromatid pairs are syntelically attached to one pole. Aurora B activity is therefore required for the correction of syntelic attachments. The drug used in these experiments also inhibits aurora A, but other evidence indicates that inhibition of this kinase is unlikely to affect chromosome attachments. Kindly provided by Michael A. Lampson and Tarun M. Kapoor. From Lampson, M.A. *et al.*: *Nat. Cell Biol.* 2004, **6**:232–237, with permission.

the kinetochore, phosphorylation of these components is reversed and attachment is stabilized. We know little about the kinetochore tension sensor and how it influences aurora B function.

Aurora B also has other effects on incorrect syntelic attachments. In cultured vertebrate cells, it stimulates movement of syntelic sister-chromatid pairs to the spindle pole before their weak microtubule attachments are rearranged to the correct amphitelic form. Aurora B may promote poleward kinetochore movement by stimulating plus-end depolymerization at the kinetochore, which is known to produce a poleward force, as described later in this chapter. Aurora B may act in part by phosphorylating the kinesin-13 protein MCAK, a catastrophe factor that helps drive plus-end depolymerization at kinetochores (see Figure 6-6).

Merotelic attachments are processed by multiple mechanisms

Most merotelic attachments involve both sister kinetochores, such that one is attached to two poles while the other is attached to just one (Figure 6-26). These attachments are likely to generate kinetochore tension, resulting in stable microtubule attachments. It is therefore unclear how these incorrect attachments are sensed or corrected.

In a merotelic attachment with two chromatids, two of the attachments—the two that connect each sister to the opposite pole—are correct, and only one attachment needs to be removed to generate a correct amphitelic attachment. In most cases, the two correctly attached fibers contain more microtubules than the one incorrectly attached fiber, indicating that some mechanism—perhaps depending on tension—increases the microtubule content of the correctly attached fibers. In some cells, the single incorrect attachment is lost before anaphase. The mechanisms that trigger detachment are not clear, although the presence of fewer microtubules in the incorrect fiber may mean that it is less stable and more likely to detach spontaneously. Because of the back-to-back geometry of the kinetochores, the incorrectly attached microtubule is essentially coming from behind and may be unable to make a strong interaction with the side of the kinetochore. Detachment mechanisms requiring aurora B and plus-end depolymerization by kinesin-13 may also be involved.

Some merotelic attachments involving both chromosomes are left uncorrected when the cell enters anaphase (see Figure 6-26). When the sisters separate, the sister that is attached only to one pole moves correctly toward that pole, while the other (merotelic) sister moves correctly to the other pole along the thicker kinetochore fiber. The incorrect fiber remains attached but simply extends to accompany the sister to the opposite pole. This behavior may depend on a greater poleward force generated by the multiple microtubule attachment sites of the correct fiber. These forces are described in sections 6-11 and 6-12.

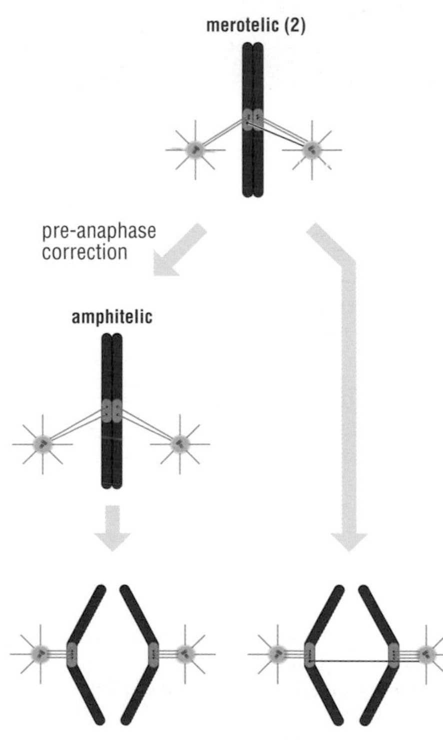

Figure 6-26 Processing of merotelic attachments As shown on the left, merotelic attachments involving both sister chromatids are sometimes corrected before anaphase by the loss of the incorrectly attached fiber (pink), perhaps because this fiber is less stable than the correctly attached fiber. In some cases, as shown on the right, anaphase occurs before correction, but the merotelically attached sister still tends to go to the correct pole because poleward forces at the correct attachment site (which usually contains more microtubules) are greater than those at the incorrect attachment site.

Hauf, S. and Watanabe, Y.: **Kinetochore orientation in mitosis and meiosis.** *Cell* 2004, **119**:317–327.

Hauf, S. *et al.*: **The small molecule Hesperadin reveals a role for Aurora B in correcting kinetochore-microtubule attachment and in maintaining the spindle assembly checkpoint.** *J. Cell Biol.* 2003, **161**:281–294.

Lampson, M.A. *et al.*: **Correcting improper chromosome-spindle attachments during cell division.** *Nat. Cell Biol.* 2004, **6**:232–237.

Lan, W. *et al.*: **Aurora B phosphorylates centromeric MCAK and regulates its localization and microtubule depolymerization activity.** *Curr. Biol.* 2004, **14**:273–286.

Shang, C. *et al.*: **Kinetochore protein interactions and their regulation by the Aurora kinase Ipl1p.** *Mol. Biol. Cell* 2003, **14**:3342–3355.

Tanaka, T.U. *et al.*: **Evidence that the Ipl1-Sli15 (Aurora kinase-INCENP) complex promotes chromosome bi-orientation by altering kinetochore-spindle pole connections.** *Cell* 2002, **108**:317–329.

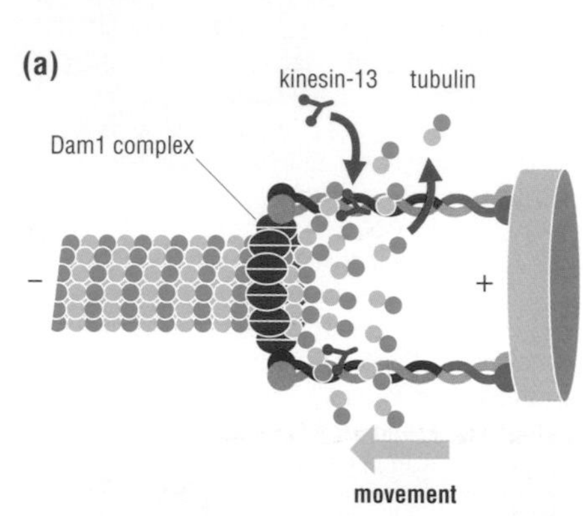

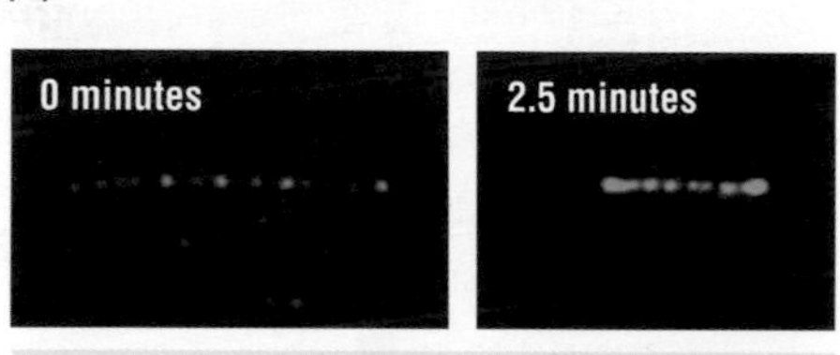

Figure 6-27 Poleward force generation by the kinetochore **(a)** Poleward force is generated by microtubule plus-end depolymerization at the kinetochore. A likely mechanism of force generation is that the outward curling of the depolymerizing microtubule protofilaments pushes against the microtubule collar provided by the Dam1 complex (see Figure 6-14). Depolymerization is stimulated by MCAK and other members of the kinesin-13 family. **(b)** In this experiment, purified microtubules were coated in fluorescently labeled Dam1 ring complexes, which are distributed randomly as green spots along the microtubule (left image). Brief treatment of these microtubules with purified MCAK caused depolymerization at both ends (right image). Dam1 proteins accumulated at the ends of the shortened microtubule, suggesting that depolymerization pushed the ring complexes inward from the ends. Panel (b) kindly provided by Stefan Westermann and Georjana Barnes. From Westermann, S. *et al.*: *Mol. Cell* 2005, **17**:277–290.

Multiple forces act on chromosomes in the spindle

Three major forces act on chromosomes to move them on the animal spindle in metaphase and anaphase. The first is a poleward force generated by the kinetochore, which pulls the chromosome along the microtubule track toward the spindle pole. This force helps drive the oscillations of sister-chromatid pairs during metaphase, and also provides much of the force that moves separated sisters toward the spindle poles in anaphase. A second major poleward force is generated by *microtubule flux*, whereby the microtubule tracks are dismantled at their minus ends, pulling the microtubule track itself and its attached chromatid towards the pole. This force is responsible for generating much of the tension between sister kinetochores in metaphase, and also helps move the separated sisters poleward in anaphase. The third force is the *polar ejection force*, which is generated by non-kinetochore microtubules and pushes chromosome arms away from the spindle poles. This force helps align sister-chromatid pairs at the spindle midzone in metaphase and is inactivated in anaphase.

The kinetochore is a major source of poleward force

Most of the poleward force generated by the kinetochore is not due to conventional motor proteins but to the process of plus-end depolymerization. The mechanism of force generation is not yet fully understood, but one attractive possibility comes from evidence of a collar at the kinetochore that holds the microtubule in place while still allowing plus-end polymerization and depolymerization (see section 6-5). When depolymerization occurs, the outward curling of the microtubule protofilaments could generate a force that pushes against the collar, thereby pushing the collar—and its attached kinetochore—toward the spindle pole (Figure 6-27a). Support for this proposal comes from studies of the yeast Dam1 complex, which is thought to form the microtubule collar, as Dam1 rings are pushed away from the ends of depolymerizing microtubules *in vitro* (Figure 6-27b).

Plus-end depolymerization—and therefore poleward force—is governed by numerous regulatory proteins. The most important are catastrophe factors of the kinesin-13 family, including MCAK (see section 6-10).

Motor proteins are thought to make only minor contributions to poleward forces at the kinetochore in metaphase and anaphase. In prometaphase, the kinetochore contains the minus-end-directed dynein motor, which contributes to the poleward movement of mono-oriented chromosomes along the sides of microtubules (see Figure 6-22). Dynein is lost from kinetochores after bi-orientation occurs, and in vertebrate cells inhibition of dynein has little effect on chromosome movements in metaphase or anaphase. Kinetochores also contain at least one other motor protein, CENP-E, a kinesin-7 that is plus-end directed and therefore suited for moving kinetochores away from the pole. However, kinetochores do not contribute significantly to the movement of bi-oriented chromatid pairs away from the pole. As mentioned earlier (see sections 6-5 and 6-9), CENP-E seems to be involved in microtubule attachment and congression of mono-oriented chromatid pairs in prometaphase.

Microtubule flux generates poleward force

In animal cells, but probably not in yeast, another poleward force is superimposed on the kinetochore-based forces. This force is generated by **microtubule flux**, which depends on the active transport of the entire microtubule toward the spindle pole, where it is depolymerized at its minus end. In metaphase, loss of tubulin at the minus end is balanced by polymerization

Definitions

microtubule flux: flow of tubulin within microtubules from the spindle midzone to the poles, as a result of loss of tubulin from the minus ends at the poles. It is usually accompanied by addition of tubulin at the plus ends.

polar ejection force: the repulsive effect of microtubules growing out from the spindle poles and interacting with chromosome arms. It tends to force chromosome arms away from the spindle poles.

References

Kapoor, T.M. and Compton, D.A.: **Searching for the middle ground: mechanisms of chromosome alignment during mitosis.** *J. Cell Biol.* 2002, **157**:551–556.

Maddox, P. *et al.*: **Direct observation of microtubule dynamics at kinetochores in *Xenopus* extract spindles: implications for spindle mechanics.** *J. Cell Biol.* 2003, **162**:377–382.

Maiato, H. *et al.*: ***Drosophila* CLASP is required for the incorporation of microtubule subunits into fluxing kinetochore fibres.** *Nat. Cell Biol.* 2005, **7**:42–47.

Miyamoto, D.T. *et al.*: **The kinesin Eg5 drives poleward microtubule flux in *Xenopus laevis* egg extract spindles.** *J. Cell Biol.* 2004, **167**:813–818.

Salmon, E.D.: **Microtubules: a ring for the depolymerization motor.** *Curr. Biol.* 2005, **15**:R299–R302.

Westermann, S. *et al.*: **Formation of a dynamic kinetochore–microtubule interface through assembly of the Dam1 ring complex.** *Mol. Cell* 2005, **17**:277–290.

of new tubulin subunits at the plus end; thus microtubule length remains relatively constant. However, when a fraction of the tubulin subunits in a cell are labeled with fluorescent tags, they can be observed in the microscope to move from the spindle midzone to the poles (Figure 6-28a).

We do not fully understand how microtubules are pulled toward the poles to drive flux. Motor proteins of the kinesin-5 family help drive interpolar microtubules toward the spindle poles, probably by cross-linking antiparallel microtubules in the spindle midzone (see Figure 6-7). These motors may also transport kinetochore microtubules poleward, but the mechanism is unclear. One intriguing possibility is that flux is not driven entirely by conventional motor proteins. Instead, minus-end depolymerization by kinesin-13 proteins at the spindle pole may pull the microtubule poleward by pushing against a collar around the minus end, much like the mechanism underlying poleward force generation at the kinetochore.

Microtubule flux pulls the kinetochore poleward because the microtubule is attached to the kinetochore in a way that resists that force. In other words, a moving microtubule could not pull a kinetochore poleward if that microtubule simply slid through the microtubule-attachment site without resistance. We can illustrate this concept by considering the metaphase spindles of *Xenopus* embryonic cells (Figure 6-28b). The kinetochore microtubule plus ends in these spindles are usually in a polymerizing state that results in tubulin addition at a rate that matches the rate of tubulin depolymerization at the minus end. Kinetochores are therefore relatively stationary in these cells during metaphase, but microtubule flux is nevertheless generating a considerable pulling force on the kinetochores, resulting in the high tension that is such an important indicator of bi-orientation. The generation of this pulling force by flux depends on friction or resistive force in the microtubule-attachment sites of the kinetochore. Movement of the microtubule through the Dam1 ring, for example, may be restricted by proteins that bind the sides of the plus end.

When sister chromatids separate in anaphase, the amount of tension in the kinetochores drops and the rate of plus-end polymerization decreases below that of minus-end depolymerization. The resistive friction in the attachment site then allows flux to move the chromosomes toward the pole. Under these low-tension conditions, the kinetochore eventually switches to a depolymerizing state, in which the outward curling of the plus end prevents the microtubule from being pulled out of the attachment site. Chromosome movement then depends on the combined effects of microtubule flux and depolymerization at the plus ends (see Figure 6-28b).

A polar ejection force is generated by chromosome arms

Whereas the kinetochore tends to be pulled toward the spindle pole in metaphase, chromosome arms tend to be pushed away by a force called the **polar ejection force** or polar wind. When chromosomes are close to the spindle pole, some of the polar ejection force may result simply from the dense barrage of growing microtubule plus ends pushing everything in their path away from the pole. However, studies in many vertebrate cell types suggest that most of the polar ejection force is generated by Kid, a plus-end-directed chromokinesin of the kinesin-10 family, which links non-kinetochore microtubules to chromosome arms. Kid is destroyed in anaphase through ubiquitination by APC^{Cdc20} (see section 3-10), thereby removing the polar ejection force and allowing separated sisters to move unopposed toward the poles.

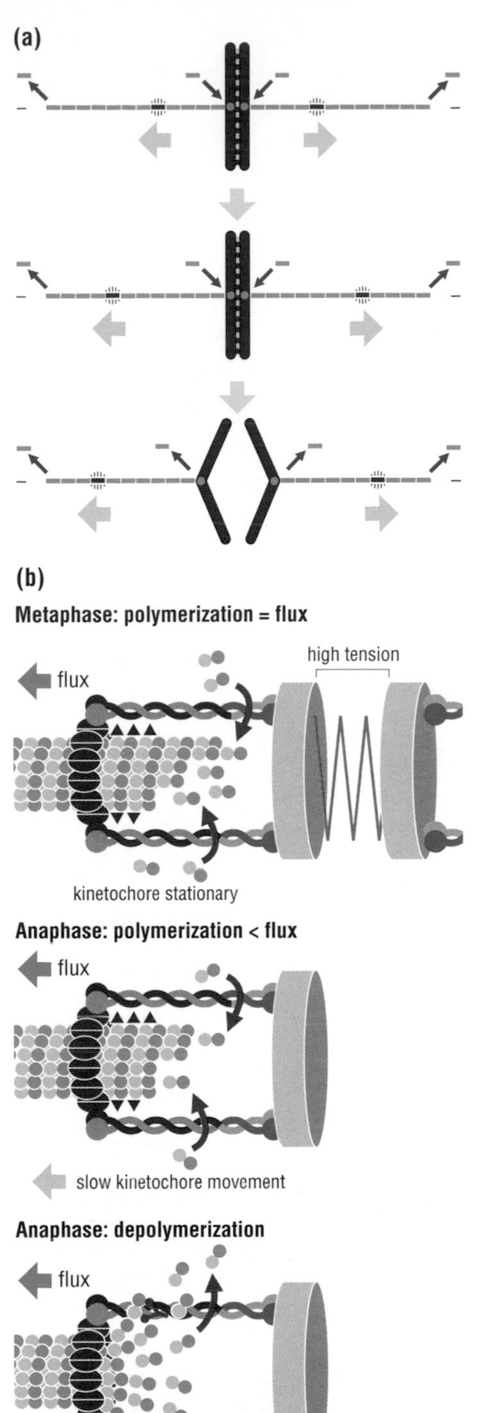

Figure 6-28 Microtubule flux in metaphase and anaphase (a) In this simplified view, addition of tubulin (green boxes) at the microtubule plus end is accompanied by dissociation of tubulin from the minus end. If cells are engineered to contain low concentrations of fluorescently tagged tubulin (red), the tagged proteins are incorporated into spindle microtubules and can be seen moving poleward, revealing that the entire microtubule is moving toward the pole. After sister-chromatid separation in anaphase (bottom panel), depolymerization at the kinetochore (see Figure 6-27) cooperates with flux to generate rapid poleward movement. In embryonic cells of *Xenopus* and *Drosophila*, flux is rapid and makes a major contribution to anaphase chromosome movement. In vertebrate somatic cells and in yeast, flux is relatively slow or nonexistent, and plus-end depolymerization at the kinetochore is the major determinant of poleward movement in anaphase. **(b)** Kinetochore behavior in mitotic spindles of *Xenopus* embryonic cells, where flux is particularly rapid. In metaphase, microtubule flux toward the pole does not move the kinetochore appreciably because plus-end polymerization equals minus-end depolymerization. Nevertheless, there is considerable tension at the kinetochore because poleward microtubule movement is resisted by friction in the kinetochore attachment site, perhaps because regulatory proteins bind the sides of the plus end as shown here (pink triangles). At the beginning of anaphase, the rate of polymerization decreases, resulting in a net movement of the kinetochore that is equal to the rate of flux minus the rate of polymerization. The kinetochore microtubule plus end typically switches to a depolymerization state in anaphase, and the kinetochore then moves more rapidly poleward—at a rate equal to the sum of flux and depolymerization. Panel (b) adapted from Maddox, P. *et al.*: *J. Cell Biol.* 2003, **162**:377–382.

Chromosome oscillations in prometaphase are generated by changes in the state of kinetochores

In vertebrate somatic cells, the initial attachment of sister chromatids to the spindle in prometaphase results in dramatic oscillations of the chromatids toward and away from the spindle pole. During these oscillations, chromosomes tend to spend more time moving away from the poles than toward them, contributing to chromosome congression. How is this biased movement toward the spindle equator achieved?

Numerous lines of evidence, particularly from studies of mono-oriented prometaphase chromosomes, suggest that kinetochores—like microtubules—shift rapidly between two distinct states: the depolymerization state, a state of poleward force generation in which the kinetochore pulls the chromatid pair toward the pole; and the polymerization state, in which the plus end is polymerizing and no active force generation is occurring at the kinetochore (Figure 6-29). When a sister-chromatid pair is attached to only one pole, the shift to the polymerization state allows the polar ejection force to push the chromosome away from the pole. As described in section 6-11, the microtubule attachment site generates friction that resists movement of the kinetochore along the microtubule. Thus, the polymerization state is also called the resistive state.

In some cell types, bi-oriented sister-chromatid pairs also exhibit oscillatory behavior in prometaphase and metaphase. Here again, oscillations are thought to result when kinetochores switch between polymerizing and depolymerizing states, with the added complexity that the behavior of the two sister kinetochores must somehow be coordinated. When a bi-oriented chromatid pair is moving toward one pole, the leading kinetochore is in the poleward force-generating state. The lagging kinetochore is usually in the polymerization state, pulled by its sister and pushed from behind by polar ejection forces.

The switch between the two kinetochore states is thought to be controlled, at least in part, by the level of tension within microtubule-attachment sites at the kinetochore. Kinetochores tend to switch to the resistive polymerization state when experiencing high tension and then switch to poleward force generation when tension is reduced. When a mono-oriented chromatid pair is approaching a spindle pole, for example, the attached kinetochore experiences increasing tension as the chromosome arms encounter the opposing polar ejection force. This kinetochore then switches to the polymerization state, allowing the ejection force to move the chromatid pair away from the pole.

One can imagine how tension-regulated switching mechanisms could promote chromosome congression (Figure 6-30). When a bi-oriented chromatid pair is moving toward one pole, for example, the leading kinetochore switches to polymerization as it meets increasing polar ejection forces. The lagging kinetochore then switches to poleward force generation and pulls

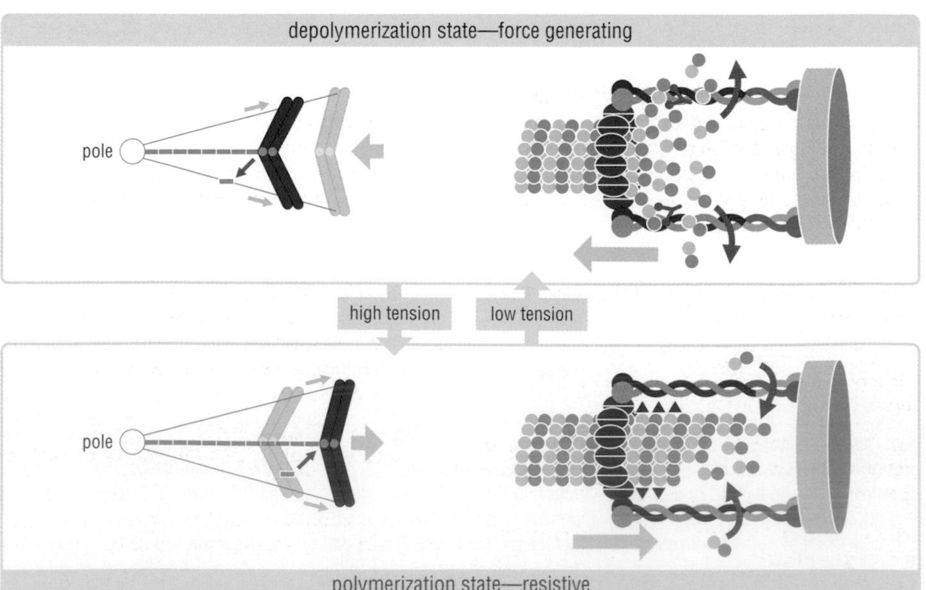

Figure 6-29 Changes in the state of kinetochores cause oscillations of chromatid pairs on the spindle A mono-oriented chromatid pair oscillates near a spindle pole in prometaphase as a result of switches in the behavior of the kinetochore. When the kinetochore is in the poleward force-generating state, the chromatids move toward the pole. Chromosome arms are pushed away from the pole by polar ejection forces (blue arrows), causing tension at the kinetochore. High tension is thought to switch the kinetochore to a polymerization or resistive state, in which it no longer generates force but moves away from the spindle pole as a result of polar ejection forces. In the diagrams on the left, the kinetochore fiber is shown as a thick green line, with tubulin dissociation and binding at the kinetochore coupled with movement. Non-kinetochore microtubules producing the polar ejection force are shown as thin green lines. For simplicity, microtubule flux is not included in this scheme. If present, it would exert a poleward force that would reduce the effects of the polar ejection force.

Figure 6-30 **Polar ejection forces in chromosome congression (a)** A bi-oriented chromatid pair moves toward one spindle pole, its leading kinetochore in the poleward force-generating state. **(b)** Polar ejection forces (blue arrows) near the pole trigger the switch to polymerization in the leading kinetochore. The lagging kinetochore switches to the poleward state, pulling the chromo-somes back toward the other pole. **(c)** When the chromosomes are near the opposite pole, polar ejection forces again increase tension and trigger the switch in kinetochore behaviors. **(d)** Eventually, chromosomes tend to gather at the center of the spindle, where polar ejection forces are equal and minimal. For simplicity, microtubule flux is not shown.

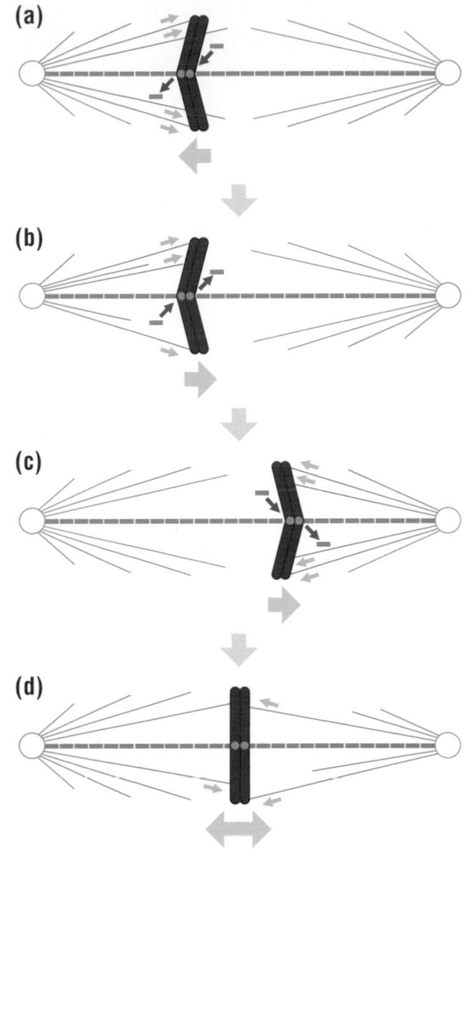

the chromatids toward the opposite pole, where polar ejection forces again increase tension and trigger the shift to polymerization. Repeated episodes of this behavior should lead to the accumulation of sister chromatids at the spindle equator, where polar ejection forces are equal and minimal. Although appealing in its simplicity, this model is not supported fully by studies of Kid, the chromokinesin (kinesin-10) that generates most of the polar ejection force (see section 6-11). Inhibition of Kid abolishes chromosome oscillations in prometaphase but does not prevent chromosome congression. Thus, although it remains likely that polar ejection forces contribute to congression, they are not strictly essential. Other mechanisms—involving a balance of poleward forces—are likely to be more important.

Microtubule flux may promote chromosome congression

Kinetochore force generation and microtubule flux, the two main mechanisms for generating poleward forces, are likely to influence chromosome congression. At present, however, there is no clear understanding of how they might do this. One proposal is that the poleward force generated at the kinetochore by microtubule flux is proportional to the length of the kinetochore fiber. If this were the case, then the flux-dependent forces pulling sister kinetochores toward opposite poles would be maximal and equal at the spindle equator, resulting in chromosome congression. Our knowledge of the mechanisms that generate microtubule flux is limited, however, and it is not clear that the force generated by flux is greater for longer microtubules. Nevertheless, it remains probable that congression is driven by multiple force gradients—involving both flux and polar ejection, for example (Figure 6-31). It is also possible that kinetochores can somehow sense their position on the spindle and are able to maximize poleward force generation—by increasing the polymerization rate, for example—when they are positioned at the middle. Such mechanisms could result from gradients of microtubule-regulatory proteins in the spindle.

At metaphase, spindle assembly is complete. Sister chromatids, pulled in opposite directions by poleward forces, await the signal to separate. The molecular basis and regulation of this anaphase signal will be discussed in Chapter 7.

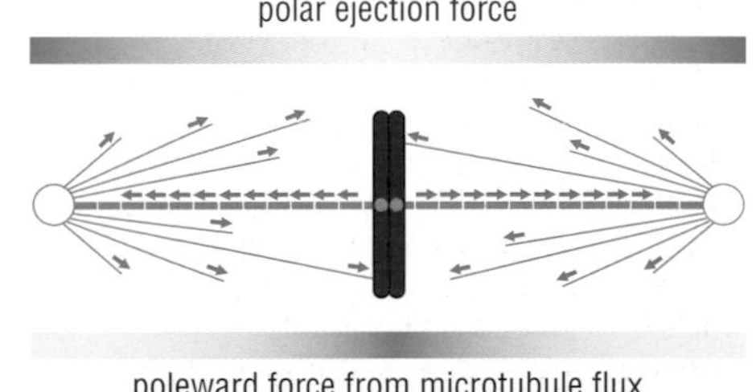

Figure 6-31 **Force gradients driving chromosome congression** One proposed mechanism for chromosome congression is that it results from a balance of polar ejection forces, which are maximal at the poles, and poleward forces driven by microtubule flux, which are proportional to microtubule length and therefore minimal near the poles. Chromosomes congregate at the spindle center, where these two forces act equally on the two sister chromatids. Adapted from Kapoor, T.M. and Compton, D.A.: *J. Cell Biol.* 2002, **157**:551–556.

References

Cimini, D. *et al.*: **Anaphase spindle mechanics prevent mis-segregation of merotelically oriented chromosomes.** *Curr. Biol.* 2004, **14**:2149–2155.

Hays, T.S. *et al.*: **Traction force on a kinetochore at metaphase acts as a linear function of kinetochore fiber length.** *J. Cell Biol.* 1982, **93**:374–389.

Kapoor, T.M. and Compton, D.A.: **Searching for the middle ground: mechanisms of chromosome alignment during mitosis.** *J. Cell Biol.* 2002, **157**:551–556.

Maddox, P. *et al.*: **Direct observation of microtubule dynamics at kinetochores in *Xenopus* extract spindles: implications for spindle mechanics.** *J. Cell Biol.* 2003, **162**:377–382.

Ostergren, G.: **The mechanism of co-orientation in bivalents and multivalents. The theory of pulling.** *Hereditas* 1951, **37**:85–156.

Rieder, C.L. and Salmon, E.D.: **The vertebrate cell kinetochore and its roles during mitosis.** *Trends Cell Biol.* 1998, **8**:310–318.

Salmon, E.D.: **Microtubules: a ring for the depolymerization motor.** *Curr. Biol.* 2005, **15**:R299–R302.

7

The Completion of Mitosis

In the final stages of mitosis the duplicated chromosomes are separated, pulled to opposite poles of the cell and packaged in identical daughter nuclei. These events are initiated by activation of the anaphase-promoting complex and by inactivation of mitotic cyclin–Cdk1 complexes.

7-0 Overview: The Completion of Mitosis

The final events of mitosis occur in anaphase and telophase

The events of early mitosis—spindle assembly and the preparation of the chromosomes for separation—bring the cell to metaphase. The bi-oriented sister chromatid pairs wait at the spindle midzone, pulled in opposite directions by intense poleward forces acting primarily at the kinetochores (see section 6-12). The events of late mitosis then lead to separation and segregation of the sister chromatids, resulting in a pair of daughter nuclei that each possess a complete and accurate copy of the genome. Late mitosis is also the time at which the cell-cycle control system is reset to the state in which it will enter G1.

The first stage of late mitosis is **anaphase A**, which is triggered by the destruction of sister-chromatid cohesion. Sister chromatids separate, allowing spindle forces to pull them to opposite ends of the spindle. To increase the distance between the chromosome sets, the spindle poles themselves move farther apart during **anaphase B**. Finally, in telophase, the spindle is dismantled, the two sets of segregated chromosomes are packaged into individual daughter nuclei, and chromatin returns to its relatively uncondensed interphase state (Figure 7-1). Cytokinesis, which we describe in Chapter 8, then divides the cell itself to complete cell division.

We have only a limited understanding of the mechanical events of anaphase and telophase—particularly the processes underlying spindle disassembly and packaging of chromosomes into new nuclei. Much of the research in this field is directed instead toward unveiling the mechanisms by which the cell-cycle control system governs progression through late mitosis, and these mechanisms form the dominant theme of this chapter. As we will see, the general outlines of late mitotic regulation are similar in all eukaryotes, although some species, budding yeast in particular, use some regulatory strategies that are not used by other organisms.

The metaphase-to-anaphase transition is initiated by ubiquitination and destruction of regulatory proteins

The initiation of sister-chromatid separation defines a major cell-cycle regulatory checkpoint called the metaphase-to-anaphase transition (see sections 3-0, 3-10 and 5-2). The basic principles underlying the control of this transition can be summarized as follows. As the cell reaches metaphase, robust inhibitory mechanisms prevent sister-chromatid separation and inhibit the events of late mitosis. When sister-chromatid bi-orientation is achieved, these braking mechanisms are removed by the ubiquitin-protein ligase APC^{Cdc20} (see section 3-10), which ubiquitinates proteins and targets them for destruction, thereby unleashing irreversible progression into anaphase and out of mitosis (see Figure 7-1). The most important targets of APC^{Cdc20} are the protein *securin*, whose destruction leads to the loss of sister-chromatid cohesion, and mitotic cyclins, whose destruction causes Cdk inactivation. This inactivation allows phosphatases to dephosphorylate Cdk targets, which stimulates the completion of late mitotic events. Cdk inactivation also resets the cell-cycle control system to a state of low Cdk activity, preparing the cell for the activation of Cdks that drive entry into the next cell cycle (see Figure 3-1).

How does securin destruction lead to sister-chromatid separation? Before anaphase, securin binds and inhibits a protease called *separase*, which is in turn responsible for cleaving Scc1, one of the subunits in the cohesin complexes that hold sister chromatids together (see section 5-8). The ubiquitination and destruction of securin liberates separase, which then cleaves Scc1 into two fragments. Cleavage of Scc1 causes dismantling of the cohesin complex, disrupting sister-chromatid cohesion and allowing sisters to separate.

Definitions

anaphase A: the stage in mitosis in which sister chromatids separate and move to opposite poles of the mitotic spindle. It follows metaphase.

anaphase B: the stage in mitosis in which the poles of the spindle move further apart. It follows **anaphase A**.

References

Harper, J.W. *et al.*: **The anaphase-promoting complex: it's not just for mitosis any more.** *Genes Dev.* 2002, **16**:2179–2206.

Morgan, D.O.: **Regulation of the APC and the exit from mitosis.** *Nat. Cell Biol.* 1999, **1**:E47–E53.

Murray, A.W. *et al.*: **The role of cyclin synthesis and degradation in the control of maturation-promoting factor activity.** *Nature* 1989, **339**:280–286.

Nasmyth, K.: **Segregating sister genomes: the molecular biology of chromosome separation.** *Science* 2002, **297**:559–565.

Peters, J.M.: **The anaphase-promoting complex: proteolysis in mitosis and beyond.** *Mol. Cell* 2002, **9**:931–943.

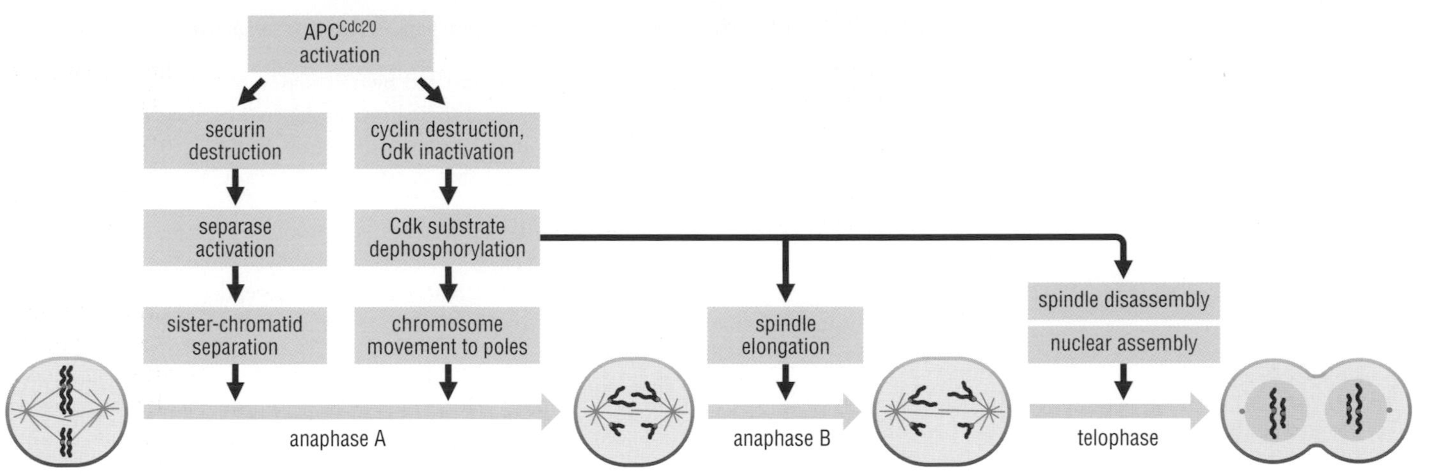

Figure 7-1 The completion of mitosis In anaphase A, separation of sister chromatids triggers their movement to the spindle poles, after which the poles themselves move apart in anaphase B. In telophase, disassembly of the spindle and the assembly of two daughter nuclei completes mitosis. All these events are initiated by the activation of APC^{Cdc20}, which triggers the ubiquitination and destruction of two key regulatory proteins. First, destruction of securin unleashes the protease separase, which cleaves a cohesin subunit and thereby disrupts sister-chromatid cohesion. Second, destruction of mitotic cyclins leads to dephosphorylation of numerous Cdk substrates, which is required for all the main events of late mitosis. Chromosome movement in anaphase A depends in large part on spindle forces already present in metaphase, but dephosphorylation of Cdk substrates also contributes to the control of these movements.

Dephosphorylation of Cdk targets drives the events of late M phase

After sister chromatids have separated, dephosphorylation of proteins that have been phosphorylated by Cdks is the major regulatory mechanism driving the completion of mitosis. It is easy to imagine why this might be so: some of the steps of late mitosis may result, at least in part, from the direct reversal of events in early mitosis. If phosphorylation of a protein promotes chromosome condensation, for example, then dephosphorylation of the same protein might be expected to be required for decondensation. Most late mitotic events, however, are unlikely to result simply from direct reversal of earlier events. Nuclear envelope assembly, for example, is a complex, multi-step process that depends on regulatory components that are not involved in mitotic entry but are activated in late mitosis, directly or indirectly, by the removal of Cdk-dependent phosphorylation. Some late mitotic processes also depend on regulation by other protein kinases or by the destruction of regulatory proteins by the APC.

Dephosphorylation of Cdk substrates is also required for two other late M-phase events: cytokinesis (see Chapter 8) and the assembly of prereplicative complexes at replication origins, which prepares the cell for the next S phase (see section 4-4).

APC^{Cdc20} initiates Cdk inactivation

Dephosphorylation of Cdk substrates in late mitosis is due primarily to the inactivation of Cdks, which allows phosphatases in the cell to act unopposed. The major mechanism of Cdk inactivation is the APC-dependent destruction of S and M cyclins. The S cyclins, such as cyclin A of animal cells, are generally destroyed earlier in mitosis than M cyclins, such as cyclin B. The destruction of M cyclins is typically the final and most critical step in Cdk inactivation. The importance of this step is illustrated by experiments with cells containing mutant M cyclins that are not recognized by the APC (see section 3-10). When APC^{Cdc20} is activated and sister chromatids separate, the mutant cyclins remain stable and Cdk1 activity remains high, resulting in cell-cycle arrest in anaphase.

The protein phosphatases that dephosphorylate Cdk targets in late mitosis are not well understood, particularly in multicellular organisms. Some Cdk substrate dephosphorylation may simply be catalyzed by general phosphatases whose activities do not vary in the cell cycle. In budding yeast, dephosphorylation of several Cdk targets depends on the activation in late mitosis of a specific phosphatase called Cdc14. It remains unclear, however, if specific phosphatases like Cdc14 are important for the regulation of late mitosis in other organisms.

In most cell types the APC activator protein Cdc20 is replaced by the alternative activator Cdh1 in late mitosis (see section 3-10). APC^{Cdh1} is not essential for mitotic progression but is required primarily for the continued destruction of cyclins—and thus Cdk inactivity—in G1. APC^{Cdh1} also ubiquitinates several regulatory proteins that are not targeted by APC^{Cdc20}, and destruction of these proteins may provide a nonessential contribution to the control of late M-phase events. In the embryonic cells of frogs and flies, however, Cdh1 is not present. APC^{Cdc20} alone is therefore capable of driving all proteolytic events required for progression through the embryonic cell cycle.

7-1 Initiation of Anaphase: Activation of the APC

APC^{Cdc20} activation in early mitosis is essential for anaphase to occur

Anaphase and the final events of mitosis are initiated by destruction of the separase inhibitor securin and the S and M cyclins. Destruction of these proteins depends on their ubiquitination by the APC, in association with its activating subunit Cdc20 (see section 3-10). Activation of APC^{Cdc20} is thus the central regulatory event that initiates the events of late mitosis.

In animal cells, APC^{Cdc20} is thought to be activated in late prophase or early prometaphase. Studies of the times during mitosis at which APC targets are destroyed provide evidence for its early mitotic activation. In *Drosophila* and vertebrates, APC^{Cdc20}-dependent destruction of the major S-phase cyclin, cyclin A, begins just after nuclear envelope breakdown at the beginning of prometaphase (Figure 7-2a). Destruction of securin and the major mitotic cyclin, cyclin B, begins abruptly after the sister chromatids have all been correctly bi-oriented on the mitotic spindle at the beginning of metaphase (Figure 7-2b). These observations indicate that the cell's population of APC^{Cdc20}, or at least some fraction of it, is activated early in mitosis, resulting in cyclin A destruction, whereas the destruction of securin and cyclin B is delayed until metaphase.

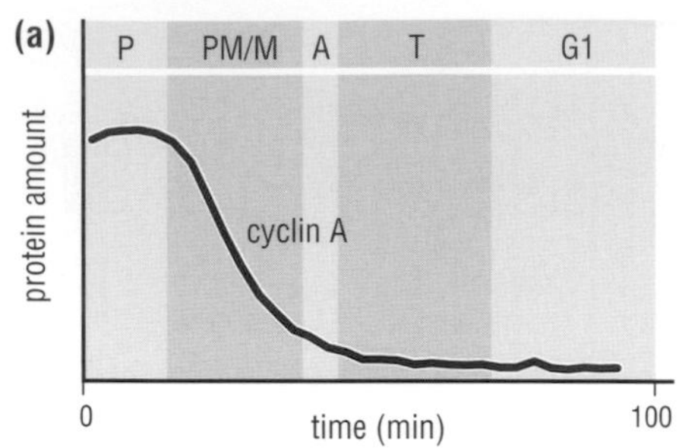

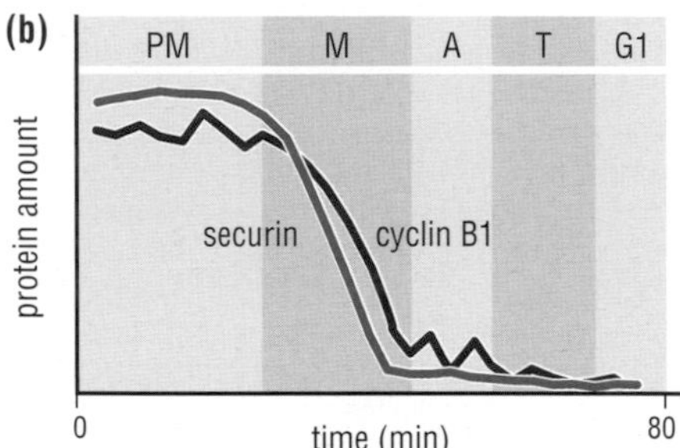

Figure 7-2 Mitotic protein destruction in human cells The precise timing of APC-dependent protein destruction was determined by injecting cultured human cells in late G2 with fluorescently tagged targets of the APC. The amount of each fluorescent protein in the injected cell was then measured at frequent intervals as the cells progressed through the stages of mitosis (PM, prometaphase; M, metaphase; A, anaphase; T, telophase). **(a)** The destruction of fluorescently tagged cyclin A began almost immediately after nuclear envelope breakdown at the beginning of prometaphase. The onset of metaphase was not determined in this experiment. **(b)** Securin (blue) and mitotic cyclin B1 (purple) were tagged with different fluorescent labels and injected into the same cell. The destruction of both proteins began just after the last sister-chromatid pair was aligned on the metaphase plate and was completed just before the onset of anaphase. Panel (a) adapted from den Elzen, N. and Pines, J.: *J. Cell Biol.* 2001, **153**:121–136. Panel (b) adapted from Hagting, A. *et al.*: *J. Cell Biol.* 2002, **157**:1125–1137.

What causes this delay? In somatic animal cells, the answer is the *spindle checkpoint system*, one of the regulatory systems that arrest progression through major cell-cycle transitions if some essential process has not been completed (see section 1-3). The spindle checkpoint system monitors the attachment of kinetochores to the spindle. In the absence of correct bi-oriented chromosome attachment, components of this system inhibit the activity of APC^{Cdc20} toward securin and mitotic cyclins, delaying their destruction until all sister-chromatid pairs have been correctly aligned. If the spindle checkpoint system is inhibited in cultured mammalian cells, securin and cyclin B are destroyed prematurely in prometaphase. It is not clear how the spindle checkpoint system prevents the destruction of these APC targets while allowing cyclin A to be destroyed.

Despite its clear importance in the timing of anaphase in somatic animal cells, the spindle checkpoint system is absent in frog and *Drosophila* embryonic cells, and mutations that inhibit it have little effect on the timing of anaphase in budding yeast. How anaphase is triggered at the correct time in these cells is not well understood. The delayed timing of securin and mitotic cyclin destruction may involve complex changes in APC phosphorylation, localization and association with other regulators.

Phosphorylation promotes APC^{Cdc20} activation in early mitosis

The activation of APC^{Cdc20} in prophase and prometaphase is controlled by several mechanisms (Figure 7-3). The most important and universal of these is phosphorylation of core subunits of the APC by mitotic Cdks, which promotes binding of the activating subunit Cdc20. In yeast, metazoans, and probably all other eukaryotes, several subunits of the APC are phosphorylated at multiple sites during mitosis. Numerous lines of evidence suggest that mitotic cyclin–Cdk1 is responsible for most of this phosphorylation. In human cells, phosphorylation of the APC is first seen in the nucleus in late prophase, just after the activation and nuclear import of cyclin B1–Cdk1 and before nuclear envelope breakdown (see section 5-6). Cdk1-dependent phosphorylation of the APC increases its affinity for Cdc20 *in vitro*, whereas mutation of a subset of Cdk1 phosphorylation sites in the APC decreases Cdc20 binding and delays APC activation. Together these observations suggest that Cdk1 is an important activator of APC^{Cdc20}. As

References

den Elzen, N. and Pines, J.: **Cyclin A is destroyed in prometaphase and can delay chromosome alignment and anaphase.** *J. Cell Biol.* 2001, **153**:121–136.

Hagting, A. *et al.*: **Human securin proteolysis is controlled by the spindle checkpoint and reveals when the APC/C switches from activation by Cdc20 to Cdh1.** *J. Cell Biol.* 2002, **157**:1125–1137.

Hansen, D.V. *et al.*: **Plk1 regulates activation of the anaphase promoting complex by phosphorylating and triggering $SCF^{\beta TrCP}$-dependent destruction of the APC inhibitor Emi1.** *Mol. Biol. Cell* 2004, **15**:5623–5634.

Harper, J.W. *et al.*: **The anaphase-promoting complex: it's not just for mitosis any more.** *Genes Dev.* 2002, **16**:2179–2206.

Kraft, C. *et al.*: **Mitotic regulation of the human anaphase-promoting complex by phosphorylation.** *EMBO J.* 2003, **22**:6598–6609.

Peters, J.M.: **The anaphase-promoting complex: proteolysis in mitosis and beyond.** *Mol. Cell* 2002, **9**:931–943.

Rudner, A.D. and Murray, A.W.: **Phosphorylation by Cdc28 activates the Cdc20-dependent activity of the anaphase-promoting complex.** *J. Cell Biol.* 2000, **149**:1377–1390.

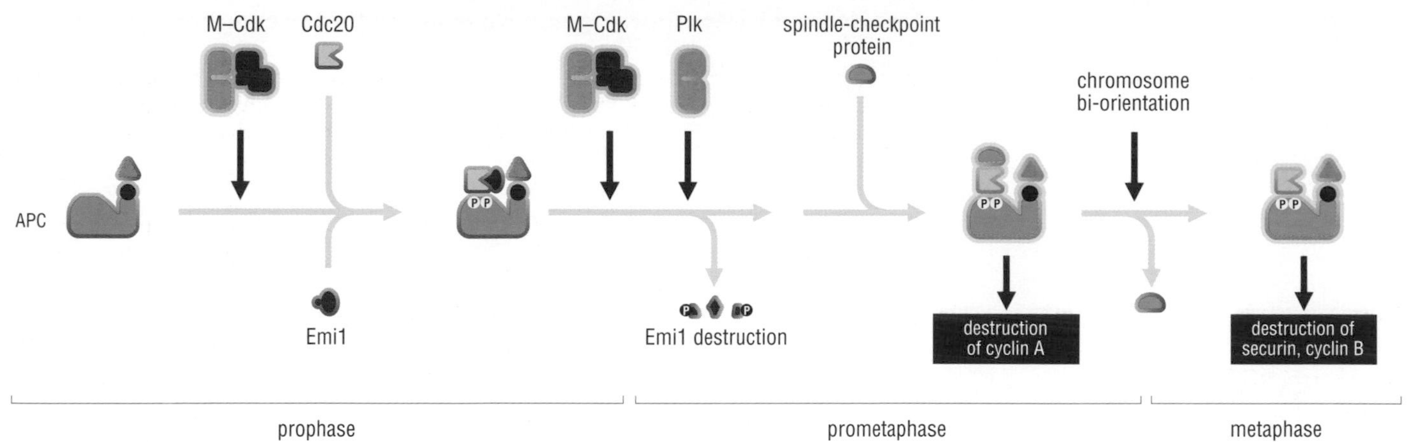

discussed in section 3-11, this regulatory interaction allows Cdk1 to set the stage for its own inactivation, thereby providing the framework for an autonomous oscillator that lies at the core of the mitotic control system.

In yeast and somatic metazoan cells (but not in embryonic cells), APC activation is also driven by increased synthesis of Cdc20 during mitosis. This results from a transient increase in the expression of the gene for Cdc20 at the beginning of mitosis, followed by decreased expression after metaphase.

In vertebrate cells, a protein called Emi1 provides an additional level of regulation. In G2 and early mitosis, Emi1 binds and inhibits Cdc20, probably by blocking its interaction with APC targets. In late prophase, Emi1 is phosphorylated by both the Polo-like protein kinase Plk, which is activated in early mitosis as a result of Cdk1 activation (see section 5-7), and by mitotic cyclin–Cdk complexes. Phosphorylation targets Emi1 to the ubiquitin-protein ligase $SCF^{\beta\text{-TrCP1}}$ (see section 3-9), resulting in its destruction in prometaphase. Thus, together with mitotic Cdks, Plk helps stimulate APC^{Cdc20} activation in prometaphase, resulting in the destruction of cyclin A. Emi1 levels rise again at the end of G1 and help inhibit APC^{Cdh1}. In *Drosophila*, an Emi1-related protein called Rca1 is involved primarily in the control of APC^{Cdh1} and not APC^{Cdc20}, so its general role in controlling early mitotic APC activity is not clear. Budding and fission yeasts do not seem to contain an Emi1-like protein.

APC^{Cdc20} activity may also be controlled by its intracellular location. In early *Drosophila* embryos, the bulk of mitotic cyclin B remains stable as the syncytial nuclei race through mitosis (see section 2-4). However, a small fraction of cyclin B, located specifically around the mitotic spindle, is degraded during each mitosis, presumably allowing sufficient local Cdk1 inactivation to promote the events of late mitosis. This localized cyclin destruction is thought to result from the concentration of APC^{Cdc20} on spindle microtubules. The regulation of APC^{Cdc20} localization thus provides a mechanism for promoting its interaction with some substrates while preventing or delaying access to others.

The control of APC activity in many cell types also depends on the alternative APC activator Cdh1. Cdh1 is inhibited by Cdk-dependent phosphorylation in early mitosis (see section 3-10). Thus, Cdk inactivation in metaphase leads to activation of APC^{Cdh1} at the beginning of anaphase, resulting in continued destruction of the targets of APC^{Cdc20}. APC^{Cdh1} also promotes the destruction of additional proteins in anaphase, such as the mitotic protein kinases Plk and aurora A (see section 5-7), which are not recognized by APC^{Cdc20}.

APC^{Cdc20} is thought to be inactivated during anaphase, primarily because Cdk inactivation leads to APC dephosphorylation (see section 3-10). APC dephosphorylation does not, however, hinder the activation of APC^{Cdh1}. One of the targets of APC^{Cdh1} is Cdc20, providing another mechanism for shutting down APC^{Cdc20} activity in anaphase.

In most cell types APC^{Cdc20} is sufficient for the destruction of mitotic cyclins. In budding yeast, however, APC^{Cdh1} makes an important contribution. The interesting and apparently unique features of the regulation of late mitosis in budding yeast will be described in section 7-5.

Figure 7-3 **Mechanisms controlling APC^{Cdc20} activation in vertebrate cells** Activation of the APC depends primarily on post-translational mechanisms and not on changes in the amount of core APC enzyme, which remains constant during the cell cycle. In most if not all eukaryotes, phosphorylation of APC core subunits in late prophase by M–Cdks enhances APC affinity for the activating subunit Cdc20. Increased production of Cdc20 in mitosis also helps drive its binding. In vertebrate cells, APC^{Cdc20} is initially inhibited by association with the protein Emi1. In late prophase and early prometaphase, M–Cdks and Plk phosphorylate Emi1, thereby triggering its destruction by $SCF^{\beta\text{-TrCP1}}$ and unleashing APC activity toward early mitotic substrates such as cyclin A. APC^{Cdc20} activity toward metaphase substrates such as securin and cyclin B is initially restrained in prometaphase by components of the spindle checkpoint system. In some way these components inhibit APC^{Cdc20} activity toward metaphase substrates without blocking activity toward cyclin A. One possibility, as shown here, is that spindle checkpoint proteins bind to all the Cdc20 in the cell but do not affect its activity toward cyclin A; another possibility is that checkpoint proteins associate only with a localized subpopulation of Cdc20 that is responsible for securin destruction. Soon after bi-orientation of sister chromatids on the spindle, spindle checkpoint components are removed from Cdc20, and full APC^{Cdc20} activation leads to destruction of securin and cyclin B.

Unattached kinetochores generate a signal that prevents anaphase

Accurate sister-chromatid segregation demands that all sister-chromatid pairs be properly attached to both spindle poles before they are separated. This dependence is ensured in most cells by a regulatory system—the **spindle checkpoint system**—that monitors the attachment of sister chromatids to the spindle and allows anaphase to occur only after bi-orientation has been achieved. The basic features of the spindle checkpoint system are clear. During prometaphase, incorrectly attached kinetochores generate a signal, sometimes called the wait anaphase signal, that inhibits APC^{Cdc20}, thereby preventing securin destruction and sister-chromatid separation. Remarkably, a single unattached kinetochore in the cell can block anaphase. If that kinetochore is destroyed with a laser, anaphase occurs rapidly. The unattached kinetochore is therefore thought to be the source of a diffusible signal that blocks sufficient APC^{Cdc20} to prevent securin destruction.

As discussed earlier (section 7-1), the spindle checkpoint system determines the time at which sister-chromatid separation occurs in somatic animal cells. Anaphase normally begins some fixed period (about 20 minutes in somatic vertebrate cells) after the last sister-chromatid pair becomes correctly bi-oriented on the spindle. Inhibition of spindle checkpoint function—by inhibitory antibody injection or gene mutation, for example—results in premature securin destruction and sister-chromatid separation. In the mouse, deletion of spindle checkpoint components causes death of the early embryo.

In budding yeast, the spindle checkpoint system is not essential for normal cell-cycle progression but becomes important when spindle assembly fails. Yeast cells with spindle checkpoint mutations are viable and undergo anaphase with normal timing—such mutations are lethal only in the presence of spindle damage. The embryonic cells of frogs and flies seem to lack the checkpoint entirely: normal oscillations in Cdk and APC activities occur in the presence of drugs or mutations that disrupt spindle assembly. In these cell types, therefore, the timing of sister-chromatid separation must be determined by other mechanisms.

The molecular framework of the spindle checkpoint system is coming into focus. The key components were first identified in screens for budding yeast mutants that do not delay in mitosis when treated with drugs that inhibit spindle assembly (Figure 7-4). The proteins identified in these screens (Mad1, 2, 3 and Bub1, 3), as well as additional proteins identified by other means (notably Mps1), are essential for spindle checkpoint function in yeast, and highly conserved homologs are also required for this function in metazoans (Figure 7-5). Most of these proteins are bound to unattached kinetochores and then released when bipolar attachment occurs, suggesting that they are directly involved in the generation of the wait anaphase signal at the kinetochore.

(a) wild type, no treatment

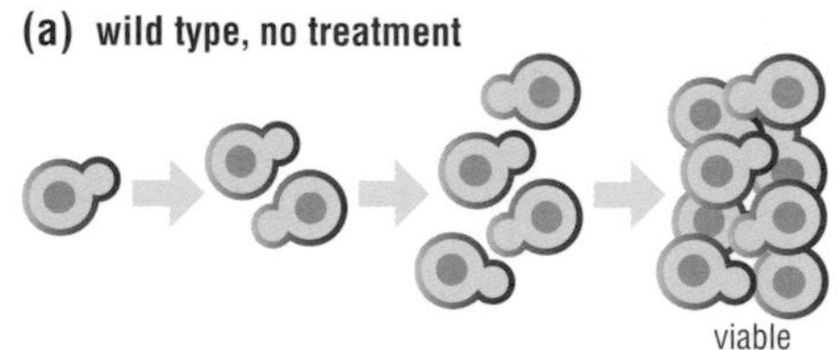

(b) wild type + benomyl

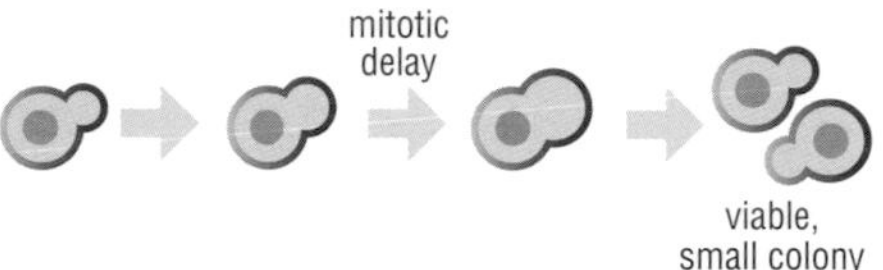

(c) *mad2* mutant + benomyl

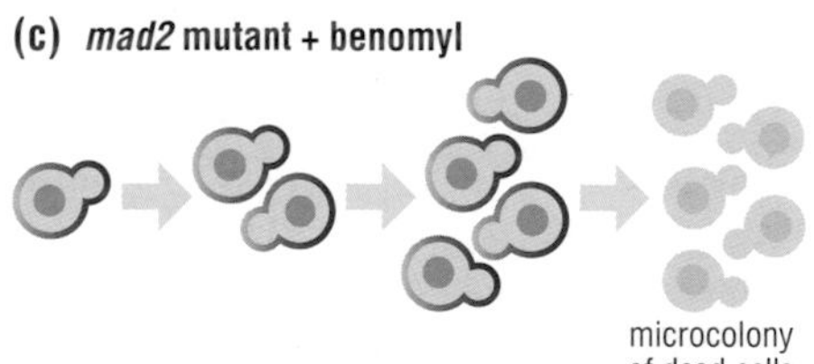

(d) *mad2* mutant, no treatment

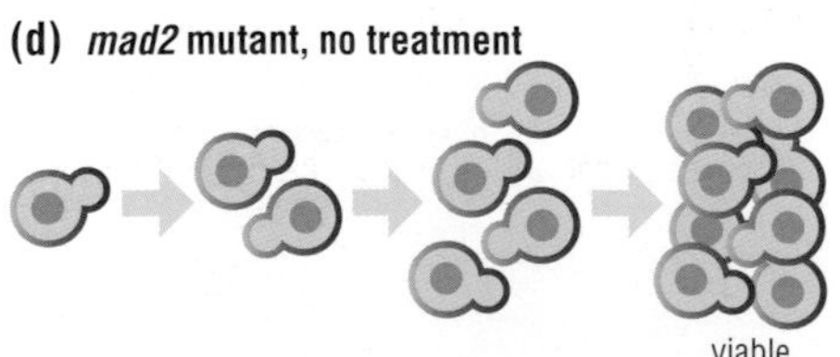

Figure 7-4 Behavior of yeast cells carrying mutations in spindle checkpoint components **(a)** A wild-type yeast cell on a culture plate divides normally to generate progeny that eventually produce a colony on the plate. **(b)** If cells are treated with low doses of the microtubule-destabilizing drug benomyl, spindle assembly is slower. The spindle defect causes a delay in mitotic progression, and }a small colony results. **(c)** If a cell carries a mutation in a spindle checkpoint component such as Mad2, mitotic progression is not inhibited in the presence of benomyl and cell division occurs at the normal rate. The spindle defect in these cells causes severe errors in chromosome segregation, however, so that cells eventually die after forming a microcolony. Sensitivity to benomyl is therefore a useful method for the identification of spindle checkpoint mutants. **(d)** Cells carrying a spindle checkpoint mutation do not display major defects in the timing of mitosis in the absence of spindle defects. This is not true in vertebrate somatic cells, where checkpoint defects result in premature anaphase and lethality.

Definitions

spindle checkpoint system: regulatory system that restrains progression through the metaphase-to-anaphase transition until all sister-chromatid pairs have been bi-oriented correctly on the mitotic spindle. It is also called the spindle assembly checkpoint or SAC.

References

Hoyt, M.A. *et al.*: ***S. cerevisiae*** **genes required for cell cycle arrest in response to microtubule function.** *Cell* 1991, **66**:507–517.

Li, R. and Murray, A.W.: **Feedback control of mitosis in budding yeast.** *Cell* 1991, **66**:519–531.

Li, X. and Nicklas, R.B.: **Mitotic forces control a cell-cycle checkpoint.** *Nature* 1995, **373**:630–632.

Musacchio, A. and Hardwick, K.G.: **The spindle checkpoint: structural insights into dynamic signalling.** *Nat. Rev. Mol. Cell Biol.* 2002, **3**:731–741.

Pinsky, B.A. and Biggins, S.: **The spindle checkpoint: tension versus attachment.** *Trends Cell Biol.* 2005, **15**:486–493.

Waters, J.C. *et al.*: **Localization of Mad2 to kinetochores depends on microtubule attachment, not tension.** *J. Cell Biol.* 1998, **141**:1181–1191.

Yu, H.: **Regulation of APC-Cdc20 by the spindle checkpoint.** *Curr. Opin. Cell Biol.* 2002, **14**:706–714.

The spindle checkpoint monitors defects in microtubule attachment and kinetochore tension

The function of the spindle checkpoint system is to block anaphase until correct bi-orientation has occurred. How does the system detect bipolar attachment of chromosomes? One possibility is that it monitors the attachment of microtubule ends to the kinetochore, and blocks anaphase when microtubule binding is absent or incomplete. Alternatively, the system could monitor the amount of tension at the kinetochore and delay anaphase when no tension is detected, as this is an indirect indication that bi-orientation has not occurred (see section 6-10). There is experimental evidence for both these possibilities, but their relative importance remains unclear.

The importance of microtubule attachment in checkpoint function is illustrated by the behavior of spindle checkpoint components at the kinetochore. Mad2, for example, is found on all unattached kinetochores during prometaphase and then disappears from each kinetochore when it becomes attached to the spindle (Figure 7-6). When only one of the two kinetochores in a pair is attached, Mad2 binding is reduced at the attached kinetochore but maintained on the unattached one. Thus, the attachment of microtubules, even in the absence of tension, results in a partial loss of Mad2 from kinetochores. Given that Mad2 dissociation from the kinetochore is likely to be an indicator of checkpoint inactivation, these observations argue that the checkpoint system can monitor microtubule attachment.

A role for tension sensing in the spindle checkpoint was first suggested by elegant studies of insect spermatocytes. In these cells, the presence of mono-oriented chromosomes normally blocks anaphase, but if a fine glass microneedle is used to pull the mono-oriented chromosome away from the nearby spindle pole, thereby generating tension at the kinetochore, anaphase begins. Tension can therefore shut off the inhibitory checkpoint signal. The possibility that lack of tension generates an inhibitory signal is also supported by the observation that some spindle checkpoint components (BubR1, for example) remain localized at kinetochores when they become attached to microtubules but are not under tension.

The relative importance of tension and microtubule attachment is obscured by the fact that they are interdependent. As discussed in section 6-10, the tension that results from correct chromosome bi-orientation is thought to increase the stability of microtubule attachment and, in animal cells, to cause an increase in the number of microtubules attached to the kinetochore; conversely, low tension reduces the strength of attachment. Because of this relationship between tension and attachment, it is difficult to be sure if tension has direct effects on spindle checkpoint function or whether it acts indirectly by changing microtubule attachment.

One simple possibility is that the spindle checkpoint system senses the presence of unoccupied microtubule-binding sites on the kinetochore. In the absence of tension, these sites are only partly or transiently occupied by microtubules—even in budding yeast kinetochores containing a single binding site. This unstable form of microtubule attachment depends on kinetochore phosphorylation by the protein kinase aurora B (see section 6-10). These weakly occupied and phosphorylated microtubule-binding sites could serve as sites at which spindle checkpoint components bind the kinetochore and generate the wait anaphase signal. When bi-orientation occurs, the high-affinity occupation of all microtubule-binding sites would displace spindle checkpoint components, thereby stopping production of the inhibitory wait anaphase signal and allowing anaphase to occur.

Although the precise kinetochore defects that are sensed by the spindle checkpoint system remain unclear, we are beginning to achieve some understanding of the mechanisms by which spindle checkpoint components at the kinetochore generate a wait anaphase signal. These mechanisms will be discussed in section 7-3.

Components of the Spindle Checkpoint System

Metazoans	*S. cerevisiae*	Comments
Mad1	Mad1	coiled-coil protein, binds Mad2 at kinetochore
Mad2	Mad2	binds Mad1 and Cdc20 independently
BubR1	Mad3	binds Cdc20 and Bub3; BubR1 (but not Mad3) contains an inactive kinase domain
Bub1	Bub1	protein kinase, binds Bub3
Bub3	Bub3	WD40 repeats, binds BubR1 and Bub1
Mps1	Mps1	protein kinase
CENP-E	–	kinesin-7 motor protein, binds BubR1
Zw10	–	forms complex with Rod
Rod	–	forms complex with Zw10

Figure 7-5 Table of spindle checkpoint system components This is a selective list of proteins known to be essential for spindle checkpoint function. Members of the Mad and Bub families, as well as Mps1, are clearly implicated in spindle checkpoint function in all eukaryotes. Other proteins listed here are required for checkpoint function only in metazoans, and clear homologs are not present in yeast.

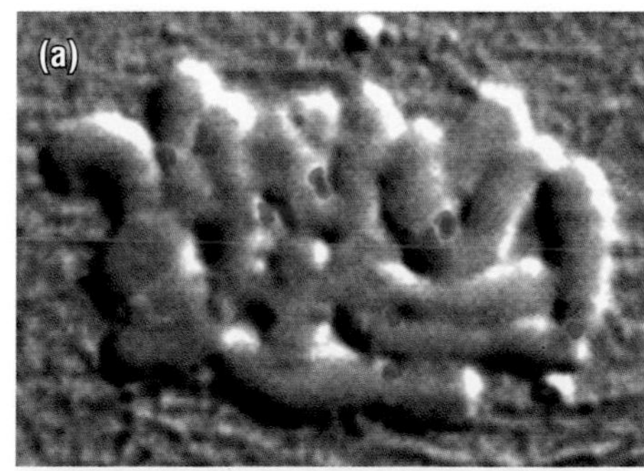

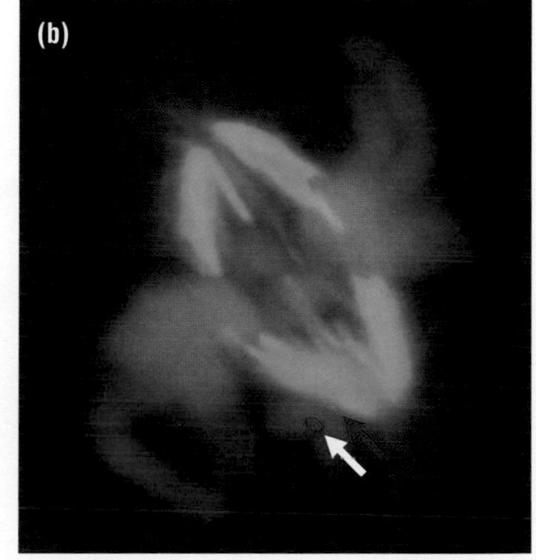

Figure 7-6 Spindle checkpoint component Mad2 at unattached kinetochores (a) A metaphase mammalian cell was treated with the spindle-depolymerizing drug nocodazole. The resulting loss of microtubules resulted in the appearance of Mad2 protein (pink) at the kinetochores of all sister-chromatid pairs. **(b)** This photograph shows a mammalian cell in late prometaphase (DNA is stained blue and microtubules green). Most sister chromatids are bioriented at the metaphase plate, but one pair is mono-oriented near the bottom spindle pole. The unattached kinetochore that faces away from the pole contains Mad2 (stained in pink, indicated by white arrow), whereas the attached kinetochore that faces the pole has very weak Mad2 staining (red arrow). Kindly provided by Jennifer C. Waters. From Waters, J.C. *et al.*: *J. Cell Biol.* 1998, **141**:1181–1191.

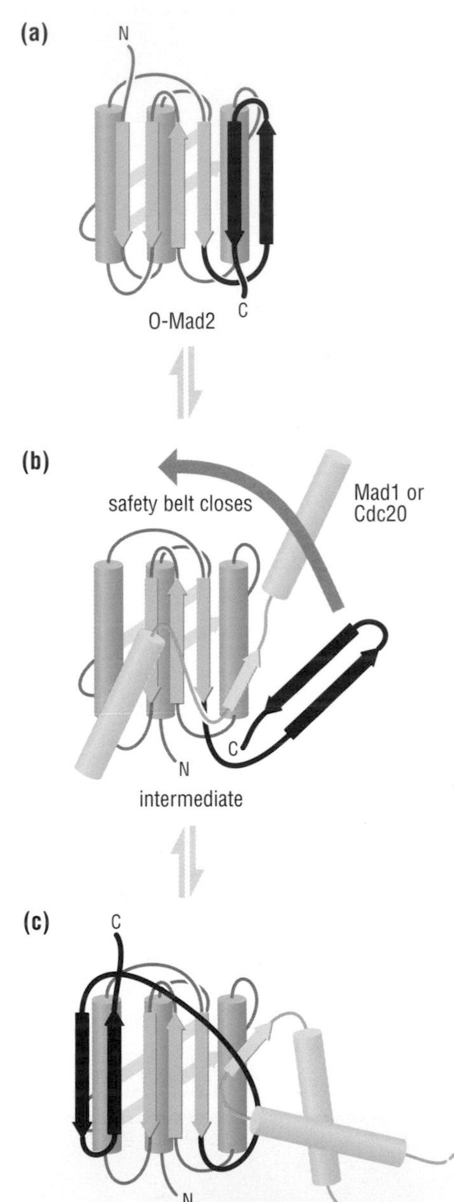

Unattached kinetochores catalyze the formation of inhibitory signaling complexes

A remarkable feature of the spindle checkpoint system is that a single unattached kinetochore can generate a signal that blocks the onset of anaphase—the wait anaphase signal. The current hypothesis is that unattached kinetochores act as catalysts for the generation of a diffusible inhibitory signal that inactivates APC^{Cdc20} activity toward its major metaphase targets securin and M cyclin (see section 7-1). This diffusible signal is thought to be composed of proteins that bind tightly to Cdc20 and block its function as an APC activator.

The spindle checkpoint component Mad2 provides the best-understood example of how an unattached or poorly attached kinetochore might generate an inhibitory signal. In the presence of spindle defects, Mad2 binds directly and tightly to Cdc20 and inhibits its function. Under these conditions, some fraction of Mad2 is also found at unattached kinetochores, where some of it seems to be rapidly coming on and off the kinetochore. These and other observations led to an intriguing hypothesis: that the unattached kinetochore acts like an enzyme that binds transiently to Mad2 to catalyze a change in its shape, or conformation, releasing it to bind and inhibit Cdc20, thus producing a diffusible inhibitor of APC. According to this hypothesis, one unattached kinetochore modifies enough Mad2 to inhibit most Cdc20 in the cell—or at least the Cdc20 that is responsible for the destruction of securin and M cyclin.

But how does an unattached kinetochore modify Mad2 conformation to make it into a Cdc20 inhibitor? The likely answer is based on the interactions between Mad2 and its binding partners. During prometaphase, a fraction of Mad2 is stably bound to another spindle checkpoint protein called Mad1 at unattached kinetochores. A separate fraction of Mad2 is bound to and inhibits Cdc20. Interestingly, Mad2 uses the same site to bind Mad1 and Cdc20 and can therefore bind only one of the two proteins at a time. Structural studies reveal that Mad2 exists in two distinct conformations (Figure 7-7). When bound to Mad1 or Cdc20, Mad2 assumes a closed conformation (called C-Mad2) in which its carboxy-terminal region is wrapped, or closed, around a narrow region of Mad1 or Cdc20 like a safety belt. When free of a binding partner, Mad2 exists in an open conformation (O-Mad2), in which the safety belt is withdrawn and held tightly against the side of Mad2. Binding of O-Mad2 to a partner does not occur efficiently until a conformational change has loosened the safety belt so that it can be repositioned around the partner.

This conformational change is proposed to occur as a result of interaction with the C-Mad2–Mad1 complexes at unattached kinetochores. There is evidence that the C-Mad2 subunit in the Mad1 complex can dimerize with a soluble O-Mad2 protein and, most importantly, promote its interaction with Cdc20. On the basis of this evidence, it has been suggested that the Mad2–Mad1 complex at unattached kinetochores catalyzes the formation of Mad2–Cdc20 complexes as shown in Figure 7-8. The C-Mad2 protein in the Mad1 complex binds a soluble O-Mad2 protein, inducing a conformational change that loosens the safety

Figure 7-7 Alternative conformations of the Mad2 protein (a) In this view of the unbound open conformation (O-Mad2), the carboxy-terminal safety belt of Mad2 (pink) is held tightly against the right side of the protein. **(b)** Mad2 binding to a ligand (yellow, for example Mad1 or Cdc20) must begin with a conformational change that loosens the safety belt, allowing it to be wrapped around the binding protein. **(c)** In the closed conformation (C-Mad2), the safety belt wraps around the binding partner and interacts with a different region of Mad2. O-Mad2 and C-Mad2 are sometimes known as N1-Mad2 and N2-Mad2, respectively. Adapted from De Antoni, A. *et al.*: *Curr. Biol.* 2005, **15**:214–225.

References

De Antoni, A. *et al.*: **The Mad1/Mad2 complex as a template for Mad2 activation in the spindle assembly checkpoint.** *Curr. Biol.* 2005, **15**:214–225.

Luo, X. *et al.*: **The Mad2 spindle checkpoint protein has two distinct natively folded states.** *Nat. Struct. Mol. Biol.* 2004, **11**:338–345.

Musacchio, A. and Hardwick, K.G.: **The spindle checkpoint: structural insights into dynamic signalling.** *Nat Rev Mol Cell Biol.* 2002, **3**:731–741.

Nasmyth, K.: **How do so few control so many?** *Cell* 2005, **120**:739–746.

Rieder, C.L. *et al.*: **Mitosis in vertebrate somatic cells with two spindles: implications for the metaphase/anaphase transition checkpoint and cleavage.** *Proc. Natl Acad. Sci. USA* 1997, **94**:5107–5112.

Vanoosthuyse, V. and Hardwick, K.G.: **Bub1 and the multilayered inhibition of Cdc20–APC/C in mitosis.** *Trends Cell Biol.* 2005, **15**:231–233.

Xia, G. *et al.*: **Conformation-specific binding of p31^{comet} antagonizes the function of Mad2 in the spindle checkpoint.** *EMBO J.* 2004, **23**:3133–3143.

Yu, H.: **Regulation of APC–Cdc20 by the spindle checkpoint.** *Curr. Opin. Cell Biol.* 2002, **14**:706–714.

belt, allowing it to become wrapped around Cdc20. The resulting C-Mad2–Cdc20 complex is released from the kinetochore, leaving the C-Mad2–Mad1 complex at the kinetochore to repeat the cycle. A particularly appealing feature of this hypothesis is that the soluble C-Mad2–Cdc20 complexes can also interact with free O-Mad2 proteins to generate more C-Mad2–Cdc20, resulting in a feedback loop that rapidly amplifies the inhibitory wait anaphase signal.

Unattached kinetochores also produce inhibitory complexes containing Cdc20 bound to the spindle checkpoint proteins BubR1 and Bub3. Formation of this complex synergizes with the Mad2–Cdc20 complex in some way to suppress APC function fully. Several other spindle checkpoint proteins are also involved in APC inhibition. Two of these, Bub1 and Mps1, are protein kinases, and phosphorylation of several checkpoint proteins, including Mad1, occurs in the presence of spindle defects. Bub1 also phosphorylates Cdc20, providing an additional mechanism of Cdc20 inhibition that helps suppress APC activation. Detailed exploration of these and other phosphorylation events is likely to unveil additional layers in the generation of the wait anaphase signal.

Mad2–Cdc20 and BubR1–Bub3–Cdc20 complexes both interact with the APC to block, or fail to promote, the formation of active APC^{Cdc20}. For reasons that are still not known, these inhibitory complexes do not inhibit the ubiquitination and destruction of the S-phase cyclin A by APC^{Cdc20} in prometaphase animal cells (see section 7-1). One possibility is that cyclin A destruction depends on a distinct subpopulation of APC^{Cdc20} that is sequestered in a location inaccessible to spindle checkpoint proteins. Alternatively, the binding of these proteins may inhibit the interaction of Cdc20 with metaphase substrates but not with cyclin A.

Another interesting and mysterious feature of the wait anaphase signal is that it does not diffuse throughout the whole cell. If two mammalian cells are fused to produce a single cell with two separate spindles, an unattached kinetochore on one spindle delays anaphase on that spindle but not on the other. Mechanisms therefore exist to restrict the movement or activity of inhibitory checkpoint complexes: perhaps they associate in some way with the spindle or are inactivated by proteins in the cytosol.

The spindle checkpoint signal is rapidly turned off once kinetochores are attached

Almost immediately after the last sister-chromatid pair is bi-oriented on the mammalian spindle, the destruction of securin and cyclin B begins, indicating that APC^{Cdc20} is no longer inhibited (see Figure 7-2b). Inhibitory checkpoint complexes must therefore disappear rapidly once kinetochores are correctly attached. The sudden drop in their numbers results, at least in part, from a decrease in their rate of production at the kinetochore. Attachment to the spindle is quickly followed by the disappearance of many checkpoint proteins from the kinetochore—in part because the minus-end-directed motor protein dynein transports them to the centrosome.

The reduced production of spindle checkpoint complexes after bi-orientation may depend specifically on a Mad2-binding protein called p31^{comet} (also called CMT2). *In vivo*, p31^{comet} antagonizes Mad2 function and, when overexpressed, promotes mitotic progression in the presence of spindle defects. *In vitro*, p31^{comet} binds the C-Mad2 conformation. Activation of p31^{comet} after bi-orientation might therefore block the formation of additional Mad2–Cdc20 complexes.

In principle, decreased production of checkpoint complexes alone will not result in rapid inactivation of the checkpoint signal unless other mechanisms exist to dismantle those complexes quickly. One simple possibility is that inhibitory checkpoint complexes are intrinsically unstable and fall apart soon after their formation. Additional factors, perhaps including protein-remodeling enzymes, might also be required for the rapid disassembly of these complexes.

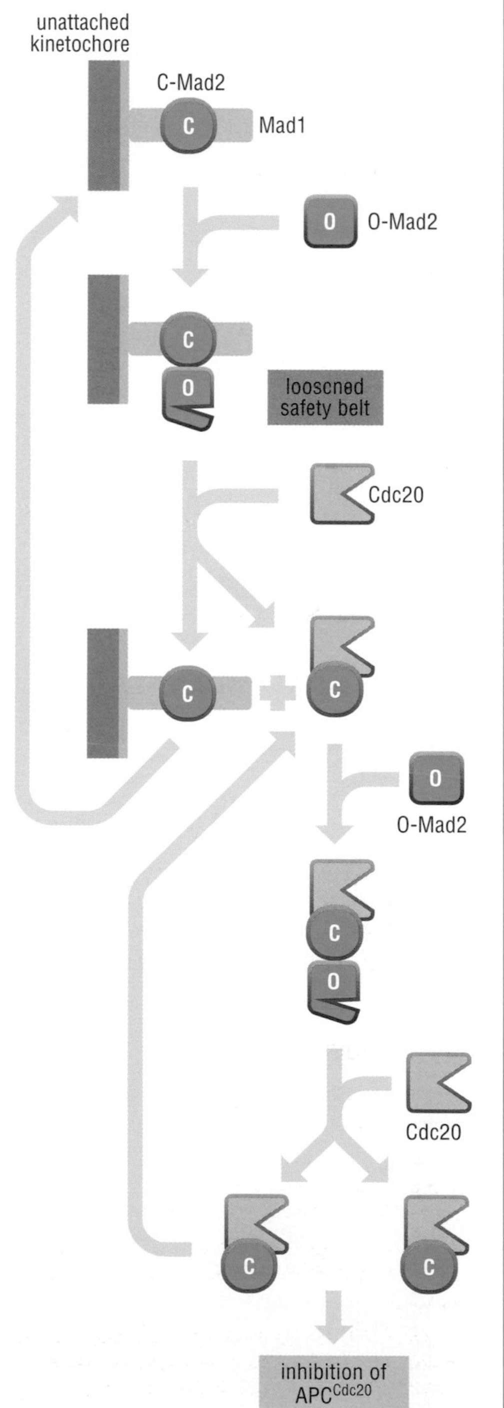

Figure 7-8 A proposed scheme for the generation of the wait anaphase signal at unattached kinetochores Unattached kinetochores are occupied by a stable complex of Mad1 and C-Mad2. Although two copies of each protein are present in the complex, single subunits are shown here for clarity. C-Mad2 dimerizes with soluble O-Mad2, triggering a conformational change that loosens the safety belt and allows binding to Cdc20. The resulting C-Mad2–Cdc20 complex dissociates and then serves as a catalyst for the production of more C-Mad2–Cdc20 complexes.

Securin and Separase

Name	Vertebrates	*Drosophila*	*S. cerevisiae*	*S. pombe*
separase	separase	SSE + THR	Esp1	Cut1
securin	PTTG	PIM	Pds1	Cut2

Figure 7-9 Table of names for separase and securin in different species Separase is divided into a carboxy-terminal protease domain and an amino-terminal regulatory domain. In *Drosophila*, these two domains seem to be found in separate but tightly associated proteins: SSE (containing the protease domain) and THR (containing the regulatory domain).

Separase is inhibited before anaphase by securin

Once bi-orientation has occurred and the spindle checkpoint system has been inactivated, sister-chromatid separation and anaphase can occur. Like so many other events in the cell cycle, chromatid separation is an all-or-none, irreversible process. It is also highly synchronous: all sister-chromatid pairs tend to separate at the same time. These features reflect the presence of an underlying regulatory system generating an abrupt and overwhelming stimulus that acts rapidly on all sister pairs. The stimulus is thought to be a wave of activity of the protease **separase**, which cleaves Scc1, a component of the cohesin complexes that hold the sisters together (see Figure 7-1). Before anaphase, separase is inhibited by the protein **securin** (Figure 7-9), which binds tightly to separase and blocks its active site. The first step toward anaphase occurs when APC^{Cdc20} catalyzes the ubiquitination of securin, thus targeting it for destruction and releasing free separase.

In budding yeast, inactivating mutations in either Cdc20 or the APC result in cell-cycle arrest before anaphase, demonstrating the importance of APC^{Cdc20} for sister-chromatid separation. Cells lacking both Cdc20 and securin, however, do undergo sister-chromatid separation (Figure 7-10). Thus, securin is the only APC^{Cdc20} target whose destruction is essential for sister separation in budding yeast.

Securin has, however, an unusual regulatory relationship with separase: as well as inhibiting separase it is also required for the protease to become active. Fission yeast or *Drosophila* cells lacking securin do not undergo premature sister-chromatid separation (which would be expected if securin were simply an inhibitor of separase) but instead fail to separate sisters at all. The molecular explanation for these observations is that the binding of securin to separase is first required to promote separase function, perhaps because securin binding helps separase achieve an active folded state or helps it reach the appropriate location in the cell. Once securin has performed this positive role, however, it remains bound as an inhibitor until it is destroyed (Figure 7-11).

Securin is not an essential activator of separase in all species, however. Budding yeast cells lacking securin show no major defect in the timing or accuracy of sister-chromatid separation, except at high temperature (see Figure 7-10). Loss of securin also has minor effects in mice and in human cells.

In vertebrate cells Cdk1 inhibits separase by phosphorylation

Although securin destruction is thought to be required for separase activation in all eukaryotes, it may not always be sufficient; other mechanisms also help govern the timing of separase activation and Scc1 cleavage. In vertebrate cells, the activity of separase is inhibited not only by securin but also by its phosphorylation by Cdk1. Activation of separase is therefore promoted by Cdk1 inactivation as well as by securin destruction. APC^{Cdc20} ubiquitinates both securin and cyclin B in vertebrate cells, resulting in their simultaneous destruction during metaphase (see Figure 7-2b), and so Cdk1 is inactivated at the same time as separase is released from securin.

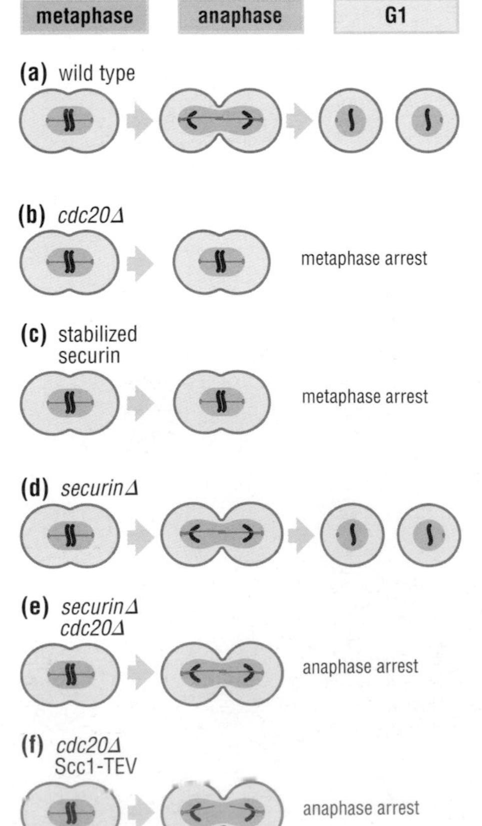

Figure 7-10 Analysis of securin and separase function in budding yeast Studies of various yeast mutants reveal the basic principles underlying the control of sister-chromatid separation. **(a)** In wild-type yeast, APC^{Cdc20} activation leads to sister-chromatid separation and the completion of mitosis. **(b)** Cells lacking Cdc20 (*cdc20Δ*) do not activate the APC at mitosis and fail to degrade securin, thereby preventing separase activation and sister-chromatid separation. **(c)** A mutant securin that is not recognized by APC^{Cdc20} is not destroyed during mitosis, and sister-chromatid separation does not occur. Securin destruction is therefore required for sister separation, because separase cannot be activated. **(d)** Budding yeast cells lacking securin do not display any major mitotic defects, indicating that the timing of separase activation or cohesin cleavage is not determined solely by securin destruction but also by other mechanisms. That is, securin destruction is required but not sufficient for sister separation. **(e)** Cells lacking both Cdc20 and securin do separate sister chromatids. This result argues that *cdc20Δ* mutants arrest in metaphase solely because they fail to degrade securin. These cells do not complete mitosis, however, because cyclin destruction is required for late mitotic events. **(f)** A mutant cell was constructed in which the cohesin subunit Scc1 was engineered to contain a short amino-acid motif that is recognized by a viral protease called TEV protease. Cells were arrested in metaphase by the removal of Cdc20. The cells were then induced to express the TEV protease, which cleaved Scc1. Scc1 cleavage alone was sufficient to trigger anaphase-like chromosome movements despite the absence of APC activity. The extended anaphase B spindle was not completely normal in these cells, however, and mitotic exit did not occur, because APC^{Cdc20} and separase are required for normal anaphase spindle function and the completion of late mitotic events.

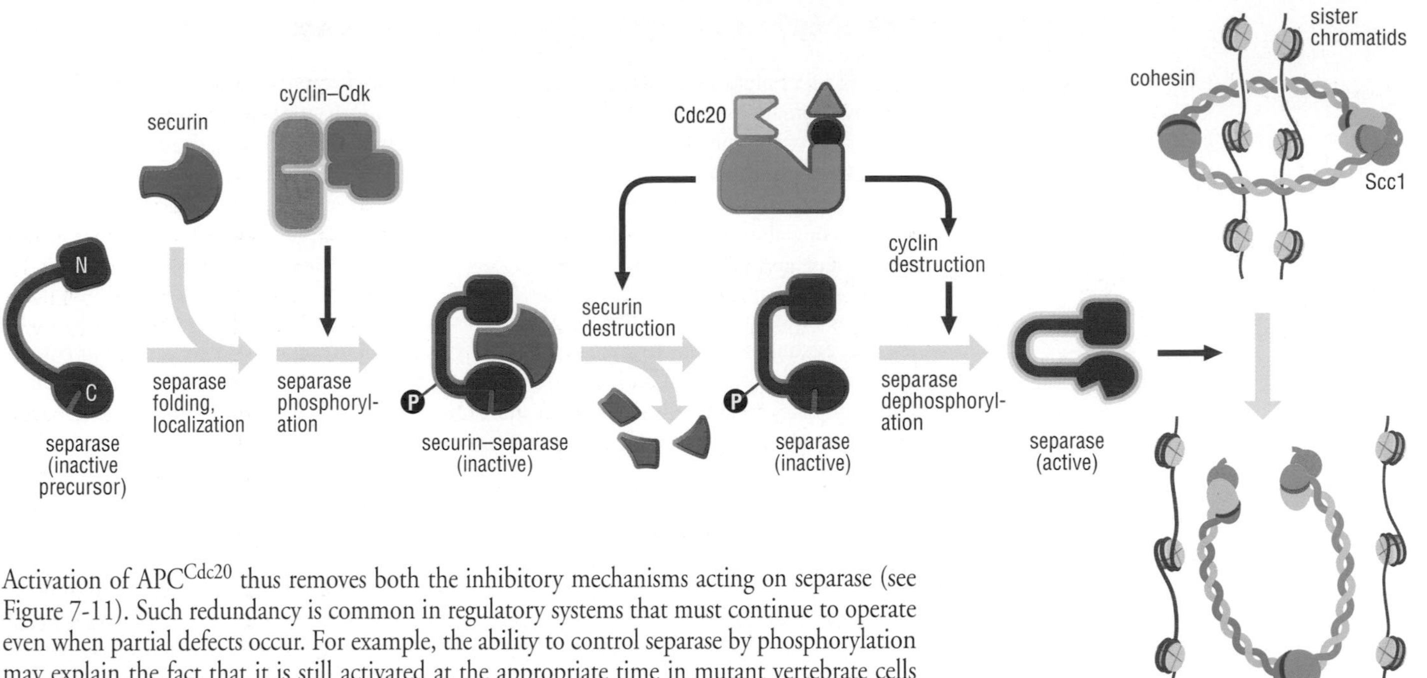

Figure 7-11 **Activation of separase and initiation of sister-chromatid separation** Binding of securin to the protease separase has both positive and negative effects on the separase. First, it is thought to promote the proper folding and localization of separase. Second, it binds tightly to both termini of separase and thereby blocks separase activation. In metazoan cells, separase activity is also suppressed by Cdk-dependent phosphorylation. Activation of APC^{Cdc20} triggers destruction of securin and mitotic cyclins, thereby reversing both inhibitory mechanisms that restrain separase activity. Activation of separase seems to involve an interaction between its amino-terminal regulatory domain and its carboxy-terminal protease domain. Active separase cleaves the Scc1 subunit of the cohesin complex, thereby initiating sister-chromatid separation and anaphase.

Activation of APC^{Cdc20} thus removes both the inhibitory mechanisms acting on separase (see Figure 7-11). Such redundancy is common in regulatory systems that must continue to operate even when partial defects occur. For example, the ability to control separase by phosphorylation may explain the fact that it is still activated at the appropriate time in mutant vertebrate cells lacking securin.

The general importance of separase regulation by phosphorylation remains uncertain. There is much evidence that separase can be activated in frog egg extracts and other animal cells in the presence of considerable Cdk1 activity—for example, when the cells contain mutant forms of cyclin B that cannot be degraded. One possible explanation is that cyclin B–Cdk1 complexes are less effective inhibitors of separase than cyclin A–Cdk1. Indeed, *Drosophila* cells expressing a nondegradable cyclin A mutant arrest in metaphase with unseparated sister chromatids, whereas cells expressing nondegradable cyclin B arrest after sister-chromatid separation. Cyclin A–Cdk1 complexes, which are normally inactivated before metaphase, might therefore be the more important inhibitors of separase activity.

The importance of separase phosphorylation in yeast is also unclear. Under certain experimental conditions, sister separation can occur in the presence of high levels of Cdk1 activity; for example, cells lacking Cdc20 and securin separate sister chromatids without destroying any of the mitotic cyclins (see Figure 7-10). Nevertheless, it is possible that phosphorylation makes some contribution to anaphase control that is critical under some conditions—particularly in the absence of securin.

In human cells, the destruction of securin and cyclin B begins in early metaphase and requires about 20 minutes (see Figure 7-2b). Sister-chromatid separation then occurs abruptly. It remains unclear how the gradual destruction of securin and cyclin B is translated into the sudden wave of Scc1 proteolysis that presumably causes rapid and synchronous sister separation. Additional regulatory components or interactions—perhaps involving positive feedback—are likely to exist in this control system. One additional regulatory mechanism may be phosphorylation of the separase target Scc1. In budding yeast, phosphorylation of Scc1 by the protein kinase Plk is required for efficient cleavage. Similar mechanisms may exist in metazoans, where Plk helps promote cohesin release from chromosome arms in early mitosis (see section 5-10).

Definitions

securin: protein that binds and inhibits the protease **separase**, thereby blocking the onset of sister-chromatid separation. It is also required in some cells for the normal folding or localization of separase.

separase: protease that initiates sister-chromatid separation by cleaving Scc1, a subunit of the cohesin complex that holds sister chromatids together before anaphase.

References

Nasmyth, K.: **Segregating sister genomes: the molecular biology of chromosome separation.** *Science* 2002, **297**:559–565.

Nasmyth, K. and Haering, C.H.: **The structure and function of smc and kleisin complexes.** *Annu. Rev. Biochem.* 2005, **74**:595–648.

Shirayama, M. *et al.*: **APC^{Cdc20} promotes exit from mitosis by destroying the anaphase inhibitor Pds1 and cyclin Clb5.** *Nature* 1999, **402**:203–207.

Stemmann, O. *et al.*: **Dual inhibition of sister chromatid separation at metaphase.** *Cell* 2001, **107**:715–726.

Thornton, B.R. and Toczyski, D.P.: **Securin and B-cyclin/CDK are the only essential targets of the APC.** *Nat. Cell Biol.* 2003 **5**:1090–1094.

Uhlmann, F.: **Chromosome cohesion and separation: from men and molecules.** *Curr. Biol.* 2003, **13**:R104–R114.

Uhlmann, F. *et al.*: **Cleavage of cohesin by the CD clan protease separin triggers anaphase in yeast.** *Cell* 2000, **103**:375–386.

Cdk inactivation in mitosis in budding yeast is not due to APC^{Cdc20} alone

The events of anaphase and telophase that follow sister-chromatid separation depend on the dephosphorylation of Cdk substrates. In most organisms this is triggered primarily by APC^{Cdc20}, which targets mitotic cyclins for destruction and thereby inactivates all major Cdk activities. Budding yeast, however, employs additional regulatory mechanisms that do not seem to be conserved in other species. In this section we review the unique features of late mitotic control in budding yeast, so that in the rest of the chapter we can see their importance in anaphase and telophase in this species.

The first unusual feature is that the APC activator Cdh1 is important in the completion of mitosis. Second, Cdk inactivation and mitotic completion in budding yeast depend in part on the Cdk inhibitor protein Sic1 (see sections 3-6 and 3-13). Finally, the dephosphorylation of Cdk substrates in late mitosis depends not only on Cdk inactivation but also on the specific activation of a protein phosphatase called Cdc14.

APC^{Cdc20} alone cannot drive mitosis to completion in budding yeast. It triggers the destruction of only about half of the mitotic cyclin Clb2 in the cell, and although this is essential for mitotic exit and will trigger spindle disassembly, it is not sufficient for all late mitotic events, such as the assembly of prereplicative complexes at replication origins (see section 4-4). Complete destruction of Clb2 requires the activation of APC^{Cdh1} in late mitosis. In budding yeast, therefore, Cdh1 has assumed a more critical function in late mitosis than in other species.

Cdh1 is activated by dephosphorylation of sites phosphorylated by Cdk1 (see section 3-10), and in most species this is brought on by APC^{Cdc20}-dependent destruction of mitotic cyclins and the inactivation of their partner Cdk1. In budding yeast, however, partial APC^{Cdc20}-dependent destruction of Clb2 is not sufficient to activate APC^{Cdh1}. Cdh1 dephosphorylation requires the APC^{Cdc20}-dependent destruction of the S-phase cyclin Clb5 (Figure 7-12). Thus, Clb5–Cdk1 is particularly important for the phosphorylation of Cdh1 in yeast.

Clb5 destruction also promotes dephosphorylation of the Cdk inhibitor Sic1, leading to an increase in its concentration in late mitosis (see section 3-6). Sic1 and APC^{Cdh1} then collaborate to complete the inactivation of Clb2–Cdk1 and the exit from mitosis. Deletion of either Sic1 or Cdh1 has only minor effects on Clb2–Cdk1 inactivation because each protein alone is sufficient for this process. Deletion of Cdh1 does have some minor effects because APC^{Cdh1} triggers the destruction of several proteins, such as the kinase Plk and the microtubule-binding protein Ase1, which help control late mitotic events.

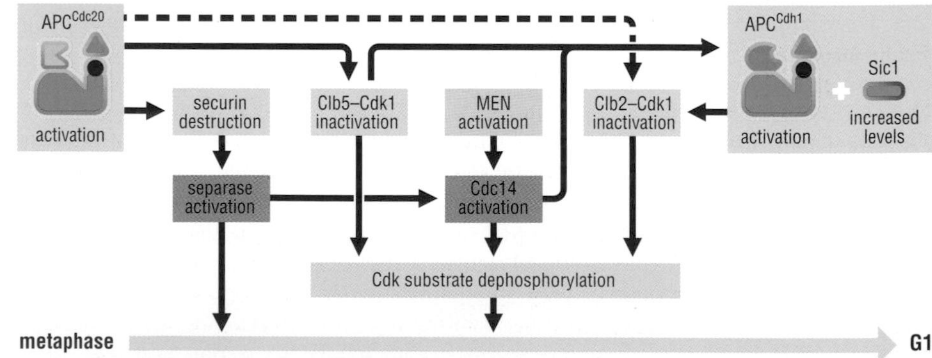

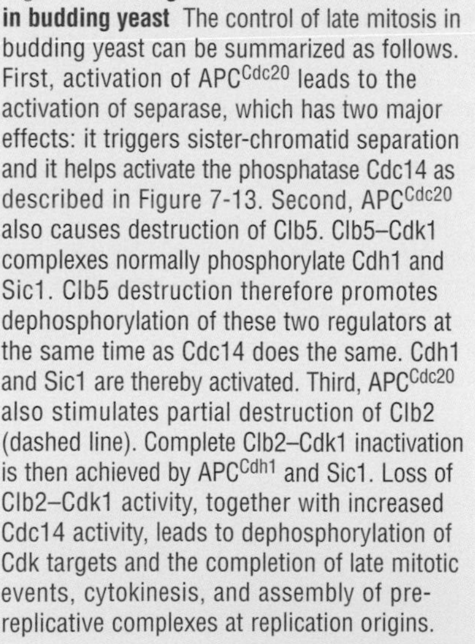
Figure 7-12 Regulation of late mitotic events in budding yeast The control of late mitosis in budding yeast can be summarized as follows. First, activation of APC^{Cdc20} leads to the activation of separase, which has two major effects: it triggers sister-chromatid separation and it helps activate the phosphatase Cdc14 as described in Figure 7-13. Second, APC^{Cdc20} also causes destruction of Clb5. Clb5–Cdk1 complexes normally phosphorylate Cdh1 and Sic1. Clb5 destruction therefore promotes dephosphorylation of these two regulators at the same time as Cdc14 does the same. Cdh1 and Sic1 are thereby activated. Third, APC^{Cdc20} also stimulates partial destruction of Clb2 (dashed line). Complete Clb2–Cdk1 inactivation is then achieved by APC^{Cdh1} and Sic1. Loss of Clb2–Cdk1 activity, together with increased Cdc14 activity, leads to dephosphorylation of Cdk targets and the completion of late mitotic events, cytokinesis, and assembly of pre-replicative complexes at replication origins.

References

D'Amours, D. and Amon, A.: **At the interface between signaling and executing anaphase—Cdc14 and the FEAR network.** *Genes Dev.* 2004, **18**:2581–2595.

Jaspersen, S.L. *et al.*: **A late mitotic regulatory network controlling cyclin destruction in *Saccharomyces cerevisiae*.** *Mol. Biol. Cell* 1998, **9**:2803–2817.

Morgan, D.O.: **Regulation of the APC and the exit from mitosis.** *Nat. Cell Biol.* 1999, **1**:E47–E53.

Shou, W. *et al.*: **Exit from mitosis is triggered by Tem1-dependent release of the protein phosphatase Cdc14 from nucleolar RENT complex.** *Cell* 1999, **97**:233–244.

Stegmeier, F. *et al.*: **Separase, polo kinase, the kinetochore protein Slk19, and Spo12 function in a network that controls Cdc14 localization during early anaphase.** *Cell* 2002, **108**:207–220.

Wäsch, R. and Cross, F.: **APC-dependent proteolysis of the mitotic cyclin Clb2 is essential for mitotic exit.** *Nature* 2002, **418**:556–562.

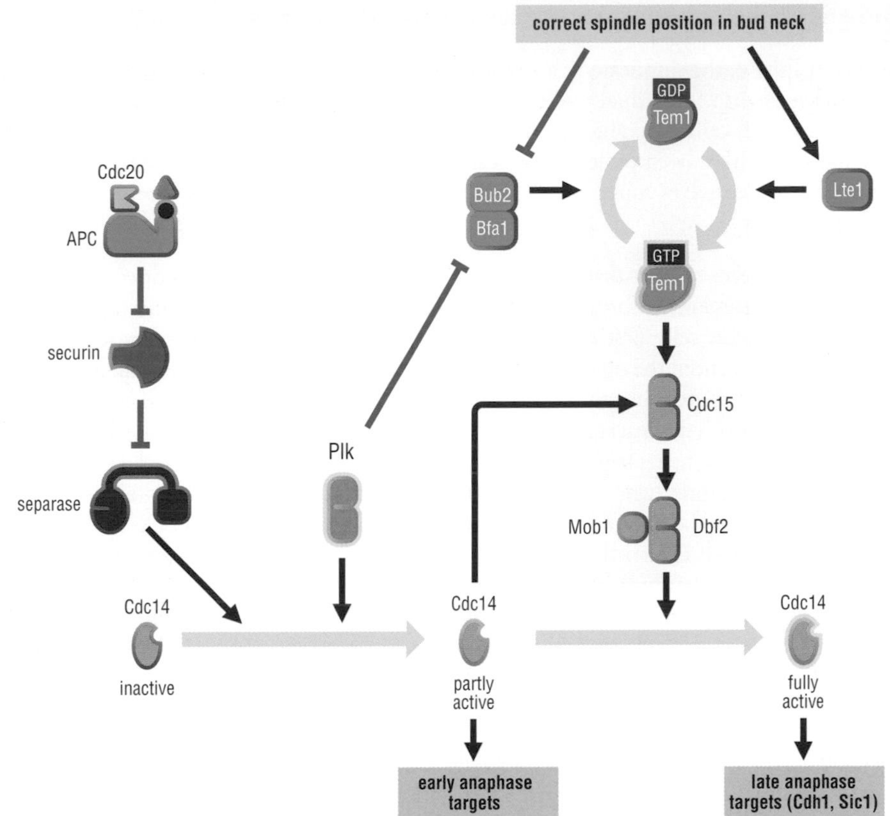

Figure 7-13 **Activation of the phosphatase Cdc14 in budding yeast** The activity of Cdc14 is controlled by two regulatory pathways. First, in early anaphase, separase collaborates with other proteins, including Plk, to initiate partial Cdc14 activation and dephosphorylation of a subset of early-anaphase Cdc14 targets. Second, later in anaphase, a cascade of regulatory proteins called the mitotic exit network (MEN; highlighted in pale yellow) stimulates the complete activation of Cdc14 and dephosphorylation of late-anaphase targets, including Cdh1 and Sic1. The MEN is composed of three major components: at the top lies the GTPase Tem1, which activates the protein kinase Cdc15, which in turn stimulates another protein kinase called Dbf2–Mob1. Dbf2–Mob1 triggers complete Cdc14 activation by unknown mechanisms. Like other small GTPases (Ran, for example; see section 6-8), Tem1 is active when bound to GTP, and its function is controlled by proteins that alter its nucleotide-binding state. It is stimulated by the guanine-nucleotide exchange factor Lte1 and inactivated by a GTPase-activating protein composed of two subunits, Bub2 and Bfa1. The mechanisms that activate the MEN are not clear, but it may be activated in part by Plk, which phosphorylates and inhibits Bub2–Bfa1, thereby promoting Tem1 activity. The MEN is also partly activated when the early-anaphase form of Cdc14 dephosphorylates the protein kinase Cdc15. Interestingly, activation of Tem1 also depends on the correct insertion of the mitotic spindle in the bud neck, thereby ensuring that mitosis is completed only when the chromosomes are correctly distributed between the mother cell and daughter bud.

The protein phosphatase Cdc14 is required to complete mitosis in budding yeast

The activation of Cdh1 and Sic1 by dephosphorylation is not simply the result of decreased Cdk activity but is also due to the protein phosphatase Cdc14, whose activity increases after anaphase (see Figure 7-12). Cdc14 catalyzes the removal of phosphates from specific Cdk1 substrates and thus promotes the dephosphorylation of Cdh1 and Sic1 at the same time as Cdks are inactivated. Cdc14 also acts on other Cdk1 substrates whose dephosphorylation is important for the control of late mitotic events.

Before anaphase, Cdc14 is sequestered in the nucleolus by an interaction with the nucleolar protein Net1/Cfi1, which is also thought to render it inactive. During anaphase, Cdc14 is liberated from the nucleolus and distributed, in active form, throughout the nucleus and cytoplasm—where it has access to its numerous targets. Cdc14 is activated by two regulatory mechanisms—one acting in early anaphase and the other slightly later (Figure 7-13). In early anaphase, Cdc14 is partly activated by separase—by an unknown mechanism that also involves the protein kinase Plk and several other proteins (sometimes called the fourteen early-anaphase release (FEAR) network). Thus, by causing separase activation, APC^{Cdc20} promotes both sister-chromatid separation and partial activation of a phosphatase that dephosphorylates Cdk targets. Partly activated Cdc14 is thought to be particularly important for the dephosphorylation of Cdk substrates involved in early-anaphase changes in spindle and chromosome behavior.

This form of Cdc14 cannot fully activate Cdh1 and Sic1, however, and thus cannot drive complete Clb2–Cdk1 inactivation and the completion of mitosis. Cdc14 is fully activated by an additional regulatory pathway called the mitotic exit network (MEN). The MEN is centered on a small GTPase called Tem1, which initiates a pathway that activates Cdc14 (see Figure 7-13). There is evidence that Plk and the partly active form of Cdc14 are involved in activating the MEN.

It is not clear whether other eukaryotes have similar regulatory mechanisms. Cdc14-related phosphatases have been identified in other species, but their functions are generally not limited to late mitosis, and they may be involved in the dephosphorylation of many Cdk targets at different cell-cycle stages. It is also not known whether Cdc14-related phosphatases in other species are activated by separase or by a MEN-like regulatory network.

7-6 Control of Anaphase Events

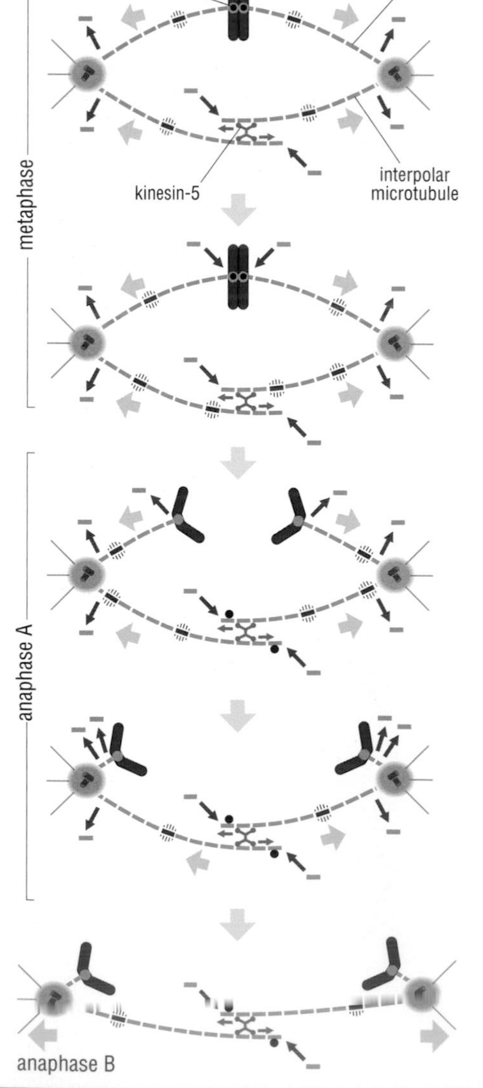

The anaphase spindle segregates the chromosomes

During anaphase, the duplicated chromosomes are moved to opposite poles of the cell and then packaged into individual nuclei during telophase, the last stage of mitosis. The first part of anaphase, called anaphase A, begins with sister-chromatid separation and continues until the sisters have been pulled to opposite poles of the spindle. During anaphase B, the spindle poles themselves move apart, further increasing the distance between the two sets of chromosomes.

Elegant experiments suggest that chromosome movement in anaphase A is driven, at least in part, by the same poleward forces that act on kinetochores in metaphase (see section 6-11). In metaphase vertebrate cells, destruction of one sister kinetochore with a laser beam results in the immediate movement of the other sister toward its closer pole. Similarly, in metaphase-arrested budding yeast it is possible to trigger poleward chromosome movements by artificially inducing the proteolytic cleavage of cohesin alone (see Figure 7-10f). Two major forces act on chromosomes in both metaphase and anaphase A (see section 6-11): the first is the kinetochore-generated poleward force, whereby chromosome movement is coupled to microtubule plus-end depolymerization at the kinetochore; and the second is poleward microtubule flux, whereby the microtubule itself is carried toward the pole by depolymerization at its minus end (Figure 7-14). The relative contributions of these two forces vary in different cell types. Kinetochore-generated forces are particularly important in yeast (which do not have microtubule flux) and in mammalian somatic cells (in which flux occurs but is slow), whereas microtubule flux is a major determinant of chromosome movement in the embryos of *Drosophila* and *Xenopus* (see section 6-11).

Spindle elongation in anaphase B results mainly from the activities of plus-end-directed bipolar kinesins (of the kinesin-5 family) that cross-link interpolar microtubules (see Figure 6-7). These kinesins help drive flux in metaphase and anaphase A by pushing interpolar microtubules toward the poles, where they depolymerize at their minus ends. In anaphase B, minus-end depolymerization is stopped, allowing the kinesin-5 motors in the midzone to push the spindle poles apart (see Figure 7-14). In some cells, minus-end-directed dynein motors anchored to astral microtubules at the cell cortex also contribute by pulling spindle poles apart (see Figure 6-7).

In metaphase, a polar ejection force helps promote chromosome congression (see section 6-11). To allow the movement of chromosomes toward the spindle poles, this ejection force is inactivated in anaphase—as a result of the proteolytic destruction or relocalization of the chromokinesin motors associated with chromosome arms. For example, one of these motors, Kid, is targeted for destruction by APC^{Cdc20} in late metaphase.

Dephosphorylation of Cdk targets governs anaphase spindle behavior

Although experimental disruption of sister-chromatid cohesion alone can initiate chromosome segregation, normal chromosome movements in anaphase A and B also depend on regulated changes in the behavior of proteins that govern microtubule behavior and chromosome attachment to the spindle. The activities of many of these proteins are controlled in a cell-cycle-dependent fashion by their Cdk-dependent phosphorylation in early mitosis and subsequent dephosphorylation in anaphase.

Figure 7-14 Spindle forces in anaphase A and B Chromosomes are pulled toward the spindle poles by two major forces: plus-end depolymerization at the kinetochores and transport of microtubules toward the poles (microtubule flux). Flux requires the addition of tubulin at the plus end and the equivalent removal of tubulin from the minus end. Flux is illustrated here by the poleward movement of tubulin subunits labeled with a red fluorescent tag (see Figure 6-28). The spindle in this diagram is representative of a *Drosophila* embryonic spindle, in which flux occurs at a rapid rate and is the predominant force-generating mechanism in the metaphase spindle. After the loss of sister-chromatid cohesion at the beginning of anaphase, sister chromatids move toward the poles as a result of a combination of microtubule flux and plus-end depolymerization at the kinetochore. When anaphase A is complete, minus-end depolymerization is inactivated and flux ceases. This allows the poles to be pushed outward by kinesin-5 motor proteins that cross-link overlapping plus ends of interpolar microtubules. The aurora B–INCENP complex (black dots) is transferred from the kinetochore in metaphase to interpolar plus ends in anaphase, allowing it to help govern microtubule behavior in the spindle midzone.

The importance of Cdk substrate dephosphorylation in anaphase can be seen in studies of cells in which that dephosphorylation is blocked experimentally. In animal cells, for example, expression of a nondegradable cyclin B mutant, which is no longer recognized by APC^{Cdc20}, does not prevent sister-chromatid separation but results in abnormal anaphase A chromosome movements and defects in anaphase B spindle elongation (Figure 7-15). A similar conclusion can be drawn from studies in budding yeast. As described in Figure 7-10f, artificial induction of cohesin cleavage in metaphase-arrested yeast cells lacking Cdc20 results in sister-chromatid separation and movement of chromosomes toward the spindle poles. Detailed analysis of these cells, however, reveals defects in anaphase A chromosome movements and anaphase B spindle elongation, often resulting in broken spindles. Artificial activation of the phosphatase Cdc14 in these cells (see section 7-5) restores normal anaphase spindle behavior, supporting once again the importance of Cdk substrate dephosphorylation.

An intriguing candidate for regulation by Cdks in anaphase is the complex of the mitotic protein kinase aurora B and its binding partner INCENP (see section 5-7). Aurora B–INCENP is found on kinetochores in metaphase, but is transferred to the spindle midzone in anaphase (see Figure 7-14). Here it forms part of a large complex of proteins (sometimes called passenger proteins) that helps stabilize the overlapping plus ends of interpolar microtubules. In animal cells, transfer of aurora B–INCENP to the spindle midzone is blocked in the presence of non-degradable cyclin B mutants, suggesting that dephosphorylation of some Cdk target is required for this transfer. The Cdk target may be the INCENP protein itself. In budding yeast, INCENP (called Sli15 in this species) is a Cdk1 substrate whose dephosphorylation in early anaphase stimulates its association with the spindle midzone and helps promote the stabilization and elongation of the spindle. Other Cdk1 substrates in yeast, including the microtubule-binding proteins Ase1 and Fin1, are also dephosphorylated in early anaphase and help stabilize the anaphase B spindle.

The transfer of aurora B from the kinetochore to the spindle midzone may also be required for normal anaphase A chromosome movements. As described elsewhere (section 6-10), aurora B in the kinetochore is thought to destabilize microtubule attachments when there is little tension across the kinetochores. This helps prevent stable attachments in the absence of bi-orientation. When sister chromatids separate in anaphase, however, the loss of tension might be expected to activate aurora B and destabilize microtubule attachments—at a time when these attachments need to be extremely stable. The problem seems to be solved, at least in part, by the removal of aurora B from the kinetochore in anaphase. If dephosphorylation of Cdk targets, and thus the removal of aurora B, is prevented, weakened microtubule–kinetochore attachments result in defective anaphase A chromosome movements.

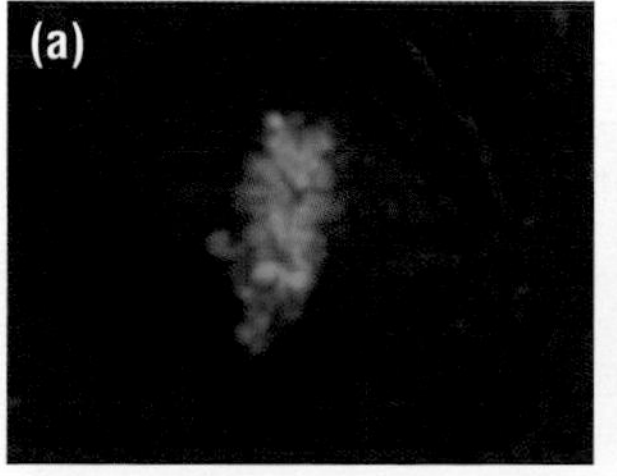

stable cyclin A

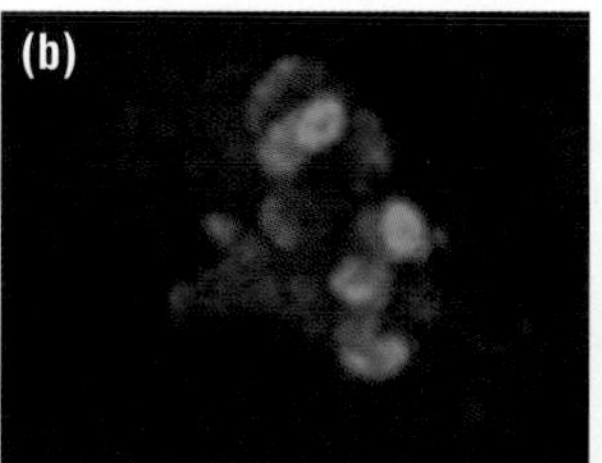

stable cyclin B

stable cyclin B3

Figure 7-15 **Anaphase defects in the presence of nondegradable cyclin mutants** *Drosophila* embryos in G2 of cycle 14 were induced to express nondegradable mutant versions of each of the three major mitotic cyclins. In each case, expression of one stable cyclin did not block destruction of the other two cyclin subtypes. The results show that different cyclins block different mitotic events, arguing that each cyclin–Cdk complex targets a specific array of protein substrates. **(a)** A stable mutant form of cyclin A causes a metaphase arrest (chromosomes are shown in green). One likely explanation for this result, as discussed in section 7-4, is that separase is inhibited by cyclin A–Cdk-dependent phosphorylation in these cells. **(b)** Nondegradable cyclin B causes an arrest in anaphase A. Anaphase B does not occur, and chromosomes oscillate back and forth without coming to rest at the poles. Dephosphorylation of cyclin B–Cdk targets is therefore required for the completion of normal anaphase. **(c)** A stable form of cyclin B3 causes an arrest at the end of anaphase B, and spindle disassembly and other late M-phase events do not occur. Photographs kindly provided by Devin Parry and Patrick O'Farrell. From Parry, D.H. and O'Farrell, P.H.: *Curr. Biol.* 2001, **11**:671–683.

References

Higuchi, T. and Uhlmann, F.: **Stabilization of microtubule dynamics at anaphase onset promotes chromosome segregation.** *Nature* 2005, **433**:171–176.

Kwon, M. and Scholey, J.M.: **Spindle mechanics and dynamics during mitosis in *Drosophila*.** *Trends Cell Biol.* 2004, **14**:194–205.

Parry, D.H. and O'Farrell, P.H.: **The schedule of destruction of three mitotic cyclins can dictate the timing of events during exit from mitosis.** *Curr. Biol.* 2001, **11**:671–683.

Parry, D.H. *et al.*: **Cyclin B destruction triggers changes in kinetochore behavior essential for successful anaphase.** *Curr. Biol.* 2003, **13**:647–653.

Pereira, G. and Schiebel, E.: **Separase regulates INCENP-Aurora B anaphase spindle function through Cdc14.** *Science* 2003, **302**:2120–2124.

Wheatley, S.P. *et al.*: **CDK1 inactivation regulates anaphase spindle dynamics and cytokinesis *in vivo*.** *J. Cell Biol.* 1997, **138**:385–393.

(a)

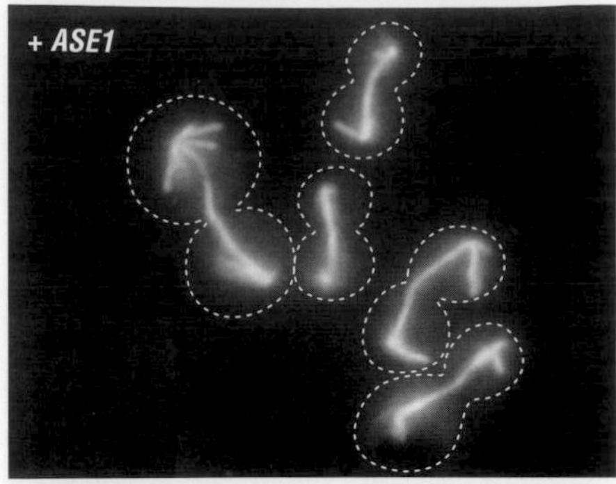

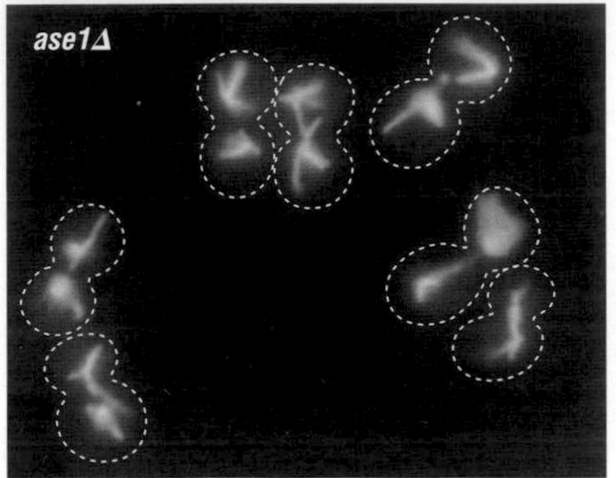

(b)

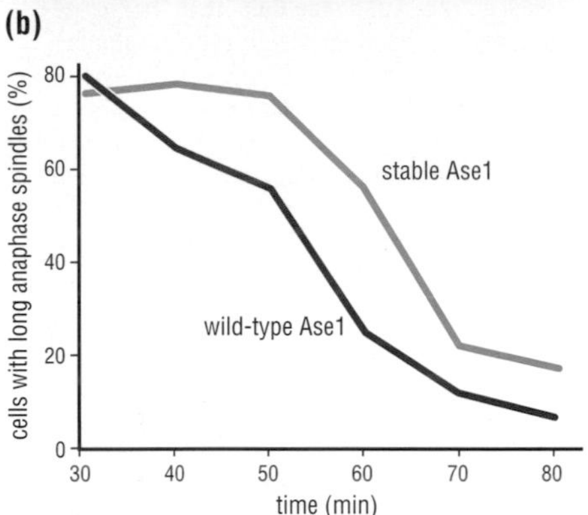

Dephosphorylation of Cdk substrates drives the final steps of mitosis

After anaphase B, the two sets of segregated chromosomes are held near the poles of the extended spindle. In telophase, these chromosomes are packaged into a pair of daughter nuclei and the cell thereby completes mitosis. The major events of telophase are disassembly of the mitotic spindle, decondensation of the chromosomes, and assembly of a nucleus around each chromosome set—except in yeast, in which the nuclear envelope remains intact in mitosis and nuclear division occurs during cytokinesis.

The major mechanism driving telophase in all eukaryotes is dephosphorylation of Cdk substrates, primarily as a result of cyclin destruction triggered by APC^{Cdc20} and the inactivation of Cdks (see sections 7-0 and 7-1). The events of telophase do not occur if Cdk inactivation is prevented, or (in budding yeast) if activation of the phosphatase Cdc14 is blocked. APC^{Cdh1}, which replaces APC^{Cdc20} after anaphase, also contributes to the control of late mitosis in some cells by promoting the destruction of regulatory proteins involved in spindle disassembly or other processes.

Another important participant in telophase is a protein called Cdc48, a member of a family of ATPases that use energy derived from ATP hydrolysis to remodel protein conformation. Cdc48 is required for spindle disassembly, chromosome decondensation and nuclear envelope assembly, possibly because it modifies the structure of proteins directly involved in these processes. In addition, Cdc48 is required for the targeting of some ubiquitinated proteins to the proteasome for destruction, which may indicate that protein destruction is important for late mitotic events.

Spindle disassembly is the central event of telophase

In all eukaryotes, the major event in the completion of mitosis is disassembly of the mitotic spindle. Mitotic increases in microtubule dynamics are reversed, kinetochores are detached from microtubules, and the centrosomes and spindle pole bodies return to their interphase states. The molecular basis of disassembly is not understood in any detail, but it is likely that it is due largely to dephosphorylation of Cdk1 targets and the targets of other protein kinases that drive mitotic changes in centrosome and microtubule behavior. The minus-end cross-linking protein NuMA, for example, is a Cdk substrate whose dephosphorylation in late mitosis promotes its dissociation from spindle poles.

In some cell types at least, late mitotic activation of APC^{Cdh1} helps promote spindle disassembly. In budding yeast, APC^{Cdh1}-dependent destruction of the microtubule-stabilizing protein Ase1 contributes to spindle disassembly (Figure 7-16). In somatic animal cells, APC^{Cdh1}-

Figure 7-16 The APC helps promote spindle disassembly in budding yeast These experiments investigated the regulation of Ase1, a protein that is thought to bind and stabilize microtubule plus ends in the anaphase spindle of budding yeast. Ase1 is ubiquitinated by APC^{Cdh1} in late mitosis, resulting in its destruction. **(a)** Yeast cells were arrested in late anaphase with a temperature-sensitive mutation in a component of the mitotic exit network (*cdc15^{ts}*), and microtubules were labeled with a fluorescent antibody. Normally, this arrest results in cells with long anaphase spindles (top panel). In cells lacking Ase1, however (bottom panel), the late anaphase arrest results in broken spindles, so that the mother and bud each contain a single spindle pole with short astral microtubules. Ase1 is therefore required for stability of the late anaphase spindle. **(b)** The importance of Ase1 destruction was tested by analyzing the timing of spindle disassembly in cells expressing a nondestructible mutant Ase1 protein. Cells were arrested in late anaphase with a *cdc15^{ts}* mutation and then released from the arrest, and the number of cells with long anaphase spindles was measured. Cells expressing stabilized Ase1 protein (green) disassemble their spindles more slowly than cells with wild-type Ase1 (blue). These results, combined with the results shown in panel (a), argue that APC-dependent destruction of Ase1 is required for rapid spindle disassembly in late mitosis. This spindle defect is not lethal, explaining in part why cells lacking Cdh1 are viable. Photographs kindly provided by David Pellman. Reprinted, with permission, from Juang, Y.L. *et al.*: *Science* 1997, **275**:1311–1314.

References

Antonin, W. *et al.*: **The integral membrane nucleoporin pom121 functionally links nuclear pore complex assembly and nuclear envelope formation.** *Mol. Cell* 2005, **17**:83–92.

Cao, K. *et al.*: **The AAA-ATPase Cdc48/p97 regulates spindle disassembly at the end of mitosis.** *Cell* 2003, **115**:355–367.

Hetzer, M. *et al.*: **Distinct AAA-ATPase p97 complexes function in discrete steps of nuclear assembly.** *Nat. Cell Biol.* 2001, **3**:1086–1091.

Hetzer, M.W. *et al.*: **Pushing the envelope: structure, function, and dynamics of the nuclear periphery.** *Annu. Rev. Cell Dev. Biol.* 2005, **21**:347–380.

Juang, Y.L. *et al.*: **APC-mediated proteolysis of Ase1 and the morphogenesis of the mitotic spindle.** *Science* 1997, **275**:1311–1314.

Walther, T.C. *et al.*: **RanGTP mediates nuclear pore complex assembly.** *Nature* 2003, **424**:689–694.

Wozniak, R. and Clarke, P.R.: **Nuclear pores: sowing the seeds of assembly on the chromatin landscape.** *Curr. Biol.* 2003, **13**:R970–R972.

dependent destruction of the protein kinases Plk and aurora A may help promote dephosphorylation of their targets in the spindle (see section 5-7). Destruction of the kinesin-7 motor protein CENP-E at kinetochores (see section 6-5) could aid in microtubule detachment. Reformation of the nuclear envelope also contributes to spindle disassembly by blocking the microtubule-stabilizing effect of the chromosomes (see section 6-8).

In animal cells, a large number of interpolar microtubules are left between the spindle poles after anaphase B. These microtubules become disconnected from the centrosomes and form a tight bundle of antiparallel microtubules, called the *central spindle*, which continues to exist until the end of cell division. As discussed in Chapter 8, the central spindle serves as an organizing center for several regulatory proteins that control cytokinesis.

Another major event in telophase is the reversal of chromosome condensation. Although dephosphorylation of Cdk1 targets is clearly important in this process, little is known about how decondensation is timed and coordinated with other late mitotic events. In vertebrate cells, most chromosome decondensation occurs after reassembly of the nuclear envelope.

Nuclear envelope assembly begins around individual chromosomes

After anaphase in vertebrate cells, a new nuclear envelope is constructed around the chromosomes gathered at each centrosome. As discussed in section 6-7, the nuclear envelope is a multi-layered structure whose major components are the double nuclear membrane, the nuclear pore complexes and the underlying nuclear lamina. Each of these components is dismantled in early mitosis and reconstructed in late mitosis.

At the beginning of prometaphase, nuclear membranes are fragmented and partly absorbed into the endoplasmic reticulum. In late mitosis this process is reversed, beginning with the association of small nuclear membrane vesicles with the surface of each chromosome as a result of direct binding interactions between inner nuclear membrane proteins and proteins on the surface of the chromosomes (Figure 7-17). At the same time, specific subunits of the nuclear pore complex also associate with chromatin and then bind other nuclear pore complex proteins associated with nuclear membrane vesicles. Membrane vesicles accumulate and then fuse laterally with each other, first enclosing small clusters of chromosomes and eventually encapsulating the entire set. In parallel, nuclear pore complexes assemble into their final interphase form, and additional inner nuclear membrane proteins are added to the growing envelope. The nuclear transport machinery soon establishes the correct interphase localization of nuclear and cytoplasmic proteins. Nuclear lamins are imported and assembled into a nuclear lamina and the nuclear envelope assumes its interphase size and shape.

Like other events of late mitosis, nuclear envelope formation is likely to involve dephosphorylation of Cdk targets. Mitotic cyclin–Cdk complexes phosphorylate numerous protein components of the nuclear pore complex, inner nuclear membrane and nuclear lamina, and dephosphorylation of these components in late mitosis is thought to be required for nuclear envelope assembly.

The small GTPase Ran, which is associated with mitotic chromosomes (see section 6-8), is required for the early steps in nuclear envelope assembly—particularly the recruitment of nuclear pore complex components and nuclear membrane vesicles to the chromosome surface. Although RanGTP is localized near the chromosomes throughout mitosis, it does not stimulate nuclear envelope assembly until Cdk inactivation occurs in late mitosis, presumably because dephosphorylation of Cdk targets is also required. Nuclear envelope assembly is thus triggered by a coincidence of temporal signals (dephosphorylation of Cdk targets in telophase) and spatial signals (RanGTP in the vicinity of chromosomes). As in mitotic spindle assembly, the effects of RanGTP on nuclear envelope assembly depend on its ability to trigger the dissociation of nuclear envelope proteins from importin (see section 6-8).

The completion of mitosis leaves the cell with a pair of genetically identical daughter nuclei in a shared cytoplasm. In most cell types, the complex events of cytokinesis then distribute these nuclei into a pair of daughter cells, as will be described in Chapter 8.

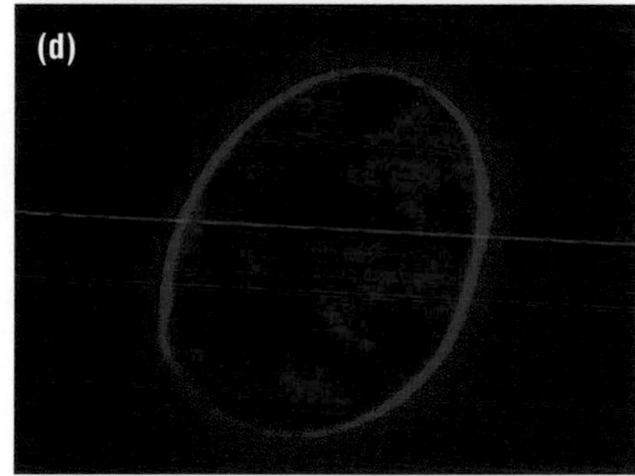

Figure 7-17 Nuclear envelope assembly in *Xenopus* embryo extracts The steps in nuclear envelope assembly can be reconstituted in a cell-free system, as shown here. **(a)** In these experiments, chromosomes (labeled with a blue dye) were prepared from demembranated *Xenopus* sperm and added to a *Xenopus* egg extract in interphase. Nuclear membrane vesicles in the extract were labeled with a red dye. **(b)** Vesicles are first seen accumulating on the surface of a chromosome, after which they become connected by a tubular network **(c)**, eventually fusing into a complete nuclear envelope **(d)**. Photographs kindly provided by Martin Hetzer and Iain Mattaj. Reprinted, with permission, from Hetzer, M. *et al.*: *Nat. Cell Biol.* 2001, **3**:1086–1091.

8

Cytokinesis

After the formation of daughter nuclei in mitosis, cytoplasmic division (cytokinesis) divides the cell itself into two daughter cells, each with identical chromosomes and, in most cases, equal amounts of other cellular components.

8-0 Overview: Cytokinesis

Cytokinesis distributes daughter nuclei into separate cells

When mitosis is complete, the two new nuclei and other cellular components are distributed into a pair of daughter cells, each with a single nucleus, single centrosome, and a roughly equal share of cytoplasmic macromolecules and organelles. This final stage in cell division is called cytokinesis. When viewed through a microscope, cytokinesis appears remarkably different in different organisms (Figure 8-1). Budding yeast divides by forming a bud that grows throughout the cell cycle and detaches from the mother cell after mitosis, whereas fission yeast (and plants) divide by constructing a new cell wall, or *septum*, at the midpoint of the cell after mitosis. Animal cells, unburdened by a rigid cell wall, divide by pulling their membranes inward after anaphase, forming a *cleavage furrow* that pinches the cell in half. Despite these outward differences, cytokinesis in all these species is based, at least in part, on similar mechanisms and molecular components.

Cytokinesis depends on a contractile ring and membrane deposition

In most eukaryotes (with the exception of higher plants), cytokinesis depends on an apparatus called the *contractile ring*, which is attached to the inside face of the cell membrane at the site of cell division and contains contractile bundles of actin and the motor protein myosin II. In animal cells, gradual contraction of this ring after anaphase pulls the membrane inward, resulting in a cleavage furrow that encircles the cell and eventually divides it in two. A similar actin–myosin-based ring helps direct the inward migration of new cell membrane and cell wall during cytokinesis in yeast.

Cytokinesis is not simply a matter of pulling the cell membrane inward. New membrane (and cell wall in yeast) must also be added at the site of division to provide the increased membrane surface area that is required. Membrane and cell-wall deposition occurs in parallel with contraction of the actin–myosin ring, so that membranes grow at a rate that matches the inward movement of the contractile ring.

The cleavage plane is positioned between the daughter nuclei

To ensure that each daughter cell receives a single copy of the genome, the plane of division is positioned between the two segregated sets of chromosomes. This positioning is achieved in most cases by one of two strategies (Figure 8-1).

In yeast, the site of cleavage is determined before mitosis. The cell then uses the mitotic spindle to position the segregated chromosomes on opposite sides of this site. In budding yeast, for example, the bud appears in late G1 and grows throughout the cell cycle, and in anaphase the spindle is positioned within the bud neck, thereby placing a set of chromosomes in each of the cell's progeny. In fission yeast, the cleavage site is marked at the center of the cell before mitosis by accumulation of proteins in a medial ring (see Figure 6-1), and the mitotic spindle poles are positioned on opposite sides of this mark.

In animal cells, the site of cleavage is not determined until anaphase, when the mitotic spindle sends a signal to the cell cortex at the spindle equator, triggering the formation of a cleavage furrow between the segregated chromosomes. In many cells, the spindle is centered in the cell, so that cytokinesis divides the cell into roughly equal halves. Some divisions, however, are asymmetric: the spindle is positioned closer to one end of the cell, resulting in daughter cells of unequal size—and often with different developmental fates.

References

Albertson, R. *et al.*: **Membrane traffic: a driving force in cytokinesis.** *Trends Cell Biol.* 2005, **15**:92–101.

Balasubramanian, M.K. *et al.*: **Comparative analysis of cytokinesis in budding yeast, fission yeast and animal cells.** *Curr. Biol.* 2004, **14**:R806–R818.

Burgess, D.R. and Chang, F.: **Site selection for the cleavage furrow at cytokinesis.** *Trends Cell Biol.* 2005, **15**:156–162.

Field, C. *et al.*: **Cytokinesis in eukaryotes: a mechanistic comparison.** *Curr. Opin. Cell Biol.* 1999, **11**:68–80.

Glotzer, M.: **Animal cell cytokinesis.** *Annu. Rev. Cell Dev. Biol.* 2001, **17**:351–386.

Glotzer, M.: **The molecular requirements for cytokinesis.** *Science* 2005, **307**:1735–1739.

Rappaport, R.: *Cytokinesis in Animal Cells* (Cambridge University Press, Cambridge, 1996).

Robinson, D.N. and Spudich, J.A.: **Towards a molecular understanding of cytokinesis.** *Trends Cell Biol.* 2000, **10**:228–237.

The mechanism by which the spindle determines the division plane in animal cells is perhaps the most enduring mystery in cytokinesis. Cleavage signals are sent to the cell cortex by microtubules of the spindle, but the source and molecular basis of these signals are not clear and probably vary in different cell types. In some cells, for example, astral microtubules from the spindle poles are thought to carry a furrow-stimulating signal that is somehow focused between the centrosomes. In other cells, a cleavage signal is sent to the cell cortex from the spindle midzone or *central spindle*, an array of antiparallel microtubules that is derived, at least in part, from the interpolar microtubules of the anaphase spindle. The central spindle contains numerous regulatory proteins and may have multiple functions in cytokinesis: it not only helps to determine the site of cleavage but it may also help to govern the inward movement of the cleavage furrow and the membrane deposition that occurs there.

As the cleavage furrow deepens, the contractile ring eventually meets the central spindle, which is then compacted into a structure called the *midbody*, which contains bundles of antiparallel microtubules and a dense protein matrix at its midline. The midbody is eventually severed or dismantled to allow the final separation of the daughter cells. Like so many aspects of cytokinesis, we know little about the mechanisms that complete the division process.

The timing of cytokinesis is coordinated with the completion of mitosis

Cytokinesis must not occur until chromosome segregation is completed. The final steps of cell division are therefore initiated by mechanisms that are closely linked to the completion of mitosis. In particular, it seems that a major signal for the initiation of cytokinesis is dephosphorylation of Cdk substrates in anaphase. In general, cytokinesis does not occur if Cdk1 inactivation is prevented (by expression of nondegradable cyclin, for example). Numerous regulatory proteins help determine the timing of cytokinesis, but we understand little about their mechanisms of action or how they are governed by the cell-cycle control system.

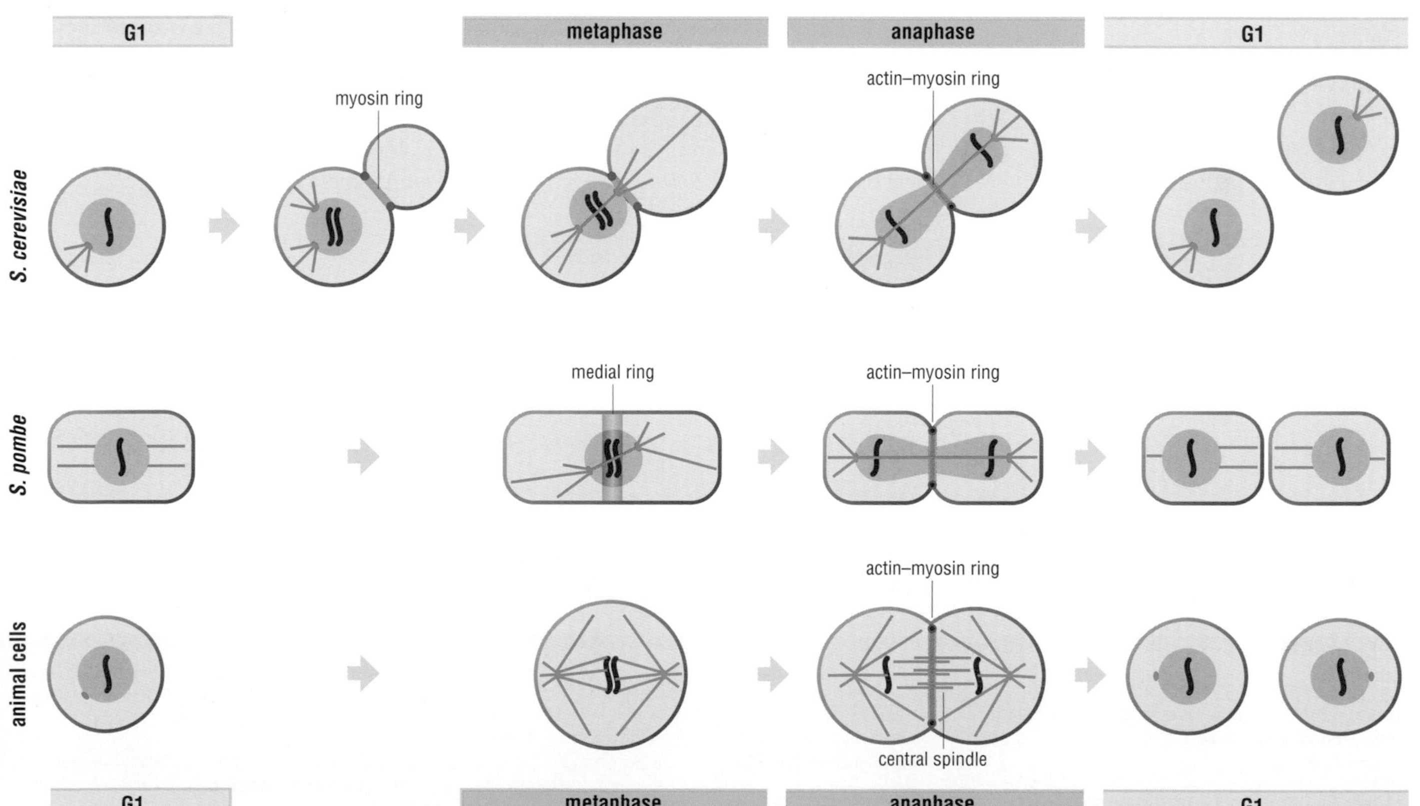

Figure 8-1 Positioning the plane of cell division The general features of cytokinesis are shown for the budding yeast *S. cerevisiae*, the fission yeast *S. pombe*, and animal cells. In the two yeasts, the site of cell division is determined before anaphase (see text). The anaphase spindle is then positioned so that the segregated chromosomes are on opposite sides of the predetermined cleavage plane. In animal cells, the site of cell division is not determined until anaphase, when the mitotic spindle sends a cleavage-inducing signal to the cell cortex.

8-1 The Actin–Myosin Ring

Bundles of actin assemble at the site of division

Cytokinesis in most eukaryotes (except plants) occurs by a centripetal mechanism: that is, the cell membrane and wall grow inward at the site of division, resulting in a **cleavage furrow** that gradually pinches the mother cell in two. This cleavage process is driven by a ring of proteins, the **contractile ring**, that surrounds the cell equator beneath the cell cortex and bisects the axis of chromosome segregation. The key components of this ring are the filamentous protein actin and the motor protein myosin II (Figure 8-2), which form contractile bundles whose contraction causes the ring to shrink and pull the membrane inward.

Actin is a major cytoskeletal protein that polymerizes to form long chains, called **microfilaments** or F-actin, arranged in a two-stranded helix (Figure 8-3). Like microtubules (see section 6-1), actin filaments are polar: one end (called the barbed end, analogous to the microtubule plus end) grows more rapidly than the other (the pointed end, analogous to the microtubule minus end). Actin filaments are thinner and more flexible than microtubules and can be organized into a wide variety of structures—ranging from highly organized linear arrays in skeletal muscle to a variety of cross-linked two- and three-dimensional networks and gels that help control cell shape and motility. The versatility of actin filaments is attributable to the numerous associated regulatory proteins that control their nucleation, stability and organization.

Force is generated in the contractile ring by non-muscle myosin II

Actin filaments serve as the tracks for motor proteins called **myosins**, which generally travel toward the actin barbed end. The force for most contractile processes—including cytokinesis—is generated by movements along actin by various types of the motor protein myosin II. Myosin II has multiple subunits, including two tightly associated large subunits called heavy chains. The long carboxy-terminal tail domains of the two heavy chains are wrapped around each other to form a coiled-coil structure. The amino-terminal ends of these proteins are globular head domains with ATP-dependent motor activity (see Figure 8-3).

Large numbers of these two-headed myosin II molecules are assembled into higher-order filaments (called thick filaments, to contrast them with the thin filaments of actin). Contractile behavior depends on the fact that these myosin II filaments are bipolar: the motor domains at each end are oriented in opposite directions (see Figure 8-3). As a result, a myosin II filament can pull the pointed ends of two actin filaments inward from opposite directions. If the barbed ends of those actin filaments are anchored (in the cell cortex, for example), then the two anchor points will be brought closer together when the actin–myosin bundle contracts.

Components of the Contractile Ring

Protein family	Mammals	*Drosophila*	*C. elegans*	*S. cerevisiae*	*S. pombe*
actin	actin	actin	ACT-5	Act1	Act1
myosin II heavy chain	myosin II	Zipper	NMY-2	Myo1	Myo2, Myp2
myosin essential light chain	EMLC	Mlc-c	?	Mlc1	Cdc4
myosin regulatory light chain	RMLC	Spaghetti squash	MLC-4	Mlc2	Rlc1
formin	Dia1	Diaphanous (Dia)	CYK-1	Bni1, Bnr1	Cdc12
profilin	profilin	Chickadee	PFN-1	Pfy1	Cdc3
cofilin	ADF/cofilin	Twinstar	UNC-60A	Cof1	Cof1

Figure 8-2 Table of proteins in the actin–myosin ring in cytokinesis

Definitions

cleavage furrow: during cytokinesis, the furrow that forms around the equator of a dividing animal cell and eventually divides it in two.

contractile ring: ring of proteins, including contractile assemblies of actin and **myosin**, that forms at the site of cleavage in dividing animal cells. Its gradual contraction pinches the cell in two.

microfilament: long helical polymer of two chains of actin monomers wound around each other. It is a component of the cytoskeleton.

myosin: motor protein that associates with **microfilaments** and moves along them. The form of myosin found in **contractile rings** is non-muscle myosin II.

References

Alberts, B. *et al.*: *Molecular Biology of the Cell* 4th ed. (Garland Science, New York, 2002).

Glotzer, M.: **The molecular requirements for cytokinesis.** *Science* 2005, **307**:1735–1739.

Matsumura, F.: **Regulation of myosin II during cytokinesis in higher eukaryotes.** *Trends Cell Biol.* 2005, **15**:371–377.

Robinson, D.N. and Spudich, J.A.: **Towards a molecular understanding of cytokinesis.** *Trends Cell Biol.* 2000, **10**:228–237.

Schroeder, T.E.: **Cytokinesis: filaments in the cleavage furrow.** *Exp. Cell Res.* 1968, **53**:272–276.

Zigmond, S.H.: **Formin-induced nucleation of actin filaments.** *Curr Opin Cell Biol.* 2004, **16**:99–105.

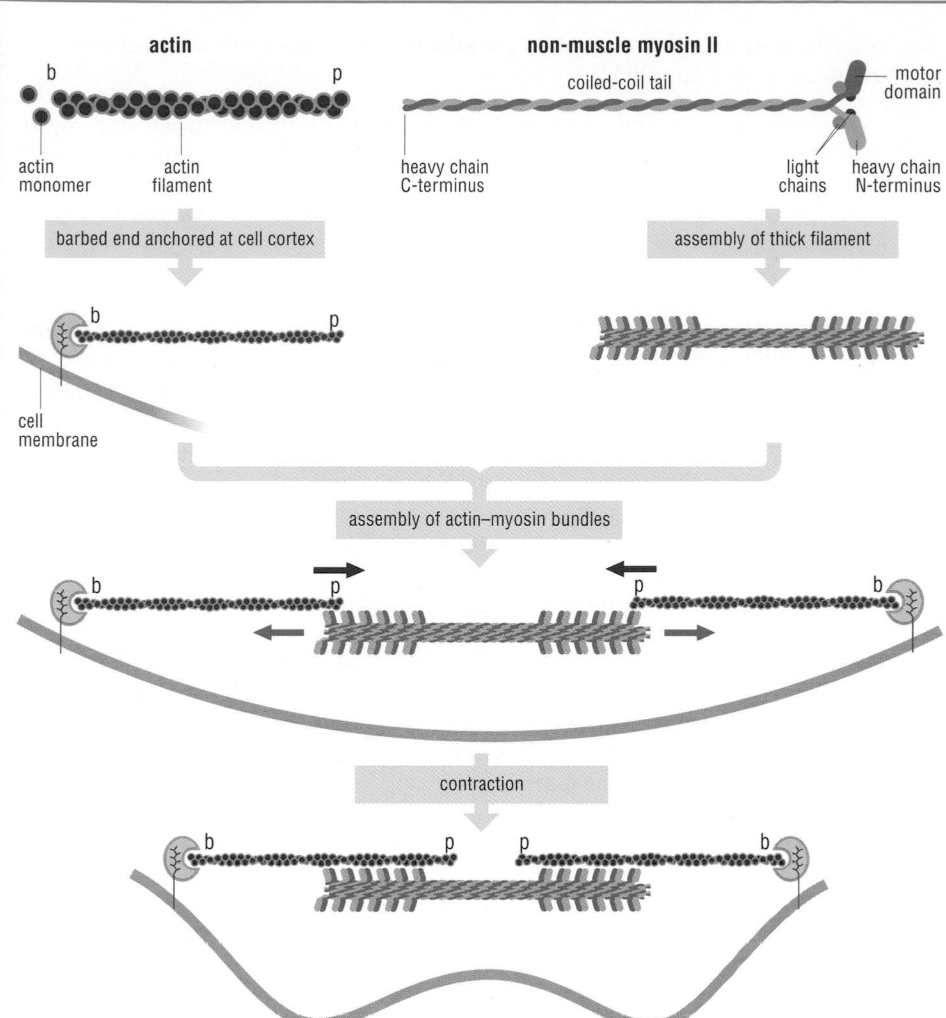

Figure 8-3 The actin–myosin ring As shown on the left, an actin filament is a polar, two-stranded helical polymer. The ends of the filament are called barbed (b) and pointed (p) on the basis of their appearance in the electron microscope when they are decorated with head domains of myosin II. During cytokinesis, the barbed ends of some filaments are anchored at the cell cortex by unknown proteins. As shown on the right, non-muscle myosin II contains two heavy chains whose elongated tail domains intertwine to form a coiled-coil structure. The globular motor domains of the heavy chains interact with two myosin light chains. The tail domains of multiple myosin II molecules interact to form thick filaments with motor domains exposed on the surface. The thick filament is polar, with the motor domains at one end oriented in the opposite direction to the motor domains at the other. Bundles of actin and myosin II are assembled at the cell cortex in the orientation shown here. Contraction occurs when the myosin II motors of the thick filament move toward the barbed ends of the actin. As a result, the anchor points on the cell cortex are brought closer together, resulting in the puckering of the cell membrane.

The myosin II in the contractile ring is a specialized version called non-muscle or cytoplasmic myosin II. Each of the two motor domains of non-muscle myosin II associates with two light chains—the essential myosin light chain and the regulatory myosin light chain. Phosphorylation of the regulatory light chain is a major mechanism by which contractile ring function is controlled.

Actin filament formation depends on formins

In principle, the appearance of actin filaments at the site of division could result either from the recruitment of preexisting filaments from elsewhere or from the generation of new filaments locally. Both possibilities seem to be true, but the latter is probably more important. Actin filament formation at the cleavage site depends primarily on a small group of proteins called formins (see Figure 8-2). Formins nucleate the assembly of new actin filaments from soluble actin monomers. Interestingly, formins interact with the barbed end of actin filaments while still promoting the addition of actin monomers at the same end. The addition of actin monomers to a formin-capped barbed end requires another protein called profilin, which binds to actin monomers and helps load them onto the filament end.

Actin and myosin II in the contractile ring do not form the highly organized arrays of interleaved filaments seen in skeletal muscle. Instead, the filaments of the contractile ring are more loosely organized in actin–myosin bundles that are roughly perpendicular to the spindle axis (Figure 8-4). During the gradual ring contraction that occurs in cytokinesis, these dynamic actin–myosin networks are dismantled and rearranged, so that the overall thickness of the contractile ring does not change significantly as its diameter shrinks. The removal of actin filaments from the shrinking contractile ring depends in part on an actin-destabilizing protein called cofilin (see Figure 8-2).

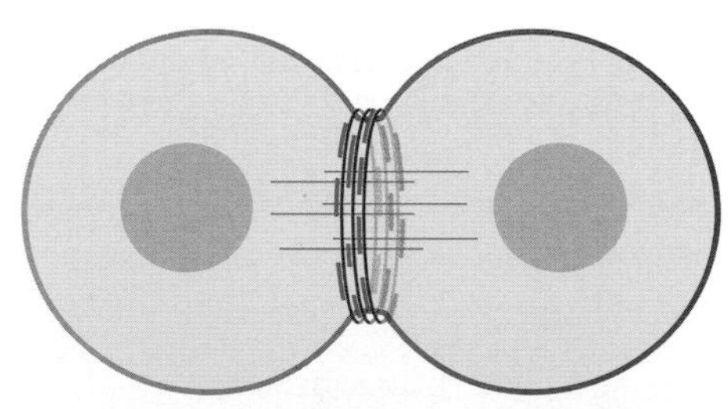

Figure 8-4 The orientation of actin–myosin bundles in the contractile ring Individual actin–myosin bundles are arranged circumferentially around the ring. As they contract, the ring pulls the membrane inward to form the cleavage furrow.

Contractile ring function depends on accessory factors whose importance varies in different species

The actin–myosin ring is a central component of the cytokinesis machinery, and every aspect of its behavior is governed by a host of other proteins. This section provides an overview of the major classes of proteins involved in contractile ring function; their species-specific functions will be addressed in greater detail later in the chapter. Our discussion here focuses on the two most important aspects of contractile ring behavior: its assembly at the correct location and its contraction at the correct time.

Contractile ring positioning and assembly are not well understood, but we are beginning to identify some of the proteins that guide this process. These include the septins, a small family of related proteins that are essential for cytokinesis in *S. cerevisiae* and are important, but not always essential, for cell division in other species. Septins are GTPases that assemble into large complexes and filaments. In budding yeast, septins form a ring at the future bud site in late G1 and are required for the formation of the actin–myosin ring and for the deposition of new cell-wall material at the bud neck. In animal cells, septins are required for normal cytokinesis and co-localize with actin filaments at the cleavage furrow. We do not understand the molecular basis of septin function nor the purpose of their GTPase activity, although one possibility is that septins serve as structural scaffolds on which the cytokinesis apparatus is organized.

Contractile ring organization in animal cells also depends partly on a protein called anillin. Anillin is a multidomain protein that associates with actin, myosin II and septins. In *Drosophila*, anillin is required for the normal localization of septins, and loss of anillin function results in partial cytokinesis defects. In fission yeast, a distantly related protein called Mid1 forms the medial ring and is essential for the normal positioning of the contractile ring (Figure 8-5). Like the septins, anillin and its relatives may help form a structural linkage between different components of the contractile ring.

The IQGAP proteins are another family that interact with actin filaments and are required for normal contractile ring formation, at least in yeast. In budding yeast, the protein Iqg1 appears at the bud neck in anaphase, just before actin, and is required for actin recruitment to that site. In fission yeast, the related Rng2 protein is required for the normal organization of actin filaments at the septum.

Proteins that Organize and Regulate the Contractile Ring

Protein family	Mammals	*Drosophila*	*C. elegans*	*S. cerevisiae*	*S. pombe*
septins	Sept1-10	Peanut, Sep1, 2, 4, 5	UNC-59, UNC-61	Cdc3, 10, 11, 12, Sep7	Spn1-7
anillin	anillin	Scraps	anillin	?	Mid1, Mid2
IQGAP	–	–	–	Iqg1/Cyk1	Rng2
Rho	RhoA	Rho1	RhoA	Rho1	Rho1
RhoGEF	Ect2	Pebble	LET-21	Rom1, Rom2	Gef1, Scd1
Rho-activated kinase	ROCK	dROK	LET-502	–	–
Citron kinase	Citron-K	Citron kinase	Citron kinase	–	–
myosin light-chain kinase	MLCK	MLCK	MLCK	–	–
myosin phosphatase	MYPT	dMYPT	MEL-11	–	–

Figure 8-5 Table of proteins that control the assembly and function of the actin–myosin ring

References

Balasubramanian, M.K. *et al.*: **Comparative analysis of cytokinesis in budding yeast, fission yeast and animal cells.** *Curr. Biol.* 2004, **14**:R806–R818.

Bement, W.M. *et al.*: **A microtubule-dependent zone of active RhoA during cleavage plane specification.** *J. Cell Biol.* 2005, **170**:91–101.

Field, C.M. *et al.*: **Characterization of anillin mutants reveals essential roles in septin localization and plasma membrane integrity.** *Development* 2005, **132**:2849–2860.

Glotzer, M.: **The molecular requirements for cytokinesis.** *Science* 2005, **307**:1735–1739.

Kinoshita, M. *et al.*: **Self- and actin-templated assembly of mammalian septins.** *Dev. Cell* 2002, **3**:791–802.

Matsumura, F.: **Regulation of myosin II during cytokinesis in higher eukaryotes.** *Trends Cell Biol.* 2005, **15**:371–377.

Matsumura, F. *et al.*: **Specific localization of serine 19 phosphorylated myosin II during cell locomotion and mitosis of cultured cells.** *J. Cell Biol.* 1998, **140**:119–129.

Prokopenko, S.N. *et al.*: **A putative exchange factor for Rho1 GTPase is required for initiation of cytokinesis in *Drosophila*.** *Genes Dev.* 1999, **13**:2301–2314.

Straight, A.F. *et al.*: **Anillin binds nonmuscle myosin II and regulates the contractile ring.** *Mol. Biol. Cell* 2005, **16**:193–201.

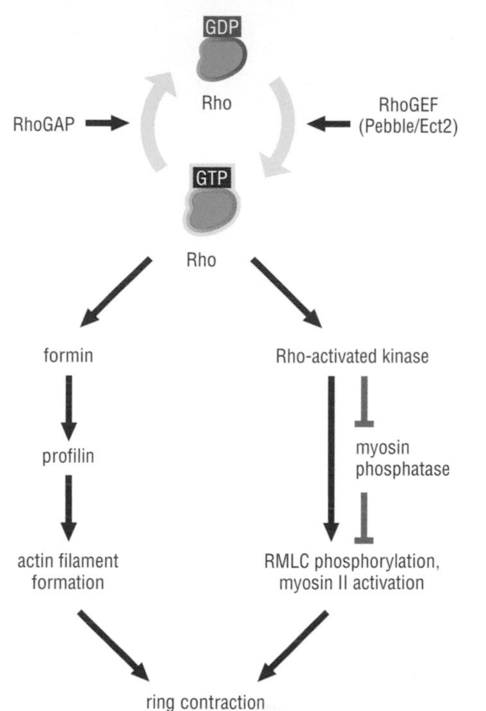

Figure 8-6 Control of actin–myosin ring function in animal cells In its GTP-bound state, Rho is thought to promote actin–myosin ring assembly and contraction by multiple mechanisms, including the two shown here. By stimulating formin activity, RhoGTP promotes actin nucleation and filament assembly at the cleavage furrow. By stimulating the Rho-activated kinase and inhibiting myosin phosphatase, RhoGTP promotes the phosphorylation of the regulatory myosin light chain (RMLC), resulting in an increase in myosin II activity. Other protein kinases, not shown here, also contribute to the control of RMLC phosphorylation. These include the Rho-activated Citron kinase and the myosin light-chain kinase (MLCK) (see Figure 8-5).

Contraction of the actin–myosin ring is regulated by activation of myosin II

The mechanisms that initiate contraction of the actin–myosin ring are best understood in animal cells, in which contraction is regulated primarily by changes in the phosphorylation state of the regulatory myosin light chain (RMLC). Non-muscle myosin II is activated by phosphorylation of a pair of serine residues near the amino terminus of the RMLC (serines 18 and 19 in vertebrate RMLC). RMLC phosphorylation is not only required for the ATP-dependent motor activity of myosin II but may also help promote the assembly of myosin II into the bipolar filaments that are required for the formation of contractile actin–myosin bundles. In animal cells, phosphorylation of these sites increases at the end of mitosis, particularly in myosin II at the cleavage furrow. In *Drosophila* embryos, mutation of RMLC phosphorylation sites inhibits contractile ring formation and cytokinesis.

RMLC phosphorylation is not seen in yeast, and other mechanisms seem to be responsible for triggering contraction of the actin–myosin ring. One potentially important mechanism in fission yeast depends on the protein Rng3, which binds and activates myosin II during cytokinesis.

The GTPase Rho controls actin and myosin behavior at the cleavage site

A major regulator of the contractile ring in animal cells is the protein Rho, a small GTPase of the Ras/Ran family (see section 6-8). In its active GTP-bound state, Rho interacts with multiple targets at the cleavage furrow, including proteins that affect both the assembly and the contraction of the actin–myosin ring (Figure 8-6).

Rho–GTP promotes actin filament formation by binding to formins and stimulating their ability to promote actin nucleation and growth. This function of Rho seems to be conserved in yeast as well as in higher eukaryotes. In animal cells but not yeast, Rho also interacts with multiple targets to increase RMLC phosphorylation at the two activating sites, thereby stimulating the assembly and motility of myosin II. These targets include the Rho-activated kinase, which phosphorylates the activating sites on the RMLC. Rho-activated kinase also phosphorylates the regulatory subunit of a myosin phosphatase, thereby decreasing phosphatase activity and further enhancing RMLC phosphorylation.

Rho is therefore a key regulator of contractile ring formation and contraction in animal cells. What, then, controls the activity of Rho? As is usually the case for small GTPases, Rho is stimulated by a specific guanine-nucleotide exchange factor (RhoGEF) and inhibited by a GTPase-activating protein (RhoGAP). Interestingly, the major RhoGEF, called Pebble in *Drosophila* and Ect2 in mammals, is localized to the site of cleavage and required for cytokinesis (Figure 8-7). An intriguing possibility is that RhoGEF is somehow recruited to the cleavage site and activated by the signals that determine the timing and position of cleavage furrow formation.

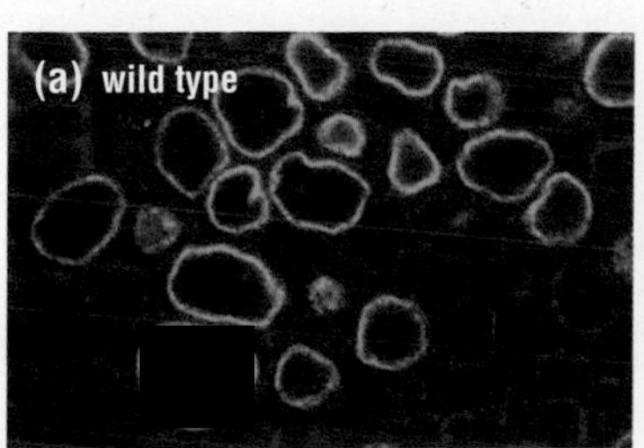

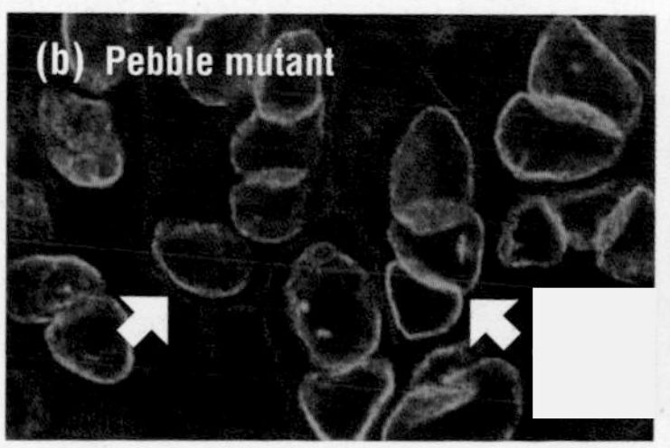

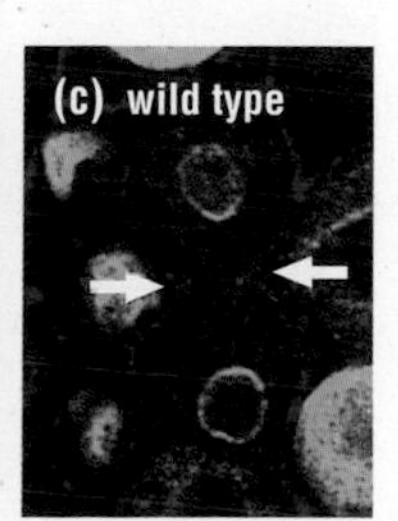

Figure 8-7 Control of cytokinesis by the RhoGEF Pebble in the *Drosophila* embryo **(a)** A wild-type embryo in interphase after mitosis 14 was labeled to mark cell membranes (red) and nuclear envelopes (green), showing that each cell contains a single nucleus. **(b)** In an embryo with a mutation in the RhoGEF Pebble, defects in cytokinesis result in the appearance of cells containing multiple nuclei. **(c)** The Pebble protein was labeled (red) in a wild-type embryo undergoing cytokinesis. Cell membranes are blue and nuclear envelopes green. The Pebble protein can be seen at the cortex of the cleavage furrow. Kindly provided by Sergei N. Prokopenko and Hugo J. Bellen. From Prokopenko, S.N. *et al.*: *Genes Dev.* 1999, **13**:2301–2314.

8-3 Membrane and Cell Wall Deposition at the Division Site

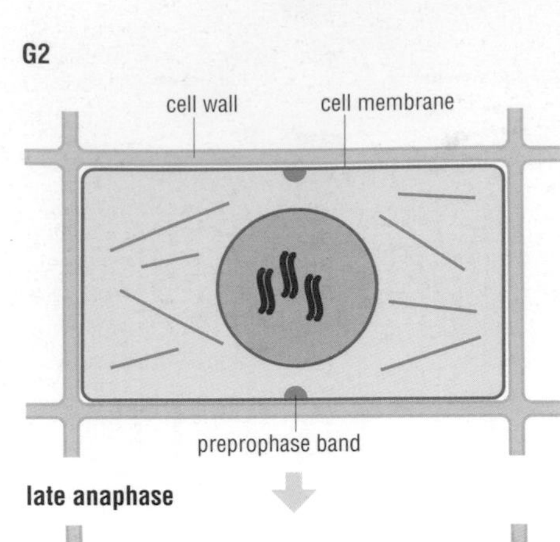

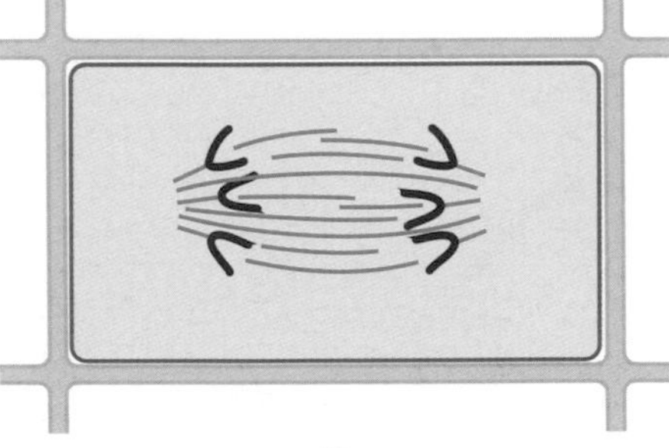

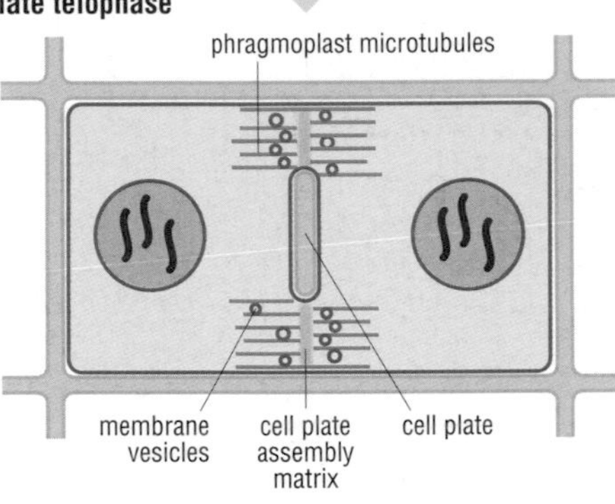

Membrane deposition is required during cytokinesis

Cytokinesis in yeast and animal cells depends on extensive remodeling of the cell membrane at the cleavage furrow. The inward movement of the furrow generally results in an increase in the surface area of the cell membrane, and the additional membrane seems to be provided by the insertion of new membrane at the cleavage site. In yeast, new cell wall materials are also deposited at this site, resulting in the process of septation—the formation of a new cell wall, or **septum**, between the daughter cells.

Addition of new membrane occurs by the fusion of membrane vesicles with the plasma membrane near the inner edge of the cleavage furrow. These small vesicles originate in the Golgi apparatus and are targeted to the plasma membrane at the cleavage site by components of the secretory apparatus, including members of the syntaxin family of vesicle-targeting proteins.

In most cell types, microtubules provide the tracks along which membrane vesicles are transported to the site of cleavage. The importance of microtubules for membrane delivery is particularly apparent in the cells of higher plants, in which cytokinesis is entirely a process of membrane and wall deposition—without any need for a contractile ring. Cytokinesis in these cells is directed by an organelle called the **phragmoplast**, an array of microtubules, derived from the anaphase spindle, whose plus ends are embedded in a protein matrix along the cell midline (Figure 8-8). Membrane vesicles carrying the raw materials for cell wall synthesis travel along these microtubules to the center of the phragmoplast, where they promote the deposition of a new membrane and cell wall to form the cell plate. Unlike the case in yeast and animal cells, where all new membrane is added to the preexisting cell membrane, new membrane formation in plants starts in the center of the cell and spreads outward until it meets the plasma membrane.

The importance of microtubules in membrane addition is also illustrated by studies of *Xenopus* embryonic cells. These cells contain a specialized microtubule array, called the furrow microtubule array, that forms at the inner edge of the cleavage furrow (Figure 8-9). It is required for membrane addition at the furrow and provides the tracks along which membrane vesicles are carried to the site of membrane fusion. The furrow microtubule array may be a specialized structure for enhancing membrane addition in the large and rapidly dividing cells of animal embryos, in which the spindle is often quite distant from the cell membrane. In smaller and more slowly dividing somatic cells, astral microtubules and microtubules of the central spindle are thought to provide a similar function.

Deposition of new membrane in cytokinesis is less extensive in yeast, in which the amount of new membrane needed represents a small fraction of total membrane surface area. In budding yeast, for example, most new membrane addition occurs throughout the cell cycle in the

Figure 8-8 Cytokinesis in a higher plant cell Just before mitosis, a band of microtubules and actin, called the preprophase band, forms around the cell at its midline. The preprophase band disappears as the cell reaches metaphase, but its position at the cortex remains marked and will determine the future site of division. After anaphase, the microtubules of the spindle form the phragmoplast. These microtubules serve as tracks on which membrane vesicles are carried to the middle of the cell from the Golgi apparatus. These vesicles contain the various glycoproteins and other components that will form the new cell wall. Vesicle fusion results in the formation of a disc-shaped membrane compartment called the cell plate, which expands outward until it eventually contacts and fuses with the cell membrane, thereby separating the daughter cells. A new cell wall is then completed between the daughter cell membranes.

Definitions

midbody: large protein complex, derived from the spindle midzone, that is involved in the final stages of cell separation in dividing animal cells.

phragmoplast: organelle in a dividing plant cell upon which the new cell membranes and cell walls between the two daughter cells are constructed. It corresponds to the central spindle of animal cells.

septum: the extracellular wall that forms between two daughter cells in fungi during cell division.

References

Albertson, R. *et al.*: **Membrane traffic: a driving force in cytokinesis.** *Trends Cell Biol.* 2005, **15**:92–101.

Cabib, E.: **The septation apparatus, a chitin-requiring machine in budding yeast.** *Arch. Biochem. Biophys.* 2004, **426**:201–207.

Danilchik, M.V. *et al.*: **Requirement for microtubules in new membrane formation during cytokinesis of *Xenopus* embryos.** *Dev. Biol.* 1998, **194**:47–60.

Gromley, A. *et al.*: **Centriolin anchoring of exocyst and SNARE complexes at the midbody is required for secretory-vesicle-mediated abscission.** *Cell* 2005, **123**:75–87.

Jürgens, G.: **Plant cytokinesis: fission by fusion.** *Trends Cell Biol.* 2005, **15**:277–283.

Straight, A.F. and Field, C.M.: **Microtubules, membranes and cytokinesis.** *Curr. Biol.* 2000, **10**:R760–R770.

VerPlank, L. and Li, R.: **Cell cycle-regulated trafficking of Chs2 controls actomyosin ring stability during cytokinesis.** *Mol. Biol. Cell* 2005, **16**:2529–2543.

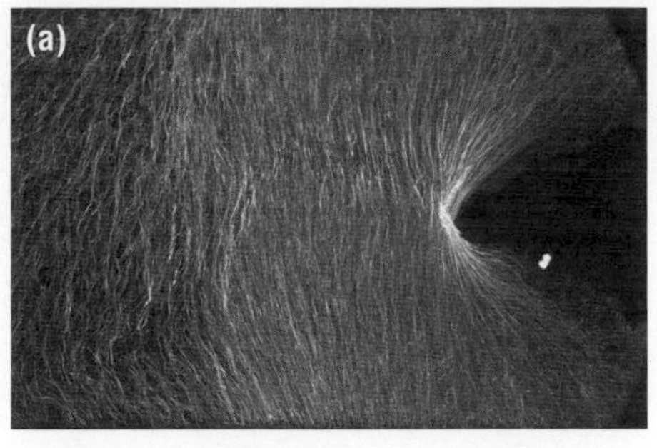

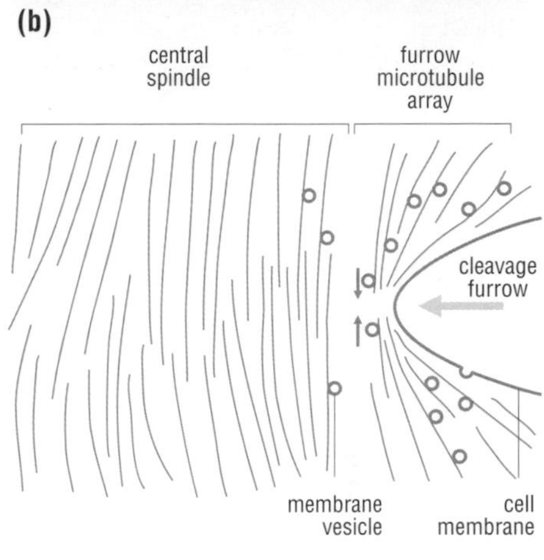

Figure 8-9 Microtubule behavior in the cleaving *Xenopus* embryo (a) Microtubule structure was analyzed near the beginning of the first division of a *Xenopus* zygote. A dense array of microtubules—the furrow microtubule array (FMA)—can be seen at the leading edge of the cleavage furrow (on the right of the photograph). The overlapping microtubules of the spindle midzone lie beneath the FMA. **(b)** This schematic shows how the FMA is thought to direct the transport of membrane vesicles for fusion to the cell membrane at the cleavage furrow. For simplicity, the contractile ring that lies at the cleavage furrow is not shown. Photograph kindly provided by Michael Danilchik and Kay Larkin. From Danilchik, M.V. *et al.*: *Dev. Biol.*1998, **194**:47–60.

growing bud, and only a small amount is required at the bud neck in the final stages of division (Figure 8-10). In these cells, as in those of metazoans, new membrane is provided by Golgi-derived vesicles. Unlike in metazoans, however, vesicle transport to the bud neck does not require microtubules but instead depends on myosin-dependent transport along actin filaments.

Transport of vesicles to the cleavage site also delivers proteins that help organize the contractile ring and serve other functions in cell separation. In budding yeast, for example, Golgi-derived vesicles that fuse with the bud-neck membrane contain the transmembrane enzyme chitin synthase, Chs2, which synthesizes chitin, the complex polysaccharide that forms the primary septum between mother and daughter cells (see Figure 8-10).

The connection between daughter cells is finally severed when the inwardly moving membranes contact and fuse with each other. This poorly understood process requires the removal of the contractile ring and, in animal cells, depends on the construction of a large protein complex called the **midbody** at the division site. The final membrane fusion event depends on machinery like that involved in membrane fusion events in the secretory pathway.

Membrane addition occurs in parallel with actin–myosin contraction

Membrane deposition generally occurs in parallel with the contraction of the actin–myosin ring, so that new membrane is added at a rate that matches the rate at which the ring moves inward. It is not clear how membrane addition and ring contraction are coordinated with each other. One simple possibility is that the two processes are independent but occur in parallel because they are both triggered by the same upstream regulatory mechanism. Dephosphorylation of Cdk targets, for example, may be an important mechanism for initiating both processes.

Another possibility is that the membrane deposition machinery and the contraction machinery are physically coupled in some way, so that the progression of each process depends on the other. There is good evidence for this in both budding and fission yeasts, in which inhibition of membrane vesicle delivery or cell wall synthesis causes defects in ring contraction. In animal cells, however, there is less evidence for direct coupling, and it seems that membrane addition and ring contraction are at least partly independent. Inhibition of actin polymerization or Rho activation, for example, blocks the formation of the contractile ring in *Xenopus* embryonic cells but does not affect membrane insertion. Conversely, inhibition of membrane insertion (by depolymerization of microtubules in the furrow microtubule array, for example) does not prevent contractile ring formation and the initiation of contraction—although full inward contraction of the ring cannot occur without the addition of new membrane.

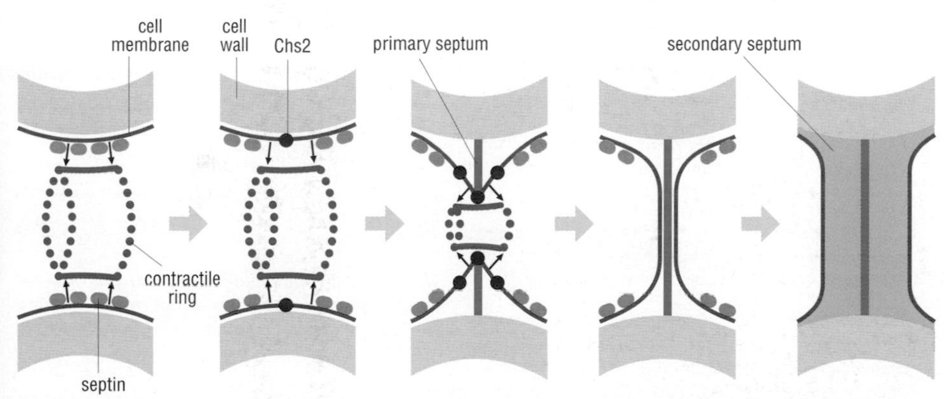

Figure 8-10 Septation in budding yeast These diagrams represent cross-sections of the bud neck during the final stages of cytokinesis. A ring of septins (gold) lies beneath the cell membrane, while unknown proteins (black arrows) link the actin–myosin ring (blue) to the membrane. Delivery of membrane vesicles (not shown) results in the appearance of Chs2 (red), a transmembrane chitin synthase, in the cell membrane adjacent to the contractile ring. As the actin–myosin ring begins to contract, Chs2 constructs a primary septum (green) behind the inwardly moving membranes, which eventually fuse to generate two separate cells with a primary septum between them. Other chitin synthases then construct a thick secondary septum. Separation of mother and daughter occurs when chitinases digest the primary septum (not shown). Interestingly, Chs2 and the actin–myosin ring are not absolutely essential for cytokinesis in budding yeast; in their absence, an abnormal but effective septum is constructed by the chitin synthases that normally synthesize the secondary septum. Adapted from Cabib, E.: *Arch. Biochem. Biophys.* 2004, **426**:201–207.

8-4 The Positioning and Timing of Cytokinesis in Yeast

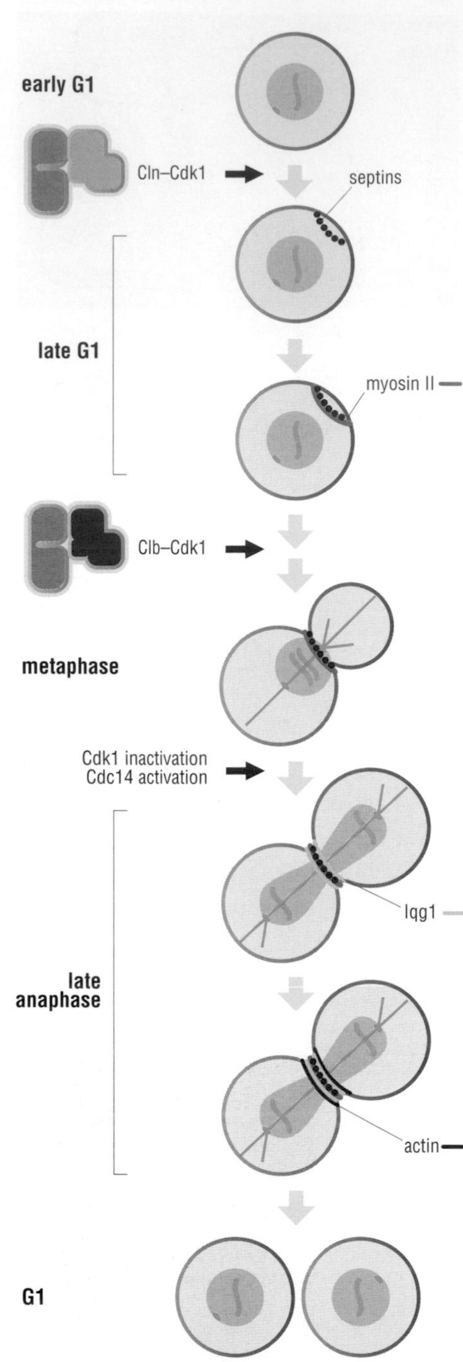

Figure 8-11 **Actin–myosin ring assembly in the budding yeast *S. cerevisiae*** In late G1, activation of Cln–Cdk1 triggers the formation of a septin ring at the future bud site. Septins then promote the recruitment of myosin II as budding is initiated. As S phase progresses, the nucleus is positioned with the daughter-bound spindle pole body in the bud neck; the completion of anaphase therefore leads to insertion of this spindle pole body, and its associated set of chromosomes, in the daughter, while the opposite spindle pole body remains in the mother. The second major phase in contractile ring formation occurs in late anaphase, when Iqg1 localizes to the bud neck and recruits actin to form the actin–myosin ring. These final steps in ring assembly may be promoted by the dephosphorylation of Cdk1 targets as well as other regulatory mechanisms.

Preparations for cytokinesis in budding yeast begin in late G1

Cytokinesis must occur in the right place (between the segregated chromosomes) and at the right time (after mitosis). Tight regulation of the positioning and timing of cytokinesis is therefore critical for the successful completion of cell division. In this section we describe how these regulatory mechanisms are thought to operate in yeast, where the site of cleavage is determined before mitosis and the spindle poles are then positioned on opposite sides of the predetermined division site. In section 8-5 we discuss the positioning of the cleavage plane in animal cells.

In the budding yeast *S. cerevisiae*, the cleavage site is the narrow neck between the mother cell and daughter bud. Preparations for budding begin in late G1, about 15 minutes before the bud first emerges, when a broad ring of septins and other proteins appears at the future bud site (Figure 8-11). A ring of myosin II, whose formation is dependent on septins, forms at this site as the new bud emerges. These early steps in the preparation for cytokinesis are triggered by G1/S–Cdk activities that rise in late G1.

As the bud grows during S phase and early mitosis, the correct positioning of the division plane is achieved by movement of the nucleus to the bud neck. Astral microtubules from one spindle pole body are captured by proteins that carry the microtubule plus ends to the bud tip along actin filaments, thereby pulling that spindle pole body to the bud neck. Astral microtubules from the other spindle pole body become anchored to the cell cortex opposite the bud, further ensuring that the nucleus is positioned with the metaphase spindle parallel to the cell's axis. When the spindle elongates in anaphase, the two sets of chromosomes are therefore pulled to opposite sides of the division plane. At this point, the assembly of the contractile ring is completed by the addition of actin to the myosin II that is already at the bud neck. This process is dependent on the IQGAP protein Iqg1 (see section 8-2) and numerous other regulators, including a formin (Bni1) and the GTPase Rho. Contraction of the actin–myosin ring, accompanied by the deposition of new cell wall and membrane (see Figure 8-10), then results in the completion of cytokinesis.

In budding yeast, the regulatory mechanisms that trigger the completion of cytokinesis after mitosis are not clear. Cytokinesis is coupled with the completion of mitosis because both processes depend, at least in part, on the dephosphorylation of Cdk1 targets—which is the result of cyclin destruction and activation of the phosphatase Cdc14 (see section 7-5). Some events in cytokinesis may also require the activation of other regulatory components, including the polo-like protein kinase Plk. The mitotic exit network (MEN), whose primary function is to activate the phosphatase Cdc14 (see section 7-5), may also have separate Cdc14-independent functions in the control of late events in cytokinesis. Much remains to be learned about the regulatory connections that govern the timing and order of the various steps in contractile ring assembly and contraction, as well as the deposition of membrane and wall components at the bud neck.

Fission yeast uses the nucleus to mark the division site in early mitosis

In the fission yeast *S. pombe*, the future site of cell division is marked in early mitosis by the appearance of a broad band of the protein Mid1 at the midline of the cell (Figure 8-12). The position of this medial ring is determined by the position of the nucleus, which is generally found at the cell center (Figure 8-13). It is not clear how the nucleus determines the location of the Mid1 ring. Mid1 is known to shuttle in and out of the nucleus and is largely concentrated

References

Daga, R.R. and Chang, F.: **Dynamic positioning of the fission yeast cell division plane.** *Proc. Natl Acad. Sci. USA* 2005, **102**:8228–8232.

Feierbach, B. and Chang, F.: **Cytokinesis and the contractile ring in fission yeast.** *Curr. Opin. Microbiol.* 2001, **4**:713–719.

Segal, M. and Bloom, K.: **Control of spindle polarity and orientation in *Saccharomyces cerevisiae*.** *Trends Cell Biol.* 2001, **11**:160–166.

Tolliday, N. *et al.*: **Assembly and regulation of the cytokinetic apparatus in budding yeast.** *Curr. Opin. Microbiol.* 2001, **4**:690–695.

Wolfe, B.A. and Gould, K.L.: **Split decisions: coordinating cytokinesis in yeast.** *Trends Cell Biol.* 2005, **15**:10–18.

Wu, J.Q. *et al.*: **Spatial and temporal pathway for assembly and constriction of the contractile ring in fission yeast cytokinesis.** *Dev. Cell* 2003, **5**:723–734.

Figure 8-12 **Actin–myosin ring positioning in the fission yeast *S. pombe*** Before mitosis, the interphase microtubule array positions the nucleus at the center of the cell. Mid1 protein forms a diffuse band around the cell adjacent to the nucleus, thereby marking the future division site. Several other proteins, including myosin II, also accumulate in diffuse patches near the cell midline during early mitosis. After Cdk1 inactivation and the onset of anaphase, the various ring proteins, now including actin, coalesce to form an actin–myosin contractile ring. As the ring contracts, the deposition of new membrane and cell wall results in septation, the formation of the septum between daughter cells.

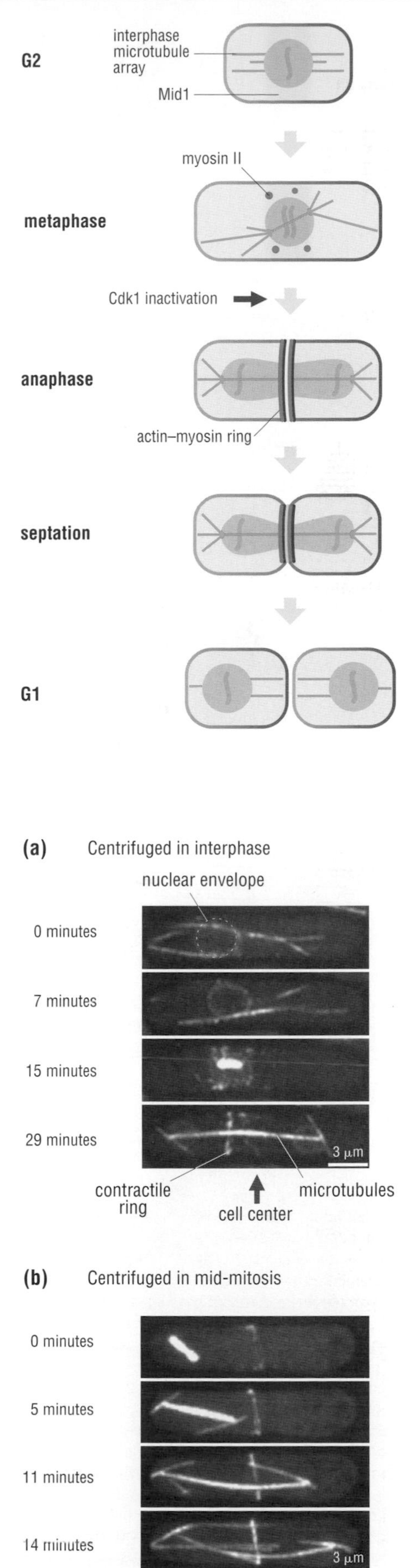

inside the nucleus during interphase. At the onset of mitosis, the protein kinase Plk promotes the export of Mid1 from the nucleus, and its localization at the nearby cell cortex might simply result from its high local concentration near the nucleus.

Numerous contractile ring components, including myosin II, actin, the formin Cdc12, and the IQGAP protein Rng2, gather in patches near the medial ring in mitosis and are required for cytokinesis. After anaphase, these components form a contractile ring structure that contracts inwards in concert with the deposition of new cell wall and membrane between the daughter nuclei (see Figure 8-12).

The positioning of the segregated chromosomes on opposite sides of the medial ring depends on the centering of the nucleus in the cell (see Figure 8-13). How is this achieved? The interphase nucleus is cradled in a meshwork of antiparallel microtubule bundles that run the length of the cell, and whose minus ends are nucleated by multiple organizing centers associated with the nuclear membrane. The plus ends of these microtubules push outward against the cell tips, and the balance of these opposing forces centers the nucleus in the cell. In mitosis, the interphase microtubule array is reorganized and the duplicated spindle pole bodies nucleate the spindle inside the nucleus. Other microtubule structures form at the site of the contractile ring and help maintain its position at the cell center. During anaphase, the spindle extends the entire length of the cell, thereby placing the two nuclei at opposite ends of the cell as cleavage begins (see Figure 8-12).

As in other cell types, dephosphorylation of Cdk1 targets seems to be a major trigger for cytokinesis in fission yeast. Blocking cyclin destruction prevents the assembly and contraction of the actin–myosin ring and the deposition of the new cell wall. Little is known about the Cdk1 targets whose dephosphorylation is required, or how they might influence the contractile ring or septation.

Septum formation in fission yeast is also regulated by a small GTPase called Spg1, which is related to Tem1 of budding yeast and governs a regulatory network (called the septation initiation network or SIN) whose components are related to those of the mitotic exit network that controls the phosphatase Cdc14 in budding yeast (see section 7-5). Unlike Tem1, however, Spg1 is not required for the completion of mitosis but is required solely for ring contraction and deposition of the new septum. By unknown mechanisms, Cdk1 inactivation in late mitosis is thought to trigger changes at the cleavage site that allow Spg1 to initiate the final steps of cytokinesis.

Figure 8-13 **Positioning the contractile ring in *S. pombe*** A yeast strain was constructed in which myosin II, the nuclear envelope, and microtubules were each genetically tagged with a fluorescent protein. Cells were treated in interphase with a drug that depolymerizes microtubules and were then lightly centrifuged in an agar matrix. **(a)** In the cell shown here, centrifugation forced the nucleus away from the center of the cell (to the left in this case). The drug was washed out, allowing reformation of microtubules, but the nucleus remained off-center as the cell entered mitosis. At the 15-minute time point, a short and intensely labeled metaphase spindle is visible inside the off-center nucleus, and diffuse patches of myosin II are gathering over the nucleus. By the 29-minute point these patches have coalesced into a contractile ring. The contractile ring forms around the nucleus, not at the cell center, showing that nuclear position determines the site of cleavage. **(b)** This cell, which lacks the nuclear envelope fluorescent tag, was in mid-mitosis when centrifuged. The metaphase spindle marks the nuclear position at the far left end of the cell. Despite the off-center position of the nucleus, the contractile ring forms at the cell center because its position had been determined at the onset of mitosis, before centrifugation. Photographs kindly provided by Rafael R. Daga and Fred Chang. From Daga, R.R. and Chang, F.: *Proc. Natl Acad. Sci. USA* 2005, **102**:8228–8232.

8-5 The Positioning and Timing of Cytokinesis in Animal Cells

Signals from the mitotic spindle determine the site of cleavage in animal cells

The correct positioning and timing of cytokinesis are determined in animal cells by a mechanism quite different from that in yeast. In animal cells the anaphase spindle determines the site of cell division by inducing cleavage at a site midway between the spindle poles. Not only does this ensure that the two sets of chromosomes (and centrosomes) are placed in separate cells, but it also helps guarantee that cytokinesis does not occur until the spindle has finished segregating the chromosomes.

Evidence that the spindle induces cleavage furrow formation came originally from micromanipulation experiments with the embryos of invertebrates and frogs. In these cells, premature cleavage furrows can be induced by moving the spindle closer to the cell surface (Figure 8-14), and multiple cleavage furrows can be induced in the same cell if the spindle is pushed with a needle to various locations at the cell cortex. Cleavage induction by the spindle is blocked when microtubules are depolymerized with chemicals, arguing that microtubules or their associated proteins are somehow generating or transporting molecular signals that trigger furrow formation at the cell cortex.

How does the spindle stimulate furrow formation? Three current hypotheses of furrow induction are illustrated in Figure 8-15. The first is the astral stimulation hypothesis (see Figure 8-15a), which postulates that astral microtubules from the spindle poles carry a furrow-inducing signal to the cell cortex, where signals from two poles are somehow focused into a ring at the spindle equator. This hypothesis originated in a variety of classic experiments with invertebrate embryos, including work showing that an ectopic cleavage furrow (called a Rappaport furrow) can form between the asters of two separate spindles in the same cell. Even a single microtubule aster can induce a cleavage furrow in anaphase cells under some conditions, suggesting that the induction signal does not require interactions between microtubules from two poles.

A second possibility, called the central spindle hypothesis (see Figure 8-15b), is that the cleavage furrow is induced by a positive stimulus that originates not in the spindle poles but in the **central spindle**, the bundle of interpolar microtubules that forms at the spindle midzone in anaphase and contains numerous potential signaling molecules. Abundant evidence from many species supports this idea. In cultured mammalian cells, for example, placing a physical block between the spindle midzone and the cortex prevents furrow formation. Mutations that disrupt central spindle formation block furrow formation in *Drosophila* (but not in *C. elegans*). In *Drosophila* meiotic cells or in cells with mutations that inhibit centrosome function, furrow formation occurs normally in the absence of astral microtubules.

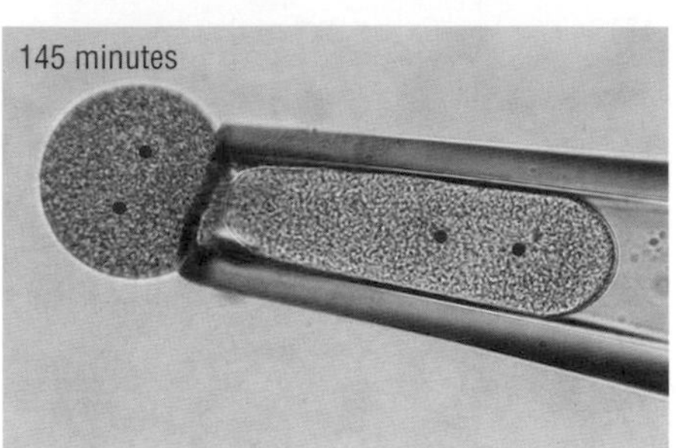

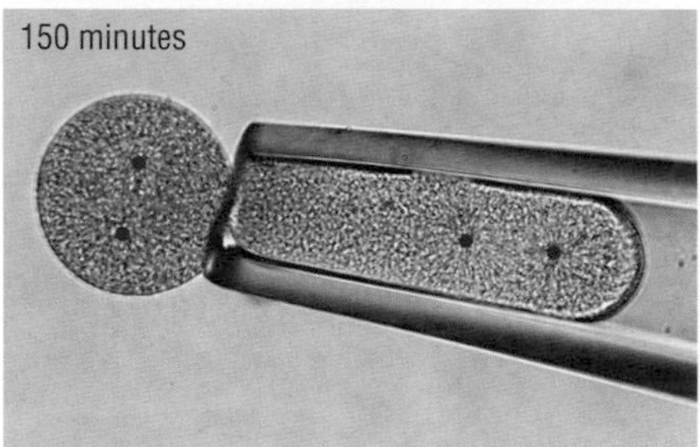

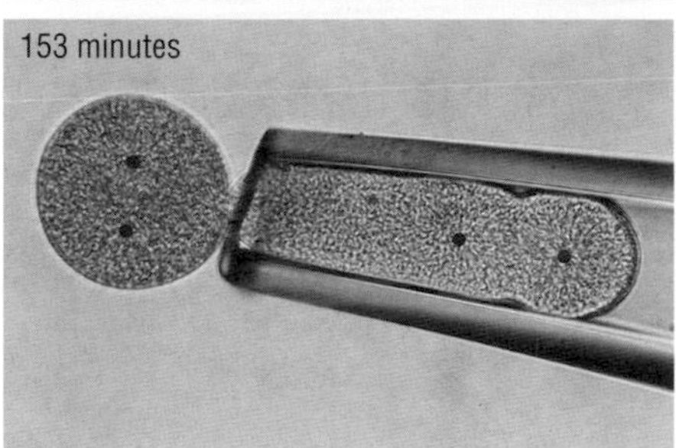

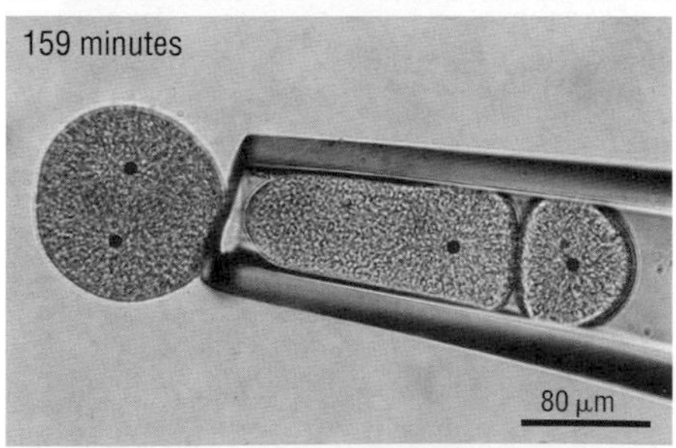

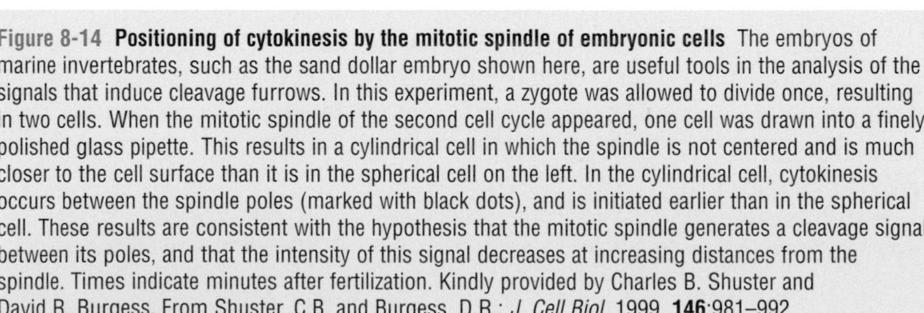

Figure 8-14 Positioning of cytokinesis by the mitotic spindle of embryonic cells The embryos of marine invertebrates, such as the sand dollar embryo shown here, are useful tools in the analysis of the signals that induce cleavage furrows. In this experiment, a zygote was allowed to divide once, resulting in two cells. When the mitotic spindle of the second cell cycle appeared, one cell was drawn into a finely polished glass pipette. This results in a cylindrical cell in which the spindle is not centered and is much closer to the cell surface than it is in the spherical cell on the left. In the cylindrical cell, cytokinesis occurs between the spindle poles (marked with black dots), and is initiated earlier than in the spherical cell. These results are consistent with the hypothesis that the mitotic spindle generates a cleavage signal between its poles, and that the intensity of this signal decreases at increasing distances from the spindle. Times indicate minutes after fertilization. Kindly provided by Charles B. Shuster and David R. Burgess. From Shuster, C.B. and Burgess, D.R.: *J. Cell Biol.* 1999, **146**:981–992.

Definitions

central spindle: the structure that forms in late mitosis in a dividing animal cell from interpolar microtubule remnants of the central part of the mitotic spindle and various associated proteins. It eventually becomes the midbody.

References

Bringmann, H. and Hyman, A.A.: **A cytokinesis furrow is positioned by two consecutive signals.** *Nature* 2005, **436**:731–734.

Burgess, D.R. and Chang, F.: **Site selection for the cleavage furrow at cytokinesis.** *Trends Cell Biol.* 2005, **15**:156–162.

Canman, J.C. *et al.*: **Determining the position of the cell division plane.** *Nature* 2003, **424**:1074–1078.

Glotzer, M.: **The molecular requirements for cytokinesis.** *Science* 2005, **307**:1735–1739.

Guse, A. *et al.*: **Phosphorylation of ZEN-4/MKLP1 by aurora B regulates completion of cytokinesis.** *Curr. Biol.* 2005, **15**:778–786.

Mishima, M. *et al.*: **Cell cycle regulation of central spindle assembly.** *Nature* 2004, **430**:908–913.

Rappaport, R.: *Cytokinesis in Animal Cells* (Cambridge University Press, Cambridge, 1996).

Shuster, C.B. and Burgess, D.R.: **Parameters that specify the timing of cytokinesis.** *J. Cell Biol.* 1999, **146**:981–992.

Yüce, O. *et al.*: **An ECT2–centralspindlin complex regulates the localization and function of RhoA.** *J. Cell Biol.* 2005, **170**:571–582.

A third hypothesis, the astral relaxation hypothesis, postulates that astral microtubules generate a negative signal that increases cortical relaxation close to the poles (see Figure 8-15c). According to this proposal, active actin–myosin bundles are distributed throughout the cell cortex, and inhibition of their contraction near the spindle poles results in a gradient of contractile activity that is highest at the midpoint between poles. There is some experimental evidence for this idea. In the embryos of *C. elegans*, and to a lesser extent in cultured mammalian cells, defects in astral microtubule formation can lead to multiple ectopic cleavage furrows and widespread contractile activity, suggesting that microtubules normally inhibit contractility at the cell cortex in these cell types.

No single proposal explains all the observations, and thus the positioning of the cleavage furrow is likely to be determined by some combination of these mechanisms and perhaps others—with variations in the importance of different mechanisms in different cell types. In the embryonic cells of *C. elegans*, for example, there is considerable evidence that astral stimulation and central spindle mechanisms are both important. A fuller understanding of these mechanisms is likely to arise from the complete identification and analysis of the signaling molecules that are involved in furrow positioning.

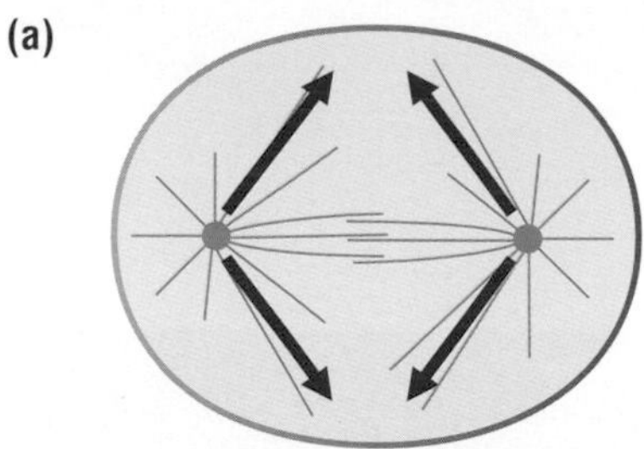

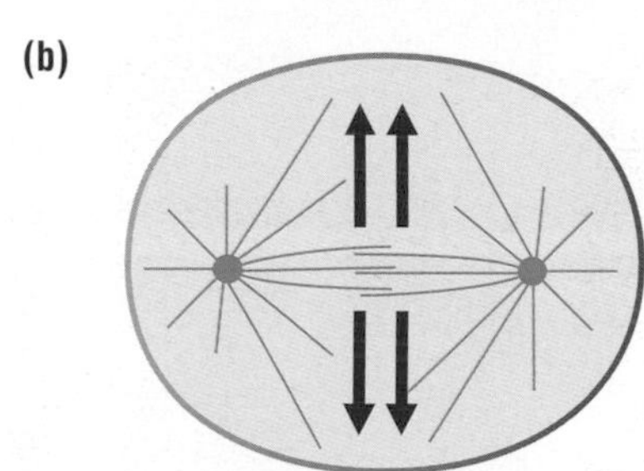

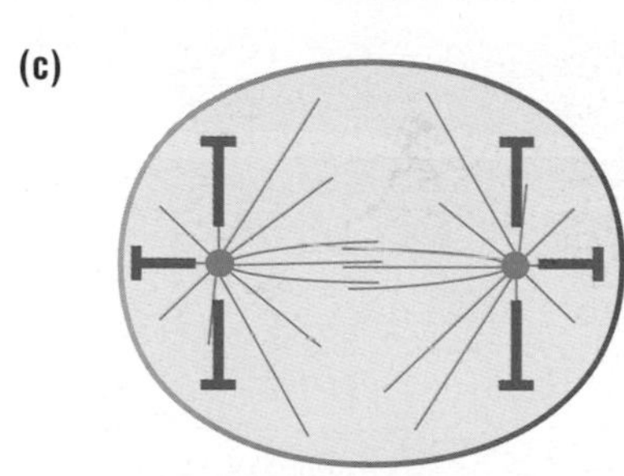

Figure 8-15 Alternative hypotheses for cleavage signal generation by the mitotic spindle The three main hypotheses to explain the ability of the mitotic spindle to induce cleavage between its poles are shown here. These hypotheses are not mutually exclusive, and it is likely that some cell types use several mechanisms to position the cleavage plane. **(a)** Astral microtubules carry a cleavage-stimulating signal to the cell cortex, and this signal is somehow focused at the spindle equator. **(b)** The central spindle generates a cleavage-inducing signal that is focused at the nearest site on the cell cortex. **(c)** Astral microtubules inhibit contractile activity at the cell cortex. The astral relaxation signal is minimal at the spindle equator, resulting in cleavage.

Multiple regulatory components at the central spindle help control cytokinesis

Clues to the molecular basis of cleavage furrow positioning—and the control of cytokinesis in general—may be found in studies of the structure, assembly, and function of the central spindle. Although poorly understood at present, it is becoming clear that this structure fulfills multiple functions in animal cell cytokinesis—in the control of cleavage furrow positioning, the delivery of membrane vesicles to the cleavage furrow, and in the formation of the midbody structure that is required for the final steps of division. We have little understanding of the mechanisms underlying any of these functions, but the molecular composition of the central spindle is coming into focus.

The core of the central spindle is an antiparallel bundle of microtubules with a dense protein matrix at its center. A key component of this structure is a plus-end-directed kinesin-6 motor called MKLP-1 (named Pavarotti in *Drosophila* and ZEN-4 in *C. elegans*). MKLP-1 forms a tight complex with a protein called CYK-4, and together the two proteins cross-link antiparallel microtubules of the central spindle. Another microtubule-bundling protein called Prc1 also contributes. The CYK-4 protein may also provide a regulatory link between the central spindle and regulation of the GTPase Rho, the key regulator of contractile ring function (see section 8-2). In some cell types, CYK-4 binds and activates the RhoGEF Pebble/Ect2, thereby contributing to the stimulation of furrow formation. Surprisingly, the CYK-4 protein also contains GTPase-activating protein (GAP) activity, which would be expected to inhibit Rho, but the function of this activity remains unclear.

The central spindle contains several other potential regulatory proteins, called passenger proteins, that are located at the kinetochores during metaphase but then appear at the central spindle at the onset of anaphase. These proteins include the two mitotic protein kinases Plk and aurora B (see section 5-7). Aurora B is essential for the completion of cytokinesis and acts in part by phosphorylating and thereby promoting the function of MKLP-1.

Cytokinesis is coordinated with mitosis by the spindle and Cdk1 inactivation

The link between the anaphase spindle and furrow formation helps ensure that cytokinesis occurs at the correct time. Cytokinesis also depends on the inactivation of Cdk1 and the dephosphorylation of Cdk1 targets, and this provides an additional mechanism for regulating its timing. Cleavage furrow formation is generally prevented by the expression or injection of nondegradable forms of mitotic cyclins. Two components of the central spindle—MKLP-1 and Prc1—are inhibited by Cdk-dependent phosphorylation before anaphase and are then activated by dephosphorylation in anaphase, providing one mechanism by which the timing of furrow formation is controlled. Numerous additional substrates of Cdk1 and other mitotic kinases are also likely to be involved.

8-6 Specialization of Cytokinesis in Animal Development

Cytokinesis can be blocked or incomplete in some stages of development

In most somatic cells, the events of the chromosomal cell cycle—duplication and segregation of the genome—are inevitably accompanied by growth and cytokinesis, which duplicate and divide the cell itself. This coupling of mitosis and cytokinesis is not seen in all cell types, however. During animal development, for example, certain features of cytokinesis, and its relationship to mitosis, can be modified dramatically. These variations on the cytokinetic theme are particularly striking during the development of insects, such as *Drosophila*, but are also seen in mammals and other species.

The first variation on cytokinesis is simply nuclear division without cell division. After fertilization in *Drosophila*, for example, the embryo undergoes 13 rounds of chromosome duplication and mitosis in the complete absence of cytokinesis, resulting in the formation of a giant syncytium containing thousands of nuclei (see section 2-4) (Figure 8-16). A less dramatic form of this process is seen in some mammalian cells, such as megakaryocytes, which accumulate multiple nuclei by undergoing multiple mitotic cycles without division. It is not known how any of these cell types shut down the cytokinesis machinery.

A second variation of cytokinesis is incomplete cytokinesis. In the *Drosophila* ovary, production of the oocyte begins when a germline stem cell undergoes a series of four nuclear divisions to produce 16 cells—one oocyte supported by 15 nurse cells. Each of these four nuclear divisions is followed by incomplete cytokinesis, in which the contraction of the cleavage furrow is arrested before cleavage is complete. An actin-based structure then forms at the incomplete division site, resulting in openings, called ring canals, between daughter cells. These canals allow the rapid transport of large amounts of material from nurse cells to the developing oocyte.

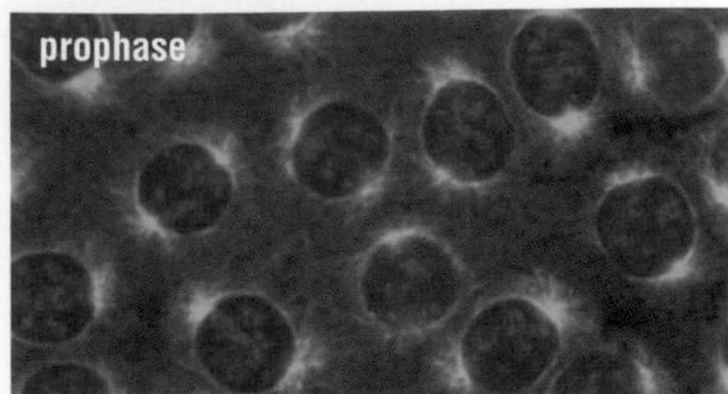

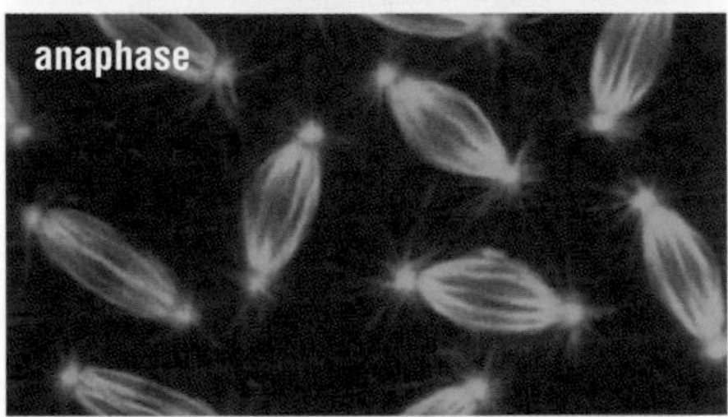

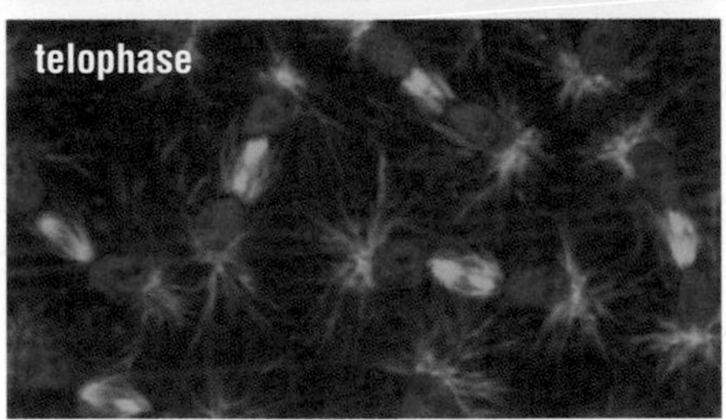

Figure 8-16 **Mitosis without cytokinesis in the *Drosophila* embryo** The extremely rapid (~8-minute) divisions of the early *Drosophila* embryo occur in the absence of cytokinesis, resulting in the formation of a syncytium. In these images, fluorescently tagged microtubules (green) and chromosomes (blue) reveal the tightly packed nuclei typical of syncytial cycles 11 and 12. Courtesy of Barbara Fasulo and William Sullivan.

Cellularization is a specialized form of cytokinesis

Another major variation on cytokinesis is cellularization, the process by which the thousands of nuclei in the syncytial embryo of *Drosophila* are packaged into individual cells (see section 2-4). Preparations for this process begin after the ninth mitosis, when the nuclei migrate to the cortex of the cell. During the next three nuclear divisions (cycles 10–12), partial invaginations, or pseudocleavage furrows, appear in the cell membrane between the nuclei after each mitosis but then recede upon entry into the next. Finally, after mitosis 13, cellularization occurs, whereby cleavage furrows descend between the nuclei and then contract to surround each nucleus with a complete cell membrane (Figure 8-17).

Cellularization and normal cytokinesis share many features and molecular components, and both depend on membrane deposition and actin–myosin ring contraction. These two processes occur in parallel during normal cytokinesis but occur in sequence during cellularization: addition of new membrane drives the invagination of the cleavage furrow, after which the actin–myosin ring contracts at the base of the cell.

Membrane deposition is particularly extensive during cellularization and leads to a 25-fold increase in the surface area of plasma membrane. As in normal cytokinesis, new membrane is added by the fusion of membrane vesicles that originate from Golgi bodies, which in cellularizing *Drosophila* embryos are in the form of numerous unstacked cisternae that are concentrated near the outer or apical surface of the cells. Early in cellularization, membrane addition occurs primarily at the apical face. In the later stages, membrane is added to the apical–lateral membrane in the upper region of the cleavage furrow (see Figure 8-17).

References

Lecuit, T.: **Junctions and vesicular trafficking during *Drosophila* cellularization.** *J. Cell Sci.* 2004, **117**:3427–3433.

Lecuit, T. and Wieschaus, E.: **Polarized insertion of new membrane from a cytoplasmic reservoir during cleavage of the *Drosophila* embryo.** *J. Cell Biol.* 2000, **150**:849–860.

Mazumdar, A. and Mazumdar, M.: **How one becomes many: blastoderm cellularization in *Drosophila melanogaster*.** *BioEssays* 2002, **24**:1012–1022.

Papoulas, O. *et al.*: **The golgin Lava lamp mediates dynein-based Golgi movements during *Drosophila* cellularization.** *Nat. Cell Biol.* 2005, **7**:612–618.

Royou, A. *et al.*: **Reassessing the role and dynamics of nonmuscle myosin II during furrow formation in early *Drosophila* embryos.** *Mol. Biol. Cell* 2004, **15**:838–850.

Sisson, J.C. *et al.*: **Lava lamp, a novel peripheral Golgi protein, is required for *Drosophila melanogaster* cellularization.** *J. Cell Biol.* 2000, **151**:905–918.

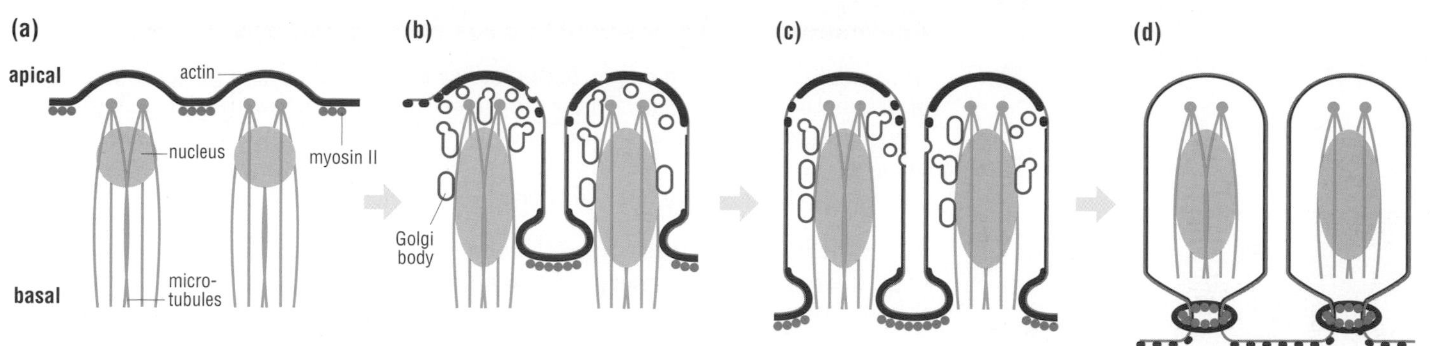

Figure 8-17 **Cellularization in the *Drosophila* embryo** **(a)** After mitosis of cycle 13, apically located centrosomes (which are duplicated in late mitosis in these cells) nucleate inverted baskets of microtubules that descend over the nuclei. Actin–myosin bundles form at the cell cortex between the nuclei. **(b, c)** Cleavage furrows, with actin–myosin bundles at their leading edges, descend along the microtubules. These early stages in cellularization are thought to be driven by the insertion of new membrane vesicles, which originate in small Golgi bodies transported upward on the microtubules. Initially, membrane insertion occurs at the apical (outer) surface but then shifts later to the apical–lateral membrane near the top of the furrow. **(d)** Cleavage furrows descend past the base of the nuclei, and actin–myosin bundles contract in a ring structure that pinches off the bottoms of the cells, thereby generating the cellular blastoderm. A thin cytoplasmic channel initially remains to connect the cells with the interior of the embryo.

As in cytokinesis, microtubules provide the tracks along which new membrane makes its way to the site of insertion during cellularization. The centrosomes associated with each nucleus lie just beneath the embryo's surface and nucleate microtubule arrays that surround each nucleus and descend into the embryo (see Figure 8-17). With assistance from minus-end-directed dynein motors, Golgi bodies are transported along these microtubules from the interior of the cell to the apical surface (Figure 8-18).

What provides the force that moves the cleavage furrow inward? Bundles of actin and myosin II are concentrated in the pseudocleavage furrows of cycles 10–12 and at the leading edge of the furrows that separate nuclei during cellularization. The motor activity of myosin II is not required for furrow invagination, however, arguing that actin–myosin contraction does not drive this process. Instead, it has been proposed that the insertion of membrane vesicles at the apical surface somehow generates a force that pushes the furrow inward. Actin may provide a structural framework that guides this process.

Once the furrow is deep enough to surround the nucleus, contraction of the actin–myosin ring is essential for completing the packaging of nuclei into individual cells. The formation and contraction of the actin–myosin ring involve many of the molecules and mechanisms that drive conventional cytokinesis. In addition to actin and myosin II, cellularization requires Rho, the formin Diaphanous, anillin, and septins. As in cytokinesis, dephosphorylation of Cdk targets is required for the formation of pseudocleavage furrows and for cellularization.

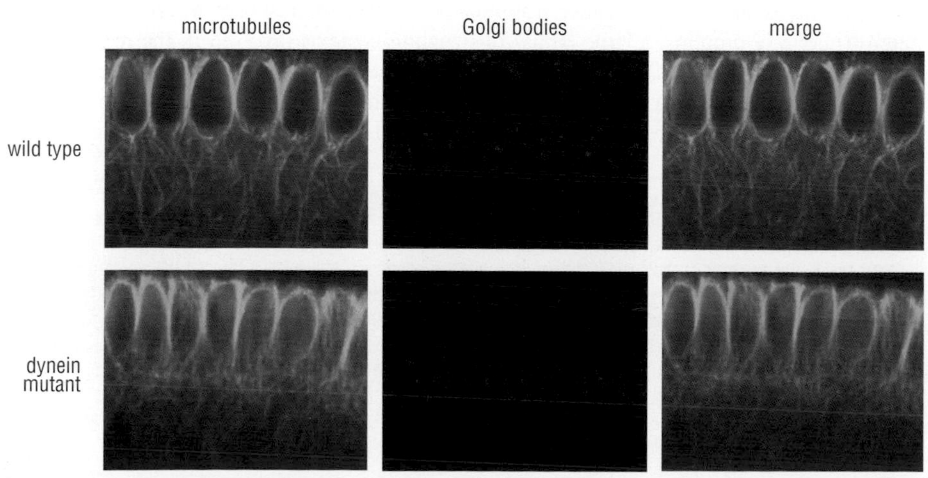

Figure 8-18 **Membrane transport during cellularization** These images are cross-sections of the surface of the *Drosophila* embryo early in cellularization, with the surface of the embryo at the top. Images in the left column show the networks of microtubules (green) that surround each nucleus, as diagrammed in Figure 8-17. In the center column, a Golgi protein called Lava lamp is stained red to label the punctate Golgi bodies. The right column provides a merged image of microtubules and Golgi. In wild-type embryos (top row), Golgi bodies are found beneath the microtubule array and at the apical cell surface. In embryos with defects in the microtubule-dependent motor protein dynein (bottom row), Golgi bodies are no longer found at the apical surface. These and various other experiments argue that dynein helps transport the Golgi bodies along microtubules to the apical surface. Mutation of dynein also results in severe defects in membrane deposition and furrow invagination. Kindly provided by John C. Sisson. From Papoulas, O. *et al.*: *Nat. Cell Biol.* 2005, **7**:612–618.

Asymmetric spindle positioning leads to daughter cells of unequal sizes

Some cell divisions are asymmetric, in that the two daughter cells do not inherit an equal share of all cytoplasmic components. During early animal development, two daughter cells may have different developmental fates, and this is often due to the unequal distribution of specific proteins, called fate determinants, at cell division. Typically, these proteins are localized at one end of the mother cell before division, so that only one daughter inherits them.

Asymmetric cell division may also result in daughter cells of unequal sizes. The first division of the *C. elegans* zygote, for example, results in one daughter cell that is about 25% larger than the other (Figure 8-19) and gives rise to most of the cells in the adult. Similarly, during the development of the nervous system in *Drosophila*, neural precursor cells called neuroblasts divide asymmetrically, resulting in one cell that is 75% larger than the other and gives rise to many more progeny (Figure 8-20).

Differently sized daughter cells result from asymmetric positioning of the cleavage plane, which is due to changes in the position of the mitotic spindle. In *C. elegans*, the spindle is centered in the zygote until anaphase B and then elongates asymmetrically: the anterior pole stays in about the same position while the posterior pole migrates nearer to the posterior cortex (see Figure 8-19). Cleavage at the spindle equator then results in a larger anterior cell. In *Drosophila* neuroblasts, spindle asymmetry is even more striking. During anaphase, microtubules from one spindle pole (the basal pole) shorten, while those from the apical pole lengthen. The basal spindle pole also shrinks dramatically in size and nucleates fewer microtubules. Interpolar microtubules from the basal pole are also shorter than those from the apical pole (see Figure 8-20). The result is a shift in the position of the central spindle toward the base of the cell, which in turn directs cleavage closer to the base.

Unequal forces on the poles underlie asymmetric spindle positioning

The position of the mitotic spindle is determined primarily by interactions between the cell cortex and the astral microtubules. Forces acting on these microtubules pull the spindle poles away from each other, toward the cell periphery. In the *C. elegans* zygote, asymmetric spindle positioning results from an imbalance in these forces: those pulling on the posterior spindle pole are greater than those pulling on the anterior pole (Figure 8-21). The molecular basis of the pulling forces in these cells is not clear, but a likely possibility is that astral microtubules are reeled in by minus-end-directed dynein motors anchored to the cell membrane (see Figure 6-7). Another intriguing possibility is that plus-end depolymerization at the cell cortex generates a pulling force, perhaps in the same way that plus-end depolymerization generates a poleward force at the kinetochore (see section 6-11).

The asymmetry of forces at the two spindle poles in the *C. elegans* zygote depends on a group of proteins called the Par proteins. Soon after fertilization, different Par proteins become focused at each end of the cell, thereby defining the anterior and posterior poles. The polarized localization of these proteins then leads to the asymmetric distribution of both fate determinants and the spindle-regulatory proteins that orient the spindle along the anterior–posterior axis and enhance astral pulling forces at the posterior pole.

The Par proteins are thought to govern spindle behavior, at least in part, by locally regulating heterotrimeric G proteins, a large family of signaling proteins involved in the regulation of countless cellular processes. As their name suggests, heterotrimeric G proteins are composed of three subunits (Gα, Gβ and Gγ), where the Gα subunit is a GTPase related to Ran (see section 6-8). G proteins are inactive in the trimeric GDP-bound state, but binding of GTP

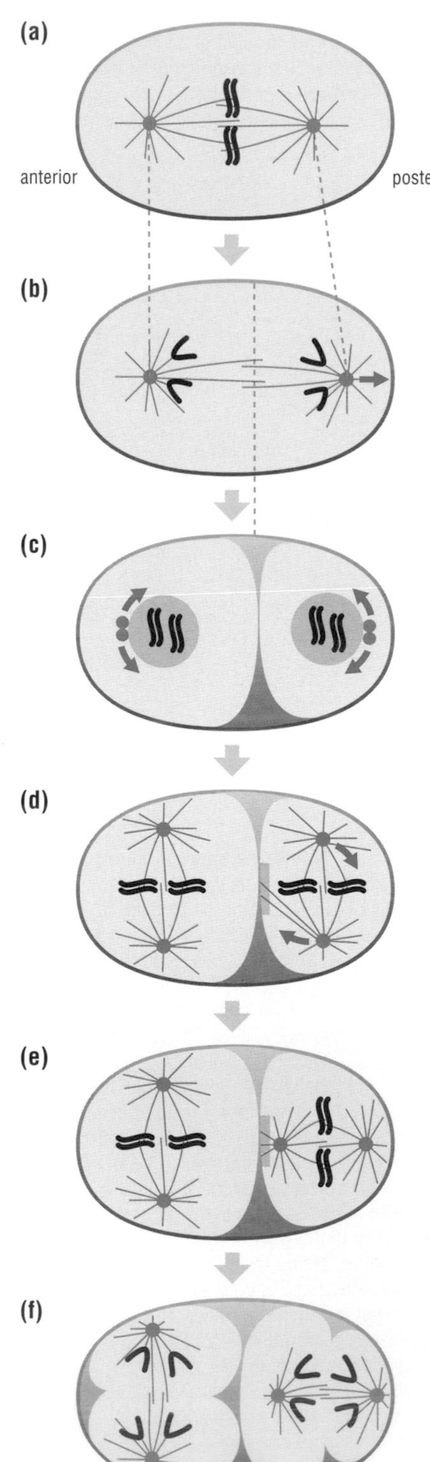

Figure 8-19 Asymmetry in the first divisions of the *C. elegans* embryo (a) In metaphase of the first division, the spindle is centered between the anterior and posterior poles of the zygote. **(b)** During anaphase B, the anterior spindle pole moves little but the posterior pole moves toward the posterior end of the cell, generating an asymmetric spindle. **(c)** Asymmetric cleavage occurs at the midpoint between the spindle poles, generating one cell that is about 25% larger than the other. **(d)** Initially, the spindles of both daughter cells are oriented as shown, perpendicular to the anterior–posterior axis. In the posterior cell, however, astral microtubules from one of the spindle poles are thought to be captured by proteins at the anterior cortex (yellow), where dynein motors or other mechanisms promote rotation of the spindle pole toward the anterior. **(e)** The spindle in the posterior cell rotates 90°, so that it is perpendicular to its original axis. **(f)** The second cleavages occur midway between the spindle poles as usual.

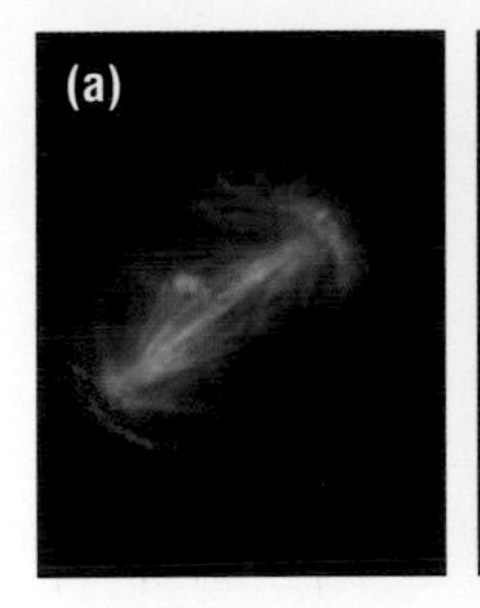

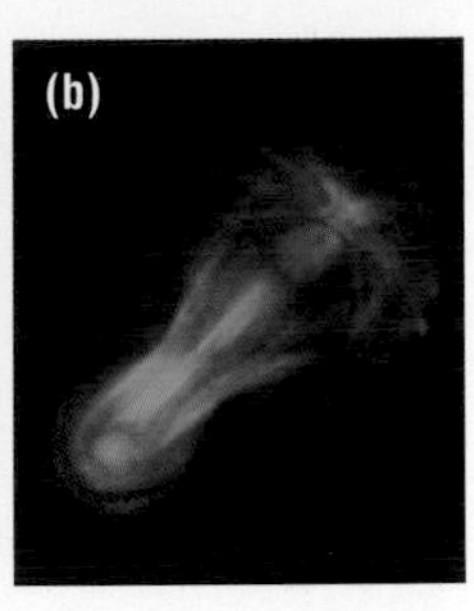

Figure 8-20 Asymmetric division in a *Drosophila* neuroblast These images of a dividing neuroblast show the mitotic spindle (green), chromosomes (blue), and the protein Miranda (red), an important fate determinant that is localized at the basal pole. **(a)** In early anaphase, the spindle is symmetrically positioned between the apical and basal poles of the cell. **(b)** During anaphase, the basal spindle pole nucleates fewer and shorter microtubules than the apical pole, resulting in a basal shift in the position of the central spindle—and therefore a basal shift in the cleavage plane. Kindly provided by Silvia Bonaccorsi. From Giansanti, M.G. *et al.*: *Development* 2001, **128**:1137–1145.

causes dissociation of the trimer, allowing the free Gα (and the Gβγ dimer in some cases) to activate cell processes. G-protein activity is governed by various proteins that influence GTP binding, hydrolysis and subunit association. In the *C. elegans* zygote, the key Gα proteins in spindle positioning are called GOA-1 and GPA-16. Their activity is controlled by several proteins, including two known as GPR-1 and GPR-2, which trigger the dissociation of the heterotrimeric G-protein complex.

How do these parts fit together into a regulatory system that enhances astral pulling forces at the posterior pole of the *C. elegans* zygote? The current hypothesis is that GOA-1 and GPA-16 stimulate astral pulling forces at both poles but are more active at the posterior. The activating GPR proteins are preferentially localized at the posterior pole, providing one mechanism by which Gα activity is enhanced at that pole. The Par proteins are required for the polarized localization of GPR proteins.

Par proteins and heterotrimeric G proteins are also important for asymmetric spindle behavior in the *Drosophila* neuroblast. The striking differences in spindle pole size and microtubule length at the two poles in these cells are likely to depend on the asymmetric distribution of proteins controlling microtubule nucleation and stability. The underlying regulatory components and mechanisms remain mysterious.

The orientation of cell division is controlled by the mitotic spindle

The position of the mitotic spindle can be used to determine the orientation of division as well as its symmetry. In the second divisions of the early *C. elegans* embryo, for example, the mitotic spindle in one daughter cell is rotated by 90° before cytokinesis, resulting in a cleavage plane perpendicular to that in the other cell (see Figure 8-19). Similar spindle movements seem be used by all metazoans to orient the plane of division in particular cell types, thereby allowing the correct positioning of cells and their progeny in multicellular tissues. In the *C. elegans* zygote, spindle rotation is thought to depend, at least in part, on cortical dynein motors that pull the astral microtubules of one spindle pole toward the anterior of the cell. Other factors thought to influence spindle rotation are the shape of the cell and the localization of other microtubule-regulating proteins at specific cortical sites.

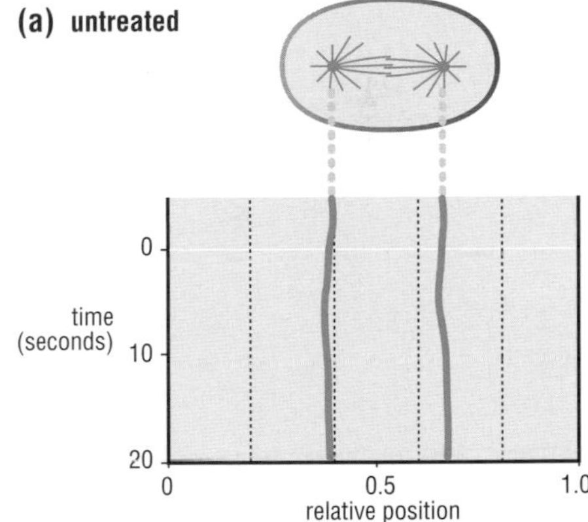

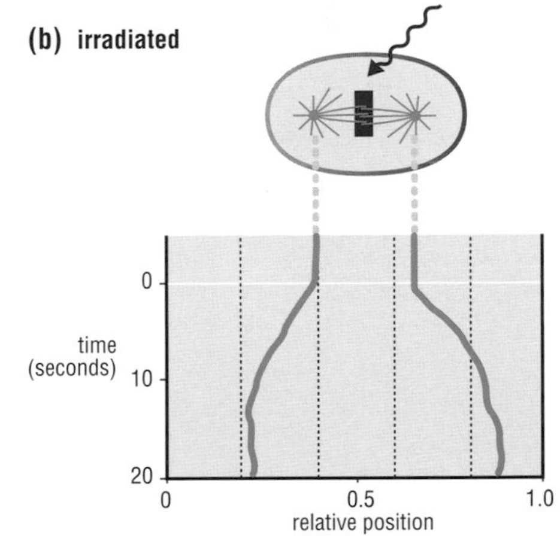

Figure 8-21 The importance of astral microtubule forces in spindle positioning in *C. elegans* In this experiment, the distance between spindle poles was measured over a brief period just after the beginning of anaphase. The graphs show the relative positions of the two poles during the experiment. **(a)** In a normal embryo, the distance between poles during early anaphase was relatively constant during the time course of the experiment. **(b)** In this embryo, the central spindle was destroyed by intense irradiation at the zero time point. The spindle poles then moved apart. The posterior pole moved more rapidly and farther than the anterior pole, revealing an asymmetry in the astral pulling forces acting on the two spindle poles. Adapted from Grill, S.W. *et al.*: *Nature* 2001, **409**:630–633.

References

Betschinger, J. and Knoblich, J.A.: **Dare to be different: asymmetric cell division in *Drosophila, C. elegans* and vertebrates.** *Curr. Biol.* 2004, **14**:R674–R685.

Giansanti, M.G. *et al.*: **The role of centrosomes and astral microtubules during asymmetric division of *Drosophila* neuroblasts.** *Development* 2001, **128**:1137–1145.

Gönczy, P.: **Mechanisms of spindle positioning: focus on flies and worms.** *Trends Cell Biol.* 2002, **12**:332–339.

Grill, S.W. and Hyman, A.A.: **Spindle positioning by cortical pulling forces.** *Dev. Cell* 2005, **8**:461–465.

Grill, S.W. *et al.*: **Polarity controls forces governing asymmetric spindle positioning in the *Caenorhabditis elegans* embryo.** *Nature* 2001, **409**:630–633.

Labbé, J.C. *et al.*: **PAR proteins regulate microtubule dynamics at the cell cortex in *C. elegans*.** *Curr. Biol.* 2003, **13**:707–714.

Tsou, M-F.B. *et al.*: **PAR-dependent and geometry-dependent mechanisms of spindle positioning.** *J. Cell Biol.* 2003, **160**:845–855.

Wodarz, A.: **Molecular control of cell polarity and asymmetric cell division in *Drosophila* neuroblasts.** *Curr. Opin. Cell Biol.* 2005, **17**:475–481.

9

Meiosis

Meiosis is a specialized form of nuclear division that generates nuclei carrying half the normal complement of chromosomes. The meiotic program involves many of the same mechanisms as the mitotic cell cycle, but also includes several unique features that allow two rounds of chromosome segregation without an intervening S phase.

9-0 Overview: Meiosis

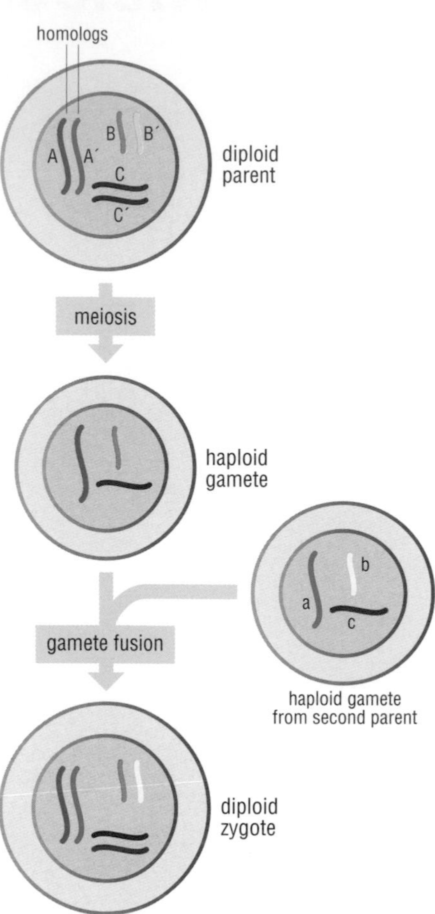

Figure 9-1 The sexual reproductive cycle The diploid cell at the top contains three different chromosomes, each with two homologs. The meiotic program reduces the complement of chromosomes by half, resulting in haploid cells, or gametes, containing one homolog of each chromosome. Meiosis can generate any combination of homologs in the gametes; only one combination is shown here. Fusion with a gamete from another parent regenerates the diploid state and completes the cycle. For clarity, this diagram does not include the effects of recombination on chromosome structure in the gametes.

Sexual reproduction is based on the fusion of haploid cells

Most eukaryotes reproduce sexually: cells from two parents fuse to generate a single cell from which a new organism develops. To avoid the chromosome complement of the offspring doubling at each generation, sexual reproduction depends on a specialized nuclear division called **meiosis** or the **meiotic program**, which reduces the chromosome complement by half during the formation of reproductive cells. In diploid organisms, such as most animals, non-reproductive cells contain two slightly different copies—called homologs—of each type of chromosome: one inherited from each parent. The cells produced by meiosis are haploid, containing a single homolog of each chromosome. These haploid cells often differentiate into specialized reproductive cells called **gametes**—for example, eggs and sperm. The reproductive cycle is completed when gametes from two parents fuse to form a diploid zygote, which carries a combination of parental chromosomes (Figure 9-1).

The meiotic program involves two rounds of chromosome segregation

The meiotic program (Figure 9-2) begins with a round of chromosome duplication—meiotic S phase—that gives rise to sister-chromatid pairs that are tightly linked by cohesins (see section 5-8). The cell then contains two copies of each parental homolog, or a total of four chromatids for each type of chromosome. Two rounds of chromosome segregation—*meiosis I* and *meiosis II*—then distribute these four chromatids into four haploid nuclei.

The unique feature of meiosis is that it results not only in the segregation of sister chromatids but also in the segregation of homologous chromosomes. Homolog segregation occurs in the first meiotic division and involves the same principles that govern sister-chromatid segregation in mitosis. First, the homologs are attached to each other, and second, they are aligned on a spindle with each homolog oriented toward the opposite spindle pole. Separation of homologs then allows them to be pulled by the spindle to opposite ends of the cell. The second meiotic division then segregates the sister-chromatid pairs by mechanisms that are essentially the same as those used in mitosis.

Homologous recombination is an important feature of meiosis

Physical linkage between homologs is essential for their accurate segregation in meiosis I. In most organisms the linkages are formed in meiotic prophase, just before the first division, by interactions between complementary DNA sequences in the two homologs. In most organisms, these interactions depend on a process known as *homologous recombination*, which occurs as follows. Double-strand breaks are introduced at numerous locations along the chromosomes, and single-stranded DNA segments originating at these breaks then interact with complementary sequences in their homologs. In some cases, the result is a reciprocal DNA exchange or *crossover*: that is, the DNA of one chromatid becomes continuous with the DNA of its homolog, and vice versa (see Figure 9-2), resulting in a strong physical linkage between the homologs. At most recombination sites, however, the interaction leads simply to repair of the DNA break, no crossing-over of chromatids and no lasting connection between homologs at the site of breakage. In the context of meiosis, this repair process is called a *noncrossover*.

The events of meiotic prophase are accompanied by chromosomal changes that are visible under the microscope. In early prophase, chromosomes are compacted and homologs gradually

Definitions

aneuploid: containing an abnormal number of chromosomes.

chiasma (plural **chiasmata**)**:** X-shaped chromosomal structure that is seen in the microscope at sites of crossing-over between homologous chromosomes at the end of meiotic prophase.

gamete: specialized haploid reproductive cell, for example egg and sperm in animals, that fuses with another gamete to form a diploid zygote in sexual reproduction.

meiosis: see **meiotic program**.

meiotic program: specialized nuclear division that occurs during formation of the gametes in sexually reproducing diploid organisms and generates haploid nuclei carrying a single homolog of each chromosome. The term **meiosis**, although originally used to describe only the events of meiotic chromosome segregation, is also used to describe the meiotic program as a whole.

References

John, B.: *Meiosis* (Cambridge University Press, New York, 1990).

Page, S.L. and Hawley, R.S.: **Chromosome choreography: the meiotic ballet.** *Science* 2003, **301**:785–789.

Roeder, G.S.: **Meiotic chromosomes: it takes two to tango.** *Genes Dev.* 1997, **11**:2600–2621.

Zickler, D. and Kleckner, N.: **Meiotic chromosomes: integrating structure and function.** *Annu. Rev. Genet.* 1999, **33**:603–754.

Figure 9-2 **The meiotic program** The diploid cell shown here has only one chromosome, represented by two homologs. Meiotic S phase results in two sister chromatids per homolog, linked tightly by cohesins along their entire lengths (black dots). In meiotic prophase, DNA recombination results in crossovers that become visible as chiasmata after the synaptonemal complex dissolves. During biorientation in meiosis I, homolog pairs are held together only by cohesion along sister-chromatid arms distal to the chiasmata. Loss of this cohesion therefore results in the separation of homologs in anaphase I. The first meiotic division is sometimes called a reductional division, because the resulting cells contain only one copy of each parental homolog and are therefore reduced to a genetically haploid state. Sister chromatids are segregated in meiosis II, which is sometimes called an equational division because the number of homologs is not changed.

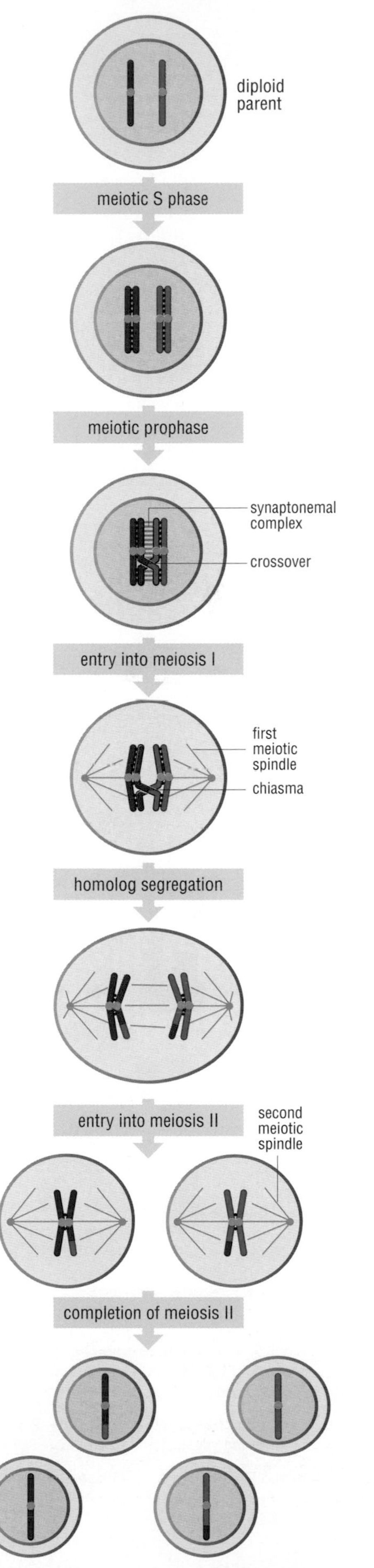

migrate to positions near each other in the nucleus. The two homologs of each chromosome become aligned in parallel in a process called homolog *pairing*, which depends in part on the DNA interactions described above. Pairing is followed by *synapsis*: homologs are brought closer together at sites of recombination and linked tightly along their entire length by a protein scaffold called the *synaptonemal complex*. Late in prophase this complex dissolves and crossovers become visible in the microscope as X-shaped structures called **chiasmata** (singular **chiasma**). Each homolog pair contains at least one chiasma, ensuring that all homolog pairs are linked as the cell begins the first meiotic division.

After meiotic prophase, homolog pairs become bi-oriented on the first meiotic spindle. Because each homolog contains two sister chromatids, there are specialized mechanisms to ensure that both kinetochores of a sister-chromatid pair attach to the same spindle pole and the homologous sister pair attaches to the opposite pole.

Chiasmata hold homologs together in meiosis I only because the sister-chromatid arms distal to the chiasmata (that is, on the side away from the centromere) are linked by cohesins (see Figure 9-2). The removal of cohesins from chromosome arms in anaphase I therefore separates the homologs, allowing their segregation to opposite poles of the first meiotic spindle. The spindle is disassembled and the cell proceeds directly to the second meiotic division.

Although homologs are separated, cohesion between sister-chromatid pairs is maintained at the centromeres throughout the first meiotic division. Sister chromatids therefore remain linked and can be bi-oriented on the second meiotic spindle. Loss of centromeric cohesion triggers sister-chromatid separation and segregation in anaphase II, just as it does in mitotic anaphase.

Meiosis is an important source of genetic variation. During the first meiotic division, a random assortment of homologs is distributed to the daughter nuclei (see Figure 9-1). Thus, in organisms with many chromosomes, the haploid gametes may contain any one of a large number of possible combinations of maternal and paternal homologs. Crossovers between homologs also generate new combinations of parental genes.

Defects in meiosis lead to aneuploidy

Errors in meiotic chromosome segregation are rare in most organisms but more common in humans, where they can lead to gametes with abnormal numbers of chromosomes—known as **aneuploid** cells. Several percent of human oocytes and sperm are aneuploid, and the frequency of errors in females increases with age. Aneuploid zygotes generally fail to survive, and meiotic segregation errors probably account for nearly half of the spontaneous abortions, or miscarriages, that are relatively common in the first trimester of pregnancy. In some cases, aneuploid gametes give rise to a viable but partly defective embryo. Embryos with three copies of chromosome 21, for example, develop into children with Down syndrome, a condition associated with mental retardation and altered physical appearance.

In this chapter we describe the major events of meiosis and how they are regulated. We begin with early meiosis and then discuss the events of meiotic prophase and the mechanism of homologous recombination. The remainder of the chapter addresses the mechanisms by which homologs and sister chromatids are segregated during the meiotic divisions.

The meiotic program is controlled at multiple checkpoints

Progression through the stages of the meiotic program is controlled at several checkpoints that are roughly equivalent to the major control points of the mitotic cell cycle (Figure 9-3). As in the mitotic cycle, these transitions are regulated by combinations of gene regulatory factors, cyclin–Cdk complexes and the APC.

In multicellular organisms, entry into the meiotic program occurs only in a small population of cells, called the germ line, which are the precursors of non-proliferating haploid gametes. The rest of the cells—the somatic cells—reproduce only by the mitotic cell cycle and cannot enter meiosis. In sexually reproducing unicellular organisms, however, all diploid cells possess the ability to undergo meiosis to generate haploid offspring. Diploid budding yeast, for example, switch on the meiotic program when nutrients become scarce, resulting in the formation of haploid spores (see section 2-1). Because the initiation of meiosis and spore formation are easily controlled and analyzed in the laboratory, the mechanisms that control meiotic entry in budding yeast are particularly well understood and are the primary focus of this section.

The transcription factor Ime1 initiates the budding yeast meiotic program

A gene regulatory factor called Ime1 triggers increased expression of a large number of genes that promote the early events of the meiotic program (Figure 9-4). These genes encode the protein machinery required for meiotic DNA synthesis and homolog recombination. Activation of Ime1 requires two coincident regulatory signals. First, a genetic signal must indicate that the cell is diploid—clearly, meiosis should not be initiated in a cell that is already haploid. Second, a nutritional signal should indicate that the cell is experiencing severe shortages of important nutrients. These two signals influence Ime1 activation by multiple mechanisms that act on both *IME1* expression and Ime1 protein activity.

Entry into the meiotic program is also coordinated with the mechanisms that control entry into the mitotic cycle. As discussed in Chapter 3 (section 3-13), entry into the mitotic cycle at Start occurs when the concentration of the G1 cyclin, Cln3, rises to a threshold that triggers expression of the G1/S cyclins Cln1 and Cln2. The activation of G1 and G1/S cyclins is tightly regulated by nutritional conditions, so that starvation—and the resulting reduction in growth rate—represses the production of these cyclins and thereby blocks Start. Thus, entry into the mitotic cycle is inhibited by the same conditions that stimulate entry into meiosis.

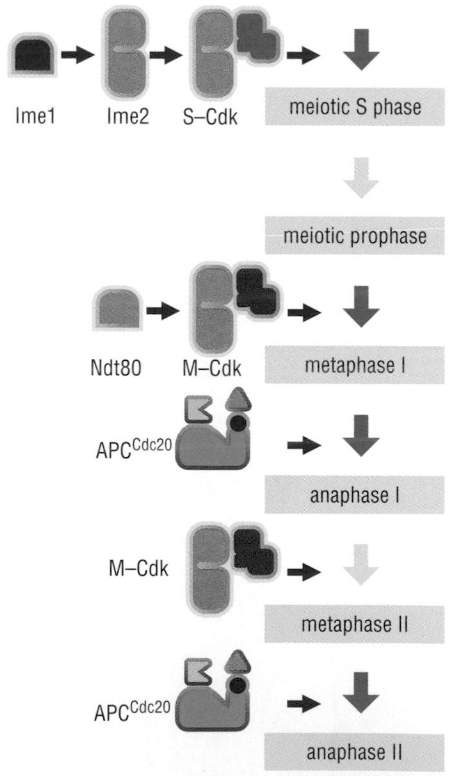

Figure 9-3 Control of the meiotic program This general scheme shows the four major regulatory transitions at which progression through the meiotic program is controlled (dark blue arrows). S–Cdk activity, driven by the gene regulatory factor Ime1 in yeast, triggers meiotic S phase and the interactions between homologs in meiotic prophase. A second gene regulatory protein, Ndt80, stimulates the M–Cdk activity that drives entry into meiosis I, as described later in this chapter. As in the mitotic cycle, initiation of anaphase in both meiotic divisions is triggered by APC^{Cdc20}. Progress through any of these four checkpoints is influenced by external factors. In budding yeast, entry into the meiotic program occurs only in starved diploid cells, and entry into meiosis I is blocked by DNA damage or incomplete recombination. Progression through these checkpoints is governed by similar mechanisms in vertebrates.

References

Benjamin, K.R. *et al.*: **Control of landmark events in meiosis by the CDK Cdc28 and the meiosis-specific kinase Ime2.** *Genes Dev.* 2003, **17**:1524–1539.

Chu, S. *et al.*: **The transcriptional program of sporulation in budding yeast.** *Science* 1998, **282**:699–705.

Honigberg, S.M.: **Ime2p and Cdc28p: co-pilots driving meiotic development.** *J. Cell Biochem.* 2004, **92**:1025–1033.

Honigberg, S.M. and Purnapatre, K.: **Signal pathway integration in the switch from the mitotic cell cycle to meiosis in yeast.** *J. Cell Sci.* 2003, **116**:2137–2147.

Marston, A.L. and Amon, A.: **Meiosis: cell-cycle controls shuffle and deal.** *Nat. Rev. Mol. Cell Biol.* 2004, **5**:983–997.

Primig, M. *et al.*: **The core meiotic transcriptome in budding yeasts.** *Nat. Genet.* 2000, **26**:415–423.

Vershon, A.K. and Pierce, M.: **Transcriptional regulation of meiosis in yeast.** *Curr. Opin. Cell Biol.* 2000, **12**:334–339.

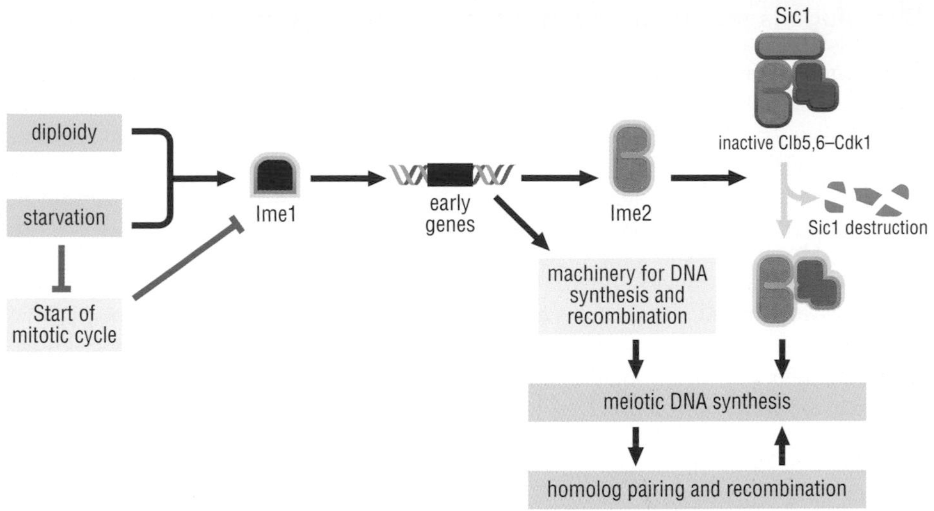

Figure 9-4 Regulatory circuit controlling the onset of meiotic S phase in budding yeast The meiotic program is initiated by the gene regulatory factor Ime1, which is activated only when two conditions are met: first, the cell must be diploid, and second, it must be exposed to low levels of nutrients. These two signals act by various mechanisms to increase expression of *IME1* and also to increase the activity of the Ime1 protein, which then triggers increased expression of the early meiotic genes. One of these genes encodes the protein kinase Ime2, which triggers the activation of Clb5,6–Cdk1 complexes by phosphorylating the Cdk inhibitor Sic1. Clb5,6–Cdk1 complexes are directly responsible for initiating meiotic DNA synthesis and may also help to activate the recombination machinery. DNA synthesis and recombination are dependent on each other, as inhibition of either process causes defects in the other.

Conversely, entry into the mitotic cycle inhibits entry into meiosis. By mechanisms that remain unclear, the G1/S cyclins Cln1 and Cln2 inhibit expression of *IME1* and the nuclear localization of the Ime1 protein. Robust mechanisms therefore exist to ensure that the mitotic and meiotic programs cannot be initiated in the same cell.

Entry into the meiotic program is driven by the protein kinase Ime2

Progression through Start in the mitotic cell cycle is triggered by Cln1,2–Cdk1 complexes, whose primary function is to phosphorylate and thereby stimulate the destruction of the Cdk inhibitor Sic1 (see section 3-6; also described in more detail in Chapter 10). Destruction of Sic1 unleashes Clb5,6–Cdk1 complexes, which then stimulate the initiation of DNA replication. The same general scheme holds true in the meiotic program, with one major difference. As described above, the G1/S cyclins Cln1 and Cln2 are not required for the meiotic program—in fact, they are inhibitors of meiosis. Their function is performed by a meiosis-specific protein kinase called Ime2.

Ime2 is encoded by one of the early meiotic genes whose expression is stimulated by Ime1. The rise in Ime2 activity in early meiosis is required for the initiation of meiotic DNA replication, primarily because Ime2 phosphorylates Sic1, thereby triggering its destruction (see Figure 9-4). Sic1 protein is stabilized in *ime2* mutants, but the deletion of Sic1 allows meiotic DNA synthesis to occur in the absence of Ime2.

Meiotic DNA replication is initiated by complexes of Cdk1 and Clb5 or Clb6. Unlike DNA replication in the mitotic cell cycle, however, meiotic DNA replication cannot occur in the absence of Clb5 and Clb6, indicating that other B-type cyclins (primarily Clb1 in meiotic cells) cannot trigger meiotic S phase (see section 4-5).

The basic mechanisms of DNA synthesis are similar in mitotic and meiotic cells. In budding yeast, initiation of DNA synthesis occurs at the same replication origins in both cell programs, and DNA synthesis is catalyzed by the same enzymes. Pre-replicative complexes containing Orc and Mcm proteins (see section 4-3) are required for origin function in meiotic S phase, and phosphorylation of these proteins by Cdks is likely to block re-replication in the meiotic cell as it does in mitosis.

Despite the similarities to premitotic S phase, however, it is clear that meiotic chromosome duplication is a specialized process that is uniquely integrated into the meiotic program. Meiotic S phase in all eukaryotes is several times longer than premitotic S phase, in part because meiotic S phase includes preparations for the homolog interactions that immediately follow DNA synthesis. Meiotic S phase is shortened, for example, by mutation of the enzyme Spo11, which generates the double-strand DNA breaks that are required for recombination. Conversely, delaying DNA replication—by deleting replication origins, for example—results in delays in the formation of the DNA breaks required for recombination. This and other evidence suggests that meiotic homologous recombination is somehow coupled to meiotic DNA synthesis. The mechanisms underlying this coupling are not known.

9-2 Homologous Recombination in Meiosis

Homologous recombination is a central feature of meiotic prophase

During meiotic prophase, homologous chromosomes—each represented by a sister-chromatid pair—become linked together in preparation for their segregation in the first meiotic division. These linkages generally depend on **homologous recombination**, in which regions of complementary DNA on the two homologs interact with each other. In this section we describe the basic mechanics of homologous recombination in meiosis, as a foundation for subsequent discussions of meiotic prophase and its regulation.

In most organisms, meiotic DNA recombination is thought to occur as illustrated in Figure 9-5. There are two main outcomes: either a **noncrossover** event that helps promote homolog pairing but does not generate a lasting interhomolog connection, or a **crossover** event that yields the strong interhomolog connection that is so important for homolog segregation in meiosis I (and in which the reciprocal exchange of large segments of homologous chromosomes occurs, resulting in genetic recombination). Only a small fraction of recombination events lead to crossovers. The noncrossover/crossover decision is rigorously controlled to ensure that each pair of homologs has at least one crossover and that crossovers tend to be distributed far apart from each other on the chromosomes.

Meiotic recombination begins when both strands of the DNA double helix are cleaved at many locations by the enzyme Spo11. The exposed DNA 5′ ends are trimmed back, or resected, resulting in single-stranded 3′ overhangs that are then coated by *recombinases* called Rad51 and Dmc1. The resulting protein–DNA filaments, called strand-transfer complexes, have the remarkable ability to invade another double-stranded DNA and scan it for complementary regions of DNA sequence. In meiosis, a single strand from one homolog invades a chromatid of the other homolog; recombination between identical sequences on sister chromatids is suppressed. On locating a homologous sequence, the invading filament pairs with the complementary strand, displacing the other strand of the DNA duplex to form a single-stranded displacement loop known as a D-loop (see Figure 9-5).

Whether a crossover or a noncrossover will be made is thought to be determined at this step. At the large number of sites that are designated to form noncrossovers, DNA repair enzymes extend the 3′ end of the invading strand on the homologous template, but the invading strand eventually dissociates and returns to the chromatid from which it came. DNA repair then fills in the gaps to generate an intact chromosome that is no longer attached to its homolog.

At the few sites that are designated to become crossovers, a different sequence of events unfolds. The first detectable step in this sequence is the stabilization, by unknown means, of the interaction between the invading single strand and its homolog—resulting in a structure called a single-end invasion. Extension of the single strand then leads to expansion of the D-loop, which interacts with the second single-stranded end generated at the original double-strand break. Further DNA synthesis and ligation fills in the gaps, and the final result is a DNA structure called a double Holliday junction. A specific pair of DNA cleavage reactions, followed by DNA repair, transforms this structure into a crossover (see Figure 9-5).

These molecular events are accompanied by chromosome movements and structural changes that provide further insights into the important problems of how crossover sites are assembled and controlled. We discuss the progression of meiotic chromosome changes in sections 9-3 and 9-4.

Definitions

crossover: in meiosis, a **homologous recombination** event that results in the reciprocal exchange of DNA between two homologs.

homologous recombination: interaction between broken and intact homologous DNA molecules that promotes repair, exchange and pairing.

noncrossover: in meiosis, a **homologous recombination** event that results in repair of a double-strand DNA break without formation of a **crossover**.

References

Allers, T. and Lichten, M.: **Differential timing and control of noncrossover and crossover recombination during meiosis.** *Cell* 2001, **106**:47–57.

Bishop, D.K. and Zickler, D.: **Early decision: meiotic crossover interference prior to stable strand exchange and synapsis.** *Cell* 2004, **117**:9–15.

Hollingsworth, N.M. and Brill, S.J.: **The Mus81 solution to resolution: generating meiotic crossovers without Holliday junctions.** *Genes Dev.* 2004, **18**:117–125.

Hunter, N. and Kleckner, N.: **The single-end invasion: an asymmetric intermediate at the double-strand break to double-Holliday junction transition of meiotic recombination.** *Cell* 2001, **106**:59–70.

Schwacha, A. and Kleckner, N.: **Interhomolog bias during meiotic recombination: meiotic functions promote a highly differentiated interhomolog-only pathway.** *Cell* 1997, **90**:1123–1135.

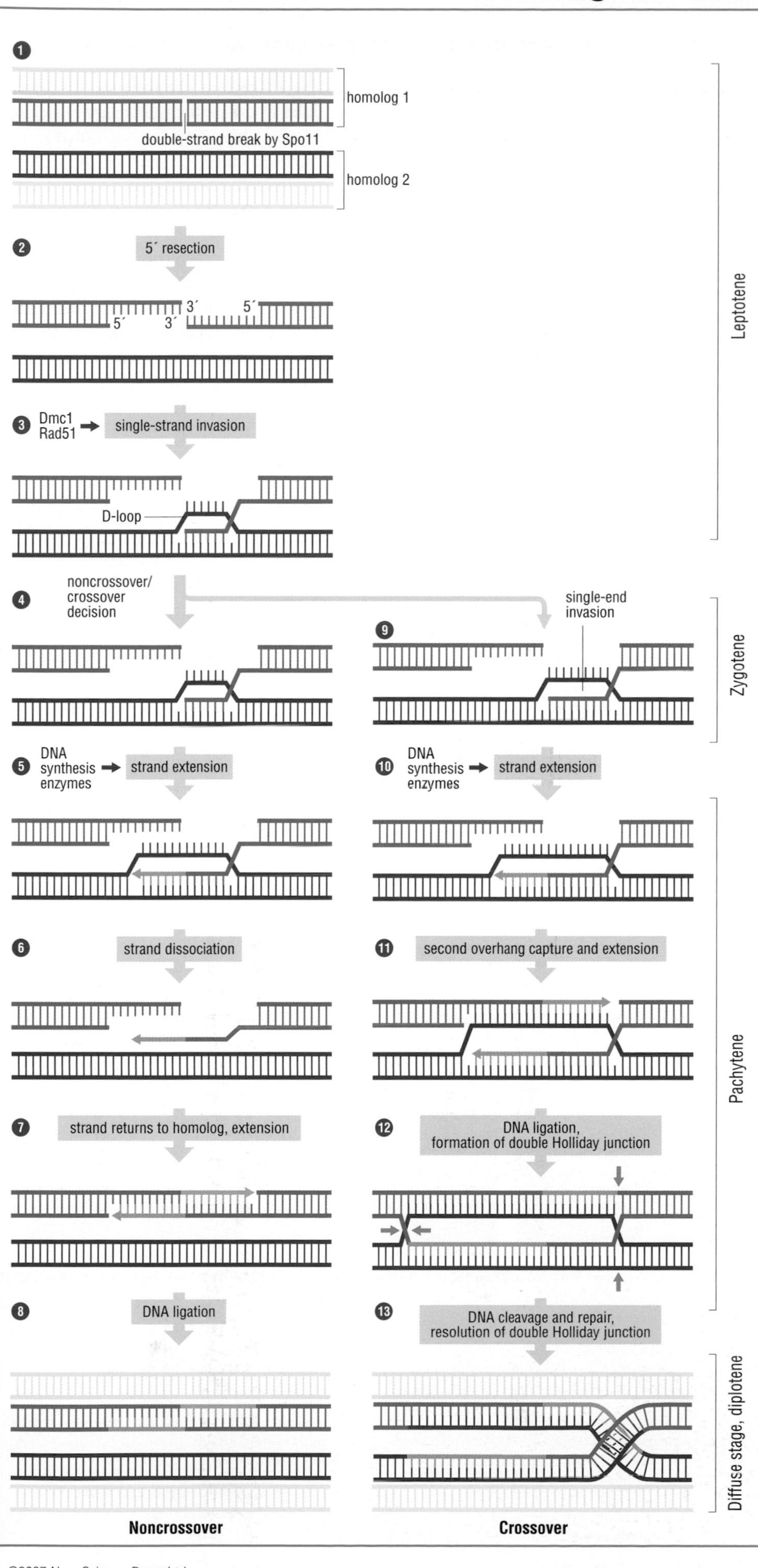

Figure 9-5 **Homologous recombination in meiotic prophase** These diagrams illustrate the current model of the major meiotic recombination mechanism. **(1)** The process begins with a double-strand break catalyzed by Spo11, which is related to the topoisomerase family of DNA-processing enzymes. **(2)** Resection of 5′ ends generates single-stranded 3′ overhangs about 300 nucleotides long. **(3)** These ends are coated with proteins called Rad51 and Dmc1, which are structurally and functionally related to the bacterial protein RecA. These proteins promote the recombination reaction and are therefore called recombinases. The recombinase–DNA complex invades a homologous chromatid and pairs with complementary sequence on one strand, displacing the other strand to form a D-loop. **(4)** By unknown mechanisms, sites of recombination are designated as either a noncrossover or crossover event. **(5)** Most recombination sites are directed toward a noncrossover event, which begins with extension of the invading single strand on the homologous template. **(6)** The invading single strand eventually detaches from the homolog. **(7)** The dissociated strand returns to the homolog from which it came and is further extended using that chromosome as template. **(8)** DNA ligation seals the single-stranded nicks, generating an intact chromosome. If the two homologs are not identical in sequence in this region, some of the sequence in the invading strand will have been changed to that of the homolog during extension of the 3′ end, a phenomenon known as gene conversion. **(9)** At the few sites selected to form crossovers, the invading single strand is converted to a stable complex called a single-end invasion. **(10)** The single strand is extended on the homologous template. **(11)** The D-loop captures the other single-stranded overhang at the original double-strand break. DNA synthesis then fills in the gaps. **(12)** DNA ligation generates a complex DNA structure in which the two DNA molecules are physically linked by exchange of a part of a DNA strand from each duplex. The region of exchange is bounded on each side by an X-shaped structure called a Holliday junction, and the whole region is called a double Holliday junction. **(13)** Asymmetric cleavage of the two Holliday junctions (green arrows), followed by DNA ligation, leads to resolution of the double Holliday junction and formation of a crossover. The major cytological stages of meiotic prophase, as described in section 9-3, are indicated along the right. Other recombination mechanisms, not shown here, also contribute to the formation of crossovers in some species. Fission yeast in particular employ a distinct pathway, involving an enzyme called Mus81, that generates crossovers by a mechanism that might not involve double Holliday junctions. Similar mechanisms may generate a small fraction of the crossovers in other species.

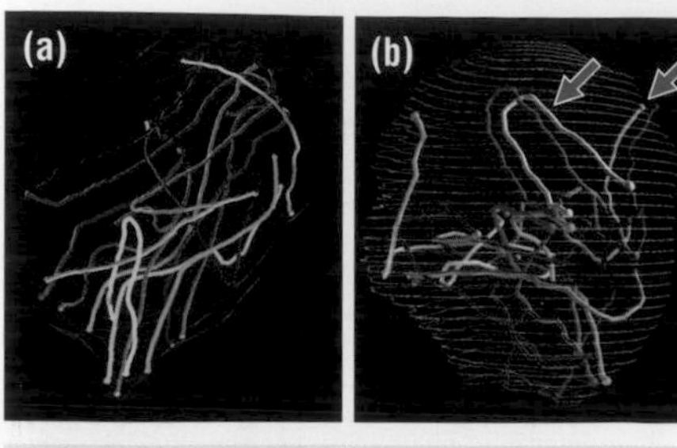

Figure 9-6 Early steps in homolog pairing Meiotic chromosomes of the fungus *Sordaria macrospora*, in which the homolog axes are labeled with a fluorescent protein. Multiple images of the same meiotic nucleus were used to reconstruct a three-dimensional image in which each chromosome is labeled with a different color. Seven homolog pairs are present. **(a)** In early leptotene the tangle of chromosomes reveals no obvious homolog interactions. **(b)** In mid-leptotene some homolog pairs are partly aligned (indicated by arrows). Photographs kindly provided by Denise Zickler. From Tessé, S. *et al.*: *Proc. Natl Acad. Sci. USA* 2003, **100**:12865–12870.

Stages of meiotic prophase are defined by cytological landmarks

Meiotic DNA recombination is accompanied by striking changes in chromosome structure, which are readily apparent in the microscope and are used to define the four major stages of meiotic prophase: *leptotene*, *zygotene*, *pachytene* and *diplotene*. In **leptotene** the duplicated sister chromatids first condense into distinct, thread-like structures. Homolog **pairing** also occurs during leptotene, in which the homolog axes are aligned with each other roughly 400 nm apart. In **zygotene**, homologs are brought even closer together, to a distance of about 100 nm, in a process called *synapsis*. An array of protein filaments called the *synaptonemal complex* is assembled between the homologs. **Pachytene** is the stage at which the synaptonemal complex is complete and homologs are tightly linked along their entire lengths. At the end of pachytene, the synaptonemal complex is disassembled and the chromatin dramatically decondenses (often called the diffuse stage between pachytene and diplotene). The chromosomes become highly condensed in **diplotene**, and crossovers become apparent as chiasmata. Diplotene is followed by entry into the first meiotic division: the spindle poles separate and form a spindle on which the homolog pairs become bi-oriented. This early stage in meiosis I is sometimes called **diakinesis**.

Homolog pairing occurs in two successive stages

One of the most fascinating problems in meiosis is how homologs find each other among the tangle of chromosomes in the nucleus. The first step in homolog pairing occurs early in leptotene, when the chromosomes undergo dramatic changes in position within the nucleus and homologs begin to interact (Figure 9-6). This early stage in pairing is probably related to a poorly understood somatic pairing process by which homologs associate in the non-meiotic cells of some species. The molecular basis of early pairing is not clear, but it is likely to involve direct interactions between complementary DNA sequences in long chromatin loops that extend from the two homologs. Breaks in the DNA are not required, and so these DNA interactions probably involve unstable interactions between intact or partly unwound double helices. In some species, interactions between homologous centromeres or other specific pairing regions also contribute to pairing.

The second major step in homolog pairing occurs in the middle of leptotene, when homologs become aligned about 400 nm apart in a process known as presynaptic alignment. This alignment occurs throughout the genome in some organisms, such as higher plants, but only occurs in specific segments in others, such as the mouse. Mutant analyses and microscopy in various organisms suggest that presynaptic alignment depends on the formation of double-strand

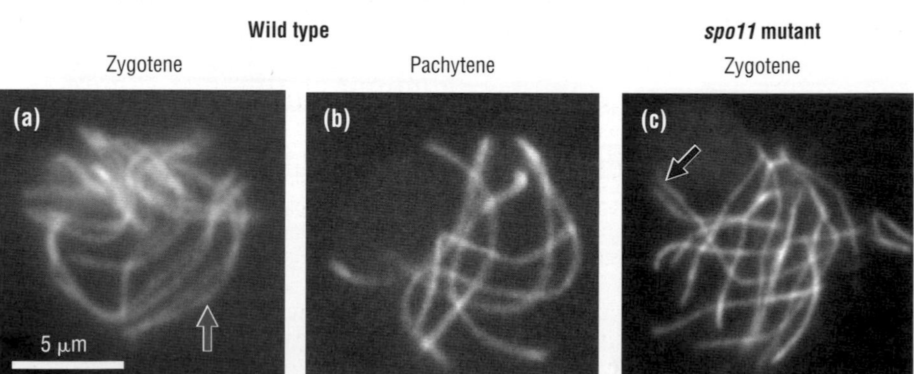

Figure 9-7 Homolog pairing defects in a *spo11* mutant As in Figure 9-6, these images represent fluorescently labeled meiotic chromosomes of the fungus *Sordaria macrospora*. **(a)** In wild-type early zygotene cells, parts of some homolog pairs display presynaptic alignment (blue arrow) while other regions have initiated synapsis. **(b)** In pachytene, synapsis is complete and homolog pairs are fused into single units by the synaptonemal complex. **(c)** In a *spo11* mutant, however, homolog pairing is severely inhibited. No presynaptic alignment is seen at the time point at which pairing is complete in wild-type cells (not shown), and a limited amount of pairing (red arrow) is found at later time points. No synaptonemal complex forms in the *spo11* mutant, and these cells will go on to have severe segregation errors in meiosis I. Photographs kindly provided by Denise Zickler. From Storlazzi, A. *et al.*: *Genes Dev.* 2003, **17**:2675–2687.

Definitions

diakinesis: stage in meiosis following **diplotene**, in which the meiotic spindle is formed and pairs of homologous chromosomes become oriented on the spindle.

diplotene: stage in meiotic prophase following **pachytene**, in which chromosomes condense and chiasmata become apparent.

leptotene: first stage in meiotic prophase, in which the pairing of homologous chromosomes occurs.

pachytene: stage in meiotic prophase following **zygotene**, in which homologous chromosome pairs are tightly linked by the synaptonemal complex.

pairing: (in meiosis) the initial interaction of homologous chromosomes with each other in early meiotic prophase.

zygotene: stage in meiotic prophase following **leptotene**, in which the synaptonemal complex begins to form between paired homologous chromosomes.

References

Albini, S.M. and Jones, G.H.: **Synaptonemal complex spreading in *Allium cepa* and *A. fistulosum*. I. The initiation and sequence of pairing.** *Chromosoma* 1987, **95**:324–338.

Bishop, D.K. and Zickler, D.: **Early decision: meiotic crossover interference prior to stable strand exchange and synapsis.** *Cell* 2004, **117**:9–15.

Börner, G.V. *et al.*: **Crossover/noncrossover differentiation, synaptonemal complex formation, and**

breaks and single-strand transfer complexes (see section 9-2). In most organisms, including fungi and vertebrates, prevention of double-strand break formation by mutation of Spo11 causes major defects in presynaptic alignment—as well as defects in synapsis and homolog segregation (Figure 9-7).

Analysis of leptotene chromosome structure provides a number of clues to the molecular events underlying presynaptic alignment. Leptotene chromosomes are thin, thread-like structures composed of long chromatin loops radiating outward from a central protein axis (Figure 9-8a). Each homolog is composed of two sister chromatids, but the sisters are so tightly linked that only a single protein axis is apparent—although individual sister axes can be detected in cells with defects in sister-chromatid cohesion. In some species the homolog axis of leptotene chromosomes is visible in electron micrographs as a dark axial element (Figure 9-8b). Light and electron microscopy also reveal that large particles containing recombination proteins (Spo11 and Dmc1, for example) associate with the axial element in leptotene (see Figure 9-8b). These particles, or nodules, are thought to reflect the formation of double-strand breaks and strand-transfer complexes on DNA loops emanating from the homolog axis.

Double-strand breaks do not occur randomly along the chromosomes but are focused at preferred sites, or recombination hotspots, that are generally found in chromatin that is relatively accessible to proteins. Recombination is thought to occur in the chromatin loops away from the homolog axis. The presence of recombination complexes on the homolog axis therefore implies that the loops containing double-strand breaks are folded back and linked to the axis by the recombination machinery. At the end of leptotene, the large nodules seen in electron micrographs are no longer associated with a single homolog axis but are positioned between the two aligned homolog axes (Figure 9-8c). These nodules can be associated with interhomolog fibers, or bridges, that link the two axes. Light microscopy also reveals the presence of interaxis bridges containing recombination proteins, and formation of bridges is blocked by mutations in these proteins.

These observations support the hypothesis that recombination complexes are initially anchored on the chromosome axis in leptotene, where they eventually interact with complementary DNA loops radiating from the homologous chromosome. This leads to interaction between the strand-transfer complex and the homologous DNA (see Figure 9-5). The recombination machinery might then act as a winch that reels in the chromatin loops of the other homolog, resulting in the formation of interaxis bridge complexes that establish presynaptic alignment.

In zygotene, homolog axes are brought even closer together, to a distance of about 100 nm (Figure 9-8d). This close alignment is accompanied by assembly of the synaptonemal complex between the homologs, as we discuss in section 9-4.

(a)

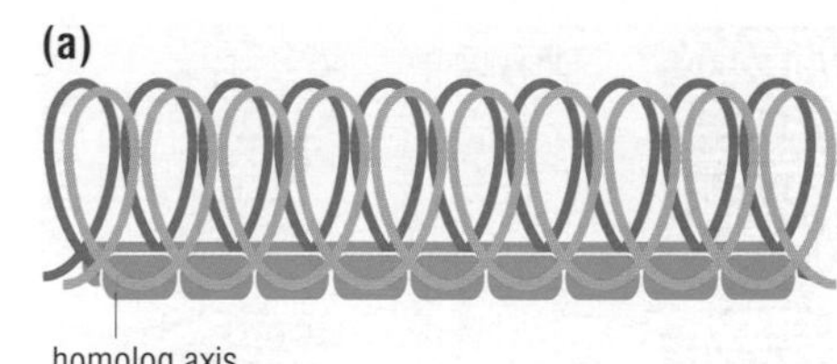

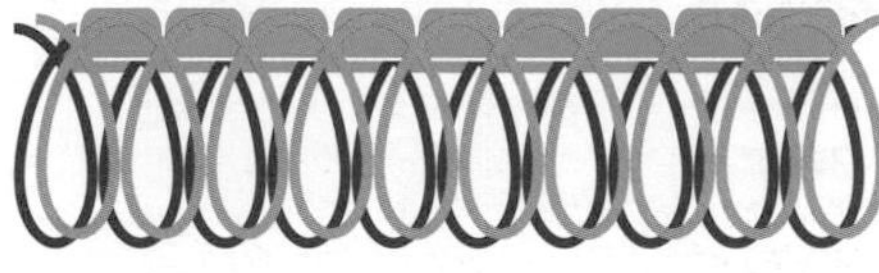

(b) Leptotene

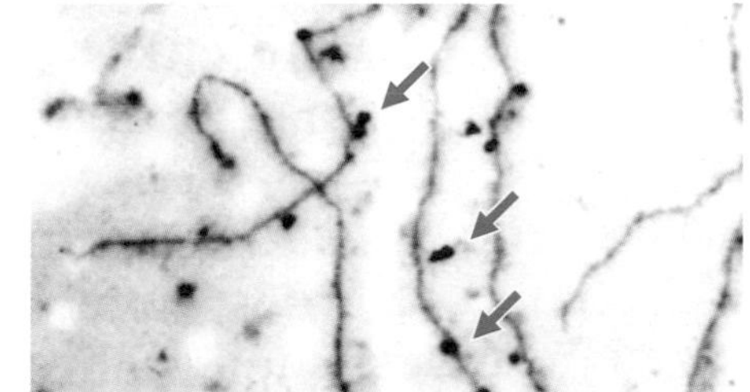

(c) Leptotene/zygotene

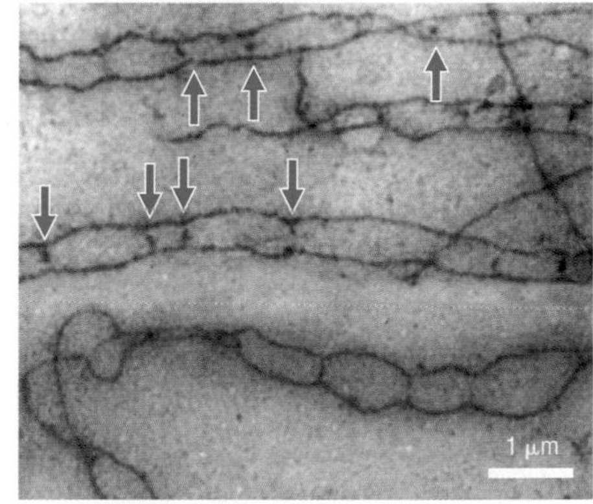

(d) Zygotene

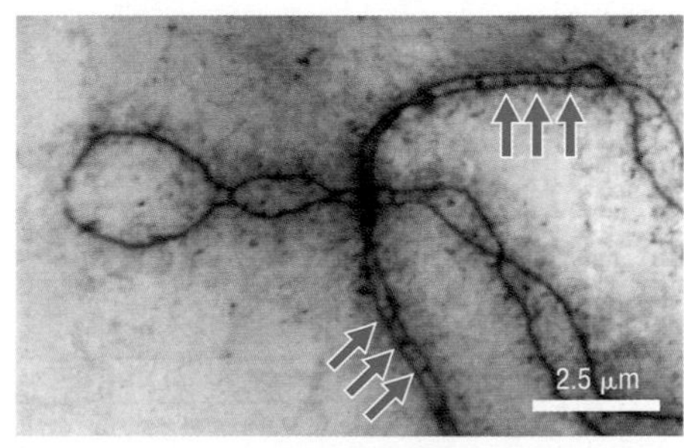

Figure 9-8 Electron microscopic analysis of chromosome structure in leptotene and early zygotene **(a)** Diagram of the general structure of homologs in early meiotic prophase, with the two sister chromatids of each homolog linked by a central protein axis. **(b)** In this electron micrograph, leptotene homologs from tomato plant cells are decorated with dark particles, such as those indicated by arrows, that are directly connected to chromosome axes. **(c)** At the end of leptotene in onion plant cells, nodules lie between the aligned homologs, and bridge structures can sometimes be seen (arrows). **(d)** In zygotene, homologs converge toward one another in regions containing nodules (arrows), setting the stage for synaptonemal complex assembly. Other studies (not shown) indicate that the nodules in these images contain recombination proteins such as Dmc1. Photographs kindly provided by Jim Henle and Nancy Kleckner. Panel (b) from Stack, S.M. and Anderson, L.K.: *Am. J. Bot.* 1986, **73**:264–281. Panels (c) and (d) from Albini, S.M. and Jones, G.H.: *Chromosoma* 1987, **95**:324–338.

regulatory surveillance at the leptotene/zygotene transition of meiosis. *Cell* 2004, **117**:29–45.

Gerton, J.L. and Hawley, R.S.: **Homologous chromosome interactions in meiosis: diversity amidst conservation.** *Nat. Rev. Genet.* 2005, **6**:477–487.

Kauppi, L. *et al.*: **Where the crossovers are: recombination distributions in mammals.** *Nat. Rev. Genet.* 2004, **5**:413–424.

MacQueen, A.J. *et al.*: **Chromosome sites play dual roles to establish homologous synapsis during meiosis in *C. elegans*.** *Cell* 2005, **123**:1037–1050.

Stack, S.M. and Anderson, L.K.: **Two-dimensional spreads of synaptonemal complexes from solanaceous plants. II. Synapsis in *Lycopersicon esculentum* (tomato).** *Am. J. Bot.* 1986, **73**:264–281.

Storlazzi, A. *et al.*: **Meiotic double-strand breaks at the interface of chromosome movement, chromosome remodeling, and reductional division.** *Genes Dev.* 2003, **17**:2675–2687.

Tessé, S. *et al.*: **Localization and roles of Ski8p protein in *Sordaria* meiosis and delineation of three mechanistically distinct steps of meiotic homolog juxtaposition.** *Proc. Natl Acad. Sci. USA* 2003, **100**:12865–12870.

Tsubouchi, T. and Roeder, G.S.: **A synaptonemal complex protein promotes homology-independent centromere coupling.** *Science* 2005, **308**:870–873.

Zickler, D. and Kleckner, N.: **Meiotic chromosomes: integrating structure and function.** *Annu. Rev. Genet.* 1999, **33**:603–754.

9-4 Chiasma Formation in Late Meiotic Prophase

(a)

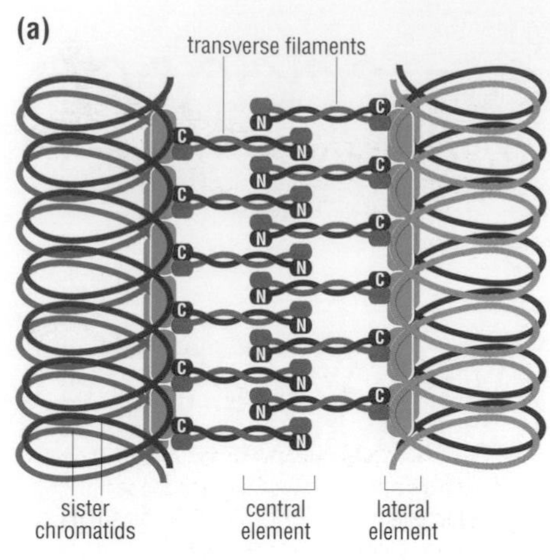

(b)

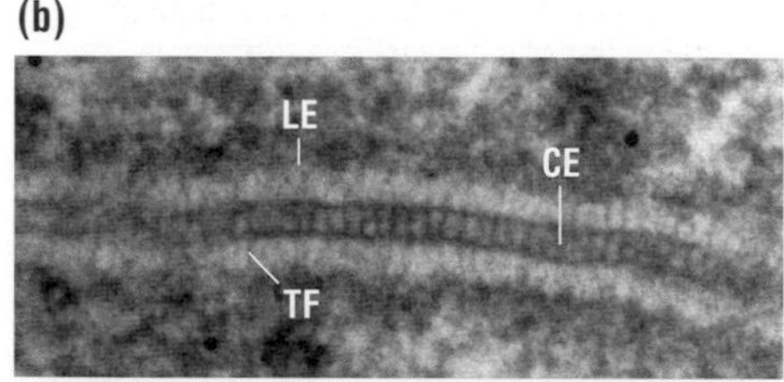

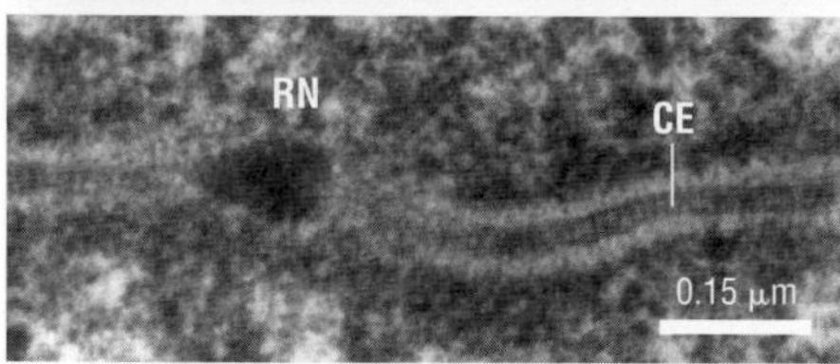

(c)

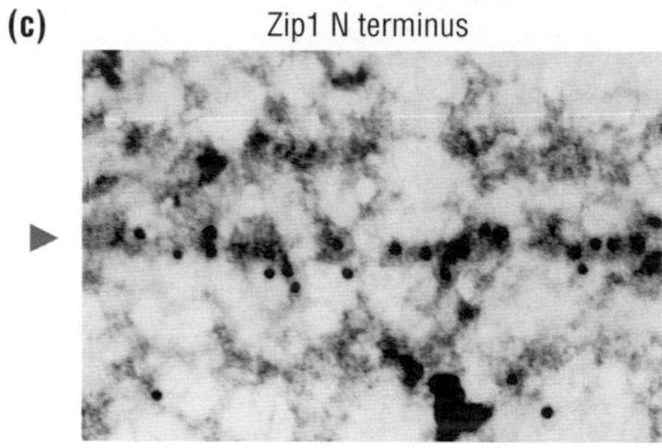

A small number of recombination sites are selected for crossover formation in zygotene

By the end of leptotene, homologs are closely aligned by large numbers of interaxis bridges that are thought to correspond to sites where DNA recombination has been initiated. A major regulatory decision then occurs. As described earlier (section 9-2), a small number of recombination sites—perhaps one to four in a typical human chromosome—are selected to become stable interactions leading to crossovers. Most recombination sites remain in a relatively unstable state that leads to noncrossover events, and by the end of pachytene the homologous DNAs are no longer connected at these sites. Thus, rather surprisingly, the function of most recombination events is not to generate DNA recombinants but to promote homolog interactions.

The designation of crossover sites is an important regulatory event because it determines the number of chiasmata that will hold the homologs together at the end of prophase. The number of crossovers is regulated to achieve two outcomes. First, it is essential that every homolog pair is connected by at least one chiasma, to ensure that homolog bi-orientation can be achieved in meiosis I. Second, in most organisms crossovers are limited in number and well spaced along the homologs by a phenomenon called crossover interference, whereby crossovers tend to suppress the formation of other crossovers in their vicinity. The mechanisms underlying the control of the noncrossover/crossover decision are not clear, but are likely to involve positive signals that promote crossovers at some sites and negative signals that spread outward from future crossover sites to block crossover formation in nearby regions.

Crossover sites nucleate the synaptonemal complex in some species

After presynaptic alignment of the homologs in leptotene (see section 9-3), the homologs are brought closer to one another in zygotene by a process called **synapsis**, which depends on the assembly of a protein scaffold called the **synaptonemal complex (SC)** between the homologs, bringing them to a distance of 100 nm apart. In some species, assembly of the synaptonemal complex begins near future sites of crossover formation at the time that these sites become committed to the crossover pathway. In budding yeast, several proteins are known to form complexes at future crossover sites in early zygotene, and mutations in these proteins block both crossover formation and synaptonemal complex assembly, so it is likely that the two processes are coupled. This coupling is not seen in all organisms, however: there are far more sites of synaptonemal complex nucleation than crossovers in many species, and synaptonemal complex formation is often seen to begin at the telomeres, which do not form crossovers.

Figure 9-9 The synaptonemal complex (a) Transverse filaments of the synaptonemal complex contain coiled-coil dimers of Zip1 (in yeast) affixed to the chromosome axes (the lateral elements of the complex). **(b)** Electron microscopy of synaptonemal complexes from the beetle *Blaps cribrosa* reveals the three major components of the complex: lateral elements (LE), central element (CE), and transverse filaments (TF). In the bottom image, a late recombination nodule (RN) lies across the top of the complex. **(c)** In these electron microscopic images of the budding yeast synaptonemal complex, the termini of the Zip1 protein are labeled with 12 nm gold beads (black dots). In the top image, the amino-terminal domain is labeled, revealing that this part of the protein is found along the central element. In the bottom image, labeling of the carboxy-terminal domain suggests that this region is found along the lateral elements. Photographs in panel (b) kindly provided by Karin Schmekel. Top photograph from Schmekel, K. and Daneholt, B.: *Chromosome Res.* 1998, **6**:155–159. Photographs in panel (c) kindly provided by Carole Rogers and Shirleen Roeder. From Dong, H. and Roeder, G.S.: *J. Cell Biol.* 2000, **148**:417–426.

Definitions

SC: see **synaptonemal complex**.

synapsis: (of chromosomes) the close linkage of two homologous chromosomes along their lengths during zygotene of meiotic prophase.

synaptonemal complex (SC): protein structure that links a pair of homologous chromosomes along their length in meiotic prophase.

References

Bishop, D.K. and Zickler, D.: **Early decision: meiotic crossover interference prior to stable strand exchange and synapsis.** *Cell* 2004, **117**:9–15.

Blat, Y. *et al.*: **Physical and functional interactions among basic chromosome organizational features govern early steps of meiotic chiasma formation.** *Cell* 2002, **111**:791–802.

Börner, G.V. *et al.*: **Crossover/noncrossover differentiation, synaptonemal complex formation, and regulatory surveillance at the leptotene/zygotene transition of meiosis.** *Cell* 2004, **117**:29–45.

Dong, H. and Roeder, G.S.: **Organization of the yeast Zip1 protein within the central region of the synaptonemal complex.** *J. Cell Biol.* 2000, **148**:417–426.

Fung, J.C. *et al.*: **Imposition of crossover interference through the nonrandom distribution of synapsis initiation complexes.** *Cell* 2004, **116**:795–802.

John, B.: *Meiosis* (Cambridge University Press, New York, 1990).

The major structural component of the synaptonemal complex is a double row of protein filaments that bridge the gap between homologs and are known as transverse filaments (Figure 9-9). Transverse filaments are composed primarily of a protein called Zip1 in budding yeast and Scp1 in mammals. These proteins form dimers containing a central rod-like coiled-coil region flanked at each end by globular domains: two amino-terminal domains at one end and two carboxy-terminal domains at the other. The carboxy-terminal domains interact with the chromosomes (the lateral elements of the synaptonemal complex), while the amino-terminal regions interact at a structure called the central element, midway between the two homologs (see Figure 9-9). The composition of the central element is not clear, but it could simply represent the increased density of amino-terminal head domains of the transverse filaments.

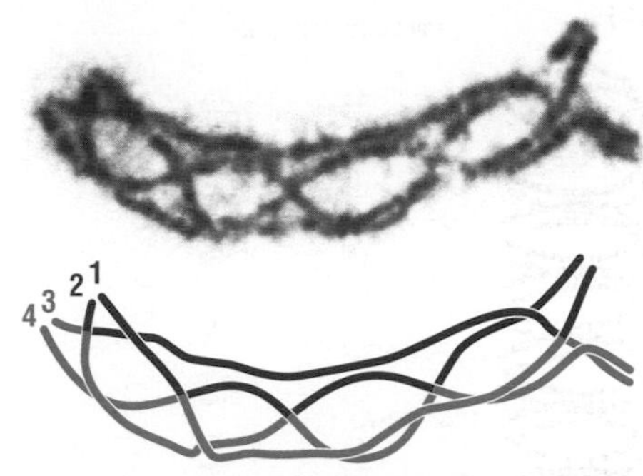

Figure 9-10 **Chiasmata** The top image shows an electron micrograph of a grasshopper homolog pair, or bivalent, with five chiasmata, stained to reveal the proteins of the chromosome axes. As shown in the bottom diagram, this bivalent includes exchanges involving all four chromatids. From Blat, Y. *et al.*: *Cell* 2002, **111**:791–802. Photograph taken from John, B.: *Meiosis* (Cambridge University Press, New York, 1990).

Electron micrographs of the synaptonemal complex reveal that large particles, called early recombination nodules, lie between the homologs as the complex assembles. These nodules are similar in size and number to the recombination complexes that are seen in leptotene and that largely disappear by the end of zygotene. It is likely, but unproven, that they represent all recombination sites or perhaps just the sites of noncrossover events. A small number of nodules mature into much larger late recombination nodules that lie atop the synaptonemal complex and correspond to sites of crossover formation (see Figure 9-9).

What is the function of the synaptonemal complex? It may provide a stable structural foundation for the maturation of crossovers—formation of double Holliday junctions, for example—or for the reorganization of the homolog axes at these sites. Mutations in transverse element proteins greatly reduce the frequency of recombination, and it seems likely that these proteins act in part by providing binding sites for the various enzymes that form the late recombination nodules. By keeping the homologs tightly linked along their entire lengths, the synaptonemal complex may also prevent different homolog pairs from becoming tangled together.

A synaptonemal complex is not present in all organisms. The fission yeast *S. pombe*, for example, does not construct a synaptonemal complex, and crossovers in this organism clearly form in the absence of extensive structural assemblies between the homologs.

Chiasmata appear in diplotene

After the diffuse stage that occurs after pachytene, the homologs are condensed once again in diplotene, and chiasmata can be seen where crossovers have formed. Electron microscopy reveals several important features of chromosome structure at this stage (Figure 9-10). First, the protein axes of the chromatids are continuous across exchange points, which makes it likely that crossover formation does not simply result in the exchange of DNA but also in the exchange of the chromatin axes that organize that DNA. Second, these images reveal that the axes of the sister chromatids are no longer fused into a single unit but are distinct, indicating that partial loss of sister cohesion has occurred. Sister-chromatid cohesion is completely absent at the chiasmata. Consistent with these observations, evidence from light microscopy indicates that cohesin protein concentrations are greatly reduced along chromatid arms and are lost at chiasmata in late prophase, presumably preparing the homologs for more rapid separation at the end of metaphase I. This partial loss of cohesin depends on condensin (see section 5-9). Condensin is thought to recruit the kinase Plk, which then phosphorylates cohesin—much like the mechanism underlying the loss of cohesins from chromosome arms in vertebrate mitosis (see section 5-10).

Roeder, G.S.: **Meiotic chromosomes: it takes two to tango.** *Genes Dev.* 1997, **11**:2600–2621.

Schmekel, K. and Daneholt, B.: **Evidence for close contact between recombination nodules and the central element of the synaptonemal complex.** *Chromosome Res.* 1998, **6**:155–159.

Schmekel, K. *et al.*: **Organization of SCP1 protein molecules within synaptonemal complexes of the rat.** *Exp. Cell Res.* 1996, **226**:20–30.

Yu, H.G. and Koshland, D.: **Chromosome morphogenesis: condensin-dependent cohesin removal during meiosis.** *Cell* 2005, **123**:397–407.

Zickler, D. and Kleckner, N.: **Meiotic chromosomes: integrating structure and function.** *Annu. Rev. Genet.* 1999, **33**:603–754.

Meiosis I is initiated by M–Cdk activity

The first major regulatory transition of the meiotic program occurs in late G1, when the diploid cell becomes committed to meiotic S phase, as described earlier in this chapter (see section 9-1). The second major transition occurs at entry into the first meiotic division or **meiosis I**. Progression through this checkpoint leads to the separation of centrosomes or spindle pole bodies (already duplicated during meiotic S phase and prophase) and assembly of the first meiotic spindle. In animal cells the nucleus breaks down, allowing the highly condensed homolog pairs to become attached to spindle microtubules. These events are similar to the events of early mitosis and, like those events, are triggered by the abrupt activation in late prophase of M–Cdk complexes. For this reason, entry into meiosis I is sometimes called the meiotic G2/M transition, although this term is not ideal because it implies, somewhat inaccurately, that meiotic prophase is analogous to premitotic G2.

Entry into the first meiotic division of animal cells is controlled in diplotene

Studies of vertebrate oocytes indicate that entry into the first meiotic division depends on M–Cdk activation. In many species, oocytes progress through meiotic S phase and meiotic prophase but then arrest at the end of meiotic prophase, in late diplotene, with highly condensed homolog pairs linked by chiasmata. These so-called immature oocytes can remain arrested in diplotene for many years—or even decades in humans. Eventually, arrested oocytes are induced to enter meiosis I—called oocyte maturation—by hormonal signals that activate M–Cdk. In frogs, for example, progesterone stimulates oocyte maturation through signaling pathways that lead to activation of cyclin B–Cdk1 complexes. Indeed, these complexes were once called maturation-promoting factor because of their ability to stimulate entry into meiosis I when injected into immature oocytes (see section 2-3).

Ndt80 and Cdk1 promote entry into the meiotic divisions of budding yeast

In budding yeast, entry into meiosis I is triggered by a gene regulatory factor called Ndt80 (see section 9-1), which stimulates the expression of numerous middle meiotic genes that encode proteins required for the two meiotic divisions. These proteins include the B-type cyclin Clb1, which is the major partner for Cdk1 during the meiotic divisions (its close relative Clb2 is less important in meiosis). Active complexes of Clb1 and Cdk1 are directly responsible for triggering the events of the first meiotic division.

The Cdk1-dependent meiotic entry checkpoint is not well defined in budding yeast and seems to have different features from the equivalent transition in vertebrates. When Ndt80 or Cdk1 is inactivated by mutation in budding yeast, meiotic cells display a prolonged delay in late pachytene with intact synaptonemal complexes but eventually arrest in the diffuse chromatin

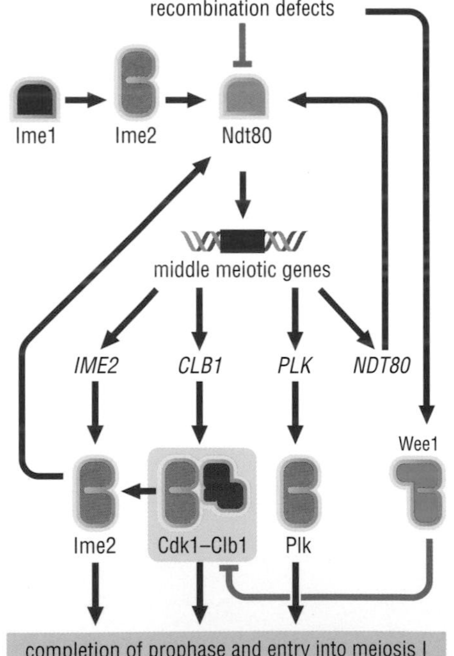

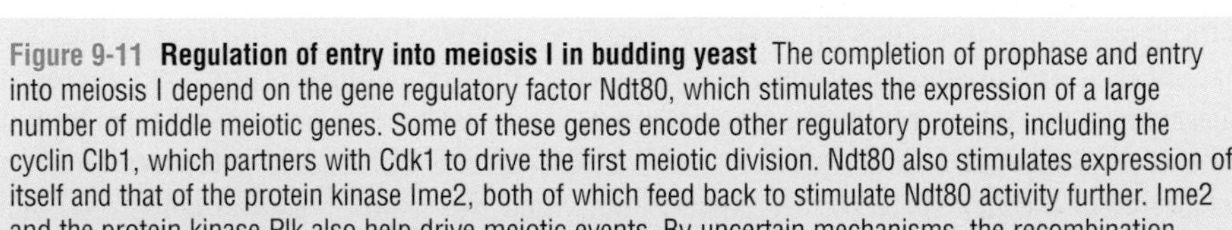

Figure 9-11 **Regulation of entry into meiosis I in budding yeast** The completion of prophase and entry into meiosis I depend on the gene regulatory factor Ndt80, which stimulates the expression of a large number of middle meiotic genes. Some of these genes encode other regulatory proteins, including the cyclin Clb1, which partners with Cdk1 to drive the first meiotic division. Ndt80 also stimulates expression of itself and that of the protein kinase Ime2, both of which feed back to stimulate Ndt80 activity further. Ime2 and the protein kinase Plk also help drive meiotic events. By uncertain mechanisms, the recombination checkpoint system blocks activation of Ndt80 and Cdk1 in response to defects in recombination.

Definitions

meiosis I: the first meiotic division, the stage in the meiotic program that includes assembly of the first meiotic spindle, bi-orientation of paired homologs on the spindle, and homolog separation and segregation.

recombination checkpoint system: in meiotic cells, a regulatory system that blocks entry into **meiosis I** when recombination fails.

stage that follows pachytene in this organism (a highly condensed diplotene chromosome state is not observed in budding yeast). Ndt80 mutants also display defects in the resolution of double Holliday junctions into crossovers. Ndt80 and Cdk1 are therefore required not simply for entry into meiosis I, as in vertebrates, but are also needed for normal rates of crossover formation and disassembly of the synaptonemal complex.

The regulatory system controlling entry into meiosis I includes circuitry that generates switch-like behavior (Figure 9-11). Ndt80 stimulates the expression of its own gene, providing the potential for positive feedback. In addition, the protein kinase Ime2, discussed earlier as a key regulator of the G1/S transition in meiosis (see section 9-1), also contributes to the control of entry into meiosis I. Ime2 activity rises in early meiosis I and is required for the normal accumulation and activity of Ndt80. Ndt80 also stimulates *IME2* expression, giving rise to another potential source of positive feedback. The ability of Ime2 to stimulate Ndt80 also provides a potentially important mechanism by which entry into meiosis I is coupled to progression through G1/S.

Ndt80 also stimulates expression of the gene encoding the polo-like kinase, Plk, of budding yeast (see section 5-7). Plk is activated in late pachytene and is required for the completion of crossover formation and, as mentioned earlier (see section 9-4), for the partial loss of cohesin from chromosome arms. Plk also regulates kinetochore behavior in meiosis I, as described later in this chapter.

Recombination defects block entry into meiosis I

When a meiotic cell initiates recombination but then fails to complete the process successfully, the cell is able to detect the problem and block further progression. When the recombinase Dmc1 is mutated in budding yeast, for example, double-strand breaks form normally but subsequent single-strand invasion is defective (see section 9-2); under these conditions, cells display a prolonged delay in pachytene and eventually arrest before entry into meiosis I. In several animal species, mutation of the recombination machinery also leads to a meiotic prophase arrest and, in some cases, eventually causes programmed cell death by apoptosis. In all cases, the meiotic cells do not attempt to segregate homologs that are partly recombined and therefore not properly connected.

The response to recombination defects is mediated by a **recombination checkpoint system** that is analogous to the checkpoint systems controlling mitotic cell-cycle progression (see section 1-3). This system depends on components of the same DNA damage response system that blocks the mitotic cell cycle in response to double-strand breaks or other abnormalities in DNA structure (described in Chapter 11). It is therefore likely that the meiotic prophase arrest that results from recombination defects is triggered by the presence of DNA breaks or single-stranded regions that are normal intermediates in the recombination process. Consistent with this possibility, prophase arrest is not triggered in yeast by mutations in Spo11, the enzyme responsible for the initial double-strand breaks in the DNA. Instead, *spo11* mutants proceed to meiosis I but then fail to segregate the homologs correctly because no interhomolog linkages exist, as we discuss later in this chapter.

Recombination defects block entry into meiosis I by preventing the activation of M–Cdk complexes. Budding yeast has at least two mechanisms for inhibiting M–Cdk. First, recombination defects somehow trigger activation of Wee1, the protein kinase that phosphorylates Cdk1 at its inhibitory tyrosine (see section 3-3). Second, recombination defects reduce the levels and activity of Ndt80, thereby inhibiting the synthesis of cyclins and other proteins required for meiosis I.

References

Benjamin, K.R. *et al.*: **Control of landmark events in meiosis by the CDK Cdc28 and the meiosis-specific kinase Ime2.** *Genes Dev.* 2003, **17**:1524–1539.

Clyne, R.K. *et al.*: **Polo-like kinase Cdc5 promotes chiasmata formation and cosegregation of sister centromeres at meiosis I.** *Nat. Cell Biol.* 2003, **5**:480–485.

Lee, B.H. and Amon, A.: **Role of Polo-like kinase CDC5 in programming meiosis I chromosome segregation.** *Science* 2003, **300**:482–486.

Leu, J.Y. and Roeder, G.S.: **The pachytene checkpoint in *S. cerevisiae* depends on Swe1-mediated phosphorylation of the cyclin-dependent kinase Cdc28.** *Mol. Cell* 1999, **4**:805–814.

Roeder, G.S. and Bailis, J.M.: **The pachytene checkpoint.** *Trends Genet.* 2000, **16**:395–403.

Tung, K.S. *et al.*: **The pachytene checkpoint prevents accumulation and phosphorylation of the meiosis-specific transcription factor Ndt80.** *Proc. Natl Acad. Sci. USA* 2000, **97**:12187–12192.

Vershon, A.K. and Pierce, M.: **Transcriptional regulation of meiosis in yeast.** *Curr. Opin. Cell Biol.* 2000, **12**:334–339.

Xu, L. *et al.*: **NDT80, a meiosis-specific gene required for exit from pachytene in *Saccharomyces cerevisiae*.** *Mol. Cell Biol.* 1995, **15**:6572–6581.

9-6 Chromosome Attachment in Meiosis I

Homolog pairs are bi-oriented on the first meiotic spindle

As in mitosis, accurate chromosome segregation in meiosis I requires that homologs are bi-oriented on the spindle, with each homolog attached to the opposite pole. Unlike the mitotic cell, however, the cell entering meiosis I faces a unique problem: each homolog contains a pair of sister chromatids, and both must be attached to the same pole. How is this achieved? The answer is not known in molecular detail, but clues are beginning to emerge. In *Drosophila* males, electron microscopy indicates that the two sister kinetochores are fused together in meiosis I, suggesting that they behave as a single unit capable of attachment to only one spindle pole. In budding yeast, detailed analysis of meiotic spindle structure (Figure 9-12) suggests that a single kinetochore microtubule is attached to each homolog—indicating either that one sister kinetochore is silenced or that the two sister kinetochores are fused into a single microtubule-binding unit.

In budding yeast, monopolar attachment of the two sister kinetochores depends on a group of proteins called **monopolins**. Mutation of any one of these proteins (Mam1, Csm1, Lrs4) results in bi-orientation of sister chromatids in meiosis I, leading to failures in homolog segregation. The mechanism by which monopolins govern kinetochore behavior is not known, but they are found at the kinetochore during meiosis I and are likely to be involved in the silencing of one kinetochore or the fusion of two kinetochores into a single functional unit. They are absent from kinetochores in meiosis II, allowing sister-chromatid kinetochores to attach to opposite poles in the second division.

One of the monopolins, Lrs4, is sequestered in the nucleolus before meiosis I but is released in late pachytene and re-localized to the kinetochores, where it is thought to recruit the monopolin Mam1. The release of Lrs4 from the nucleolus is triggered by the protein kinase Plk. Plk is therefore required for Mam1 localization at the kinetochore and for accurate homolog segregation—as well as being required for crossover formation, as mentioned earlier (see section 9-5).

Correct chromosome attachment in meiosis, as in mitosis, is thought to be achieved by a trial-and-error process whereby only the correct orientation is stabilized and allows progression to anaphase (see section 6-10). Most homolog pairs achieve the correct bipolar orientation early in meiosis I, and the kinetochore tension that results from this orientation leads to stabilization of attachment. Some homolog pairs may initially be attached with an incorrect orientation, but these incorrect attachments do not generate tension at the kinetochore (Figure 9-13) and are therefore not locked in place.

Homolog bi-orientation depends on cohesion of sister-chromatid arms

Accurate chromosome alignment in meiosis I requires connections between homolog pairs. In most species this linkage depends on the chiasmata that result from crossovers between homologs. Chiasmata can link the homologs only because sister chromatids are tightly attached

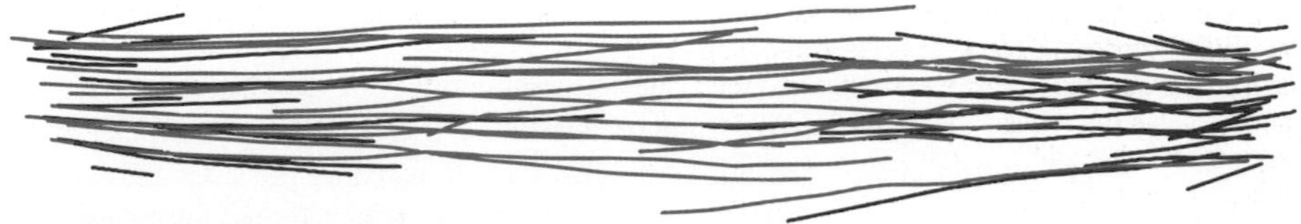

Figure 9-12 Microtubules of the first meiotic spindle in budding yeast Multiple electron microscopic images were used to reconstruct the path of every microtubule in a meiosis I spindle. From its length, it is likely that this spindle was from a cell in early anaphase I. Blue indicates long microtubules that overlap in the spindle midzone and are therefore likely to be interpolar microtubules (which are far more abundant in meiotic spindles than they are in mitotic spindles). Each spindle pole also nucleates about 16 short microtubules (red) that do not overlap with microtubules from the other pole. These microtubules probably represent kinetochore microtubules—one for each of the 16 homologs. Courtesy of Mark Winey.

Definitions

monopolins: group of proteins that localize to the kinetochores of sister chromatids in meiosis I and somehow promote the attachment of both kinetochores to the same spindle pole.

References

Bickel, S.E. *et al.*: **The sister-chromatid cohesion protein ORD is required for chiasma maintenance in *Drosophila* oocytes.** *Curr. Biol.* 2002, **12**:925–929.

Clyne, R.K. *et al.*: **Polo-like kinase Cdc5 promotes chiasmata formation and cosegregation of sister centromeres at meiosis I.** *Nat. Cell Biol.* 2003, **5**:480–485.

Goldstein, L.S.: **Kinetochore structure and its role in chromosome orientation during the first meiotic division in male *D. melanogaster*.** *Cell* 1981, **25**:591–602.

Koehler, K.E. *et al.*: **Spontaneous X chromosome MI and MII nondisjunction events in *Drosophila melanogaster* oocytes have different recombinational histories.** *Nat. Genet.* 1996, **14**:406–414.

Lee, B.H. and Amon, A.: **Role of Polo-like kinase CDC5 in programming meiosis I chromosome segregation.** *Science* 2003, **300**:482–486.

Marston, A.L. and Amon, A.: **Meiosis: cell-cycle controls shuffle and deal.** *Nat. Rev. Mol. Cell Biol.* 2004, **5**:983–997.

Nilsson N.O. and Säll, T.: **A model of chiasma reduction**

along their arms distal to the chiasmata: that is, on the side of the chiasmata away from the kinetochores. Cohesion of sister-chromatid arms is therefore essential for accurate homolog segregation and must be maintained until metaphase I. As mentioned earlier (see section 9-4), some cohesion is removed in meiotic prophase, but sufficient cohesion remains to support normal homolog bi-orientation in meiosis I.

Effective homolog linkage depends on the positioning of chiasmata along the chromosomes (Figure 9-14a). If homologs are linked by a single chiasma that is very close to the end of the chromosome (Figure 9-14b), then the small amount of arm cohesion distal to that chiasma may not be sufficient to hold the homologs together before metaphase, and errors in segregation may occur. Conversely, a single chiasma that is too close to the centromere might not be resolved effectively in meiosis I because sister-chromatid cohesion is maintained near the centromere until meiosis II. For these reasons, crossover formation is suppressed near telomeres and centromeres.

Crossover interference—the ability of one crossover to inhibit the formation of other crossovers in the vicinity (see section 9-4)—may enhance the effectiveness of homolog linkage in some species. The amount of arm cohesion may be insufficient, for example, when two crossovers occur close to each other between the same chromatids (Figure 9-14c). On the other hand, many species—including the fission yeast *S. pombe*—do not have crossover interference and achieve accurate segregation despite having very large numbers of crossovers in each homolog pair. In these cases the total amount of arm cohesion, although scattered over multiple regions, is clearly sufficient to prevent premature homolog separation.

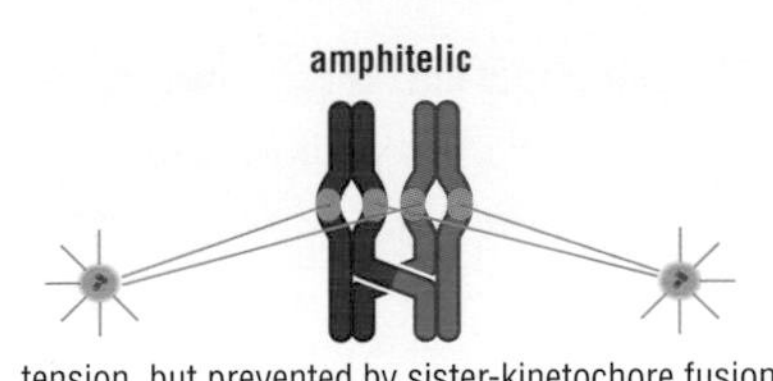

tension, but prevented by sister-kinetochore fusion

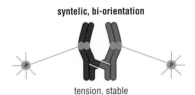

syntelic, no bi-orientation

no tension, unstable

syntelic, bi-orientation

tension, stable

Figure 9-13 Homolog attachment to the first meiotic spindle Because each homolog contains two sister kinetochores, there are, in principle, numerous incorrect ways in which the homologs can attach to the spindle (see also Figure 6-24). Some incorrect forms of attachment—like the amphitelic attachment shown at top—are prevented by modifying the two sister kinetochores to form a single unit that can attach to only one pole. Two forms of attachment can then occur. The incorrect form (center), in which bi-orientation has not occurred and tension is not generated, is thought to be unstable. The correct bi-oriented attachment (bottom) generates tension that leads to stabilization of attachment.

Homolog linkage does not involve chiasmata in some species

Although crossover formation is critical for homolog linkage and segregation in most species, there are a few cases in which homologs are linked by mechanisms other than chiasmata. This achiasmate segregation is seen, for example, in the oocytes of the silk moth, in which homolog pairing and synapsis occur despite the absence of recombination. The synaptonemal complex in these cells is not disassembled before metaphase I, so that homologs remain in close contact until they are separated in anaphase I. Achiasmate segregation can also occur in *Drosophila* meiosis. In females, an occasional achiasmate homolog pair is tolerated because heterochromatin proteins continue to connect homologs after the synaptonemal complex has disintegrated. In *Drosophila* males there is no recombination or synaptonemal complex assembly, and it is likely that other chromatin proteins are responsible for homolog interactions. In none of these cases is it known how homolog linkages are released in anaphase I.

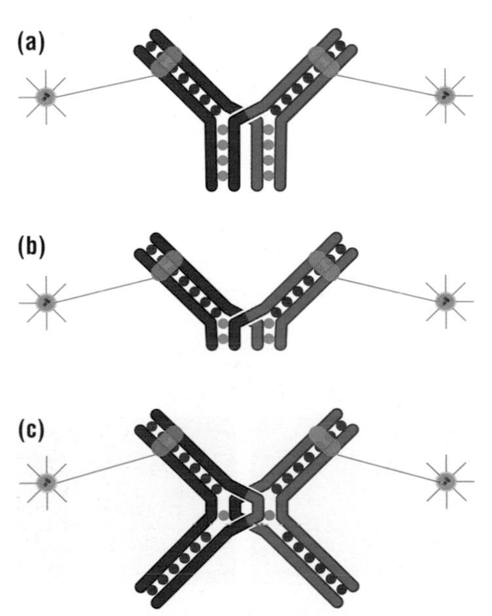

Figure 9-14 Variations in the amount of homolog linkage in meiosis I Accurate homolog segregation in meiosis I occurs only if the homologs are linked by sufficient amounts of sister-chromatid cohesion distal to chiasmata—that is, on the side of the chiasma opposite the centromere. **(a)** If a single chiasma is present at the center of the chromosome arm, the homologs will be held together by the abundant amount of cohesion distal to the chiasma (orange dots). **(b)** If a single chiasma is too close to the end of the chromosome, then the homologs are held together by only a short region of sister cohesion. This small amount of cohesion may not withstand the pulling forces of the spindle, resulting in premature separation. **(c)** Insufficient homolog linkage can also result if two crossovers are very close to each other on the same two sister chromatids.

of closely formed crossovers. *J. Theor. Biol.* 1995, **173**:93–98.

Page, S.L. and Hawley, R.S.: **Chromosome choreography: the meiotic ballet.** *Science* 2003, **301**:785–789.

Petronczki, M. *et al.*: **Un ménage à quatre: the molecular biology of chromosome segregation in meiosis.** *Cell* 2003, **112**:423–440.

Toth, A. *et al.*: **Functional genomics identifies monopolin: a kinetochore protein required for segregation of homologs during meiosis I.** *Cell* 2000, **103**:1155–1168.

Winey, M. *et al.*: **Three-dimensional ultrastructure of *Saccharomyces cerevisiae* meiotic spindles.** *Mol. Biol. Cell* 2005, **16**:1178–1188.

Yokobayashi, S. and Watanabe, Y.: **The kinetochore protein Moa1 enables cohesion-mediated monopolar attachment at meiosis I.** *Cell* 2005, **123**:803–817.

Meiotic Subunits of the Cohesin Complex

Mitotic cohesin subunit	Meiotic cohesin subunits		
	S. cerevisiae	*S. pombe*	Vertebrates
Smc1			SMC1β
Smc3			
Scc1	Rec8	Rec8	Rec8
Scc3		Rec11	STAG3

Figure 9-15 Table of meiosis-specific cohesin subunits Meiotic chromosomes may contain two forms of some subunits. In *S. pombe*, Rec11 only partly replaces the mitotic subunit Scc3, and in mammals the mitotic Smc1 and Scc3 subunits are only partly replaced with Smc1β and STAG3, respectively. There is also evidence in some species that Scc1 is not completely replaced by Rec8 in meiosis. See Figure 5-20 for a listing of the mitotic cohesin subunits.

Loss of sister-chromatid arm cohesion initiates anaphase I

At metaphase I, the pulling forces of the meiotic spindle are opposed only by sister-chromatid arm cohesion distal to the chiasmata. Anaphase I is triggered by the removal of cohesin from chromosome arms, which is achieved in most species by cleavage of cohesin by the protease separase (see section 7-4). Cohesin at the centromere is protected from cleavage in meiosis I, thereby allowing homologs to separate while sister chromatids remain linked until meiosis II.

In most organisms, meiosis-specific cohesin subunits completely or partly replace their mitotic counterparts (Figure 9-15). Most importantly, the key regulatory subunit Scc1 (see Figure 7-11) is replaced in meiosis with Rec8. In most species, homolog separation is triggered by Rec8 cleavage along chromosome arms, which is catalyzed by separase. As in mitosis, separase is activated when APC^{Cdc20} triggers the destruction of the separase inhibitor securin (see section 7-4). Mutations that inactivate separase or prevent APC-dependent securin destruction have been shown to block homolog segregation in numerous organisms, including *C. elegans*, mice (Figure 9-16) and budding yeast (Figure 9-17).

In some organisms, cohesin removal from sister-chromatid arms in meiosis I may involve other mechanisms. Inhibition of the APC does not block the first meiotic division in frog oocytes, for example. Cohesin must therefore be removed from chromosome arms in these cells by other mechanisms—perhaps involving cohesin phosphorylation, as occurs in vertebrate mitosis (see section 5-10).

As discussed earlier in this chapter (see section 9-4), some loss of chromatid arm cohesion occurs late in meiotic prophase, probably as a result of the partial removal of cohesins from arms but not centromeres. It is not clear when decatenation of sister chromatids occurs, but it is likely that this process also contributes to the resolution of sisters in meiotic prophase—as it does in mitosis (see section 5-8).

The spindle checkpoint system helps control anaphase I

In mitotic cells the spindle checkpoint system delays sister-chromatid separation in response to defects in chromosome attachment to the spindle (see section 7-2). This system also operates in meiosis I. When homologs are not attached correctly to the first meiotic spindle, the spindle checkpoint system generates a signal that blocks APC^{Cdc20} activity and thereby prevents securin destruction and the loss of sister-chromatid cohesion. As in mitotic cells, the spindle checkpoint system is thought to help determine the normal timing of anaphase in meiosis I, at least in mammalian cells. Defects in spindle checkpoint components such as Mad2 cause premature segregation of homologs in mouse oocytes, for example.

The spindle checkpoint system is partly activated in meiotic cells carrying homolog pairs that are not linked by chiasmata. This can be seen in mutant cells lacking Spo11, the enzyme that catalyzes the first step in recombination. In these mutants, homologs remain unlinked and are therefore distributed randomly in meiosis I (see Figure 9-17). Activation of the spindle checkpoint system in *spo11* mutant cells delays the destruction of securin, but this does not prevent anaphase I because loss of sister cohesion is not required to separate homologs that are not linked in the first place. In the presence of unlinked homologs, the checkpoint system presumably responds either to the lack of tension in the kinetochores of unlinked homologs or to the poor microtubule attachment that results from that low tension (see sections 6-10 and 9-6).

(a) Wild-type securin

metaphase I | anaphase I

securin fluorescence (0–140)

Time after nuclear breakdown (hours) (0–14)

polar body

DNA

(b) Stable securin mutant

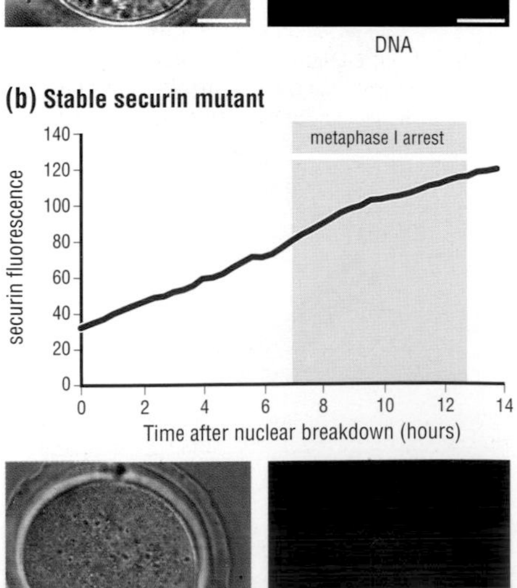

Figure 9-16 Securin destruction is required for meiotic anaphase I in mouse oocytes **(a)** A mouse oocyte was injected with fluorescently tagged securin protein, and the amount of securin was measured as the oocyte progressed through meiosis I. Starting 8 hours after nuclear envelope breakdown, securin levels dropped and anaphase I occurred. As shown in the photographs below the graph, one set of homologs was packaged in the first polar body (see section 2-3), whereas the other set remained in the oocyte as it entered meiosis II. **(b)** In this experiment an oocyte was injected with a mutant form of securin lacking the sequence that targets the protein to APC^{Cdc20}. Securin destruction was prevented, and the oocyte arrested in metaphase I. Homolog segregation in these cells therefore requires securin destruction, which presumably allows separase to cleave chromatid arm cohesins. Photographs kindly provided by Mary Herbert. Adapted, with permission, from Herbert, M. *et al.*: *Nat. Cell Biol.* 2003, **5**:1023–1025.

Centromeric cohesin is protected from cleavage in meiosis I

Cohesion is maintained at centromeres in meiosis I because the cohesins linking sister centromeres are protected from separase. The mechanisms underlying centromeric protection involve the meiosis-specific cohesin subunit Rec8, as revealed in studies of a mutant yeast strain in which meiotic cells are engineered to express the mitotic Scc1 cohesin subunit in place of Rec8 (see Figure 9-17). After meiotic S phase, sister-chromatid cohesion seems normal in these cells, indicating that Scc1 is fully functional in this respect. During anaphase I, however, Scc1 cleavage occurs at centromeres as well as at chromatid arms, revealing that protection from cleavage at the centromeres is conferred by some unique feature of Rec8.

Centromeric Rec8 is protected from cleavage in meiosis I by a protein called Sgo1 (MEI-S332 in *Drosophila*), which is localized at the centromeres in meiosis I and associates with Rec8. In fission and budding yeasts, mutation of Sgo1 allows centromeric Rec8 cleavage to occur during the first meiotic division, resulting in a loss of centromeric cohesion that leads to random sister-chromatid segregation in the second meiotic division (see Figure 9-17). Sgo1 disappears from the cell in the second meiotic division, thereby exposing Rec8 to cleavage in anaphase II. We know little about the mechanisms that govern Sgo1 localization to the centromere or its destruction after meiosis I.

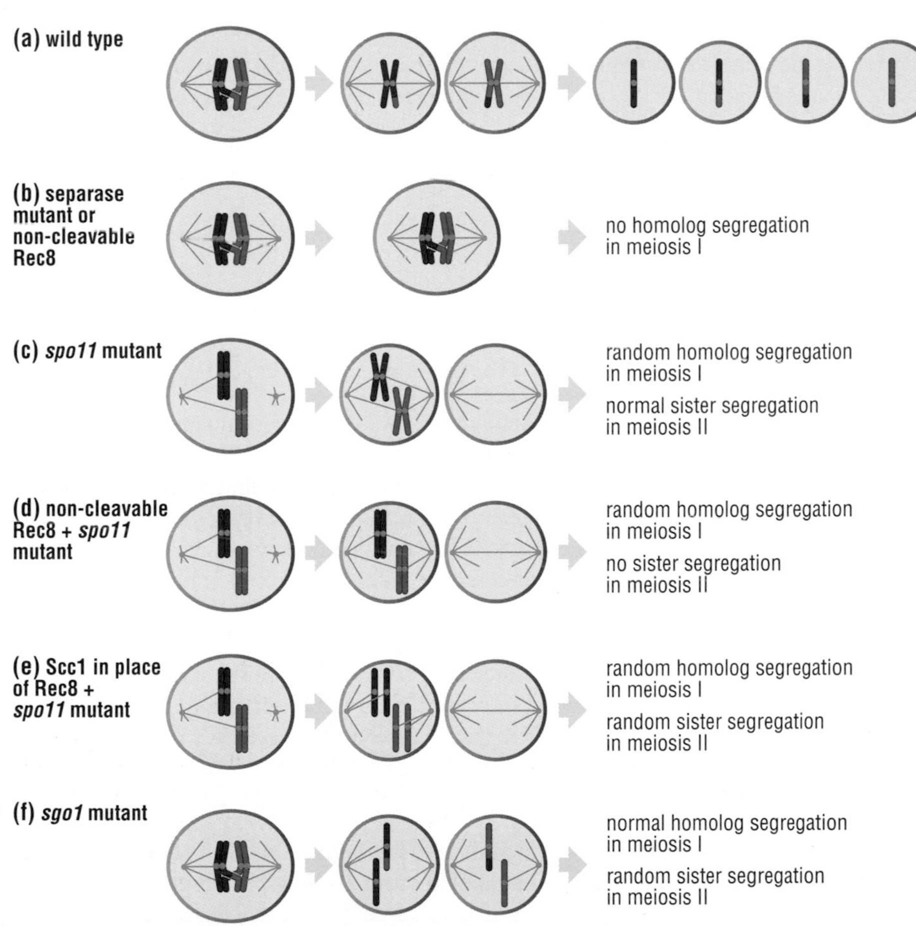

Figure 9-17 Mechanisms controlling homolog segregation in budding yeast Analysis of various mutant yeast strains provides a variety of insights into the mechanisms that control sister-chromatid cohesion in meiosis. For simplicity, these diagrams show only one homolog pair. **(a)** In wild-type cells, the two meiotic divisions generate four haploid gametes. **(b)** Mutation of separase, or mutation of the separase cleavage sites in the cohesin subunit Rec8, prevents the loss of cohesin from sister-chromatid arms and therefore blocks homolog separation and segregation. **(c)** In cells lacking the enzyme Spo11, recombination does not occur and homologs are not linked by chiasmata. Homologs therefore attach to either spindle pole and segregate randomly, with both going to one pole (as shown here) or each to separate poles (not shown). Securin destruction is delayed in these cells by the spindle checkpoint system, but cohesion is eventually lost from chromatid arms and sisters are segregated normally in meiosis II. **(d)** In cells carrying both a *spo11* mutation and a version of Rec8 that is resistant to separase, homologs segregate randomly in meiosis I because, in the absence of chiasmata, there is no need to remove chromatid arm cohesion. In meiosis II, however, cohesion remains intact and sisters cannot separate. **(e)** In this experiment meiotic cells were engineered to produce the mitotic cohesin subunit Scc1 instead of Rec8. Because Rec8 has complex functions in recombination as well as sister cohesion, clearer results also required mutation of *SPO11*. In anaphase I, cohesion was lost both on chromatid arms and at centromeres, and separated sisters therefore attached incorrectly to the second meiotic spindle, resulting in random segregation. Scc1 is therefore not protected from cleavage in the first division. **(f)** In cells lacking Sgo1, homologs segregate normally in meiosis I. Because centromeric cohesion is also lost in meiosis I, the sister chromatids separate prematurely and segregate randomly in meiosis II.

References

Buonomo, S.B. *et al.*: **Disjunction of homologous chromosomes in meiosis I depends on proteolytic cleavage of the meiotic cohesin Rec8 by separin.** *Cell* 2000, **103**:387–398.

Herbert, M. *et al.*: **Homologue disjunction in mouse oocytes requires proteolysis of securin and cyclin B1.** *Nat. Cell Biol.* 2003, **5**:1023–1025.

Homer, H.A. *et al.*: **Mad2 prevents aneuploidy and premature proteolysis of cyclin B and securin during meiosis I in mouse oocytes.** *Genes Dev.* 2005, **19**:202–207.

Kitajima, T.S. *et al.*: **The conserved kinetochore protein shugoshin protects centromeric cohesion during meiosis.** *Nature* 2004, **427**:510–517.

Petronczki, M. *et al.*: **Un ménage à quatre: the molecular biology of chromosome segregation in meiosis.** *Cell* 2003, **112**:423–440.

Revenkova, E. *et al.*: **Cohesin SMC1 beta is required for meiotic chromosome dynamics, sister chromatid cohesion and DNA recombination.** *Nat. Cell Biol.* 2004, **6**:555–562.

Shonn, M.A. *et al.*: **Spindle checkpoint component Mad2 contributes to biorientation of homologous chromosomes.** *Curr. Biol.* 2003, **13**:1979–1984.

Watanabe, Y.: **Sister chromatid cohesion along arms and at centromeres.** *Trends Genet.* 2005, **21**:405–412.

Meiosis I is followed by meiosis II

The segregation of homologs in the first meiotic division is followed immediately by entry into **meiosis II**. As in mitosis, the central function of meiosis II is to segregate sister chromatids, and the events of meiosis II are therefore closely related to those of mitosis—and are regulated by similar mechanisms.

Homolog segregation in anaphase I is followed by disassembly of the first meiotic spindle. Soon thereafter, spindle pole bodies or centrosomes duplicate and separate to begin the construction of the second meiotic spindle. Sister-chromatid kinetochores, no longer fused as they are in meiosis I, are attached to the spindle with the correct bi-orientation.

The final regulatory checkpoint of the meiotic program occurs at the second metaphase-to-anaphase transition. As in mitosis, APC^{Cdc20}-dependent destruction of securin leads to the activation of separase, which then cleaves the Rec8 subunit of cohesin complexes at the sister centromeres—which are no longer protected as they are in meiosis I. Sister-chromatid separation and segregation occur, and exit from meiosis II then completes the meiotic program.

Wild type
+ Wee1
DNA
Cy3-dUTP

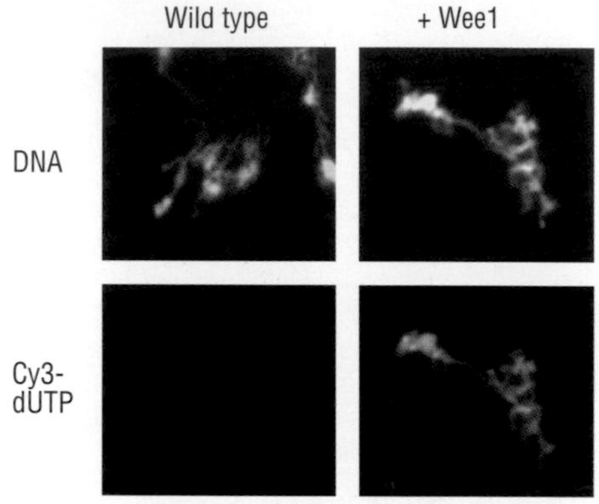

Figure 9-18 Inhibition of Cdk1 triggers DNA synthesis after meiosis I *Xenopus* oocytes were induced to mature with progesterone. When the oocytes entered meiosis I (as judged by nuclear envelope breakdown, which is visible on the cell surface), they were chilled and crushed to produce a concentrated metaphase I extract with high levels of cyclin B–Cdk1 activity. Sperm chromosomes were added to the extract to monitor chromosome behavior, and the extract was then warmed to room temperature to allow progress out of metaphase I and into meiosis II. The top images in this figure show the condensed chromosomes of metaphase II. To monitor the synthesis of DNA, a fluorescently tagged nucleotide (Cy3-dUTP) was also added. In normal extracts (left), no fluorescence is seen on the chromosomes because there is normally no DNA synthesis between the two meiotic divisions. If a small amount of the protein kinase Wee1 is added, however, DNA synthesis does occur (right). These results indicate that the small amount of Cdk1 activity that is normally present after meiosis I is required to suppress DNA synthesis before the next division. Photographs kindly provided by Keita Ohsumi. Reprinted, with permission, from Iwabuchi, M. *et al.*: *EMBO J.* 2000, **19**:4513–4523.

Partial Cdk1 inactivation occurs after meiosis I

The mechanisms driving the completion of meiosis I are related to but distinct from those controlling the completion of mitosis. As in mitosis, spindle disassembly must occur at the end of meiosis I, so that the two sets of sister-chromatid pairs are segregated on separate new spindles in meiosis II. On the other hand, some events of late mitosis in animal cells—nuclear reformation and chromosome decondensation, for example—are not strictly required to prepare for the second division, and in many cases these events do not occur or occur only partly.

There is one major event that occurs after mitosis but must never occur after meiosis I: DNA synthesis. Robust mechanisms exist to suppress the assembly of pre-replicative complexes (preRCs; see section 4-4) after meiosis I, thereby preventing the abundant Cdk activity of meiosis II from driving a new round of DNA replication.

How are some but not all late mitotic events triggered during the completion of meiosis I? The answer probably lies in Cdk1 regulation. As described in Chapter 7 (see section 7-0), the events of late mitosis are triggered primarily by inactivation of S–Cdks and M–Cdks, which leads to dephosphorylation of Cdk1 targets. After meiosis I, however, Cdk1 inactivation is incomplete. In *Xenopus* oocytes, for example, a low but significant amount of Cdk1 activity remains between the meiotic divisions. If complete Cdk1 inactivation is forced upon these cells by addition of the Cdk1 inhibitory kinase Wee1 (see section 3-3), the oocyte undergoes DNA synthesis after the first meiotic division (Figure 9-18). It therefore seems likely that partial Cdk1 inactivation at the end of meiosis I results in dephosphorylation of only a subset of Cdk1 targets. Targets involved in spindle disassembly, for example, may be more effectively dephosphorylated, particularly if phosphatases acting specifically on these proteins are generally more active. Components of the preRC, however, must remain phosphorylated after meiosis I, thereby blocking origin resetting.

The partial inactivation of Cdk1 after meiosis I in frog oocytes is due, at least in part, to mechanisms that suppress the cyclin-ubiquitination activity of APC^{Cdc20}. A complex network of regulatory proteins, including protein kinases called Mos and Rsk, is responsible for

Definitions

meiosis II: the second meiotic division, during which the sister chromatids are segregated on the second meiotic spindle.

References

Buonomo, S.B. *et al.*: **Division of the nucleolus and its release of CDC14 during anaphase of meiosis I depends on separase, SPO12, and SLK19.** *Dev. Cell* 2003, **4**:727–739.

Iwabuchi, M. *et al.*: **Residual Cdc2 activity remaining at meiosis I exit is essential for meiotic M-M transition in *Xenopus* oocyte extracts.** *EMBO J.* 2000, **19**:4513–4523.

Izawa, D. *et al.*: **Fission yeast Mes1p ensures the onset of meiosis II by blocking degradation of cyclin Cdc13p.** *Nature* 2005, **434**:529–533.

Marston, A.L. *et al.*: **The Cdc14 phosphatase and the FEAR network control meiotic spindle disassembly and chromosome segregation.** *Dev. Cell* 2003, **4**:711–726.

Marston, A.L. and Amon, A.: **Meiosis: cell-cycle controls shuffle and deal.** *Nat. Rev. Mol. Cell Biol.* 2004, **5**:983–997.

Rauh, N.R. *et al.*: **Calcium triggers exit from meiosis II by targeting the APC/C inhibitor XErp1 for degradation.** *Nature* 2005, **437**:1048–1052.

Tunquist, B.J. and Maller, J.L.: **Under arrest: cytostatic factor (CSF)-mediated metaphase arrest in vertebrate eggs.** *Genes Dev.* 2003, **17**:683–710.

restraining the activity of APC^{Cdc20} after meiosis I. These proteins may also suppress the activity of the Myt1 kinase that inhibits Cdk1 (see section 3-3). In fission yeast, and perhaps in other organisms as well, a protein called Mes1 interacts with the Cdc20 subunit and thus reduces APC^{Cdc20} activity during meiosis I, ensuring that M-cyclin levels decline only partly between the meiotic divisions.

The completion of meiosis I is also stimulated by the activation of phosphatases that dephosphorylate a subset of Cdk1 substrates. In budding yeast, late mitotic events depend on the phosphatase Cdc14, which is activated by separase (see section 7-5). The same regulatory mechanism is important in late meiosis I, in which mutation of Cdc14 causes defects in disassembly of the spindle (Figure 9-19). In *Xenopus* oocytes, exit from meiosis I seems to occur normally even if cyclin B destruction is prevented by inhibition of the APC, and meiotic exit may be driven in these cells by activation of a phosphatase that targets Cdk1 substrates.

Surprisingly, there does not seem to be a major regulatory checkpoint between the two meiotic divisions. Entry into meiosis I represents a commitment to both meiotic divisions, regardless of whether the first division is carried out successfully or not. In various yeast mutants—including *spo11* and *cdc14*, for example—cells enter meiosis II despite major defects in the segregation of homologs in meiosis I (see Figures 9-17 and 9-19). This lack of control may reflect the fact that entry into meiosis II is not governed by a robust molecular switch—such as total Cdk1 inactivation and reactivation—like the one that triggers entry into meiosis I.

The meiotic program is coordinated with gametogenesis

After meiosis II, haploid nuclei are packaged into cells that differentiate into specialized cell types, such as spores in yeast or gametes (sperm and eggs) in animals. The formation of these cells depends on the coordination of the meiotic program with cytokinesis and differentiation.

In budding yeast, each of the four haploid nuclei within the mother cell is packaged inside a tough spore wall that is built around the nuclear envelope, starting at each spindle pole body. The mother cell then collapses around the four spores, forming an ascus.

In most animals, spermatogenesis also involves specialized forms of division and differentiation. Partial cytokinesis occurs after both meiotic divisions, resulting in multiple spermatids that are joined by cytoplasmic bridges as they complete their transformation into spermatozoa.

During vertebrate oogenesis, the first meiotic division is followed by an asymmetric division in which one set of homologs is packaged into a tiny cellular remnant called a polar body (see section 2-3 and Figure 9-16), which will eventually degenerate. The oocyte enters meiosis II and in most vertebrates remains arrested at metaphase II until fertilization. The metaphase II arrest in frog eggs (called a cytostatic factor arrest) is due primarily to inhibition of APC^{Cdc20} by an inhibitory protein, Erp1, that is related to the Emi1 protein that restrains APC^{Cdc20} before mitosis (see section 7-1). Fertilization triggers the destruction of Erp1 and thus the activation of APC^{Cdc20}, allowing the completion of meiosis II. One set of chromosomes is again pinched off into a polar body, leaving the egg with a single haploid chromosome set that fuses with the haploid sperm nucleus. The diploid zygote then enters the first embryonic cell division—and thereby returns to the mitotic cell cycle.

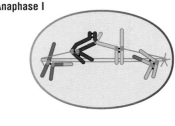

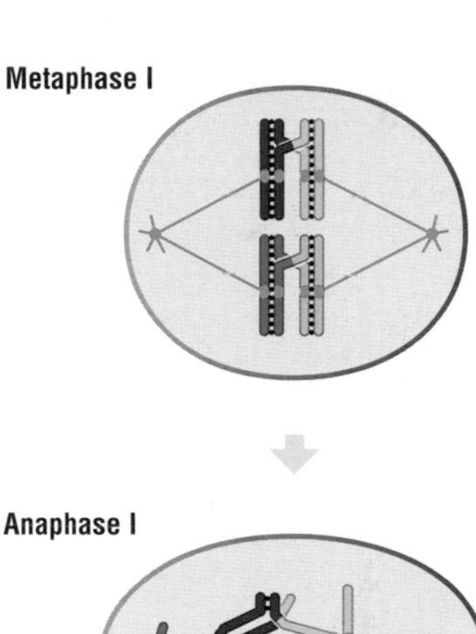

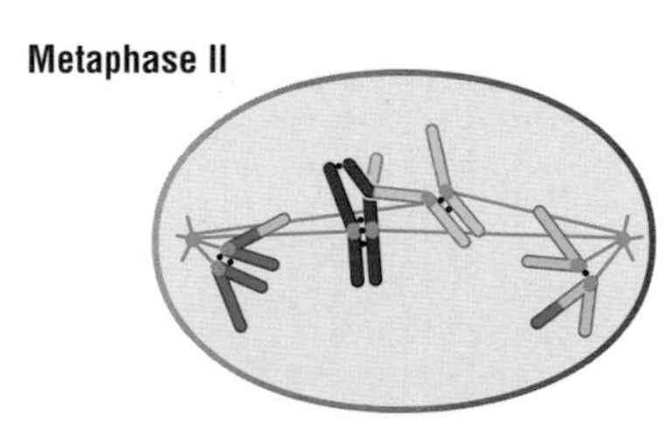

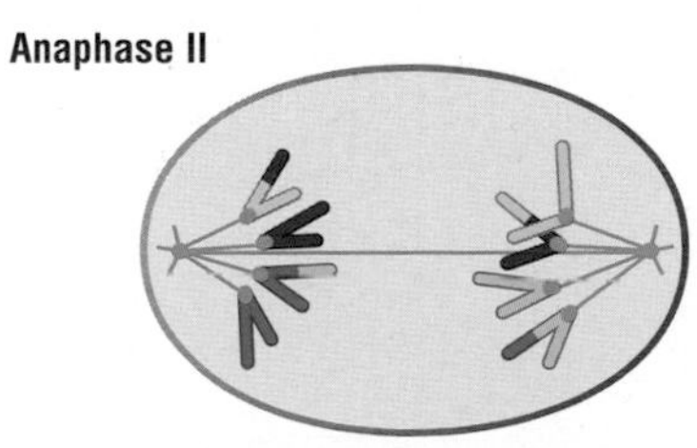

Figure 9-19 The phosphatase Cdc14 is required for disassembly of the first meiotic spindle in budding yeast This diagram summarizes the complex phenotypes seen in meiotic budding yeast cells with a mutation in the phosphatase Cdc14, which normally helps promote the completion of mitosis by dephosphorylating Cdk1 targets—and by stimulating APC^{Cdh1} (see section 7-5). In the absence of Cdc14, the spindle is not disassembled after meiosis I, but the cell continues into meiosis II, apparently using the same spindle. Numerous defects occur, however. Some homolog pairs (blue) are segregated normally in the first division but fail to segregate in the second division, perhaps because it is difficult for these sister chromatids to attach to the distant opposite pole in time. Some homolog pairs, however, fail to segregate completely in the first division because the *cdc14* mutation delays the loss of cohesion in some chromosome regions. These non-segregated homologs, which remain in the middle of the spindle, are more successful in bipolar spindle attachment in meiosis II, thereby allowing sister segregation to opposite poles. Adapted from Marston, A.L. *et al.*: *Dev. Cell* 2003, **4**:711–726.

10

Control of Cell Proliferation and Growth

In late G1, activation of G1/S– and S–Cdks triggers an irreversible commitment to a new cell cycle. Activation of these kinases is tightly regulated by extracellular factors, so that cells proliferate only when needed. In most proliferating cell populations, cell division is coordinated with cell growth, ensuring that cell size remains constant.

10-0 Overview: Control of Cell Proliferation and Growth

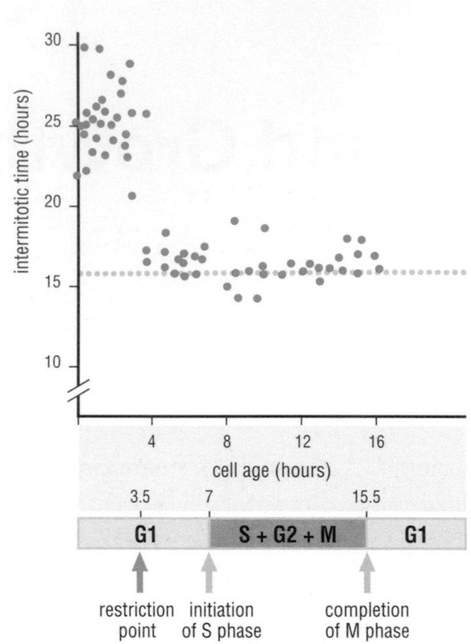

Figure 10-1 Commitment to cell division in cultured mammalian cells A population of mouse 3T3 fibroblast cells was cultured in standard medium containing serum, which provides the mitogens and growth factors required for proliferation and growth. Culture dishes were placed on a microscope heated to the optimal temperature for cell growth and proliferation (37 °C) and equipped with a video camera. After 24 hours of video recording, the culture medium was replaced with serum-free medium for 1 hour, after which the serum was added back and cells were filmed for an additional 48 hours. The video record of the 24 hours before serum removal was used to determine the time of birth for all cells in the population. The age of every cell at the beginning of serum starvation was calculated (plotted along the horizontal axis). The time of the next mitosis was then determined for every cell, allowing calculation of cell-cycle length (plotted along the vertical axis). Each dot represents a single cell. Average cell-cycle time for these cells in serum-containing medium is 15.5 hours (dashed yellow line). The results reveal that young cells, less than 3.5 hours old, undergo a prolonged cell-cycle delay of about 8 hours when deprived of serum for 1 hour. Other experiments showed that this delay occurs in G1. Once cells are more than 3.5 hours old, however, they do not experience any G1 delay and divide at the normal time. These elegant studies suggest that cells are able to respond to changes in extracellular factors only during the first few hours of G1. After this point—called Start or the restriction point in mammals—the cells complete cell division on time regardless of culture conditions. In addition, the lengthy delay after a brief serum starvation suggests that cells exit from the cell cycle into a quiescent state, G0, from which it takes hours to return. The timeline underneath the graph indicates the average length of the various cell-cycle stages in these cells. Adapted from Zetterberg, A. and Larsson, O.: *Proc. Natl Acad. Sci. USA* 1985, **82**:5365–5369.

Cell proliferation is controlled at a checkpoint in late G1

When environmental conditions are ideal, yeast and other unicellular organisms reproduce as rapidly as possible, progressing from one cell cycle to the next without delay. This unrestrained cell proliferation is also seen in the cleavage divisions of the early animal embryo, or when cultured mammalian cells are provided with unlimited space and resources in a plastic dish. Such unbridled cell proliferation is rare in nature, however. In multicellular organisms, the rate of cell division is tightly controlled to ensure that new cells are produced only when needed. During early animal development, for example, rates of cell division vary widely in different regions of the embryo, providing one mechanism by which the sizes of different organs are determined. In the adult animal, cell division is limited strictly to those tissues in which new cells are required to maintain organ size or for other functions.

The rate of cell division is governed by a combination of intracellular and extracellular factors. Yeast proliferation is limited primarily by external nutrient levels, which can vary widely in the wild. Animal cells, by contrast, are usually bathed in constant high concentrations of nutrients, and their proliferation depends on both tissue-specific genetic programming and signals provided by extracellular proteins produced by other cells. In early embryos, for example, extracellular and intracellular developmental patterning signals influence the rate of cell division in different regions. Extracellular proteins that stimulate cell division are known as *mitogens*.

The rate of cell division in most cell types is governed at Start, the major cell-cycle checkpoint in late G1 (see section 1-3). As cells progress through the cell cycle, they monitor various aspects of their external environment and internal programming. When the appropriate nutrients and extracellular mitogens are present, and their internal program permits, G1 cells pass through the Start checkpoint and enter a new cell cycle. When conditions are unfavorable, progression through Start is delayed or blocked, thereby reducing the rate of division. In some cases, as in cultured mammalian cells deprived of mitogens (Figure 10-1), cells exit from the cell cycle entirely and enter a specialized resting state called G0 (G-zero). The length and stability of this quiescent state vary widely in different tissues. In adult tissues in which new cells are rarely produced, as in the nervous system, cells may remain in G0 for the lifetime of the organism.

Start represents an irreversible commitment to cell-cycle entry, after which the cell will complete the entire cell cycle even if environmental conditions become unfavorable (see Figure 10-1).

Progression through Start depends on an irreversible wave of Cdk activity

As we have seen in previous chapters (see, for example, section 3-13), the central molecular event underlying Start is the activation of G1/S cyclin–Cdk complexes—Cln1–Cdk1 and Cln2–Cdk1 in budding yeast; cyclin E–Cdk2 in metazoans (Figure 10-2). These Cdks have several important functions. In most species they directly initiate the early events of the cell cycle—notably the duplication of the centrosome or spindle pole body (SPB) and the initiation of budding in yeast (see Chapter 6). Directly or indirectly, they also trigger the initiation of DNA replication. In *Drosophila*, and perhaps some mammalian cells, cyclin E–Cdk2 seems to be a direct regulator of replication origin firing. In yeast and most mammalian cells, DNA replication is initiated by the S-phase Cdks, and G1/S–Cdks trigger DNA synthesis indirectly—by causing the destruction of proteins that inhibit S–Cdks (see Chapter 4).

Start is controlled by a regulatory system containing positive feedback loops and other control mechanisms that generate highly switch-like, irreversible Cdk activation—resulting in irreversible entry into the cell cycle (see section 3-13). The robustness of this system ensures that it continues to operate effectively when important components fail.

Progression through Start requires changes in gene expression

Several regulatory events are integrated to bring on the wave of Cdk activity that catapults the cell into a new cell cycle. A central event is the activation in G1 of gene regulatory proteins that increase the expression of the genes encoding the G1/S cyclins and other cell-cycle components. These transcription factors are activated, at least in part, by G1 cyclin–Cdk complexes. In principle, therefore, the circuitry that controls Start can be reduced to a series of three cyclin–Cdk complexes (G1–, G1/S– and S–Cdks), each of which helps activate the next in the series (see Figure 10-2).

Mitogens and other regulatory factors influence the rate of cell division by controlling the components that govern progression through Start. In budding yeast, for example, mating pheromone arrests haploid cells in G1 in preparation for mating by activating a protein that inhibits G1– and G1/S–Cdks. In mammalian cells, mitogens stimulate cell proliferation by increasing the production of G1–Cdks or by acting on other pathways that influence gene regulatory proteins or G1/S–Cdks. Genetic changes that deregulate these mitogenic signaling pathways are often associated with diseases of excess cell proliferation, such as cancer.

Cell division is often coordinated with cell growth

Cell growth is the process by which a cell increases its size—by synthesizing the proteins, membranes, organelles and other components that make up the bulk of cell mass. Like cell division, cell growth is controlled by a combination of intrinsic programming and extracellular signals. The key growth-regulating factors are external nutrient concentrations (in yeast) and extracellular *growth factors* produced by other cells (in animals).

In a proliferating cell population, maintenance of cell size requires that a cell double in mass during each cell cycle. To coordinate growth and division, the signaling pathways controlling cell growth are generally linked in some way to those controlling cell division. In yeast, for example, the rate of cell division depends on the rate of cell growth, so that cells divide only when a threshold growth rate is achieved.

The coordination of cell growth and division in multicellular organisms is poorly understood. Growth and division are clearly coordinated in cell populations in which cell size must be maintained, but the mechanism is not yet clear. In contrast, the two processes are often unlinked in tissues in which cell size must be changed. Cell growth occurs in the absence of cell division to produce large cells, such as neurons and oocytes, whereas the early cleavage divisions of the embryo occur in the absence of growth, producing smaller and smaller cells at each division. Both during embryonic development and in the adult animal, cell growth and cell division are also governed by more global regulatory mechanisms that determine the sizes of organs and organisms. The molecular mechanisms underlying the control of cell, tissue and organism size are among the most fascinating and mysterious problems in biology.

This chapter focuses on the molecular mechanisms governing the rate of cell division. In Chapters 3 and 4 we described the basic features of the regulatory systems that drive progression through Start and initiate chromosome duplication in S phase. In the first sections of this chapter (10-1 to 10-5) we discuss these mechanisms in more detail, with an emphasis on the control of G1/S gene expression and Cdk activation at Start in budding yeast and animal cells. In sections 10-6 to 10-9 we describe how progression through Start in animal cells is governed by external influences, such as mitogens, and by intrinsic developmental programming. Next, in sections 10-10 to 10-13, we turn to the problem of how cell growth is controlled and how it is coordinated with cell division. Finally, in section 10-14, we discuss briefly how the number of cells in a population can be governed in part through regulated changes in the rate of cell death.

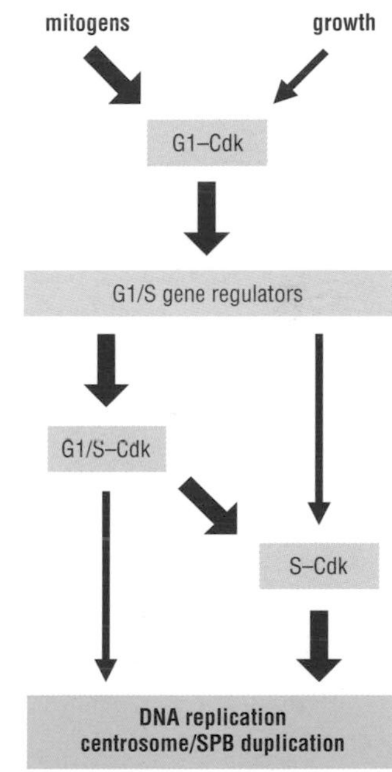

Figure 10-2 Regulatory scheme governing cell proliferation In most cells, G1 cyclin–Cdk complexes trigger the activation of gene regulatory proteins that stimulate the expression of numerous G1/S genes, including genes encoding G1/S cyclins, S cyclins and the proteins that carry out the events of the early cell cycle. The resulting waves of G1/S– and S–Cdk activities result in irreversible commitment to cell division. This is a highly simplified representation of a complex and robust regulatory system whose composition and behavior vary greatly among different species and cell types.

References

Baserga, R.: *The Biology of Cell Reproduction* (Harvard University Press, Cambridge, MA, 1985).

Hall, M.N. *et al.* (eds): *Cell Growth: Control of Cell Size* (Cold Spring Harbor Laboratory Press, Cold Spring Harbor, 2004).

Pardee, A.B.: **G1 events and regulation of cell proliferation.** *Science* 1989, **246**:603–608.

Zetterberg, A. and Larsson, O.: **Kinetic analysis of regulatory events in G1 leading to proliferation or quiescence of Swiss 3T3 cells.** *Proc. Natl Acad. Sci. USA* 1985, **82**:5365–5369.

The gene regulatory proteins SBF and MBF drive expression of Start-specific genes in yeast

In most cell types, entry into the cell cycle is accompanied by marked increases in the expression of a large number of genes. In budding yeast, genome-wide analyses have identified about 200 of these so-called G1/S genes, which encode two major classes of proteins: one comprises cell-cycle regulatory components, including the G1/S cyclins Cln1 and Cln2 and the S cyclins Clb5 and Clb6 (see section 4-5), and the other contains components of the cell-cycle machinery that are responsible for carrying out the events of late G1 and S phase—such as the enzymes that replicate the DNA. Thus, the overall effect of increased gene expression is to increase the Cdk activities that drive passage through Start and to increase the synthesis of the proteins that carry out early cell-cycle events.

In Chapter 3 we briefly described the regulatory systems that drive G1/S gene expression and Cdk activation in budding yeast growing in abundant nutrients. In this and the following section we describe these mechanisms in more detail, after which we discuss, in section 10-3, how a specific external signal—mating pheromone—can act on this system to prevent cell division.

Most G1/S gene expression in budding yeast is controlled by a pair of related gene regulatory proteins, SBF (SCB-binding factor) and MBF (MCB-binding factor), which interact, respectively, with the SCB and MCB control sequences in the promoters of G1/S genes (see section 3-12). SBF and MBF each contain at least two protein subunits, including one subunit responsible for sequence-specific DNA binding (called Swi4 in SBF and Mbp1 in MBF) and a second subunit (Swi6) that is present in both complexes.

SBF and MBF are activated by Cln3–Cdk1 at Start

The activation of SBF is best understood. During late mitosis and G1, Swi4 and Swi6 bind to each other in the nucleus to form SBF, which binds to SCB sequences on G1/S gene promoters (Figure 10-3). DNA-bound SBF initially fails to activate G1/S gene transcription, however, because it is bound by an inhibitory protein called Whi5. SBF is activated by the dissociation of Whi5, which is caused by its phosphorylation by the G1 Cln3–Cdk1 complex (see Figure 10-3). Abundant evidence suggests that Cln3 is the major initiator of SBF (and MBF) activation.

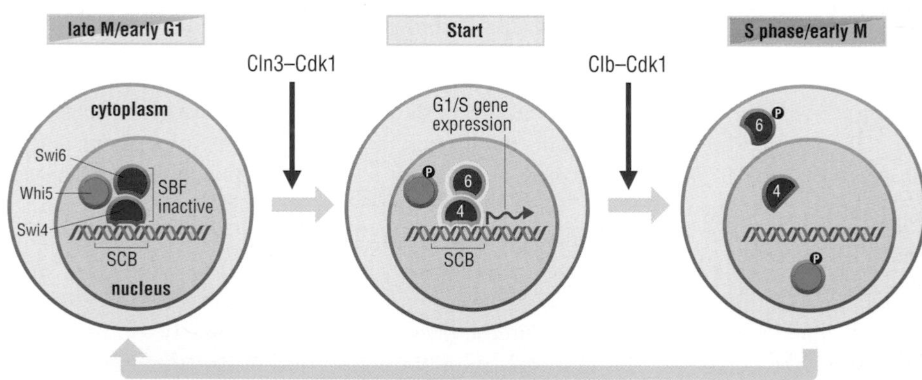

Figure 10-3 Proposed mechanism for the regulation of SBF activity during the cell cycle SBF is composed of two subunits, Swi4 and Swi6. In early G1, SBF is bound to the promoters of G1/S genes but is inactivated by bound Whi5. Cln3–Cdk1 triggers SBF activation, and thus G1/S gene expression, by phosphorylating Whi5, which then dissociates from the complex. On entry into S phase, Clb–Cdk1 complexes inactivate SBF by phosphorylating one of its subunits, Swi6, thereby promoting the disassembly of SBF. Swi4 no longer binds DNA, and Swi6 is exported from the nucleus.

References

Bean, J.M. *et al.*: **Coherence and timing of cell cycle start examined at single-cell resolution.** *Mol. Cell* 2006, **21**:3–14.

Breeden, L.L.: **Periodic transcription: a cycle within a cycle.** *Curr. Biol.* 2003, **13**:R31–R38.

Costanzo, M. *et al.*: **CDK activity antagonizes Whi5, an inhibitor of G1/S transcription in yeast.** *Cell* 2004, **117**:899–913.

de Bruin, R.A. *et al.*: **Cln3 activates G1-specific transcription via phosphorylation of the SBF bound repressor Whi5.** *Cell* 2004, **117**:887–898.

Futcher, B.: **Transcriptional regulatory networks and the yeast cell cycle.** *Curr. Opin. Cell Biol.* 2002, **14**:676–683.

MacKay, V.L. *et al.*: **Early cell cycle box-mediated transcription of *CLN3* and *SWI4* contributes to the proper timing of the G_1-to-S transition in budding yeast.** *Mol. Cell Biol.* 2001, **21**:4140–4148.

Pramila, T. *et al.*: **Conserved homeodomain proteins interact with MADS box protein Mcm1 to restrict ECB-dependent transcription to the M/G1 phase of the cell cycle.** *Genes Dev.* 2002, **16**:3034–3045.

Wang, H. *et al.*: **Recruitment of Cdc28 by Whi3 restricts nuclear accumulation of the G1 cyclin–Cdk complex to late G1.** *EMBO J.* 2004, **23**:180–190.

Wittenberg, C. and Reed, S.I.: **Cell cycle-dependent transcription in yeast: promoters, transcription factors, and transcriptomes.** *Oncogene* 2005, **24**:2746–2755.

In cells with reduced levels of Cln3, G1/S gene expression and Start are greatly delayed, whereas overproduction of Cln3 accelerates cell-cycle entry. Phosphorylation of Swi6 may also contribute to SBF activation.

Among the targets of SBF and MBF are the genes for the G1/S cyclins Cln1 and 2, and so Cln3-dependent activation of these transcription factors leads to increased expression of Cln1 and 2. These cyclins, in complexes with Cdk1, also phosphorylate Whi5 and might thereby help activate SBF. Although this could potentially form a positive feedback loop, it does not seem to do so in practice, because disabling this loop—by inhibiting Cln1 and Cln2 activities, for example—does not affect the timing of SBF-dependent G1/S gene expression.

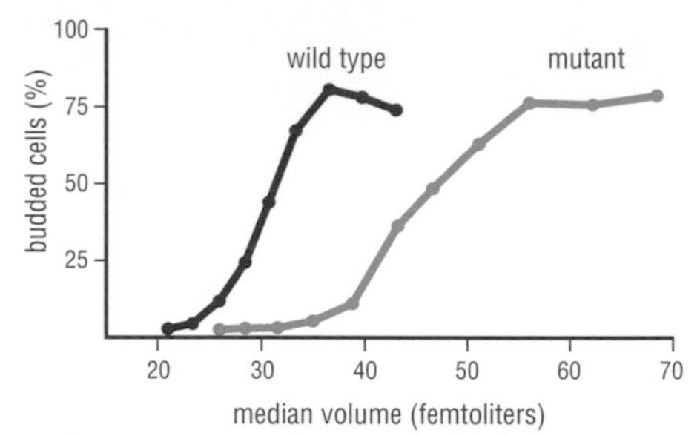

Figure 10-4 Synchronicity of Start in budding yeast An asynchronous population of yeast was subjected to centrifugal elutriation, a technique in which a specialized centrifuge is used to separate cells on the basis of their size. The smallest cells isolated by this method are newborn daughter cells, which have a volume of about 15 femtoliters. In the experiment shown here, newborn cells were observed as they progressed through G1 and initiated budding, which provides a useful marker of passage through Start. Wild-type cells (blue line) began to bud at an average volume of about 30 femtoliters, and the steepness of the curve reveals that passage through Start was highly synchronous. The green line shows results with a mutant yeast strain in which the ECB regulatory sequences upstream of the *CLN3* and *SWI4* genes were deleted. These mutations prevent the small increases in *CLN3* and *SWI4* expression that normally occur in G1, resulting in a delay in the onset of Start until cells have reached a larger volume. In addition, progression through Start is less synchronous in the mutant than in wild-type cells (the green curve is less steep than the blue one), revealing that ECB-dependent *CLN3* and *SWI4* expression is required for switch-like progression through Start. Adapted from MacKay, V.L. *et al.*: *Mol. Cell Biol.* 2001, **21**:4140–4148.

Small changes in the amount of Cln3 help trigger cell-cycle entry

The activity of Cln3–Cdk1 is a major determinant of G1/S gene activation and thus progression through Start. What regulates Cln3–Cdk1 activity? Numerous mechanisms are involved. To begin with, the concentration of Cln3 protein is regulated by external nutrient levels and intracellular metabolism, which seem to influence the rate at which *CLN3* mRNA is translated. The amount of Cln3 in the cell is therefore thought to serve as an important indicator of cell growth rate, as we describe in more detail later in this chapter.

Increased transcription of both *CLN3* and *SWI4* in late mitosis and early G1 leads to a small rise in the levels of both proteins in G1. *CLN3* and *SWI4* belong to a large group of genes whose expression increases in late mitosis (see section 3-12); in proliferating cells, expression of *CLN3* rises about threefold. The expression of these genes is controlled at a site in the promoter called the early cell cycle box (ECB), and suppressing *CLN3* and *SWI4* expression by deleting the ECB delays the onset of G1/S gene expression (Figure 10-4). This indicates that passage through Start depends in part on the small increase in Cln3 and Swi4 in early G1.

We do not yet understand how the relatively small changes in Cln3 concentration lead to the dramatic, switch-like upswing in G1/S gene expression that underlies Start. One source of switch-like behavior may lie in the control of *CLN3* and *SWI4* transcription by the ECB sequence, as mutation of this sequence not only delays cell-cycle entry but also decreases its synchronicity in the population (see Figure 10-4). Another mechanism might depend on the large number of Cdk1 phosphorylation sites in the Whi5 protein. If Whi5 were inactivated only when fully phosphorylated, then the response to small increases in Cln3–Cdk1 activity would be ultrasensitive (see section 3-7): that is, Whi5 inactivation would not occur at low levels of Cln3–Cdk1 activity, at which only partial Whi5 phosphorylation occurs, but would occur suddenly when Cln3–Cdk1 activity rose beyond some high threshold that triggered full Whi5 phosphorylation.

Because the major targets of Cln3–Cdk1, such as Whi5, are inside the nucleus, it is likely that the nuclear concentration of Cln3–Cdk1, and not its total cellular concentration, is the critical determinant of cell-cycle entry. Nuclear localization is controlled, at least in part, by a protein called Whi3, which anchors many of the Cln3–Cdk1 complexes in the cytoplasm during much of the cell cycle. In late G1, Cln3–Cdk1 is released from Whi3—by an unknown molecular mechanism—and its level inside the nucleus increases.

SBF and MBF are inactivated in S phase by Clb–Cdk1 complexes

In contrast to the three Cln cyclins, the yeast S- and M-cyclins Clb1–6 are negative regulators of SBF and MBF, and thus of G1/S gene expression. As discussed in section 3-13, the suppression of G1/S expression by S– and M–Cdks has an important role in generating the oscillations in G1/S-Cdk activity. SBF inactivation occurs in S phase when Swi4 and Swi6 dissociate from each other and Swi6 is exported to the cytoplasm. Although Swi4 remains in the nucleus, its DNA binding activity is lost—apparently because binding to Swi6 is required to unmask the DNA-binding site on Swi4. SBF inactivation is triggered by the increase in Clb–Cdk1 activities in S phase, and one inactivation mechanism may be the direct phosphorylation of both Swi4 and Swi6 (see Figure 10-3). Phosphorylation of Swi6 on Ser 160, for example, obscures a nearby nuclear localization signal, thereby promoting Swi6 accumulation in the cytoplasm. The molecular effect of Swi4 phosphorylation is not known, but an appealing possibility is that it disrupts Swi4 binding to Swi6.

G1/S–Cdks promote activation of S–Cdks

In budding yeast, a major consequence of the initiation of G1/S gene expression is the increased production of both G1/S cyclins and S cyclins (Figure 10-5). The G1/S cyclins Cln1 and Cln2 immediately form active complexes with Cdk1. The S cyclins Clb5 and Clb6 also bind Cdk1, but these S–Cdks are held in the inactive state by the Cdk inhibitor Sic1, which is abundant in G1 cells and specifically inhibits Clb–Cdk1 complexes but not Cln–Cdk1 complexes (see section 3-6). The major function of the G1/S–Cdks is to trigger the activation of the S–Cdks by promoting the destruction of Sic1.

Because of the high levels of Sic1 in G1, a stockpile of inactive Clb5,6–Cdk1–Sic1 complexes accumulates as the cell approaches S phase. Sic1 is then phosphorylated by the rising wave of Cln1,2–Cdk1 activity. Phosphorylation of Sic1 at multiple sites triggers its destruction, thereby unleashing the Clb5,6–Cdk1 complexes and triggering the onset of chromosome duplication (see section 4-5). Clb5,6–Cdk1 complexes also phosphorylate Sic1, providing the potential for a positive feedback loop whereby Clb5,6–Cdk1 complexes can help promote their own activation (see Figure 10-5).

Multisite phosphorylation of Sic1 generates switch-like S–Cdk activation

The destruction of phosphorylated Sic1 depends on its ubiquitination by SCF, a ubiquitin-protein ligase that collaborates with the ubiquitin-conjugating enzyme Cdc34 to promote the destruction of several cell-cycle regulatory proteins. SCF can recognize multiple targets using substrate-specific adaptor subunits called F-box proteins, which contain one domain (the F-box) that interacts with the SCF core and another that interacts with a specific phosphorylated protein target (see section 3-9). Sic1 ubiquitination requires the F-box protein Cdc4, which binds specifically to phosphorylated Sic1 and recruits it to the SCF active site.

The binding of phosphorylated Sic1 to Cdc4 is an intriguing example of how multiple phosphorylated sites interact with a single binding site on Cdc4. High-resolution structure analysis indicates that Cdc4 contains a single binding site for a single phosphorylated residue. High-affinity binding of Cdc4 occurs when the phosphorylation site on the target protein is found within a specific local sequence context: hydrophobic residues must be present just upstream of the phosphorylation site. Several SCF^{Cdc4} targets, including mammalian cyclin E

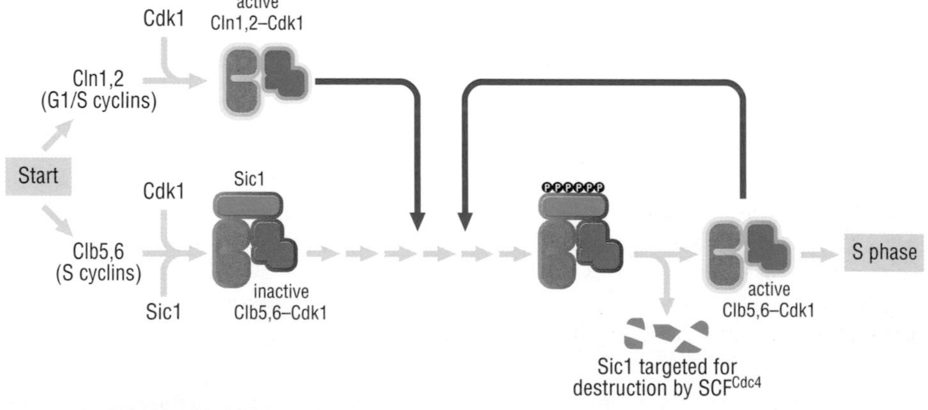

Figure 10-5 Mechanisms controlling S–Cdk activation in budding yeast Passage through Start depends on increased expression of G1/S and S cyclins, resulting in the formation of G1/S–Cdk and S–Cdk complexes. S–Cdks are initially bound and inhibited by Sic1, but multisite phosphorylation of Sic1 by the rising wave of G1/S–Cdk activity results in Sic1 destruction. Activated S–Cdks then initiate DNA replication.

References

Huang, J.N. *et al.*: **Activity of the APC^{Cdh1} form of the anaphase-promoting complex persists until S phase and prevents the premature expression of Cdc20p.** *J. Cell Biol.* 2001, **154**:85–94.

Klein, P. *et al.*: **Mathematical modeling suggests cooperative interactions between a disordered polyvalent ligand and a single receptor site.** *Curr. Biol.* 2003, **13**:1669–1678.

Mendenhall, M.D.: **An inhibitor of $p34^{CDC28}$ protein kinase activity from *Saccharomyces cerevisiae*.** *Science* 1993, **259**:216–219.

Nash, P. *et al.*: **Multisite phosphorylation of a CDK inhibitor sets a threshold for the onset of DNA replication.** *Nature* 2001, **414**:514–521.

Orlicky, S. *et al.*: **Structural basis for phosphodependent substrate selection and orientation by the SCF^{Cdc4} ubiquitin ligase.** *Cell* 2003, **112**:243–256.

Schwob, E. *et al.*: **The B-type cyclin kinase inhibitor $p40^{SIC1}$ controls the G1 to S transition in *S. cerevisiae*.** *Cell* 1994, **79**:233–244.

Verma, R. *et al.*: **Phosphorylation of Sic1p by G1 Cdk required for its degradation and entry into S phase.** *Science* 1997, **278**:455–460.

and yeast Gcn4, contain a single phosphorylation site with these ideal properties for recognition by Cdc4, and phosphorylation at this one site results in rapid protein ubiquitination and destruction.

Sic1, by contrast, does not contain such a site. Instead, its nine or so Cdk-dependent phosphorylation sites are each suboptimal, low-affinity Cdc4-binding sites. Detailed analysis of these sites led to the remarkable finding that Sic1 must be phosphorylated at a minimum of six sites to be recognized efficiently by SCF^{Cdc4} and targeted for destruction. This finding suggested that the multiple low-affinity sites on Sic1 can, by a mechanism as yet unknown, lead to sufficient occupation of the Cdc4-binding site to allow Sic1 ubiquitination.

Why are multiple low-affinity sites used on Sic1 instead of the single high-affinity site found on other SCF^{Cdc4} targets? The answer may lie in the stimulus–response relationship between rising Cln1,2–Cdk1 activity and Sic1 destruction (Figure 10-6a). The requirement for multiple phosphorylation events results in an ultrasensitive response (see section 3-7). At low levels of Cln1,2–Cdk1 activity, Sic1 is phosphorylated at only a few sites and is not destroyed: the system thereby filters out random, low-level stimuli. As Cln1,2–Cdk1 activity reaches a higher level, however, Sic1 phosphorylation reaches the multisite threshold for Cdc4 binding and destruction. When combined with the positive feedback loops that help control G1/S- and S–Cdks, this system should, in principle, provide a robust, switch-like activation of S–Cdks (Figure 10-6b).

The importance of this regulatory mechanism is revealed by analysis of yeast strains carrying mutant forms of Sic1. If Sic1 is engineered to contain a single high-affinity Cdc4-binding site instead of multiple low-affinity sites, destruction of Sic1 occurs in an asynchronous fashion throughout G1, as expected if random fluctuations in Cln1,2–Cdk1 activity lead to low levels of Sic1 phosphorylation (Figure 10-6c). These cells also tend to have defects in replication origin resetting because S–Cdk activity rises prematurely and prevents normal assembly of pre-replicative complexes at origins (see section 4-4). A long-term outcome of these problems is chromosome instability and lethality. Such defects are even more severe in cells lacking Sic1 entirely. Robust, switch-like S–Cdk activation in late G1 is therefore critical for successful chromosome inheritance in cell division.

G1/S- and S–Cdks collaborate to inactivate APC^{Cdh1} after Start

Another key regulatory event in late G1 is the inactivation of the ubiquitin-protein ligase APC^{Cdh1}, which is responsible for the ubiquitination and destruction of mitotic cyclins and other important targets in late mitosis and throughout G1 (see section 3-10). Inactivation of APC^{Cdh1} is required to allow the renewed accumulation of these targets as cells enter S phase and approach mitosis.

APC^{Cdh1} is inactivated by Cdk1-dependent phosphorylation of its Cdh1 subunit. Phosphorylation and inactivation of APC^{Cdh1} are thought to be initiated at Start by Cln1,2–Cdk1 complexes, which are not targeted for destruction by APC^{Cdh1}. Complete APC^{Cdh1} inactivation also requires the activation in early S phase of Clb5,6–Cdk1 complexes. Clb5 and 6 are thought to be partly resistant to APC^{Cdh1}-mediated destruction (they are targeted primarily by APC^{Cdc20}), but inactivation of APC^{Cdh1} may be required for them to accumulate to the high levels required for S phase.

(a)

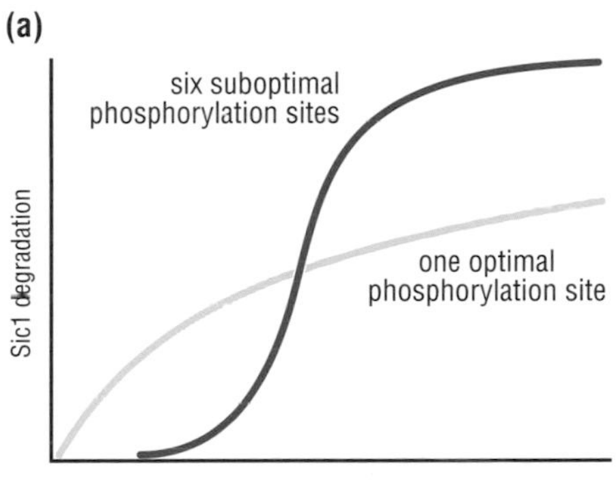

(b) six suboptimal phosphorylation sites

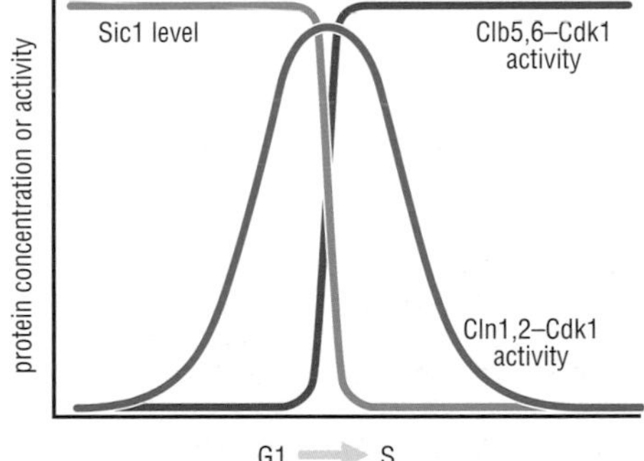

(c) one optimal phosphorylation site

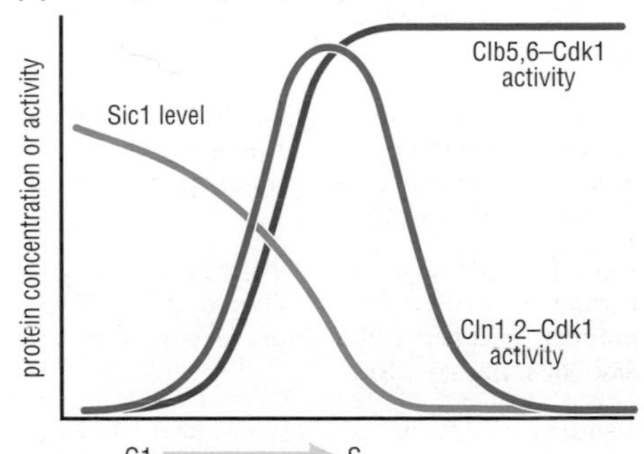

Figure 10-6 Multisite phosphorylation of Sic1 results in an ultrasensitive response (a) These curves portray the theoretical effects of increasing Cln1,2–Cdk1 activity on Sic1 degradation, when degradation requires one or six separate phosphorylation events. Theoretically, if Sic1 phosphorylation at a single optimal site is sufficient for degradation (yellow curve), Sic1 degradation will display a hyperbolic response (see section 3-7) as Cln1,2–Cdk1 activity increases. If six sites must be phosphorylated (red curve), the result is an ultrasensitive system that responds poorly to low levels of Cln1,2–Cdk1 activity but responds more abruptly at higher stimulus levels. **(b)** In the cell, the ultrasensitivity that results from multisite phosphorylation converts the rise in Cln1,2–Cdk1 activity to a switch-like drop in Sic1 levels and rise in Clb5,6–Cdk1 activity. **(c)** If Sic1 is mutated to contain a single optimal recognition site for Cdc4 binding, Sic1 destruction and Clb5,6–Cdk1 activation occur prematurely and less abruptly, resulting in premature and asynchronous entry into S phase. Adapted from Nash, P. *et al.*: *Nature* 2001, **414**:514–521.

Yeast mating factors induce cell-cycle arrest in G1

Yeast cells do not require extracellular mitogens to stimulate their proliferation, and they divide as rapidly as possible when nutrients are abundant. There is one point in the yeast life cycle, however, at which an extracellular peptide controls the rate of cell division. This occurs during mating, when haploid yeast cells arrest in G1 in response to peptide pheromones called mating factors (see section 2-1). In comparison with the mitogenic signaling pathways of animal cells, the mechanism by which yeast mating factors govern progression through Start is relatively simple and well understood. The mating factor signaling pathway therefore provides a useful illustration of how extracellular factors can influence cell proliferation.

Laboratory strains of budding yeast proliferate indefinitely in the haploid state under ideal culture conditions (see section 2-1). When a haploid cell of one mating type (for example an α cell) encounters a haploid cell of the other mating type (an **a** cell), however, mating factors produced by each cell (α-factor and **a**-factor, respectively) bind receptors on the surface of the other cell. Receptor activation in both cases initiates intracellular signaling pathways that lead to cell-cycle arrest in G1, as well as to changes in gene expression and cell morphology that lead to mating, cell fusion and the formation of a diploid zygote.

Mating factors induce cell-cycle arrest by causing the inhibition of all three Cln–Cdk1 complexes in G1 cells. The expression of G1/S genes, including *CLN1* and *CLN2*, is prevented, Start does not occur, and the cell is arrested in G1. Inhibition of Cdk1 function is achieved primarily by a protein called Far1, which is activated by mating factor, binds specifically to Cln–Cdk1 complexes—but not Clb–Cdk1 complexes—and inhibits their activity.

Far1 has multiple functions in proliferating and arrested cells

In normally proliferating cells, the concentration of Far1 increases in late mitosis and remains high throughout G1. Two mechanisms are involved. First, *FAR1* gene expression increases in late mitosis. Second, the stability of Far1 is increased in G1. Between late G1 and mitosis, Far1 is phosphorylated at Ser 87 by cyclin–Cdk1 complexes, resulting in SCF^{Cdc4}-dependent ubiquitination and destruction. Inactivation of Cdks in late mitosis therefore results in Far1 stabilization.

Although Far1 is present in G1 it does not inhibit Cdk activity unless the cell is treated with mating factor. Mating factor triggers activation of the protein kinase Fus3, which phosphorylates

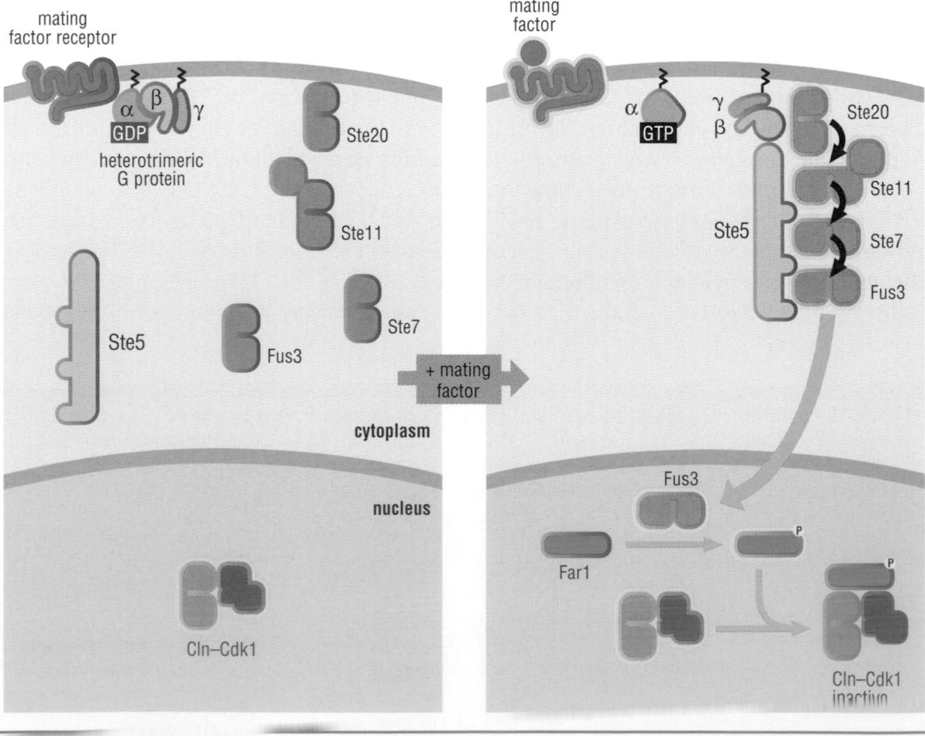

Figure 10-7 Inhibition of Start by mating factor signaling In the absence of mating factor, the inactive receptor is associated with a heterotrimeric G protein, in which the α subunit is bound by GDP. Activation of the receptor by mating factor leads to replacement of GDP with GTP, resulting in release of the βγ complex. This complex then promotes the assembly at the membrane of a signaling complex containing Ste20 and the scaffold protein Ste5, leading to activation of a MAP kinase cascade containing Ste11, Ste7 and Fus3. Fus3 phosphorylates and activates Far1, which inhibits Cln–Cdk1 activity and prevents Start. This simplified diagram does not include several features of the system, including the ability of Fus3 to promote changes in gene expression and the ability of Far1 to regulate the actin cytoskeleton at the cell tip.

Far1 at Thr 306, thereby allowing Far1 to bind and inhibit Cln–Cdk1 complexes. Not only does this block Start, but it also prevents Cln–Cdk1 complexes from phosphorylating Far1 and triggering its destruction.

Far1 also interacts with and inhibits the guanine-nucleotide exchange factor Cdc24. In proliferating cells, destruction of Far1 in late G1 liberates Cdc24, which then activates the small GTPase Cdc42 to promote actin rearrangements at the bud site. In cells treated with mating factor, some Far1 protein travels with Cdc24 to the cell membrane to stimulate the formation of the mating projection, or schmoo, that becomes the site of fusion with the other cell.

Far1 phosphorylation is triggered by a G-protein signaling pathway

How does mating factor trigger the activation of Fus3 and the consequent phosphorylation of Far1? The signaling process begins at the mating factor receptor, a large protein that spans the cell membrane seven times (Figure 10-7). The cytoplasmic face of the receptor is coupled to a heterotrimeric G protein (see section 8-7). These proteins, like the related Ras-like small GTPases, undergo conformational changes in response to guanine-nucleotide binding and hydrolysis, and are used as molecular switches in many signaling pathways. The G protein associated with the mating factor receptor contains three subunits: an α subunit (Scg1) that binds guanine nucleotides and a βγ subunit pair (Ste4 and Ste18). In the absence of mating factor, the three subunits are bound together and the Gα subunit is occupied by GDP. Binding of mating factor alters the shape of its receptor, resulting in the release of GDP by the Gα subunit. Gα then binds GTP from the cytoplasm, which causes a conformational change that triggers release of the Gβγ pair.

Once released from the Gα subunit, the primary function of the Gβγ pair is to activate a cascade of kinase activities that ultimately results in the activation of Fus3. Fus3 is a member of a large family of protein kinases, called the MAP kinases or extracellular signal-regulated kinases (ERKs), which are found in a wide range of signaling pathways in all eukaryotes. Typically, members of the MAP kinase family are controlled by a signaling module called a MAP kinase cascade, a series of three protein kinases each of which phosphorylates and thereby activates the next in the series. Fus3, like other MAP kinases, is found at the bottom of such a series (see Figure 10-7). Fus3 is activated by the protein kinase Ste7, which is activated by the protein kinase Ste11. The three kinases of this cascade are all bound next to each other on a scaffold protein, Ste5, which thereby promotes a productive interaction between the three kinases while insulating them from related kinases involved in other functions.

After its release by the activated mating factor receptor, the Gβγ complex recruits Ste5—and its associated kinases—to the cell membrane. At the same time, Gβγ also binds another protein kinase, Ste20, that is likewise found at the cell membrane. Ste20 is an activator of Ste11. By bringing Ste5 and Ste20 together, Gβγ allows Ste20 to activate Ste11, leading ultimately to activation of the final kinase in the series, Fus3 (see Figure 10-7).

Active Fus3 travels to the nucleus, where it phosphorylates a fraction of the Far1 molecules, thereby allowing them to inhibit Cln–Cdk1 complexes (some Far1–Cdc24 complexes leave the nucleus to help initiate schmoo formation, as noted above). Fus3 also phosphorylates gene regulatory proteins, resulting in the increased expression of genes required for the mating process.

References

Chang, F. and Herskowitz, I.: **Identification of a gene necessary for cell cycle arrest by a negative growth factor of yeast: FAR1 is an inhibitor of a G1 cyclin, CLN2.** *Cell* 1990, **63**:999–1011.

Dohlman, H.G. and Thorner, J.W.: **Regulation of G protein-initiated signal transduction in yeast: paradigms and principles.** *Annu. Rev. Biochem.* 2001, **70**:703–754.

Gartner, A. *et al.*: **Pheromone-dependent G_1 cell cycle arrest requires Far1 phosphorylation, but may not involve inhibition of Cdc28-Cln2 kinase, *in vivo*.** *Mol. Cell Biol.* 1998, **18**:3681–3691.

Henchoz, S. *et al.*: **Phosphorylation- and ubiquitin-dependent degradation of the cyclin-dependent kinase inhibitor Far1p in budding yeast.** *Genes Dev.* 1997, **11**:3046–3060.

Peter, M. *et al.*: **FAR1 links the signal transduction pathway to the cell cycle machinery in yeast.** *Cell* 1993, **73**:747–760.

van Drogen, F. *et al.*: **MAP kinase dynamics in response to pheromones in budding yeast.** *Nat. Cell Biol.* 2001, **3**:1051–1059.

Winters, M.J. *et al.*: **A membrane binding domain in the ste5 scaffold synergizes with Gβγ binding to control binding to control localization and signaling in pheromone response.** *Mol. Cell* 2005, **20**:21–32.

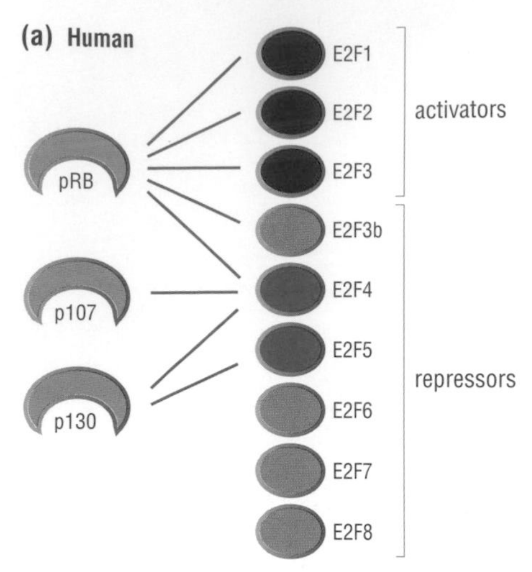

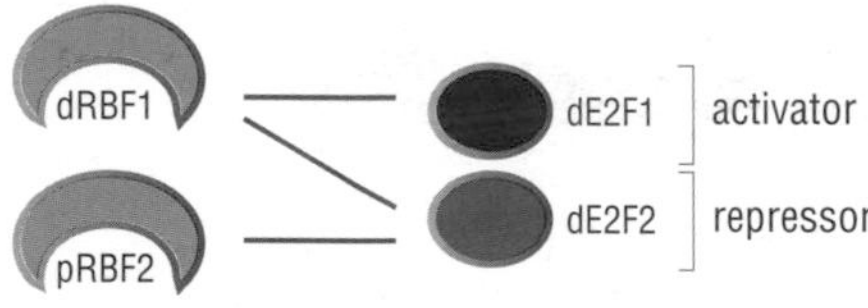

Figure 10-8 E2F proteins and their interactions with members of the pRB family **(a)** In human cells, the three activator E2Fs, E2F1–3, stimulate G1/S gene expression and thus promote cell-cycle entry. Before Start, they are inhibited by pRB. Other members of the E2F family are repressors of gene expression. The best understood of these, E2F4 and E2F5, interact with p107 and p130 to form complexes that actively inhibit G1/S gene expression during G0 and G1. The functions of other E2F family members are not clear. E2F3b is a version of E2F3 that is encoded by a unique mRNA derived from the *E2F3* gene; E2F3b is a repressor protein that associates with pRB in some types of quiescent cells. E2F6, 7 and 8 are also repressors but are not regulated by association with pRB-related proteins, and their function in the cell cycle is uncertain. All E2F proteins except E2F7 and 8 also interact with a DP protein, either DP-1 or DP-2 (not shown here). **(b)** In *Drosophila*, the activator E2F (dE2F1) and repressor E2F (dE2F2) are regulated by two pRB-related proteins, dRBF1 and dRBF2. Each E2F protein interacts with a DP protein (dDP, not shown).

E2F transcription factors help control G1/S gene expression in animals

Progression through Start in animals, as in yeast, depends in most cell types on the induction of G1/S gene expression by regulatory proteins. The most important of these regulators are members of the E2F family, which govern the expression in late G1 of a large number of genes—perhaps more than a thousand in human cells. Two critical G1/S gene products are the G1/S cyclin, cyclin E, and the S cyclin, cyclin A. As in yeast, therefore, an important function of increased G1/S gene expression is to promote the activation of Cdks that trigger cell-cycle entry and DNA synthesis.

Of the E2F family members encoded in the human genome (Figure 10-8a), E2F1–5 are most clearly implicated in the control of G1/S gene expression and will be the focus of our discussion. Each associates with a second subunit, DP-1 or DP-2, to form a heterodimeric transcription factor that binds to a specific DNA sequence in the promoters of G1/S target genes (see section 3-12). The activities of these factors are controlled by interactions with proteins of the pRB family. Here and in section 10-5 we review the major features of E2F function and regulation by pRB proteins, after which we describe, in sections 10-6 to 10-8, how external mitogens control these proteins and thereby govern cell proliferation.

Stimulation of G1/S gene expression results from a combination of increased gene activation and decreased gene repression

The five major human E2F proteins are divided into two functional groups: the activator E2Fs and repressor E2Fs (see Figure 10-8a). The activator E2Fs—E2F1, 2 and 3—act primarily as transcriptional activators whose interaction with target gene promoters in late G1 and early S phase increases gene expression. Experimental overexpression of activator E2Fs in mammalian cells generally promotes entry into S phase, whereas decreased function of activator E2Fs reduces the expression of many E2F target genes and inhibits cell proliferation. In contrast, the repressor E2Fs—E2F4 and 5—are transcriptional repressors that bind to G1/S gene promoters in quiescent (G0) cells and inhibit gene expression; increased gene expression at Start therefore depends on the inactivation of these repressor E2Fs. Mutation of repressor E2Fs generally results in an increase in E2F-dependent gene expression in quiescent cells.

The two branches of the E2F family therefore act antagonistically to control G1/S gene expression. This principle is best illustrated by genetic studies in *Drosophila*, which express only one activator E2F (dE2F1) and one repressor E2F (dE2F2) (Figure 10-8b). dE2F1 is a transcriptional activator whose overexpression stimulates S-phase entry. Mutation of dE2F1 inhibits G1/S gene expression and cell proliferation. Mutation of dE2F2 results in the increased expression of some G1/S genes but has relatively minor effects on cell proliferation. When a mutation in dE2F1 is combined with a mutation in dE2F2, near-normal cell proliferation is restored (Figure 10-9), despite the fact that the expression of some of these G1/S genes now occurs throughout the cell cycle rather than being limited to late G1 and S phase.

These results in flies have several important implications. First, they suggest that a balance of E2F-dependent activation and repression normally governs G1/S gene expression and cell proliferation. As a result, when the activator E2F is inhibited by mutation, the repressor E2F is unopposed and G1/S gene expression is markedly inhibited. Second, they show that the two dE2Fs are required for normal oscillations in the expression of G1/S genes and that high basal expression occurs in their absence—perhaps because loss of the repressor E2F allows high expression even if the activator E2F is not present. Third, the ability of cells to divide normally

References

Attwooll, C. *et al.*: **The E2F family: specific functions and overlapping interests.** *EMBO J.* 2004, **23**:4709–4716.

Blais, A. and Dynlacht, B.D.: **Hitting their targets: an emerging picture of E2F and cell cycle control.** *Curr. Opin. Genet. Dev.* 2004, **14**:527–532.

Classon, M. and Harlow, E.: **The retinoblastoma tumour suppressor in development and cancer.** *Nat. Rev. Cancer* 2002, **2**:910–917.

Dimova, D.K. and Dyson, N.J.: **The E2F transcriptional network: old acquaintances with new faces.** *Oncogene* 2005, **24**:2810–2826.

Frolov, M.V. *et al.*: **Functional antagonism between E2F family members.** *Genes Dev.* 2001, **15**:2146–2160.

Trimarchi, J.M. and Lees, J.A.: **Sibling rivalry in the E2F family.** *Nat. Rev. Mol. Cell Biol.* 2002, **3**:11–20.

in the absence of both E2Fs indicates that E2F-dependent oscillations in gene expression are not essential for cell proliferation: high basal levels of gene expression seem to be sufficient.

It is likely, however, that E2F-dependent mechanisms are required for robust, reliable Start control in *Drosophila* and other animals. In mice, for example, mutation of individual activator E2Fs results in tissue-specific defects in cell proliferation, whereas mutation of repressor E2Fs causes defects in the ability of some cell types to exit from the cell cycle and differentiate. Thus, the finely tuned balance of positive and negative E2F activities is an important feature of the regulatory networks governing cell number and differentiation in animal tissues.

Although E2F proteins are most clearly implicated in the control of G1/S gene expression, they also contribute to the control of other cell-cycle stages. Many E2F target genes in human cells are involved in mitosis, and mutations in some E2F proteins result in mitotic defects. In *Drosophila* wing imaginal discs, and probably in other tissues as well, the activator dE2F1 is a key regulator of the G2/M transition. A major target of dE2F1 is the gene for the phosphatase Cdc25 (see section 3-3), whose increased expression is required for entry into mitosis.

E2F function is regulated by pRB proteins

The activities of the E2F proteins are regulated to ensure that G1/S genes are expressed only at the appropriate time in the cell cycle—and are repressed in non-proliferating cells. The function of the major E2Fs is controlled primarily by binding to the pRB proteins, which are sometimes called the pocket proteins because they contain an E2F-binding pocket. Mammals contain three pRB-related proteins—pRB, p107 and p130—and *Drosophila* contains two, dRBF1 and dRBF2. Each pRB protein interacts with specific members of the E2F family (see Figure 10-8) to inhibit the expression of E2F target genes in quiescent cells: in mammals, pRB inhibits the activator E2Fs, whereas p107 and p130 act as co-repressors for the repressor E2Fs. Re-entry into the cell cycle from quiescence therefore requires removal of the inhibitory effects of the pRB proteins.

pRB-related proteins inhibit E2F-dependent gene expression by at least two mechanisms. First, and most importantly, pRB proteins bound to repressor E2Fs interact with nucleosome-modifying enzymes (such as histone deacetylases) and chromatin-remodeling complexes (such as the Swi/Snf complex) (see section 4-9). By recruiting these factors to the promoters of E2F-responsive genes, pRB proteins stimulate the local formation of a chromatin structure that inhibits gene expression. Second, one member of the pRB family, pRB itself, inhibits E2F-dependent gene expression by binding the transcriptional activation domain at the carboxy terminus of the activator E2Fs, thereby blocking their action.

pRB proteins provide braking mechanisms that are not essential for regulating cell-cycle progression but are involved primarily in the control of cell-cycle exit and in maintaining the quiescent state. Mutation of all three pRB proteins in mouse cells, or mutation of pRB proteins in *Drosophila*, results in small decreases in the length of G1 in proliferating cell populations growing under ideal conditions. Progression directly from mitosis to Start occurs with only minor defects when E2F-dependent gene expression is not restrained by pRB proteins. Cells lacking pRB proteins, however, have profound defects in their ability to exit the cell cycle in response to extracellular signals or stresses such as DNA damage. Mutations affecting the pRB-mediated braking mechanism are important in human health. Almost all human cancers are associated with some defect in this pathway, as discussed in Chapter 12.

(a) Larval growth

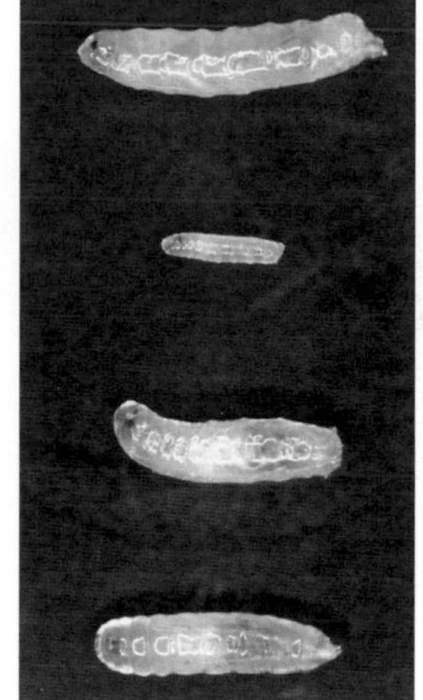

(b) Cyclin E expression

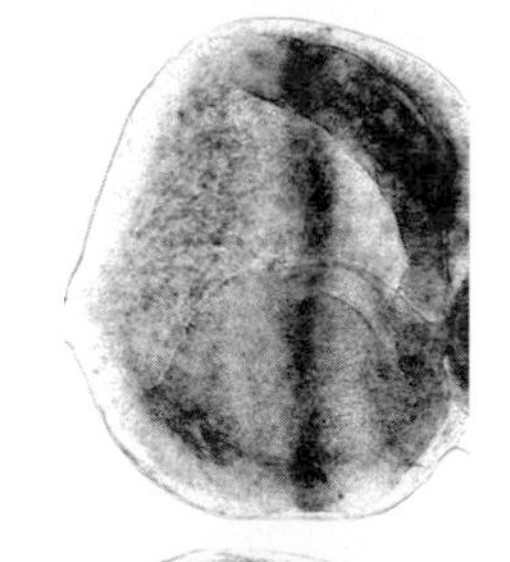

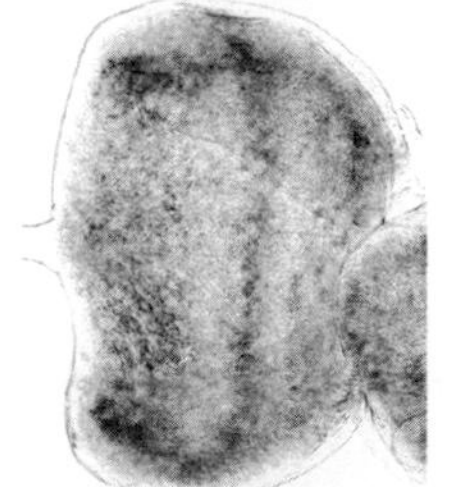

Figure 10-9 Antagonistic functions of the two E2F homologs in *Drosophila* **(a)** In *Drosophila*, mutation of the activator dE2F1 results in defects in cell proliferation and tissue growth, as revealed here by the small size of the mutant fly larva. Mutation of dE2F2 suppresses the defects of the dE2F1 mutant, suggesting that these defects result from unchecked repression of G1/S genes by dE2F2. **(b)** This experiment demonstrates the minor effects of E2F mutations on the expression of the G1/S gene cyclin E in the eye imaginal disc, a flat sheet of cells in the larva that is the precursor of the adult eye (see section 2-4). Cyclin E mRNA is stained black. In wild-type embryos (top panel), it is found throughout the disc but is particularly focused in a vertical line of proliferating cells adjacent to the morphogenetic furrow, which passes anteriorly (left) across the cells of the disc as it develops (see section 2-4). Mutation of dE2F1 causes a significant drop in cyclin E expression (not shown), whereas mutation of dE2F2 has little effect (middle panel). Mutation of both dE2Fs together also results in only minor defects (bottom panel). Thus, the reduced expression in the absence of the activator dE2F1 can be restored by mutation of the repressor dE2F2. Photographs kindly provided by Maxim Frolov. From Frolov, M.V. *et al.*: *Genes Dev.* 2001, **15**:2146–2160.

G1/S gene expression at Start involves the replacement of repressor E2Fs with activator E2Fs

Before Start in animal cells, expression of the G1/S genes is inhibited by the binding of repressor E2F proteins. Increased G1/S gene expression at Start depends on the removal of the repressor E2Fs from G1/S gene promoters and their replacement at those promoters by activator E2Fs. This depends on the inactivation of pRB proteins, which, as we saw in section 10-4, inhibit G1/S gene expression in quiescent or G1 cells by binding E2F proteins. Mitogens promote entry into the cell cycle by activating cyclin–Cdk complexes that phosphorylate pRB proteins, causing their dissociation from E2Fs and enabling G1/S gene expression.

Much of our understanding of the regulation of E2F and pRB proteins comes from studies of cultured mammalian cells. When these cells are deprived of mitogens, they exit from the cell cycle in G1 and enter the quiescent G0 state. E2F-dependent gene expression is inhibited, primarily because gene promoters are occupied by repressor E2Fs (E2F4 and 5), bound by the pRB family members p107 or p130 and their associated chromatin-remodeling complexes (see section 10-4). Activator E2Fs are present at very low levels (if at all) in quiescent cells, because the genes encoding these proteins are also repressed by E2F4 and E2F5. The small population of activator E2Fs in quiescent cells is inhibited by interactions with pRB.

After the mitogenic stimulation of quiescent cells, the dissociation of pRB proteins from E2Fs leads to a shift in the composition of E2F proteins at G1/S gene promoters (Figure 10-10). First, repressor E2Fs and their associated chromatin-modifying enzymes dissociate from these promoters, thereby enabling gene expression. Even in the absence of activator E2Fs, removal of repressor E2Fs can result in the derepression, or higher basal transcription, of some G1/S genes. Second, dissociation of pRB from activator E2Fs, coupled with increased production of activator E2Fs, triggers the binding of activator E2Fs to gene promoters, greatly stimulating G1/S gene expression. Genes encoding activator E2Fs are also E2F-responsive, resulting in a potential positive feedback loop that enhances the production of activator E2Fs and further stimulates G1/S gene expression.

Phosphorylation of pRB proteins releases E2F

The key to the regulation of G1/S genes at Start is that the phosphorylation of pRB proteins results in their dissociation from E2Fs (see Figure 10-10). As mitogen-stimulated mammalian cells enter a new cell cycle from quiescence, the initial phosphorylation of pRB proteins is catalyzed primarily by G1 cyclin–Cdk complexes—cyclin D and its partners Cdk4 and Cdk6. Cyclin D–Cdk complexes are activated in response to mitogens, and thus provide one of the main links between mitogens and the cell-cycle control system.

The function of G1 cyclin–Cdk complexes in progression through Start is illustrated by the effects of their inhibition. In many mammalian cell lines, specific inhibition of cyclin D–Cdk activity—by overexpression of the Cdk-inhibitory protein $p16^{INK4a}$, for example (see section 3-6)—results in G1 delay or arrest. Inhibition of cyclin D–Cdk has little effect in cells lacking pRB proteins, however. This shows that the central function of cyclin D–Cdk complexes is to remove the inhibitory effects of pRB family members. The pRB proteins serve primarily as a braking mechanism to help induce exit from the cell cycle, and cyclin D–Cdk complexes may have evolved as one major mechanism by which these brakes are released. If the brakes are not present in the first place, however, no mechanism is required to release them.

References

Attwooll, C. *et al.*: **The E2F family: specific functions and overlapping interests.** *EMBO J.* 2004, **23**:4709–4716.

Classon, M. and Harlow, E.: **The retinoblastoma tumour suppressor in development and cancer.** *Nat. Rev. Cancer* 2002, **2**:910–917.

Kozar, K. *et al.*: **Mouse development and cell proliferation in the absence of D-cyclins.** *Cell* 2004, **118**:477–491.

Rayman, J.B. *et al.*: **E2F mediates cell cycle-dependent transcriptional repression *in vivo* by recruitment of an HDAC1/mSin3B corepressor complex.** *Genes Dev.* 2002, **16**:933–947.

Rubin, S.M. *et al.*: **Structure of the Rb C-terminal domain bound to E2F1-DP1: a mechanism for phosphorylation-induced E2F release.** *Cell* 2005, **123**:1093–1106.

Sherr, C.J. and Roberts, J.M.: **CDK inhibitors: positive and negative regulators of G1-phase progression.** *Genes Dev.* 1999, **13**:1501–1512.

Takahashi, Y. *et al.*: **Analysis of promoter binding by the E2F and pRB families *in vivo*: distinct E2F proteins mediate activation and repression.** *Genes Dev.* 2000, **14**:804–816.

Trimarchi, J.M. and Lees, J.A.: **Sibling rivalry in the E2F family.** *Nat. Rev. Mol. Cell Biol.* 2002, **3**:11–20.

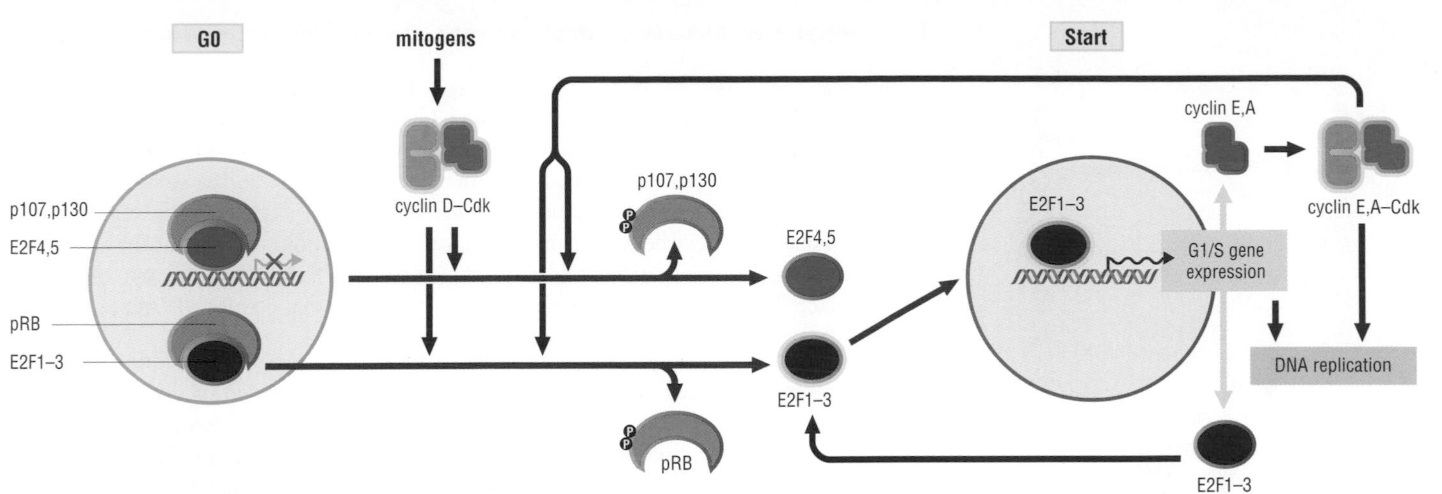

Figure 10-10 Control of E2F-dependent gene expression at mammalian Start In quiescent cells (G0/G1), the promoters of G1/S genes are repressed by complexes of repressor E2Fs (blue ovals) and the pRB family members p107 or p130 (which recruit chromatin-modifying enzymes, not shown). Small amounts of activator E2Fs (red ovals) may be present at low levels in quiescent cells but are inactivated by pRB binding. Treatment with mitogens results in Cdk-dependent phosphorylation of pRB family members, thereby promoting their dissociation from E2Fs. Repressor E2Fs on the DNA are then replaced with pRB-free activator E2Fs, resulting in activation of G1/S gene expression. This leads to increased expression of activator E2Fs and the cyclins E and A, which then feed back to enhance the levels and activity of activator E2Fs.

Some cell types in mammals, and almost all in flies, can divide normally in the absence of cyclin D–Cdk activities. In *Drosophila*, mutation of cyclin D or Cdk4 (the sole partner for cyclin D in this species) has remarkably little effect in any tissue, suggesting that cyclin D–Cdk4 complexes are not required for the proliferation of any major cell type, although they are required for cell growth. The same is true for several cell types in mice lacking Cdk4 or specific D-type cyclins; indeed, even mice lacking all three D-type cyclins survive for most of embryogenesis. There are at least two possible explanations for these results. First, it seems likely that, in some cell types, pRB proteins are either not present or have relatively weak effects on cell proliferation, rendering cyclin D unnecessary. Alternatively, in some cells the inhibitory effects of pRB proteins are overcome by mechanisms that are independent of cyclin D. In particular, it is clear that other cyclin–Cdk complexes, notably cyclin E–Cdk2, can also phosphorylate pRB proteins, as we discuss next.

Multiple mechanisms of E2F activation provide robust regulation of Start

In many mammalian cell types, cyclin D–Cdk catalyzes only partial pRB phosphorylation and E2F activation. Complete E2F activation occurs when the activity of the G1/S–Cdk, cyclin E–Cdk2, rises in late G1 and completes the phosphorylation of pRB proteins (see Figure 10-10). Cyclin E–Cdk2 thus makes a major contribution to the stimulation of E2F-dependent gene expression, and in cells lacking cyclin D–Cdk activity, cyclin E–Cdk2 is likely to be the primary activator of this gene expression. This also seems to be the case in *Drosophila*, where cyclin E–Cdk2 (unlike cyclin D–Cdk4) is essential for cell-cycle entry.

S-phase cyclin A–Cdk complexes can also phosphorylate pRB proteins and are thought to maintain this phosphorylation as the cell progresses through S phase and into mitosis. There is also evidence that cyclin A–Cdk complexes phosphorylate activator E2Fs and thereby inhibit their activities in S phase, thus providing a mechanism for generating the transient pulse of E2F activity that is seen in late G1.

Despite the abundant evidence for the central regulatory role of cyclin E–Cdk2 in progression through Start, the loss of cyclin E or Cdk2 in mouse cells has surprisingly minor effects on cell proliferation and even on the viability of the whole animal. It is therefore likely that there is some redundancy in this regulatory system, such that other complexes—perhaps a combination of cyclin D–Cdk4 and cyclin A–Cdk1—can carry out the functions of cyclin E–Cdk2 in its absence. As we have seen, many other important regulators, such as cyclin D–Cdk and members of the E2F and pRB families, are also not essential for cell proliferation. Thus, the regulatory system driving Start in animal cells is remarkably robust and can withstand the failure of seemingly critical components.

10-6 Mitogenic Signaling in Animal Cells

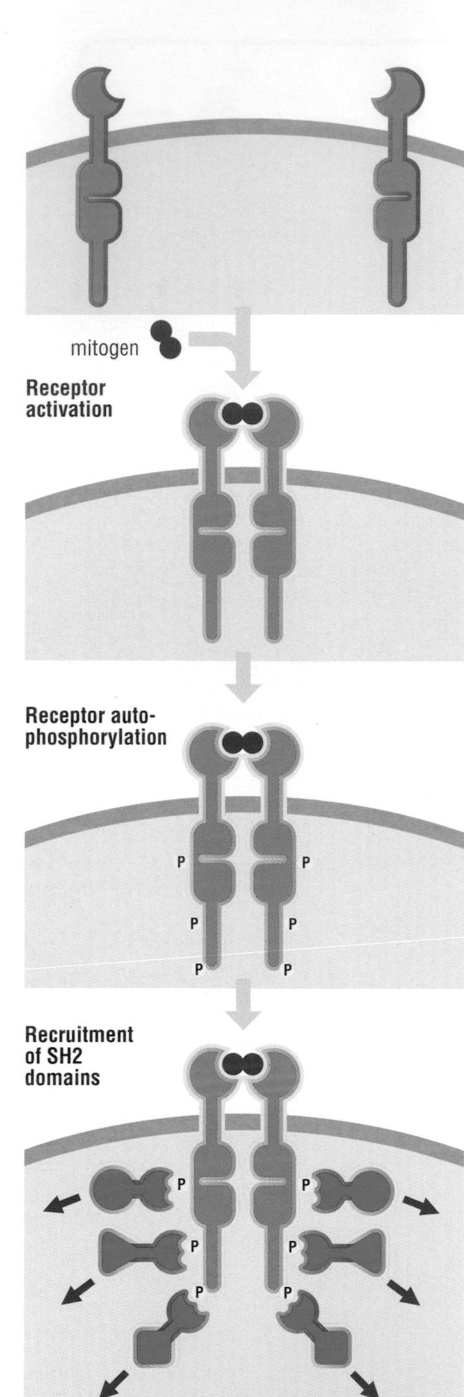

Figure 10-11 **Origin of the mitogenic signal at the cell membrane** The binding of mitogens causes dimerization of their receptors at the cell surface. In most cases, dimerization is due simply to cross-linking of two receptors by a bivalent mitogen (as shown here), whereas in others a monovalent mitogen causes conformational changes that lead to receptor dimerization. The intracellular protein kinase domains of the receptors then phosphorylate tyrosine residues at several positions within the receptors themselves. Each phosphotyrosine binds to a specific SH2 domain linked to a signaling molecule, resulting in the assembly of giant signaling complexes at the cell membrane.

Extracellular mitogens control the rate of cell division in animals

Unlike yeast cells, the cells of multicellular organisms divide only when the organism as a whole needs new cells to build or maintain tissues. In general, entry into a new cell cycle occurs in animals only when the cell is exposed to the appropriate extracellular **mitogens**. Mitogens are generally soluble peptides or small proteins secreted by neighboring cells, or they can be insoluble components of the extracellular matrix. Some are highly specific regulators of division in one cell type, whereas others have much broader actions throughout the body. The best understood mitogens include platelet-derived growth factor (PDGF) and epidermal growth factor (EGF), both of which are soluble polypeptides that control the rate of division in many different cell types.

Entry into the animal cell cycle at Start is triggered by the activation of G1–, G1/S– and S–Cdks, coupled with the activation of E2F-dependent gene expression (see section 10-5). In this and the next two sections we address the complex and poorly understood mechanisms by which extracellular mitogens promote Cdk activation. First, we focus here on the earliest steps in mitogenic signaling. In sections 10-7 and 10-8 we move downstream to the signaling events that trigger the sequential activation of G1– and G1/S–Cdks.

Activated mitogen receptors recruit signaling complexes to the cell membrane

Peptide and protein mitogens such as PDGF and EGF stimulate the cell by binding to transmembrane receptor proteins, which transmit the mitogenic signal across the cell membrane by activating a protein kinase on the intracellular side (Figure 10-11). The intracellular regions of many mitogen receptors contain a protein kinase catalytic domain that phosphorylates the hydroxyl group on tyrosine residues. Mitogen binding to the extracellular part of the receptor causes the dimerization of two receptor molecules. This activates the protein kinase domains, which then phosphorylate multiple tyrosine residues within the cytoplasmic domains of the receptors themselves.

The phosphorylation of tyrosine residues creates new binding sites on the receptor that recruit intracellular signaling proteins. Phosphotyrosines are binding sites for members of the SH2 family of protein domains, with different SH2 domains specifically binding phosphotyrosines in different sequence contexts. Thus, each phosphotyrosine on the activated receptor interacts with a signaling protein containing a specific SH2 domain, resulting in decoration of the receptor with various signaling proteins (see Figure 10-11). In this way, a single activated receptor can initiate multiple signaling pathways.

Ras and Myc are components of many mitogenic signaling pathways

Many mitogenic signaling pathways begin with the activation of the small GTPase Ras at the cell membrane. Like other small GTPases such as Ran (see section 6-8), Ras can exist in an inactive GDP-bound state or an active GTP-bound state. In quiescent cells, Ras is found primarily in the inactive GDP-bound form. When mitogen receptors are activated, specific phosphotyrosines on these receptors interact with an SH2-containing protein called Grb2, which then binds the protein Sos. Sos is a guanine-nucleotide exchange factor for Ras. Its recruitment to the receptor at the cell membrane is thought to bring it close to Ras, which it activates by stimulating it to exchange its GDP for GTP (Figure 10-12).

Definitions

mitogen: extracellular molecule that stimulates cell proliferation.

References

Adhikary, S. and Eilers, M.: **Transcriptional regulation and transformation by Myc proteins.** *Nat. Rev. Mol. Cell Biol.* 2005, **6**:635–645.

Alberts, B. *et al.*: *Molecular Biology of the Cell* 4th ed. (Garland Science, New York, 2002).

Cantley, L.C.: **The phosphoinositide 3-kinase pathway.** *Science* 2002, **296**:1655–1657.

Shaulian, E. and Karin, M.: **AP-1 as a regulator of cell life and death.** *Nat. Cell Biol.* 2002, **4**:E131–E136.

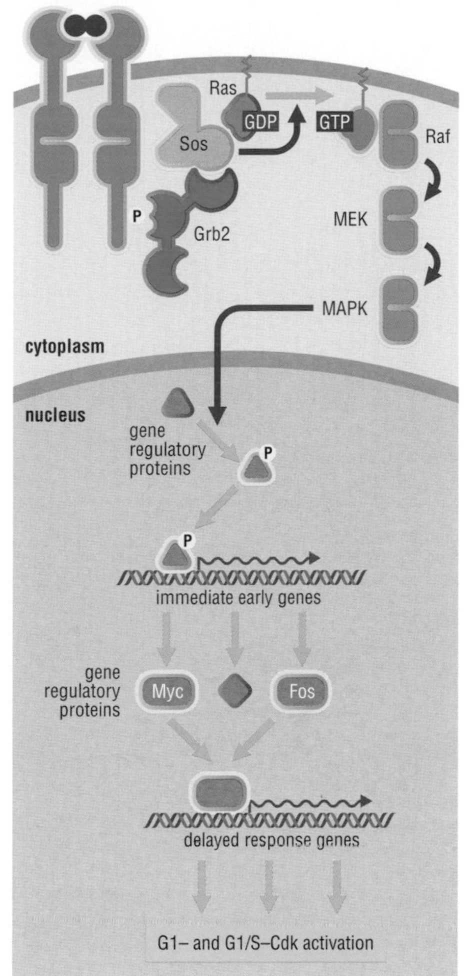

Figure 10-12 Mitogenic signaling by the Ras pathway Activated mitogen receptors bring the guanine-nucleotide exchange factor Sos to the membrane, resulting in the local activation of membrane-bound Ras. Ras recruits and activates the protein kinase Raf, thereby triggering the activation of a MAP kinase cascade. The final kinase in this cascade, MAP kinase, is transported to the nucleus, where it activates gene regulatory proteins that increase transcription of a series of immediate early genes. Some of these genes encode additional gene regulatory proteins (such as Myc and Fos) that activate more genes (including genes encoding G1 cyclins), eventually leading to the activation of G1– and G1/S–Cdks and G1/S gene expression.

Ras promotes cell division by a variety of downstream pathways, but the most important is the stimulation of a three-component MAP kinase cascade similar to that in the yeast mating pheromone pathway (see section 10-3). The first kinase in this cascade, Raf, is activated at the cell membrane by binding to activated Ras. Raf phosphorylates and activates the second kinase, MEK, which then activates the third kinase, MAP kinase. MAP kinase relays the mitogenic signal to the nucleus, where it probably phosphorylates multiple targets. These include gene regulatory proteins that are activated by the phosphorylation and induce the expression of a set of genes that are called the immediate early genes because their expression is the earliest transcriptional event after mitogen stimulation. One of these gene regulatory proteins is serum-response factor (SRF), which is directly phosphorylated by MAP kinase.

One of the most important immediate early genes encodes the transcription factor Fos. Increased production of Fos, in combination with various post-translational mechanisms, promotes the assembly and activation of a transcription factor complex called AP-1. AP-1 is a dimer of Fos (or a Fos-related protein) and a member of the Jun family of gene regulatory proteins. AP-1 triggers the expression of a second wave of genes called the delayed response genes (see Figure 10-12). These encode proteins such as the G1 cyclin, cyclin D1, providing a link to the activation of G1–Cdks and thus the stimulation of cell-cycle entry.

Another immediate early gene product is the protein Myc, whose levels in the cell rise after mitogenic stimulation and remain high throughout the cell cycle. Increased Myc concentrations result in part from increased *MYC* expression and in part from stabilization of Myc protein, which depends on its phosphorylation by mitogen-stimulated protein kinases.

Myc interacts with various related proteins to form gene regulatory complexes that increase the expression of a large number of target genes. Some of these genes encode cell-cycle regulatory molecules such as cyclin D2 and Cdk4, providing a mechanism whereby Myc may promote passage through Start. In addition, many Myc target genes encode regulators of cell growth and metabolism, and there is considerable evidence from many different cell types that Myc is a major promoter of cell growth—indeed, this may be its major function, as we discuss later in this chapter.

Activation of PI3 kinase helps promote mitogenesis

In addition to activating the Ras–MAP kinase pathway, activated mitogen receptors interact with another signaling protein complex to activate a second signaling pathway (Figure 10-13). One subunit of this complex, p85, binds the activated receptor through an SH2 domain. The other subunit, p110, contains an enzyme called phosphoinositide-3-kinase, or PI3 kinase. PI3 kinase catalyzes the phosphorylation of a hydroxyl at the 3-position on the inositol ring of various forms of phosphatidylinositol, a phospholipid in the cell membrane. During mitogenic signaling, for example, PI3 kinase catalyzes the conversion of phosphatidylinositol 4,5-bisphosphate (PIP_2) to phosphatidylinositol 3,4,5-trisphosphate (PIP_3). Thus, the recruitment of PI3 kinase to the cell membrane by activated receptors results in the local formation of PIP_3 in the membrane. This reaction may be enhanced by activated Ras, which also binds the p110 subunit of PI3 kinase and may stimulate its activity.

PIP_3 in the cell membrane then serves as a docking site for a number of signaling proteins that contain a protein domain called the pleckstrin-homology (PH) domain. One of these is the protein kinase Akt (also called PKB), whose activity is stimulated when it binds PIP_3. Akt then stimulates mitogenic pathways in various ways, some of which may involve the direct control of G1–Cdk activity. It also helps stimulate cell growth, and, in some cell types, promotes cell survival, as we discuss later in this chapter.

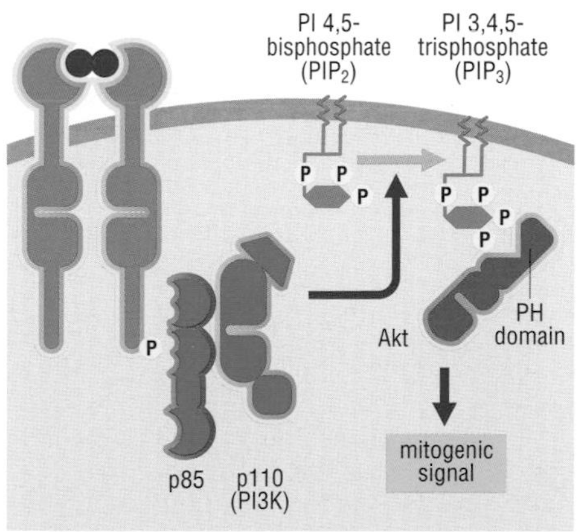

Figure 10-13 Mitogenic signaling by the PI3 kinase pathway Activated mitogen receptors interact with the two-subunit PI3 kinase complex, which phosphorylates phosphatidylinositol (PI) at the 3-position on the inositol ring. In most signaling pathways, the key function of PI3 kinase is to catalyze the conversion of PI 4,5-bisphosphate (PIP_2) to PI 3,4,5-trisphosphate (PIP_3) in the cell membrane. PIP_3 acts as a binding site for the PH domains of signaling proteins, such as the protein kinase Akt.

Mitogenic signaling pathways lead to activation of cyclin D–Cdk complexes

The ultimate function of mitogenic signaling pathways is to trigger activation of the Cdks that initiate the early events of the cell cycle. How are the pathways discussed in section 10-6 linked to the activation of these Cdks? In some cells at least, mitogenic signals directly activate the G1–Cdk, cyclin D–Cdk. As discussed earlier (see section 10-5), cyclin D–Cdk activity helps initiate E2F-dependent gene expression, cyclin E–Cdk2 activation and progression through Start.

Mitogens influence the activation of cyclin D–Cdk in several ways (Figure 10-14). To begin with, they stimulate expression of the cyclin D gene, which is one of the delayed response genes discussed earlier (see section 10-6). Cyclin D levels remain elevated throughout the cell cycle as long as mitogens are present, suggesting that cyclin D gene expression is responsive to mitogenic activity regardless of cell-cycle stage.

The stimulation of cyclin D production by mitogens requires activation of the Ras–MAP kinase pathway, which leads to the activation of at least two gene regulatory proteins that promote cyclin D gene expression. First, AP-1 (see section 10-6) directly promotes the expression of the gene for cyclin D1 in some cell types. Second, Myc triggers modest increases in the expression of the genes for cyclin D2 and Cdk4. The general importance of these mechanisms remains unclear, however, and a great deal remains to be learned about the regulation of cyclin D production.

The assembly of active cyclin D–Cdk complexes requires not only increased levels of cyclin D protein but also cofactors such as the Cip/Kip family proteins p27 and p21, which activate cyclin D–Cdk4 but inhibit other cyclin–Cdk complexes such as cyclin E–Cdk2 (see section 3-6). In addition, there is evidence that the assembly of cyclin D–Cdk4–p27 complexes requires an assembly factor—as yet unidentified—that is activated by Ras and the MAP kinase cascade (see Figure 10-14).

Mitogens control cyclin D–Cdk localization and destruction

The few cyclin D–Cdk complexes in quiescent cells are inhibited by the phosphorylation of cyclin D on Thr 286, which promotes the nuclear export and cytoplasmic destruction of the cyclin D protein (see Figure 10-14). This inhibitory phosphorylation is catalyzed by glycogen synthase kinase (GSK3β), which is highly active in quiescent cells. Mitogenic stimulation reduces GSK3β activity—by a mechanism that depends on the PI3 kinase–Akt pathway described earlier (see section 10-6). Cyclin D is dephosphorylated, thereby promoting the accumulation of stable cyclin D–Cdk complexes in the nucleus—where their major targets, the pRB proteins, are found.

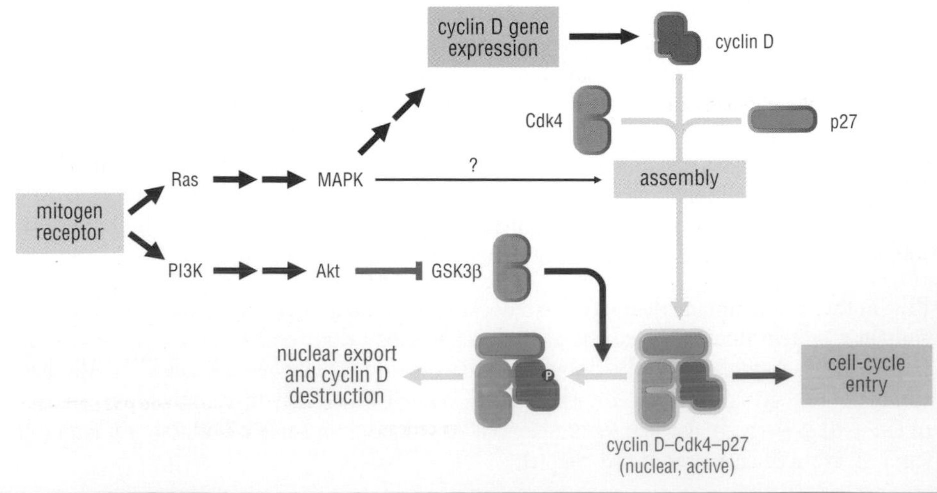

Figure 10-14 Mitogenic control of G1–Cdk activity Mitogens increase G1–Cdk activity by multiple mechanisms. Activation of the MAP kinase (MAPK) cascade results in increased cyclin D gene expression and also promotes the assembly of cyclin D–Cdk4–p27 complexes. Mitogens also act through PI3 kinase (PI3K) to inhibit the protein kinase GSK3β, which normally stimulates the nuclear export and destruction of cyclin D; its inhibition by mitogens therefore enhances cyclin D–Cdk activity in the nucleus.

GSK3β also phosphorylates and inhibits other important signaling components, including the transcription factors AP-1 and Myc. Mitogenic inhibition of GSK3β therefore helps promote the expression of genes that are targets of AP-1 and Myc.

Mitogens and anti-mitogens control the concentrations of Cdk inhibitor proteins

Cyclin D–Cdk function is suppressed in many cell types by members of the INK4 family of Cdk inhibitors (see section 3-6), which bind specifically to the cyclin D partners Cdk4 and Cdk6 (but not Cdk2 and Cdk1) and block cyclin binding. One INK4 protein in particular—the $p15^{INK4b}$ protein—seems to be an important inhibitor of cyclin D–Cdk activation. In some cell types, expression of the $p15^{INK4b}$ gene is stimulated during quiescence by the gene regulatory protein Miz1. When these cells are treated with mitogens, the Myc protein binds to Miz1 and inhibits its effects, thereby reducing $p15^{INK4b}$ levels and unleashing Cdk4 or Cdk6.

Cell proliferation in many tissues is governed not only by external mitogens but also by **anti-mitogens**, which act through various mechanisms to inhibit passage through Start. One such anti-mitogen is transforming growth factor β (TGF-β), which binds to specific cell-surface receptors to initiate anti-mitogenic signals. In many cell types TGF-β treatment triggers a marked increase in the intracellular concentration of $p15^{INK4b}$. The effects of TGF-β are mediated by gene regulatory proteins called the Smad proteins, which increase $p15^{INK4b}$ production in at least two different ways. First, Smad proteins collaborate with the repressor E2Fs, E2F4 and 5, to inhibit the production of Myc, thereby removing its inhibitory effects on Miz1. Second, Smad proteins interact positively with Miz1 and further enhance $p15^{INK4b}$ production (Figure 10-15).

In TGF-β-treated cells, the ability of $p15^{INK4b}$ to block the formation of cyclin D–Cdk complexes not only inhibits their kinase activity but also prevents these complexes from sequestering p27 and p21. These proteins are therefore free to act as Cdk inhibitors and inhibit cyclin E–Cdk2 and cyclin A–Cdk2 (see section 3-6). TGF-β also stimulates expression of the p21 gene, thereby promoting further inhibition of cyclin E–Cdk2 and cyclin A–Cdk2.

Despite the clear involvement of cyclin D–Cdk in cell proliferation control, mice lacking all three cyclin D proteins are capable of near-normal responses to mitogens in many cell types (see section 10-5). Mitogens must therefore act, at least in part, by cyclin D-independent mechanisms to influence the activation of cyclin E–Cdk2 and cyclin A–Cdk2. We will discuss some of these mechanisms in section 10-8.

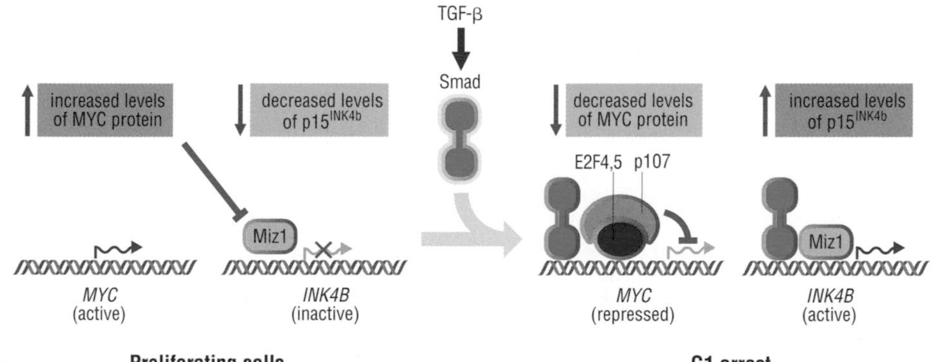

Figure 10-15 Anti-mitogenic actions of TGF-β In proliferating cells (left), high levels of Myc protein bind and inhibit the gene regulatory protein Miz1, thereby blocking activation of the gene encoding $p15^{INK4b}$. Treatment of cells with TGF-β stimulates the activation of Smad regulatory proteins, which interact with E2F4, 5 and p107 to repress expression of *MYC*. The decrease in Myc protein then releases Miz1, which together with Smads triggers activation of *INK4B*.

Definitions

anti-mitogen: extracellular molecule that inhibits cell proliferation.

References

Adhikary, S. and Eilers, M.: **Transcriptional regulation and transformation by Myc proteins.** *Nat. Rev. Mol. Cell Biol.* 2005, **6**:635–645.

Diehl, J.A. *et al.*: **Glycogen synthase kinase-3β regulates cyclin D1 proteolysis and subcellular localization.** *Genes Dev.* 1998, **12**:3499–3511.

Sherr, C.J. and McCormick, F.: **The RB and p53 pathways in cancer.** *Cancer Cell* 2002, **2**:103–112.

Sherr, C.J. and Roberts, J.M.: **CDK inhibitors: positive and negative regulators of G1-phase progression.** *Genes Dev.* 1999, **13**:1501–1512.

Shi, Y. and Massagué, J.: **Mechanisms of TGF-beta signaling from cell membrane to the nucleus.** *Cell* 2003, **113**:685–700.

10-8 Activation of G1/S- and S–Cdk Complexes in Animal Cells

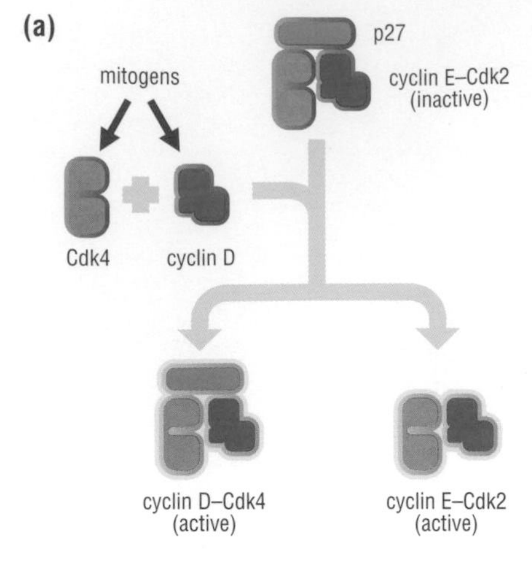

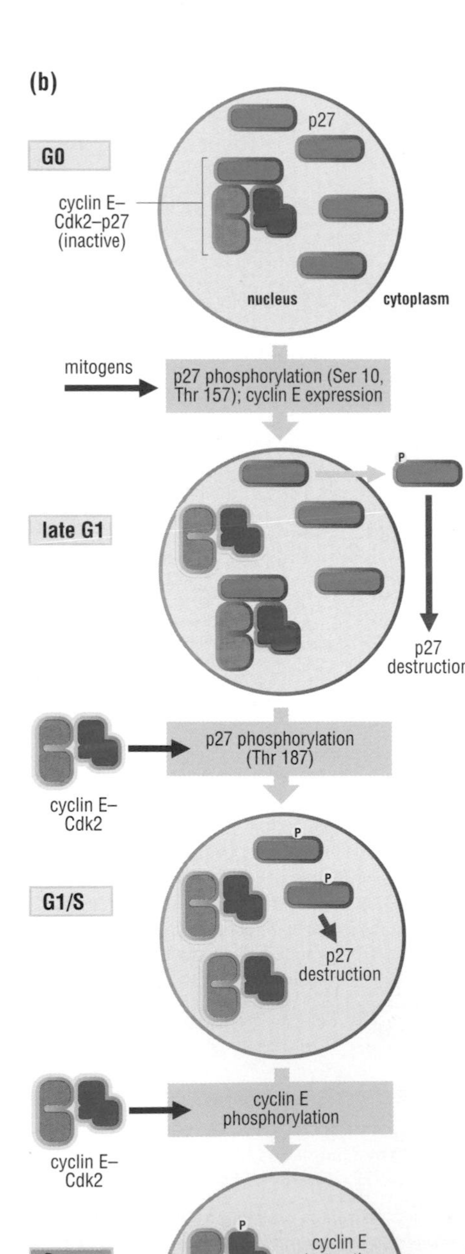

G1/S–Cdk activation at Start depends on removal of the inhibitor p27

As in yeast, Start in multicellular organisms is driven primarily by a G1/S–Cdk (cyclin E–Cdk2) that collaborates with an S–Cdk (cyclin A–Cdk2) to promote robust and irreversible commitment to cell-cycle entry and S phase. As described in section 10-5, activation of these Cdks depends in part on the ability of G1–Cdks to initiate E2F-dependent gene expression, which leads to enhanced production of G1/S and S cyclins. In this section we describe in more detail the molecular mechanisms by which mitogens promote G1/S– and S–Cdk activation at Start.

G1/S– and S–Cdk activity during entry into the cell cycle is governed in part by specific Cdk-inhibitory proteins: p27 in mammalian cells and Dacapo in *Drosophila* (see section 3-6). Like the yeast Sic1 protein described earlier (see section 10-2), these inhibitors help suppress the activity of these Cdks in G1. Unlike Sic1, however, they inhibit G1/S–Cdk activity as well as S–Cdk activity. Removal of these inhibitors is therefore important for both cyclin E–Cdk2 and cyclin A–Cdk2 activation.

After mitogenic stimulation of a quiescent cell, p27 is inactivated by a variety of mechanisms. First, the rising levels of cyclin D–Cdk complexes help inactivate p27—essentially by taking it out of circulation. As noted in section 10-7, Cdk inhibitors of the p27 (Cip/Kip) family do not inhibit cyclin D–Cdk complexes. On the contrary, these inhibitors are required for the productive formation of active cyclin D–Cdk complexes. Thus, when cyclin D levels rise after mitogenic stimulation, many p27 molecules become associated with cyclin D–Cdk complexes, which prevents them from binding and inhibiting cyclin E–Cdk2 (Figure 10-16a).

p27 is also inactivated during cell-cycle entry by proteolytic destruction, which is triggered by at least two mechanisms (Figure 10-16b). The first depends on a ubiquitin-protein ligase called KPC, which resides in the cytoplasm. In quiescent cells, p27 (like its Cdk targets) is located inside the nucleus. After mitogenic stimulation, protein kinases (including Akt) phosphorylate p27 at multiple sites (including Ser 10 and Thr 157 in human cells), thereby promoting p27 export from the nucleus to the cytoplasm. This leads to ubiquitination of p27 by KPC and its consequent destruction in the proteasome. As cells progress into late G1 and early S phase, a second proteolytic mechanism is brought into play: p27 is phosphorylated at a different residue (Thr 187) by rising cyclin E–Cdk2 and cyclin A–Cdk2 activities. This generates a high-affinity binding site for the nuclear ubiquitin-protein ligase SCF^{Skp2} (Figure 10-17), which ubiquitinates p27 and thereby triggers its destruction by proteasomes in the nucleus.

Given the importance of p27 in the activation of cyclin D–Cdk complexes, it remains unclear how these complexes can remain active if p27 is destroyed upon cell-cycle entry. Although this issue remains unresolved, one possibility is that a fraction of p27 remains associated with cyclin

Figure 10-16 Regulation of p27 during cell-cycle entry The Cdk inhibitor p27 is controlled by two mechanisms, which are diagrammed separately here for clarity. **(a)** After mitogenic stimulation, increasing numbers of cyclin D–Cdk4 complexes associate with p27, thereby reducing the amount of p27 available for binding to cyclin E–Cdk2. **(b)** Mitogenic stimulation promotes p27 phosphorylation at multiple sites, resulting in its export from the nucleus, ubiquitination in the cytoplasm and destruction by cytoplasmic proteasomes. In late G1, cyclin E–Cdk2 phosphorylates p27 at an additional site (Thr 187), triggering its ubiquitination by SCF and destruction by nuclear proteasomes. Eventually, in early S phase, cyclin E–Cdk2 and cyclin A–Cdk2 trigger the destruction of cyclin E, in part through the phosphorylation of cyclin E on Thr 380 and other sites.

References

Hao, B. *et al.*: **Structural basis of the Cks1-dependent recognition of $p27^{Kip1}$ by the SCF^{Skp2} ubiquitin ligase.** *Mol. Cell* 2005, **20**:9–19.

Hsu, J.Y. *et al.*: **E2F-dependent accumulation of hEmi1 regulates S phase entry by inhibiting APC^{Cdh1}.** *Nat. Cell Biol.* 2002, **4**:358–366.

Hwang, H.C. and Clurman, B.E.: **Cyclin E in normal and neoplastic cell cycles.** *Oncogene* 2005, **24**:2776–2786.

Kamura, T. *et al.*: **Cytoplasmic ubiquitin ligase KPC regulates proteolysis of $p27^{Kip1}$ at G1 phase.** *Nat. Cell Biol.* 2004, **6**:1229–1235.

Sherr, C.J. and Roberts, J.M.: **CDK inhibitors: positive and negative regulators of G_1-phase progression.** *Genes Dev.* 1999, **13**:1501–1512.

Welcker, M. *et al.*: **Multisite phosphorylation by Cdk2 and GSK3 controls cyclin E degradation.** *Mol. Cell* 2003, **12**:381-392.

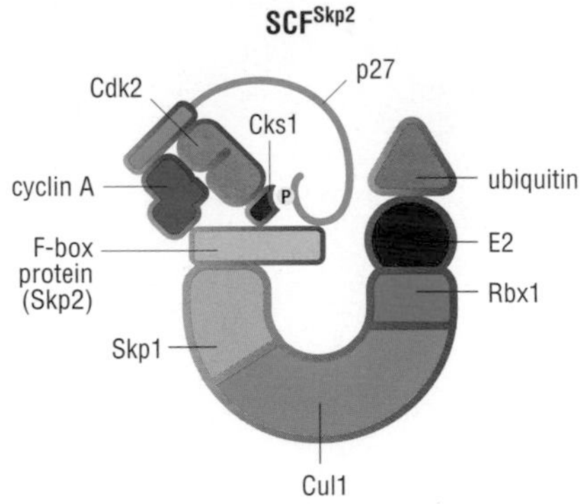

Figure 10-17 Ubiquitination of phosphorylated p27 by SCF^{Skp2} This diagram of the interaction of phosphorylated p27 and SCF^{Skp2} is based on several different structures derived from X-ray crystallography, including two structures shown in previous figures (see Figures 3-16 and 3-24). In the ubiquitin-protein ligase SCF^{Skp2}, the F-box protein that provides substrate specificity is Skp2. Another SCF subunit, Rbx1, associates with the E2 enzyme that provides the ubiquitin for conjugation to the target. p27 contains two major segments: an amino-terminal half that binds and inhibits cyclin A–Cdk2 as shown in previous figures (see Figure 3-16), plus an unstructured carboxy-terminal loop that is not involved in Cdk inhibition and contains the phosphorylated threonine (Thr 187) that is required for p27 destruction. The binding of SCF^{Skp2} to phosphorylated p27 depends on a small protein called Cks1, which holds the SCF–p27 complex together by interacting with three different components: Cdk2, Skp2 and the region of p27 that contains phosphorylated Thr 187. Interestingly, Cks1 is best known as a protein that interacts with Cdks and contributes to Cdk substrate targeting (see section 3-5). SCF–p27 complex formation is also promoted by interactions between p27 and Skp2 and between cyclin A and Skp2. According to this model, SCF^{Skp2} presents a loop in p27 for ubiquitination by the E2–ubiquitin conjugate.

D–Cdk complexes and is resistant to destruction. In addition, cyclin D–Cdk activation in cycling cells may depend on the other major Cip/Kip protein, p21 (see section 3-6), whose levels generally increase during cell-cycle entry.

p27 is thus an important contributor to the regulation of Cdks at Start, both as a stimulus for cyclin D–Cdk assembly and as an inhibitor of cyclin E–Cdk2 and cyclin A–Cdk2. It is not an essential regulator, however. Mice lacking p27 (or mice lacking both p27 and p21) have severe defects in cyclin D–Cdk activity and high cyclin E–Cdk2 activity but do not display major problems in cell-cycle progression or development, apart from increases in the number of cells in many tissues. In *Drosophila*, mutation of the p27-related inhibitor Dacapo has greater effects, resulting in defects in the ability of many cell types of the embryo to exit from the cycle into a quiescent state.

Early in S phase, cyclin E is destroyed by an SCF-dependent mechanism (see Figure 10-16b). Phosphorylation of cyclin E at multiple sites (primarily Thr 380) targets the protein to the F-box protein Cdc4 to allow its ubiquitination by SCF. Cyclin E is thought to be phosphorylated in late G1 by cyclin E–Cdk2 complexes, raising the question of how cyclin E can accumulate in the first place if it promotes its own destruction. As in other oscillating systems (see section 3-11), this system must include additional features—which may depend on other phosphorylation sites in cyclin E—that delay cyclin E destruction until after its levels have risen to those needed to carry out its important functions at Start.

Activation of cyclin E–Cdk2 at Start also depends on changes in Cdk2 phosphorylation. Members of the Wee1 family of protein kinases phosphorylate inhibitory sites in the active site of Cdk2, and activation of Cdk2 in late G1 requires the removal of inhibitory phosphorylation by the phosphatase Cdc25A (see section 3-3). We will see in Chapter 11 that Cdc25A has important functions in the control of cell-cycle progression after DNA damage.

Cyclin A–Cdk2 activation is promoted in part by APC inhibition

In most mammalian cells, activation of cyclin A–Cdk2 is required for the initiation of DNA replication (see section 4-6). As in yeast, activation of this S–Cdk complex depends, at least in part, on G1/S–Cdk activity. Much remains to be learned, however, about the mechanisms that control the timing of cyclin A–Cdk2 activation.

The formation of cyclin A–Cdk2 is initiated in late G1, when G1–Cdks and G1/S–Cdks drive the E2F-dependent expression of the cyclin A gene. Like cyclin E–Cdk2, cyclin A–Cdk2 complexes are partly inhibited by p27, which must therefore be removed to allow their complete activation. Cyclin A–Cdk2 activation also depends on dephosphorylation by Cdc25A.

Cyclin A concentration in the cell is determined in part by its degradation rate (Figure 10-18). During G1, APC^{Cdh1} targets cyclin A for destruction. The accumulation of active cyclin A–Cdk2 complexes in S phase therefore requires inactivation of APC^{Cdh1}, which occurs in at least two ways. First, as in yeast, G1/S– and S–Cdks collaborate to phosphorylate Cdh1 and thereby block its function. The ability of cyclin A–Cdk2 to promote cyclin A stabilization provides a potential positive feedback loop. Second, APC^{Cdh1} is shut off in late G1 by an E2F-dependent increase in the production of Emi1, a protein that binds and inhibits Cdh1 (Emi1 also contributes to the timing of APC activation in mitosis, as described in section 7-1).

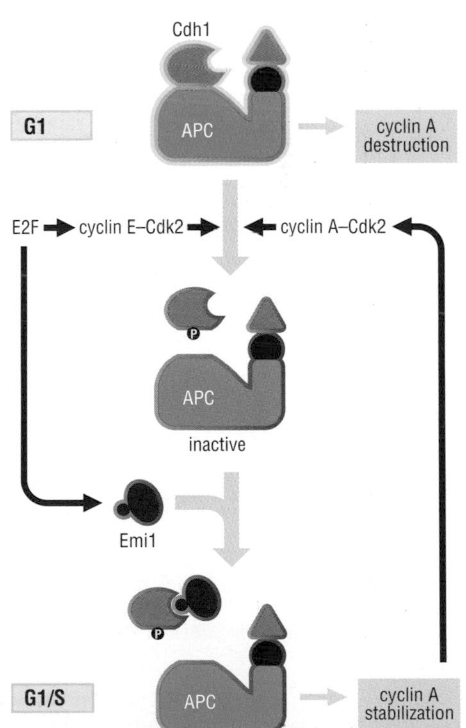

Figure 10-18 Control of cyclin A stability during cell-cycle entry Cyclin A is targeted for destruction by APC^{Cdh1} in G1. Mitogens promote inactivation of Cdh1 by stimulating its phosphorylation and by triggering increased production of Emi1, which binds and inhibits Cdh1.

10-9 Developmental Control of Cell Proliferation

Developmental signals limit cell division to specific embryonic regions

During embryogenesis, developmental signals establish the basic body plan and direct the patterns of cell movement and differentiation that form the tissues. These signals also govern rates of cell division, ensuring that developing tissues are provided with new cells at the correct time and place.

The developmental control of cell proliferation is best understood in *Drosophila*, and this will be our focus here. The regulation of cell proliferation in the fly embryo involves several mechanisms that are distinct from the yeast and mammalian examples described earlier in this chapter. First, extracellular mitogens are often not involved; instead, the control of cell number depends primarily on intracellular, tissue-specific gene regulatory proteins that stimulate the proliferation of specific cells at the appropriate time. Second, the rate of cell division in the *Drosophila* embryo is sometimes controlled at unusual cell-cycle stages. In some cell populations, proliferation is governed at the entry to the cell cycle, where the concentration of the G1/S–cyclin, cyclin E, is the key rate-limiting factor. In others, cell-cycle progression is controlled at the entry to mitosis—through regulation of the phosphatase Cdc25 (called String in *Drosophila*), which dephosphorylates and thereby activates mitotic cyclin B–Cdk1 (see section 3-3).

Embryonic divisions are limited by depletion of key cell-cycle regulators

The *Drosophila* zygote is a giant cell containing stockpiles of nutrients and other materials provided by the mother (see section 2-4). The first 13 cell cycles are rapid and synchronous and occur in the absence of growth, cytokinesis or detectable gap phases. This results in a syncytium containing large numbers of nuclei in a shared cytoplasm. These nuclei migrate to the cell surface and, after the 13th mitosis, are packaged into individual cells. At this stage, about 2.5 hours after fertilization, the 6,000 cells of the embryo lie in a single superficial layer, the cellular blastoderm.

During the first 13 cell cycles, all cell-cycle regulatory components are synthesized from maternal mRNAs, and there is no need for embryonic cells to transcribe any of their own genes. After mitosis 13, however, several essential maternal mRNAs, including that for Cdc25, are abruptly degraded, and continued development depends on zygotic transcription. Maternal Cdc25 protein declines dramatically after cycle 13. As Cdc25 is essential for mitotic Cdk1 activation, its disappearance results in an arrest in G2 throughout the embryo (Figure 10-19).

Entry into mitosis 14 and all subsequent mitoses depends on synthesis of new Cdc25 in the embryonic cells. Expression of *Drosophila cdc25* is controlled by a promoter region that includes binding sites for a wide variety of tissue-specific gene regulatory proteins. This complex

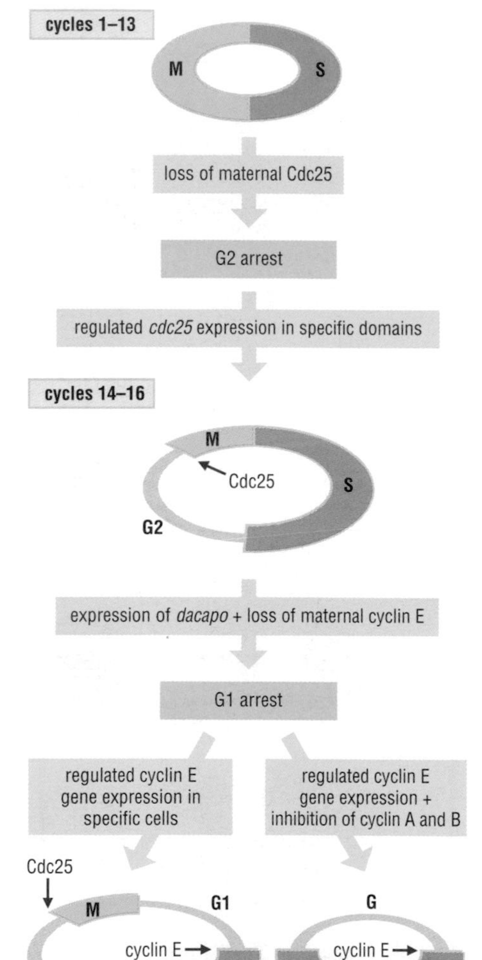

Figure 10-19 Variations in cell-cycle structure in the *Drosophila* embryo The rapid cell cycles of the early embryo are supported by abundant maternal supplies of mRNAs encoding key cell-cycle regulators. By the end of cycle 13, maternal mRNA encoding Cdc25 is exhausted, leading to a G2 arrest. Further cell-cycle progression requires zygotic production of Cdc25, which occurs in specific regions at specific times. Loss of maternal mRNA encoding cyclin E, coupled with synthesis of the Cdk inhibitor Dacapo, leads to a G1 arrest in most cell types after cycle 16. A few embryonic cell types, such as neuroblasts, initiate cell cycles that depend on newly produced cyclin E. Other embryonic cells, such as the larval precursor cells, initiate pulsatile cyclin E production while suppressing mitotic cyclin synthesis, thereby triggering endoreduplication cycles that increase cell ploidy. The imaginal cells that will form adult structures remain arrested in G1 until the larva hatches, after which they begin to proliferate by conventional four-phase cell cycles that are coordinated with cell growth.

References

Edgar, B.A. and O'Farrell, P.H.: **The three postblastoderm cell cycles of *Drosophila* embryogenesis are regulated in G2 by string.** *Cell* 1990, **62**:469–480.

Edgar, B.A. *et al.*: **Transcriptional regulation of string (cdc25): a link between developmental programming and the cell cycle.** *Development* 1994, **120**:3131–3143.

Jones, L. *et al.*: **Tissue-specific regulation of *cyclin E* transcription during *Drosophila melanogaster* embryogenesis.** *Development* 2000, **127**:4619–4630.

Knoblich, J.A. *et al.*: **Cyclin E controls S phase progression and its down-regulation during *Drosophila* embryogenesis is required for the arrest of cell proliferation.** *Cell* 1994, **77**:107–120.

Lee, L.A. and Orr-Weaver, T.L.: **Regulation of cell cycles in *Drosophila* development: intrinsic and extrinsic cues.** *Annu. Rev. Genet.* 2003, **37**:545–578.

Lehman, D.A. *et al.*: ***Cis*-regulatory elements of the mitotic regulator, string/Cdc25.** *Development* 1999, **126**:1793–1803.

Liu, T.H. *et al.*: **Transcription of the *Drosophila* CKI gene *dacapo* is regulated by a modular array of *cis*-regulatory sequences.** *Mech. Dev.* 2002, **112**:25–36.

Meyer, C.A. *et al.*: ***Drosophila* p27^{Dacapo} expression during embryogenesis is controlled by a complex regulatory region independent of cell cycle progression.** *Development* 2002, **129**:319–328.

promoter enables the timing of Cdc25 synthesis, and thus the rate of entry into mitosis, to vary in different regions of the embryo (Figure 10-20). *cdc25* expression in cycle 14 occurs in distinct pulses, so that Cdc25 protein levels rise abruptly in late G2 and decline after mitosis. Because the key G1/S regulator, cyclin E, continues to be produced in constant large quantities from maternal mRNA during cycles 14 to 16, entry into S phase is not restricted and cells progress directly to S phase after mitosis. These cycles are therefore S–G2–M cycles in which the rate of proliferation is controlled by the timing of *cdc25* expression at the G2/M transition (see Figure 10-19). Cycles 14–16 take about 5–6 hours, so the first 16 cell cycles in the developing fly occupy a total of about 8–9 hours out of the 24 hours it takes to complete embryogenesis (see Figure 2-13).

After cycle 16, changes in several regulatory components trigger a prolonged G1 arrest in many cell types. Most importantly, maternal mRNA encoding cyclin E is degraded, and maternal cyclin E protein levels decline. In addition, the Cdk inhibitor protein Dacapo (see section 3-6) is first synthesized at this time and helps suppress cyclin E–Cdk activity. Cell proliferation beyond cycle 16 depends on new cyclin E synthesis by the embryonic cells. The promoter of *CycE*, like that of *cdc25*, is responsive to a variety of tissue-specific gene regulatory proteins, thereby allowing pulses of *CycE* expression, and thus entry into the cell cycle, to be triggered in a precise spatiotemporal pattern.

For many cells, the division of cycle 16 is the last embryonic cell division, and many of the cellular rearrangements of gastrulation occur in the absence of further division. A small number of cell types, however, continue to divide using newly synthesized cyclin E (see Figure 10-19). Cells of the developing nervous system, for example, contain gene regulatory proteins that trigger *CycE* expression, allowing their continued proliferation during late embryogenesis.

Many cells in the embryo are destined to form larval tissues (see section 2-4). After cycle 16, these cells enter endoreduplication cycles (see section 1-2) in which repeated S phases occur in the absence of M phases (see Figure 10-19). These cycles require the pulsatile expression of *CycE* (see section 4-6), which depends on the interaction of tissue-specific gene regulatory proteins, as well as E2F, with the *CycE* promoter. Endoreduplication also requires the suppression of mitotic cyclin synthesis. In most larval precursor cells, repeated endoreduplication generates marked increases in cell ploidy, which helps support the rapid growth of these cells after the larva hatches.

The imaginal cells in the embryo, which will eventually produce the tissues of the adult fly (section 2-4), remain arrested in G1 after cycle 16 and do not begin to proliferate until after the larva hatches and begins to feed, triggering mitogenic signals. These cells proliferate by a four-phase cell cycle that is governed in late G1 by the transient expression of *CycE* under the control of E2F and other factors (see Figure 10-19). Imaginal cell-cycle progression can also be controlled in G2 by changes in the production of Cdc25. Unlike the embryonic divisions, imaginal cell divisions in the larva are accompanied by cell growth, and in most cases growth and division are coordinated to maintain a roughly constant cell size. Each imaginal cell type is present in small numbers in the embryo (typically 10–50 cells). In the larva they grow and proliferate, typically forming sheets of thousands of cells called imaginal discs, which are the precursors of adult structures such as eyes and wings.

The coordination of growth and division in imaginal cell populations depends on complex interactions between the regulatory pathways driving cell division and those driving cell growth. We discuss this important and poorly understood problem in the next sections of this chapter.

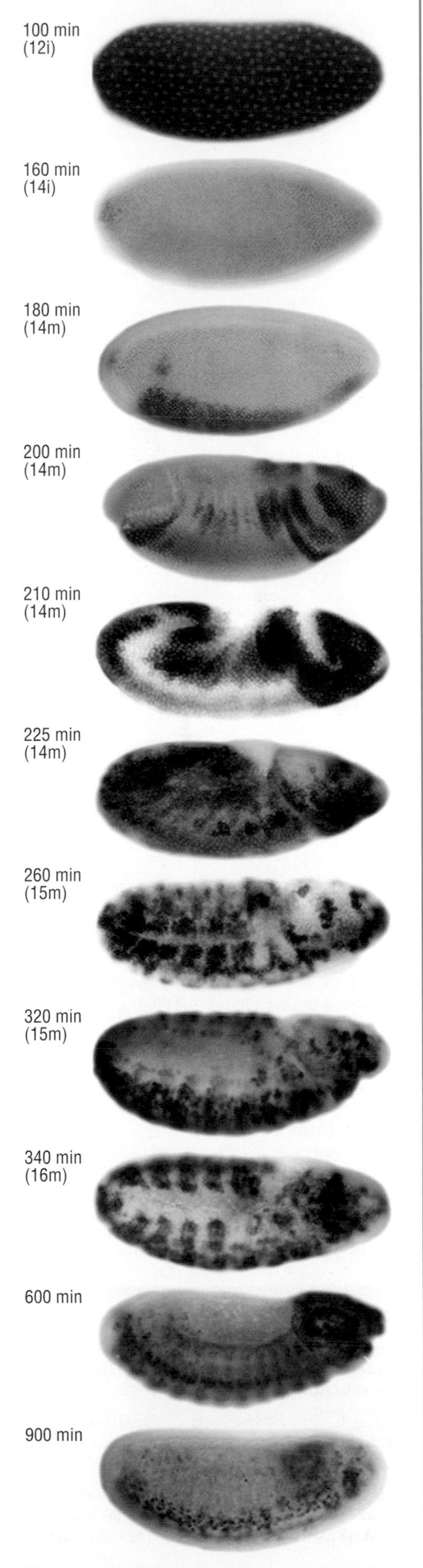

Figure 10-20 Patterns of *cdc25 (string)* expression in the fly embryo These images of the *Drosophila* embryo indicate the level of *cdc25* mRNA (dark color) at the indicated times after fertilization. The cell-cycle number and stage are indicated in parentheses (i, interphase; m, mitosis). In the early syncytial embryo (top image, cycle 12 interphase), maternal *cdc25* mRNA is abundant throughout the embryo. After cycle 13, however, maternal *cdc25* mRNA is degraded (second image, cycle 14 interphase), and progression through mitoses 14 to 16 and beyond depends on bursts of zygotic *cdc25* mRNA production, which occur in specific spatiotemporal patterns that generate the appropriate cell numbers in different cell populations. The bottom two images show embryos at later stages, when most cells are arrested in G1 of cycle 17 but some cells, primarily neural cells, express *cdc25* and divide. Some G1 cells in these later stages also retain residual levels of *cdc25* mRNA from the previous mitosis. Photographs kindly provided by Bruce Edgar. From Edgar, B.A. *et al.*: *Development* 1994, **120**:3131–3143.

10-10 Overview: Coordination of Cell Division and Cell Growth

Cell division and cell growth are separate processes

In many animals, such as the fly and the frog, the fertilized egg is a large, well-stocked cell that is subdivided rapidly during early embryogenesis into thousands of cells. Cell growth does not occur in these divisions, and so average cell size decreases while the size of the embryo remains constant. When the larva begins to feed, the influx of nutrients triggers cell and organism growth; that is, an increase in the net amount of proteins and other cellular components. The cells of some tissues—most larval tissues of the fly, for example—then grow without dividing, resulting in an increase in cell size. In other tissues, such as the fly imaginal discs, cell growth is accompanied by cell division, and the relative rates of the two processes determine the size of the cells in each tissue. In those tissues in which cell size remains constant, division and growth are coordinated to ensure that cell size doubles with each division. Cell growth and cell division are therefore distinct processes that are often, but not always, coordinated.

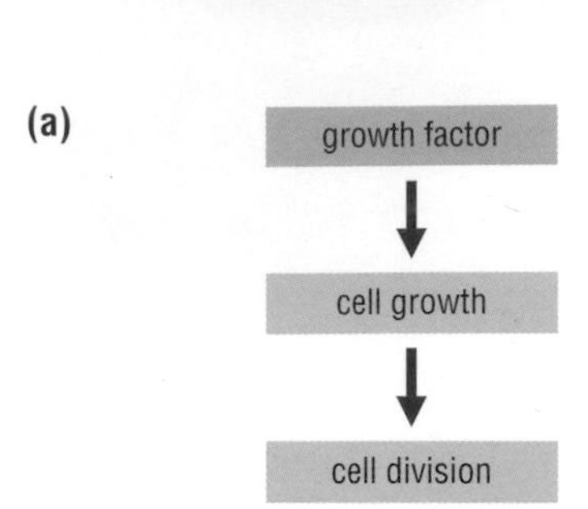

Cell growth is regulated by extracellular nutrients and growth factors

Many eukaryotic cells monitor the levels of various nutrients in the environment and adjust their rates of growth and metabolism accordingly. If the concentration of important nutrients decreases, cells conserve precious resources by decreasing the rates of various synthetic processes—thereby reducing the rate of cell growth. These mechanisms are particularly important in unicellular eukaryotes such as yeast, which must be able to survive in the face of drastic changes in nutrient levels in their natural environment.

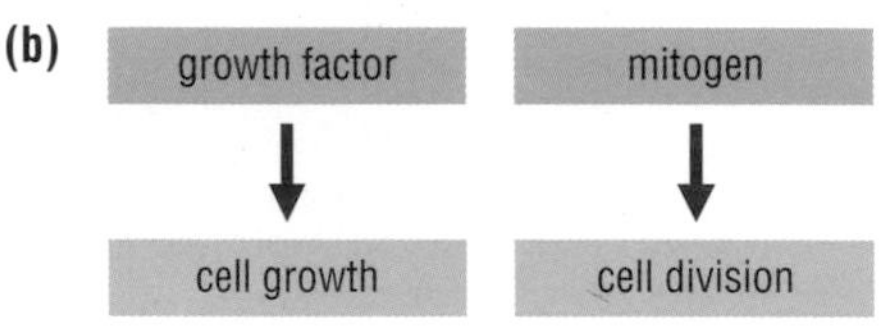

In multicellular organisms, cell growth—like cell division—occurs only when necessary for the benefit of the organism. Nutrient concentrations tend to remain constant in the intercellular environment, and cell growth is instead controlled by a combination of internal, cell-type-specific genetic programming and extracellular proteins, called **growth factors**, produced by other cells.

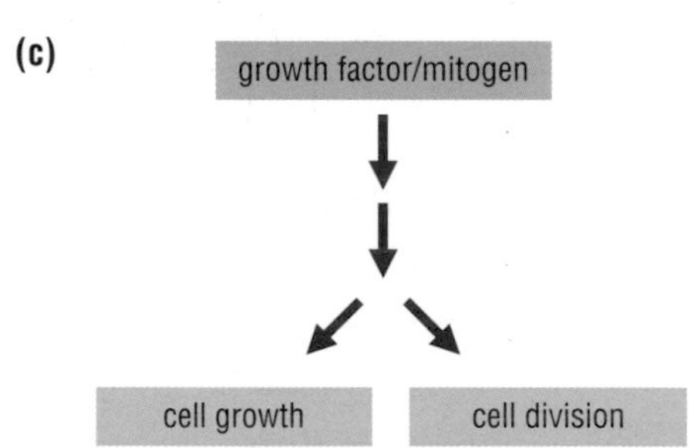

Figure 10-21 Three general mechanisms for coordinating cell growth and division (a) The rate of cell division depends on cellular growth rate, thereby ensuring that division occurs only when growth is sufficient. **(b)** In some animal cells, growth and division are controlled by separate growth factors and mitogens, respectively, acting through independent signaling pathways that stimulate constant rates of growth and division. **(c)** A single extracellular factor can stimulate both growth and division by stimulating a signaling pathway that branches to promote both processes. These three mechanisms are not mutually exclusive, and many animal cell types are likely to use some combination of all three.

Cell growth and division are coordinated by multiple mechanisms

How are cell growth and division coordinated in proliferating cell populations, thereby ensuring that cell size doubles in each division and average cell size remains constant? There is evidence for at least three major mechanisms, each of differing importance in different cell types (Figure 10-21). It is likely that most cells employ a combination of these mechanisms.

Coordination of growth and division is achieved in many cell types by making cell division depend on cell growth; that is, the cell cycle is allowed to progress only when cell growth rate is sufficient to support the doubling of cell size during each cell cycle (Figure 10-21a). In yeast and some animal cells, experimental inhibition of cell growth—by removal of nutrients or growth factors, for example—inhibits cell-cycle progression, usually by causing a delay before Start in G1. The dependence of division on growth results from links between the cell's metabolic machinery and the cell-cycle control system. In budding yeast, for example, changes in the overall rate of protein synthesis seem to influence the activities of G1– and G1/S–Cdks. These effects are rapid: a sudden drop in growth rate (caused by a decline in external nutrients, for example) leads to an immediate cell-cycle delay, ensuring that normal cell size is maintained even in the face of rapidly changing environmental conditions.

In some mammalian cells, division does not seem to depend on growth; instead, growth factors and mitogens act through independent, parallel signaling pathways to promote balanced rates

Definitions

growth factor: extracellular factor that stimulates cell growth (an increase in cell mass). This term is sometimes used incorrectly to describe a factor that stimulates cell proliferation, for which the correct term is mitogen.

References

Brooks, R.F.: **Variability in the cell cycle and the control of cell proliferation** in *The Cell Cycle*. John, P.C.L. ed. (Cambridge University Press, Cambridge, 1981), 35–61.

Conlon, I. and Raff, M.: **Differences in the way a mammalian cell and yeast cells coordinate cell growth and cell-cycle progression.** *J. Biol.* 2003, **2**:7.

Conlon, I. and Raff, M.: **Size control in animal development.** *Cell* 1999, **96**:235–244.

Hall, M.N. *et al.* (eds): *Cell Growth: Control of Cell Size* (Cold Spring Harbor Laboratory Press, Cold Spring Harbor, 2004).

Jorgensen, P. and Tyers, M.: **How cells coordinate growth and division.** *Curr. Biol.* 2004, **14**:R1014–R1027.

Mitchison, J.M.: **Growth during the cell cycle.** *Int. Rev. Cytol.* 2003, **226**:165–258.

Rupeš, I.: **Checking cell size in yeast.** *Trends Genet.* 2002, **18**:479–485.

of growth and division (Figure 10-21b). In other words, growth and division are correlated but not coupled to each other in these cells. In this case, the maintenance of cell size in a proliferating population depends on the maintenance of constant concentrations of the extracellular growth factor and mitogen.

This strategy of correlating growth and division can work only if cell growth rate is constant regardless of cell size: that is, big and small cells must add an equal amount of mass per unit time. We can illustrate this concept by considering what happens when cell division accidentally results in daughter cells of unequal size (Figure 10-22). If the rates of division and growth are constant, then random variations in daughter cell size will be corrected—over a period of several cell generations—so that the offspring of abnormally sized daughter cells eventually return to the mean size in the population, even if there is no communication between growth rate and the cell cycle. Studies of some mammalian cell types indicate that growth rate does not vary significantly with cell size. Thus, in these cells at least, a constant cell size can be achieved simply by using constant extracellular growth factor and mitogen concentrations to drive constant rates of growth and division, respectively.

Finally, in many animal cell types there is a third mechanism for coordinating cell growth and division (Figure 10-21c). In these cases, a single extracellular protein acts both as a growth factor and as a mitogen by triggering intracellular signaling pathways that stimulate both growth and division. As described earlier in this chapter, mitogens often stimulate cell division by activating the small GTPase Ras, the transcription factor Myc, and the kinase Akt. In many cell types these proteins can also stimulate cell growth. Thus, activation of these signaling molecules by a single extracellular protein triggers a coordinated increase in the rates of both growth and division.

The size of a cell depends on its genomic content

One of the most mysterious features of cell-size control is that, within a given species, the size of a cell depends on its chromosome content. In a wide range of eukaryotes, haploid cells are smaller than their diploid counterparts, and experimental increases in ploidy generally result in increases in cell size. The polyploid cells that result from endoreduplication (see section 1-2) are generally very large. This relationship between nuclear DNA content and cell size has been well established for many decades, but we do not have a good mechanism to explain it. One possibility is that chromosome content determines cell growth rate, so that cells with more chromosomes grow faster relative to cell division and therefore become larger. The solution to this problem is likely to be found as we develop a better understanding of growth regulation and the connections between growth and the cell cycle.

In the following sections of this chapter we provide a basic overview of the coordination of growth and division in yeast and animal cells. We begin with a description of the mechanisms by which external nutrients and growth factors stimulate cell growth.

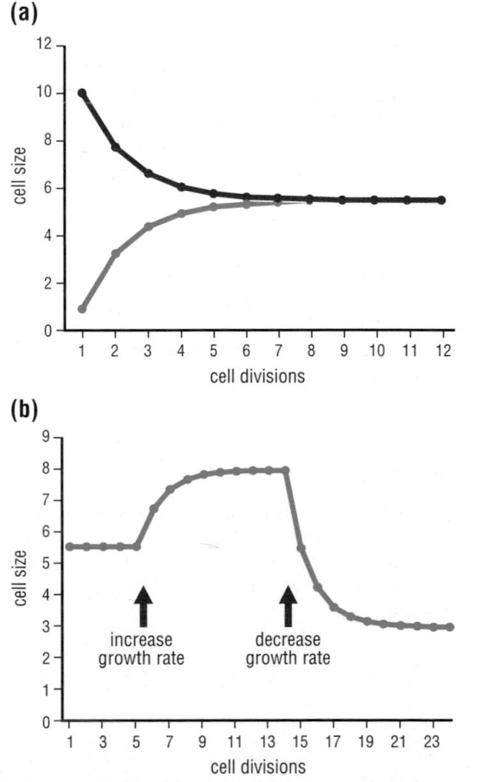

Figure 10-22 The maintenance of cell size at constant rates of growth and division These diagrams reveal that constant cell size can be maintained by separate growth factors and mitogens, as in Figure 10-21b, if cell growth rate does not vary with cell size. **(a)** Consider a theoretical cell population in which average cell size at birth is 5.5 mass units. The growth rate of all cells in this population is constant, so that each cell grows 5.5 mass units per cell cycle—resulting in a mass of 11 units at division. Now imagine that a cell in this population undergoes an unequal division, resulting in one daughter cell (blue curve) of 10 mass units and another (green curve) of 1 mass unit. If the growth rate for both daughters remains constant at 5.5 mass units per cell cycle, then the offspring of these asymmetric daughter cells will—over a period of several cell generations—return to the normal average birth size of 5.5 units. For this to be true, cells of all sizes must have the same growth rate of 5.5 units per cell cycle. If larger cells grow faster than small cells (as in yeast), then unequal daughter cells will never return to the same size, and some kind of coupling mechanism between growth and division (Figure 10-21a) is required to maintain constant cell size. **(b)** Consider a cell growing at the rate of 5.5 mass units per cell cycle as in panel (a). Now imagine that this cell is treated at division 5 with an extracellular growth factor that increases growth rate to 8 units per cell cycle. The cell will gradually reach a larger constant cell size of 8 units. The same result could be achieved if the cell were exposed to a reduced concentration of mitogen, resulting in an increase in cell-cycle time (not shown). If the concentration of growth factor is sharply reduced at division 14 (reducing growth rate to 3 mass units per cell cycle in this example), cell size will decline to a lower set point. Thus, constant cell size can be achieved at various rates of growth and division. Adapted from Conlon, I. and Raff, M.: *J. Biol.* 2003, **2**:7 and Brooks, R.F.: In: *The Cell Cycle* John, P.C.L. ed. (Cambridge University Press, Cambridge, 1981), 35–61.

Cell growth rate is determined primarily by the rate of protein synthesis

Many eukaryotic cells are able to adjust their rate of growth in response to changes in various regulatory factors, including the concentration of nutrients or growth factors in the environment. Given that the growth of a cell depends on a remarkably complex array of metabolic processes, it is not surprising that growth-promoting factors have a wide range of effects inside the cell—from increasing the uptake of raw materials to stimulating the incorporation of these materials into proteins, membranes and other macromolecules. Because most of the cell's dry mass is protein, the major determinant of cell growth is the rate of protein synthesis, and for this reason the protein synthetic machinery—particularly the ribosome—is a key destination in all growth-regulatory pathways. In this section we briefly review the major mechanisms of cell-growth regulation, with an emphasis on the control of protein synthesis.

Figure 10-23 Growth control by insulin-like growth factors Insulin-like growth factors (IGFs), including the insulin-related peptides of *Drosophila*, activate cell-surface receptors. Unlike the conventional mitogen receptors described earlier in this chapter (Figure 10-11), the IGF receptor is a dimer in the absence of its ligand, and IGF binding to two separate sites triggers a conformational change that causes activation and phosphorylation of the internal kinase domains. This leads to binding of a phosphotyrosine-binding protein called IRS, which recruits the two-subunit PI3 kinase (see Figure 10-13). PI3 kinase produces PIP_3 from PIP_2 in the membrane, and the increased PIP_3 recruits the protein kinases PDK1 and Akt. Akt is thought to inhibit a GTPase-activating protein complex called Tsc1–Tsc2, resulting in increased GTP binding by the small GTPase Rheb. By mechanisms that remain unclear, Rheb–GTP stimulates the nutrient-sensitive protein kinase TOR, which stimulates protein synthesis and other metabolic processes. Two well established targets of TOR are the protein kinase S6K and the eukaryotic initiation factor binding protein 4E-BP. Other targets are thought to mediate the effects of TOR on ribosome synthesis and other processes. S6K activation also depends directly on PDK1, but it remains unclear whether the actions of PDK1 depend simply on PI3 kinase as shown here: in *Drosophila*, analysis of various mutants indicates that S6K activation requires PDK1 and TOR, but not PI3 kinase or Akt.

Extracellular nutrients and growth factors stimulate cell growth by activating the protein kinase TOR

Cells monitor the levels of the many types of nutrients in the environment and adjust their metabolism accordingly, leading to changes in growth rate. Nutrient-sensing systems are particularly important in controlling the growth of unicellular organisms such as yeast, which experience marked changes in external nutrient concentrations. Acute nutrient responses are less necessary for cells in multicellular animals, in which extracellular nutrient concentrations tend to be kept constant by physiological mechanisms. In this case, growth is governed primarily by extracellular growth factors.

Central to the control of growth in response to both nutrients and growth factors is a protein kinase called TOR, which increases the rate of protein synthesis and thus promotes growth (Figure 10-23). In budding yeast, TOR is activated (by unknown mechanisms) in response to increasing levels of nutrients and is required for the stimulation of growth that occurs under high-nutrient conditions. In *Drosophila* and mammals, TOR is similarly activated in response to nutrients but can also be activated by growth factors, and loss of TOR function results in decreased cell growth and cell size.

Several other protein kinases are sensitive to nutrient levels and help regulate cell metabolism and growth. In budding yeast, the most important of these is the cAMP-dependent protein kinase (PKA). Mutant analyses in yeast provide evidence that PKA is a downstream target of TOR and is responsible, at least in part, for carrying out the positive effects of TOR on protein synthesis.

TOR affects cell growth mainly by stimulating protein synthesis

TOR stimulates growth through a variety of mechanisms, including increased uptake of amino acids from the environment, increased expression of genes encoding metabolic enzymes, and inhibition of protein degradation. Its primary growth-promoting effect is the stimulation of

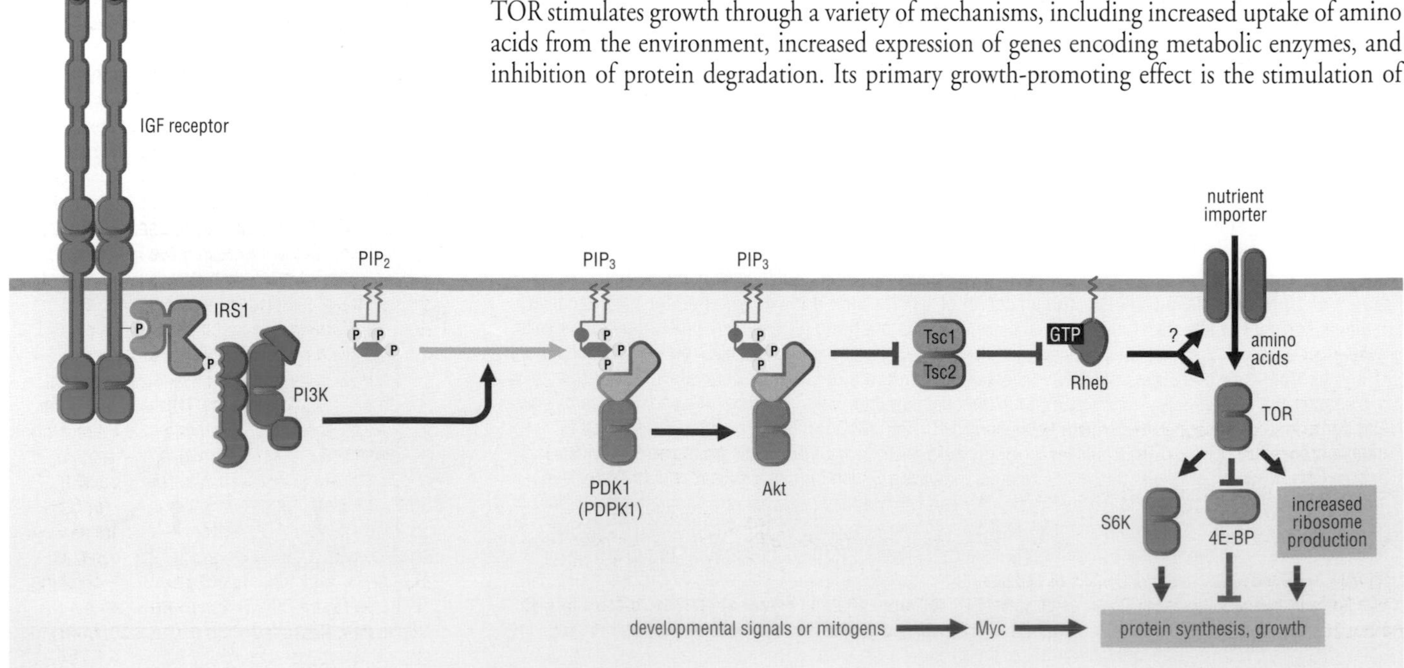

protein synthesis, chiefly through increasing the production of ribosomes. Activation of TOR results in an increase in the expression of genes for ribosomal proteins and ribosomal RNAs (rRNAs), as well as genes encoding proteins required for ribosome assembly. In budding yeast, TOR stimulates the expression of these genes by activating several gene regulatory proteins, including the proteins Fhl1 and Sfp1, which interact with ribosomal gene promoters.

In animal cells, TOR promotes protein synthesis not just by stimulating ribosome synthesis but also by several other mechanisms (see Figure 10-23). One is the regulation of the eukaryotic initiation factor 4E (eIF-4E), an important component of a regulatory complex that helps initiate the translation of most mRNAs. The activity of eIF-4E is inhibited by interaction with a small protein called 4E-binding protein (4E-BP). When TOR is activated, 4E-BP is phosphorylated at multiple sites, liberating eIF-4E and thus stimulating translation. TOR also helps activate a protein kinase called ribosomal protein S6 kinase, which phosphorylates a ribosomal subunit called S6, thereby enhancing the translation of a major subset of mRNAs. Other signaling pathways also contribute to the activation of S6 kinase independently of TOR, and we consider these pathways next.

Growth factors stimulate protein synthesis through the activation of PI3 kinase

The best understood of the extracellular growth factors are the members of the insulin-like family of proteins, including insulin itself and the related polypeptides IGF-I and IGF-II. Studies in *Drosophila* and cultured mammalian cells have begun to provide clues about the signaling pathways through which these factors stimulate growth (see Figure 10-23 and Figure 10-24). These pathways, like the mitogenic pathways described earlier (see section 10-6), begin with the autophosphorylation of the IGF receptor at several tyrosine residues. This leads to the recruitment of a phosphotyrosine-binding protein (called IRS in mammals and Chico in *Drosophila*), which is then phosphorylated on tyrosines by the active receptor. Phosphorylated IRS binds the PI3 kinase complex (see section 10-6), which catalyzes the formation of PIP_3 in the membrane. PIP_3 acts primarily to recruit the protein kinases PDK1 and Akt to the membrane. PDK1 phosphorylates and helps activate Akt, which then has multiple effects, including the stimulation of a small GTPase called Rheb, which acts through TOR to stimulate protein synthesis and cell growth. PDK1 can also phosphorylate the S6 kinase directly, contributing to its activation (see Figure 10-23).

External growth factors and internal developmental signals also influence the rate of cell growth through signaling pathways involving Ras, Myc and other proteins described earlier as components of mitogenic pathways (see section 10-6). Myc, for example, directly regulates the expression of genes for metabolic enzymes and other proteins involved in cell growth. It also stimulates ribosome synthesis by directly activating genes encoding ribosomal proteins and rRNAs. The stimulation of Myc thus results in a wide range of growth-promoting signals. Interestingly, Myc has considerably more growth-promoting targets than mitogenic targets, perhaps indicating that its major function in many cell types is to promote growth, not division. In those cells whose division depends on growth (see Figure 10-21a), the effects of Myc on cell proliferation may be an indirect effect of its growth-promoting actions.

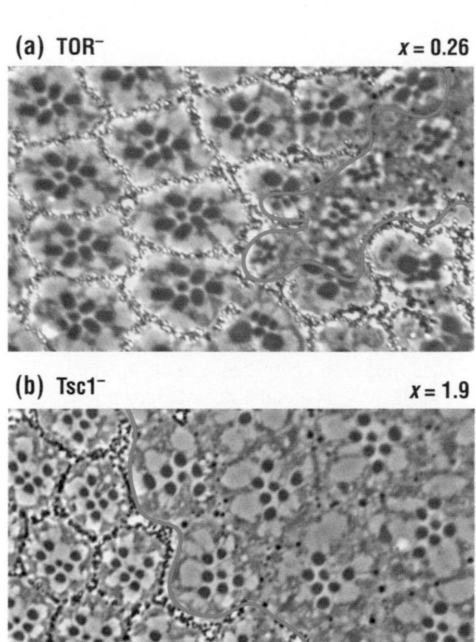

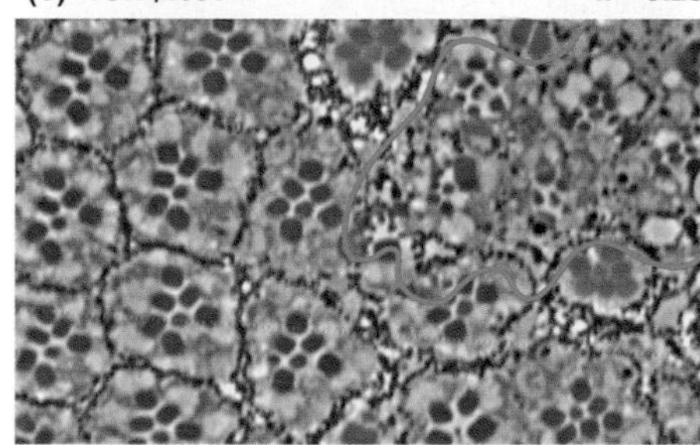

Figure 10-24 Analysis of growth control in the *Drosophila* eye These microscope images show cross-sections of the developing *Drosophila* compound eye, constructed of numerous clusters of photoreceptor cells (seen here as groups of 6–8 grey nuclei). In these experiments the effects of specific gene mutations were analyzed by creating mixed populations of wild-type and mutant cells (genetic mosaics), as follows: early in the growth of the eye imaginal disc, the desired gene was deleted in a few random cells, which then proliferated to form clones of mutant cells that gave rise to several mutant photoreceptor cell clusters. The blue line in these images indicates the boundary between yellow wild-type cells on the left and grey mutant cells on the right. **(a)** If the gene encoding TOR is deleted, mutant cells are much smaller than wild type (about 0.26 times the size of wild-type cells, as indicated by the *x*-value), so TOR is required for normal cell growth. **(b)** Mutation of the signaling component Tsc1 (see Figure 10-23) results in cells 1.9 times larger than wild type, revealing that Tsc1 is an inhibitor of cell growth. **(c)** Cells with mutations in both TOR and Tsc1 are as small as cells with mutations in TOR alone, as expected if Tsc1 inhibits cell growth by inhibiting TOR (see Figure 10-23). Kindly provided by Duojia Pan. Reprinted, with permission, from Gao, X. *et al.*: *Nat. Cell Biol.* 2002, **4**:699–704.

References

Edgar, B.A. and Orr-Weaver, T.L.: **Endoreplication cell cycles: more for less.** *Cell* 2001, **105**:297–306.

Gao, X. *et al.*: **Tsc tumour suppressor proteins antagonize amino-acid–TOR signalling.** *Nat. Cell Biol.* 2002, **4**:699–704.

Hall, M.N. *et al.* (eds): *Cell Growth: Control of Cell Size* (Cold Spring Harbor Laboratory Press, Cold Spring Harbor, 2004).

Jorgensen, P. and Tyers, M.: **How cells coordinate growth and division.** *Curr. Biol.* 2004, **14**:R1014–R1027.

Martin, D.E. and Hall, M.N.: **The expanding TOR signaling network.** *Curr. Opin. Cell Biol.* 2005, **17**:158–166.

Martin, D.E. *et al.*: **TOR regulates ribosomal protein gene expression via PKA and the Forkhead transcription factor FHL1.** *Cell* 2004, **119**:969–979.

Oskarsson, T. and Trumpp, A.: **The Myc trilogy: lord of RNA polymerases.** *Nat. Cell Biol.* 2005, **7**:215–217.

10-12 Coordination of Cell Growth and Division in Yeast

Yeast cell growth and division are tightly coupled

Yeast cells can encounter sudden large changes in the levels of nutrients in their environment—and must respond quickly with widely varying rates of growth. Yeast cell size, however, changes relatively little in different conditions. It is therefore clear that robust mechanisms exist in yeast to coordinate cell growth and division, so that cell size is maintained within a narrow range in the face of much greater variations in growth rate.

The coordination of yeast cell growth and division is achieved primarily by directly coupling cell division to cell growth (see Figure 10-21a): progression through the yeast cell cycle depends on cell growth. In the budding yeast *S. cerevisiae* this regulatory coupling is thought to involve an array of regulatory interactions between the protein synthetic machinery and the regulators of G1/S gene expression at Start. Although our knowledge of this system is limited, numerous lines of evidence suggest that it operates, at least in outline, as shown in Figure 10-25. Two key features can be discerned in this scheme. First, cell growth leads to the accumulation of a regulatory protein (probably the G1 cyclin Cln3) that triggers Start when its concentration reaches some threshold. Second, as we discuss later in this section, it is possible to reset the threshold required for cell-cycle entry, providing a mechanism for adjusting cell size.

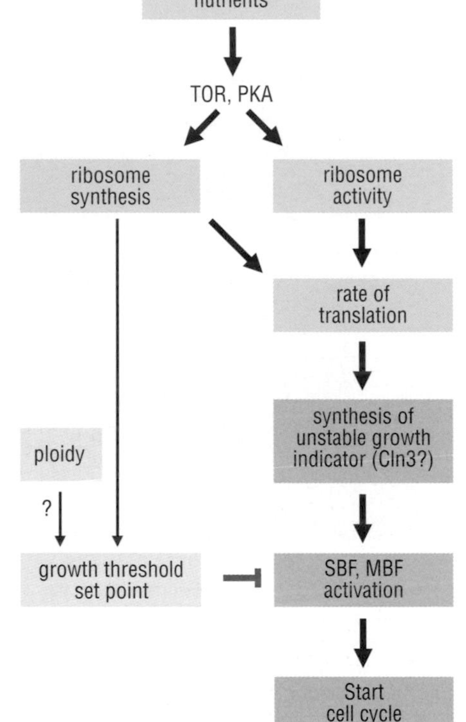

Figure 10-25 Possible mechanisms for coupling growth and division in budding yeast cells This is a speculative overview of the mechanisms by which growth pathways may influence cell-cycle progression. As described in section 10-11, nutrients stimulate growth by promoting ribosome synthesis and translational activity, which leads to an increased rate of protein synthesis and thus to an increased growth rate. Higher translation rates lead to increased concentrations of unstable regulatory proteins such as the G1 cyclin Cln3. When Cln3 levels reach some threshold, the gene regulatory proteins SBF and MBF are activated. This threshold can be adjusted by nutrients or other factors, resulting in a different cell size. One mechanism by which the threshold is reset may depend on the rate of ribosome synthesis, as shown here. By unknown mechanisms, an increased rate of ribosome synthesis suppresses SBF and MBF activities, so that more Cln3 is required to trigger Start. Higher cell ploidy may also result in larger cells because this also somehow reduces SBF and MBF activation.

Yeast cells monitor translation rates as an indirect indicator of cell size

There is abundant evidence that progression through Start in budding yeast occurs only when a cell grows to some minimal size threshold. But how can a yeast cell measure its size? The most likely answer is that size itself is not the relevant factor; instead, the overall rate of protein translation, which is generally proportional to cell volume in yeast, serves as an indirect indicator of size. Only when some threshold translation rate is achieved is the cell allowed to progress through Start.

How, then, does a change in the cell's translation rate control progression through Start? Our current view is that cell-cycle progression is triggered by a threshold concentration of a highly unstable regulatory protein, whose levels in the cell are determined primarily by its rate of translation. The major candidate for this growth-sensing regulator is the G1 cyclin Cln3, a highly unstable protein whose concentration is clearly a critical determinant in cell-cycle progression (see section 10-1).

Translation of the *CLN3* mRNA is regulated by the growth rate of the cell in such a way that small changes in the rate of general protein synthesis result in disproportionately large changes in the rate of Cln3 synthesis. The 5′ noncoding region of the *CLN3* mRNA contains a short upstream open reading frame (uORF) that hinders the progress of ribosomes. When growth rates are high, ribosomes are abundant in the cell and are capable of bypassing the uORF, resulting in high rates of *CLN3* translation. When ribosome numbers decrease under nutrient-poor conditions, the uORF slows the translational apparatus more drastically, resulting in decreased rates of Cln3 synthesis. By these and other mechanisms, the rate of protein synthesis directly influences the concentration of Cln3 and thus progression through Start.

Nutrients also control the amount of Cln3 protein by other mechanisms. Increased amounts of glucose, for example, activate the cAMP-dependent protein kinase PKA (see section 10-11), which initiates a signaling pathway leading to the activation of gene regulatory proteins that stimulate *CLN3* expression. These nutrients therefore stimulate passage through Start at the same time as they promote cell growth, providing another mechanism for coordinating growth and division.

In the fission yeast *S. pombe*, cell growth regulates cell division at the G2/M transition. Changes in the growth rate of fission yeast cells lead to changes in the length of G2, suggesting that entry into mitosis cannot occur until the cell achieves some minimal growth rate. In this organism, cell growth rate influences the activation of the major mitotic cyclin–Cdk complex, Cdc13–Cdk1 (see section 5-2), whose activity is regulated by inhibitory phosphorylation by the kinase Wee1 and dephosphorylation by the phosphatase Cdc25 (see section 3-3). The translation of both *CDC25* and *CDC13* mRNAs is highly sensitive to changes in the general rate of protein synthesis, probably as a result of regulatory sequences in the 5′ noncoding regions of the mRNAs. In addition, increased concentrations of nutrients decrease Wee1 activity by stimulating two protein kinases, Cdr1 and Cdr2, that phosphorylate and thereby inhibit

Wee1. Thus, the concentrations and activities of several key Cdk1 activators are coupled to cell growth in fission yeast, in much the same way that Cln3 levels are tied to growth in budding yeast.

Growth thresholds are rapidly adjustable

Yeast cell size is not always constant: small changes in the average size of cells in a population can occur under certain conditions. This implies that the growth threshold for cell-cycle progression can be reset: larger cells must have a higher threshold, for example. The resetting of growth thresholds is best illustrated by the behavior of yeast cells in different nutrient conditions. Cells growing in nutrient-rich medium do not just grow faster but are also slightly larger than cells in nutrient-poor medium (Figure 10-26). At first glance this might seem surprising: if cells growing in high nutrient concentrations have higher ribosome concentrations and thus higher translation rates, then one might expect that they would achieve the threshold growth rate at a smaller volume, not a larger one. The solution to this puzzle is that high nutrient concentrations, in addition to stimulating growth, also act through separate signaling pathways to slightly inhibit cell-cycle progression, thereby delaying division until cells are larger. In other words, nutrient levels can reset the threshold growth rate, or Cln3 concentration, at which Start is triggered (see Figure 10-25).

In budding yeast, high nutrient levels increase the size threshold by delaying the activation of Cln1,2–Cdk1. In the presence of a nutrient-rich environment, activation of PKA suppresses the transcription of *CLN1* and *CLN2*. Nutrients may govern Cln1,2–Cdk1 activation through their ability to stimulate the rate of ribosome synthesis. Mutations in genes required for ribosome synthesis result in small cells whose size no longer increases in high nutrient concentrations. From these and other lines of evidence, it has been proposed that mechanisms exist to monitor rates of ribosome synthesis, and that increased rates of synthesis somehow inhibit the gene regulatory proteins SBF and MBF that are required for most G1/S gene expression (see section 10-1), thereby delaying Start until cells grow larger. Little is known as yet about the molecular basis of these regulatory interactions, but the exploration of these mechanisms should reveal much about the control of cell size.

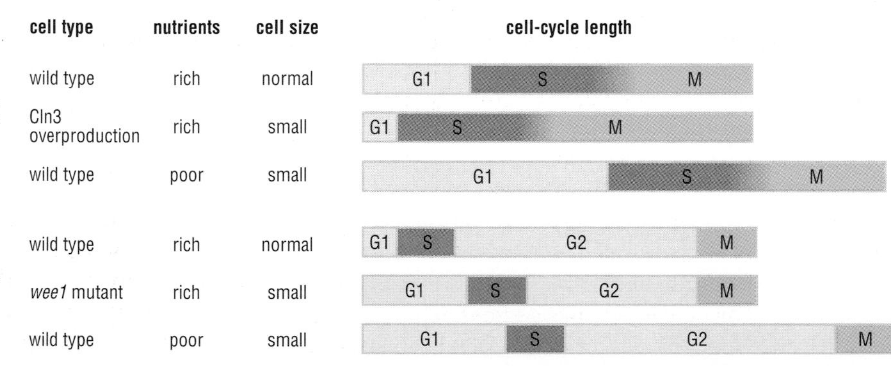

Figure 10-26 Size control in budding and fission yeasts Features of cell-size control are revealed by studies of yeast mutants. In budding yeast, overexpression of *CLN3* causes cells to pass Start at a smaller size—resulting in a shortened G1. These cells have a normal cell-cycle length, however, because later cell-cycle stages are extended—suggesting that additional growth-sensing mechanisms delay progression through M phase of budding yeast (recall from section 1-2 that S and M phases overlap in budding yeast, and there is no distinct G2/M checkpoint). In fission yeast, where growth governs progression through G2/M, a *wee1* mutation causes entry into mitosis at an abnormally small size, resulting in a shorter G2. G1 is lengthened, suggesting that progression through Start is responsive to growth when the G2/M control is lost. Changes in nutrient conditions have effects on both cell growth rate and cell-cycle time. As shown here, cells in poor-nutrient conditions grow slowly and therefore take longer to double in size and divide; low nutrient levels also reset the growth threshold so that cells progress through the cell cycle at a smaller size. *CLN3*-overexpressing budding yeast cells or *wee1* mutant fission yeast cells are smaller than wild-type cells at all cell-cycle stages—despite having roughly the same cell-cycle length. For this to be so, small mutant cells and large wild-type cells must take the same amount of time to double in size. It follows that small cells must grow at a slower rate (that is, they must add less total mass per unit time) than large cells. Indeed, it is known that small yeast cells grow more slowly than large yeast cells.

References

Daga, R.R. and Jimenez, J.: **Translational control of the Cdc25 cell cycle phosphatase: a molecular mechanism coupling mitosis to cell growth.** *J. Cell Sci.* 1999, **112**:3137–3446.

Hall, M.N. *et al.* (eds): *Cell Growth: Control of Cell Size* (Cold Spring Harbor Laboratory Press, Cold Spring Harbor, 2004).

Jorgensen, P. *et al.*: **Systematic identification of pathways that couple cell growth and division in yeast.** *Science* 2002, **297**:395–400.

Jorgensen, P. and Tyers, M.: **How cells coordinate growth and division.** *Curr. Biol.* 2004, **14**:R1014–R1027.

Mitchison, J.M.: **Growth during the cell cycle.** *Int. Rev. Cytol.* 2003, **226**:165–258.

Nurse, P.: **Genetic control of cell size at cell division in yeast.** *Nature* 1975, **256**:547–551.

Polymenis, M. and Schmidt, E.V.: **Coupling of cell division to cell growth by translational control of the G_1 cyclin *CLN3* in yeast.** *Genes Dev.* 1997, **11**:2522–2531.

Rupeš, I.: **Checking cell size in yeast.** *Trends Genet.* 2002, **18**:479–485.

Growth and division are coordinated by multiple mechanisms in animal cells

The coordination of growth and division is achieved in animal cells by various combinations of the general mechanisms described earlier (see Figure 10-21). As in yeast, division is coupled to growth in many animal cell types, such that entry into a new cell cycle depends on reaching some threshold growth rate or cell size (see Figure 10-21a). In other animal cell types, division does not seem to depend on growth, and cell size may be maintained by constant levels of growth factors and mitogens acting through independent signaling pathways (see Figure 10-21b). Finally, there is abundant evidence that some extracellular factors act as both growth factors and mitogens, stimulating intracellular signals that promote both growth and division (see Figure 10-21c).

Division depends on growth in many animal cell types

In many cultured vertebrate cell lines, the length of G1, and thus entry into the cell cycle, seems to depend on some threshold rate of protein synthesis or cell size. Early studies of mouse fibroblasts, for example, revealed that cell size after mitosis is highly variable, and that smaller G1 cells tend to require more time and accumulation of mass before they enter the next cell cycle. Similarly, if cultured avian erythroblasts or mouse fibroblasts are treated with low concentrations of DNA synthesis inhibitors, S phase is prolonged by a few hours but the cells continue to grow at a normal rate. These cells complete mitosis at an abnormally large size and enter the next cell cycle more quickly than untreated cells, suggesting that the length of G1 can be shortened to compensate for increased cell size.

Coupling of division and growth are clearly apparent in studies of *Drosophila* development. The initiation of endoreduplication cycles in larval tissues depends on the cell growth that results when the larva begins to feed. Experimental inhibition of growth in these cells inhibits entry into these cell cycles, and this delay can be abolished by overproduction of growth-promoting signaling proteins of the IGF pathway. There are also hints of growth-dependent division in the mitotic cycles of the developing wing imaginal disc. Flies carrying mutations in the IGF pathway (such as mutations in S6 kinase; see section 10-11) display defects in cellular growth rate and have smaller cells. Progression through all stages of the cell cycle is also slowed in these cells, suggesting that the reduced growth rate results in a reduced rate of division.

Additional evidence for a connection between growth and division comes from studies of imaginal cells in flies engineered to overexpress various growth promoters (such as PI3 kinase or Rheb; see section 10-11). These cells display increased growth rates, increased size, and an increased rate of cell-cycle entry at the G1/S transition (Figure 10-27), suggesting that growth rate might influence cell-cycle entry. A likely possibility is that growth rate influences the cellular concentration of cyclin E, which is the critical determinant of G1/S progression in both endoreduplicating larval cells and wing imaginal cells.

The overproduction of growth promoters shortens G1 in wing imaginal cells but does not affect overall cell-cycle length because G2 is lengthened (see Figure 10-27). Progression into mitosis seems to be controlled by separate factors acting through the phosphatase Cdc25 (see section 10-9). It remains unclear whether growth rate can also influence Cdc25 levels, and thus G2/M progression, under certain conditions. If it can, this could explain the general extension of all cell-cycle stages in cells carrying defects in S6 kinase, as noted earlier.

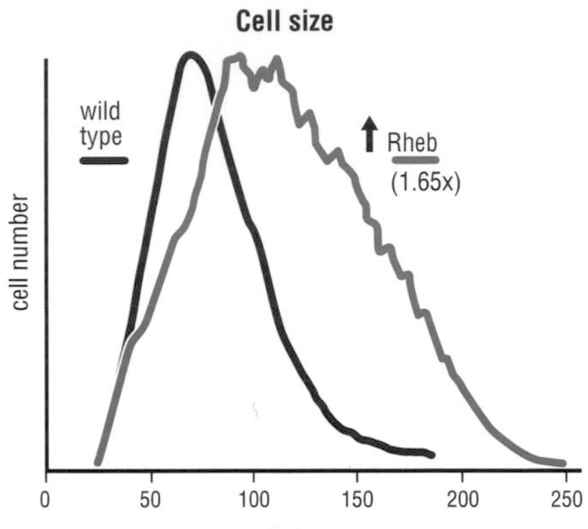

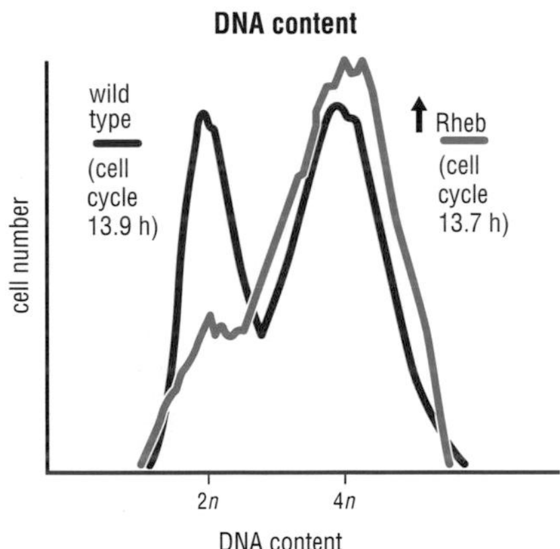

Figure 10-27 Evidence for the coupling of growth and cell-cycle progression in *Drosophila* These experiments were performed to test the effects of overexpression of the growth promoter Rheb (see section 10-11) on cell size and cell-cycle progression. Cells of the *Drosophila* wing imaginal disc were engineered to express high levels of the Rheb protein, and the wing disc was then dissociated into individual cells and analyzed by flow cytometry to determine cell size and DNA content (see section 2-6). Overproduction of Rheb resulted in a larger cell size and also caused a decrease in the number of cells with a G1 DNA content, suggesting that the stimulation of growth caused an increase in progression through Start. Cell-cycle time was not changed by Rheb, however, indicating that the shortening of G1 was compensated for by a lengthening of later cell-cycle stages. Similar changes occur after overexpression of numerous other growth promoters, including Myc and Ras. These proteins increase the levels of cyclin E, which might therefore be responsible for the increased rate of cell-cycle entry. Adapted from Saucedo, L.J. *et al.*: *Nat. Cell Biol.* 2003, **5**:566–571.

References

Conlon, I. and Raff, M.: **Differences in the way a mammalian cell and yeast cells coordinate cell growth and cell-cycle progression.** *J. Biol.* 2003, **2**:7.

Conlon, I.J. *et al.*: **Extracellular control of cell size.** *Nat. Cell Biol.* 2001, **3**:918–921.

Dolznig, H. *et al.*: **Evidence for a size-sensing mechanism in animal cells.** *Nat. Cell Biol.* 2004, **6**:899–905.

Edgar, B.A. and Nijhout, H.F.: **Growth and cell cycle control in *Drosophila*.** in *Cell Growth: Control of Cell Size.* Hall, M.N. *et al.* eds (Cold Spring Harbor, Cold Spring Harbor Laboratory Press, 2004), 23–83.

Jorgensen, P. and Tyers, M.: **How cells coordinate growth and division.** *Curr. Biol.* 2004, **14**:R1014–R1027.

Killander, D. and Zetterberg, A.: **A quantitative cytochemical investigation of the relationship between cell mass and initiation of DNA synthesis in mouse fibroblasts *in vitro*.** *Exp. Cell Res.* 1965, **40**:12–20.

We saw earlier (see section 10-12) that in budding yeast it is possible to reset the growth threshold required for entry into the cell cycle, resulting in changes in average cell size. In animal cells, control of this threshold is also likely to be an important mechanism for changing the size at which cells enter the cycle. As in yeast, growth rate itself may help control the growth threshold. *Drosophila* imaginal cells with higher growth rates are also larger than usual. Similarly, in avian erythroblasts, a sudden increase in growth rate (achieved by changing the balance of growth factor and mitogen stimulation) leads to a larger cell size at the beginning of the next cell cycle.

Animal cell growth and division are sometimes controlled independently

In some animal cell types, cell growth and division seem to be completely independent of each other: growth is stimulated by a growth factor through one signaling pathway and division is stimulated by a mitogen acting through a separate pathway (see Figure 10-21b). In cultured mouse fibroblast lines, for example, EGF is a potent mitogen but a poor growth factor, whereas IGF-I is a more effective growth factor and a poor mitogen.

In cultured rat Schwann cells, glial growth factor (GGF) stimulates cell division but not growth, whereas IGF-I is primarily a growth factor. When these cells are cultured in a fixed concentration of IGF-I, increasing the concentration of GGF stimulates cell division without affecting growth rate, resulting in smaller cells (Figure 10-28). The change in cell size occurs over several generations. Thus, a sudden change in growth rate does not result in a change in cell size in the next cell cycle, as would be expected if cell-cycle entry depended on some threshold growth rate. These cells are therefore different from the avian erythroblasts discussed earlier, in which cell size can be changed within a single cell cycle because division is coupled to growth.

In many cultured mammalian cell lines, a single extracellular factor will stimulate both growth and division. Indeed, most of the best-known growth factors, including PDGF, EGF and IGF-I, are both mitogens and growth factors for many cell types, although a combination of different factors may be needed to achieve an optimal balance of the two processes. Multifunctional factors promote both growth and division by triggering signaling pathways that begin at a single receptor and branch out to stimulate the cell-growth machinery on one branch and the cell-cycle machinery controlling Start on the other. Several major signaling components stimulate both growth and division in certain mammalian cells. The protein kinase Akt, the GTPase Ras, and the gene regulatory factor Myc are all thought to possess both growth-promoting and mitogenic activities. These multifunctional regulators lie near the top of the signaling network, and their activation by a single upstream receptor brings about the coordinated stimulation of both growth and division.

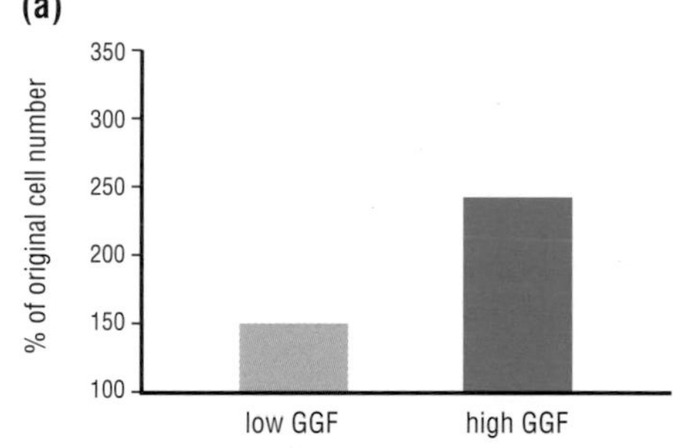

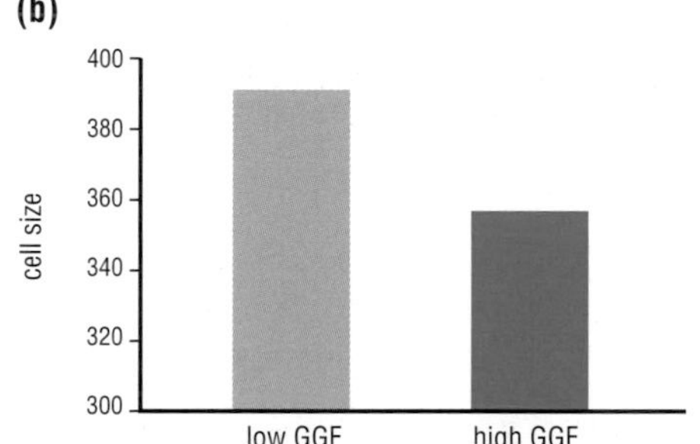

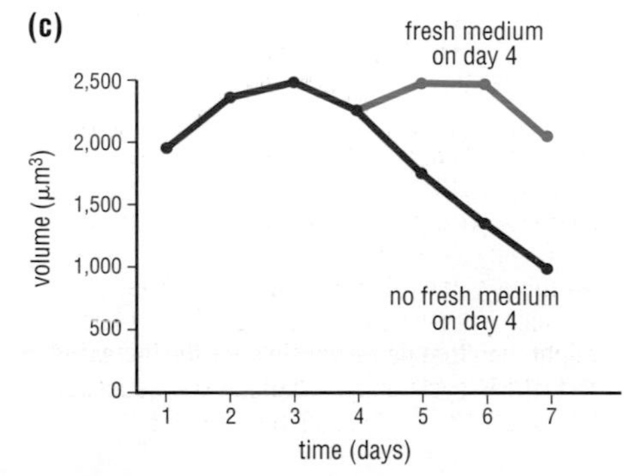

Figure 10-28 Separate control of growth and division in mammalian cells Cultured Schwann cells from rat were treated with the mitogen GGF (which has no growth factor activity) and the growth factor IGF-I (which is a very weak mitogen). **(a)** Cells growing in a constant IGF-I concentration were treated with two different concentrations of GGF. Higher GGF levels promoted increased cell numbers. **(b)** The increase in cell number at high GGF concentration was accompanied by a decrease in cell size. Thus, a simple change in the ratio of mitogen to growth factor resulted in a change in cell size. **(c)** When Schwann cells are grown for long periods in serum-containing culture medium, growth factors are depleted more rapidly than mitogens. Cell size declines (red line) unless the medium is replaced with fresh medium (blue line). Competition for limited amounts of growth factors can therefore have a significant impact on cell size. Panels (a) and (b) adapted from Conlon, I.J. *et al.*: *Nat. Cell Biol.* 2001, **3**:918–921; panel (c) adapted from Conlon, I. and Raff, M.: *J. Biol.* 2003, **2**:7.

Reis, T. and Edgar, B.A.: **Negative regulation of dE2F1 by cyclin-dependent kinases controls cell cycle timing.** *Cell* 2004, **117**:253–264.

Saucedo, L.J. *et al.*: **Rheb promotes cell growth as a component of the insulin/TOR signalling network.** *Nat. Cell Biol.* 2003, **5**:566–571.

Zetterberg, A. *et al.*: **The relative effects of different types of growth factors on DNA replication, mitosis, and cellular enlargement.** *Cytometry* 1984, **5**:368–375.

10-14 Control of Cell Death

Animal cell numbers are determined by a balance of cell birth and death

The number of cells in a tissue, and thus the size of that tissue, is determined not only by the rate of cell birth by division but also by the rate of cell death. Cell death in most tissues is not a passive process but is achieved by a complex and tightly scripted death program called **apoptosis**. Apoptosis results from the irreversible activation of a group of intracellular proteases and nucleases, which digest various components of the cell, including the nuclear lamina, parts of the cytoskeleton, and the DNA of the chromosomes. Apoptotic cells shrink and come loose from their neighbors, and eventually break apart into membrane-bound fragments that are engulfed by neighboring cells or extruded from the tissue.

Most animal cells contain the molecular machinery required for apoptosis and are therefore prepared to destroy themselves. Whether a cell lives or dies depends on developmental cues and other signals that promote or inhibit cell death depending on the needs of the organism. In many growing tissues, increasing the cell number depends on the local production of extracellular proteins called **survival factors**, which suppress apoptosis in that tissue. Survival factors bind to cell-surface receptors and initiate intracellular signaling pathways that block the activation of the apoptotic program. The best-understood survival signaling mechanisms involve proteins that are also used in mitogenic and growth-promoting signaling pathways; thus, mitogens and growth factors often promote cell survival at the same time as they stimulate division and growth.

Survival factors suppress the mitochondrial pathway of apoptosis

Survival signaling pathways act by suppressing the intrinsic or mitochondrial pathway of apoptosis, as illustrated in Figure 10-29. This apoptotic mechanism begins with the release from mitochondria of several proteins, including cytochrome c, one of the components of the mitochondrial electron-transport chain. Once in the cytoplasm, cytochrome c interacts with a protein called Apaf1, and together the two proteins form a complex that binds and activates a protease called caspase-9. Caspase-9 (called the initiator caspase) activates additional members of the caspase family (the effector caspases), resulting in an overwhelming, irreversible wave of protease activity that commits the cell to death by apoptosis. Other proteins released from the mitochondria promote apoptosis by other means.

Release of cytochrome c from mitochondria, and thus the initiation of the intrinsic apoptotic pathway, is controlled by a large family of proteins called the Bcl-2 family. Members of this family all contain regions of sequence homology called Bcl-2 homology (BH) domains. Each member of the Bcl-2 family is classified into one of three subfamilies according to the number of BH domains it contains and whether it stimulates or inhibits apoptosis. The first subfamily comprises the multidomain proapoptotic proteins, including two proteins called Bax and Bak, which contain several BH domains and bind directly to mitochondrial membranes, where they trigger cytochrome c release and thus apoptosis. The second subfamily comprises the multidomain antiapoptotic proteins, including Bcl-2 itself and Bcl-X_L, which inhibit apoptosis primarily by binding to antiapoptotic proteins such as Bax and Bak. The third subfamily is the BH3-only proapoptotic proteins, a large group that includes the proteins Bad, Bim, Bid and others. Members of the BH3-only subfamily promote apoptosis by one of two mechanisms. Some BH3-only proteins bind and inhibit the multidomain antiapoptotic proteins (such as Bcl-2), thereby enabling apoptosis. Others bind and stimulate the multidomain proapoptotic

Definitions

apoptosis: regulated cell death in which activation of specific proteases and nucleases leads to death characterized by chromatin condensation, protein and DNA degradation, loss of plasma membrane lipid asymmetry and disintegration of the cell into membrane-bound fragments.

survival factor: extracellular factor that inhibits cell death by apoptosis.

References

Adams, J.M.: **Ways of dying: multiple pathways to apoptosis.** *Genes Dev.* 2003, **17**:2481–2495.

Chittenden, T.: **BH3 domains: intracellular death-ligands critical for initiating apoptosis.** *Cancer Cell* 2002, **2**:165–166.

Danial, N.N. and Korsmeyer, S.J.: **Cell death: critical control points.** *Cell* 2004, **116**:205–219.

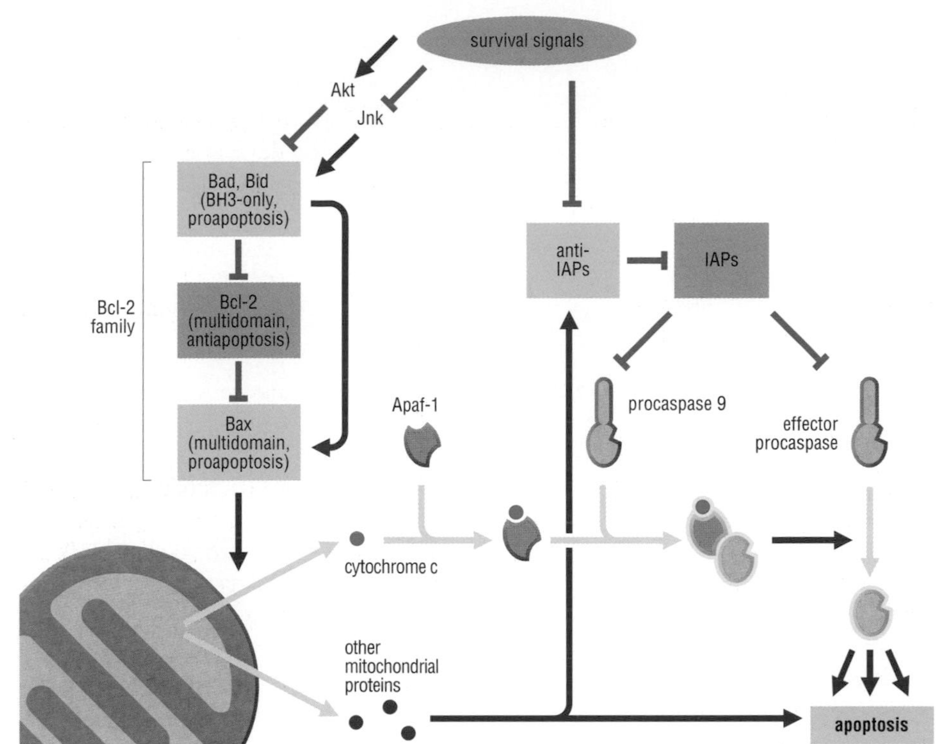

Figure 10-29 Suppression of the intrinsic apoptotic pathway by survival signals Activation of the intrinsic apoptotic pathway begins with the mitochondrial release of various proteins, including cytochrome c, which triggers the activation of a cascade of caspases that carry out the proteolysis of numerous proteins in the cell. The release of cytochrome c is controlled by members of the Bcl-2 protein family, which can be divided into three subfamilies, whose members control each other as shown here. Apoptosis can also be regulated by IAPs that inhibit caspase activity and anti-IAPs that block this inhibition. Survival factors and other antiapoptotic signals suppress apoptosis primarily by reducing the activity of the BH3-only members of the Bcl-2 family. In *Drosophila*, survival factors act by inhibiting anti-IAPs.

proteins (such as Bax), thereby directly activating apoptosis (see Figure 10-29). The BH3-only proapoptotic proteins are generally the key targets of the regulatory pathways that control apoptosis in mammalian cells.

The rate of apoptosis is also regulated in some cell types—particularly in *Drosophila*—by proteins called IAPs (inhibitors of apoptosis), which bind and thereby inhibit the activity of the caspases that drive apoptosis. The survival of many cells depends on the presence of IAPs in the cytoplasm. When the intrinsic apoptotic pathway is triggered, mitochondria release proteins called anti-IAPS, which bind IAPs and thereby block their effects, allowing caspases to trigger cell death. In *Drosophila*, signals that control the rate of apoptosis often act by changing the levels of IAPs or anti-IAPs.

In mammalian cells, survival factors act primarily by inhibiting the concentrations or activities of BH3-only proteins (see Figure 10-29). The best characterized survival signaling mechanism involves the protein kinase Akt. We saw earlier in this chapter that activation of Akt by PI3 kinase is used in some cell types to promote cell division (see section 10-6) or growth (see section 10-11). Akt also serves as a key promoter of cell survival in many cell types by catalyzing the phosphorylation of Bad, a member of the BH3-only subfamily. Phosphorylation of Bad blocks its ability to promote apoptosis. Some survival factors act by inhibiting another protein kinase called Jnk, which otherwise stimulates expression of the gene encoding Bim, another BH3-only protein. In *Drosophila*, survival signals generally act by triggering the phosphorylation of anti-IAPs, thereby blocking their effects on IAPs and thus inhibiting caspase function.

DNA damage and other stresses can trigger apoptosis

Apoptosis is often used to remove cells that are potentially harmful to the organism. Cells that suffer severe DNA damage, for example, are destroyed by apoptosis; they might otherwise acquire mutations that would enable them to grow and proliferate uncontrollably. If left unchecked these cells can give rise to a tumor that threatens the organism's survival. For similar reasons, cell death can also be triggered in cells that produce excessive amounts of mitogenic signaling proteins, such as Myc or E2F1. DNA damage and other stresses generally trigger apoptosis by activating a gene regulatory protein called p53, which triggers apoptosis in part by stimulating the expression of BH3-only proteins. We describe this and the many other cellular responses to DNA damage in Chapter 11.

11

The DNA Damage Response

Cell-cycle progression is blocked when the chromosomal DNA is damaged, thereby ensuring that potentially lethal genetic errors are not passed on to the cell's offspring. This important safeguard mechanism is founded on a complex network of regulatory proteins that sense many forms of genomic damage and send inhibitory signals to the cell-cycle control system.

11-0 Overview: The DNA Damage Response

The DNA damage response helps maintain the genome

The information encoded in our DNA must be maintained faithfully through countless cell divisions. This is achieved in part by the remarkable accuracy of the molecular machines that carry out chromosome duplication in S phase and segregation in M phase. The accuracy of these machines is not enough, however, to ensure the integrity of the genome. Every cell, dividing or not, must also be equipped to prevent the potentially harmful effects of DNA damage. DNA is a large and complex entity and is subject to a variety of chemical changes that are either spontaneous or catalyzed by the chemicals and radiation that bombard every cell. These alterations can lead to gene mutations that cripple essential cellular processes or, in the case of multicellular organisms, alter the cell's behavior in a way that threatens the survival of the organism. To avoid this problem, most damaged DNA is repaired before it is replicated or segregated, and thus alterations in gene sequence are only rarely passed on to a cell's offspring.

DNA damage can take many forms, ranging from subtle changes in nucleotide base structure to breaks in both strands of the double helix, and it can occur at all phases of the cell cycle. All cells possess sensor proteins that scan the genome, detect DNA damage and recruit specialized enzymes to repair it. If the DNA is extensively damaged and not easily repaired, the damage sensors trigger a more extensive response called the **DNA damage response**. Signaling pathways are activated that transmit the damage signal to a variety of effector proteins, some of which trigger increased production of DNA repair enzymes. Other effectors inhibit the cell-cycle control system, thereby blocking cell-cycle progression. This branch of the DNA damage response is sometimes called the DNA damage checkpoint. If the damage is repaired, the cell-cycle block is lifted and cell proliferation resumes.

In some cases the damage is particularly extensive and cannot be repaired. The response to irreparable DNA damage varies in different organisms. In yeast and other unicellular species, each cell is an individual organism, the survival of which depends on continued proliferation even in the face of severe damage. Thus, the cell-cycle arrest that occurs after DNA damage in these cells is not permanent, and yeast cells with irreparably damaged DNA eventually resume proliferation despite the risks. Multicellular organisms, however, are social communities of cells in which the survival of the organism as a whole is more important than the survival of an individual cell. Unrepaired DNA damage in one cell can lead to mutations that cause its uncontrolled proliferation or other behavior that could kill the organism. Thus, the response to irreparable damage in animal cells is often a permanent cell-cycle arrest or, in many cases, the death of the cell by apoptosis.

ATR and ATM are conserved protein kinases at the heart of the DNA damage response

In all eukaryotes the DNA damage response is centered on a pair of related protein kinases called ATR and ATM, whose sequence and function have been well conserved in evolution (Figure 11-1). Early in the response, these proteins bind to the chromosomes at sites of DNA damage, together with accessory proteins that provide platforms on which damage-response components and DNA repair complexes are assembled. Association of ATR or ATM with damaged DNA leads to their phosphorylation of regulatory proteins and ultimately triggers the activation of two other protein kinases called Chk1 and Chk2, which are known as the effector kinases of the damage response. These initiate signaling pathways that inhibit cell-cycle progression and stimulate the expression of large numbers of genes encoding proteins involved in DNA repair.

Definitions

DNA damage response: cellular response to extensive DNA damage, in which DNA repair is promoted and cell-cycle progression is delayed.

References

Bakkenist, C.J. and Kastan, M.B.: **Initiating cellular stress responses.** *Cell* 2004, **118**:9–17.

Fry, R.C. *et al.*: **Genome-wide responses to DNA-damaging agents.** *Annu. Rev. Microbiol.* 2005, **59**:357–377.

Melo, J. and Toczyski, D.: **A unified view of the DNA-damage checkpoint.** *Curr. Opin. Cell Biol.* 2002, **14**:237–245.

Nyberg, K.A. *et al.*: **Toward maintaining the genome: DNA damage and replication checkpoints.** *Annu. Rev. Genet.* 2002, **36**:617–656.

Rouse, J. and Jackson, S.P.: **Interfaces between the detection, signaling, and repair of DNA damage.** *Science* 2002, **297**:547–551.

Wahl, G.M. and Carr, A.M.: **The evolution of diverse biological responses to DNA damage: insights from yeast and p53.** *Nat. Cell Biol.* 2001, **3**:E277–E286.

Weinert, T.A. and Hartwell, L.H.: **The RAD9 gene controls the cell cycle response to DNA damage in *Saccharomyces cerevisiae*.** *Science* 1988, **241**:317–322.

Zhou, B.B. and Elledge, S.J.: **The DNA damage response: putting checkpoints in perspective.** *Nature* 2000, **408**:433–439.

Animal cells, but not yeast, possess an additional DNA damage response that enables permanent cell-cycle arrest or cell death when damage cannot be repaired. This response is based on a gene regulatory protein called p53, which is activated by ATR or ATM and triggers the increased expression of numerous target genes involved in DNA repair, cell-cycle arrest and apoptosis (see Figure 11-1).

The importance of the DNA damage response for the survival of multicellular organisms is illustrated by the human diseases that result when components of the response are missing. Because so much spontaneous DNA damage occurs in every cell, a defective DNA damage response inevitably leads to the accumulation of damaged DNA, which eventually generates mutations that lead to inappropriate cell behavior. Thus, mutations in DNA damage-response components such as ATR, ATM, Chk1, Chk2 and p53 generally result in an increased sensitivity to DNA damage and an increased likelihood of developing cancer—as discussed in Chapter 12.

Replication defects trigger a DNA damage response

One of the consequences of certain types of DNA damage is the stalling of replication forks at the damaged site when the DNA is replicated. Stalled replication forks generate abnormal DNA structures that are sensed by the DNA damage response, resulting in the recruitment and activation of ATR. This leads to a complex response that blocks cell-cycle progression, prevents the firing of other replication origins, and stabilizes the stalled replication fork so that it can safely restart when the damage has been repaired.

The cell does not enter mitosis if DNA replication is not completed successfully, as we saw in earlier chapters (see section 5-1). This important feature of cell-cycle control is based on the ability of stalled replication forks to trigger a DNA damage response, which sends inhibitory signals to the M–Cdks that promote entry into mitosis. Any treatment that produces stalled replication forks and blocks replication—such as chemical inhibition of nucleotide synthesis or DNA polymerase—will induce the DNA damage response and halt cell-cycle progression.

This chapter provides a brief review of the DNA damage response, with an emphasis on the effects of DNA damage on the cell-cycle regulatory mechanisms discussed in earlier chapters. We begin with a discussion of the most common forms of DNA damage and then make our way through the regulatory pathways that detect this damage and send inhibitory signals to the cell-cycle control system.

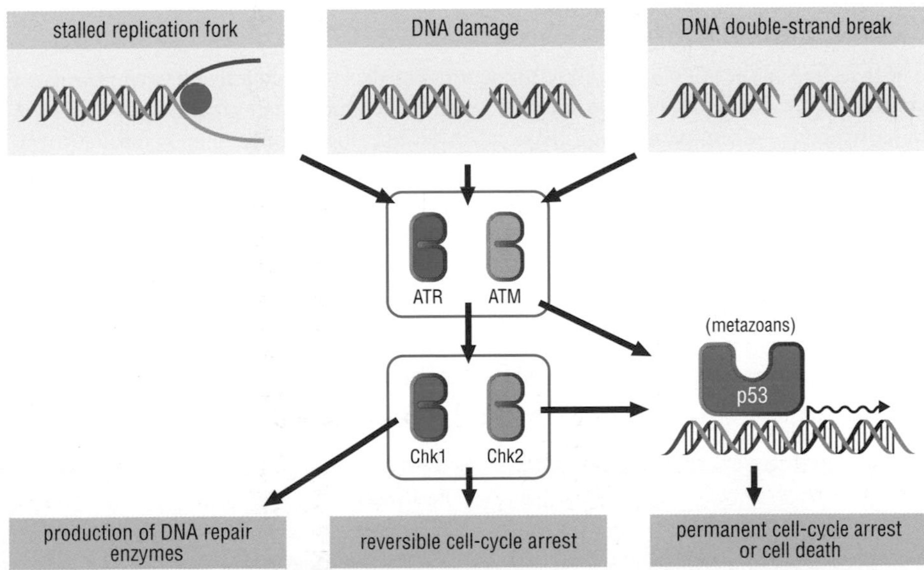

Figure 11-1 The DNA damage response DNA damage of various types leads to the recruitment of the protein kinases ATR or ATM to the damaged site. With assistance from numerous accessory proteins (not shown), ATR and ATM generate a damage signal that leads to activation of the protein kinases Chk1 and Chk2. These kinases then trigger increased expression of large numbers of genes encoding DNA repair enzymes and other repair proteins; they also block further progression through the cell cycle. In multicellular organisms, the gene regulatory protein p53 is also activated during the damage response, resulting in the increased expression of genes that promote long-term cell-cycle arrest or apoptosis.

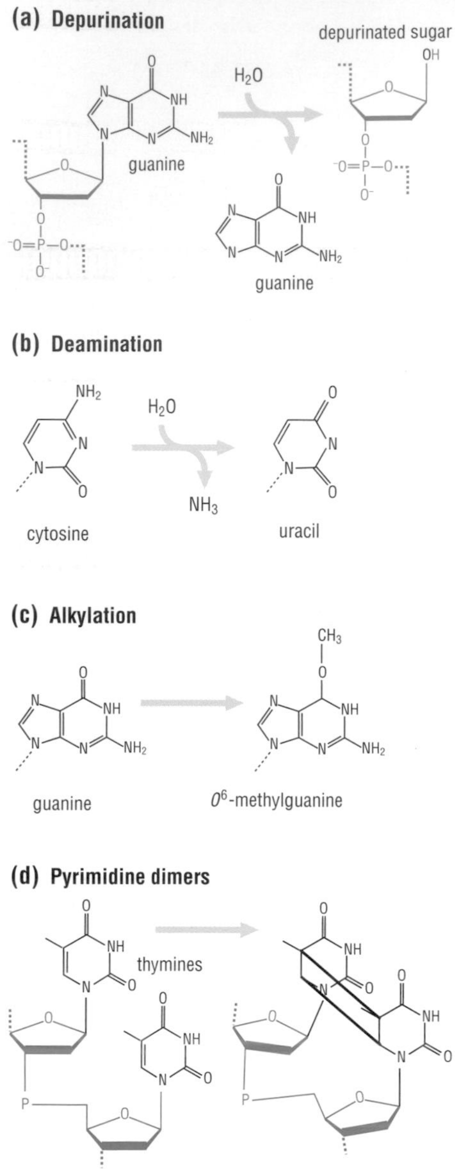

11-1 Detection and Repair of DNA Damage

DNA can be damaged in many ways

DNA is a very stable molecule, but it is still subject to chemical reactions that can lead to potentially harmful gene mutations. Under normal conditions the nucleotides in DNA are continually being modified by spontaneous hydrolysis and oxidation. Such reactions lead to several types of damage: depurination by hydrolysis of the bond connecting guanine or adenine bases to the nucleotide (Figure 11-2a); deamination by hydrolytic attack on amino groups in some bases (particularly cytosine, which is converted to uracil) (Figure 11-2b); and alkylation, the modification of oxygens and nitrogens in bases by reactive metabolites, resulting, for example, in the addition of methyl groups (Figure 11-2c). These and many other changes are remarkably common, affecting thousands of nucleotides in every cell each day.

Environmental factors also contribute to DNA damage. Ultraviolet (UV) radiation from sunlight causes the covalent cross-linking of adjacent pyrimidine bases—producing thymine dimers, for example, which interfere with replication (Figure 11-2d). Nucleotide structure can also be modified by environmental chemicals such as the carcinogen benzopyrene, which is a large hydrocarbon that covalently attaches to bases in DNA, forming a bulky adduct that distorts the DNA helix. The commonly used experimental mutagen methyl methanesulfonate (MMS) generates mutations by methylating certain reactive sites on some bases. Methylated guanine, for example, can mispair during DNA synthesis with thymine instead of cytosine, resulting in a point mutation on the new strand.

These alterations in nucleotide structure usually affect just one DNA strand at a given site, but both strands of the DNA double helix can also be broken—generally as a result of ionizing radiation such as X-rays or exposure to chemicals such as bleomycin. This creates a lesion known as a **double-strand break**. Breaks are particularly harmful, not simply because they can fragment chromosomes but also because the DNA repair machinery can accidentally fuse exposed DNA ends from different chromosomes, resulting in chromosome rearrangements.

Base and nucleotide excision repair systems repair nucleotide damage

Armies of DNA repair enzymes constantly scan the chromosomes and detect all forms of DNA damage with remarkable speed and sensitivity. Simple nucleotide alterations on only one strand of the DNA, such as those mentioned above, are repaired easily because the undamaged strand is available to provide the correct sequence information. Thus, repair of this type of damage is achieved readily by removing the damaged portion of the DNA strand and resynthesizing it correctly, using the undamaged DNA strand as the template.

The detection and repair of altered nucleotide structure depends primarily on two major repair systems (Figure 11-3). One is **base excision repair**, which finds relatively minor alterations in base structure—deamination and base methylation, for example—and repairs them. Components of this system are thought to scan the DNA by flipping each base out of the helix to check it for abnormalities. When an altered base is found, it is removed from the DNA backbone. The base-free strand is then processed to remove the sugar-phosphate backbone at the site of the damaged nucleotide. The same happens to depurinated nucleotides. New nucleotides are then added by DNA polymerase, using the undamaged strand as a template. The nick in the DNA strand is sealed by DNA ligase.

Nucleotide excision repair is responsible for the detection and repair of major modifications that alter the conformation of the double helix. These bulky lesions include pyrimidine dimers

(a) Depurination

(b) Deamination

(c) Alkylation

(d) Pyrimidine dimers

Figure 11-2 Examples of nucleotide damage **(a)** Hydrolytic cleavage can result in loss of the base from purine nucleotides. **(b)** Several bases contain amino groups that are susceptible to hydrolysis. **(c)** Metabolic byproducts and environmental chemicals alkylate various positions on DNA bases. **(d)** Ultraviolet radiation leads to the formation of bonds between adjacent pyrimidine bases.

Definitions

base excision repair: mechanism of DNA repair in which a nucleotide with a damaged or missing base is excised from the DNA and replaced with an undamaged one, using the undamaged DNA strand as a template.

double-strand break: type of DNA damage in which a DNA molecule is broken across both strands.

non-homologous end joining: mechanism for repairing double-strand breaks in DNA in which the broken ends are rejoined directly, usually with the loss of nucleotides at the join.

nucleotide excision repair: DNA repair pathway in which major structural defects, such as thymine dimers, are excised along with a stretch of 12–30 nucleotides surrounding the site of damage, and the damaged strand is resynthesized using the undamaged strand as the template.

References

Friedberg, E.C. *et al.*: *DNA Repair and Mutagenesis.* 2nd ed. (ASM Press, Washington, D.C., 2005).

Hoeijmakers, J.H.: **Genome maintenance mechanisms for preventing cancer.** *Nature* 2001, **411**:366–374.

Sancar, A. *et al.*: **Molecular mechanisms of mammalian DNA repair and the DNA damage checkpoints.** *Annu. Rev. Biochem.* 2004, **73**:39–85.

Figure 11-3 **Repair of nucleotide damage** Two major mechanisms are used to remove and replace damaged nucleotides. **(a)** In base excision repair, a damaged base (deaminated cytosine, for example) is detected and removed from the DNA backbone by specific repair enzymes, after which the backbone itself is excised. DNA polymerase then fills in the missing nucleotide, using the undamaged strand as a template. DNA ligase completes the repair by sealing the nicked backbone. **(b)** Nucleotide excision repair is used to repair bulky lesions in the DNA (a thymine dimer, for example). A stretch of the damaged strand on each side of the lesion is removed by nucleases and helicases and then replaced with a new strand by DNA polymerase and DNA ligase.

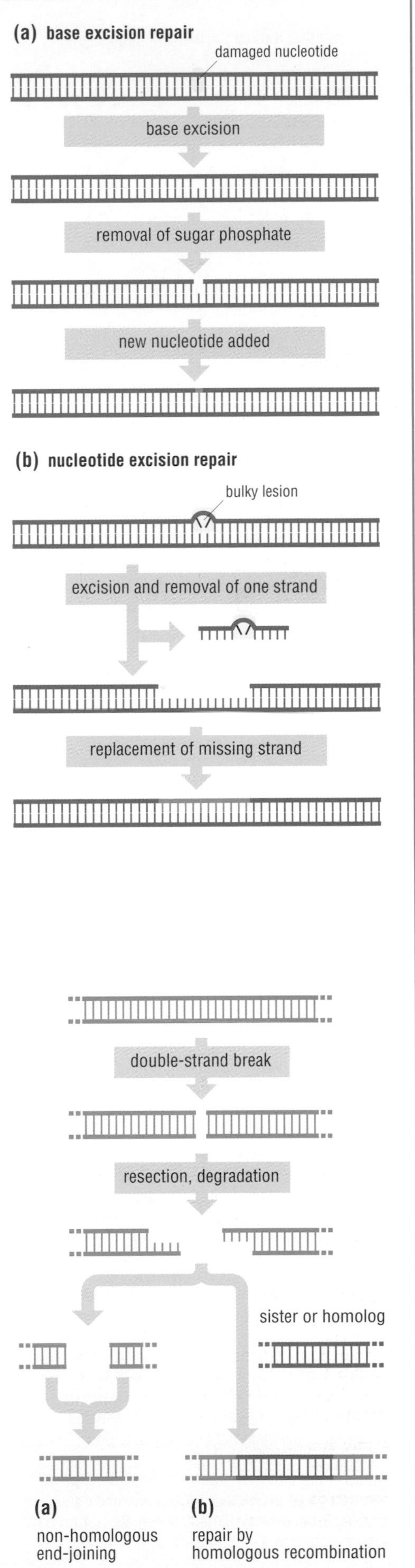

or DNA alkylated by large chemicals such as benzopyrene. The nucleotide excision repair machinery scans the DNA in search of major helical distortions and then, using nucleases and DNA helicases, removes a short stretch of the damaged strand. The undamaged strand is then used as a template to synthesize a new strand, thereby restoring the original sequence.

Double-strand breaks are repaired by two main mechanisms

One way of repairing double-strand breaks is by a process called **non-homologous end joining**, in which the two broken ends are simply rejoined by DNA ligases (Figure 11-4). This approach is not ideal, however, because nucleotides are usually lost at the repair site—generally because the exposed ends of double-strand breaks are resected and degraded by nucleases before being rejoined. Non-homologous end joining is a common repair mechanism during G1 in mammalian cells, where a small loss of sequence can be tolerated because so much mammalian DNA does not encode proteins. It is less frequent in eukaryotes that have relatively little noncoding DNA—yeast and flies, for example.

More accurate repair of double-strand breaks can be achieved by homologous recombination between the broken chromosome and a homologous sequence in a sister chromatid or homologous chromosome (see Figure 11-4). We discussed the molecular mechanism of homologous recombination in section 9-2, in the context of recombination between homologs in meiotic prophase. As we saw, meiotic recombination often leads to noncrossover events that repair the broken chromosome without generating a lasting connection between homologs. Similar mechanisms are used in somatic cells to repair double-strand breaks. In G2 cells, for example, sequences in a broken sister chromatid are usually repaired by recombination with undamaged sequences in the other sister. Single-stranded DNA from the damaged chromatid invades the sister helix and base-pairs with the complementary strand. As in meiotic recombination, invasion depends on recombinases, primarily Rad51 (see section 9-2). Extension of the invading strand on the sister template then replaces the sequence across the double-strand break with that of the sister chromatid, after which the extended strand returns to the original chromosome to complete the repair.

In diploid cells, double-strand break repair by recombination with a homologous chromosome can lead to loss of heterozygosity in a gene. For example, a wild-type gene on the damaged chromosome might be replaced with a mutant form of the gene from the homolog, resulting in a potentially harmful homozygous mutant state. To avoid this problem, the preferred template for repair in somatic cells is an identical sister chromatid, and mechanisms exist to promote recombination between sisters and suppress recombination between homologs. Sister-chromatid cohesion, for example, enhances recombination between sisters. In addition, as we describe in section 11-2, double-strand break repair by recombination is activated only during cell-cycle stages in which a sister chromatid is available. In other stages of the cell cycle, double-strand breaks are usually repaired by non-homologous end joining, even if a homolog is available for recombinational repair.

Figure 11-4 **Repair of double-strand breaks** When a double-strand break occurs in DNA, the exposed ends are often resected by nucleases, resulting in single-stranded overhangs. **(a)** In some cases, these damaged ends are processed to blunt ends and rejoined by non-homologous end joining, resulting in a new DNA molecule that is lacking several nucleotides of the original sequence. **(b)** A more accurate repair is achieved by homologous recombination. As described in section 9-2, a single-stranded end from the broken DNA invades a homologous sequence in a sister chromatid or homologous chromosome. The invading strand is extended along the homologous template, returns to the original chromosome and acts as template for the repair of the other broken strand. This results in a repaired DNA molecule in which the damaged region is replaced with sequence from the sister or homolog.

11-2 The DNA Damage Response: Recruitment of ATR and ATM

Alternative Names for DNA Damage Response Components

Name	*S. cerevisiae*	*S. pombe*	Vertebrates
Sensor kinases			
ATR	Mec1	Rad3	ATR
ATM	Tel1	Tel1	ATM
ATR regulatory subunit			
ATRIP	Ddc2/Lcd1	Rad26	ATRIP
Effector kinases			
Chk1	Chk1	Chk1/Rad27	Chk1
Chk2	Rad53	Cds1	Chk2
MRN complex			
Mre11	Mre11	Rad32	Mre11
Rad50	Rad50	Rad50	Rad50
Nbs1	Xrs2	Nbs1	Nbs1
9-1-1 (PCNA-like) complex			
Rad9	Ddc1	Rad9	Rad9
Hus1	Mec3	Hus1	Hus1
Rad1	Rad17	Rad1	Rad1
Rad17–RFC complex			
Rad17	Rad24	Rad17	Rad17
Rfc2-5	Rfc2-5	Rfc2-5	Rfc2-5
Adaptors and mediators			
BRCT	Rad9	Crb2/Rhp1	BRCA1 53BP1 MDC1/NFBD1
claspin	Mrc1	Mrc1	claspin

Figure 11-5 Table of alternative names for DNA damage response components The proteins of the BRCT and claspin groups have not been well conserved in evolution and are referred to by their species-specific names.

ATR is required for the response to multiple forms of damage

Many forms of DNA damage are repaired quickly and do not trigger a DNA damage response leading to cell-cycle arrest. Some damage, however, is particularly extensive or difficult to repair—for example, when a sister chromatid is not available for recombinational double-strand break repair, or when double-strand breaks are accompanied by extensive nucleotide alterations. In these cases a specialized damage response is initiated by recruitment of one or both of the protein kinases ATR and ATM to the site of damage. These kinases activate damage responses by phosphorylating various proteins that also gather at the damaged site (Figure 11-5).

ATR is required for the response to many different forms of DNA damage, including nucleotide damage, stalled replication forks, and double-strand breaks. ATM is specialized for the response to double-strand breaks.

How is it possible for ATR to recognize so many different types of DNA damage? The likely answer is that ATR specifically recognizes tracts of single-stranded DNA (Figure 11-6). Single-stranded DNA is formed, for example, during nucleotide excision repair of UV-induced thymine dimers and in some types of base excision repair. Mutations in nucleotide excision repair pathways prevent the ATR-dependent response to UV damage, suggesting that the generation of single-stranded DNA during processing of the damaged site is required for the response. The ATR response to defects that disrupt DNA replication is also likely to depend on the single-stranded DNA that accumulates at stalled replication forks.

Single-stranded DNA is usually coated by the single-strand binding protein RPA, which has been discussed previously in the context of DNA replication (see section 4-1). ATR recruitment to single-stranded DNA probably involves an interaction with RPA, because mutations in RPA block the ATR-dependent damage response (Figure 11-7). The interaction of ATR with the complex of single-stranded DNA and RPA depends, at least in part, on the direct binding of RPA to an ATR-associated adaptor subunit called ATRIP. Cells with mutations in ATRIP have the same damage-response defects as those with mutations in ATR, demonstrating the central importance of ATRIP in ATR function.

The cell is generally more sensitive to DNA damage in S phase than it is in G1. Certain forms of minor DNA damage, such as methylation and UV-induced thymine dimers, trigger little or no ATR response when they occur in G1, perhaps because they are repaired by the time that

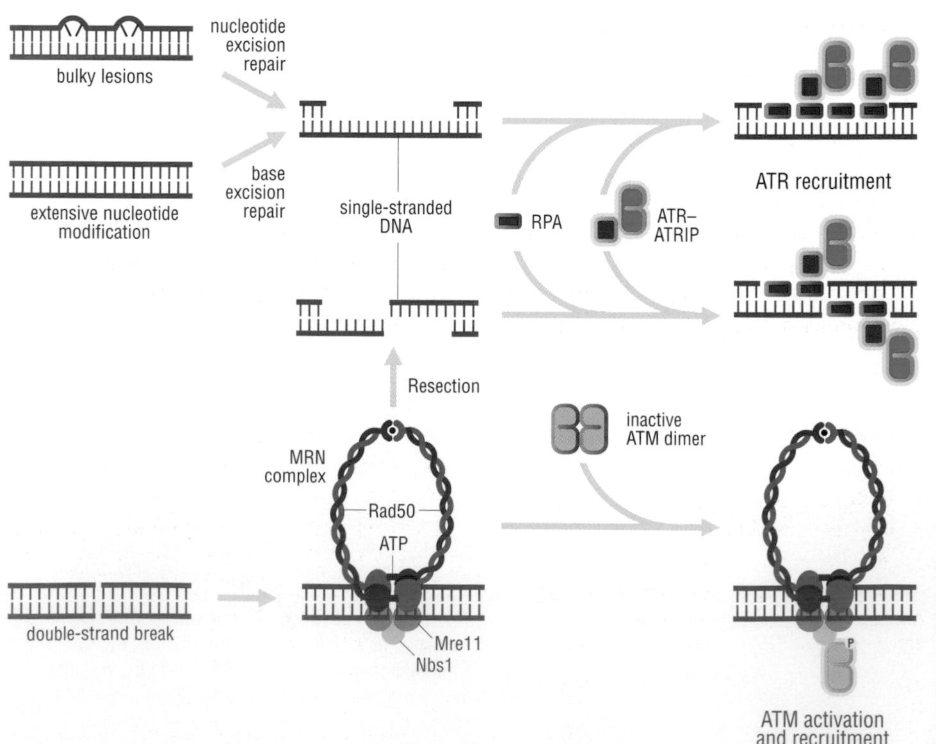

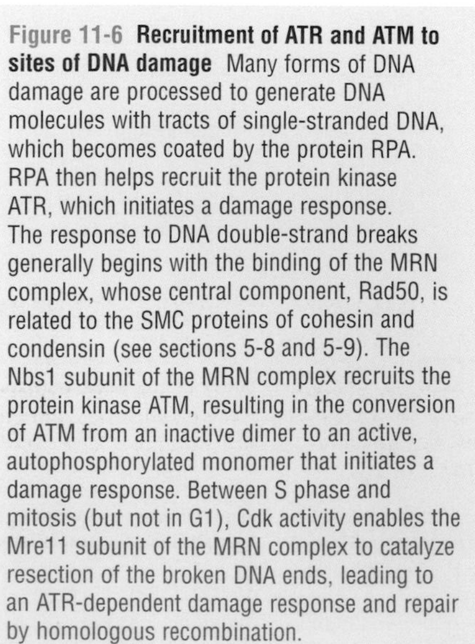

Figure 11-6 Recruitment of ATR and ATM to sites of DNA damage Many forms of DNA damage are processed to generate DNA molecules with tracts of single-stranded DNA, which becomes coated by the protein RPA. RPA then helps recruit the protein kinase ATR, which initiates a damage response. The response to DNA double-strand breaks generally begins with the binding of the MRN complex, whose central component, Rad50, is related to the SMC proteins of cohesin and condensin (see sections 5-8 and 5-9). The Nbs1 subunit of the MRN complex recruits the protein kinase ATM, resulting in the conversion of ATM from an inactive dimer to an active, autophosphorylated monomer that initiates a damage response. Between S phase and mitosis (but not in G1), Cdk activity enables the Mre11 subunit of the MRN complex to catalyze resection of the broken DNA ends, leading to an ATR-dependent damage response and repair by homologous recombination.

the DNA begins to be replicated, or because minor repairs, such as base excision repair, do not always generate large tracts of single-stranded DNA. During S phase, however, these types of damage do activate ATR and stimulate a damage response, probably because replication forks are delayed or stalled at damaged sites, resulting in extensive formation of single-stranded DNA.

The intrinsic kinase activity of ATR does not seem to change on its recruitment to DNA. Instead, it is thought that the binding of active ATR to sites of damage promotes its phosphorylation of target proteins that are also recruited to those sites.

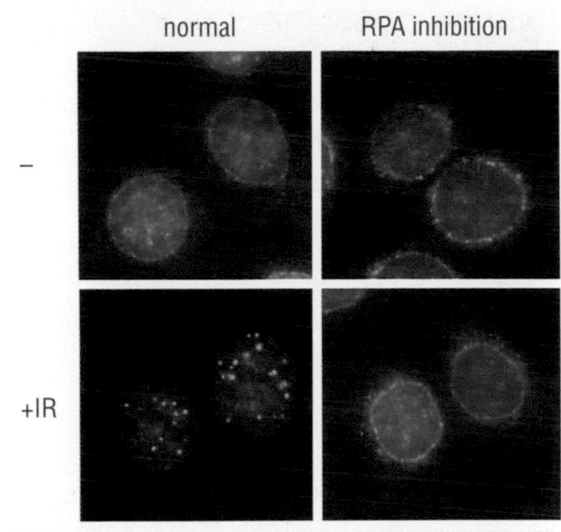

Figure 11-7 RPA-dependent recruitment of ATR to sites of DNA damage Cultured human cells were either left untreated (top panels) or treated with ionizing radiation (+IR, bottom panels) to generate double-strand breaks, which in these experiments were resected to trigger an ATR response. The location of ATR was revealed by staining with antibodies. As seen in the upper left image, ATR is normally localized throughout the nucleus; irradiation leads to the appearance of ATR foci at sites of damage. In the right column, cells were treated with a short interfering RNA (siRNA; see section 2-5) that inhibits the synthesis of RPA in the cell. The resulting decrease in RPA protein prevented the recruitment of ATR to DNA damage sites and blocked the damage response. In other experiments, the ATRIP regulatory subunit of ATR was found to bind directly to RPA on single-stranded DNA. These and other results argue that the ATR–ATRIP complex is recruited to sites of DNA damage by a direct interaction with RPA-coated DNA, as shown in Figure 11-6. Kindly provided by Stephen J. Elledge. From Zou, L. and Elledge, S.J.: *Science* 2003, **300**:1542–1548.

ATM is specialized for the response to unprocessed double-strand breaks

The kinase ATM is required primarily for the response to double-strand breaks. Mutation of the gene for ATM in humans results in the disease *ataxia telangiectasia*, which is characterized by, among other things, a greatly reduced ability to repair radiation-induced double-strand breaks—and an increased risk of developing cancer. ATM is recruited to sites of double-strand break formation, where it phosphorylates effector molecules that carry out the damage response.

The ATM response to double-strand breaks depends on a trimeric complex of three proteins: Mre11, Rad50 and Nbs1 (called Xrs2 in yeast; see Figure 11-6). This complex, called the MRN complex in humans, assembles at double-strand breaks immediately after their formation and helps hold the two ends together. Mutations in genes encoding subunits of the MRN complex block the ATM-dependent damage response in yeast, and partial defects in the Nbs1 subunit reduce the response of human cells to double-strand breaks. ATM interacts with the Nbs1 subunit of the MRN complex, which results not only in recruitment of ATM to the site of damage but also in its conversion from an inactive dimer into a monomer with protein kinase activity. ATM activation is accompanied by, and may depend on, autophosphorylation of the kinase. Active ATM initiates the damage response by phosphorylating target proteins that are found both at the damage site and in the nucleoplasm.

The cellular response to double-strand breaks also depends partly on the ATR kinase. Under certain conditions, double-strand breaks undergo 5′ to 3′ resection of one strand, resulting in single-stranded DNA tracts that provide the signal to recruit ATR (see Figure 11-6). Resection of double-strand breaks is thought to be catalyzed primarily by nuclease activity within the Mre11 subunit of the MRN complex, and inhibition of this enzyme blocks the ATR response to this damage.

The response to double-strand breaks varies during the cell cycle, primarily because the ability of the MRN complex to catalyze resection of double-strand breaks is in some way promoted by Cdk activity. When Cdk activity is high between S phase and mitosis, enhanced resection of double-strand breaks leads not only to an ATR response but also generates the single-stranded overhang that initiates homologous recombination—at a stage in the cell cycle when the ideal repair template, a sister chromatid, is available (see section 11-1). When Cdk activity is low during G1, double-strand break resection is suppressed, thereby preventing the ATR response and suppressing recombinational repair. The MRN complex, together with other repair proteins, then directs repair of the double-strand break by non-homologous end joining.

References

Abraham, R.T.: **Cell cycle checkpoint signaling through the ATM and ATR kinases.** *Genes Dev.* 2001, **15**:2177–2196.

Bakkenist, C.J. and Kastan, M.B.: **DNA damage activates ATM through intermolecular autophosphorylation and dimer dissociation.** *Nature* 2003, **421**:499–506.

Bakkenist, C.J. and Kastan, M.B.: **Initiating cellular stress responses.** *Cell* 2004, **118**:9–17.

Garber, P.M. *et al.*: **Damage in transition.** *Trends Biochem. Sci.* 2005, **30**:63–66.

Ira, G. *et al.*: **DNA end resection, homologous recombination and DNA damage checkpoint activation require CDK1.** *Nature* 2004, **431**:1011–1077.

Jazayeri, A. *et al.*: **ATM- and cell cycle-dependent regulation of ATR in response to DNA double-strand breaks.** *Nat. Cell Biol.* 2006, **8**:37–45.

Lee, J.-H. and Paull, T.T.: **ATM activation by DNA double-strand breaks through the Mre11-Rad50-Nbs1 complex.** *Science* 2005, **308**:551–554.

Lisby, M. *et al.*: **Choreography of the DNA damage response: spatiotemporal relationships among checkpoint and repair proteins.** *Cell* 2004, **118**:699–713.

Zou, L. and Elledge, S.J.: **Sensing DNA damage through ATRIP recognition of RPA-ssDNA complexes.** *Science* 2003, **300**:1542–1548.

Protein complexes assemble at DNA damage sites to coordinate DNA repair and the damage response

The binding of ATR and ATM to sites of DNA damage is accompanied by the recruitment of numerous other proteins to the surrounding DNA. Together these components form large multiprotein complexes that help recruit and coordinate the enzymes that repair the DNA. These complexes also bind and activate two additional protein kinases called Chk1 and Chk2, which transmit the damage signal to components of the cell-cycle control system, leading to delays in cell-cycle progression.

A PCNA-like complex is required for the ATR-mediated damage response

One of the complexes that is recruited to certain sites of DNA damage is the 9-1-1 complex, which is required for the ATR-mediated DNA damage response and also seems to promote the processing of damage by repair proteins. The three subunits of the 9-1-1 complex (see Figure 11-5) are related in sequence to the subunits of PCNA, the sliding clamp that is loaded onto DNA at the primer–template junction and binds DNA polymerase (see section 4-1). On the basis of this sequence homology, the 9-1-1 complex is thought to form a ring around damaged DNA. The association of the 9-1-1 complex with DNA depends on a second large complex called Rad17–RFC, which is a modified form of the eukaryotic clamp loader, RFC (see section 4-1). The largest subunit of Rad17–RFC is Rad17, which is related to Rfc1, the largest subunit of RFC. The other four subunits of both complexes are Rfc2–5. As predicted by the similar structures, Rad17–RFC is required for loading of the 9-1-1 complex onto damaged DNA (Figure 11-8). Biochemical studies indicate that, as with the replication clamp, the 9-1-1 complex is loaded by Rad17–RFC at junctions between single- and double-stranded DNA, which are found at sites of DNA damage.

The 9-1-1 complex and ATR–ATRIP are recruited independently to sites of DNA damage (see Figure 11-8). Once on the DNA they probably interact with each other, resulting in the phosphorylation of several subunits of both the 9-1-1 and Rad17–RFC complexes. The function of this phosphorylation is not clear, but it may help recruit additional components.

Adaptor proteins link DNA damage to activation of Chk1 and Chk2

ATR and ATM initiate damage responses in part by phosphorylating, and thereby activating, the protein kinases Chk1 and Chk2. These are recruited to the damaged DNA by **adaptor** or **mediator proteins** that present them to ATR or ATM for phosphorylation (Figure 11-9).

(a) wild type, no damage

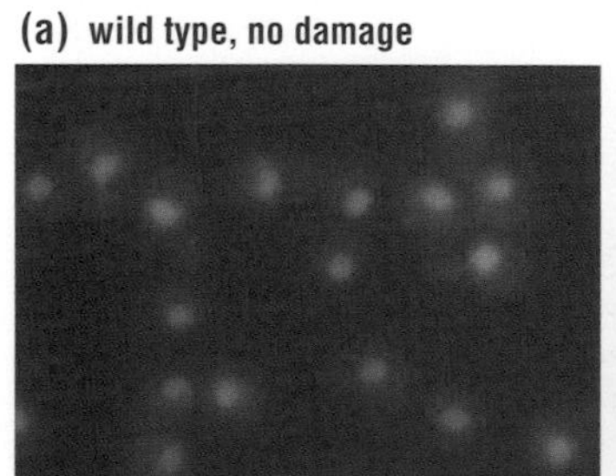

(b) wild type plus damage

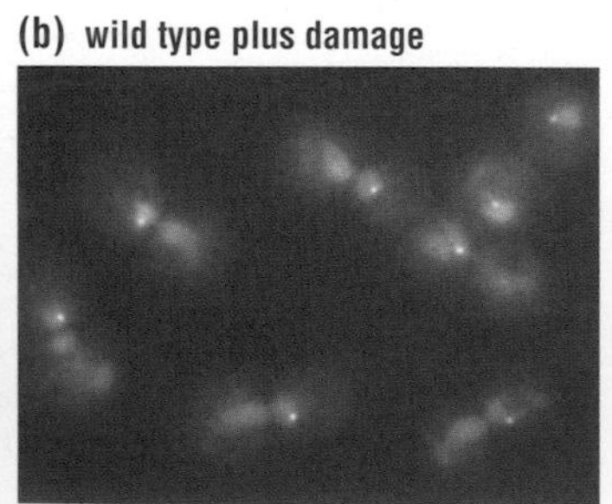

(c) Rad17–RFC mutant plus damage

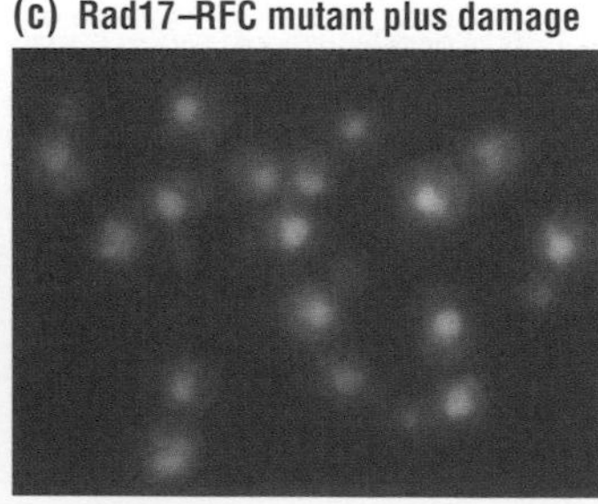

(d) ATR mutant plus damage

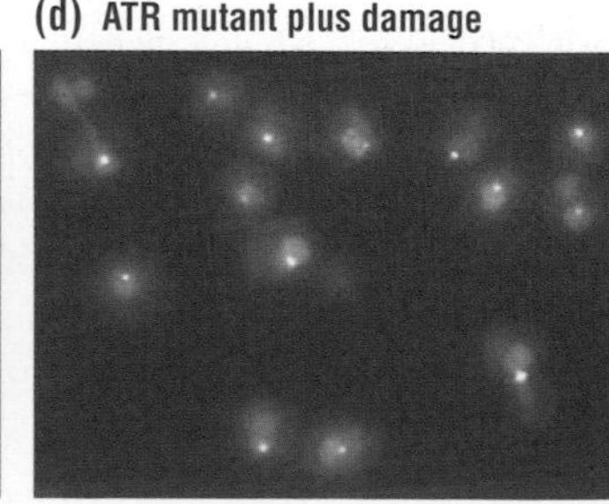

Figure 11-8 **Recruitment of the 9-1-1 complex to sites of DNA damage** In these experiments, one subunit of the budding yeast 9-1-1 complex was tagged with a fluorescent protein to allow its detection by microscopy. The cells were also engineered so that addition of a specific sugar to the culture medium stimulated an enzyme that caused a single double-strand break in one chromosome. Resection of this break then initiated an ATR response. **(a)** In wild-type cells lacking DNA damage, the 9-1-1 complex is diffusely localized in cell nuclei (each patch of fluorescence corresponds to a nucleus). **(b)** When the DNA damage response is triggered by a single double-strand break, the 9-1-1 complex is focused at the damage site, resulting in the appearance of a single fluorescent dot. **(c)** The 9-1-1 complex does not localize to the damage site in the absence of the Rad17–RFC complex. **(d)** The 9-1-1 complex localizes normally to the damage site in the absence of ATR. Similar experiments (not shown) revealed that localization of ATR to sites of damage does not depend on the 9-1-1 complex, indicating that these two complexes are recruited independently to the damage. Courtesy of Justine Melo and David Toczyski.

Definitions

adaptor protein: a protein that links two other proteins together in a regulatory network or protein complex; also called a **mediator protein**.

mediator protein: see **adaptor protein**.

References

Bartek, J. and Lukas, J.: **Chk1 and Chk2 kinases in checkpoint control and cancer.** *Cancer Cell* 2003, **3**:421–429.

Celeste, A. *et al.*: **Histone H2AX phosphorylation is dispensable for the initial recognition of DNA breaks.** *Nat. Cell Biol.* 2003, **5**:675–679.

Gilbert, C.S. *et al.*: **Budding yeast Rad9 is an ATP-dependent Rad53 activating machine.** *Mol. Cell* 2001, **8**:129–136.

Glover, J.N. *et al.*: **Interactions between BRCT repeats and phosphoproteins: tangled up in two.** *Trends Biochem. Sci.* 2004, **29**:579–585.

Huyen, Y. *et al.*: **Methylated lysine 79 of histone H3 targets 53BP1 to DNA double-strand breaks.** *Nature* 2004, **432**:406–411.

Melo, J.A. *et al.*: **Two checkpoint complexes are independently recruited to sites of DNA damage *in vivo*.** *Genes Dev.* 2001, **15**:2809–2821.

Pellicioli, A. and Foiani, M.: **Signal transduction: how rad53 kinase is activated.** *Curr. Biol.* 2005, **15**:R769–R771.

Sanders, S.L. *et al.*: **Methylation of histone H4 lysine 20 controls recruitment of Crb2 to sites of DNA damage.**

The Rad9 protein of budding yeast is the best understood of these adaptors. After the binding of ATR–ATRIP, 9-1-1 and Rad17–RFC to damage sites, Rad9 is also recruited, possibly as a result of interactions with ATR, the phosphorylated 9-1-1 complex and local chromatin proteins. After its binding, Rad9 is extensively phosphorylated by ATR (see Figure 11-9). Some of these phosphorylations trigger Rad9 self-association, leading to the assembly of Rad9 oligomers on the chromosome. Phosphorylation of Rad9 on other residues generates binding sites for Chk2, which is thereby recruited to the damage site and activated by ATR. Rad9-associated Chk2 also phosphorylates itself, further contributing to its activation. Activated Chk2 is released from Rad9 to pursue its targets in the cell-cycle control system (see Figure 11-9). Rad9 is also required for the activation of Chk1 after DNA damage, but little is known about the molecular basis of Chk1 activation.

Several proteins distantly related to yeast Rad9 are thought to serve as adaptors in the response to DNA damage in human cells. The most prominent are BRCA1, 53BP1 and MDC1 (see Figure 11-5), which are recruited to damaged DNA and phosphorylated by ATM or ATR. The precise molecular function of these human adaptor proteins is not clear, but they seem to provide a structural framework for the assembly of DNA repair machinery and for the activation of Chk1 or Chk2.

Adaptor proteins are recruited to sites of DNA damage in part by interactions with modified chromatin proteins. In mammalian cells, double-strand breaks trigger ATM-dependent phosphorylation of histone H2A.X, a variant form of histone H2A (see section 4-9), along large tracts of the chromosome near the damage site. Yeast does not contain histone H2A.X, but damage similarly causes local phosphorylation of specific residues in histone H2A. Phosphorylation of histones is thought to generate binding sites for a protein domain called the BRCT domain, which is found in most adaptor proteins, including yeast Rad9 and the three major human adaptors (see Figure 11-5).

Binding of adaptor proteins to phosphorylated histone H2A.X contributes to the damage response, but it is not essential because loss of H2A.X does not completely abolish the response or the recruitment of some adaptors. H2A.X deletion does reduce the rate of DNA repair, however, which argues that local assembly of repair enzymes depends on histone phosphorylation. Histone phosphorylation may also promote the repair of double-strand breaks by stimulating a local increase in the concentration of cohesin between sister chromatids (see section 5-8), thereby ensuring that the broken DNA is closely associated with a template for recombination.

Histone methylation also contributes to the assembly of damage response components on the chromosome. Several adaptor proteins, including human 53BP1, fission yeast Crb2 and budding yeast Rad9, contain a domain, called a Tudor domain, that interacts with methylated lysine residues in histone tails (see section 4-9). 53BP1 binds methylated Lys 79 of histone H3, whereas Crb2 binds methylated Lys 20 of histone H4, and loss of these interactions as a result of mutation leads to defects in the DNA damage response. Surprisingly, histone methylation at these sites is not limited to regions surrounding DNA damage but occurs throughout the chromosomes. An intriguing possibility is that DNA damage somehow alters local chromatin structure and exposes the methylated histones, thereby generating binding sites for the damage response proteins.

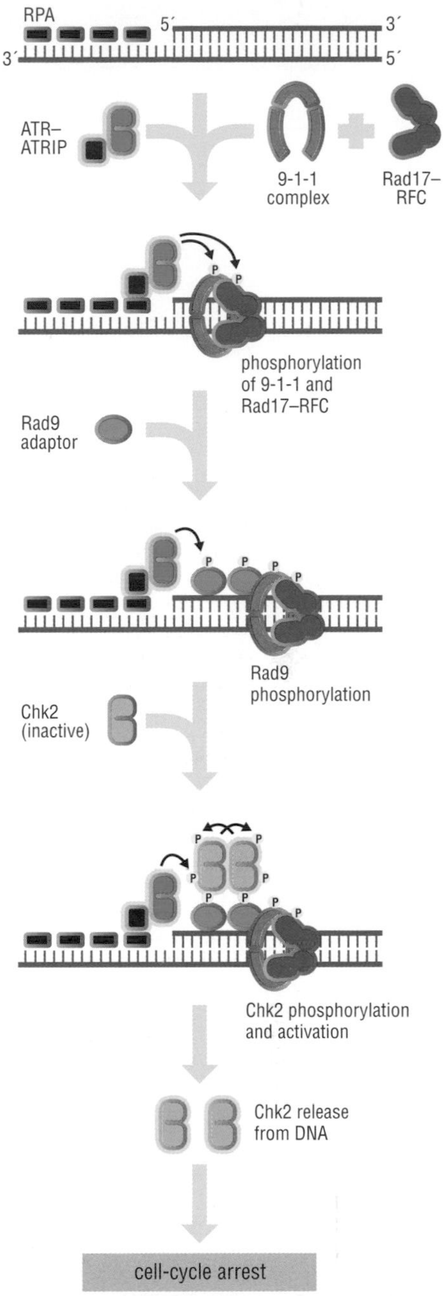

Figure 11-9 Steps in the activation of effector kinases after DNA damage The importance of adaptor proteins is illustrated here by the ATR-dependent response to a resected double-strand break in budding yeast. This process begins with the independent recruitment of ATR–ATRIP, which binds RPA on single-stranded DNA, and the 9-1-1 complex, which is probably loaded on or near the adjacent 5′-recessed DNA structure. ATR then phosphorylates components of the 9-1-1 complex. The adaptor protein Rad9 forms oligomers that associate with the damage site, possibly through interactions with ATR, the phosphorylated 9-1-1 complex or modified histones. ATR phosphorylates Rad9, thereby creating binding sites on Rad9 for the kinase Chk2. Chk2 is then phosphorylated by ATR, and also phosphorylates itself, resulting in its activation and dissociation from the complex.

Cell 2004, **119**:603–614.

Ström, L. *et al.*: **Postreplicative recruitment of cohesin to double-strand breaks is required for DNA repair.** *Mol. Cell* 2004, **16**:1003–1015.

Stucki, M. *et al.*: **MDC1 directly binds phosphorylated histone H2AX to regulate cellular responses to DNA double-strand breaks.** *Cell* 2005, **123**:1213–1226.

Sweeney, F.D. *et al.*: ***Saccharomyces cerevisiae* Rad9 acts as a Mec1 adaptor to allow Rad53 activation.** *Curr. Biol.* 2005, **15**:1364–1375.

Unal, E. *et al.*: **DNA damage response pathway uses histone modification to assemble a double-strand break-specific cohesin domain.** *Mol. Cell* 2004, **16**:991–1002.

Vidanes, G.M. *et al.*: **Complicated tails: histone modifications and the DNA damage response.** *Cell* 2005, **121**:973–976.

p53 is responsible for long-term inhibition of cell proliferation in animal cells

In a multicellular organism, one cell with severely damaged DNA can be a substantial threat to the organism as a whole. It is therefore to the benefit of the organism to prevent badly damaged cells from proliferating, either by arresting them permanently in G1 or by removing them entirely by apoptosis. In animal cells, one regulatory protein, known as p53, lies at the heart of this important response to DNA damage.

The importance of p53 in the response to DNA damage and other stresses is best illustrated by the fact that p53 is the single most frequently mutated protein in cancer, being inactivated in at least half of all cases. In the absence of p53 function, the response to DNA damage is deficient, and the resulting accumulation of gene mutations greatly enhances the likelihood that cancer will develop. p53 is therefore called a tumor suppressor, as we discuss in Chapter 12.

p53 is a gene regulatory protein that binds directly to the promoters of its target genes and alters the rate at which their transcription is initiated. In most cases the expression of target genes is stimulated, and the overall result of p53 activation is increased production of proteins that inhibit cell-cycle progression or stimulate apoptosis. p53 represses the transcription of some target genes, notably those encoding inhibitors of apoptosis. The result of p53 action is therefore either cell-cycle arrest or cell death, depending on the cell type and other factors.

The vast majority of studies of p53 have focused on mouse and human cells. A p53-related gene regulatory factor has been identified in *Drosophila* and is required for the stimulation of cell death in response to DNA damage. Unlike the p53 of higher animals, however, *Drosophila* p53 does not stimulate cell-cycle arrest after damage.

p53 is of central importance in the response to DNA damage and other cellular stresses, and its activation can cause the death of the cell. It is therefore subject to an unusually large array of regulatory modifications that ensure it is present and active only when necessary. Most of these modifications increase its concentration or its intrinsic gene regulatory activity, or both, when DNA damage occurs.

The major regulators of p53 include Mdm2, p300 and ARF

The primary structure of p53 is illustrated schematically in Figure 11-10. Like many transcriptional regulators, p53 has a DNA-binding domain that recognizes specific sequences in the regulatory regions of the genes it controls, and a separate amino-terminal transcriptional activation region that interacts with the transcriptional machinery. Together with a regulatory region at the carboxy-terminal end of p53, the transcriptional activation domain is the target of an array of regulatory proteins that catalyze the ubiquitination, phosphorylation and acetylation of specific amino acids.

The major regulator of p53 is Mdm2, an E3 ubiquitin-protein ligase that ubiquitinates several lysine residues near the carboxyl terminus of p53, thereby targeting it for destruction in the proteasome (see section 3-9). Mice lacking Mdm2 die early in embryonic development, apparently because excessive amounts of p53 accumulate and block the proliferation of many cell types; mutation of p53 in these mice prevents the lethal effects of the Mdm2 mutation. In the absence of DNA damage, Mdm2 associates with p53 and keeps its concentration at a minimum. When DNA damage occurs, numerous mechanisms reduce the activity of Mdm2, thereby stabilizing p53. Mdm2 binds to the amino-terminal region of p53, which contains the transcriptional activation domain (see Figure 11-10). By interacting with this domain, Mdm2 inhibits the intrinsic gene regulatory activity of p53 as well as promoting its destruction.

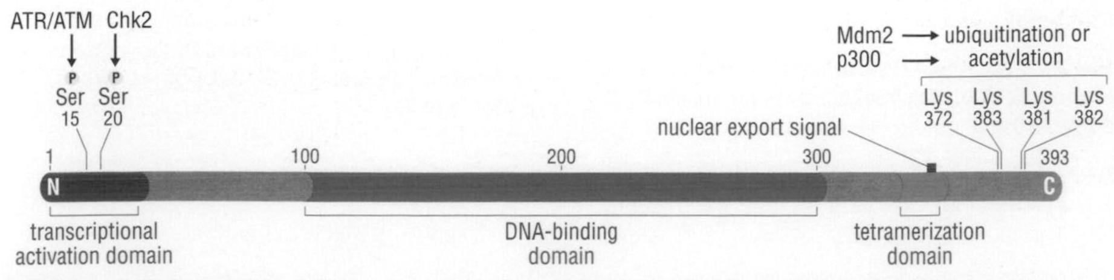

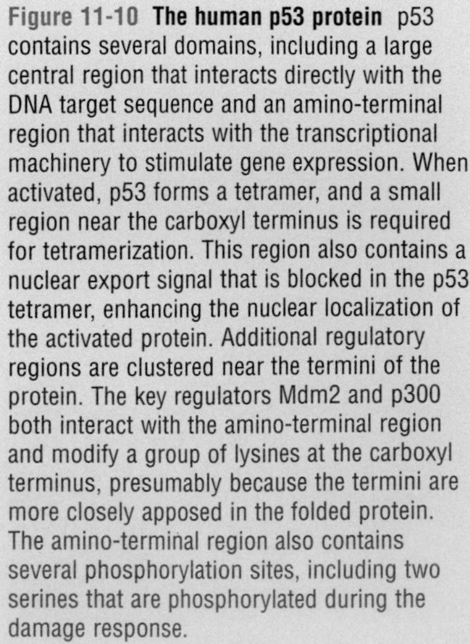

Figure 11-10 The human p53 protein p53 contains several domains, including a large central region that interacts directly with the DNA target sequence and an amino-terminal region that interacts with the transcriptional machinery to stimulate gene expression. When activated, p53 forms a tetramer, and a small region near the carboxyl terminus is required for tetramerization. This region also contains a nuclear export signal that is blocked in the p53 tetramer, enhancing the nuclear localization of the activated protein. Additional regulatory regions are clustered near the termini of the protein. The key regulators Mdm2 and p300 both interact with the amino-terminal region and modify a group of lysines at the carboxyl terminus, presumably because the termini are more closely apposed in the folded protein. The amino-terminal region also contains several phosphorylation sites, including two serines that are phosphorylated during the damage response.

The function of p53 is also regulated by the protein p300, which contains histone acetyltransferase activity. p300 associates with p53 during the DNA damage response and helps promote local gene expression by acetylating histones, thereby generating a more open chromatin structure (see section 4-9). p300 also acetylates p53 itself, at the same lysines that are ubiquitinated by Mdm2 in the absence of damage (see Figure 11-10). Acetylation of these lysines blocks their ubiquitination, thus further ensuring the stabilization of p53 during the damage response.

Another important p53 regulator is the protein ARF, which binds Mdm2 and inhibits p53 degradation. This protein is not central to the DNA damage response, but mutant cells lacking ARF tend to have blunted p53 responses because of high Mdm2 activity. ARF has functions in the response of the cell to imbalances in mitogenic signaling, as discussed later in this chapter in the context of stress responses (see section 11-8).

Damage-response kinases phosphorylate p53 and Mdm2

The stability and activity of p53 are regulated in large part by protein kinases and phosphatases that control the phosphorylation of a remarkably large number of residues in p53 and Mdm2. The function of many of these phosphorylation events remains obscure, but the key regulatory modifications in the DNA damage response seem clear. One of the most rapid and best established early events is phosphorylation of Mdm2 at Ser 395. This inhibits the association of Mdm2 and p53 and thus stabilizes p53. This phosphorylation is probably catalyzed by ATM, and perhaps ATR. The same kinases also phosphorylate p53 itself on Ser 15 in the activation domain. This both inhibits Mdm2 binding and increases p300 binding and acetylation, thereby increasing the stability and gene regulatory activity of p53. Finally, the effector kinase Chk2 (and probably Chk1 as well) phosphorylates p53 at serine 20, which also reduces Mdm2 binding and helps stabilize p53 (see Figure 11-10).

The nuclear localization of p53 is also regulated. In the absence of damage, a nuclear export signal near the carboxyl terminus (see Figure 11-10) ensures that p53 is kept out of the nucleus, thereby preventing its association with target genes. When p53 is stabilized and activated after DNA damage, it forms a tetrameric complex in which the nuclear export signal is blocked, ensuring that the active tetramer is retained in the nucleus. The DNA damage response signal therefore employs multiple overlapping mechanisms to ensure the rapid and robust activation of p53-dependent gene expression (Figure 11-11).

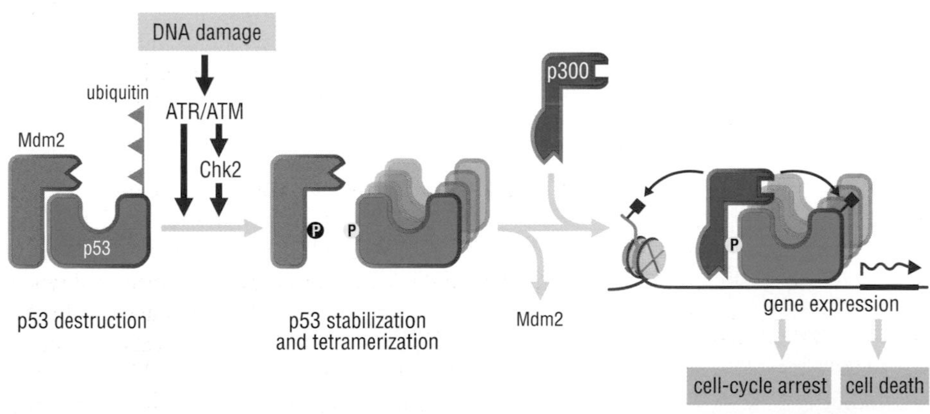

Figure 11-11 Activation of p53-dependent gene expression by DNA damage In the absence of DNA damage, Mdm2 ubiquitinates p53, thereby promoting its destruction by the proteasome. The small amount of p53 in undamaged cells is further restrained by its export from the nucleus. After DNA damage, phosphorylation of Mdm2 and p53 disrupts their association, resulting in stabilization and activation of p53. Tetramerization of active p53 blocks its nuclear export, further increasing its concentration in the nucleus. Phosphorylation of p53 also enhances its interactions with transcriptional proteins, including the histone acetylase p300. p300 associates with p53 and promotes the acetylation of histones and of p53 itself, both of which act to increase p53-dependent gene expression.

References

Brooks, C.L. and Gu, W.: **Ubiquitination, phosphorylation and acetylation: the molecular basis for p53 regulation.** *Curr. Opin. Cell Biol.* 2003, **15**:164–171.

Harris, S.L. and Levine, A.J.: **The p53 pathway: positive and negative feedback loops.** *Oncogene* 2005, **24**:2899–2908.

Levine, A.J.: **p53, the cellular gatekeeper for growth and division.** *Cell* 1997, **88**:323–331.

Vogelstein, B. *et al.*: **Surfing the p53 network.** *Nature* 2000, **408**:307–310.

Vousden, K.H. and Lu, X.: **Live or let die: the cell's response to p53.** *Nat. Rev. Cancer* 2002, **2**:594–604.

Wahl, G.M. and Carr, A.M.: **The evolution of diverse biological responses to DNA damage: insights from yeast and p53.** *Nat. Cell Biol.* 2001, **3**:E277–E286.

Yee, K.S. and Vousden, K.H.: **Complicating the complexity of p53.** *Carcinogenesis* 2005, **26**:1317–1322.

11-5 Effects of DNA Damage on Progression through Start

DNA damage blocks cell-cycle progression at multiple points

The survival of any organism is more likely if its cells are prevented from dividing with damaged DNA. One immediate effect of severe DNA damage is therefore cell-cycle arrest, caused by the activated protein kinases of the DNA damage response system acting on the cell-cycle control system to block progression through all major cell-cycle transitions. The effects of DNA damage at each of these transitions varies in different species: arrest in G1 is the principal effect of DNA damage in mammals, whereas a delay in progression through mitosis is more important in yeast. In this section we will discuss the effects of DNA damage on progression through Start, and then turn in later sections to the effects of damage on DNA replication and mitosis.

DNA damage has minor effects on progression through Start in budding yeast

In budding yeast, most forms of DNA damage cause little or no response in G1—depending on the type and severity of damage. Extensive amounts of nucleotide methylation or cross-linking trigger a brief G1 delay that depends on ATR and Chk2, partly because Chk2 phosphorylates the gene regulatory factor Swi6, thereby delaying expression of the G1/S cyclins Cln1 and Cln2 (see section 10-1). Typically, however, minor nucleotide damage in G1 is either repaired rapidly or stimulates a damage response only when the cell reaches S phase, when replication forks encounter the damage and initiate an ATR-dependent response—as discussed later in this chapter (section 11-6).

After a double-strand break in a G1 yeast cell, the MRN complex and ATM are recruited to the break but there is little cell-cycle response. As mentioned earlier in this chapter (see section 11-2), processing and recombinational repair of double-strand breaks are suppressed in G1, and non-homologous end joining is not well developed in yeast: as a result, a double-strand break in G1 yeast cells is often left unrepaired, leading to potentially lethal chromosome damage on progression through the cell cycle.

DNA damage in vertebrate cells triggers a robust G1 arrest

Extensive DNA damage in G1 vertebrate cells results in a strong and often irreversible block to cell-cycle entry. The G1 damage response can be divided into two phases (Figure 11-12). The first is the rapid response, which occurs within minutes of the damage and is mediated by changes in the phosphorylation state of key cell-cycle regulators. The second part of the response is the delayed or maintenance phase, which involves the activation of p53 and the increased expression of regulatory proteins governing progression through Start.

Vertebrate G1 cells are particularly responsive to double-strand breaks, which trigger an ATM-dependent damage response leading to activation of the effector kinase Chk2. A key target for Chk2 is the phosphatase Cdc25A, one of the three members of the Cdc25 family of Cdk-activating phosphatases (see section 5-4). Cdc25A normally contributes to the activation of cyclin E–Cdk2 and cyclin A–Cdk2 at Start by dephosphorylating an inhibitory tyrosine residue in Cdk2. Phosphorylation of Cdc25A by Chk2, however, targets the protein for ubiquitination by the ubiquitin-protein ligase SCF (coupled with the F-box protein β-TRCP; see section 3-9). The DNA damage response thereby promotes the destruction of Cdc25A, allowing inhibitory phosphorylation to accumulate on Cdk2 and prevent its activation at Start.

References

Bakkenist, C.J. and Kastan, M.B.: **Initiating cellular stress responses.** *Cell* 2004, **118**:9–17.

Bartek, J. and Lukas, J.: **Mammalian G1- and S-phase checkpoints in response to DNA damage.** *Curr. Opin. Cell Biol.* 2001, **13**:738–747.

Chipuk, J.E. *et al.*: **PUMA couples the nuclear and cytoplasmic proapoptotic function of p53.** *Science* 2005, **309**:1732–1735.

Jazayeri, A. *et al.*: **ATM- and cell cycle-dependent regulation of ATR in response to DNA double-strand breaks.** *Nat. Cell Biol.* 2006, **8**:37–45.

Lisby, M. *et al.*: **Choreography of the DNA damage response: spatiotemporal relationships among checkpoint and repair proteins.** *Cell* 2004, **118**:699–713.

Vousden, K.H. and Lu, X.: **Live or let die: the cell's response to p53.** *Nat. Rev. Cancer* 2002, **2**:594–604.

Wahl, G.M. and Carr, A.M.: **The evolution of diverse biological responses to DNA damage: insights from yeast and p53.** *Nat. Cell Biol.* 2001, **3**:E277–E286.

Yee, K.S. and Vousden, K.H.: **Complicating the complexity of p53.** *Carcinogenesis* 2005, **26**:1317–1322.

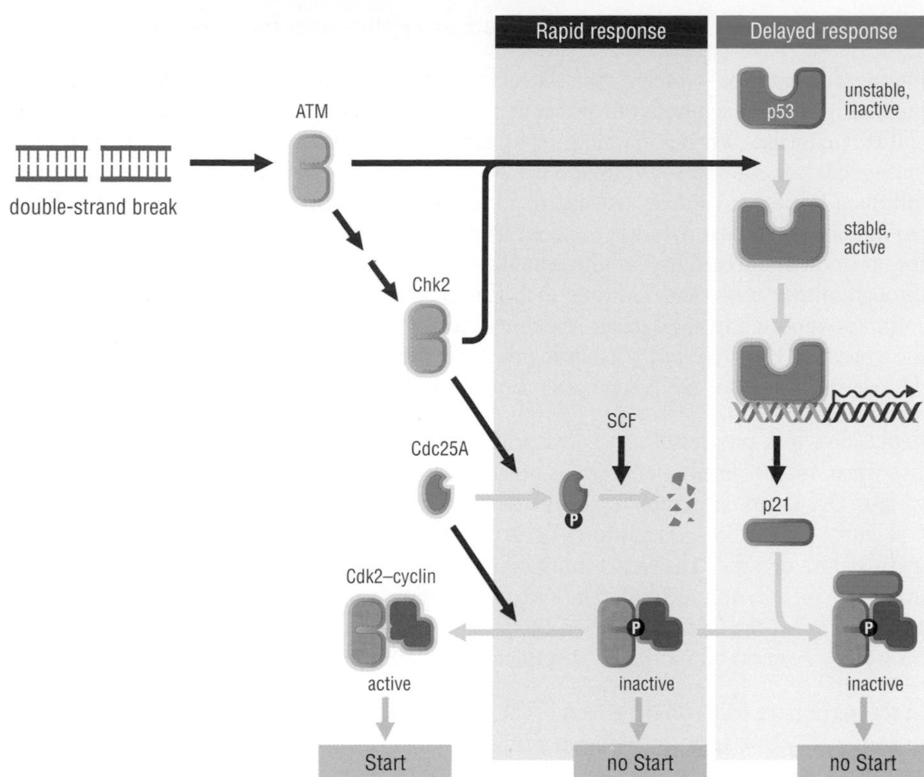

Figure 11-12 Inhibition of progression through Start after DNA damage in human cells Double-strand breaks in G1 lead to activation of ATM and of Chk2, which phosphorylates Cdc25A, targeting it for ubiquitination and destruction. As a result, inhibitory phosphorylation accumulates on Cdk2 and blocks progression through Start. ATM activation also causes the stabilization and activation of p53 (see section 11-4), resulting in increased expression of the gene encoding the Cdk inhibitor p21, which further suppresses Cdk activities and helps maintain long-term cell-cycle arrest.

ATR seems to be a less critical component of the DNA damage response in G1. As mentioned earlier (section 11-2), double-strand breaks are usually not resected during G1 and do not trigger a secondary ATR response as they do in later cell-cycle stages. In many vertebrate cell types, minor nucleotide damage (methylation or UV-induced thymine dimers, for example) only triggers an ATR response during S phase and not during G1. There is evidence, however, that an ATR-dependent response can occur during G1 in some cell types, and this response is thought to be mediated primarily through the activation of Chk1.

ATM and ATR initiate a delayed damage response by stabilizing and activating p53, as described earlier (see section 11-4). Increased p53 levels lead to changes in the expression of numerous p53 target genes, which eventually promote prolonged cell-cycle arrest. A key target of p53 is the gene encoding the Cdk inhibitor p21 (see section 3-6). Activation of p53 typically leads to increased expression of p21, which binds and inhibits the cyclin–Cdk2 complexes required for passage through Start (see Figure 11-12).

In many cell types, the long-term response to DNA damage and p53 activation is cell death by apoptosis. Several of the p53 target genes encode proteins that promote apoptosis or inhibit signaling by survival factors. Chief among these is the gene encoding the protein PUMA, a proapoptotic BH3-only member of the Bcl-2 family of apoptosis regulators (see section 10-14). In addition, p53 itself is thought to possess BH3-like activity that allows it to promote cell death through direct activation of proapoptotic members of the Bcl-2 family.

p53 has different effects in different cell types

The response to p53 activation varies markedly in different cell types, ranging from a reversible, p21-dependent cell-cycle arrest to cell death. The mechanisms underlying the variation in the p53 response are not well understood, but several potential explanations are taking shape. Survival signaling pathways may be particularly intense in some cells, resulting in suppression of the apoptotic response to p53. There may be cell-specific differences in co-activators, chromatin structure, and other regulatory molecules that affect some p53 target genes and not others. Similarly, phosphorylation of p53 at certain sites promotes expression of specific subsets of target genes, indicating that cell-specific differences in the activity of p53 kinases could affect the pattern of gene expression in different cell types. p53 thus lies at the heart of an immensely complex network of regulatory interactions, and its precise effects in different cells vary as a result of differences in the concentrations and behaviors of the components of this network.

11-6 Effects of DNA Damage at Replication Forks

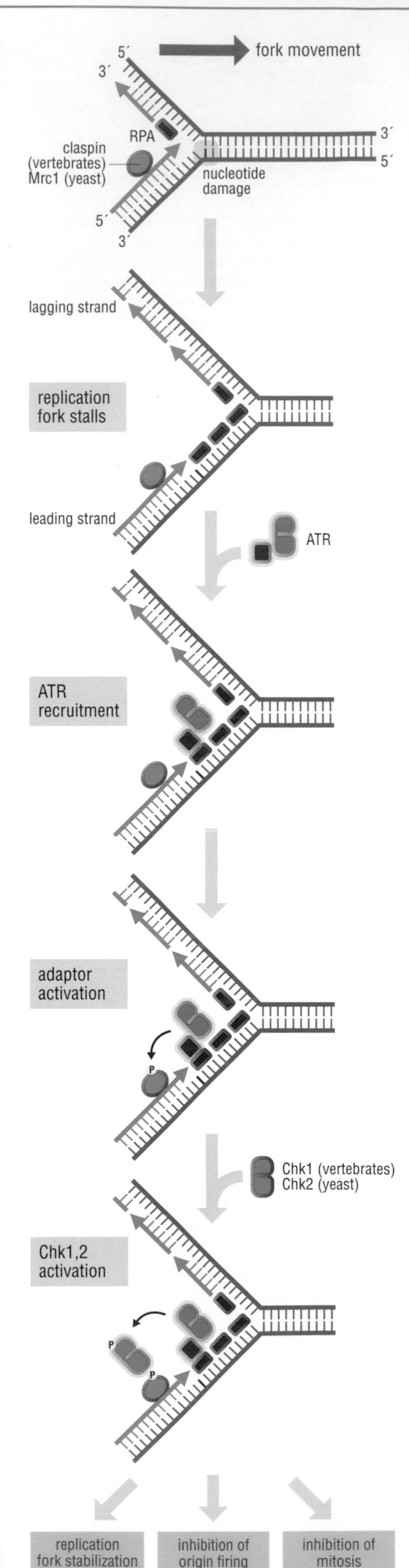

A DNA damage response is initiated at replication forks during S phase

During S phase, the proteins that assemble at the replication fork serve not only as the machinery of DNA synthesis but also as important and highly sensitive sensors of many forms of DNA damage. When a traveling replication fork encounters damaged DNA—alkylated nucleotides, for example—its progress can be delayed or even blocked. The stalled replication fork then initiates a complex damage response that has at least three major outcomes. First, replication origins that have not yet been used are prevented from firing, thereby preventing the initiation of further DNA synthesis until the damage has been repaired. Second, progression through mitosis is blocked, ensuring that damaged chromosomes are not segregated. Finally, and most importantly, the damage response system helps govern the structure and function of the replication fork machinery, stabilizing the fork when it stalls at sites of damage and allowing the safe resumption of DNA synthesis when damage has been repaired.

Replication fork progression is stalled or delayed not only as a result of DNA damage but also when the fork encounters various other obstacles, such as tightly packed chromatin and various DNA–protein structures. Forks can also stall when replication enzymes fail. In the laboratory, for example, stalled forks occur after depletion of nucleotides (by treatment with hydroxyurea or other chemicals) or inhibition of DNA polymerase activity (by chemical inhibitors such as aphidicolin). The system that responds to DNA synthesis inhibitors was originally called the DNA replication checkpoint, but it is no longer considered to be distinct from the system that responds to DNA damage or other stresses at forks.

In this section we focus on the effects of the damage response at the replication fork itself, and in section 11-7 we turn to the mechanisms by which the response blocks origin firing and mitotic progression.

ATR is the key initiator of the response to stalled replication forks

The protein kinase ATR, with its subunit ATRIP, is the central component of the response to stalled replication forks (Figure 11-13). ATR–ATRIP, 9-1-1 and Rad17–RFC complexes are all recruited to stalled replication forks, although the last two are not essential for the response. The damage signal is initiated by ATR-dependent activation of either Chk1 (in vertebrates) or Chk2 (in yeast). Phosphorylation of these kinases depends on the adaptor protein claspin in vertebrates and the related protein Mrc1 in yeast (see Figure 11-5). Unlike the adaptor proteins described earlier in this chapter (see section 11-3), which are recruited to the DNA only after damage occurs, claspin and Mrc1 are present at the replication fork throughout S phase and have functions in DNA synthesis as well as in the damage response.

As discussed earlier in this chapter (see section 11-2), ATR tends to be activated by DNA damage that leads to the formation of tracts of single-stranded DNA coated with RPA. It is therefore likely that single-stranded DNA–RPA is the signal that initiates a damage response at stalled replication forks. Single-stranded DNA can be generated in a number of ways. In some cases, the Mcm helicase continues unwinding the DNA after the polymerase machinery has stalled. In other cases, synthesis of the leading and lagging strands is uncoupled because the polymerase on one strand continues operating while the other is blocked.

The DNA damage response stabilizes the replication fork

The S-phase damage response is usually reversible, particularly in yeast. When damage is repaired or polymerase function is restored, stalled replication forks resume DNA synthesis. Restarting of DNA synthesis might seem at first glance to be a straightforward process. In fact, stalled replication forks are highly unstable structures and the cell uses a great deal of energy

Figure 11-13 The response to stalled replication forks When a replication fork encounters DNA damage—an alkylated nucleotide, for example—the polymerase machinery is blocked on the damaged strand, resulting in a stalled replication fork. Single-stranded DNA accumulates at the stalled fork, probably because DNA synthesis continues briefly on the other strand (although the precise DNA structure at stalled forks is likely to vary with different types of damage). The single-stranded DNA, coated with RPA, recruits ATR. The adaptor proteins claspin or Mrc1 promote phosphorylation and thus activation of the kinase Chk1 (in vertebrates) or Chk2 (in yeast).

in stabilizing forks so that they can be restarted without generating replication errors. Fork stabilization is dependent on the DNA damage response. Indeed, studies of damage-response mutants in yeast indicate that stabilization of replication forks might be the only function of the S-phase damage response that is essential for cell survival after DNA damage.

The importance of replication fork stabilization is best illustrated by the problems that arise when it does not occur. In budding yeast, mutations in damage response components prevent fork stabilization, leading to a complex and poorly understood process called replication fork collapse. In *chk2* mutants, for example, the DNA at collapsed forks forms a variety of abnormal structures, including extensive single-stranded regions and reversed replication forks (Figure 11-14). Double-strand breaks can also occur. If a replication fork stalls at a nick in one strand of the DNA, for example, a double-strand break can result if the fork collapses. Many of the abnormal DNA structures at collapsed forks are difficult to repair or are susceptible to degradation, breakage or recombination.

How does the damage response prevent the collapse of replication forks? At least two mechanisms are involved. First, ATR and Chk2 phosphorylate components of the replication machinery, apparently locking them in place on the DNA. In budding yeast, for example, mutations in ATR result in loss of polymerases from stalled forks, whereas defects in Chk2 cause the loss of the Mcm helicase. Second, the damage response inhibits the activity of recombination enzymes that process certain DNA structures at stalled forks and thus generate mutations and chromosomal rearrangements.

A wide range of mechanisms are used to help stalled or collapsed replication forks resume DNA synthesis. In those cases where the fork has stalled at nucleotide damage, specialized but error-prone DNA polymerases called translesion polymerases may be recruited to the fork. These simply continue synthesis across the damaged nucleotide. For double-strand breaks at stalled forks, a more complex repair mechanism uses the newly replicated and undamaged sister chromatid as a template for recombinational repair (see section 11-1); this type of repair can, however, lead to chromosomal rearrangements if the recombination intermediates are not processed correctly. In addition, helicases of the RecQ family (Sgs1 in budding yeast, five proteins in humans) help prepare collapsed forks for restart by unraveling abnormal DNA structures and preventing improper recombination events. Mutations in human RecQ proteins are responsible for important human diseases, including *Bloom's syndrome* and *Werner's syndrome*, which are characterized by defects in chromosome stability and increased rates of cancer.

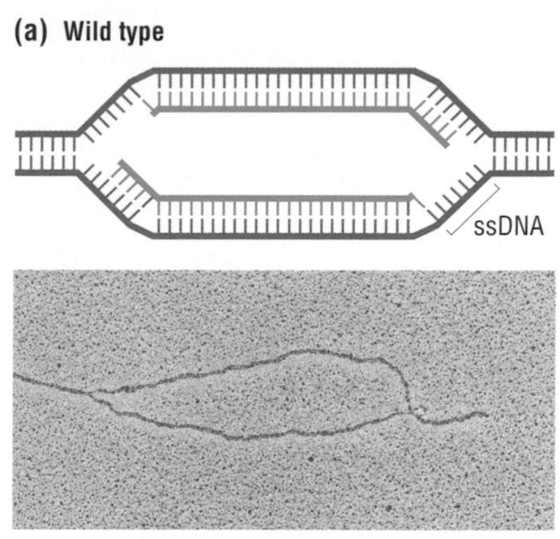

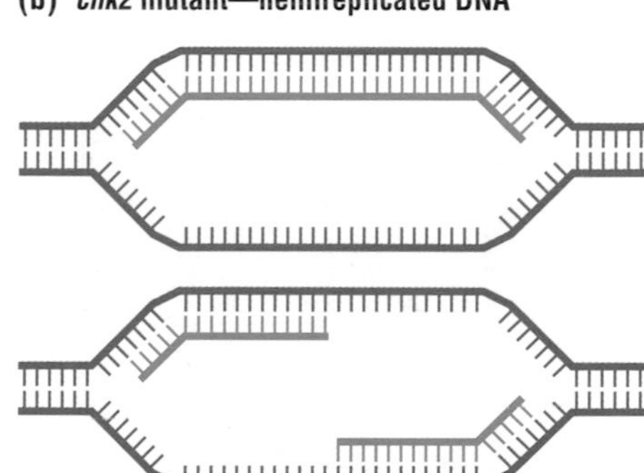

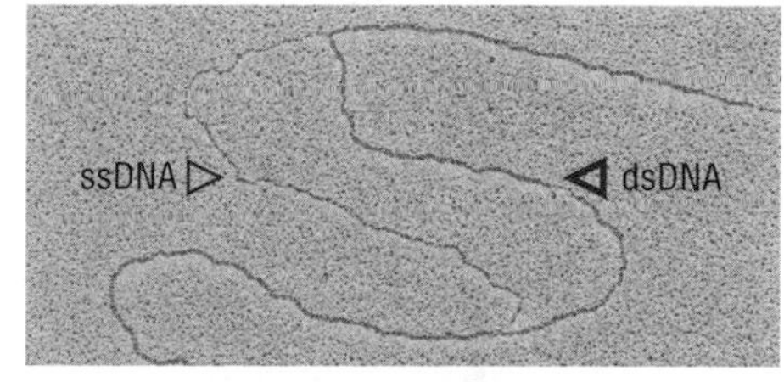

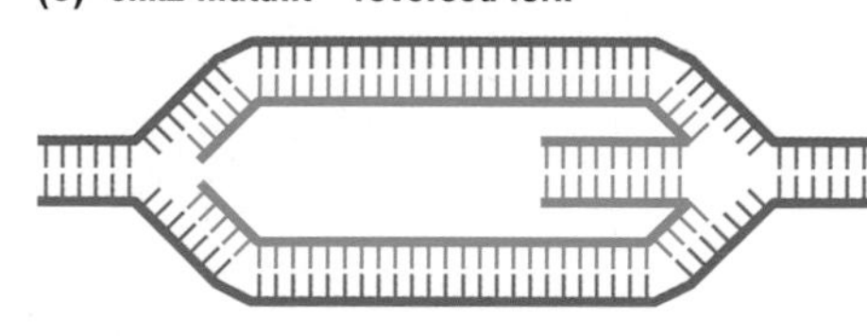

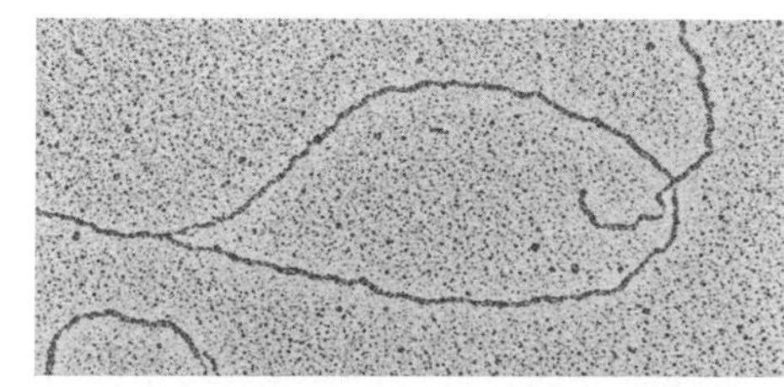

Figure 11-14 Abnormal DNA structures at stalled replication forks in yeast *chk2* mutants The structure of stalled replication forks was determined by electron microscopy of DNA isolated from yeast cells treated with the DNA synthesis inhibitor hydroxyurea. The thickness of the DNA in these images indicates whether it is single-stranded (ssDNA) or double-stranded (dsDNA). **(a)** In wild-type cells, the DNA at stalled replication forks is generally double-stranded, although it sometimes contains short single-stranded stretches (not visible in the microscope image). **(b)** In cells lacking Chk2, abnormal DNA structures are seen at stalled forks. In some cases, imbalances in the rate of synthesis on the two strands lead to hemireplicated DNA, in which large sections of single-stranded DNA are seen. This DNA is highly susceptible to processing by nucleases. **(c)** In other cases, poorly understood mechanisms transform stalled replication forks into an X-shaped structure, sometimes called a reversed fork, in which the two newly synthesized strands anneal. This structure can form a Holliday junction like that seen during meiotic recombination (see section 9-2). If this junction is resolved inappropriately by the recombination machinery, double-strand breaks can result. Thus, Chk2 is required to prevent the formation of potentially harmful DNA structures at stalled forks. Courtesy of Massimo Lopes and Marco Foiani.

References

Branzei, D. and Foiani, M.: **The DNA damage response during DNA replication.** *Curr. Opin. Cell Biol.* 2005, **17**:568–575.

Byun, T.S. *et al.*: **Functional uncoupling of MCM helicase and DNA polymerase activities activates the ATR-dependent checkpoint.** *Genes Dev.* 2005, **19**:1040–1052.

Cobb, J.A. *et al.*: **Redundancy, insult-specific sensors and thresholds: unlocking the S-phase checkpoint response.** *Curr. Opin. Genet. Dev.* 2004, **14**:292–300.

Cobb, J.A. *et al.*: **Replisome instability, fork collapse, and gross chromosomal rearrangements arise synergistically from Mec1 kinase and RecQ helicase mutations.** *Genes Dev.* 2005, **19**:3055–3069.

Desany, B.A. *et al.*: **Recovery from DNA replicational stress is the essential function of the S-phase checkpoint pathway.** *Genes Dev.* 1998, **12**:2956–2970.

Lee, J. *et al.*: **Roles of replication fork-interacting and Chk1-activating domains from Claspin in a DNA replication checkpoint response.** *Mol. Biol. Cell* 2005, **16**:5269–5282.

Lopes, M. *et al.*: **Multiple mechanisms control chromosome integrity after replication fork uncoupling and restart at irreparable UV lesions.** *Mol. Cell* 2006, **21**:15–27.

Sogo, J.M. *et al.*: **Fork reversal and ssDNA accumulation at stalled replication forks owing to checkpoint defects.** *Science* 2002, **297**:599–602.

Tercero, J.A. *et al.*: **A central role for DNA replication forks in checkpoint activation and response.** *Mol. Cell* 2003, **11**:1323–1336.

DNA damage in S phase blocks replication origin firing

Damage to DNA and defects in the replication fork during S phase trigger multiple responses, including the inhibition of replication origin firing. Damage in early S phase thereby prevents the initiation of DNA synthesis at other origins until the damage has been repaired.

Double-strand breaks block origin firing by a mechanism that depends on ATM and the MRN complex (see section 11-2). When this response is defective, as in patients with mutations in ATM or in subunits of the MRN complex, DNA synthesis is not blocked—resulting in a phenomenon called radioresistant DNA synthesis (Figure 11-15). ATM blocks origin firing by preventing the activation of cyclin–Cdk2 complexes, which are required for the initiation of DNA synthesis at origins (see section 4-6). As in the ATM-mediated response in G1 cells (see Figure 11-12), ATM acts through the activation of Chk2, which phosphorylates sites in the amino-terminal region of Cdc25A that trigger its ubiquitination by SCF (Figure 11-16). Cdc25A is destroyed, allowing inhibitory phosphorylation to accumulate on Cdk2. ATM activation in S phase also leads to phosphorylation of the protein Smc1—a component of the cohesin complex that holds sister chromatids together after replication (see section 5-8). Mutation of the ATM-dependent phosphorylation sites in Smc1 reduces the damage response, but it remains unclear how this phosphorylation helps block origin firing.

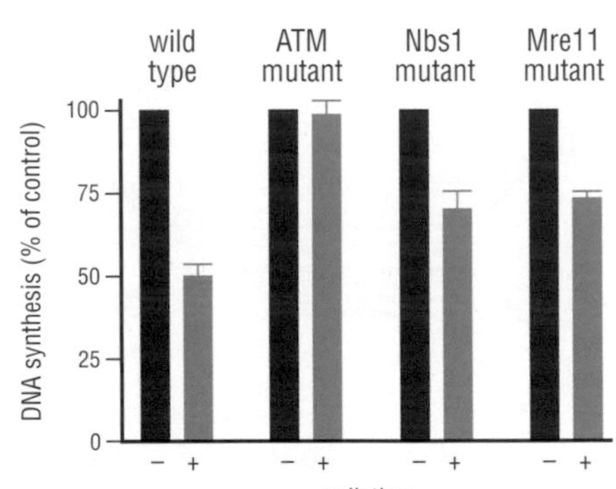

Figure 11-15 The response to double-strand breaks during S phase in human cells Several human diseases are caused by mutations in DNA damage response components, including ataxia telangiectasia (AT, caused by mutations in the gene *ATM*), Nijmegen breakage syndrome (NBS, caused by mutations in *NBS1*) and AT-like disease (ATLD, caused by mutations in *MRE11*). Cells from patients with these diseases have reduced responses to DNA damage, including damage in S phase, as shown here. In normal human cells, DNA synthesis is inhibited by about 50% one hour after treatment with ionizing radiation. In cells from patients with *ATM* mutations, this response is lost, resulting in radioresistant DNA synthesis (RDS). The response is partly lost in patients with mutations in *NBS1* or *MRE11*, encoding two subunits of the MRN complex. The mutant Nbs1 and Mre11 proteins in these patients are thought to retain some function, resulting in a partial response to DNA damage. Adapted from Falck, J. *et al.*: *Nat. Genet.* 2002, **30**:290–294.

Stalled replication forks block origin firing by an ATR-dependent mechanism. An ATR response is also triggered by resection of double-strand breaks during S phase, as discussed earlier (see section 11-2). In human cells, ATR acts primarily through the activation of Chk1, which prevents origin firing by inhibiting the phosphatase Cdc25A (see Figure 11-16). Two mechanisms are involved. First, Chk1 (like Chk2) phosphorylates Cdc25A at the amino-terminal sites that promote Cdc25A destruction. In addition, Chk1 (but not Chk2) phosphorylates Cdc25A at a site in the carboxy-terminal region that inhibits the ability of Cdc25A to interact with and dephosphorylate Cdk2. This second inhibitory mechanism is less critical in ATM-dependent DNA damage responses, which in human cells are mediated primarily by Chk2.

DNA damage in S phase may also block activation of the protein kinase Cdc7, which is required for origin firing (see section 4-7). In budding yeast, ATR-dependent activation of Chk2 leads to phosphorylation of the Dbf4 subunit of Cdc7, thereby inhibiting the ability of Cdc7 to drive origin firing. There is also evidence from studies in *Xenopus* embryo extracts that the activation of ATR leads to inhibition of Cdc7 activity.

DNA damage blocks mitotic entry in most eukaryotes

When DNA damage occurs in S phase or G2, entry into mitosis is blocked until the damage has been repaired, ensuring that the cell does not make a potentially dangerous attempt to segregate damaged chromosomes. In most eukaryotes—including fission yeast, *Drosophila* and vertebrates—the DNA damage response acts by blocking the activation of mitotic cyclin–Cdk1 complexes (see Figure 11-6).

Activation of the DNA damage response—through ATR or ATM—leads to increased Chk1 and Chk2 activities. In vertebrates these kinases phosphorylate multiple sites on the three members of the Cdc25 family of phosphatases that normally trigger mitotic Cdk1 activation (see section 5-4). As discussed above, Chk1 (but not Chk2) phosphorylates a carboxy-terminal site in Cdc25A, thereby inhibiting its ability to activate Cdk1; similar sites are probably phosphorylated in Cdc25B and Cdc25C, further blocking Cdk1 activation (see Figure 11-16).

References

Bartek, J. and Lukas, J.: **Chk1 and Chk2 kinases in checkpoint control and cancer.** *Cancer Cell* 2003, **3**:421–429.

Chan, T.A. *et al.*: **Cooperative effects of genes controlling the G_2/M checkpoint.** *Genes Dev.* 2000, **14**:1584–1588.

Costanzo, V. *et al.*: **An ATR- and Cdc7-dependent DNA damage checkpoint that inhibits initiation of DNA replication.** *Mol. Cell* 2003, **11**:203–213.

Falck, J. *et al.*: **The DNA damage-dependent intra-S phase checkpoint is regulated by parallel pathways.** *Nat. Genet.* 2002, **30**:290–294.

McGowan, C.H. and Russell, P.: **The DNA damage response: sensing and signaling.** *Curr. Opin. Cell Biol.* 2004, **16**:629–633.

Nyberg, K.A. *et al.*: **Toward maintaining the genome: DNA damage and replication checkpoints.** *Annu. Rev. Genet.* 2002, **36**:617–656.

Uto, K. *et al.*: **Chk1, but not Chk2, inhibits Cdc25 phosphatases by a novel common mechanism.** *EMBO J.* 2004, **23**:3386–3396.

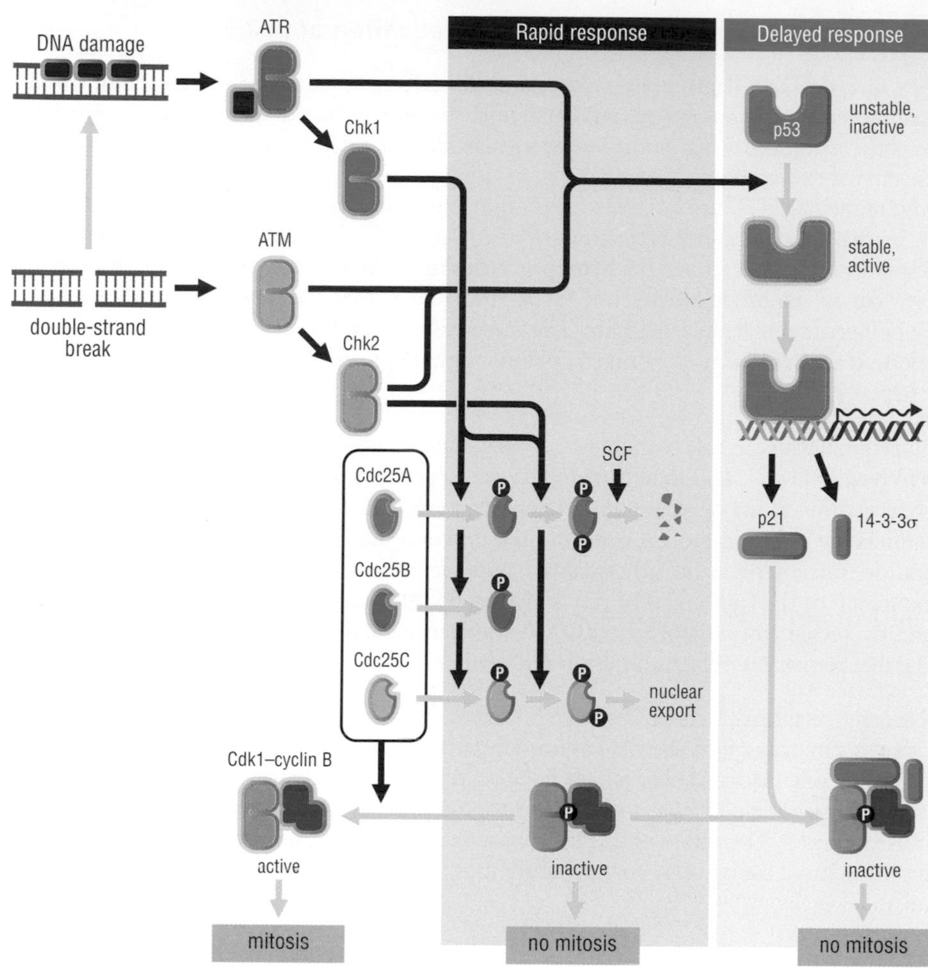

Figure 11-16 Inhibition of mitotic entry by DNA damage Activation of ATR and/or ATM during S phase or G2 leads to activation of the protein kinases Chk1 and Chk2. Chk1 and Chk2 have complex inhibitory effects on the three members of the Cdc25 phosphatase family. Chk1 phosphorylates a carboxy-terminal site in all three members, thereby inhibiting their activity. Chk1 and Chk2 phosphorylate other sites in Cdc25A, leading to its ubiquitination by SCF. Chk1 and Chk2 also phosphorylate Cdc25C at a site that inhibits its activity and its nuclear import. Activation of Cdk1–cyclin B1 is thereby blocked, preventing entry into mitosis. In the long term, p53 activation further suppresses Cdk1 activity by stimulating expression of the Cdk1 inhibitor p21 and the protein 14-3-3σ, which is thought to bind Cdk1–cyclin B1 and block its import into the nucleus. The damage response also inhibits the protein kinase Plk (not shown), another major promoter of mitotic entry.

Phosphorylation by Chk1 and Chk2 also promotes Cdc25A destruction as described earlier. Finally, as discussed in Chapter 5 (see section 5-6), Chk1 and Chk2 phosphorylate Cdc25C at another amino-terminal site (Ser 216 in human cells), blocking its activity and also its import into the nucleus—thereby reducing its ability to activate cyclin B1–Cdk1 complexes that enter the nucleus in late prophase.

In addition to these rapid inhibitory effects on Cdk1 activation, DNA damage in S phase or G2 in many human cells also promotes the long-term maintenance of cell-cycle arrest through the stabilization and activation of p53. Activated p53 enhances cell-cycle arrest by multiple mechanisms that are not well understood. As in G1 cells (see section 11-5), p53 acts in part by stimulating expression of the Cdk inhibitor p21, which binds and inhibits multiple types of cyclin–Cdk complexes (although it has a higher affinity for G1/S- and S-cyclin–Cdk2 than it does for mitotic cyclin B1–Cdk1). In some cell types p53 also promotes the expression of another protein, 14-3-3σ, which is thought to inhibit mitotic entry by preventing nuclear import of cyclin B1–Cdk1. Cells lacking p21 or 14-3-3σ or both have severe defects in the maintenance of the DNA damage response.

DNA damage blocks anaphase in budding yeast

In budding yeast, mitotic entry is not a discrete regulatory transition. In the presence of DNA damage, cells enter mitosis but arrest in metaphase. The molecular basis of this arrest is not clear, but one likely mechanism is that the damage-response kinase Chk1 phosphorylates securin, thereby preventing its ubiquitination by APC^{Cdc20} (see section 7-5). The stabilization of securin prevents the activation of separase and therefore prevents sister-chromatid separation. Because separase also helps trigger Cdk1 inactivation in yeast, securin stabilization also blocks the completion of mitosis. A robust metaphase arrest after DNA damage also requires the damage-response kinase Chk2, whose direct targets remain unclear. It remains likely that other regulators—perhaps including APC^{Cdc20}—are inhibited during the response.

Hyperproliferative signals trigger the activation of p53

As well as the DNA damage response, mammalian cells possess a related stress-response system that prevents inappropriate cell proliferation in response to excessive mitogenic stimuli. This response is revealed when normal cells, such as fibroblasts taken from a mouse embryo, are engineered to overproduce key mitogenic signaling proteins such as the gene regulatory protein Myc or the small GTPase Ras (see section 10-6). Surprisingly, overproduction of these proteins in normal cells causes cell-cycle arrest or apoptosis rather than increased proliferation. This response, sometimes termed the **hyperproliferation stress response** or oncogene checkpoint, provides an important mechanism for removing cells from the population if they display inappropriate proliferative behavior. The mechanism is also important for preventing cancer and so is also referred to as a tumor suppressor or tumor surveillance system—as we discuss in Chapter 12.

Hyperproliferative signals block cell-cycle progression or induce cell death in part by activating p53 (Figure 11-17). The underlying mechanism depends on a small protein called ARF, whose concentration in the cell increases in response to excessive mitogenic stimuli. ARF binds and inhibits the ubiquitin-protein ligase Mdm2, thereby reducing the rate of p53 ubiquitination and destruction. As in the DNA damage response, increased p53 activity leads to increased expression of the Cdk inhibitor p21 and proapoptotic proteins. Depending on the cell type and the presence of survival factors or other influences on the apoptotic machinery, the result is either permanent cell-cycle arrest or cell death.

The gene regulatory factor E2F1, which is activated by mitogenic signals (see section 10-4), can also stimulate apoptosis when overproduced. One of its targets is the gene encoding ARF, providing one mechanism by which E2F1, and thus mitogenic stimuli such as Myc, could drive p53 activation. E2F1 also stimulates apoptosis through mechanisms that are independent of ARF or p53. The ability of E2F1 to promote apoptosis may be important in limiting the proliferation of tumor cells, because deletion of the E2F1 gene increases cancer incidence in mice.

The overactivity of mitogenic signaling proteins such as Myc and Ras is believed to make an important contribution to the excessive proliferation of cancer cells. Indeed, the genes encoding these proteins were originally termed *oncogenes* because they were found to promote excessive proliferation in certain cultured cell lines. It is now clear that oncogenic proteins promote proliferation only when ARF or p53 is absent, which cripples the hyperproliferation stress response. We discuss this issue further in Chapter 12.

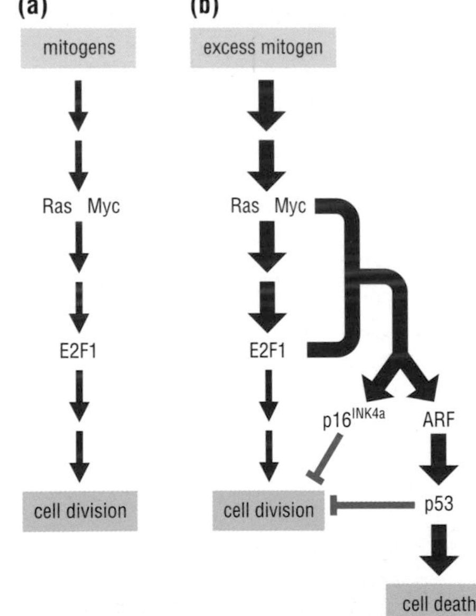

Figure 11-17 The hyperproliferation stress response in mammalian cells **(a)** In normal mammalian cells, mitogens promote cell division through various signaling proteins such as Ras and Myc (see section 10-6), whose activities lead eventually to activation of the gene regulatory factor E2F1 and related proteins (see section 10-4). **(b)** When cells are grown in culture or are engineered to contain overactive Ras, Myc or E2F1, the excessive mitogenic signal triggers increased expression of ARF and, in some cases, the Cdk inhibitor $p16^{INK4a}$. ARF activates p53, resulting in a permanent cell-cycle arrest (replicative senescence) or, under some conditions, cell death.

Imbalances in mitogenic stimuli promote replicative senescence in mouse cells

Fibroblasts taken directly from a mouse and grown in culture divide approximately 15 times and then undergo a stable cell-cycle arrest termed replicative senescence (see section 2-5). In many mouse cell types, this phenomenon results primarily from an increase in the production of ARF and the consequent increase in p53 levels. Senescence in these cells is thought to result from the nonphysiological conditions in the culture dish, in which high levels of serum seem to generate a hyperproliferation stress response like that seen when Myc or Ras are overactive (see Figure 11-17). Senescence might also result from other nonphysiological aspects of life in cell culture—such as the lack of cell–cell contacts, insufficient extracellular matrix components and inappropriate oxygen levels. Replicative senescence is not seen in mouse cells if efforts are made to culture them in a more physiological environment (see section 2-5).

The Cdk inhibitor $p16^{INK4a}$ (see section 3-6) also contributes to the cell-cycle arrest that occurs after prolonged cell culture—although its importance varies in different cell types. Although ARF is the major component required for senescence in many mouse cells, $p16^{INK4a}$ and ARF make equal contributions in other mouse cell types and in many human cells. The concentration of $p16^{INK4a}$ in the cell gradually increases as cultured cells are passaged, presumably because of the nonphysiological conditions of the culture environment.

The molecular mechanism that leads to the accumulation of ARF and $p16^{INK4a}$ in cell culture is not yet clear. The expression of both proteins is actively repressed in fibroblasts *in vivo* by chromatin-remodeling enzymes, and a gradual loss of repressive chromatin structure that occurs during cell culture may contribute to their increased expression. Numerous gene regulatory proteins that help govern *INK4A* expression have been identified, but their regulation during senescence is not well understood.

Mouse cells lacking ARF or p53 do not undergo senescence when subjected to nonphysiological culture conditions. Such cells proliferate indefinitely in culture and are therefore termed immortal (see section 2-5).

Telomere degeneration promotes cell-cycle arrest in human cells

Another important regulatory mechanism that limits cell proliferation centers on telomeres, the complex DNA–protein structures that cap the ends of chromosomes (see section 4-1). This mechanism is most clearly observed in human somatic cells, which often do not express telomerase, the enzyme primarily responsible for maintaining telomere length. In these cells, a gradual decrease in telomere length over many cell generations leads to deterioration of the protein cap on the telomere. Eventually, telomere dysfunction in human cells triggers a permanent cell-cycle arrest (which can be prevented by artificially activating telomerase expression, leading to telomere lengthening, as discussed in section 2-5). Cell-cycle arrest as a result of telomere deterioration does not normally occur in mouse cells, which generally express abundant telomerase, but does occur if telomere function is compromised (by engineering mouse cells with mutations in telomerase or telomere-capping proteins, for example).

The cell-cycle arrest brought about by telomere degeneration is dependent in large part on the DNA damage response. Exposed mammalian telomeres are interpreted as double-strand breaks, leading to the recruitment of ATM and the initiation of a p53-dependent cell-cycle arrest like that seen in response to double-strand breaks elsewhere in the genome (Figure 11-18). Exposed telomeres may also, in some cases, contain single-stranded DNA tracts that activate the ATR branch of the damage response pathway. In yeast, defective telomeres contain abundant single-stranded DNA and the damage response in these cells is mediated primarily by ATR.

In human cells (but not in mice), telomere dysfunction is also thought to promote the accumulation of $p16^{INK4a}$. Little is known about the molecular basis of this response, although it does seem to depend on activation of the DNA damage response but not on p53.

The response to telomere shortening, like the hyperproliferation stress response, is an important tumor suppressor mechanism that limits the proliferative potential of human cells. The loss of these response systems is an important step in the formation of cancer cells, as we discuss in Chapter 12.

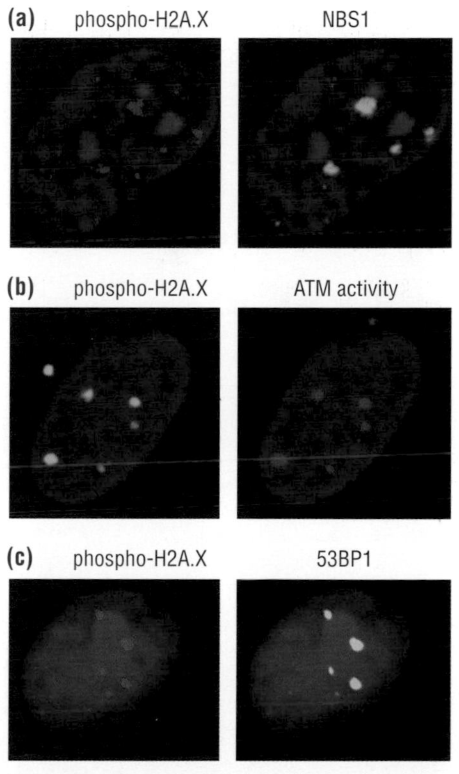

Figure 11-18 Generation of a DNA damage response in senescent human cells In these experiments, normal human fibroblasts were grown in culture until they reached senescence and stopped dividing. To determine if DNA damage was present, cells were then labeled with antibodies against various damage response components. **(a)** In this experiment, cells were labeled with antibodies against phosphorylated histone H2A.X, which is found in large quantities at sites of DNA damage in human cells (see section 11-3). In addition, the same cells were co-labeled with antibodies against Nbs1, a component of the MRN complex (see section 11-2). **(b)** These cells were co-labeled with antibodies raised against phosphorylated substrates of ATM, allowing detection of ATM activity. **(c)** These cells were co-labeled with antibodies against 53BP1, an adaptor protein (see section 11-3). Numerous DNA damage components therefore appear at discrete foci in senescent cells, suggesting that a damage response has been initiated. Other studies (not shown) revealed that these foci are found primarily at the telomeres. Photographs kindly provided by Fabrizio d'Adda di Fagagna. Reprinted, with permission, from d'Adda di Fagagna, F. *et al.*: *Nature* 2003, **426**:194–198.

Definitions

hyperproliferation stress response: cell-cycle arrest or apoptosis seen in response to excessive mitogenic signals.

References

Campisi, J.: **Senescent cells, tumor suppression, and organismal aging: good citizens, bad neighbors.** *Cell* 2005, **120**:513–522.

Celli, G.B. and de Lange, T.: **DNA processing is not required for ATM-mediated telomere damage response after TRF2 deletion.** *Nat. Cell Biol.* 2005, **7**:712–718.

d'Adda di Fagagna, F. *et al.*: **A DNA damage checkpoint response in telomere-initiated senescence.** *Nature* 2003, **426**:194–198.

d'Adda di Fagagna, F. *et al.*: **Functional links between telomeres and proteins of the DNA-damage response.** *Genes Dev.* 2004, **18**:1781–1799.

Dimova, D.K. and Dyson, N.J.: **The E2F transcriptional network: old acquaintances with new faces.** *Oncogene* 2005, **24**:2810–2826.

Sherr, C.J.: **The INK4a/ARF network in tumour suppression.** *Nat. Rev. Mol. Cell Biol.* 2001, **2**:731–737.

Sherr, C.J. and DePinho, R.A.: **Cellular senescence: mitotic clock or culture shock?** *Cell* 2000, **102**:407–410.

Takai, H. *et al.*: **DNA damage foci at dysfunctional telomeres.** *Curr. Biol.* 2003, **13**:1549–1556.

12

The Cell Cycle in Cancer

Cancer is a complex group of diseases characterized by abnormal increases in cell number. Progression of this disease is an evolutionary process driven by the gradual accumulation of gene mutations, which increase the activity of regulatory proteins that stimulate cell proliferation and decrease the activity of proteins that normally restrain it.

12-0 Overview: Cell-Cycle Defects in Cancer

Cancer cells break the communal rules of tissues

The human body is made up of populations of cells organized into tight-knit communities called tissues. Every tissue has an optimal size, which is based on the body's requirements and determined by the number and size of each cell type in that tissue. To achieve and maintain this ideal size, the growth, division, death and differentiation of every cell must be tightly controlled—which is achieved by a complex blend of intrinsic programming and extracellular factors produced by neighboring cells. The cells of each tissue also remain tightly associated or compartmentalized, preventing the cells of one tissue from invading another and disrupting its function.

Cancer is a disease in which cells no longer respond to the social signals that normally govern their behavior in a tissue. Tumor cells grow and divide when they should not, and fail to die when they should. They lose their attachments to the tissue and spread to other tissues. The goal of cancer research is to understand the molecular mechanisms that lead to this behavior, and then develop sophisticated methods to prevent or reverse it. The past few decades have seen great strides in the former, but far less progress in the latter.

Cancer progression is an evolutionary process driven by gene mutation

Like the evolution of species, the development of a tumor is based on the natural selection of new genetic traits that confer some competitive advantage on the cell (Figure 12-1). Most cancers begin when a single gene mutation—in some mitogenic regulatory pathway, for example—allows a cell to reproduce slightly more vigorously than other cells in the tissue. Tumor evolution continues as the descendants of the original mutant cell acquire additional genetic defects that allow them to overcome, one by one, the regulatory barriers that normally restrain their proliferation and survival. Because all features of cell behavior are governed by complex and robust regulatory networks, multiple gene mutations are generally necessary to break down each of the barriers that limit the ability of the developing tumor cell to proliferate and spread.

Gene mutations generally accumulate with age, and for this and other reasons the incidence of most cancers increases later in life. Most cancers are thought to arise as a result of six to ten mutations, which accumulate over a period of several decades. Some cancers, such as certain leukemias, may depend on only two mutations and can therefore arise in childhood. Others, such as prostate cancer, are likely to require as many as 30 mutations and tend to appear in old age.

The underlying foundation for all cancer progression is an abnormal increase in cell number, which can be attributed to three major changes in cell behavior. First, the cancer cell grows and divides at an inappropriate rate, either because it no longer requires stimulation by mitogens and growth factors (see Chapter 10) or because it has developed a resistance to extracellular factors that inhibit proliferation or stimulate terminal differentiation. Second, cancer cells carry mutations that allow them to survive under conditions that normally trigger cell death by apoptosis; cancer cells often display a reduced dependence on extracellular survival factors, for example. Third, the cells of most established tumors are not restrained by the telomere degeneration that normally limits the number of times a cell can divide (see section 11-8)—often because cancer cells, unlike most normal human cells, express telomerase or have other mechanisms that maintain telomere integrity.

Definitions

angiogenesis: formation of new blood vessels, often governed by extracellular angiogenic factors produced by local cells.

metastasis: in cancer biology, the spread of cells from a tumor to locations in other regions of the body, resulting in secondary tumors called metastases.

neoplasm: a tumor—a mass of cells that are proliferating at an inappropriate rate, generally as a result of mutations in mitogenic signaling pathways.

An abnormal increase in cell number leads to the formation of a tumor, or **neoplasm**: a mass of highly proliferative cells that is initially benign, or nonlethal. As solid tumors increase in size, cells in the tumor experience a lack of oxygen and nutrients. This initially limits their proliferation but also leads to the local production of extracellular angiogenic factors that promote **angiogenesis**: the formation of new blood vessels that infiltrate and nourish the tumor. Eventually, cells may escape from the primary tumor, spread to adjacent tissues and travel through the bloodstream or lymphatic system to other sites in the body. This is known as **metastasis**, and results in the formation of secondary tumors, or metastases. Metastatic tumors are often malignant, or lethal, if left untreated. The term cancer is generally reserved for malignant tumors.

New gene mutations may help promote tumor angiogenesis and metastasis, but in many cases they may not be required. Angiogenesis could result simply from natural responses to increased tissue size, and the spread of tumor cells to other tissues may depend in some cases on normal mechanisms of cell motility and invasiveness. In addition, angiogenesis and metastasis are sometimes promoted by mutations already present in the tumor cells: mutations in certain mitogenic signaling proteins are known to stimulate the production of angiogenic factors, for example, and mutations that promote cell survival could help tumor cells proliferate and spread outside the primary tumor. Thus, early mutations that increase cell proliferation can set the stage for later angiogenesis and metastasis to occur without further mutation.

Genetic instability accelerates cancer progression

Mutation is clearly a key driving force in tumor evolution. How do these mutations arise? In a small fraction of cases (about 5–10%), a tumor-promoting mutation is inherited—resulting in a familial cancer syndrome in which there is an increased incidence of cancer in specific tissues and often at an earlier age than usual. In most cases, however, cancer-promoting mutations arise spontaneously in somatic cells as a result of natural errors in the duplication, repair or segregation of DNA and chromosomes.

Environmental factors that damage DNA can increase the rate of mutation and accelerate tumor evolution. Exposure to ultraviolet radiation in sunlight, for example, triggers the formation of pyrimidine dimers (see section 11-1) and promotes skin cancer. Chemical carcinogens, like those in tobacco smoke, cause various forms of nucleotide damage that accelerate tumorigenesis in exposed tissues.

The rate of mutation can also be increased by the acquisition of defects in the cellular machinery. Most cancer cells display varying degrees of genetic instability—an increase in the rate at which DNA and chromosomes are damaged, lost or rearranged. In most cases, this genetic instability seems to arise from mutations in regulatory proteins that govern DNA repair, the DNA damage response and the behavior of chromosomes during mitosis.

This chapter focuses on the mechanisms that drive abnormal cell proliferation in cancer, with an emphasis on cancer-associated defects in the regulatory systems discussed in earlier chapters.

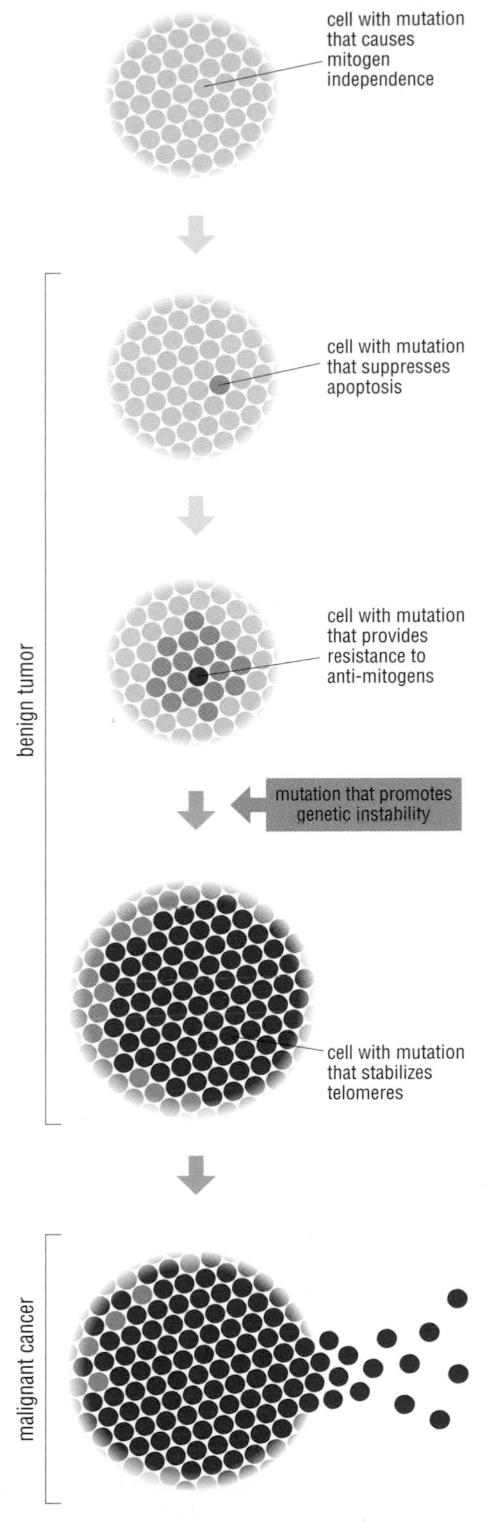

Figure 12-1 Evolution of a tumor Tumor evolution begins with a mutation that provides a cell with a competitive advantage—perhaps, as shown here, by enabling the cell to proliferate more rapidly than its neighbors. Over a period of many years, the descendants of that mutant cell acquire additional mutations that allow them to overcome the various regulatory barriers that restrain cell proliferation. In addition, mutations in genes required for genome maintenance trigger an increase in genetic instability, thereby enhancing the rate of mutation.

References

Alberts, B. *et al.*: *Molecular Biology of the Cell* 4th ed. (Garland Science, New York, 2002).

Devita, V.T. *et al.*: *Cancer: Principles and Practice of Oncology* 7th ed. (Lippincott, Williams and Wilkins, Philadelphia, 2004).

Evan, G. and Vousden, K.H.: **Proliferation, cell cycle and apoptosis in cancer.** *Nature* 2001, **411**:342–348.

Hanahan, D. and Weinberg, R.A.: **The hallmarks of cancer.** *Cell* 2000, **100**:57–70.

Nowell, P.C.: **The clonal evolution of tumor cell populations.** *Science* 1976, **194**:23–28.

Ramaswamy, S. *et al.*: **A molecular signature of metastasis in primary solid tumors.** *Nat. Genet.* 2003, **33**:49–54.

Varmus, H. and Weinberg, R. A.: *Genes and the Biology of Cancer* (Scientific American Library, New York, 1993).

Vogelstein, B. and Kinzler, K.W.: *The Genetic Basis of Human Cancer* 2nd ed. (McGraw-Hill, New York, 2002).

Weinberg, R.A.: *The Biology of Cancer* (Garland Science, New York, 2006).

Mutations in oncogenes and tumor suppressors stimulate tumor progression

Vast numbers of intracellular and extracellular proteins govern cell division, cell growth, cell death and the DNA damage response. In general, it is possible to divide these proteins into two groups: the positive regulators that promote an increase in cell number and the negative regulators that restrain it. The increase in cell number that occurs in cancer is driven by mutations that affect both these groups: mutations that cause excessive activity of positive growth regulators and mutations that reduce the activity of negative regulators.

Mutations that cause hyperactivity in a positive growth regulator tend to be genetically dominant; that is, a mutant gene on one chromosome promotes tumor formation even in the presence of a normal gene on the homologous chromosome. The mutant, cancer-promoting forms of these genes are sometimes called **oncogenes** and their normal versions are called **proto-oncogenes**.

Negative regulatory genes whose mutation promotes cancer are called **tumor suppressor genes**. A mutation in a tumor suppressor gene usually causes a loss of function that is genetically recessive, so that both copies—or alleles—of the gene must be mutated to promote tumorigenesis.

Oncogenes can be activated by many different mechanisms

Proto-oncogenes can be converted to cancer-causing oncogenes by multiple mechanisms (Figure 12-2). First, errors in DNA replication or repair can generate a single base change in the protein-coding sequence of the gene, resulting in a hyperactive protein. Many cancer cells, for example, have a single amino-acid change in the GTPase Ras, a positive regulator of many mitogenic pathways (see section 10-6). This mutation blocks the GTPase activity of Ras, thereby locking the protein in the active GTP-bound form. Similarly, point mutations in certain protein kinases, such as the tyrosine kinase Src, not only enhance enzymatic activity but also broaden substrate specificity so that the kinase phosphorylates and activates mitogenic proteins that are not normally its targets.

Another important oncogenic mechanism is gene amplification—an increase in the number of copies of a gene in the cell, which leads to the synthesis of the gene product in excessive amounts. The mitogenic gene regulatory protein Myc, for example (see section 10-6), is sometimes overproduced in cancer cells as a result of massive *MYC* amplification. Amplification occurs by many mechanisms. Errors in DNA repair or chromosome segregation can cause changes in chromosome structure or number, thereby increasing gene copy number. In addition, errors in DNA replication can lead to genetic recombination events that dramatically increase the number of gene copies.

In some cancers, chromosomal rearrangements have altered the expression of a mitogenic protein by placing its coding sequence under the control of another regulatory region. Here again, Myc provides a classic example: Burkitt's lymphoma is associated with a chromosome

Oncogenes

Gene	Function of gene product	Mechanism of activation in cancer
RAS	GTPase in mitogenic signaling	mutation that blocks GTPase
MYC	gene regulatory factor in mitogenic signaling	gene overexpression
FOS, *JUN*	gene regulatory factors in mitogenic signaling	gene overexpression
RAF	protein kinase in mitogenic signaling	mutations that activate kinase
EGFR	receptor for mitogen EGF	gene overexpression or activating mutation
PDGF	mitogen	gene overexpression
BCL2	inhibitor of apoptosis	gene overexpression
ABL	protein kinase with multiple functions	gene overexpression or activating mutation
SRC	protein kinase with multiple functions	mutations that activate kinase
PIK3CA	catalytic subunit of PI3 kinase	gene overexpression or activating mutation
AKT	protein kinase with multiple functions	gene overexpression or activating mutation

Figure 12-2 Table of representative oncogenes This partial list illustrates some of the major positive regulatory mechanisms that are over-stimulated in cancer cells. These genes are all capable of inducing a cancer-like phenotype when overexpressed in cultured mammalian cells. Countless other positive regulatory genes, not listed here, probably contribute to cancer but are not sufficient to induce the transformation of cultured cells.

Definitions

oncogene: a gene whose protein product promotes cancer, generally because mutations or rearrangements in a normal gene (the **proto-oncogene**) have resulted in a protein that is overactive or overproduced.

proto-oncogene: a gene that when dysregulated or mutated can promote malignancy (see **oncogene**). Proto-oncogenes generally regulate cell growth, cell division, cell survival, or cell differentiation.

tumor suppressor gene: a gene that encodes a protein that normally restrains cell proliferation or tumorigenesis, such that loss of the gene increases the likelihood of cancer formation.

References

Beachy, P.A. *et al.*: **Tissue repair and stem cell renewal in carcinogenesis.** *Nature* 2004, **432**:324–331.

Coussens, L.M. and Werb, Z.: **Inflammation and cancer.** *Nature* 2002, **420**:860–867.

Lowe, S.W. *et al.*: **Intrinsic tumour suppression.** *Nature* 2004, **432**:307–315.

Rowley, J.D.: **Chromosome translocations: dangerous liaisons revisited.** *Nat. Rev. Cancer* 2001, **1**:245–250.

Sherr, C.J.: **Principles of tumor suppression.** *Cell* 2004, **116**:235–246.

Vogelstein, B. and Kinzler, K.W.: **Cancer genes and the pathways they control.** *Nat. Med.* 2004, **10**:789–799.

Weir, B. *et al.*: **Somatic alterations in the human cancer genome.** *Cancer Cell* 2004, **6**:433–438.

translocation that places *MYC* under the control of immunoglobulin gene regulatory sequences, resulting in abnormally high Myc production in lymphoid cells.

Chromosome rearrangements can also generate fusion proteins with oncogenic potential. In chronic myeloid leukemia (CML), for example, a translocation between chromosomes 9 and 22 generates a hybrid chromosome—called the Philadelphia chromosome—in which two genes, *BCR* and *ABL*, are fused. The Abl protein is a protein kinase that becomes overactive when a small region at its amino terminus is deleted, apparently because this region normally suppresses kinase activity. In the Bcr–Abl fusion protein, this amino-terminal region in Abl is replaced with sequence from the Bcr protein, resulting in a hyperactive Abl protein that contributes to the development of the cancer.

Multiple mutations are required to cripple tumor suppressor genes

Loss of tumor suppressor gene function usually requires the inactivation of both alleles of the gene—which requires two separate and independent mutation events. Simple point mutations or small deletions in critical regions of the protein-coding or regulatory sequences are often the cause of inactivation of the first copy of a tumor suppressor gene. The remaining normal allele can be lost by similar gene-specific mutations, but in most cases it is lost by more general mechanisms. Errors in chromosome segregation or DNA repair, for example, can lead to complete or partial loss of the homologous chromosome carrying the normal allele. Chromosome rearrangements can result in the replacement of sequences from the normal allele with homologous regions from the mutant chromosome.

In rare cases, mutations in one allele of a tumor suppressor gene are inherited from one parent. This greatly accelerates tumor progression, as complete loss of gene function then requires mutation of only the normal copy inherited from the other parent, an event that is known as loss of heterozygosity (LOH). Recessive mutations in tumor suppressor genes are found in several familial cancers (Figure 12-3). Patients with these diseases tend to have an increased likelihood of developing cancer in specific tissues. Although familial cancer syndromes account for only a small fraction of cancers, mapping and identifying the mutant genes in these diseases has been one of the most effective means of discovering tumor suppressor genes.

Mutation of both alleles of a tumor suppressor gene is not always required for cancer progression. In some cases, loss of function of only one allele will suffice, indicating that a single intact copy of the gene is not sufficient for normal function (a condition termed haploinsufficiency). In other cases, one allele of a tumor suppressor gene can be mutated in such a way that the gene product gains the ability to interfere with the function of the normal protein. The mutant gene is therefore genetically dominant, and such mutations are called dominant-negative mutations.

Cancer can be initiated by mechanisms other than gene mutation

Although gene mutation is clearly the major driving force in tumorigenesis, not all cancer-promoting agents act by altering chromosomal gene sequences. A class of chemicals called phorbol esters, for example, enhance tumor formation by binding and activating protein kinases that promote mitogenic signaling. Some viruses can also cause tumors by mechanisms that do not involve the mutation of genes in the host cell. In a cell infected with papillomavirus, for example, the proteins E6 and E7 are expressed at high levels from the viral genome. These bind and inhibit pRB and p53, respectively, thus inactivating two major tumor suppressor proteins. The genomes of some animal retroviruses contain mutant, activated oncogenes whose overexpression in the infected cell promotes tumor formation.

Chronic viral infections and other forms of chronic tissue injury are also thought to promote tumor formation by non-mutational mechanisms. Long-term infection with hepatitis virus B or C is associated with liver cancer, in part because chronic infection triggers inflammatory and repair responses that promote liver cell proliferation. Similar responses may be responsible for the cancers that are associated with chronic infection with certain bacteria and parasites. Similarly, persistent irritation, such as that caused by long-term exposure of lung cells to asbestos fibers or tobacco smoke particles, promotes inflammatory and proliferative responses. Over the long term, these proliferative responses may provide fertile ground in which gene mutations can trigger tumor formation.

Tumor Suppressor Genes

Gene	Function of gene product	Familial cancer syndrome
p53	gene regulatory factor in stress responses	Li–Fraumeni syndrome
RB	inhibitor of G1/S gene expression	retinoblastoma
INK4A	Cdk inhibitor p16^{INK4a}	melanoma
ARF	positive regulator of p53	melanoma
APC	inhibitor of mitogenic signaling (not related to anaphase-promoting complex)	familial adenomatous polyposis coli
PTEN	antagonist of PI3 kinase	Cowden syndrome
NF1	GTPase-activating protein for Ras, mitogenic signaling	neurofibromatosis
TSC1,2	GTPase-activating protein for Rheb, growth signaling	tuberous sclerosis
DPC4/ SMAD4	gene regulatory factor in anti-mitogenic signaling	–
ATM	DNA damage response kinase	ataxia telangiectasia
NBS1	DNA repair, damage response	Nijmegen breakage syndrome
BRCA1	DNA repair, damage response	familial breast and ovarian cancer

Figure 12-3 Table of representative tumor suppressor genes This partial list illustrates some of the major negative regulatory mechanisms that are lost in cancer cells. Although the loss of tumor suppressor gene function generally results from sporadic mutations of both alleles, mutations in these genes are sometimes associated with inherited syndromes, named at right, that result in an increased likelihood of developing cancer in specific tissues. As discussed later in this chapter, two genes on this list, *INK4A* and *ARF*, overlap in the same chromosome region, and some cancers are associated with deletions that remove both genes. Adapted from Sherr, C.J.: *Cell* 2004, **116**:235–246.

12-2 Tissue Specificity in Cancer

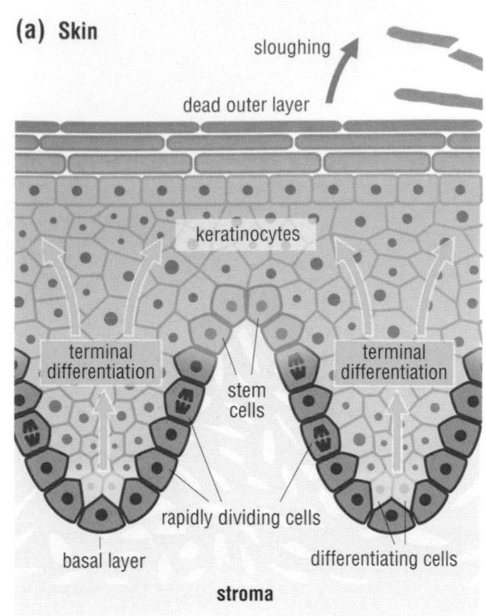

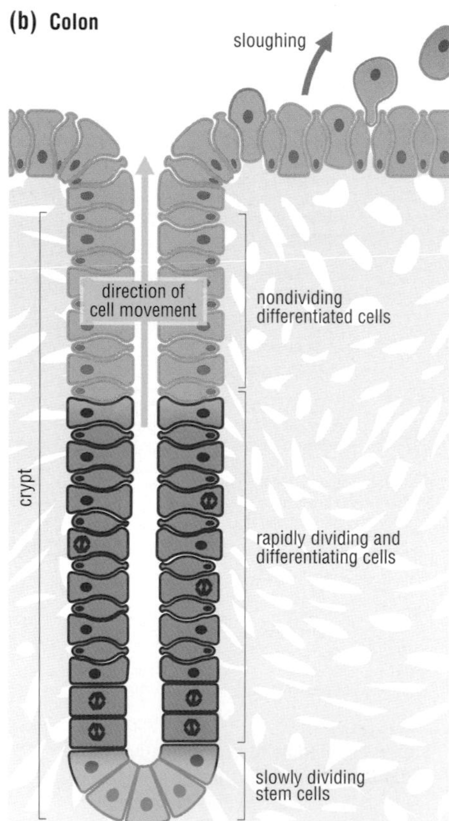

Cancers are a complex group of diseases

Cancer can arise in almost any human tissue, and even within the same tissue it can take different forms. Cancer is therefore a large and complex group of related but distinct diseases, each with unique features that depend on the biological characteristics of the cell type or tissue in which the disease originates.

Cancers are named on the basis of several factors, including the tissue of origin, the pathological features of the tumor tissue and the stage in tumor progression. **Carcinomas** are malignant tumors of epithelia—the sheets of cells that line the lung, gut, skin, reproductive tract and other organs (Figure 12-4). The vast majority (about 90%) of human cancers are carcinomas, probably because epithelial tissues tend to proliferate throughout life and are most directly exposed to chemicals and other insults that help promote tumorigenesis. **Sarcomas** are cancers of the connective tissue and muscle, and leukemias and lymphomas are cancers of the hematopoietic cells of the blood and lymphatic system (Figure 12-5).

Benign tumors are named differently from malignant tumors in the same tissue (see Figure 12-5). In the breast or colon, most benign tumors are **adenomas**—a name generally applied to benign epithelial tumors with a glandular origin. Malignant carcinomas in glandular tissues are called adenocarcinomas. Benign and malignant tumors in the same tissue do not necessarily represent different stages in the evolution of a cancer. In some tissues, such as colon, a malignant adenocarcinoma typically originates in a benign adenoma, whereas in other tissues, such as breast and liver, adenomas are rarely pre-malignant. Malignant breast tumors instead tend to originate in a different type of benign tumor called a carcinoma-*in-situ*.

In most tissues, new cells are produced by division of undifferentiated cells called stem cells. Differentiation of stem-cell progeny yields the specialized cell types—many of them incapable of division—that function in that tissue. Because stem cells and their immediate descendants retain the capacity to divide, it is often in this population that tumor cells arise. Tumor cells may undergo varying degrees of differentiation, resulting in cancer cells that appear moderately or well differentiated. In many cases, however, mutations block the differentiation process, resulting in a lack of differentiation that is called anaplasia. A related phenomenon is dysplasia, which refers to the disordered growth, disruption of tissue architecture, and variability in cell size and shape that often occurs in cancer—particularly in epithelia.

Epithelial cells are supported by a foundation of stromal tissue containing a variety of non-epithelial cell types, including fibroblasts, which secrete proteins that help govern epithelial cell growth, proliferation and movement. Carcinomas often contain large numbers of stromal fibroblasts, and tumor growth is thought to depend in part on factors produced by these cells. Epithelial cancer progression may be promoted in some cases by mutations in fibroblasts that trigger increased production of mitogens that drive epithelial cell division.

The molecular basis of tumorigenesis can vary in different tissues

Cancers of all tissues share the same antisocial patterns of proliferative and invasive behavior, which result from the gradual loss by mutation of the mechanisms that normally limit these behaviors. Because different cell types often use the same regulatory proteins to control their proliferation, a cancer-promoting mutation in the gene encoding one of these proteins can promote tumorigenesis in many different tissues. It is clear, for example, that mutations in the tumor suppressors p53 or pRB can enhance the progression of most, if not all, types of human cancer. Similarly, hyperactivating mutations in widely used mitogenic signaling proteins, such as Ras and Myc, are associated with many different tumor types.

Figure 12-4 **Examples of epithelial tissues** Carcinomas arise in the epithelial linings of many tissues, including the two tissues shown here. **(a)** The epidermal layer of the skin rests on a dermal layer of stromal connective tissue. The basal cell layer of the epidermis contains stem cells that divide to produce a population of rapidly dividing cells that differentiate into keratinocytes, which continue to divide a few times as they migrate toward the skin surface, eventually dying and falling off. Numerous carcinomas can arise in the epidermis, including basal cell carcinoma (arising in the basal cell layer) and squamous cell carcinoma (in the upper layers). The basal layer also contains specialized pigment cells called melanocytes (not shown), which can give rise to a highly metastatic carcinoma called malignant melanoma. **(b)** The epithelium that lines the colon also undergoes constant renewal by division of stem cells, which are located in invaginations called crypts. These stem cells give rise to a rapidly dividing cell population that differentiates and migrates toward the surface, where cells are shed into the gut lumen. Tumors typically arise in the rapidly dividing cell populations along the walls of the crypts, resulting in benign adenomas, or polyps, that protrude from the epithelium into the lumen of the colon.

It is also clear, however, that there are tissue-specific differences in the ability of certain mutations to promote cancer. A mutation in some mitogenic pathways, for example, may initiate tumorigenesis in only one cell type and have little effect in others. This concept is particularly well illustrated by the familial cancer syndromes, in which a mutation shared by every somatic cell in the body tends to result in cancer in only a limited subset of tissues (see Figure 12-3).

The causes of this tissue specificity are not understood in much molecular detail, but several possibilities can be imagined. To begin with, different cell types may have different regulatory systems to govern their growth, division and survival: one cell type may be much more dependent than another on a particular anti-mitogenic protein to suppress its proliferation, and thus more sensitive to the loss of that protein. Different cell types may also display different sensitivities to cancer-promoting mutations because of differences in their normal proliferative behavior. Highly proliferative cell types, such as stem cells of the colon epithelium, contain molecular systems that enable constant growth and division; such cells are more readily transformed into tumor cells than are nondividing cell types such as neurons in which these systems are irreversibly shut down. Other biological features of specific cell types are also important. In the epidermis, for example, cells achieve a terminally differentiated state and are soon lost from the surface of the skin; this acts as a protective mechanism that rapidly removes any cell that has acquired oncogenic mutations (see Figure 12-4). Mutations that specifically block epidermal cell differentiation are an important step toward skin cancer but may have little impact in other tissues. In general, therefore, a complete understanding of the molecular basis of each type of cancer requires an understanding of the anatomy and physiology of the tissue in which that cancer originates.

In sections 12-3 and 12-4 we discuss the general mechanisms that lead to excessive cell numbers in most cancers, after which we turn to the question of how genetic instability arises in tumors. Later in the chapter (section 12-8) we look at the mechanisms underlying cancer progression in a specific tissue: the epithelium of the colon.

Types of Cancers

Tissue	Benign	Malignant
Epithelia		
glands or ducts (e.g. gut, breast, lung, uterus)	adenoma carcinoma-*in-situ*	adenocarcinoma
stratified squamous tissue (e.g. skin, lung)	squamous cell papilloma	squamous cell or epidermoid carcinoma
basal cells (skin)		basal cell carcinoma
melanocytes (skin)	nevus	malignant melanoma
renal epithelium	renal tubular adenoma	renal cell carcinoma
liver cells	liver cell adenoma	hepatocellular carcinoma
Connective tissue and derivatives		
fibroblasts	fibroma	fibrosarcoma
cartilage	chondroma	chondrosarcoma
bone	osteoma	osteogenic sarcoma
striated muscle	rhabdomyoma	rhabdomyosarcoma
blood vessels	hemangioma	angiosarcoma
Bone marrow, spleen		
hematopoietic (blood-forming) cells		leukemias, erythroleukemia, myeloma, lymphomas
Nerves		
peripheral nervous system		neuroblastomas
central nervous system		gliomas, astrocytomas, medulloblastomas

Figure 12-5 Table of cancer nomenclature Note that benign tumors do not necessarily represent the precursors of malignant tumors in the same tissue.

Definitions

adenoma: benign tumor of epithelial cells, typically of glandular origin.

carcinoma: malignant tumor of epithelial tissue.

sarcoma: malignant tumor of connective tissue or muscle.

References

American Cancer Society: *Cancer Facts and Figures* (American Cancer Society, Atlanta, 2006).

Beachy, P.A. *et al.*: **Tissue repair and stem cell renewal in carcinogenesis.** *Nature* 2004, **432**:324–331.

Bhowmick, N.A. *et al.*: **Stromal fibroblasts in cancer initiation and progression.** *Nature* 2004, **432**:332–337.

Cotran, R.S. *et al.*: *Robbins Pathologic Basis of Disease* (Elsevier, Philadelphia, 1998).

Devita, V.T. *et al.*: *Cancer: Principles and Practice of Oncology* 7th ed. (Lippincott, Williams and Wilkins, Philadelphia, 2004).

Huntly, B.J. and Gilliland, D.G.: **Leukaemia stem cells and the evolution of cancer-stem-cell research.** *Nat. Rev. Cancer* 2005, **5**:311–321.

Tlsty, T.D. and Hein, P.W.: **Know thy neighbor: stromal cells can contribute oncogenic signals.** *Curr. Opin. Genet. Dev.* 2001, **11**:54–59.

American Cancer Society website: www.cancer.org

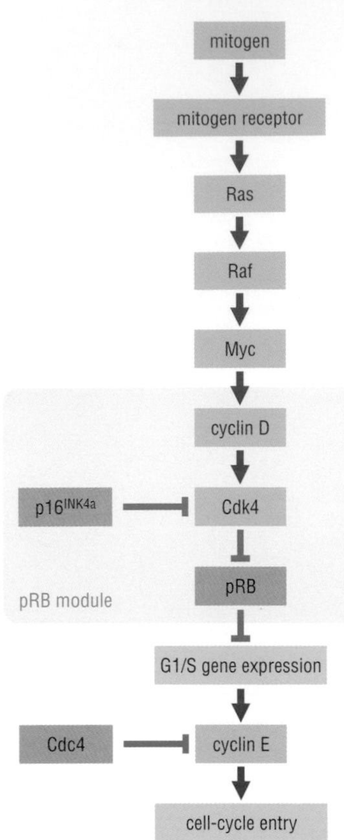

Figure 12-6 Sites of tumorigenic mutations in mitogenic signaling pathways Many mitogens act through signaling pathways that include the regulatory proteins shown here (see Chapter 10). Mutations in these proteins are associated with cancer: green components are proto-oncogene products whose overactivation promotes cancer; red components are tumor suppressor gene products whose loss contributes to cancer. For simplicity, mitogenic signaling is shown here as a simple linear pathway. In reality, many of the components in this diagram initiate a broad range of signals that govern numerous downstream targets involved in mitogenesis, growth, survival and other cellular processes. As a result, hyper-activation of Ras and Myc will each have different effects on cell behavior, and mutation of both components is more oncogenic than mutation of either component alone. This does not seem to be true for the components of the pRB regulatory module, in which the mutation of any one component is sufficient for tumorigenesis (see Figure 12-7).

Tumor cells are independent of mitogens and resistant to anti-mitogens

The cells of multicellular organisms do not proliferate unless they are instructed to do so by extracellular mitogens, growth factors and survival factors. Every tissue contains a specific blend of factors on which the cells of that tissue depend, and any cell that escapes the unique microenvironment of that tissue will not proliferate. Cancer cells have lost this dependence on extracellular signals and therefore divide, grow and survive in their absence, resulting in the increased cell numbers that are the central characteristic of this disease.

Cell proliferation in most tissues depends not only on soluble mitogens but on components of the extracellular matrix to which the cells are attached. Tumor cells acquire the ability to proliferate in the absence of these attachments, and analysis of this so-called anchorage independence is often used in the laboratory to assess the transformation of normal cells into tumor cells.

Defects in mitogenic signaling lie at the heart of uncontrolled proliferation of tumor cells. As discussed in Chapter 10, numerous mitogenic pathways govern the rate of cell division by controlling the activation of G1– and G1/S–Cdks that drive entry into the cell cycle (Figure 12-6). Dominant oncogene mutations have been found in cancer cells at nearly every step in these pathways, beginning with mutations in the genes encoding the mitogens themselves. One of the earliest oncogenes to be identified, in the genome of the tumor-inducing simian sarcoma virus, encodes a form of the mitogen PDGF, whose overproduction in infected cells contributes to tumorigenesis. It is now clear that many tumors depend on mutations that cause excessive production of mitogens by the tumor cell—resulting in a form of positive feedback known as autocrine stimulation.

Cell-surface receptors for mitogens are also common targets of oncogenic mutations. The cytoplasmic protein kinase domain that is often found in these receptors (see section 10-6) can be made hyperactive by point mutations or deletions in the extracellular domain, resulting in mitogen-independent kinase activity. Alternatively, some cancers are associated with greatly increased production of mitogen receptors, resulting in cells that can be stimulated by low levels of mitogen that normally fail to trigger division.

Many of the signaling molecules that lie downstream of mitogen receptors are also hyperactive in tumor cells. One particularly important mitogenic signaling pathway involves the activation of Ras (see section 10-6). Ras has effects on several subsequent signaling pathways, including activation of a MAP kinase cascade that begins with the protein kinase Raf. It is likely that most, if not all, human cancers have some oncogenic defect in the Ras–Raf–MAPK signaling module. Mutations that activate Ras are particularly common, occurring in about 25% of cancers.

Later steps in the mitogen-response pathway—the activation of gene regulatory proteins—are also frequent targets for dominant oncogenic mutations. As discussed in section 12-1, the transcriptional regulator Myc in particular is overproduced in many different cancers, resulting in excessive expression of its target genes—which produce proteins that help drive cell growth and division (see section 10-6).

The proliferation of many cell types is governed both by mitogens that promote division and by anti-mitogenic factors that inhibit it. Tumor cells tend not only to be independent of mitogens but also resistant to anti-mitogens. Part of this resistance may result simply from the overstimulation by positive mitogenic signals. In addition, tumor cells sometimes contain loss-of-function mutations in components of anti-mitogenic signaling pathways. Some cancer

References

Hahn, S.A. *et al.*: **DPC4, a candidate tumor suppressor gene at human chromosome 18q21.1.** *Science* 1996, **271**:350–353.

Hanahan, D. and Weinberg, R.A.: **The hallmarks of cancer.** *Cell* 2000, **100**:57–70.

Lowe, S.W. *et al.*: **Intrinsic tumour suppression.** *Nature* 2004, **432**:307–315.

Massagué, J.: **G1 cell-cycle control and cancer.** *Nature* 2004, **432**:298–306.

Sherr, C.J.: **Principles of tumor suppression.** *Cell* 2004, **116**:235–246.

Sherr, C.J. and McCormick, F.: **The RB and p53 pathways in cancer.** *Cancer Cell* 2002, **2**:103–112.

Vogelstein, B. and Kinzler, K.W.: **Cancer genes and the pathways they control.** *Nat. Med.* 2004, **10**:789–799.

Differentiation is often inhibited in tumor cells

The cells of many tissues cease dividing and enter terminally differentiated states from which they rarely return. Tumor cells often acquire mutations that inhibit differentiation, thereby allowing the cell to remain in a proliferative state that contributes to cancer progression. In many cases it seems that the same oncogenic mutations that drive proliferation also inhibit differentiation. Overexpression of Myc, for example, not only promotes the expression of genes involved in cell division and growth but also prevents the formation of gene regulatory complexes that promote differentiation. Similarly, the pRB protein interacts in some cells with specific gene regulatory proteins that stimulate differentiation. Loss of pRB therefore prevents differentiation while releasing the brakes on proliferation.

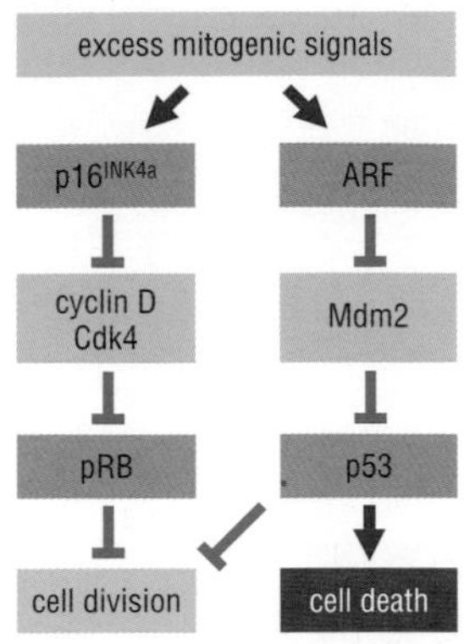

Figure 12-9 The hyperproliferation stress response in cancer Hyperproliferative signals lead to increased levels of ARF and p16^{INK4a} (see section 11-8), resulting in cell-cycle arrest or cell death through the pathways shown here (see also Figure 12-6 for additional details of p16^{INK4a} function). Mutations in components of these pathways are associated with cancer: green components are proto-oncogene products whose overactivation promotes cancer; red components are tumor suppressor gene products whose loss contributes to cancer.

Tumor cells are resistant to the hyperproliferation stress response

Overactivation of mitogenic signaling components such as Ras and Myc does not stimulate the proliferation of normal cells but instead causes either cell death or a permanent cell-cycle arrest called senescence (see section 11-8). This hyperproliferation stress response provides a mechanism by which cells with oncogenic mutations are prevented from initiating tumor formation.

The hyperproliferation stress response tends to be suppressed in cancer cells, allowing the continued proliferation of cells carrying overactive mitogenic signals. The apoptosis that normally results from Myc overproduction in some cells, for example, can be suppressed in tumor cells by oncogenic mutations that stimulate survival signals or directly inhibit the apoptotic machinery, as described above (see Figure 12-8). Another mechanism by which cancer cells bypass the hyperproliferation stress response is through inactivation of the stress-response protein ARF. Increased production of ARF in response to hyperproliferative signals normally leads to activation of p53, which triggers either cell-cycle arrest or apoptosis (see Figure 11-17). Loss of either of the tumor suppressors ARF or p53 therefore prevents this response (Figure 12-9).

In many cell types—particularly in humans—the Cdk inhibitor p16^{INK4a} also contributes to the cell-cycle arrest that occurs after hyperproliferative stress (see section 11-8). Thus, the loss of p16^{INK4a}, which occurs in many cancers (see section 12-3), helps abolish this response in some cell types. Interestingly, p16^{INK4a} and ARF are encoded by overlapping genes at the same chromosomal locus (Figure 12-10). Some cancer-associated chromosomal deletions disrupt both genes, thereby knocking out regulators of both the pRB and p53 pathways.

Loss of p53 function is a remarkably common event in tumor cells—and an event of unparalleled importance in tumor evolution, because it allows cell proliferation to continue in the face of many different forms of stress and DNA damage. This can contribute to genetic instability, as we discuss in the next three sections of this chapter.

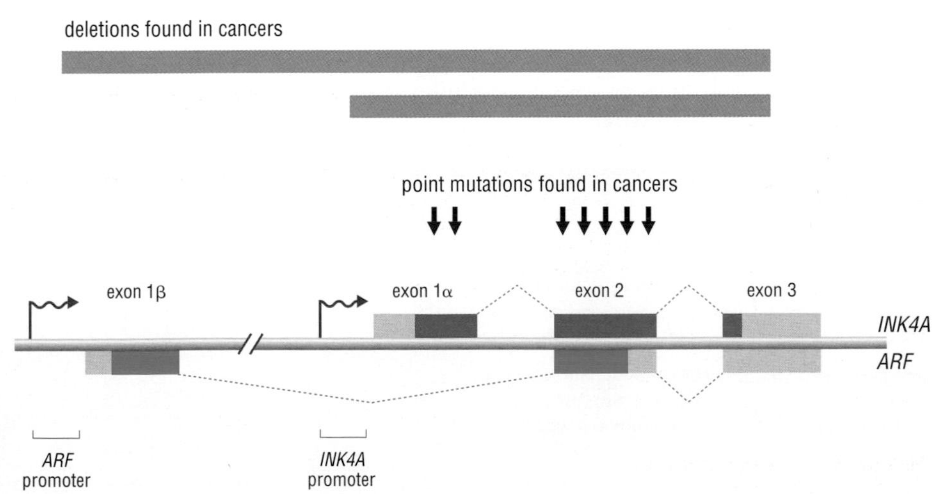

Figure 12-10 The *INK4A/ARF* locus p16^{INK4a} and ARF are encoded by a single chromosomal locus with the four exons shown here. Transcription of *INK4A* begins upstream of exon 1α, and processing of the resulting RNA leads to splicing of exons 1α, 2 and 3 as shown (the p16^{INK4a} open reading frame is colored blue). Transcription of *ARF* begins upstream of exon 1β, and the resulting RNA is spliced to exons 2 and 3 as shown. The ARF protein is encoded by the open reading frame colored green; exon 2 is used to encode both proteins. The p53-stimulatory function of ARF is thought to reside in the amino-terminal region encoded by exon 1β, and so mutation of this exon is required for loss of ARF function. Human cancers are associated with various types of mutations and deletions in the *INK4A/ARF* locus. Familial melanoma, for example, is caused by point mutations in the exons required for p16^{INK4a} function. Some cancers are associated with gene deletions that disrupt one or both genes. Cancer may also be driven in some cases by inhibition of *INK4A* promoter function by nucleotide methylation. Adapted from Haber, D.A.: *Cell* 1997, **91**:555–558.

Most cancer cells have unstable genomes

Tumor evolution is driven by the gradual accumulation of mutations. It is unlikely, however, that the rate of spontaneous mutation in most human cells is sufficient to produce the large number of defects required for progression to malignant cancer. Tumorigenesis is therefore thought to depend on the acquisition of **genetic instability**: an increase in the rate at which genes and chromosomes are mutated, lost, amplified or rearranged. There is considerable evidence that this genetic instability contributes to tumor formation—although its importance in some types of cancer remains uncertain.

Excessive amounts of DNA and chromosome damage are lethal, even to a seemingly invincible cancer cell. Thus, tumor evolution requires some optimal level of genetic instability that accelerates the rate of gene mutation without killing all cells in the population.

Defects in the DNA damage response promote genetic instability in cancer

Genetic instability is accentuated in most cancer cells by defects in the DNA damage response (see Chapter 11). In particular, the gene regulatory protein p53 is lost or defective in most, if not all, cancer cells. Other components, such as the damage-response kinase ATM, can also be mutated in some familial cancer syndromes. Cells with damage-response defects fail to undergo cell-cycle arrest or apoptosis in response to increased levels of DNA damage, which allows genetically unstable cells to continue proliferating.

The DNA damage response may be an important barrier to cancer progression even at the earliest stages of tumor formation. Cells from certain early-stage tumors display molecular markers of a DNA damage response, such as phosphorylated Chk2 and histone H2A.X (see section 11-3). DNA damage might arise early in tumor development because hyperactive mitogenic signaling leads to defects in the control of replication fork progression, resulting in replication fork defects and cell-cycle delays (see section 11-6). Thus, even in the early stages of tumor progression, there seem to be selective pressures that lead to the emergence of mutant cells lacking p53 or other components of the DNA damage response.

Genetic instability sometimes results from an increased rate of point mutation

Point mutations in specific genes are common contributors to many, if not all, types of cancer. One might therefore expect that tumorigenesis would be accelerated by increased rates of point mutation, perhaps arising out of defects in DNA repair. There is good evidence for this in only a small number of cancers, however.

About 15% of colon cancers, and a smaller fraction of other cancers, carry mutations in the enzymes responsible for a process called **mismatch repair**. These enzymes correct the occasional nucleotide-mispairing errors that occur naturally during DNA replication. In the absence of this repair system, mismatched nucleotide pairs accumulate, generating point mutations. These mutations are particularly abundant in repetitive DNA regions called microsatellites, and so the increased mutation rate due to mismatch repair defects is sometimes called **microsatellite instability (MIN)**.

Definitions

chromosomal instability (CIN): an abnormally high incidence of defects in chromosome number (numerical CIN) or chromosome structure (structural CIN).

chromosomal translocation: alteration in chromosome structure in which pieces of one chromosome become exchanged with or attached to another chromosome.

CIN: see **chromosomal instability**.

genetic instability: an abnormal increase in the rate at which genes and chromosomes are mutated, rearranged or lost. Also called genomic instability.

karyotype: the set of condensed metaphase chromosomes in a eukaryotic cell stained so that each chromosome can be uniquely identified.

microsatellite instability (MIN): an abnormally high frequency of point mutations in repetitive DNA regions, generally as a result of defects in **mismatch repair**.

MIN: see **microsatellite instability**.

mismatch repair: repair of nucleotide mismatches that are generated during DNA replication.

Two other major DNA repair systems are nucleotide excision repair, which is responsible primarily for the repair of bulky lesions such as the pyrimidine dimers caused by UV radiation, and base excision repair, which replaces damaged nucleotide bases (see section 11-1). Defects in these types of DNA repair do not seem to make a major contribution to genetic instability in cancer. An exception is a rare inherited disease called xeroderma pigmentosum, which is caused by mutations in a group of proteins that are required for nucleotide excision repair. Patients with this disease are at a high risk of developing skin cancer as a result of exposure to sunlight—as expected from a defect in the system that repairs UV-induced DNA damage.

Chromosomal instability is the major form of genetic instability

Most cancer cells are aneuploid; that is, they contain abnormal numbers of chromosomes, and in most cases these chromosomes display major structural abnormalities—particularly **chromosomal translocations**, in which large pieces of one chromosome become exchanged with or attached to another chromosome (Figure 12-11). These chromosomal alterations can contribute to cancer progression by altering the expression of proto-oncogenes or tumor suppressor genes.

Aneuploidy in cancer generally results from **chromosomal instability (CIN)**, which is an increase in the rate of gains or losses of whole chromosomes, or large portions of them. Unlike genetic instability due to an increased rate of point mutation, chromosomal instability is likely to be widespread and important in many human cancers, although different cancer types display varying degrees of instability. Some cancers, such as the small number of colon or breast cancers in which mismatch repair mechanisms are defective, have relatively stable chromosomes (see Figure 12-11). Many cancers progress through a transient period of extreme instability—when chromosome content varies widely in the tumor-cell population—followed by a relatively stable state in which many cells in the tumor have a similar abnormal chromosome content. These bursts of chromosomal instability may be triggered by the erosion of telomeres, the protective caps on the ends of chromosomes.

In the next two sections of this chapter we focus on two classes of defects that are thought to give rise to chromosomal instability, particularly in the carcinomas that make up the majority of human cancers. First, we address the mechanisms by which DNA damage—particularly double-strand breaks and the exposure of chromosome ends as a result of degenerating telomeres—can lead to instability in chromosome structure (structural CIN). Second, we discuss how defects in mitotic chromosome segregation and cytokinesis can generate instability in chromosome number (numerical CIN).

Figure 12-11 Chromosomal abnormalities in cancer cells To assess the number and structure of chromosomes in a cell, condensed sister-chromatid pairs are isolated from a metaphase cell, spread on a glass slide, and treated with a set of fluorescent labels that each interact with a specific human chromosome. Images of the 23 human chromosomes, each represented by a pair of homologs, can then be aligned and ordered as shown here. The resulting image of the cell's chromosome complement is called a **karyotype**, and the method used here is called spectral karyotyping. **(a)** This is the apparently normal karyotype of a cell line derived from a breast cancer with microsatellite instability but not chromosomal instability. **(b)** This is a karyotype from a cell line derived from a breast cancer with chromosomal instability. Extra copies of several chromosomes are present, and translocations between chromosomes are revealed by the presence of chromosomes with multiple colors. Courtesy of Kylie Gorringe, Mira Grigorova and Paul Edwards.

(a)

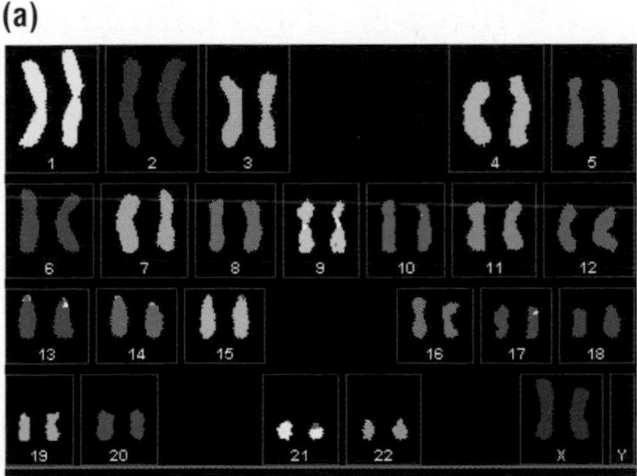

(b)

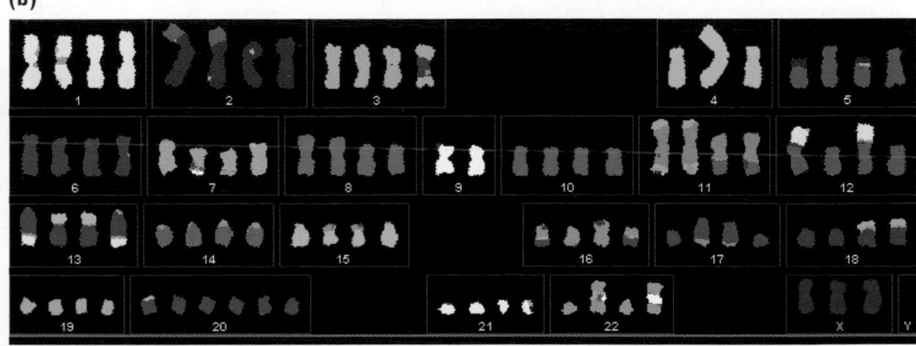

References

Abdel-Rahman, W.M. *et al.*: **Spectral karyotyping suggests additional subsets of colorectal cancers characterized by pattern of chromosome rearrangement.** *Proc. Natl Acad. Sci. USA* 2001, **98**:2538–2543.

Bartkova, J. *et al.*: **DNA damage response as a candidate anti-cancer barrier in early human tumorigenesis.** *Nature* 2005, **434**:864–870.

Gorgoulis, V.G. *et al.*: **Activation of the DNA damage checkpoint and genomic instability in human precancerous lesions.** *Nature* 2005, **434**:907–913.

Hoeijmakers, J.H.: **Genome maintenance mechanisms for preventing cancer.** *Nature* 2001, **411**:366–374.

Kastan, M.B. and Bartek, J.: **Cell-cycle checkpoints and cancer.** *Nature* 2004, **432**:316–323.

Liyanage, M. *et al.*: **Multicolour spectral karyotyping of mouse chromosomes.** *Nat. Genet.* 1996, **14**:312–315.

Pihan, G. and Doxsey, S.J.: **Mutations and aneuploidy: co-conspirators in cancer?** *Cancer Cell* 2003, **4**:89–94.

Rajagopalan, H. *et al.*: **The significance of unstable chromosomes in colorectal cancer.** *Nat. Rev. Cancer* 2003, **3**:695–701.

Storchova, Z. and Pellman, D.: **From polyploidy to aneuploidy, genome instability and cancer.** *Nat. Rev. Mol. Cell Biol.* 2004, **5**:45–54.

Defective DNA damage responses can lead to chromosomal instability

Many cancer cells undergo chromosomal rearrangements that cause large losses, amplifications or exchanges of chromosome segments (see Figure 12-11). In many cases, these gross rearrangements of chromosome structure are thought to result from defects in the system that repairs two similar forms of DNA damage: double-strand breaks in DNA and eroded telomeres, which expose the ends of a DNA molecule.

Several inherited genetic diseases in which there is a predisposition to cancer are caused by mutations in the cellular systems that respond to and repair double-strand breaks. These familial syndromes include ataxia telangiectasia, which is caused by mutations in the damage-response kinase ATM, as well as syndromes caused by mutations in BRCA1 or in the MRN complex, which are involved in the DNA damage response to double-strand breaks (see Chapter 11; see also Figure 12-3). Defects in these components not only lead to faulty DNA repair but also prevent the cell-cycle arrest or apoptosis that normally follows damage. Thus, additional DNA damage may occur because the cell attempts to replicate or segregate defective chromosomes.

How might defective double-strand break repair lead to gross chromosome rearrangements? To begin with, rearrangements can arise from failures in the repair of double-strand breaks by homologous recombination (see sections 9-2 and 11-1). Because human chromosomes contain so much repetitive DNA, a broken DNA end in a repetitive region of one chromosome can accidentally recombine with similar sequences on a non-homologous chromosome. If such recombination events are processed incorrectly as a result of defects in repair enzymes the result can sometimes be a non-reciprocal translocation in which a part of one chromosome is joined to another. Translocations can also result when broken ends from two different chromosomes are joined together by non-homologous end joining (see section 11-1).

Non-reciprocal translocations can be particularly dangerous if they produce a chromosome with two centromeres (a **dicentric chromosome**). In the next mitosis the two centromeres of one sister chromatid could become attached to opposite spindle poles, so that the chromatid is torn in two when the spindle elongates in anaphase. This would generate a pair of broken DNA ends

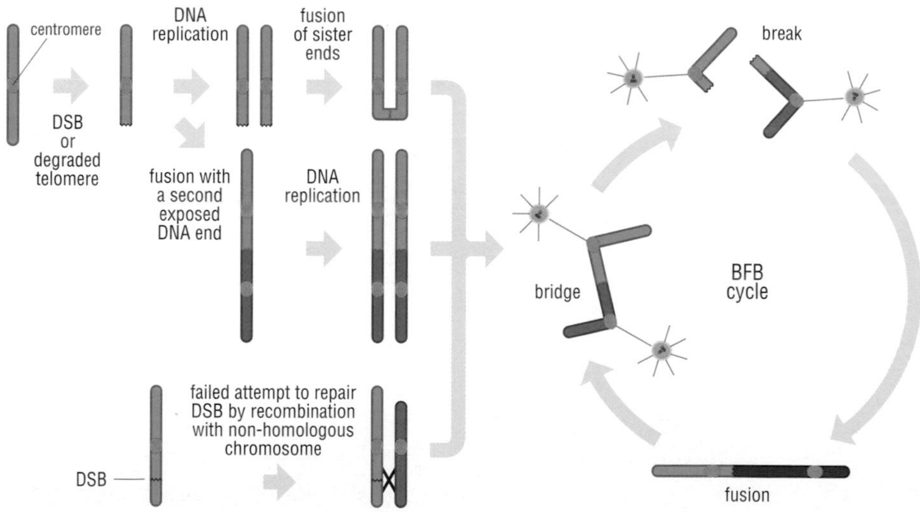

Figure 12-12 Scrambling chromosome structure with breakage-fusion-bridge cycles In the absence of a normal DNA damage response, faulty DNA repair can result in the formation of a dicentric chromosome containing two centromeres. Some mechanisms by which this can occur are shown on the left side of this diagram. A double-strand break or degraded telomere, for example, can fuse with other exposed DNA ends to form a dicentric chromosome. Alternatively, a dicentric chromosome can sometimes result from a failed attempt to repair a double-strand break by recombination with a non-homologous chromosome. If the two centromeres are attached to different spindle poles during mitosis, the chromosome will be torn apart during anaphase. One broken end can then fuse with a broken end on another chromosome to produce another dicentric chromosome (bridge formation), thereby initiating another cycle of breakage, fusion and bridge formation (the BFB cycle). The end result in most cases is non-reciprocal translocations: the transfer of a piece of one chromosome onto a different chromosome. Large sections of chromosomes can be lost or amplified during multiple rounds of this process.

Definitions

breakage-fusion-bridge (BFB) cycle: complex process that generates chromosome rearrangements. It begins when a **dicentric chromosome** is attached to both spindle poles and then torn in two in anaphase. Fusion of the broken ends can generate another dicentric chromosome that begins the process again.

dicentric chromosome: chromosome that contains two centromeres and therefore two kinetochores, usually as a result of chromosomal rearrangements.

References

Artandi, S.E. and Attardi, L.D.: **Pathways connecting telomeres and p53 in senescence, apoptosis, and cancer.** *Biochem. Biophys. Res. Commun.* 2005, **331**:881–890.

Artandi, S.E. *et al.*: **Telomere dysfunction promotes non-reciprocal translocations and epithelial cancers in mice.** *Nature* 2000, **406**:641–645.

Chin, K. *et al.*: ***In situ* analyses of genome instability in breast cancer.** *Nat. Genet.* 2004, **36**:984–988.

Feldser, D.M. *et al.*: **Telomere dysfunction and the initiation of genome instability.** *Nat. Rev. Cancer* 2003, **3**:623–627.

Kastan, M.B. and Bartek, J.: **Cell-cycle checkpoints and cancer.** *Nature* 2004, **432**:316–323.

Pihan, G. and Doxsey, S.J.: **Mutations and aneuploidy: co-conspirators in cancer?** *Cancer Cell* 2003, **4**:89–94.

Sharpless, N.E. and DePinho, R.A.: **Telomeres, stem cells, senescence, and cancer.** *J. Clin. Invest.* 2004, **113**:160–168.

that can then fuse with other broken ends to generate more translocations, leading to another round of chromosome breakage and fusion (Figure 12-12). It can be imagined that these events, known as **breakage-fusion-bridge cycles**, might continue indefinitely—generating all sorts of chromosome translocations and resulting in the loss or amplification of large chromosomal regions. Most of these chromosomal arrangements will be lethal, and so large fractions of the cell population will die. In a few rare cases, however, chromosomal rearrangements can result in a viable cell that has lost tumor suppressor genes or has amplified proto-oncogenes.

Degenerating telomeres can lead to chromosomal instability

Telomeres are the tracts of repetitive DNA at the ends of chromosomes, which are packaged in specific proteins to provide a protective cap on the end of the DNA molecule (see section 4-1). They are shortened at each round of DNA replication, but can be maintained by the enzyme telomerase, which adds new telomere DNA sequence. Telomerase is not expressed in most human somatic cells, however, and as a result their telomeres progressively shorten at each cell division. After about 25–50 divisions, telomere function can be lost, triggering a p53-dependent DNA damage response that leads either to apoptosis or to a permanent cell-cycle arrest called senescence (see section 11-8). This response is thought to act as a tumor suppressor mechanism: any cell that displays unbridled proliferation will eventually be stopped by the loss of its telomeres. The importance of this mechanism is revealed in studies of mice engineered to lack telomerase. Unlike humans, mice normally continue to express telomerase in most somatic tissues, and significant telomere shortening does not occur in proliferating mouse cells. In mice engineered to lack telomerase, however, telomeres shrink gradually over several generations. Severe telomere erosion in the later generations of these mice reduces the incidence of tumors in particular tissues, presumably by causing senescence or apoptosis. Shortened telomeres therefore seem to suppress the formation of some types of tumors.

Paradoxically, telomere degeneration can promote tumorigenesis in some tissues while preventing it in others. Later generations of mice lacking telomerase, for example, display an increased incidence of cancer in some tissues. How do we reconcile the two opposing effects of telomere dysfunction? The answer probably lies in whether or not the incipient tumor cell possesses a p53-dependent DNA damage response. The ability of telomere dysfunction to block tumor formation depends on this response. In tumor cells that carry a mutational defect in p53, cells with eroded telomeres continue to proliferate instead of undergoing apoptosis, leading to chromosomal rearrangements that accelerate tumorigenesis. These rearrangements are thought to occur when poorly protected chromosome ends, like the broken DNA ends discussed above, fuse with each other or recombine with other chromosomes, thereby initiating breakage-fusion-bridge cycles that generate drastic chromosomal rearrangements (see Figure 12-12).

The chromosomal instability that results from telomere loss in the absence of p53 seems to be temporary in many cancers. Although chromosomal rearrangements in most cells are lethal, a few cells emerge that have somehow restored expression of telomerase before chromosome damage was sufficient to kill them. By repairing the protective caps on chromosome ends, telomerase helps reduce the breakage-fusion-bridge cycles and stabilizes chromosome structure. Thus, telomere degeneration seems to trigger a transient surge in chromosomal instability that accelerates tumor progression.

The effects of telomere degeneration are particularly striking in mice engineered to lack both telomerase and p53 (Figure 12-13). These animals have an increased incidence of epithelial cell cancers that display severe chromosomal instability, as predicted if telomere degradation, when accompanied by defects in the DNA damage response, is carcinogenic. Studies with these mice may also help explain the differences in the types of cancers seen in humans compared with those in mice. Most mouse cancers are normally soft-tissue sarcomas and lymphomas that do not display extensive chromosomal instability. In mice lacking telomerase and p53, however, most of the resulting cancers are carcinomas with extensive chromosomal instability (see Figure 12-13). Loss of telomere function in mice therefore results in a shift in tumor type toward the carcinomas that are most common in humans, arguing that telomere regulation may be a key determinant of the cancer spectrum in both species.

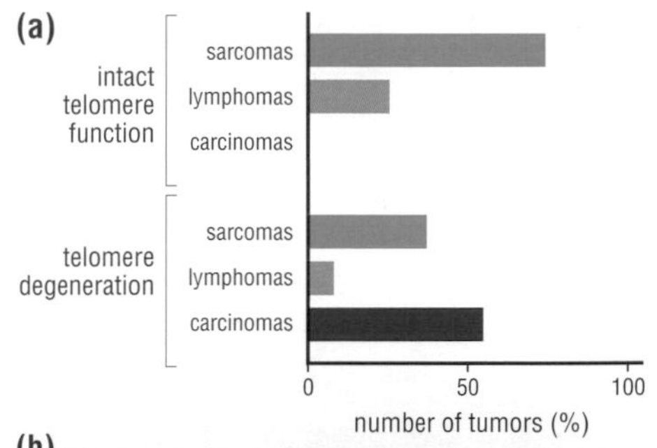

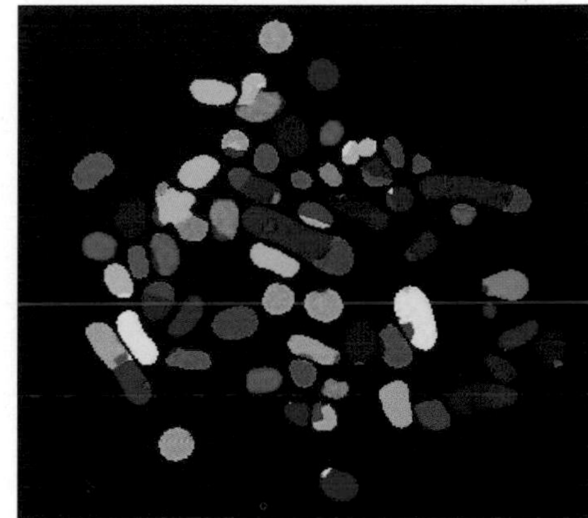

Figure 12-13 Telomere degeneration in the formation of carcinomas **(a)** Mice were engineered to contain mutations in both p53 and telomerase. In early generations, when telomeres were still intact, the p53 mutation alone resulted in the formation of various types of sarcomas and lymphomas, but no carcinomas. In later generations, when telomere degeneration became significant, a large number of carcinomas was observed. **(b)** Analysis of tumor-cell karyotypes in these mice revealed that cells from the sarcomas and lymphomas of early generations had normal or near-normal karyotypes (not shown). In contrast, spectral karyotyping revealed that cells from the carcinomas that appeared after telomere degeneration displayed abnormal karyotypes with numerous non-reciprocal translocations (as revealed by the presence of multiple colors in the same chromosome; see also Figure 12-11). These results are consistent with the idea that telomere degeneration in the absence of p53 results in chromosomal rearrangements, probably through breakage-fusion-bridge mechanisms like those shown in Figure 12-12. Photograph kindly provided by Ronald A. DePinho. From Artandi, S.E. *et al.*: *Nature* 2000, **406**:641–645, with permission.

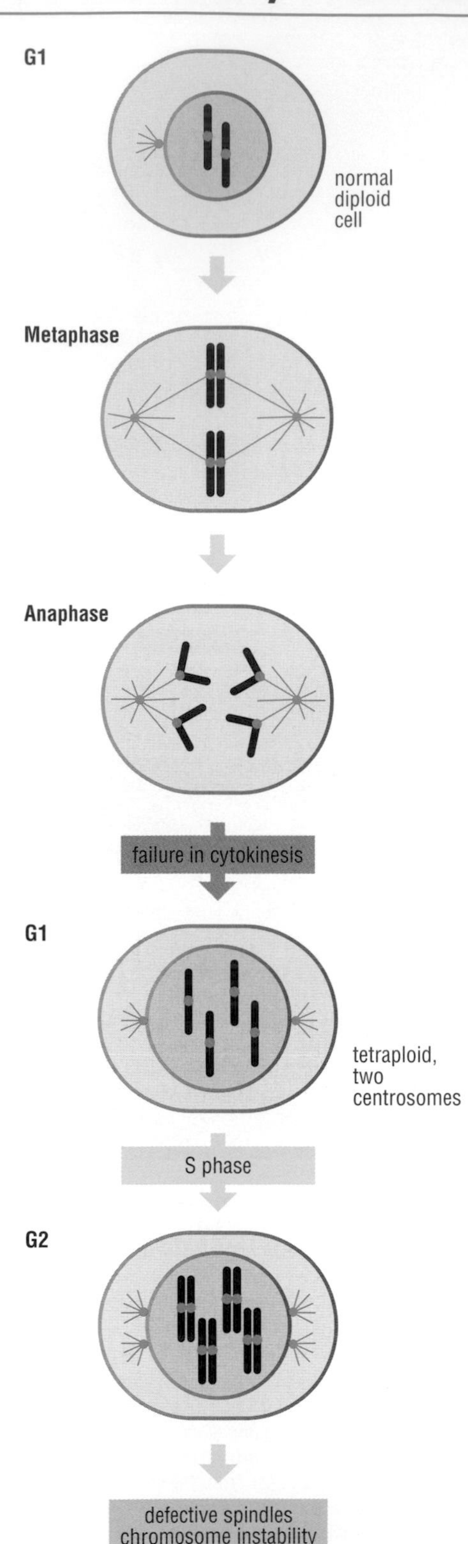

Figure 12-14 Tetraploidization as a potential source of chromosome instability If a cell duplicates its chromosomes but then fails to complete cytokinesis, the result is a cell with twice the normal number of chromosomes and centrosomes. When centrosomes are duplicated in the following S phase, multipolar spindles may result. Chromosome segregation defects will then generate a state of numerical chromosome instability in which daughter cells receive abnormal numbers of chromosomes. For clarity, the cell shown in this diagram contains homologs of one chromosome only.

Cancer cells often become aneuploid through a tetraploid intermediate

Human cancer cells usually have abnormal numbers of chromosomes, a condition called aneuploidy. In the later stages of cancer, chromosome number can increase markedly—to 60 to 90 instead of the usual 46. The most likely explanation is that cancer cells occasionally undergo a defective cell division in which their chromosome number is doubled to a **tetraploid** state, after which defects in chromosome segregation lead to decreases in chromosome number to an unstable state between diploid and tetraploid. Tumor progression could then be promoted by the shifts in gene expression that arise from the gain and loss of chromosomes.

Tetraploidy is thought to result mainly from failures in cytokinesis (Figure 12-14). In the mammalian cell cycle, progression from mitosis to G1 does not depend on the successful completion of cytokinesis, and so it is conceivable that occasional errors in cytokinesis in proliferating tumor cells generate tetraploid cells. Tetraploidization may sometimes depend on a poorly understood regulatory system that inhibits cytokinesis when chromosome segregation fails. Even minor errors in chromosome segregation, such as nondisjunction of a single chromosome, can block the completion of cytokinesis in cultured human cell lines, resulting in tetraploid G1 cells. This system may also be responsible for the tetraploidy that results when cells are forced to progress through mitosis without segregating the sister chromatids. Cells that are treated with microtubule poisons, for example, initially arrest in mitosis in response to activation of the spindle checkpoint (see section 7-2) but eventually escape the mitotic arrest and progress to G1 without segregating sister chromatids or undergoing cytokinesis.

What triggers tetraploidization in tumor cells? Mutations may not be necessary: tetraploidization occurs at a small but significant rate even in normal cells, perhaps as a response to minor defects in chromosome segregation. On the other hand, the frequency of tetraploidization may be enhanced in tumors by mutations in regulatory proteins that govern late M-phase events. Overexpression of the protein kinase aurora A (see section 5-7), for example, can cause an abortive mitosis that leads to tetraploidization of cultured mammalian cells, and aurora A is found overexpressed in many breast cancers. Mutational defects in the spindle checkpoint system are also found in many cancer cells and may cause chromosome-segregation defects that lead to tetraploidy.

In most normal human tissues, tetraploidy is rarely observed or tolerated, suggesting that mechanisms exist to detect and remove the tetraploid cells that occasionally appear. Indeed, there is evidence that a failure in cytokinesis, and thus conversion to tetraploidy, triggers a p53-dependent cell-cycle arrest in G1. Unlike normal cells, p53-deficient cells are often tetraploid or aneuploid. When p53-containing tetraploid cells are experimentally generated by overproduction of aurora A or by treatment with microtubule inhibitors, they arrest in the following G1. If p53 is inactivated, such cells enter the next cell cycle. The mechanism underlying this p53 response, sometimes called the tetraploidy checkpoint, is not known.

Cancer cells often contain excessive numbers of centrosomes

A failure of cytokinesis produces a cell that not only has double the normal complement of chromosomes but also has twice the number of centrosomes (see Figure 12-14). If a tetraploid cell progresses into a new cell cycle and duplicates its centrosomes as usual, it will then enter mitosis with four centrosomes—possibly resulting in a multipolar spindle that could cause major errors in chromosome segregation (Figure 12-15). Tetraploidy might therefore lead to a

Definitions

tetraploid: (of a cell) possessing four copies, or homologs, of each chromosome, often as a result of a failure in cytokinesis in diploid cells.

References

Fujiwara, T. *et al.*: **Cytokinesis failure generating tetraploids promotes tumorigenesis in p53-null cells.** *Nature* 2005, **437**:1043–1047.

Kops, G.J. *et al.*: **On the road to cancer: aneuploidy and the mitotic checkpoint.** *Nat. Rev. Cancer* 2005, **5**:773–785.

Margolis, R.L.: **Tetraploidy and tumor development.** *Cancer Cell* 2005, **8**:353–354.

state of considerable instability in chromosome number. In cells lacking p53 function, failure of cytokinesis and the subsequent proliferation of tetraploid cells leads not only to aneuploidy but also to major structural rearrangements in chromosomes, much like those discussed in section 12-6. Why tetraploidy promotes structural chromosome instability is not clear.

Cancer cells with high numbers of centrosomes can also arise as a result of excessive centrosome replication. In normal cells, centrosome duplication is triggered in early S phase by cyclin E–Cdk2 (see section 6-4). Some cancer cells contain mutations that cause the overproduction of cyclin E, either because of the overexpression of the gene for cyclin E or because of defects in the ubiquitin-protein ligase SCF^{Cdc4}, which targets cyclin E for destruction. These mutations also cause chromosome instability, perhaps because they promote excessive centrosome duplication.

An excess of centrosomes need not result in mitotic defects. Some cancer cells carrying high numbers of centrosomes manage to construct bipolar spindles in which multiple centrosomes are clustered at each pole. This clustering may be associated with the mechanisms that drive the formation of bipolar spindles in the absence of centrosomes (see section 6-8).

Mutations in mitotic spindle components contribute to chromosomal instability

Small changes in chromosome number usually result from errors in chromosome segregation. These errors can occur when abnormal centrosome numbers lead to multipolar mitotic spindles, as described above. Other spindle defects may also contribute to the conversion of a tetraploid (or diploid) cell to a cell with an unbalanced chromosome content. Components of the spindle checkpoint system, including Mad2 and Bub1 (see section 7-2), are partly defective in some cancer cells, resulting in premature separation of sister chromatids in cells that have not completed spindle assembly. It is likely that other aspects of spindle behavior—microtubule motor function, kinetochore attachment, sister-chromatid cohesion and resolution, to name a few—will also turn out to be defective in some cancer cells and could contribute to instability of chromosome number.

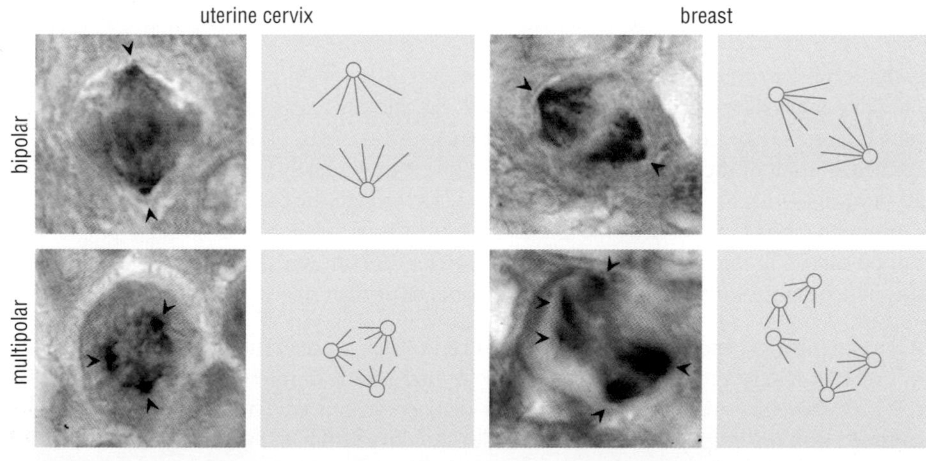

Figure 12-15 Mitotic spindle defects arising from abnormal centrosome number Tumor cells were labeled with antibodies against gamma-tubulin (brown) to reveal the number of spindle poles and the general structure of the spindle. The left panels show cells from a carcinoma-*in-situ* of the cervix, and the right panels show cells from a ductal carcinoma-*in-situ* of the breast. Normal spindles are shown in the top row, and abnormal spindles in the bottom row. From Pihan, G.A. *et al.*: *Cancer Res.* 2003, **63**:1398–1404.

Meraldi, P. *et al.*: **Aurora-A overexpression reveals tetraploidization as a major route to centrosome amplification in $p53^{-/-}$ cells.** *EMBO J.* 2002, **21**: 483–492.

Nigg, E.A.: **Centrosome aberrations: cause or consequence of cancer progression?** *Nat. Rev. Cancer* 2002, **2**:815–825.

Pihan, G.A. *et al.*: **Centrosome abnormalities and chromosome instability occur together in pre-invasive carcinomas.** *Cancer Res.* 2003, **63**:1398–1404.

Quintyne, N.J. *et al.*: **Spindle multipolarity is prevented by centrosomal clustering.** *Science* 2005, **307**:127–129.

Rajagopalan, H. and Lengauer, C.: **Aneuploidy and cancer.** *Nature* 2004, **432**:338–341.

Rajagopalan, H. *et al.*: **Inactivation of hCDC4 can cause chromosomal instability.** *Nature* 2004, **428**:77–81.

Shi, Q. and King, R.W.: **Chromosome nondisjunction yields tetraploid rather than aneuploid cells in human cell lines.** *Nature* 2005, **437**:1038–1042.

Storchova, Z. and Pellman, D.: **From polyploidy to aneuploidy, genome instability and cancer.** *Nat. Rev. Mol. Cell Biol.* 2004, **5**:45–54.

Stukenberg, P.T.: **Triggering p53 after cytokinesis failure.** *J. Cell Biol.* 2004, **165**:607–608.

There are many genetic routes to a malignant cancer

All cancers eventually acquire mutations that allow them to bypass the various proliferative and invasive barriers that normally restrain their behavior. It is clear, however, that all cancers need not take the same route to that final outcome. In one tumor, for example, hyperactivation of Ras may initiate tumorigenesis, after which loss of pRB might provide the second step; in another tumor, this order of events could be reversed. Much of this variation may depend on the tissue involved: certain mutations will be more tumorigenic in some tissues than in others, for the reasons discussed in section 12-2. Nevertheless, cancer progression can occur by remarkably different sequences of mutational events even within the same tissue or cell type.

Understanding the variable routes to a malignant cancer is simply a matter of evolutionary principles. The evolving tumor cell acquires, through a form of natural selection, mutations that provide some selective advantage in the face of environmental pressures. Mutations that activate mitogenic signaling pathways, for example, allow adaptation to the normally limited amounts of mitogenic factors. In most cases, multiple mutations in these pathways are required for the cell to arrive at complete independence of mitogens, but the order in which these mutations occur, or the precise components that are mutated, may not be important. In the same way, a mutation that activates mitogenic signaling can be selected before or after a mutation that inhibits apoptosis (in some cell types at least), as both mutations may be advantageous regardless of the order in which they occur.

In contrast, some selective pressures change as a tumor evolves, and some mutations are advantageous only at later stages of tumor evolution. Telomere degeneration, for example, requires many rounds of cell division to show an effect and is less likely to limit the proliferation of early tumor cells. Thus, telomerase reactivation is expected to be advantageous only at later stages. The same can be said for mutations that enhance angiogenesis or invasive behavior.

Some mutations that are acquired early in tumorigenesis can provide additional advantages later. Mutations that activate mitogenic signaling pathways are clearly advantageous in early stages; these same mutations can also promote angiogenesis, which becomes important later. Similarly, inactivation of p53 can provide an early advantage in overcoming the cell-cycle delay caused by excess mitogenic signaling (see section 12-4) and will also allow cells to continue proliferating when telomere degeneration occurs in later stages.

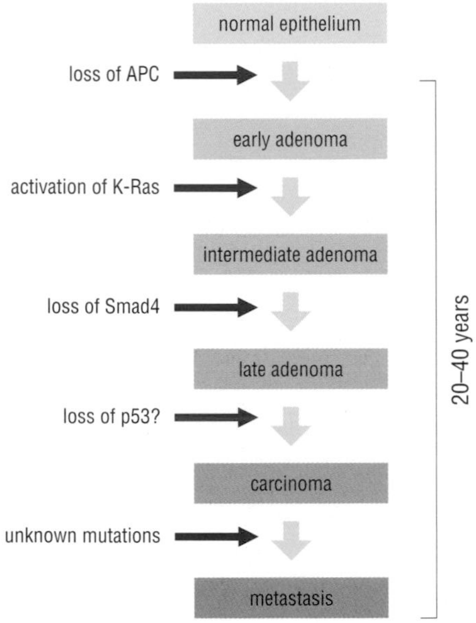

normal epithelium
loss of APC →
early adenoma
activation of K-Ras →
intermediate adenoma
loss of Smad4 →
late adenoma
loss of p53? →
carcinoma
unknown mutations →
metastasis
20–40 years

Figure 12-16 Cancer progression in the colon epithelium There are many different routes to metastatic cancer in the colon, but loss of the gene *APC* is likely to be the initiating event in most cases. The chronology of mutations that occur in later stages is variable and not well defined, but stimulation of mitogenesis by activation of K-Ras and inhibition of Smad4 are important in many cases. Loss of p53 and the appearance of chromosomal instability are also likely to be important early events, but the time at which these events occur remains controversial. See Figure 12-4 for a diagram of the colon epithelial crypts where tumors arise.

Colon cancer progression usually begins with mutations in the gene *APC*

Colon cancer is unusually well understood because it has been possible to analyze in considerable molecular detail its progression from the earliest stages through to the malignant state (Figure 12-16). It typically appears first as small adenomas, or polyps, that bud from the epithelial lining of the colon. These benign tumors can gradually progress—over a period of 20–40 years—to a malignant carcinoma that breaks through the basal membrane, invades the surrounding muscle, and finally metastasizes to other tissues. Tumors at all stages of progression can be surgically removed and their genes, proteins and chromosomes studied, giving considerable insight into the molecular defects that arise at each stage.

A large number of genes are known to be mutated in colon cancers, and it is clear that the route to cancer can vary greatly. Nevertheless, loss of one tumor suppressor gene in particular—*APC*—is thought to be the earliest event in most colon cancer. The name *APC* (not to be confused with the anaphase-promoting complex) is derived from an inherited syndrome, famil-

References

Evan, G. and Vousden, K.H.: **Proliferation, cell cycle and apoptosis in cancer.** *Nature* 2001, **411**:342–348.

Hanahan, D. and Weinberg, R.A.: **The hallmarks of cancer.** *Cell* 2000, **100**:57–70.

Kinzler, K.W. and Vogelstein, B.: **Lessons from hereditary colorectal cancer.** *Cell* 1996, **87**:159–170.

Parsons, D.W. *et al.*: **Colorectal cancer: mutations in a signalling pathway.** *Nature* 2005, **436**:792.

Rajagopalan, H. *et al.*: **The significance of unstable chromosomes in colorectal cancer.** *Nat. Rev. Cancer.* 2003, **3**:695–701.

Reya, T. and Clevers, H.: **Wnt signalling in stem cells and cancer.** *Nature* 2005, **434**:843–850.

ial adenomatous polyposis coli (FAP), that causes abundant colon polyps and a predisposition to malignant cancer. This inherited syndrome is caused by mutations in *APC*. The loss of *APC* function also occurs early in most sporadic (non-heritable) forms of colon cancer.

APC encodes an inhibitory component in a signaling pathway triggered by a mitogen called Wnt (Figure 12-17). Thus, in the absence of APC, Wnt signaling is hyperactivated and drives abnormally high rates of cell proliferation. About 85% of colon cancers bear mutations in *APC*, and the remaining tumors are thought to contain mutations that activate the Wnt signaling pathway by other means. Inherited mutations in *APC* do not generally cause cancer in tissues other than the colon, so a hyperactive Wnt pathway must be particularly advantageous in the initiation of tumorigenesis in colon epithelial cells.

The further development of colon polyps is driven by hyperactivating mutations in other mitogenic regulators, including K-Ras, a member of the Ras family of GTPases, and B-Raf, a member of the Raf family of protein kinases that lie at the top of the MAP kinase cascade (see section 10-6). These mutations presumably generate abnormal mitogenic signals that are not produced by activation of the Wnt pathway alone. Additional mutations, particularly in a gene regulatory protein called Smad4, disrupt the signaling pathway activated by the anti-mitogenic factor TGF-β (see section 10-7). Other mutations also disrupt the expression of $p16^{INK4a}$ and other regulators of the pRB pathway controlling E2F function. As is common in all cancers, therefore, the abnormal cell-cycle entry in colon cancer cells is the result of mutations in multiple regulatory systems.

Abnormally high levels of signaling by mitogenic proteins normally block cell proliferation through the activation of p53 by ARF (see section 12-4). Not surprisingly, most colon cancer cells have adapted to this block by inactivation of *ARF* early in their evolution.

Two forms of genetic instability drive colorectal cancer progression

The evolution of a colon cancer, like that of other cancer types, is likely to depend on genetic instability. About 85% of colon cancers, including those due to FAP, display considerable alterations in chromosome structure and number that probably result from chromosomal instability. The mechanisms that trigger this instability are not well defined, but are thought to include many of those discussed previously (see sections 12-6 and 12-7). In many cases of colon cancer, mutations in p53 occur around the time when a benign adenoma is transformed into a malignant carcinoma, and these mutations may coincide with telomere degeneration to promote chromosomal instability at this stage.

Interestingly, about 15% of colon cancers do not display marked chromosomal instability but have developed genetic instability in another way. This is illustrated by another familial colon cancer syndrome, hereditary nonpolyposis colorectal cancer (HNPCC), in which patients have an increased likelihood of developing colon cancer but do not develop large numbers of polyps. HNPCC is caused by mutations in any one of several enzymes responsible for DNA mismatch repair (see section 12-5), and so these patients display microsatellite instability, the form of genetic instability that results from an increased rate of point mutations. HNPCC accounts for about 2–4% of colon cancers, and about 13% of sporadic cases are probably caused by similar defects leading to microsatellite instability. Interestingly, colon cancers rarely have both microsatellite instability and chromosomal instability—just one is sufficient to boost the progression of this disease to malignancy. Thus, strikingly different routes can be taken during progression to cancer, even in the same tissue.

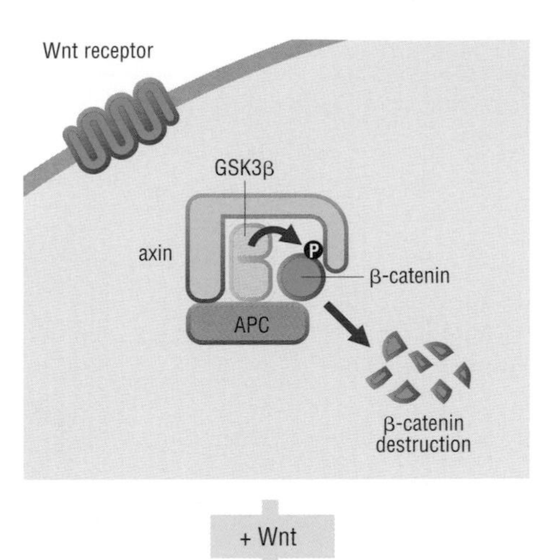

+ Wnt

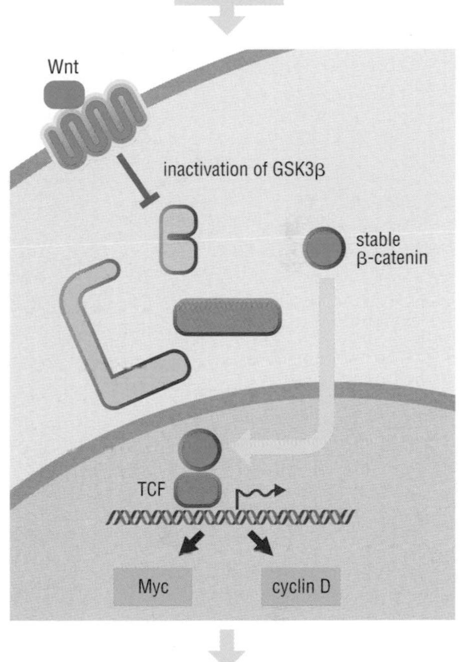

cell-cycle entry

Figure 12-17 The Wnt mitogenic signaling pathway The division of the epithelial cells of the colon is partly regulated by a mitogen called Wnt. A key component in Wnt signaling is the protein β-catenin, which stimulates the expression of genes that drive entry into the cell cycle. In the absence of Wnt (top panel), β-catenin is inhibited by association in the cytoplasm with a protein complex that includes APC and the enzyme glycogen synthase kinase-3β (GSK3β). GSK3β phosphorylates β-catenin, thereby targeting it for destruction (through the ubiquitin-protein ligase SCF, not shown). In the presence of Wnt (bottom panel), GSK3β is inactivated, which leads to dissociation of the APC complex. β-catenin is not destroyed and can carry out its mitogenic function. APC is required for the phosphorylation and destruction of β-catenin in the absence of Wnt, and thus the loss of APC, as in colon cancer, results in constitutive activation of the mitogenic β-catenin signal. For the same reasons, colon cancer is sometimes associated with mutant forms of β-catenin that are resistant to phosphorylation or destruction.

Reducing cancer mortality begins with prevention and early diagnosis

Millions of people will be killed by cancer this year. How can this death rate be slowed? The first and most effective line of defense is prevention. Some common cancers are triggered by clearly identified environmental factors, and these risk factors can be avoided. There is little doubt, for example, that tobacco smoke is an underlying cause in most cases of lung cancer, and the abolition of smoking would greatly reduce worldwide cancer deaths. It is important that we continue to identify and eradicate the preventable causes of this disease.

Cancer is lethal in most cases because it invades and colonizes tissues, eventually destroying their function. Many tumors are curable if they are found at an early stage when surgery and other methods can be used to remove or destroy them. Early diagnosis is therefore a powerful approach to reducing cancer mortality. Most early tumors do not produce any symptoms, however, and are difficult to detect. The development of more sensitive detection tools—advanced imaging methods and molecular markers, for example—that can spot tumors when they first develop is an important direction for future research. Even existing detection methods could save countless lives if they were applied more effectively.

Therapies must kill cancer cells but leave healthy cells intact

The ideal cancer treatment must achieve two goals. First, the treatment must focus specifically on tumor cells and leave normal cells unscathed. Second, all cells in the tumor must be killed, as just one survivor has the potential to seed another tumor.

The most straightforward approach to treatment is the surgical removal of a tumor. This can be very effective when metastasis has not occurred, and even in early metastatic cancer can be moderately successful if both the primary and secondary (metastatic) tumors are removed. When tumors are not accessible to surgery, they can often be destroyed non-surgically by ionizing radiation.

Treatments that damage DNA or inhibit DNA synthesis, including radiation and cytotoxic drugs, are also commonly used against cancer. Although these therapies produce side effects in normal cells, cancer cells are generally more sensitive to them. In many cases this is because the cancer cells are genetically unstable and have lost their DNA damage response—and therefore continue to progress through the cell cycle despite the damage, resulting in their death by apoptosis. Alternatively, some tumor cells not only retain their damage response but are extremely sensitive to any proapoptotic insult—presumably because proapoptotic signals are constitutively active in tumor cells with hyperactive mitogenic signaling (see section 12-4).

Some microtubule poisons, such as the drug Taxol, are effective cancer treatments that are thought to act by disrupting spindle function. Tumor cells often have a defective spindle checkpoint system (see section 7-2) and therefore attempt to progress through mitosis despite the presence of spindle defects—resulting in massive chromosome segregation errors that kill the cell.

Tumors contain a mixture of cell populations with different proliferative properties and thus different importance in therapy. Some types of tumor are thought to contain cancer stem cells that retain the capacity to proliferate indefinitely and are the major source of new cells in the tumor. Therapies that target this cell population are likely to be the most effective.

References

Alberts, B. *et al.*: *Molecular Biology of the Cell* 4th ed. (Garland Science, New York, 2002).

Dean, M. *et al.*: **Tumour stem cells and drug resistance.** *Nat. Rev. Cancer* 2005, **5**:275–284.

Devita, V.T. *et al.*: *Cancer: Principles and Practice of Oncology* 7th ed. (Lippincott, Williams and Wilkins, Philadelphia, 2004).

Ferrara, N. and Kerbel, R.S.: **Angiogenesis as a therapeutic target.** *Nature* 2005, **438**:967–974.

Hingorani, S.R. and Tuveson, D.A.: **Targeting oncogene dependence and resistance.** *Cancer Cell* 2003, **3**:414–417.

Klein, S. *et al.*: **Killing time for cancer cells.** *Nat. Rev. Cancer* 2005, **5**:573–580.

Sawyers, C.: **Targeted cancer therapy.** *Nature* 2004, **432**:294–297.

Shah, N.P. *et al.*: **Multiple BCR-ABL kinase domain mutations confer polyclonal resistance to the tyrosine kinase inhibitor imatinib (STI571) in chronic phase and blast crisis chronic myeloid leukemia.** *Cancer Cell* 2002, **2**:117–125.

A detailed understanding of the molecular basis of cancer may lead to rational and more specific cancer therapies

A great deal of effort is being devoted to the development of more rational and sophisticated approaches to cancer treatment, based on the immense body of knowledge we have acquired about the molecular mechanisms underlying cell proliferation in normal and cancerous tissues. It is hoped that if we identify the proteins that are primarily responsible for driving progression in each type of cancer, then we can find ways of specifically inhibiting these proteins and thus stop that cancer. This approach has had some success: the drug imatinib (more commonly known as Gleevec) specifically inhibits the enzymatic activity of the protein kinase Abl, which is abnormally active in chronic myeloid leukemia (CML, see section 12-1) (Figure 12-18). Imatinib has been remarkably successful in the treatment of early-stage CML, and this success has stimulated other attempts to develop specific inhibitors for enzymes that are hyperactivated in other cancers. Inhibitors of numerous other mitogenic and cell-cycle regulatory components, including protein kinases, GTPases and other signaling components, are currently being developed and tested in the treatment of various cancers, and it seems likely that some of these efforts will eventually yield useful treatments. Efforts are also being made to develop chemicals that alter other processes—apoptosis, telomere maintenance, angiogenesis and metastasis, for example—that are important in cancer biology.

Efforts to develop rational therapies may be hindered by a number of problems. First, some of the targeted enzymes are important in the control of normal cell proliferation, and inhibitors of these enzymes will therefore have undesirable side effects in normal tissues. Second, cancer cell misbehavior is rarely dependent on a single hyperactive enzyme as in early CML; instead, the malignant cancer is typically a robust proliferative and invasive machine whose behavior results from the overactivity of multiple positive regulatory components, and often the lack of negative regulation as well. Inhibition of only one component is unlikely to have much effect, and it will probably be necessary to develop cocktails of therapies that target multiple cellular pathways and are tailored specifically for each cancer type. Third, most advanced cancer cells are genetically unstable, which in some cases helps them acquire mutations that provide resistance to specific chemical inhibitors—some late-stage CML patients have developed resistance to imatinib, for example (see Figure 12-18). Fourth, many of these treatments will inhibit the proliferation of cancer cells without killing them, in part because proapoptotic or other stress response systems are often absent in these cells. Chronic drug treatment is therefore required, which provides the time for a cell in the tumor to develop resistance.

Many cancers arise through the loss of tumor suppressor gene function as well as the activation of proto-oncogenes. How might we develop therapies that restore the function of mutant tumor suppressors? One possibility is gene therapy: the replacement of a defective gene with a normal version. Cancer is, after all, a genetic disease, and the ultimate solution to genetic defects would seem to be gene repair—at least in the early stages of cancer, before genetic instability runs amok. Unfortunately, there have been only minor successes in developing effective methods for introducing genes into diseased cells, and so the great promise of gene therapy remains unfulfilled. Given the pace of scientific progress in the recent past, however, there is good reason to hope that these and other approaches, combined with effective prevention and diagnosis, will eventually succeed.

(a)

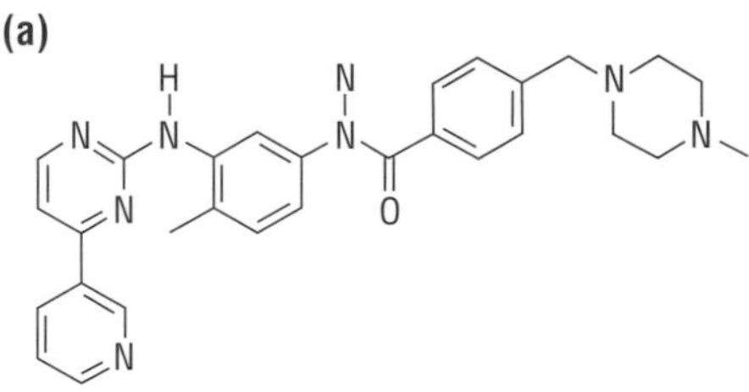

(b)

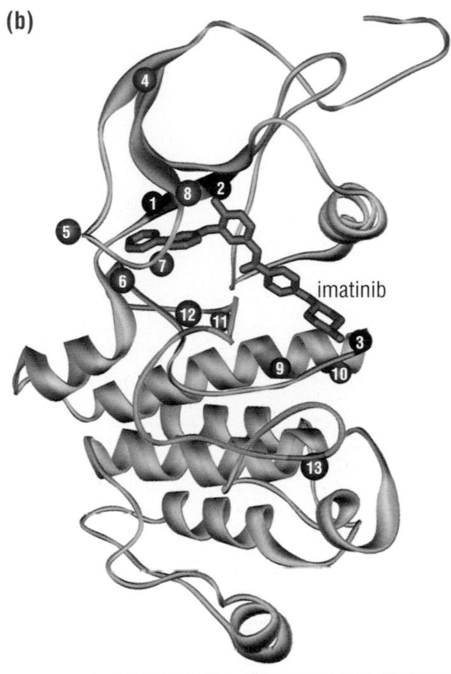

Figure 12-18 **Inhibition of the protein kinase Abl by imatinib** **(a)** Chemical structure of imatinib. **(b)** The structure of Abl (grey) bound to imatinib (orange). Like other protein kinases (see section 3-1), Abl contains an amino-terminal lobe dominated by beta-sheet structure (top) and a larger carboxy-terminal lobe composed of alpha helices (bottom). Imatinib occupies the ATP-binding site that lies between the two kinase lobes, and thus inhibits enzyme activity. In many patients, resistance to imatinib arises because the Abl protein acquires point mutations that block the effects of the drug. Some mutations (labeled 1–3, purple) directly inhibit imatinib binding. Other mutations (4–13, green) are thought to inhibit protein conformational changes that are required for the kinase to bind the drug with high affinity. From Shah, N.P. *et al.*: *Cancer Cell* 2002, **2**:117–125.

Glossary

activator subunits: two regulatory subunits, **Cdc20** and **Cdh1**, that bind to and stimulate the ubiquitin-protein ligase activity of the anaphase-promoting complex (APC). (3-10)

adaptor protein: a protein that links two other proteins together in a regulatory network or protein complex; also called a **mediator protein.** (11-3)

adenoma: benign tumor of epithelial cells, typically of glandular origin. (12-2)

alpha-satellite (alphoid) DNA: in human chromosomes, the repetitive DNA sequence found at the **centromere.** (4-12)

amphitelic attachment: see **bi-orientation.** (6-10)

anaphase: the first stage of mitotic exit, when the **sister chromatids** are segregated by the **mitotic spindle.** In most species, anaphase is divided into anaphase A, the initial movement of chromosomes to the spindle poles, and anaphase B, the movement of spindle poles away from each other. (5-0)

anaphase A: the stage in mitosis in which sister chromatids separate and move to opposite poles of the mitotic spindle. It follows metaphase. (7-0)

anaphase B: the stage in mitosis in which the poles of the spindle move further apart. It follows **anaphase A.** (7-0)

anaphase-promoting complex (APC): large, multisubunit ubiquitin-protein ligase (also known as an E3 enzyme) that catalyzes the attachment of **ubiquitin** to mitotic proteins such as M cyclin and securin, thus promoting their ubiquitin-dependent proteolysis. It is also called the cyclosome or APC/C. (3-9)

aneuploid: containing an abnormal number of chromosomes. (9-0)

angiogenesis: formation of new blood vessels, often governed by extracellular angiogenic factors produced by local cells. (12-0)

anti-mitogen: extracellular molecule that inhibits cell proliferation. (10-7)

APC: see **anaphase-promoting complex.** (3-9)

apoptosis: regulated cell death in which activation of specific proteases and nucleases leads to death characterized by chromatin condensation, protein and DNA degradation, loss of plasma membrane lipid asymmetry and disintegration of the cell into membrane-bound fragments. (10-14)

ARS: see **autonomously replicating sequence.** (4-2)

ATP-dependent nucleosome-remodeling complex: see **chromatin-remodeling complex.** (4-9)

aurora A: serine/threonine protein kinase that is activated at the beginning of M phase and is involved in centrosome function and spindle assembly. (5-7)

aurora B: serine/threonine protein kinase that is activated at the beginning of M phase and is involved in chromosome condensation, spindle assembly, attachment of kinetochores, sister-chromatid segregation and cytokinesis. (5-7)

autonomously replicating sequence (ARS): specific DNA sequence that functions as an origin of replication when transferred to a new location in a DNA plasmid or other DNA fragment. (4-2)

base excision repair: mechanism of DNA repair in which a nucleotide with a damaged or missing base is excised from the DNA and replaced with an undamaged one, using the undamaged DNA strand as template. (11-1)

bi-orientation: correct bipolar attachment of a sister-chromatid pair to the spindle, with the two kinetochores attached to opposite poles. (6-10)

bistable: able to exist stably in one of two alternative states, but cannot come to rest in an intermediate state between them. (3-7)

blastula: hollow ball of cells that results from the early **cleavage** divisions in some animal embryos. (2-3)

breakage-fusion-bridge (BFB) cycle: complex process that generates chromosome rearrangements. It begins when a **dicentric chromosome** is attached to both spindle poles and then torn in two in anaphase. Fusion of the broken ends can generate another dicentric chromosome that begins the process again. (12-6)

bromodomain: protein domain that binds to acetyl-lysine. Bromodomains are found in several non-histone chromatin proteins and interact with specific acetylated lysines on histone H3 or H4. (4-9)

CAK: see **Cdk-activating kinase.** (3-3)

carcinoma: malignant tumor of epithelial tissue. (12-2)

catastrophe: (in **microtubules**) sudden shrinkage that occurs when GTP hydrolysis occurs at the microtubule tip. (6-1)

Cdc20: activator subunit for the anaphase-promoting complex (APC), primarily responsible for promoting the destruction of proteins that control the metaphase-to-anaphase transition. (3-10)

Cdc25: protein phosphatase that activates cyclin-dependent kinases by removing phosphate from specific residues in the Cdk active site (Tyr 15 in most Cdks; also Thr 14 in animals). (3-3)

Cdh1: activator subunit for the anaphase-promoting complex (APC), primarily responsible for maintaining APC activity in late mitosis and G1. (3-10)

Cdk: see **cyclin-dependent kinase.** (1-3)

Cdk-activating kinase (CAK): protein kinase that activates cyclin-dependent kinases by phosphorylating a threonine residue (Thr 160 in human Cdk2) in the T-loop. (3-3)

Cdk inhibitor protein (CKI): protein that interacts with Cdks or Cdk–cyclin complexes to block activity, usually during G1 or in response to inhibitory signals from the environment or damaged DNA. (3-6)

cell cycle: sequence of events that leads to the reproduction of the cell. In eukaryotic cells, it includes two major phases: S phase, in which the chromosomes are duplicated, and M phase, in which the duplicated chromosomes are segregated and the cell divides. In many cells, additional gap phases separate S and M phase: G1 before S phase and G2 before M phase. (1-0)

cell-cycle control system: network of regulatory proteins that controls the timing and coordination of cell-cycle events. (1-0)

cell line: a genetically homogeneous cell population that can proliferate indefinitely in culture. It is also called an immortalized or established cell line. (2-5)

cellular blastoderm: early stage in *Drosophila* embryonic development, comprising a superficial epithelial layer of several thousand cells surrounding a yolky center. (2-4)

cellularization: in insect development, the packaging of the nuclei of the syncytial embryo into individual cells, generating the **cellular blastoderm.** (2-4)

central spindle: the structure that forms in late mitosis in a dividing animal cell from interpolar microtubule remnants of the central part of the mitotic spindle and various associated proteins. It eventually becomes the midbody. (8-5)

centriole: cylindrical array of microtubules, typically found in pairs in the **centrosomes** of animal cells. (6-3)

centromere: region of the chromosome where kinetochores are assembled and attach to the mitotic spindle. (4-12)

centrosome: large protein organelle that serves as the major microtubule-organizing center of most animal cells. It contains a pair of orthogonally oriented **centrioles,** as well as a surrounding matrix containing γ-tubulin ring complexes. (6-3)

centrosome disjunction: severing of the protein linkage between duplicated centrosomes in early mitosis. (6-6)

centrosome maturation: improvement in the microtubule-nucleating ability of the centrosome that occurs during mitosis. It is due to an increase in the number of γ-tubulin ring complexes. (6-6)

centrosome separation: the movement apart of duplicated centrosomes that follows **centrosome disjunction** in early mitosis. (6-6)

checkpoint: regulated transition point in the cell cycle, where progression to the next phase can be blocked by negative signals. This term is sometimes defined to include the signaling mechanisms that monitor cell-cycle events and transmit the information to the control system; in this book the term is used to define the transition point in the cell cycle where these mechanisms act. (1-3)

chiasma (plural **chiasmata**)**:** X-shaped chromosomal structure that is seen in the microscope at sites of crossing-over between homologous chromosomes at the end of meiotic prophase. (9-0)

chromatin: the complex of DNA and protein that forms a eukaryotic chromosome. (4-0)

chromatin-remodeling complex: large protein complex that, through release of energy by ATP hydrolysis, causes changes in **nucleosome** structure that allow rearrangements in nucleosome position or access to proteins involved in DNA transcription, repair or replication. (4-9)

chromodomain: protein domain that binds to methyl-lysine. Chromodomains are found in several non-histone chromatin proteins and interact with specific methylated lysines on histone H3 or H4. (4-9)

chromosomal instability (CIN): an abnormally high incidence of defects in chromosome number (numerical CIN) or chromosome structure (structural CIN). (12-5)

chromosomal translocation: alteration in chromosome structure in which pieces of one chromosome become exchanged with or attached to another chromosome. (12-5)

chromosome condensation: in mitosis, the structural changes that result in compaction of the chromosomes into short, thick structures. (5-9)

chromosome congression: (in animal cells) the alignment of sister-chromatid pairs at the center of the spindle in metaphase. (6-9)

CIN: see **chromosomal instability.** (12-5)

Cip/Kip: small family of **CKIs** in animal cells, including mammalian p21 (Cip1) and p27 (Kip1), that inhibit Cdk activity by interaction with both subunits of the Cdk–cyclin complex. (3-6)

CKI: see **Cdk inhibitor protein.** (3-6)

clamp loader: five-subunit protein that loads the **sliding clamp** onto the DNA during DNA replication. (4-1)

cleavage: the early cell divisions of animal embryos, which occur in the absence of growth and rapidly subdivide the large fertilized egg into thousands of smaller cells. (2-3)

cleavage furrow: during cytokinesis, the furrow that forms around the equator of a dividing animal cell and eventually divides it in two. (8-1)

cohesin: a complex of four proteins that links sister chromatids together after S phase. (5-8)

condensin: complex of five proteins that helps condense and resolve the sister chromatids in mitosis. (5-9)

contractile ring: ring of proteins, including contractile assemblies of actin and **myosin**, that forms at the site of cleavage in dividing animal cells. Its gradual contraction pinches the cell in two. (8-1)

crossover: in meiosis, a **homologous recombination** event that results in the reciprocal exchange of DNA between two homologs. (9-2)

cullin: large sequence motif found in a core subunit of some ubiquitin-protein ligases, including **SCF** and the **APC**. It interacts with the **RING finger** subunit of these enzymes. (3-9)

cyclin: positive regulatory subunit that binds and activates **cyclin-dependent kinases**, and whose levels oscillate in the cell cycle. (1-3)

cyclin-dependent kinase (Cdk): protein kinase whose catalytic activity depends on an associated **cyclin** subunit. Cyclin-dependent kinases are key components of the cell-cycle control system. (1-3)

cytokinesis: cell division, the process in late **M phase** by which the duplicated nuclei and cytoplasmic components are distributed into daughter cells by division of the mother cell. (1-1)

D-box: see **destruction box.** (3-10)

destruction box (D-box): sequence motif (RXXLXXXXN) found in many targets of the anaphase-promoting complex (APC). (3-10)

diakinesis: stage in meiosis following **diplotene**, in which the meiotic spindle is formed and pairs of homologous chromosomes become oriented on the spindle. (9-3)

dicentric chromosome: chromosome that contains two centromeres and therefore two kinetochores, usually as a result of chromosomal rearrangements. (12-6)

diploid: (of a cell) possessing two copies, or **homologs**, of each chromosome. The somatic cells of most multi-cellular organisms are diploid. (1-2)

diplotene: stage in meiotic prophase following **pachytene**, in which chromosomes condense and chiasmata become apparent. (9-3)

DNA catenation: the intertwining of sister DNA molecules that occurs when replication forks meet. (5-8)

DNA damage response: cellular response to extensive DNA damage, in which DNA repair is promoted and cell-cycle progression is delayed. (11-0)

DNA helicase: protein that moves along DNA, unwinding the double helix as it goes. DNA helicases have ATPase activity and use the energy of ATP hydrolysis to move along the DNA. The helicases involved in DNA replication in bacteria and eukaryotes are six-subunit ring structures. Many of the helicases involved in other cellular processes such as repair and recombination are monomeric. (4-3)

DNA ligase: enzyme that links the ends of two strands of DNA by catalyzing the formation of a phosphodiester bond between the 3′ hydroxyl group at the end of one fragment and the 5′ phosphate at the end of the other. (4-1)

DNA polymerase: enzyme that synthesizes new DNA by copying a single-stranded DNA template. The polymerase moves along the template and synthesizes a new strand of complementary DNA sequence by adding nucleotides, one at a time, onto the 3′ OH end of the new strand. Which nucleotide is added is determined by correct base pairing with the nucleotide template. There are many different types of DNA polymerase, each specialized for a different cellular function. (4-0)

dominant negative: refers to a mutant gene product that inhibits the function of the wild-type gene product in a genetically dominant fashion, often by interfering with the ability of the wild-type protein to interact with other proteins. (2-5)

double-strand break: type of DNA damage in which a DNA molecule is broken across both strands. (11-1)

double thymidine block: a method for synchronously arresting a mammalian cell population at the beginning of S phase. Asynchronous cells are treated with thymidine, which causes arrest throughout S phase. These cells are released from arrest to allow progression out of S phase. A second thymidine treatment is then used to arrest all cells at the beginning of the subsequent S phase. (2-6)

dynamic instability: the tendency of **microtubules** to switch between states of rapid growth and rapid shrinkage. (6-1)

endoreduplication: the repeated replication of chromosomes without accompanying mitosis or cell division. This can result in large **polytene chromosomes** consisting of many copies in parallel. (1-2)

epigenetic: inherited through mechanisms that are not dependent on DNA sequence. Known epigenetic mechanisms often concern gene regulation and are dependent on modifications of the DNA or local chromatin structure. (4-12)

euchromatin: chromatin in which DNA is packaged in such a way as to be accessible to enzymes and gene regulatory proteins. (4-0)

FACS: see **fluorescence-activated cell sorter.** (2-6)

F-box protein: any of the numerous alternative substrate-targeting and binding subunits of **SCF**, which contain an amino-acid sequence called the F-box. It mediates interaction of the protein to be ubiquitinated with core SCF subunits. (3-9)

flow cytometer: instrument through which a stream of cells is passed and their fluorescence measured in the technique of **flow cytometry.** (2-6)

flow cytometry: technique used to enumerate and analyze a sample of cells by incubating them with one or more fluorescently labeled antibodies and/or other molecules that can bind to cellular components and measuring the fluorescence intensity of each fluor for each cell. It is used to count the numbers of cells of different types, or at different stages in development or the cell cycle. (2-6)

fluorescence-activated cell sorter (FACS): a modified **flow cytometer** that sorts individual cells into different containers according to their fluorescence. (2-6)

γ-tubulin ring complex (γ-TuRC): large, multisubunit protein complex containing a ring of γ-tubulin subunits. It is thought to serve as a template for microtubule nucleation. (6-2)

γ-TuRC: see **γ-tubulin ring complex.** (6-2)

G0: a prolonged nondividing state that is reached from **G1** when cells are exposed to extracellular conditions that arrest cell proliferation. (1-1)

G1: the cell-cycle gap phase between **M phase** and **S phase.** (1-1)

G1 cyclins: cyclins that bind and activate Cdks that stimulate entry into a new cell cycle at Start; their concentration depends on the rate of cell growth or on growth-promoting signals rather than on the phase of the cell cycle. (3-2)

G1/S cyclins: cyclins that activate Cdks that stimulate progression through Start; their concentration peaks in late G1. (3-2)

G2: the cell-cycle gap phase between **S phase** and **M phase.** (1-1)

G2/M checkpoint: important regulatory transition where entry into M phase can be controlled by various factors such as DNA damage or the completion of DNA replication. (1-3)

gamete: specialized haploid reproductive cell, for example egg and sperm in animals, that fuses with another gamete to form a diploid zygote in sexual reproduction. (9-0)

gastrulation: cell movements that reorganize the blastula (**cellular blastoderm** in *Drosophila*) into an embryo with a gut surrounded by cell layers in place for the development of tissues and organs. (2-4)

genetic instability: an abnormal increase in the rate at which genes and chromosomes are mutated, rearranged or lost. Also called genomic instability. (12-5)

growth factor: extracellular factor that stimulates cell growth (an increase in cell mass). This term is sometimes used incorrectly to describe a factor that stimulates cell proliferation, for which the correct term is mitogen. (10-10)

haploid: (of a cell) possessing one copy, or **homolog**, of each chromosome. The egg and sperm cells of animals are haploid. (1-2)

heterochromatin: chromatin in which DNA is packaged in such a way as to be poorly accessible to enzymes and gene regulatory proteins. (4-0)

histone octamer: protein core of the **nucleosome**, composed of eight histone subunits (two each of histones H2A, H2B, H3 and H4). (4-9)

homolog: in sexually reproducing organisms, either of the two copies of each chromosome normally present in the **diploid** somatic cells. For each chromosome, one homolog is inherited from one parent and the other homolog from the other parent. (1-2)

homologous recombination: interaction between broken and intact homologous DNA molecules that promotes repair, exchange and pairing. It involves the invasion of the intact DNA helix by a complementary single strand originating at a double-strand break in the other DNA molecule, and leads either to reciprocal exchange of DNA sequence (a **crossover**) or to a **non-crossover** event in which repair but not exchange occurs. It occurs between duplicated homologous chromosomes in meiotic prophase, and between sister chromatids or homologous chromosomes in somatic cells. (9-2)

hydrophobic patch: small hydrophobic region on the surface of a protein. Many cyclins contain a hydrophobic patch that is based on the MRAIL sequence in the cyclin box. They interact with the **RXL motif** of Cdk substrates and inhibitors. (3-5)

hyperbolic response: response to an increasing stimulus that is initially linear but levels off as the system becomes saturated. (3-7)

hyperproliferation stress response: cell-cycle arrest or apoptosis seen in response to excessive mitogenic signals. (11-8)

hysteresis: (in the context of **bistable** signaling systems) tendency of a system to respond differently to the same stimulus depending on the initial state of the system. (3-7)

imaginal cells: cells of the *Drosophila* larva that are the precursors of adult structures. (2-4)

imaginal discs: small sheets or pouches of **imaginal cells** in the *Drosophila* embryo, which will differentiate into major structures of the adult fly. (2-4)

initiator protein: any of a complex of proteins that bind to origins of replication in DNA and initiate the unwinding of the double helix in preparation for DNA replication. (4-3)

INK4: small family of mammalian **CKIs**, including $p15^{INK4b}$ and $p16^{INK4a}$, that bind the Cdk4 and Cdk6 proteins and reduce their binding affinity for cyclin D. (3-6)

interphase: the period between the end of one **M phase** and the beginning of the next. (1-1)

karyotype: the set of condensed metaphase chromosomes in a eukaryotic cell stained so that each chromosome can be uniquely identified. (12-5)

KEN-box: sequence motif (KENXXXN) found in many targets of the anaphase-promoting complex (APC). (3-10)

kinetochore: protein complex at the centromere of a chromosome, where the microtubules of the spindle are attached during mitosis. (5-0)

kinetochore fiber: bundle of microtubules that links a chromatid kinetochore to a spindle pole. (6-9)

knock out: to render a gene inactive by disrupting it in the animal, usually by replacing most of the gene with an inactivating insertion. (2-5)

L12 helix: small alpha helix adjacent to the **T-loop** in the active site of Cdk2 (residues 147–151), which changes structure to a beta strand upon cyclin binding. (3-1)

lagging strand: in the replication of DNA, the new DNA strand that is synthesized on the template strand that runs from 5′ to 3′ into a replication fork. It is synthesized discontinuously as a series of short DNA fragments that are subsequently joined together to form a continuous DNA strand. (4-1)

leading strand: in the replication of DNA, the new DNA strand that is synthesized on the template strand that runs from 3′ to 5′ into a replication fork. It is synthesized continuously as the fork moves forward. (4-1)

leptotene: first stage in meiotic prophase, in which the pairing of homologous chromosomes occurs. (9-3)

maturation-promoting factor: see **MPF**. (2-3)

Mcm complex: multiprotein complex that functions as the helicase that unwinds DNA at replication origins and replication forks in eukaryotes. (4-3)

M cyclins: cyclins that activate Cdks necessary for entry into mitosis; their concentration rises at the approach to mitosis and peaks in metaphase. (3-2)

mediator protein: see **adaptor protein**. (11-3)

meiosis: see **meiotic program**. (9-0)

meiosis I: the first meiotic division, the stage in the meiotic program that includes assembly of the first meiotic spindle, bi-orientation of paired homologs on the spindle, and homolog separation and segregation. (9-5)

meiosis II: the second meiotic division, during which the sister chromatids are segregated on the second meiotic spindle. (9-8)

meiotic program: specialized nuclear division that occurs during formation of the gametes in sexually reproducing diploid organisms and generates haploid nuclei carrying a single homolog of each chromosome. The term **meiosis**, although originally used to describe only the events of meiotic chromosome segregation, is also used to describe the meiotic program as a whole. (9-0)

merotelic attachment: incorrect attachment of a sister-chromatid pair to the spindle, with one kinetochore attached to both poles. Merotelic attachment is often seen in a form where one sister is attached to both poles and the other sister to just one, resulting in an incorrect form of **bi-orientation**. (6-10)

metaphase: the last stage of mitotic entry, when the **sister chromatids** are fully attached to the spindle and await the signal to separate in **anaphase**. (5-0)

metaphase-to-anaphase transition: cell-cycle transition where the initiation of sister-chromatid separation can be blocked if the spindle is not fully assembled. Also called the M/G1 checkpoint, but this is not an ideal term because it does not coincide with the boundary between M phase and G1. (1-3)

metastasis: in cancer biology, the spread of cells from a tumor to locations in other regions of the body, resulting in secondary tumors called metastases. (12-0)

microfilament: long helical polymer of two chains of actin monomers wound around each other. It is a component of the cytoskeleton. (8-1)

microsatellite instability (MIN): an abnormally high frequency of point mutations in repetitive DNA regions, generally as a result of defects in **mismatch repair**. (12-5)

microtubule: long hollow polymer of tubulin subunits with two distinct ends, a **plus end** and a **minus end**, that display different polymerization behaviors. Microtubules are part of the cytoskeleton of interphase cells and form the mitotic and meiotic spindles. There are three classes of spindle microtubule. Astral microtubules have their minus ends in a spindle pole and plus ends in the cytoplasm. The plus ends often contact the cell cortex, providing a mechanism for anchoring and positioning the spindle in the cell. Interpolar microtubules interdigitate in the spindle midzone but are not directly involved in attachment to sister-chromatid kinetochores. Kinetochore microtubules connect the spindle poles and the kinetochores of sister chromatids. (6-1)

microtubule flux: flow of tubulin within microtubules from the spindle midzone to the poles, as a result of loss of tubulin from the minus ends at the poles. It is usually accompanied by addition of tubulin at the plus ends. (6-11)

midblastula transition: in some animal embryos, transition from development based primarily on maternally supplied protein and RNA to development based on transcription of embryonic genes. (2-3)

midbody: large protein complex, derived from the spindle midzone, that is involved in the final stages of cell separation in dividing animal cells. (8-3)

MIN: see **microsatellite instability**. (12-5)

minus end: the end of a **microtubule** with α-tubulin exposed. Tubulin subunits are added more slowly at this end than the other. (6-1)

mismatch repair: repair of nucleotide mismatches that are generated during DNA replication. (12-5)

mitogen: extracellular molecule that stimulates cell proliferation. (10-6)

mitosis: nuclear division, the process in early **M phase** by which the duplicated chromosomes are segregated by the mitotic spindle and packaged into daughter nuclei. (1-1)

mitosis-promoting factor: see **MPF**. (2-3)

mitotic index: fraction of cells in a population that are undergoing mitosis. (2-6)

mitotic spindle: bipolar array of microtubules, generally with a centrosome or spindle pole body at each pole, which segregates the **sister chromatids** during mitosis. (5-0)

monopolins: group of proteins that localize to the kinetochores of sister chromatids in meiosis I and somehow promote the attachment of both kinetochores to the same spindle pole. (9-6)

monotelic attachment: attachment of a sister-chromatid pair to the spindle, with one kinetochore attached to one pole and the other kinetochore unattached. (6-10)

morphogenetic furrow: in the eye **imaginal disc** of *Drosophila*, a shallow indentation that results from a wave of cell differentiation that passes anteriorly across the disc during the third larval stage. (2-4)

motor protein: any of a wide range of proteins with ATPase activity that can move along microtubules or actin filaments. Those connected with spindle assembly are the microtubule motors of the kinesin and dynein families. (6-2)

MPF: maturation-promoting factor or **mitosis-promoting factor**: active complex of Cdk1 and cyclin B, which promotes the onset of meiotic maturation in immature **oocytes** and mitosis in somatic cells. (2-3)

M phase: the cell-cycle phase during which the duplicated chromosomes are segregated and packaged into daughter nuclei (**mitosis**) and distributed into daughter cells (**cytokinesis**). (1-1)

myosin: motor protein that associates with **microfilaments** and moves along them. The form of myosin found in **contractile rings** is non-muscle myosin II. (8-1)

negative feedback oscillator: regulatory system in which a regulatory component activates its own inhibitor after a delay, resulting in oscillations in the activity of that component. (3-11)

neocentromere: chromosomal region that nucleates the formation of centromeric heterochromatin and kinetochore assembly at a position other than that of the normal centromere for that chromosome. (4-12)

neoplasm: a tumor—a mass of cells that are proliferating at an inappropriate rate, generally as a result of mutations in mitogenic signaling pathways. (12-0)

noncrossover: in meiosis, a **homologous recombination** event that results in repair of a double-strand DNA break without formation of a **crossover**. (9-2)

non-homologous end joining: mechanism for repairing double-strand breaks in DNA in which the broken ends are rejoined directly, usually with the loss of nucleotides at the join. (11-1)

nucleosome: fundamental unit of eukaryotic chromatin structure, containing about 147 bp of DNA wrapped around a **histone octamer**. (4-9)

nucleosome assembly factor: protein that binds to histones and facilitates their assembly into nucleosomes. (4-11)

nucleotide excision repair: DNA repair pathway in which major structural defects, such as thymine dimers, are excised along with a stretch of 12–30 nucleotides surrounding the site of damage, and the damaged strand is resynthesized using the undamaged strand as the template. (11-1)

Okazaki fragments: short fragments of DNA that are made during **lagging strand** synthesis. They are initiated at the replication fork, synthesized in the direction away from the direction of fork movement and subsequently joined together by **DNA ligase**. (4-1)

oncogene: a gene whose protein product promotes cancer, generally because mutations or rearrangements in a normal gene (the **proto-oncogene**) have resulted in a protein that is overactive or overproduced. (12-1)

oocyte: precursor of the haploid egg cell in animals such as frogs, flies and mammals. (2-3)

oocyte maturation: the process by which a frog **oocyte** arrested in meiotic prophase is induced by progesterone to undergo meiosis I and then arrest in metaphase of meiosis II. (2-3)

ORC: see **origin recognition complex**. (4-3)

origin recognition complex (ORC): protein complex that binds to replication origins in eukaryotes and recruits other proteins to unwind DNA and initiate DNA replication. (4-3)

pachytene: stage in meiotic prophase following **zygotene**, in which homologous chromosome pairs are tightly linked by the synaptonemal complex. (9-3)

pairing: (in meiosis) the initial interaction of homologous chromosomes with each other in early meiotic prophase. (9-3)

phragmoplast: organelle in a dividing plant cell upon which the new cell membranes and cell walls between the two daughter cells are constructed. It corresponds to the central spindle of animal cells. (8-3)

Plk: see **polo-like kinase**. (5-7)

plus end: the end of a **microtubule** with β-tubulin exposed. Tubulin subunits are added more rapidly at this end than the other. (6-1)

polar body: small cell produced at each meiotic division in **oocytes**. At division, one chromosome set remains in the oocyte and the other is discarded in the polar body, which is eventually resorbed. (2-3)

polar ejection force: the repulsive effect of microtubules growing out from the spindle poles and interacting with chromosome arms. It tends to force chromosome arms away from the spindle poles. (6-11)

polo-like kinase (Plk): serine/threonine protein kinase that is activated at the beginning of M phase and inactivated in late mitosis and G1, and is involved in a variety of mitotic processes including spindle assembly and kinetochore function, and in cytokinesis. (5-7)

polytene chromosome: giant chromosome arising from repeated rounds of DNA replication in nondividing cells. (1-2)

positive feedback: process whereby an action induces the same action. In a signaling system, for example, a component may activate itself and this can allow full activation at low stimulus levels. (3-7)

preinitiation complex: large complex of proteins that assembles at the replication origin when the origin is activated by S–Cdks and Cdc7. It includes DNA polymerases and other components that initiate DNA replication. (4-8)

pre-RC: see **prereplicative complex**. (4-3)

prereplicative complex (pre-RC): large complex of proteins, including the **origin recognition complex** and its associated proteins, including the inactive Mcm helicase, that assembles at replication origins in late mitosis and G1 and which is activated to initiate DNA replication at the origin at the beginning of S phase. (4-3)

primary cells: cells taken directly from the tissue of an intact animal. They are generally susceptible to **replicative senescence** after several generations of proliferation in culture. (2-5)

primase: specialized RNA polymerase that synthesizes a short stretch of **primer** RNA on the template at the beginning of DNA replication, thus generating a 3′ end to which nucleotides can be added by DNA polymerases. In eukaryotes, RNA primer synthesis is carried out by a polymerase α–primase complex, which has primase and DNA polymerase activity. (4-1)

primer: in DNA synthesis, a short stretch of RNA synthesized by a **primase** on a DNA template, and from which the new DNA strand is elongated by DNA polymerase. It is extended by DNA polymerases to form the new DNA. Primers are subsequently removed and filled in with DNA. (4-1)

prometaphase: the second stage of mitosis in animal and plant cells, when the nuclear envelope breaks down and the **sister chromatids** become attached to the spindle. (5-0)

prophase: the first stage of mitosis, when chromosome condensation, centrosome separation and spindle assembly begin. (5-0)

proteasome: large multisubunit enzyme complex that degrades cytosolic proteins into short peptides. (3-9)

proto-oncogene: a gene that when dysregulated or mutated can promote malignancy (see **oncogene**). Proto-oncogenes generally regulate cell growth, cell division, cell survival, or cell differentiation. (12-1)

PSTAIRE helix: alpha helix in the amino-terminal lobe of Cdks (also known as the α1 helix), which interacts with cyclin and is moved inward upon cyclin binding, resulting in reorientation of key active-site residues. The name of this helix comes from its amino-acid sequence, which is conserved among all major Cdks. (3-1)

recombination checkpoint system: in meiotic cells, a regulatory system that blocks entry into **meiosis I** when recombination fails. (9-5)

relaxation oscillator: oscillating system in which a regulatory component oscillates between two stable states, generally as a result of positive feedback. (3-11)

replication foci: localized regions of replicating DNA in the nucleus, generally seen by microscopic analysis of cells incubated with the thymidine analog bromodeoxyuridine (BrdU) and labeled with anti-BrdU antibodies. (4-2)

replication fork: the site at which DNA strands are separated and new DNA is synthesized. It is a Y-shaped structure and moves away from the site of replication initiation. Both strands of the DNA are copied at the replication fork. (4-0)

replication origin: site or region in a chromosome where DNA synthesis is initiated by unwinding of the double helix and assembly of the DNA synthetic machinery. (4-0)

replicative senescence: general term for the eventual cessation of division by **primary cells** when grown in artificial culture conditions. (2-5)

replicon clusters: regions of DNA in which a group of neighboring replication origins fire simultaneously. (4-2)

rescue: (in **microtubules**) sudden shift from shrinkage to growth that occurs when a GTP cap forms at the microtubule tip. (6-1)

RING finger: small structural domain that binds zinc ions and is a central component in a large class of ubiquitin-protein ligases, including **SCF** and the **APC**. It mediates interaction of the enzyme with the E2–ubiquitin conjugate. (3-9)

RNAi: see **RNA interference**. (2-5)

RNA interference (RNAi): mechanism by which short fragments of double-stranded RNA lead to the degradation of homologous mRNAs. Also called RNA silencing. RNAi is widely used experimentally to decrease the expression of a gene of interest to study its function. Many plant and some insect viruses have been shown to encode proteins that block RNAi. (2-5)

RXL motif: degenerate amino-acid sequence on Cdk substrates and inhibitors that interacts with the **hydrophobic patch** on the surface of cyclins. Also called a Cy motif. (3-5)

sarcoma: malignant tumor of connective tissue or muscle. (12-2)

SC: see **synaptonemal complex**. (9-4)

SCF: multisubunit ubiquitin-protein ligase that catalyzes the attachment of **ubiquitin** to a number of proteins involved in G1/S control and other processes, thus promoting their ubiquitin-dependent proteolysis. It is targeted to substrates by the associated **F-box protein**. (3-9)

S cyclins: cyclins that activate Cdks necessary for DNA synthesis; their concentrations rise and remain high during S phase, G2 and early mitosis. (3-2)

securin: protein that binds and inhibits the protease **separase**, thereby blocking the onset of sister-chromatid separation. It is also required in some cells for the normal folding or localization of separase. (7-4)

separase: protease that initiates sister-chromatid separation by cleaving Scc1, a subunit of the cohesin complex that holds sister chromatids together before anaphase. (7-4)

septum: the extracellular wall that forms between two daughter cells in fungi during cell division. (8-3)

silencing: (of chromatin) establishment of a heritable state of chromatin, known as heterochromatin, characterized by repression of gene expression and recombination and delayed replication. (4-12)

sister chromatids: pair of chromosomes that is generated by chromosome duplication in S phase. (5-0)

sister-chromatid cohesion: linkages that hold **sister chromatids** together between S phase and anaphase. (5-0)

sister-chromatid resolution: the gradual disentangling of sister chromatids before separation in anaphase, which makes them visible as distinct structures under the microscope. (5-9)

sister-chromatid segregation: the process by which separated **sister chromatids** are pulled to opposite poles of the cell for packaging in daughter nuclei. (5-0)

sister-chromatid separation: the process by which **sister-chromatid cohesion** is dissolved and **sister-chromatid** pairs dissociate at the metaphase-to-anaphase transition. (5-0)

sliding clamp: ring-shaped protein that moves along the DNA with the DNA polymerase in DNA replication and by tethering the polymerase to the DNA increases its processivity. (4-1)

SMC (structural maintenance of chromosomes) proteins: family of large proteins composed of a long coiled-coil region with a terminal ATPase domain made up of the amino and carboxyl termini of the protein, and which play a part in chromosome segregation and DNA recombination and repair. SMC proteins typically form dimers with another SMC protein to form large ring structures that may encircle chromosomes and position them with respect to each other, as well as recruiting other proteins essential for their maintenance. (5-8)

SPB: see **spindle pole body**. (6-3)

S phase: the cell-cycle phase during which DNA replication and chromosome duplication occurs. (1-1)

spindle checkpoint system: regulatory system that restrains progression through the metaphase-to-anaphase transition until all sister-chromatid pairs have been bi-oriented correctly on the mitotic spindle. It is also called the spindle assembly checkpoint or SAC. (7-2)

spindle pole body (SPB): the major microtubule-organizing center of yeast cells. In budding yeast it is a multilayered structure embedded in the nuclear envelope throughout the cell cycle. (6-3)

sporulation: the formation of spores. In yeast, refers to the formation of haploid spores from diploid cells by meiosis in conditions unfavorable for growth and proliferation. (2-1)

Start: major regulatory transition at the entry into the cell cycle in mid to late G1, also called the G1/S checkpoint or the restriction point (in animal cells). Progression past this point is prevented if cell growth is insufficient, DNA is damaged or other preparations for cell-cycle entry are not complete. Unlike cells arrested at the **G2/M checkpoint** or **metaphase-to-anaphase transition**, cells prevented from passing Start do not arrest at this point but typically exit the cell cycle into a prolonged nondividing state from which a return to the cycle is a lengthy process. (1-3)

survival factor: extracellular factor that inhibits cell death by apoptosis. (10-14)

synapsis: (of chromosomes) the close linkage of two homologous chromosomes along their lengths during zygotene of meiotic prophase. (9-4)

synaptonemal complex (SC): protein structure that links a pair of homologous chromosomes along their length in meiotic prophase. (9-4)

syncytium: multinucleate cell. (1-2)

syntelic attachment: incorrect attachment of a sister-chromatid pair to the spindle, with both kinetochores attached to the same pole. (6-10)

telomerase: specialized DNA polymerase that synthesizes the repetitive DNA sequence of **telomeres** using a built-in RNA template. Conventional replication enzymes replicate most of the DNA of the telomere. However the terminal-most repeats are added by telomerase. (4-1)

telomere: the specialized repetitive structure at the end of a eukaryotic chromosome that enables the DNA to be fully replicated and also maintains the integrity of the chromosome. It is synthesized by **telomerase**. (4-1)

telophase: the final stage of mitosis, when the spindle is disassembled and, in multicellular eukaryotes, the chromosomes decondense and the nuclear envelope reforms. (5-0)

tetrad: in yeast, the four haploid spores produced by meiosis from a single diploid cell. (2-1)

tetraploid: (of a cell) possessing four copies, or homologs, of each chromosome, often as a result of a failure in cytokinesis in diploid cells. (12-7)

T-loop: flexible loop adjacent to the active site of Cdks, named for the threonine whose phosphorylation is required for maximal activity. Sometimes called the activation loop. (3-1)

transformed cell line: **cell line** that has acquired mutations that render it independent of normal proliferation controls, and typically capable of forming tumors when injected into mice. (2-5)

treadmilling: (in **microtubules**) the addition of GTP–tubulin to the **plus end** while GDP–tubulin is dissociating from the **minus end**. It results in the net movement of tubulin subunits from the plus end to the minus end. (6-1)

tumor suppressor gene: a gene that encodes a protein that normally restrains cell proliferation or tumorigenesis, such that loss of the gene increases the likelihood of cancer formation. (12-1)

ubiquitin: a small protein that, when attached to other proteins in multiple copies, targets them to the **proteasome** for degradation. Sometimes ubiquitin tagging targets a protein to other fates such as endocytosis. (3-9)

ubiquitination: attachment of **ubiquitin** to a protein. (3-9)

ultrasensitive: property of a system that displays a sigmoidal dose–response curve because low levels of stimulus generate a poor response but higher levels generate an abrupt increase in the response. (3-7)

variant histone: any histone other than the canonical five histones H2A, H2B, H3, H4 and H1 found in **nucleosomes** in specific chromosome regions. (4-9)

Wee1: protein kinase that inhibits cyclin-dependent kinases by phosphorylating a tyrosine residue in the Cdk active site (Tyr 15 in most Cdks); a related protein kinase, Myt1, also phosphorylates this site and an adjacent threonine (Thr 14) in animals. (3-3)

zygote: in sexually reproducing organisms, the **diploid** cell produced by the fusion of two **haploid** cells, such as egg and sperm, from the parents. (1-2)

zygotene: stage in meiotic prophase following **leptotene**, in which the synaptonemal complex begins to form between paired homologous chromosomes. (9-3)

References

Abdel-Rahman, W.M., Katsura, K., Rens, W., Gorman, P.A., Sheer, D., Bicknell, D., Bodmer, W.F., Arends, M.J., Wyllie, A.H. and Edwards, P.A.: **Spectral karyotyping suggests additional subsets of colorectal cancers characterized by pattern of chromosome rearrangement.** *Proc. Natl Acad. Sci. USA* 2001, **98**:2538–2543. (12-5)

Abraham, R.T.: **Cell cycle checkpoint signaling through the ATM and ATR kinases.** *Genes Dev.* 2001, **15**:2177–2196. (11-2)

Abrieu, A., Brassac, T., Galas, S., Fisher, D., Labbe, J.C. and Doree, M.: **The Polo-like kinase Plx1 is a component of the MPF amplification loop at the G2/M-phase transition of the cell cycle in *Xenopus* eggs.** *J. Cell Sci.* 1998, **111**:1751–1757. (5-5)

Adams, I.R. and Kilmartin, J.V.: **Spindle pole body duplication: a model for centrosome duplication?** *Trends Cell Biol.* 2000, **10**:329–335. (6-3)

Adams, J.M.: **Ways of dying: multiple pathways to apoptosis.** *Genes Dev.* 2003, **17**:2481–2495. (10-14)

Adhikary, S. and Eilers, M.: **Transcriptional regulation and transformation by Myc proteins.** *Nat. Rev. Mol. Cell Biol.* 2005, **6**:635–645. (10-6, 10-7)

Akey, C.W. and Luger, K.: **Histone chaperones and nucleosome assembly.** *Curr. Opin. Struct. Biol.* 2003, **13**:6–14. (4-11)

Alberts, B., Johnson, A., Lewis, J., Raff, M., Roberts, K. and Walter, P.: *Molecular Biology of the Cell* 4th ed. (Garland Science, New York, 2002). (4-0, 4-1, 6-1, 8-1, 10-6, 12-0, 12-9)

Albertson, R., Riggs, B. and Sullivan, W.: **Membrane traffic: a driving force in cytokinesis.** *Trends Cell Biol.* 2005, **15**:92–101. (8-0, 8-3)

Albini, S.M. and Jones, G.H.: **Synaptonemal complex spreading in *Allium cepa* and *A. fistulosum*. I. The initiation and sequence of pairing.** *Chromosoma* 1987, **95**:324–338. (9-3)

Allers, T. and Lichten, M.: **Differential timing and control of noncrossover and crossover recombination during meiosis.** *Cell* 2001, **106**:47–57. (9-2)

American Cancer Society: *Cancer Facts and Figures* (American Cancer Society, Atlanta, 2006). (12-2)

Andersen, J.S., Wilkinson, C.J., Mayor, T., Mortensen, P., Nigg, E.A. and Mann, M.: **Proteomic characterization of the human centrosome by protein correlation profiling.** *Nature* 2003, **426**:570–574. (6-3)

Anderson, D.E., Losada, A., Erickson, H.P. and Hirano, T.: **Condensin and cohesin display different arm conformations with characteristic hinge angles.** *J. Cell Biol.* 2002, **156**:419–424. (5-8, 5-9)

Andrews, P.D., Ovechkina, Y., Morrice, N., Wagenbach, M., Duncan, K., Wordeman, L. and Swedlow, J.R.: **Aurora B regulates MCAK at the mitotic centromere.** *Dev. Cell* 2004, **6**:253–268. (6-10)

Antonin, W., Franz, C., Haselmann, U., Antony, C. and Mattaj, I.W.: **The integral membrane nucleoporin pom121 functionally links nuclear pore complex assembly and nuclear envelope formation.** *Mol. Cell* 2005, **17**:83–92. (7-7)

Aparicio, O.M., Stout, A.M. and Bell, S.P.: **Differential assembly of Cdc45p and DNA polymerases at early and late origins of DNA replication.** *Proc. Natl Acad. Sci. USA* 1999, **96**:9130–9135. (4-8)

Arias, E.E. and Walter, J.C.: **PCNA functions as a molecular platform to trigger Cdt1 destruction and prevent re-replication.** *Nat. Cell Biol.* 2006, **8**:84–90. (4-4)

Artandi, S.E. and Attardi, L.D.: **Pathways connecting telomeres and p53 in senescence, apoptosis, and cancer.** *Biochem. Biophys. Res. Commun.* 2005, **331**:881–890. (12-6)

Artandi, S.E., Chang, S., Lee, S.L., Alson, S., Gottlieb, G.J., Chin, L. and DePinho, R.A.: **Telomere dysfunction promotes non-reciprocal translocations and epithelial cancers in mice.** *Nature* 2000, **406**:641–645. (12-6)

Attwooll, C., Lazzerini Denchi, E. and Helin, K.: **The E2F family: specific functions and overlapping interests.** *EMBO J.* 2004, **23**:4709–4716. (10-4, 10-5)

Austin, R.J., Orr-Weaver, T.L. and Bell, S.P.: ***Drosophila* ORC specifically binds to ACE3, an origin of DNA replication control element.** *Genes Dev.* 1999, **13**:2639–2649. (4-3)

Bakkenist, C.J. and Kastan, M.B.: **DNA damage activates ATM through intermolecular autophosphorylation and dimer dissociation.** *Nature* 2003, **421**:499–506. (11-2)

Bakkenist, C.J. and Kastan, M.B.: **Initiating cellular stress responses.** *Cell* 2004, **118**:9–17. (11-0, 11-2, 11-5)

Balasubramanian, M.K., Bi, E. and Glotzer, M.: **Comparative analysis of cytokinesis in budding yeast, fission yeast and animal cells.** *Curr. Biol.* 2004, **14**:R806–R818. (8-0, 8-2)

Barr, F.A.: **Golgi inheritance: shaken but not stirred.** *J. Cell Biol.* 2004, **164**:955–958. (6-7)

Barr, F.A., Sillje, H.H. and Nigg, E.A.: **Polo-like kinases and the orchestration of cell division.** *Nat. Rev. Mol. Cell Biol.* 2004, **5**:429–440. (5-5, 5-7, 6-6)

Bartek, J. and Lukas, J.: **Chk1 and Chk2 kinases in checkpoint control and cancer.** *Cancer Cell* 2003, **3**:421–429. (11-3, 11-7)

Bartek, J. and Lukas, J.: **Mammalian G1- and S-phase checkpoints in response to DNA damage.** *Curr. Opin. Cell Biol.* 2001, **13**:738–747. (11-5)

Bartkova, J., Hořejší, Z, Koed, K., Krämer, A., Tort, F., Zieger, K., Guldberg, P., Sehested, M., Nesland, J.M., Lukas, C., Ørntoft, T., Lukas, J., Bartek, J.: **DNA damage response as a candidate anti-cancer barrier in early human tumorigenesis.** *Nature* 2005, **434**:864–870. (12-5)

Baserga, R.: *The Biology of Cell Reproduction* (Harvard University Press, Cambridge, MA, 1985). (1-0, 2-0, 2-6, 10-0)

Bayliss, R., Sardon, T., Vernos, I. and Conti, E.: **Structural basis of Aurora-A activation by TPX2 at the mitotic spindle.** *Mol. Cell 2003*, **12**:851–862. (5-7)

Beach, D., Durkacz, B. and Nurse, P.: **Functionally homologous cell cycle control genes in budding and fission yeast.** *Nature* 1982, **300**:706–709. (2-2)

Beachy, P.A., Karhadkar, S.S. and Berman, D.M.: **Tissue repair and stem cell renewal in carcinogenesis.** *Nature* 2004, **432**:324–331. (12-1, 12-2)

Bean, J.M., Siggia, E.D. and Cross, F.R.: **Coherence and timing of cell cycle start examined at single-cell resolution.** *Mol. Cell* 2006, **21**:3–14. (10-1)

Becker, P.B. and Hoerz, W.: **ATP-dependent nucleosome remodeling.** *Annu. Rev. Biochem.* 2002, **71**:247–273. (4-9, 4-11)

Bednar, J., Horowitz, R.A., Grigoryev, S.A., Carruthers, L.M., Hansen, J.C., Koster, A.J. and Woodcock, C.L.: **Nucleosomes, linker DNA, and linker histone form a unique structural motif that directs the higher-order folding and compaction of chromatin.** *Proc. Natl Acad. Sci. USA* 1998, **95**:14173–14178. (4-9)

Bell, S.P. and Dutta, A.: **DNA replication in eukaryotic cells.** *Annu. Rev. Biochem.* 2002, **71**:333–374. (4-0, 4-1, 4-3, 4-4, 4-7, 4-8)

Bell, S.P. and Stillman, B.: **ATP-dependent recognition of eukaryotic origins of DNA replication by a multiprotein complex.** *Nature* 1992, **357**:128–134. (4-3)

Bement, W.M., Benink, H.A. and von Dassow, G.: **A microtubule-dependent zone of active RhoA during cleavage plane specification.** *J. Cell Biol.* 2005, **170**:91–101. (8-2)

Benjamin, K.R., Zhang, C., Shokat, K.M. and Herskowitz, I.: **Control of landmark events in meiosis by the CDK Cdc28 and the meiosis-specific kinase Ime2.** *Genes Dev.* 2003, **17**:1524–1539. (9-1, 9-5)

Betschinger, J. and Knoblich, J.A.: **Dare to be different: asymmetric cell division in *Drosophila*, *C. elegans* and vertebrates.** *Curr. Biol.* 2004, **14**:R674–R685. (8-7)

Bettencourt-Dias, M., Rodrigues-Martins, A., Carpenter, L., Riparbelli, M., Lehmann, L., Gatt, M.K., Carmo, N., Balloux, F., Callaini, G. and Glover, D.M.: **SAK/PLK4 is required for centriole duplication and flagella development.** *Curr. Biol.* 2005, **15**:2199–2207. (6-4)

Bhowmick, N.A., Neilson, E.G. and Moses, H.L.: **Stromal fibroblasts in cancer initiation and progression.** *Nature* 2004, **432**:332–337. (12-2)

Bickel, S.E., Orr-Weaver, T.L. and Balicky, E.M.: **The sister-chromatid cohesion protein ORD is required for chiasma maintenance in *Drosophila* oocytes.** *Curr. Biol.* 2002, **12**:925–929. (9-6)

Bishop, D.K. and Zickler, D.: **Early decision: meiotic crossover interference prior to stable strand exchange and synapsis.** *Cell* 2004, **117**:9–15. (9-2, 9-3, 9-4)

Blais, A. and Dynlacht, B.D.: **Hitting their targets: an emerging picture of E2F and cell cycle control.** *Curr. Opin. Genet. Dev.* 2004, **14**:527–532. (10-4)

Blat, Y., Protacio, R., Hunter, N. and Kleckner, N.: **Physical and functional interactions among basic chromosome organizational features govern early steps of meiotic chiasma formation.** *Cell* 2002, **111**:791–802. (9-4)

Blower, M.D., Nachury, M., Heald, R. and Weis, K.: **A Rae1-containing ribonucleoprotein complex is required for mitotic spindle assembly.** *Cell* 2005, **121**:223–234. (6-8)

Bornens, M.: **Centrosome composition and microtubule anchoring mechanisms.** *Curr. Opin. Cell Biol.* 2002, **14**:25–34. (6-3)

Börner, G.V., Kleckner, N. and Hunter, N.: **Crossover/non-crossover differentiation, synaptonemal complex formation, and regulatory surveillance at the leptotene/zygotene transition of meiosis.** *Cell* 2004, **117**:29–45. (9-3, 9-4)

Bourne, Y., Watson, M.H., Hickey, M.J., Holmes, W., Rocque, W., Reed, S.I. and Tainer, J.A.: **Crystal structure and mutational analysis of the human CDK2 kinase complex with cell cycle-regulatory protein CksHs1.** *Cell* 1996, **84**:863–874. (3-5)

Bousset, K. and Diffley, J.F.: **The Cdc7 protein kinase is required for origin firing during S phase.** *Genes Dev.* 1998, **12**:480–490. (4-7)

Bowers, J.L., Randell, J.C., Chen, S. and Bell, S.P.: **ATP hydrolysis by ORC catalyzes reiterative Mcm2-7 assembly at a defined origin of replication.** *Mol. Cell* 2004, **16**:967–978. (4-3)

Branzei, D. and Foiani, M.: **The DNA damage response during DNA replication.** *Curr. Opin. Cell Biol.* 2005, **17**:568–575. (11-6)

Breeden, L.L.: **Periodic transcription: a cycle within a cycle.** *Curr. Biol.* 2003, **13**:R31–R38. (3-12, 10-1)

Bringmann, H. and Hyman, A.A.: **A cytokinesis furrow is positioned by two consecutive signals.** *Nature* 2005, **436**:731–734. (8-5)

Brooks, C.L. and Gu, W.: **Ubiquitination, phosphorylation and acetylation: the molecular basis for p53 regulation.** *Curr. Opin. Cell Biol.* 2003, **15**:164–171. (11-4)

Brooks, R.F.: **Variability in the cell cycle and the control of cell proliferation** in *The Cell Cycle* John, P.C.L. ed. (Cambridge University Press, Cambridge, 1981), 35–61. (10-10)

Brotherton, D.H., Dhanaraj, V., Wick, S., Brizuela, L., Domaille, P.J., Volyanik, E., Xu, X., Parisini, E., Smith, B.O., Archer, S.J., Serrano, M., Brenner, S.L., Blundell, T.L. and Laue, E.D.: **Crystal structure of the complex of the cyclin D-dependent kinase Cdk6 bound to the cell-cycle inhibitor p19^{INK4d}.** *Nature* 1998, **395**:244–250. (3-6)

Brown, N.R., Noble, M.E.M., Endicott, J.A. and Johnson, L.N.: **The structural basis for specificity of substrate and recruitment peptides for cyclin-dependent kinases.** *Nat. Cell Biol.* 1999, **1**:438–443. (3-4, 3-5)

Brown, N.R., Noble, M.E.M., Endicott, J.A., Garman, E.F., Wakatsuki, S., Mitchell, E., Rasmussen, B., Hunt, T. and Johnson, L.N.: **The crystal structure of cyclin A.** *Structure* 1995, **3**:1235–1247. (3-2)

Buonomo, S.B., Clyne, R.K., Fuchs, J., Loidl, J., Uhlmann, F. and Nasmyth, K.: **Disjunction of homologous chromosomes in meiosis I depends on proteolytic cleavage of the meiotic cohesin Rec8 by separin.** *Cell* 2000, **103**:387–398. (9-7)

Buonomo, S.B., Rabitsch, K.P., Fuchs, J., Gruber, S., Sullivan, M., Uhlmann, F., Petronczki, M., Toth, A. and Nasmyth, K.: **Division of the nucleolus and its release of CDC14 during anaphase of meiosis I depends on separase, SPO12, and SLK19.** *Dev. Cell* 2003, **4**:727–739. (9-8)

Burgess, D.R. and Chang, F.: **Site selection for the cleavage furrow at cytokinesis.** *Trends Cell Biol.* 2005, **15**:156–162. (8-0, 8-5)

Burke, B. and Ellenberg, J.: **Remodelling the walls of the nucleus.** *Nat. Rev. Mol. Cell Biol.* 2002, **3**:487–497. (6-7)

Byun, T.S., Pacek, M., Yee, M.C., Walter, J.C. and Cimprich, K.A.: **Functional uncoupling of MCM helicase and DNA polymerase activities activates the ATR-dependent checkpoint.** *Genes Dev.* 2005, **19**:1040–1052. (11-6)

Cabib, E.: **The septation apparatus, a chitin-requiring machine in budding yeast.** *Arch. Biochem. Biophys.* 2004, **426**:201–207. (8-3)

Campisi, J.: **Senescent cells, tumor suppression, and organismal aging: good citizens, bad neighbors.** *Cell* 2005, **120**:513–522. (11-8, 12-4)

Canman, J.C., Cameron, L.A., Maddox, P.S., Straight, A., Tirnauer, J.S., Mitchison, T.J., Fang, G., Kapoor, T.M. and Salmon, E.D.: **Determining the position of the cell division plane.** *Nature* 2003, **424**:1074–1078. (8-5)

Cantley, L.C.: **The phosphoinositide 3-kinase pathway.** *Science* 2002, **296**:1655–1657. (10-6)

Cao, K., Nakajima, R., Meyer, H.H. and Zheng, Y.: **The AAA-ATPase Cdc48/p97 regulates spindle disassembly at the end of mitosis.** *Cell* 2003, **115**:355–367. (7-7)

Cardozo, T. and Pagano, M.: **The SCF ubiquitin ligase: insights into a molecular machine.** *Nat. Rev. Mol. Cell Biol.* 2004, **5**:739–751. (3-9)

Carmena, M. and Earnshaw, W.C.: **The cellular geography of aurora kinases.** *Nat. Rev. Mol. Cell Biol.* 2003, **4**:842–854. (5-7)

Caudron, M., Bunt, G., Bastiaens, P. and Karsenti, E.: **Spatial coordination of spindle assembly by chromosome-mediated signaling gradients.** *Science* 2005 **309**:1373–1376. (6-8)

Celeste, A., Fernandez-Capetillo, O., Kruhlak, M.J., Pilch, D.R., Staudt, D.W., Lee, A., Bonner, R.F., Bonner, W.M. and Nussenzweig, A.: **Histone H2AX phosphorylation is dispensable for the initial recognition of DNA breaks.** *Nat. Cell Biol.* 2003, **5**:675–679. (11-3)

Celli, G.B. and de Lange, T.: **DNA processing is not required for ATM-mediated telomere damage response after TRF2 deletion.** *Nat. Cell Biol.* 2005, **7**:712–718. (11-8)

Chan, T.A., Hwang, P.M., Hermeking, H., Kinzler, K.W. and Vogelstein, B.: **Cooperative effects of genes controlling the G_2/M checkpoint.** *Genes Dev.* 2000, **14**:1584–1588. (11-7)

Chang, F. and Herskowitz, I.: **Identification of a gene necessary for cell cycle arrest by a negative growth factor of yeast: FAR1 is an inhibitor of a G1 cyclin, CLN2.** *Cell* 1990, **63**:999–1011. (10-3)

Cheeseman, I.M., Anderson, S., Jwa, M., Green, E.M., Kang, J., Yates, J.R. 3rd, Chan, C.S., Drubin, D.G. and Barnes, G.: **Phospho-regulation of kinetochore-microtubule attachments by the Aurora kinase Ipl1p.** *Cell* 2002, **111**:163–172. (6-10)

Cheeseman, I.M., Niessen, S., Anderson, S., Hyndman, F., Yates, J.R. 3rd, Oegema, K. and Desai, K.: **A conserved protein network controls assembly of the outer kinetochore and its ability to sustain tension.** *Genes Dev.* 2004, **18**:2255–2268. (6-5)

Chen, K.C., Calzone, L., Csikasz-Nagy, A., Cross, F.R., Novak, B. and Tyson, J.J.: **Integrative analysis of cell cycle control in budding yeast.** *Mol. Biol. Cell* 2004, **15**:3841–3862. (3-13)

Chin, K., de Solorzano, C.O., Knowles, D., Jones, A., Chou, W., Rodriguez, E.G., Kuo, W.-L., Ljung, B.-M., Chew, K., Myambo, K., Miranda, M., Krig, S., Garbe, J., Stampfer, M., Yaswen, P., Gray, J.W. and Lockett, S.J.: ***In situ* analyses of genome instability in breast cancer.** *Nat. Genet.* 2004, **36**:984–988. (12-6)

Chipuk, J.E., Bouchier-Hayes, L., Kuwana, T., Newmeyer, D.D. and Green, D.R.: **PUMA couples the nuclear and cytoplasmic proapoptotic function of p53.** *Science* 2005, **309**:1732–1735. (11-5)

Chittenden, T.: **BH3 domains: intracellular death-ligands critical for initiating apoptosis.** *Cancer Cell* 2002, **2**:165–166. (10-14)

Cho, R.J., Campbell, M.J., Winzeler, E.A., Steinmetz, L., Conway, A., Wodicka, L., Wolfsberg, T.G., Gabrielian, A.E., Landsman, D., Lockhart, D.J. and Davis, R.W.: **A genome-wide transcriptional analysis of the mitotic cell cycle.** *Mol. Cell* 1998, **2**:65–73. (3-12)

Chu, S, DeRisi, J., Eisen, M., Mulholland, J., Botstein, D., Brown, P.O. and Herskowitz, I.: **The transcriptional program of sporulation in budding yeast.** *Science* 1998, **282**:699–705. (9-1)

Cimini, D., Cameron, L.A. and Salmon E.D.: **Anaphase spindle mechanics prevent mis-segregation of merotelically oriented chromosomes.** *Curr. Biol.* 2004, **14**:2149–2155. (6-10, 6-12)

Classon, M. and Harlow, E.: **The retinoblastoma tumour suppressor in development and cancer.** *Nat. Rev. Cancer* 2002, **2**:910–917. (10-4, 10-5)

Cleveland, D.W., Mao, Y. and Sullivan, K.F.: **Centromeres and kinetochores: from epigenetics to mitotic checkpoint signaling.** *Cell* 2003, **112**:407–421. (4-12)

Clyne, R.K., Katis, V.L., Jessop, L., Benjamin, K.R., Herskowitz, I., Lichten, M. and Nasmyth, K.: **Polo-like kinase Cdc5 promotes chiasmata formation and cosegregation of sister centromeres at meiosis I.** *Nat. Cell Biol.* 2003, **5**:480–485. (9-5, 9-6)

Cobb, J.A., Schleker, T., Rojas, V., Bjergbaek, L., Tercero, J.A. and Gasser, S.M.: **Replisome instability, fork collapse, and gross chromosomal rearrangements arise synergistically from Mec1 kinase and RecQ helicase mutations.** *Genes Dev.* 2005, **19**:3055–3069. (11-6)

Cobb, J.A., Shimada, K. and Gasser, S.M.: **Redundancy, insult-specific sensors and thresholds: unlocking the S-phase checkpoint response.** *Curr. Opin. Genet. Dev.* 2004, **14**:292–300. (11-6)

Conlon, I. and Raff, M.: **Differences in the way a mammalian cell and yeast cells coordinate cell growth and cell-cycle progression.** *J. Biol.* 2003, **2**:7. (10-10, 10-13)

Conlon, I. and Raff, M.: **Size control in animal development.** *Cell* 1999, **96**:235–244. (10-10)

Conlon, I.J., Dunn, G.A., Mudge, A.W. and Raff, M.C.: **Extracellular control of cell size.** *Nat. Cell Biol.* 2001, **3**:918–921. (10-13)

Costanzo, M., Nishikawa, J.L., Tang, X., Millman, J.S., Schub, O., Breitkreuz, K., Dewar, D., Rupeš, I., Andrews, B. and Tyers, M.: **CDK activity antagonizes Whi5, an inhibitor of G1/S transcription in yeast.** *Cell* 2004, **117**:899–913. (3-12, 10-1)

Costanzo, V., Shechter, D., Lupardus, P.J., Cimprich, K.A., Gottesman, M. and Gautier, J.: **An ATR- and Cdc7-dependent DNA damage checkpoint that inhibits initiation of DNA replication.** *Mol. Cell* 2003, **11**:203–213. (11-7)

Cotran, R. S., Kumar, V. and Collins, T.: *Robbins Pathologic Basis of Disease* (Elsevier, Philadelphia, 1998). (12-2)

Coussens, L.M. and Werb, Z.: **Inflammation and cancer.** *Nature* 2002, **420**:860–867. (12-1)

Cross, F.R.: **Two redundant oscillatory mechanisms in the yeast cell cycle.** *Dev. Cell* 2003, **4**:741–752. (3-11, 3-13)

Cross, F.R., Archambault, V., Miller, M. and Klovstad, M.: **Testing a mathematical model of the yeast cell cycle.** *Mol. Biol. Cell* 2002, **13**:52–70. (3-13)

Cross, F.R., Yuste-Rojas, M., Gray, S. and Jacobson, M.D.: **Specialization and targeting of B-type cyclins.** *Mol. Cell* 1999, **4**:11–19. (4-5)

d'Adda di Fagagna, F., Reaper, P.M., Clay-Farrace, L., Fiegler, H., Carr, P., Von Zglinicki, T., Saretzki, G., Carter, N.P. and Jackson, S.P.: **A DNA damage checkpoint response in telomere-initiated senescence.** *Nature* 2003, **426**:194–198. (11-8)

d'Adda di Fagagna, F., Teo, S.H. and Jackson, S.P.: **Functional links between telomeres and proteins of the DNA-damage response.** *Genes Dev.* 2004, **18**:1781–1799. (11-8)

Daga, R.R. and Chang, F.: **Dynamic positioning of the fission yeast cell division plane.** *Proc. Natl Acad. Sci. USA* 2005, **102**:8228–8232. (8-4)

Daga, R.R. and Jimenez, J.: **Translational control of the Cdc25 cell cycle phosphatase: a molecular mechanism coupling mitosis to cell growth.** *J. Cell Sci.* 1999, **112**:3137–3446. (10-12)

D'Amours, D. and Amon, A.: **At the interface between signaling and executing anaphase—Cdc14 and the FEAR network.** *Genes Dev.* 2004, **18**:2581–2595. (7-5)

Danial, N.N. and Korsmeyer, S.J.: **Cell death: critical control points.** *Cell* 2004, **116**:205–219. (10-14)

Danilchik, M.V., Funk, W.C., Brown, E.E. and Larkin, K.: **Requirement for microtubules in new membrane formation during cytokinesis of *Xenopus* embryos.** *Dev. Biol.* 1998, **194**:47–60. (8-3)

Dean, M., Fojo, T. and Bates, S.: **Tumour stem cells and drug resistance.** *Nat. Rev. Cancer* 2005, **5**:275–284. (12-9)

De Antoni, A., Pearson, C.G., Cimini, D., Canman, J.C., Sala, V., Nezi, L., Mapelli, M., Sironi, L., Faretta, M., Salmon, E.D. and Musacchio, A.: **The Mad1/Mad2 complex as a template for Mad2 activation in the spindle assembly checkpoint.** *Curr. Biol.* 2005, **15**:214–225. (7-3)

De Bondt, H.L., Rosenblatt, J., Jancarik, J., Jones, H.D., Morgan, D.O. and Kim, S.-H.: **Crystal structure of cyclin-dependent kinase 2.** *Nature* 1993, **363**:595–602. (3-1, 3-4)

de Bruin, R.A., McDonald, W.H., Kalashnikova, T.I., Yates, J.R. 3rd and Wittenberg, C.: **Cln3 activates G1-specific transcription via phosphorylation of the SBF bound repressor Whi5.** *Cell* 2004, **117**:887–898. (3-12, 10-1)

Delattre, M. and Gönczy, P.: **The arithmetic of centrosome biogenesis.** *J. Cell Sci.* 2004, **117**:1619–1630. (6-3, 6-4)

DeLuca, J.G., Dong, Y., Hergert, P., Strauss, J., Hickey, J.M., Salmon, E.D. and McEwan, B.F.: **Hec1 and nuf2 are core components of the kinetochore outer plate essential for organizing microtubule attachment sites.** *Mol. Biol. Cell* 2005, **16**:519–531. (6-5)

den Elzen, N. and Pines, J.: **Cyclin A is destroyed in prometaphase and can delay chromosome alignment and anaphase.** *J. Cell Biol.* 2001, **153**:121–136. (7-1)

Desai, A. and Mitchison, T.J.: **Microtubule polymerization dynamics.** *Annu. Rev. Cell Dev. Biol.* 1997, **13**:83–117. (6-1)

Desany, B.A., Alcasabas, A.A., Bachant, J.B. and Elledge, S.J.: **Recovery from DNA replicational stress is the essential function of the S-phase checkpoint pathway.** *Genes Dev.* 1998, **12**:2956–2970. (11-6)

Devita, V.T., Hellman, S. and Rosenberg, S.A.: *Cancer: Principles and Practice of Oncology* 7th ed. (Lippincot, Williams and Wilkins, Philadelphia, 2004). (12-0, 12-2, 12-9)

Diehl, J.A., Cheng, M., Roussel, M.F. and Sherr, C.J.: **Glycogen synthase kinase-3β regulates cyclin D1 proteolysis and subcellular localization.** *Genes Dev.* 1998, **12**:3499–3511. (10-7)

Diffley, J.F.: **Regulation of early events in chromosome replication.** *Curr. Biol.* 2004, **14**:R778–R786. (4-4)

Dimova, D.K. and Dyson, N.J.: **The E2F transcriptional network: old acquaintances with new faces.** *Oncogene* 2005, **24**:2810–2826. (10-4, 11-8)

Dohlman, H.G. and Thorner, J.W.: **Regulation of G protein-initiated signal transduction in yeast: paradigms and principles.** *Annu. Rev. Biochem.* 2001, **70**:703–754. (10-3)

Dolznig, H., Grebien, F., Sauer, T., Beug, H. and Müllner, E.W.: **Evidence for a size-sensing mechanism in animal cells.** *Nat. Cell Biol.* 2004, **6**:899–905. (10-13)

Dominski, Z., Erkmann, J.A., Yang, X., Sanchez, R. and Marzluff, W.F.: **A novel zinc finger protein is associated with U7 snRNP and interacts with the stem-loop binding protein in the histone pre-mRNP to stimulate 3′-end processing.** *Genes Dev.* 2002, **16**:58–71. (4-10)

Dominski, Z., Yang, X.C., Kaygun, H., Dadlez, M. and Marzluff, W.F.: **A 3′ exonuclease that specifically interacts with the 3′ end of histone mRNA.** *Mol. Cell* 2003, **12**:295–305. (4-10)

Donaldson, A.D.: **The yeast mitotic cyclin Clb2 cannot substitute for S phase cyclins in replication origin firing.** *EMBO Rep.* 2000, **1**:507–512. (4-5)

Donaldson, A.D., Fangman, W.L. and Brewer, B.J.: **Cdc7 is required throughout the yeast S phase to activate replication origins.** *Genes Dev.* 1998, **12**:491–501. (4-7)

Donaldson, A.D., Raghuraman, M.K., Friedman, K.L., Cross, F.R., Brewer, B.J. and Fangman, W.L.: ***CLB5*-dependent activation of late replication origins in *S. cerevisiae*.** *Mol. Cell* 1998, **2**:173–182. (4-5)

Dong, H. and Roeder, G.S.: **Organization of the yeast Zip1 protein within the central region of the synaptonemal complex.** *J. Cell Biol.* 2000, **148**:417–426. (9-4)

Dorigo, B., Schalch, T., Kalungara, A., Duda. S., Schroeder, R.R. and Richmond, T.J.: **Nucleosome arrays reveal the two-start organization of the chromatin fiber.** *Science* 2004, **306**:1571–1573. (4-9)

Dunphy, W.G.: **The decision to enter mitosis.** *Trends Cell Biol.* 1994, **4**:202–207. (3-3)

Dyson, N.: **The regulation of E2F by pRB-family proteins.** *Genes Dev.* 1998, **12**:2245–2262. (3-12)

Edgar, B.A. and Lehner, C.F.: **Developmental control of cell cycle regulators: a fly's perspective.** *Science* 1996, **274**:1646–1652. (2-4)

Edgar, B.A. and Nijhout, H.F.: **Growth and cell cycle control in *Drosophila*** in *Cell Growth: Control of Cell Size.* Hall, M.N. *et al.* eds (Cold Spring Harbor, Cold Spring Harbor Laboratory Press, 2004), 23–83. (10-13)

Edgar, B.A. and O'Farrell, P.H.: **The three postblastoderm cell cycles of *Drosophila* embryogenesis are regulated in G2 by string.** *Cell* 1990, **62**:469–480. (10-9)

Edgar, B.A. and Orr-Weaver, T.L.: **Endoreplication cell cycles: more for less.** *Cell* 2001, **105**:297–306. (10-11)

Edgar, B.A., Lehman, D.A. and O'Farrell, P.H.: **Transcriptional regulation of string (cdc25): a link between developmental programming and the cell cycle.** *Development* 1994, **120**:3131–3143. (10-9)

Elia, A.E., Cantley, L.C. and Yaffe, M.B.: **Proteomic screen finds pSer/pThr-binding domain localizing Plk1 to mitotic substrates.** *Science* 2003, **299**:1228–1231. (5-7)

Evan, G. and Vousden, K.H.: **Proliferation, cell cycle and apoptosis in cancer.** *Nature* 2001, **411**:342–348. (12-0, 12-8)

Evans, T., Rosenthal, E.T., Youngblom, J., Distel, D. and Hunt, T.: **Cyclin: a protein specified by maternal mRNA in sea urchin eggs that is destroyed at each cleavage division.** *Cell* 1983, **33**:389–396. (3-2)

Falck, J., Petrini, J.H., Williams, B.R., Lukas, J. and Bartek, J.: **The DNA damage-dependent intra-S phase checkpoint is regulated by parallel pathways.** *Nat. Genet.* 2002, **30**:290–294. (11-7)

Feierbach, B. and Chang, F.: **Cytokinesis and the contractile ring in fission yeast.** *Curr. Opin. Microbiol.* 2001, **4**:713–719. (8-4)

Feldser, D.M., Hackett, J.A. and Greider, C.W.: **Telomere dysfunction and the initiation of genome instability.** *Nat. Rev. Cancer* 2003, **3**:623–627. (12-6)

Ferguson, A.M., White, L.S., Donovan, P.J. and Piwnica-Worms, H.: **Normal cell cycle and checkpoint responses in mice and cells lacking Cdc25B and Cdc25C protein phosphatases.** *Mol. Cell Biol.* 2005, **25**:2853–2860. (5-4)

Ferrara, N. and Kerbel, R.S.: **Angiogenesis as a therapeutic target.** *Nature* 2005, **438**:967–974. (12-9)

Ferrell, J.E. Jr: **How regulated protein translocation can produce switch-like responses.** *Trends Biochem. Sci.* 1998, **23**:461–465. (5-6)

Ferrell, J.E. Jr: **Self-perpetuating states in signal transduction: positive feedback, double-negative feedback and bistability.** *Curr. Opin. Cell Biol.* 2002, **14**:140–148. (3-8, 3-11, 5-5)

Ferrell, J.E. Jr: **Tripping the switch fantastic: how a protein kinase cascade can convert graded inputs into switch-like outputs.** *Trends Biochem. Sci.* 1996, **21**:460–466. (3-7, 3-8)

Field, C., Li, R. and Oegema, K.: **Cytokinesis in eukaryotes: a mechanistic comparison.** *Curr. Opin. Cell Biol.* 1999, **11**:68–80. (8-0)

Field, C.M., Coughlin, M., Doberstein, S., Marty, T. and Sullivan, W.: **Characterization of anillin mutants reveals essential roles in septin localization and plasma membrane integrity.** *Development* 2005, **132**:2849–2860. (8-2)

Fisher, D. and Nurse, P.: **Cyclins of the fission yeast *Schizosaccharomyces pombe*.** *Semin. Cell Biol.* 1995, **6**:73–78. (5-2)

Fisher, D.L. and Nurse, P.: **A single fission yeast mitotic cyclin B $p34^{cdc2}$ kinase promotes both S-phase and mitosis in the absence of G_1 cyclins.** *EMBO J.* 1996, **15**:850–860. (4-5)

Fisk, H.A., Mattison, C.P. and Winey, M.: **Centrosomes and tumour suppressors.** *Curr. Opin. Cell Biol.* 2002, **14**:700–705. (6-3)

Fitch, I., Dahmann, C., Surana, U., Amon, A., Nasmyth, K., Goetsch, L., Byers, B. and Futcher, B.: **Characterization of four B-type cyclin genes of the budding yeast *Saccharomyces cerevisiae*.** *Mol. Biol. Cell* 1992, **3**:805–818. (5-2)

Follette, P.J. and O'Farrell, P.H.: **Cdks and the *Drosophila* cell cycle.** *Curr. Opin. Genet. Dev.* 1997, **7**:17–22. (4-6)

Follette, P.J., Duronio, R.J. and O'Farrell, P.H.: **Fluctuations in cyclin E levels are required for multiple rounds of endocycle S phase in *Drosophila*.** *Curr. Biol.* 1998, **8**:235–238. (4-6)

Forsburg, S.L. and Nurse, P.: **Cell cycle regulation in the yeasts *Saccharomyces cerevisiae* and *Schizosaccharomyces pombe*.** *Annu. Rev. Cell Biol.* 1991, **7**:227–256. (2-1, 2-2)

Friedberg, E.C., Walker, G.C., Siede, W., Wood, R.D., Schultz, R.A. and Ellenberger, T.: *DNA Repair and Mutagenesis.* 2nd ed. (ASM Press, Washington, D.C., 2005). (11-1)

Frolov, M.V., Huen, D.S., Stevaux, O., Dimova, D., Balczarek-Strang, K., Elsdon, M. and Dyson, N.J.: **Functional antagonism between E2F family members.** *Genes Dev.* 2001, **15**:2146–2160. (10-4)

Fry, R.C., Begley, T.J. and Samson, L.D.: **Genome-wide responses to DNA-damaging agents.** *Annu. Rev. Microbiol.* 2005, **59**:357–377. (11-0)

Fujiwara, T., Bandi, M., Nitta, M., Ivanova, E.V., Bronson, R.T. and Pellman, D.: **Cytokinesis failure generating tetraploids promotes tumorigenesis in p53-null cells.** *Nature* 2005, **437**:1043–1047. (12-7)

Fung, J.C., Rockmill, B., Odell, M. and Roeder, G.S.: **Imposition of crossover interference through the nonrandom distribution of synapsis initiation complexes.** *Cell* 2004, **116**:795–802. (9-4)

Furuno, N., den Elzen, N. and Pines, J.: **Human cyclin A is required for mitosis until mid prophase.** *J. Cell Biol.* 1999, **147**:295–306. (5-3)

Futcher, B.: **Transcriptional regulatory networks and the yeast cell cycle.** *Curr. Opin. Cell Biol.* 2002, **14**:676–683. (3-12, 3-13, 10-1)

Gadde, S. and Heald, R.: **Mechanisms and molecules of the mitotic spindle.** *Curr. Biol.* 2004, **14**:R797–R805. (6-0, 6-2, 6-8, 6-9)

Gao, X., Zhang, Y., Arrazola, P., Hino, O., Kobayashi, T., Yeung, R.S., Ru, B. and Pan, D.: **Tsc tumour suppressor proteins antagonize amino-acid–TOR signalling.** *Nat. Cell Biol.* 2002, **4**:699–704. (10-11)

Garber, P.M., Vidanes, G.M. and Toczyski, D.P.: **Damage in transition.** *Trends Biochem. Sci.* 2005, **30**:63–66. (11-2)

Gartner, A., Jovanovic, A., Jeoung, D.I., Bourlat, S., Cross, F.R. and Ammerer, G.: **Pheromone-dependent G_1 cell cycle arrest requires Far1 phosphorylation, but may not involve inhibition of Cdc28-Cln2 kinase, *in vivo*.** *Mol. Cell Biol.* 1998, **18**:3681–3691. (10-3)

Gerton, J.L. and Hawley, R.S.: **Homologous chromosome interactions in meiosis: diversity amidst conservation.** *Nat. Rev. Genet.* 2005, **6**:477–487. (9-3)

Giansanti, M.G., Gatti, M. and Bonaccorsi, S.: **The role of centrosomes and astral microtubules during asymmetric division of *Drosophila* neuroblasts.** *Development* 2001, **128**:1137–1145. (8-7)

Gilbert, C.S., Green, C.M. and Lowndes, N.F.: **Budding yeast Rad9 is an ATP-dependent Rad53 activating machine.** *Mol. Cell* 2001, **8**:129–136. (11-3)

Gilbert, D.M.: **Making sense of eukaryotic DNA replication origins.** *Science* 2001, **294**:96–100. (4-2)

Girard, F., Strausfeld, U., Fernandez, A. and Lamb, N.J.C.: **Cyclin A is required for the onset of DNA replication in mammalian fibroblasts.** *Cell* 1991, **67**:1169–1179. (4-6)

Glotzer, M.: **Animal cell cytokinesis.** *Annu. Rev. Cell Dev. Biol.* 2001, **17**:351–386. (8-0)

Glotzer, M.: **The molecular requirements for cytokinesis.** *Science* 2005, **307**:1735–1739. (8-0, 8-1, 8-2, 8-5)

Glover, J.N., Williams, R.S. and Lee, M.S.: **Interactions between BRCT repeats and phosphoproteins: tangled up in two.** *Trends Biochem. Sci.* 2004, **29**:579–585. (11-3)

Goldstein, L.S.: **Kinetochore structure and its role in chromosome orientation during the first meiotic division in male *D. melanogaster*.** *Cell* 1981, **25**:591–602. (9-6)

Gönczy, P.: **Mechanisms of spindle positioning: focus on flies and worms.** *Trends Cell Biol.* 2002, **12**:332–339. (8-7)

Gorgoulis, V.G., Vassiliou, L.V., Karakaidos, P., Zacharatos, P., Kotsinas, A., Liloglou, T., Venere, M., Ditullio, R.A. Jr, Kastrinakis, N.G., Levy, B., Kletsas, D., Yoneta, A., Herlyn, M., Kittas, C. and Halazonetis, T.D.: **Activation of the DNA damage checkpoint and genomic instability in human precancerous lesions.** *Nature* 2005, **434**:907–913. (12-5)

Gottschling, D.E., Aparicio, O.M., Billington, B.L. and Zakian, V.A.: **Position effect at *S. cerevisiae* telomeres: reversible repression of Pol II transcription.** *Cell* 1990, **63**:751–762. (4-12)

Gould, K.L. and Nurse, P.: **Tyrosine phosphorylation of the fission yeast *cdc2*$^+$ protein kinase regulates entry into mitosis.** *Nature* 1989, **342**:39–45. (5-4)

Grill, S.W. and Hyman, A.A.: **Spindle positioning by cortical pulling forces.** *Dev. Cell* 2005, **8**:461–465. (8-7)

Grill, S.W., Gönczy, P., Stelzer, E.H. and Hyman, A.A.: **Polarity controls forces governing asymmetric spindle positioning in the *Caenorhabditis elegans* embryo.** *Nature* 2001, **409**:630–633. (8-7)

Gromley, A., Yeaman, C., Rosa, J., Redick, S., Chen, C.T., Mirabelle, S., Guha, M., Sillibourne, J. and Doxsey, S.J.: **Centriolin anchoring of exocyst and SNARE complexes at the midbody is required for secretory-vesicle-mediated abscission.** *Cell* 2005, **123**:75–87. (8-3)

Gunjan, A. and Verreault, A.: **A Rad53 kinase-dependent surveillance mechanism that regulates histone protein levels in *S. cerevisiae*.** *Cell* 2003, **115**:537–549. (4-10)

Guse, A., Mishima, M. and Glotzer, M.: **Phosphorylation of ZEN-4/MKLP1 by aurora B regulates completion of cytokinesis.** *Curr. Biol.* 2005, **15**:778–786. (8-5)

Habedanck, R., Stierhof, Y.D., Wilkinson, C.J. and Nigg, E.A.: **The Polo kinase Plk4 functions in centriole duplication.** *Nat. Cell Biol.* 2005, **7**:1140–1146. (6-4)

Haber, D.A.: **Splicing into senescence: the curious case of p16 and p19^{ARF}.** *Cell* 1997, **91**:555–558. (12-4)

Hagting, A., Den Elzen, N., Vodermaier, H.C., Waizenegger, I.C., Peters, J.M. and Pines, J.: **Human securin proteolysis is controlled by the spindle checkpoint and reveals when the APC/C switches from activation by Cdc20 to Cdh1.** *J. Cell Biol.* 2002, **157**:1125–1137. (7-1)

Hagting, A., Jackman, M., Simpson, K. and Pines, J.: **Translocation of cyclin B1 to the nucleus at prophase requires a phosphorylation-dependent nuclear import signal.** *Curr. Biol.* 1999, **9**:680–689. (5-6)

Hahn, S.A., Schutte, M., Hoque, A.T., Moskaluk, C.A., da Costa, L.T., Rozenblum, E., Weinstein, C.L., Fischer, A., Yeo, C.J., Hruban, R.H. and Kern, S.E.: **DPC4, a candidate tumor suppressor gene at human chromosome 18q21.1.** *Science* 1996, **271**:350–353. (12-3)

Hall, M.N., Raff, M. and Thomas, G. (eds): *Cell Growth: Control of Cell Size* (Cold Spring Harbor Laboratory Press, Cold Spring Harbor, 2004). (10-0, 10-10, 10-11, 10-12)

Hanahan, D. and Weinberg, R.A.: **The hallmarks of cancer.** *Cell* 2000, **100**:57–70. (12-0, 12-3, 12-4, 12-8)

Hansen, D.V., Loktev, A.V., Ban, K.H. and Jackson, P.K.: **Plk1 regulates activation of the anaphase promoting complex by phosphorylating and triggering SCF$^{\beta TrCP}$-dependent destruction of the APC inhibitor Emi1.** *Mol. Biol. Cell* 2004, **15**:5623–5634. (7-1)

Hao, B., Zheng, N., Schulman, B.A., Wu, G., Miller, J.J., Pagano, M., Pavletich, N.P.: **Structural basis of the Cks1-dependent recognition of p27^{Kip1} by the SCF^{Skp2} ubiquitin ligase.** *Mol. Cell* 2005, **20**:9–19. (10-8)

Hara, K., Tydeman, P. and Kirschner, M.: **A cytoplasmic clock with the same period as the division cycle in *Xenopus* eggs.** *Proc. Natl Acad. Sci. USA* 1980, **77**:462–466. (2-3)

Harel, A. and Forbes, D.J.: **Importin beta: conducting a much larger cellular symphony.** *Mol. Cell.* 2004, **16**:319–330. (6-8)

Harper, J.W. and Elledge, S.J.: **The role of Cdk7 in CAK function, a retro-retrospective.** *Genes Dev.* 1998, **12**:285–289. (3-3)

Harper, J.W., Burton, J.L. and Solomon, M.J.: **The anaphase-promoting complex: it's not just for mitosis any more.** *Genes Dev.* 2002, **16**:2179–2206. (3-10, 7-0, 7-1)

Harris, H.: *The Birth of the Cell* (Yale University Press, New Haven, 2000). (1-0)

Harris, S.L. and Levine, A.J.: **The p53 pathway: positive and negative feedback loops.** *Oncogene* 2005, **24**:2899–2908. (11-4)

Hartwell, L.H.: **Twenty-five years of cell cycle genetics.** *Genetics* 1991, **129**:975–980. (2-2)

Hartwell, L.H. and Weinert, T.A.: **Checkpoints: controls that ensure the order of cell cycle events.** *Science* 1989, **246**:629–634. (1-3)

Hartwell, L.H., Culotti, J. and Reid, B.: **Genetic control of the cell-division cycle in yeast, I. Detection of mutants.** *Proc. Natl Acad. Sci. USA* 1970, **66**:352–359. (2-2)

Hartwell, L.H, Culotti, J., Pringle, J.R. and Reid, B.J.: **Genetic control of the cell division cycle in yeast.** *Science* 1974, **183**:46–51. (2-2)

Hauf, S. and Watanabe, Y.: **Kinetochore orientation in mitosis and meiosis.** *Cell* 2004, **119**:317–327. (6-5, 6-10)

Hauf, S., Cole, R.W., LaTerra, S., Zimmer, C., Schnapp, G., Walter, R., Heckel, A., van Meel, J., Rieder, C.L. and Peters, J.M.: **The small molecule Hesperadin reveals a role for Aurora B in correcting kinetochore-microtubule attachment and in maintaining the spindle assembly checkpoint.** *J. Cell Biol.* 2003, **161**:281–294. (6-10)

Hauf, S., Roitinger, E., Koch, B., Dittrich, C.M., Mechtler, K. and Peters, J.M.: **Dissociation of cohesin from chromosome arms and loss of arm cohesion during early mitosis depends on phosphorylation of SA2.** *PLoS Biol.* 2005, **3**:e69. (5-10)

Haushalter, K.A. and Kadonaga, J.T.: **Chromatin assembly by DNA-translocating motors.** *Nat. Rev. Mol. Cell Biol.* 2003, **4**:613–620. (4-11)

Hayes, J.J. and Hansen, J.C.: **Nucleosomes and the chromatin fiber.** *Curr. Opin. Genet. Dev.* 2001, **11**:124–129. (4-9)

Hays, T.S., Wise, D. and Salmon, E.D.: **Traction force on a kinetochore at metaphase acts as a linear function of kinetochore fiber length.** *J. Cell Biol.* 1982, **93**:374–389. (6-12)

Heald, R. and McKeon, F.: **Mutations of phosphorylation sites in lamin A that prevent nuclear lamina disassembly in mitosis.** *Cell* 1990, **61**:579–589. (6-7)

Henchoz, S., Chi, Y., Catarin, B., Herskowitz, I., Deshaies, R.J. and Peter, M.: **Phosphorylation- and ubiquitin-dependent degradation of the cyclin-dependent kinase inhibitor Far1p in budding yeast.** *Genes Dev.* 1997, **11**:3046–3060. (10-3)

Henikoff, S., Furuyama, T. and Ahmad, K.: **Histone variants, nucleosome assembly and epigenetic inheritance.** *Trends Genet.* 2004, **20**:320–326. (4-11)

Herbert, M., Levasseur, M., Homer, H., Yallop, K., Murdoch, A. and McDougall, A.: **Homologue disjunction in mouse oocytes requires proteolysis of securin and cyclin B1.** *Nat. Cell Biol.* 2003, **5**:1023–1025. (9-7)

Hershko, A. and Ciechanover, A.: **The ubiquitin system.** *Annu. Rev. Biochem.* 1998, **67**:425–479. (3-9)

Hetzer, M., Meyer, H.H., Walther, T.C., Bilbao-Cortes, D., Warren, G. and Mattaj, I.W.: **Distinct AAA-ATPase p97 complexes function in discrete steps of nuclear assembly.** *Nat. Cell Biol.* 2001, **3**:1086–1091. (7-7)

Hetzer, M., Walther, T.C. and Mattaj, I.W.: **Pushing the envelope: structure, function, and dynamics of the nuclear periphery.** *Annu. Rev. Cell Dev. Biol.* 2005, **21**:347–380. (6-7, 7-7)

Higuchi, T. and Uhlmann, F.: **Stabilization of microtubule dynamics at anaphase onset promotes chromosome segregation.** *Nature* 2005, **433**:171–176. (7-6)

Hinchcliffe, E.H., and Sluder, G.: **"It takes two to tango": understanding how centrosome duplication is regulated throughout the cell cycle.** *Genes Dev.* 2001, **15**:1167–1181. (6-4)

Hinchcliffe, E.H., Li, C., Thompson, E.A., Maller, J.L. and Sluder, G.: **Requirement of Cdk2-cyclin E activity for repeated centrosome reproduction in *Xenopus* egg extracts.** *Science* 1999, **283**:851–854. (6-4)

Hingorani, S.R. and Tuveson, D.A.: **Targeting oncogene dependence and resistance.** *Cancer Cell* 2003, **3**:414–417. (12-9)

Hirano, T.: **Condensins: organizing and segregating the genome.** *Curr. Biol.* 2005, **15**:R265–R275. (5-9, 5-10)

Hoeijmakers, J.H.: **Genome maintenance mechanisms for preventing cancer.** *Nature* 2001, **411**:366–374. (11-1, 12-5)

Hollingsworth, N.M. and Brill, S.J.: **The Mus81 solution to resolution: generating meiotic crossovers without Holliday junctions.** *Genes Dev.* 2004, **18**:117–125. (9-2)

Homer, H.A., McDougall, A., Levassuer, M., Yallop, K., Murdoch, A.P. and Herbert, M.: **Mad2 prevents aneuploidy and premature proteolysis of cyclin B and securin during meiosis I in mouse oocytes.** *Genes Dev.* 2005, **19**:202–207. (9-7)

Honda, R., Lowe, E.D., Dubinina, E., Skamnaki, V., Cook, A., Brown, N.R. and Johnson, L.N.: **The structure of cyclin E1/CDK2: implications for CDK2 activation and CDK2-independent roles.** *EMBO J.* 2005, **24**:452–463. (3-4)

Honigberg, S.M.: **Ime2p and Cdc28p: co-pilots driving meiotic development.** *J. Cell Biochem.* 2004, **92**:1025–1033. (9-1)

Honigberg, S.M. and Purnapatre, K.: **Signal pathway integration in the switch from the mitotic cell cycle to meiosis in yeast.** *J. Cell Sci.* 2003, **116**:2137–2147. (9-1)

Hopfner, K.P.: **Chromosome cohesion: closing time.** *Curr. Biol.* 2003, **13**:R866–R868. (5-8)

Hopfner, K.P. and Tainer, J.A.: **Rad50/SMC proteins and ABC transporters: unifying concepts from high-resolution structures.** *Curr. Opin. Struc. Biol.* 2003, **13**:249–255. (5-8)

Hoyt, M.A., Totis, L. and Roberts, B.T.: ***S. cerevisiae* genes required for cell cycle arrest in response to microtubule function.** *Cell* 1991, **66**:507–517. (7-2)

Hsu, J.Y., Reimann, J.D., Sorensen, C.S., Lukas, J. and Jackson, P.K.: **E2F-dependent accumulation of hEmi1 regulates S phase entry by inhibiting APC^{Cdh1}.** *Nat. Cell Biol.* 2002, **4**:358–366. (10-8)

Hu, F. and Aparicio, O.M.: **Swe1 regulation and transcriptional control restrict the activity of mitotic cyclins toward replication proteins in *Saccharomyces cerevisiae*.** *Proc. Natl Acad. Sci. USA* 2005, **102**:8910–8915. (4-5)

Huang, J.N., Park, I., Ellingson, E., Littlepage, L.E. and Pellman, D.: **Activity of the APC^{Cdh1} form of the anaphase-promoting complex persists until S phase and prevents the premature expression of Cdc20p.** *J. Cell Biol.* 2001, **154**:85–94. (10-2)

Huberman, J.A. and Tsai, A.: **Direction of DNA replication in mammalian cells.** *J. Mol. Biol.* 1973, **75**:5–12. (4-2)

Hunter, N. and Kleckner, N.: **The single-end invasion: an asymmetric intermediate at the double-strand break to double-Holliday junction transition of meiotic recombination.** *Cell* 2001, **106**:59–70. (9-2)

Huntly, B.J. and Gilliland, D.G.: **Leukaemia stem cells and the evolution of cancer-stem-cell research.** *Nat. Rev. Cancer* 2005, **5**:311–321. (12-2)

Huyen, Y., Zgheib, O., Ditullio, R.A. Jr, Gorgoulis, V.G., Zacharatos, P., Petty, T.J., Sheston, E.A., Mellert, H.S., Stavridi, E.S. and Halazonetis, T.D.: **Methylated lysine 79 of histone H3 targets 53BP1 to DNA double-strand breaks.** *Nature* 2004, **432**:406–411. (11-3)

Hwang, H.C. and Clurman, B.E.: **Cyclin E in normal and neoplastic cell cycles.** *Oncogene* 2005, **24**:2776–2786. (10-8)

Ira, G., Pellicioli, A., Balijja, A., Wang, X., Fiorani, S., Carotenuto, W., Liberi, G., Bressan, D., Wan, L., Hollingsworth, N.M., Haber, J.E. and Foiani, M.: **DNA end resection, homologous recombination and DNA damage checkpoint activation require CDK1.** *Nature* 2004, **431**:1011–1077. (11-2)

Ivanov, D. and Nasmyth, K.: **A topological interaction between cohesion rings and a circular minichromosome.** *Cell* 2005, **122**:849–860. (5-8)

Iwabuchi, M., Ohsumi, K., Yamamoto, T.M., Sawada, W. and Kishimoto, T.: **Residual Cdc2 activity remaining at meiosis I exit is essential for meiotic M-M transition in *Xenopus* oocyte extracts.** *EMBO J.* 2000, **19**:4513–4523. (9-8)

Izawa, D., Goto, M., Yamashita, A., Yamano, H. and Yamamoto, M.: **Fission yeast Mes1p ensures the onset of meiosis II by blocking degradation of cyclin Cdc13p.** *Nature* 2005, **434**:529–533. (9-8)

Jacobs, H.W., Knoblich, J.A. and Lehner, C.F.: ***Drosophila* Cyclin B3 is required for female fertility and is dispensable for mitosis like Cyclin B.** *Genes Dev.* 1998, **12**:3741–3751. (5-2, 5-3)

Jackman, M., Lindon, C., Nigg, E.A. and Pines, J.: **Active cyclin B1–Cdk1 first appears on centrosomes in prophase.** *Nat. Cell Biol.* 2003, **5**:143–148. (5-6)

Jackson, P.K., Chevalier, S., Philippe, M. and Kirschner, M.W.: **Early events in DNA replication require cyclin E and are blocked by p21^{CIP1}.** *J. Cell Biol.* 1995, **130**:755–769. (4-6)

Jaspersen, S.L. and Winey, M.: **The budding yeast spindle pole body: structure, duplication, and function.** *Annu. Rev. Cell Dev. Biol.* 2004, **20**:1–28. (6-3)

Jaspersen, S.L., Charles, J.F., Tinker-Kulberg, R.L. and Morgan, D.O.: **A late mitotic regulatory network controlling cyclin destruction in *Saccharomyces cerevisiae*.** *Mol. Biol. Cell* 1998, **9**:2803–2817. (7-5)

Jaspersen, S.L., Huneycutt, B.J., Giddings, T.H. Jr, Resing, K.A., Ahn, N.G. and Winey, M.: **Cdc28/Cdk1 regulates spindle pole body duplication through phosphorylation of Spc42 and Mps1.** *Dev. Cell* 2004, **7**:263–274. (6-4)

Jazayeri, A., Falck, J., Lukas, C., Bartek, J., Smith, G.C., Lukas, J. and Jackson, S.P.: **ATM- and cell cycle-dependent regulation of ATR in response to DNA double-strand breaks.** *Nat. Cell Biol.* 2006, **8**:37–45. (11-2, 11-5)

Jeffrey, P.D., Russo, A.A., Polyak, K., Gibbs, E., Hurwitz, J., Massagué, J. and Pavletich, N.P.: **Mechanism of CDK activation revealed by the structure of a cyclin A–CDK2 complex.** *Nature* 1995, **376**:313–320. (3-4)

Jenuwein, T. and Allis, C.D.: **Translating the histone code.** *Science* 2001, **293**:1074–1080. (4-9)

Jin, J., Cardozo, T., Lovering, R.C., Elledge, S.J., Pagano, M. and Harper, J.W.: **Systematic analysis and nomenclature of mammalian F-box proteins.** *Genes Dev.* 2004, **18**:2573–2580. (3-9)

John, B.: *Meiosis* (Cambridge University Press, New York, 1990). (9-0, 9-4)

Jones, L., Richardson, H. and Saint, R.: **Tissue-specific regulation of cyclin E transcription during *Drosophila melanogaster* embryogenesis.** *Development* 2000, **127**:4619–4630. (10-9)

Jorgensen, P. and Tyers, M.: **How cells coordinate growth and division.** *Curr. Biol.* 2004, **14**:R1014–R1027. (10-10, 10-11, 10-12, 10-13)

Jorgensen, P., Nishikawa, J.L., Breitkreutz, B.J. and Tyers, M.: **Systematic identification of pathways that couple cell growth and division in yeast.** *Science* 2002, **297**:395–400. (10-12)

Juang, Y.-L., Huang, J., Peters, J.-M., McLaughlin, M.E., Tai, C.-Y. and Pellman, D.: **APC-mediated proteolysis of Ase1 and the morphogenesis of the mitotic spindle.** *Science* 1997, **275**:1311–1314. (7-7)

Jürgens, G.: **Plant cytokinesis: fission by fusion.** *Trends Cell Biol.* 2005, **15**:277–283. (8-3)

Kalab, P., Weis, K. and Heald, R.: **Visualization of a Ran-GTP gradient in interphase and mitotic *Xenopus* egg extracts.** *Science* 2002, **295**:2452–2456. (6-8)

Kamakaka, R.T. and Biggins, S.: **Histone variants: deviants?** *Genes Dev.* 2005, **19**:295–310. (4-9)

Kamura, T., Hara, T., Matsumoto, M., Ishida, N., Okumura, F., Hatakeyama, S., Yoshida, M., Nakayama, K. and Nakayama, K.I.: **Cytoplasmic ubiquitin ligase KPC regulates proteolysis of p27^{Kip1} at G1 phase.** *Nat. Cell Biol.* 2004, **6**:1229–1235. (10-8)

Kapoor, T.M. and Compton, D.A.: **Searching for the middle ground: mechanisms of chromosome alignment during mitosis.** *J. Cell Biol.* 2002, **157**:551–556. (6-9, 6-11, 6-12)

Kastan, M.B. and Bartek, J.: **Cell-cycle checkpoints and cancer.** *Nature* 2004, **432**:316–323. (12-5, 12-6)

Kauppi, L., Jeffreys, A.J. and Keeney, S.: **Where the crossovers are: recombination distributions in mammals.** *Nat. Rev. Genet.* 2004, **5**:413–424. (9-3)

Kearsey, S.E. and Cotterill, S.: **Enigmatic variations: divergent modes of regulating eukaryotic DNA replication.** *Mol. Cell* 2003, **12**:1067–1075. (4-3)

Kelly, T.J. and Brown, G.W.: **Regulation of chromosome replication.** *Annu. Rev. Biochem.* 2000, **69**:829–880. (4-1)

Kennedy, B.K., Barbie, D.A., Classon, M., Dyson, N. and Harlow, E.: **Nuclear organization of DNA replication in primary mammalian cells.** *Genes Dev.* 2000, **14**:2855–2868. (4-2)

Khodjakov, A. and Rieder, C.L.: **The sudden recruitment of gamma-tubulin to the centrosome at the onset of mitosis and its dynamic exchange throughout the cell cycle, do not require microtubules.** *J. Cell Biol.* 1999, **146**:585–596. (6-6)

Khorasanizadeh, S.: **The nucleosome: from genomic organization to genomic regulation.** *Cell* 2004, **116**:259–272. (4-9)

Killander, D. and Zetterberg, A.: **A quantitative cytochemical investigation of the relationship between cell mass and initiation of DNA synthesis in mouse fibroblasts *in vitro*.** *Exp. Cell Res.* 1965, **40**:12–20. (10-13)

Kim, K.K., Chamberlin, H.M., Morgan, D.O. and Kim, S.H.: **Three dimensional structure of human cyclin H, a positive regulator of the CDK-activating kinase.** *Nat. Struct. Biol.* 1996, **3**:849–855. (3-2)

Kimura, K., Hirano, M., Kobayashi, R. and Hirano, T.: **Phosphorylation and activation of 13S condensin by Cdc2 *in vitro*.** *Science* 1998, **282**:487–490. (5-10)

Kinoshita, K., Arnal, I., Desai, A., Drechsel, D.N. and Hyman, A.A.: **Reconstitution of physiological microtubule dynamics using purified components.** *Science* 2001, **294**:1340–1343. (6-2)

Kinoshita, M., Field, C., Coughlin, M., Straight, A. and Mitchison, T.: **Self- and actin-templated assembly of mammalian septins.** *Dev. Cell* 2002, **3**:791–802. (8-2)

Kinzler, K.W. and Vogelstein, B.: **Lessons from hereditary colorectal cancer.** *Cell* 1996, **87**:159–170. (12-8)

Kireeva, N., Lakonishok, M., Kireev, I., Hirano, T. and Belmont, A.S.: **Visualization of early chromosome condensation: a hierarchical folding, axial glue model of chromosome structure.** *J. Cell Biol.* 2004, **166**:775–785. (5-9)

Kitajima, T.S., Kawashima, S.A. and Watanabe, Y.: **The conserved kinetochore protein shugoshin protects centromeric cohesion during meiosis.** *Nature* 2004, **427**:510–517. (9-7)

Klein, P., Pawson, T. and Tyers, M.: **Mathematical modeling suggests cooperative interactions between a disordered polyvalent ligand and a single receptor site.** *Curr. Biol.* 2003, **13**:1669–1678. (10-2)

Klein, S., McCormick, F. and Levitzki, A.: **Killing time for cancer cells.** *Nat. Rev. Cancer* 2005, **5**:573–580. (12-9)

Kline-Smith, S.L. and Walczak, C.E.: **Mitotic spindle assembly and chromosome segregation: refocusing on microtubule dynamics.** *Mol. Cell* 2004, **15**:317–327. (6-2, 6-6)

Kline-Smith, S.L., Sandall, S. and Desai, A.: **Kinetochore-spindle microtubule interactions during mitosis.** *Curr. Opin. Cell Biol.* 2005, **17**:35–46. (6-5)

Knoblich, J.A., Sauer, K., Jones, L., Richardson, H., Saint, R. and Lehner, C.F.: **Cyclin E controls S phase progression and its down-regulation during *Drosophila* embryogenesis is required for the arrest of cell proliferation.** *Cell* 1994, **77**:107–120. (10-9)

Koehler, K.E., Boulton, C.L., Collins, H.E., French, R.L., Herman, K.C., Lacefield, S.M, Madden, L.D., Schuetz, C.D. and Scott Hawley, R.: **Spontaneous X chromosome MI and MII nondisjunction events in *Drosophila melanogaster* oocytes have different recombinational histories.** *Nat. Genet.* 1996, **14**:406–414. (9-6)

Kops, G.J., Weaver, B.A. and Cleveland, D.W.: **On the road to cancer: aneuploidy and the mitotic checkpoint.** *Nat. Rev. Cancer* 2005, **5**:773–785. (12-7)

Kozar, K., Ciemerych, M.A., Rebel, V.I., Shigematsu, H., Zagozdzon, A., Sicinska, E., Geng, Y., Yu, Q., Bhattacharya, S., Bronson, R.T., Akashi, K. and Sicinski, P.: **Mouse development and cell proliferation in the absence of D-cyclins.** *Cell* 2004, **118**:477–491. (10-5)

Kraft, C., Herzog, F., Gieffers, C., Mechtler, K., Hagting, A., Pines, J. and Peters, J.M.: **Mitotic regulation of the human anaphase-promoting complex by phosphorylation.** *EMBO J.* 2003, **22**:6598–6609. (7-1)

Kumagai, A. and Dunphy, W.G.: **Purification and molecular cloning of Plx1, a Cdc25-regulatory kinase from *Xenopus* egg extracts.** *Science* 1996, **273**:1377–1380. (5-5)

Kwon, M. and Scholey, J.M.: **Spindle mechanics and dynamics during mitosis in *Drosophila*.** *Trends Cell Biol.* 2004, **14**:194–205. (6-0, 6-2, 6-6, 7-6)

Labbé, J.C., Maddox, P., Salmon, E. and Goldstein, B.: **PAR proteins regulate microtubule dynamics at the cell cortex in *C. elegans*.** *Curr. Biol.* 2003, **13**:707–714. (8-7)

Lampson, M.A., Renduchitala, K., Khodjakov, A. and Kapoor, T.M.: **Correcting improper chromosome-spindle attachments during cell division.** *Nat. Cell Biol.* 2004, **6**:232–237. (6-10)

Lan, W., Zhang, X., Kline-Smith, S.L., Rosasco, S.E., Barrett-Wilt, G.A., Shabanowitz, J., Hunt, D.F., Walczak, C.E. and Stukenberg, P.T.: **Aurora B phosphorylates centromeric MCAK and regulates its localization and microtubule depolymerization activity.** *Curr. Biol.* 2004, **14**:273–286. (6-10)

Lavoie, B.D., Hogan, E. and Koshland, D.: ***In vivo* requirements for rDNA chromosome condensation reveal two cell-cycle-regulated pathways for mitotic chromosome folding.** *Genes Dev.* 2004, **18**:76–87. (5-9, 5-10)

Lawrence, C.J., Dawe, R.K., Christie, K.R., Cleveland, D.W., Dawson, S.C., Endow, S.A., Goldstein, L.S., Goodson, H.V., Hirokawa, N., Howard, J., Malmberg, R.L., McIntosh, J.R., Miki, H., Mitchison, T.J., Okada, Y., Reddy, A.S., Saxton, W.M., Schliwa, M., Scholey, J.M., Vale, R.D., Walczak, C.E. and Wordeman, L.: **A standardized kinesin nomenclature.** *J. Cell Biol.* 2004, **167**:19–22. (6-2)

Lecuit, T.: **Junctions and vesicular trafficking during *Drosophila* cellularization.** *J. Cell Sci.* 2004, **117**:3427–3433. (8-6)

Lecuit, T. and Wieschaus, E.: **Polarized insertion of new membrane from a cytoplasmic reservoir during cleavage of the *Drosophila* embryo.** *J. Cell Biol.* 2000, **150**:849–860. (8-6)

Lee, B.H. and Amon, A.: **Role of Polo-like kinase CDC5 in programming meiosis I chromosome segregation.** *Science* 2003, **300**:482–486. (9-5, 9-6)

Lee, L.A. and Orr-Weaver, T.L.: **Regulation of cell cycles in *Drosophila* development: intrinsic and extrinsic cues.** *Annu. Rev. Genet.* 2003, **37**:545–578. (2-4, 10-9)

Lee, J.-H. and Paull, T.T.: **ATM activation by DNA double-strand breaks through the Mre11-Rad50-Nbs1 complex.** *Science* 2005, **308**:551–554. (11-2)

Lee, J., Gold, D.A., Shevchenko, A., Shevchenko, A. and Dunphy, W.G.: **Roles of replication fork-interacting and Chk1-activating domains from Claspin in a DNA replication checkpoint response.** *Mol. Biol. Cell* 2005, **16**:5269–5282. (11-6)

Leffak, I.M., Grainger, R. and Weintraub, H.: **Conservative assembly and segregation of nucleosomal histones.** *Cell* 1977, **12**:837–845. (4-11)

Lehman, D.A., Patterson, B., Johnston, L.A., Balzer, T., Britton, J.S., Saint, R. and Edgar, B.A.: ***Cis*-regulatory elements of the mitotic regulator, *string/Cdc25*.** *Development* 1999, **126**:1793–1803. (10-9)

Lehner, C.F. and O'Farrell, P.H.: **The roles of *Drosophila* Cyclins A and B in mitotic control.** *Cell* 1990, **61**:535–547. (5-3)

Leidel, S. and Gönczy, P.: **Centrosome duplication and nematodes: recent insights from an old relationship.** *Dev. Cell* 2005, **9**:317–325. (6-4)

Lenart, P. and Ellenberg, J.: **Nuclear envelope dynamics in oocytes: from germinal vesicle breakdown to mitosis.** *Curr. Opin. Cell Biol.* 2003, **15**:88–95. (6-7)

Lenart, P., Rabut, G., Daigle, G., Hand, A.R., Terasaki, M. and Ellenberg, J.: **Nuclear envelope breakdown in starfish oocytes proceeds by partial NPC disassembly followed by a rapidly spreading fenestration of nuclear membranes.** *J. Cell Biol.* 2003, **160**:1055–1068. (6-7)

Leu, J.Y. and Roeder, G.S.: **The pachytene checkpoint in *S. cerevisiae* depends on Swe1-mediated phosphorylation of the cyclin-dependent kinase Cdc28.** *Mol. Cell* 1999, **4**:805–814. (9-5)

Levine, A.J.: **p53, the cellular gatekeeper for growth and division.** *Cell* 1997, **88**:323–331. (11-4)

Li, A. and Blow, J.J.: **Non-proteolytic inactivation of geminin requires CDK-dependent ubiquitination.** *Nat. Cell Biol.* 2004, **6**:260–267. (4-4)

Li, J., Meyer, A.N. and Donoghue, D.J.: **Nuclear localization of cyclin B1 mediates its biological activity and is regulated by phosphorylation.** *Proc. Natl Acad. Sci. USA* 1997, **94**:502–507. (5-6)

Li, R. and Murray, A.W.: **Feedback control of mitosis in budding yeast.** *Cell* 1991, **66**:519–531. (7-2)

Li, X. and Nicklas, R.B.: **Mitotic forces control a cell-cycle checkpoint.** *Nature* 1995, **373**:630–632. (7-2)

Li, F., Long, T., Lu, Y., Ouyang, Q. and Tang, C.: **The yeast cell-cycle network is robustly designed.** *Proc. Natl Acad. Sci. USA* 2004, **101**:4781–4786. (3-13)

Lilly, M.A. and Spradling, A.C.: **The *Drosophila* endocycle is controlled by cyclin E and lacks a checkpoint ensuring S-phase completion.** *Genes Dev.* 1996, **10**:2514–2526. (4-6)

Lindqvist, A., Kallstrom, H. and Karlsson Rosenthal, C.: **Characterisation of Cdc25B localisation and nuclear export during the cell cycle and in response to stress.** *J. Cell Sci.* 2004, **117**:4979–4990. (5-4)

Lindqvist, A., Kallstrom, H., Lundgren, A., Barsoum, E. and Rosenthal, C.K.: **Cdc25B cooperates with Cdc25A to induce mitosis but has a unique role in activating cyclin B1–Cdk1 at the centrosome.** *J. Cell Biol.* 2005, **171**:35–45. (5-4, 5-5)

Lisby, M., Barlow, J.H., Burgess, R.C. and Rothstein, R.: **Choreography of the DNA damage response: spatiotemporal relationships among checkpoint and repair proteins.** *Cell* 2004, **118**:699–713. (11-2, 11-5)

Liu, T.H., Li, L. and Vaessin, H.: **Transcription of the *Drosophila* CKI gene *dacapo* is regulated by a modular array of *cis*-regulatory sequences.** *Mech. Dev.* 2002, **112**:25–36. (10-9)

Liyanage, M., Coleman, A., du Manoir, S., Veldman, T., McCormack, S., Dickson, R.B., Barlow, C., Wynshaw-Boris, A., Janz, S., Wienberg, J., Ferguson-Smith, M.A., Schrock, E. and Ried, T.: **Multicolour spectral karyotyping of mouse chromosomes.** *Nat. Genet.* 1996, **14**:312–315. (12-5)

Lohka, M.J., Hayes, M.K. and Maller, J.L.: **Purification of maturation-promoting factor, an intracellular regulator of early mitotic events.** *Proc. Natl Acad. Sci. USA* 1988, **85**:3009–3013. (2-3)

Loog, M. and Morgan, D.O.: **Cyclin specificity in the phosphorylation of cyclin-dependent kinase substrates.** *Nature* 2005, **434**:104–108. (3-5, 4-5)

Lopes, M., Foiani, M. and Sogo, J.M.: **Multiple mechanisms control chromosome integrity after replication fork uncoupling and restart at irreparable UV lesions.** *Mol. Cell* 2006, **21**:15–27. (11-6)

Losada, A., Hirano, M. and Hirano, T.: **Cohesin release is required for sister chromatid resolution, but not for condensin-mediated compaction, at the onset of mitosis.** *Genes Dev.* 2002, **16**:3004–3016. (5-10)

Lowe, S.W., Cepero, E. and Evan, G.: **Intrinsic tumour suppression.** *Nature* 2004, **432**:307–315. (12-1, 12-3, 12-4)

Luo, X., Tang, Z., Xia, G., Wassmann, K., Matsumoto, T., Rizo, J. and Yu, H.: **The Mad2 spindle checkpoint protein has two distinct natively folded states.** *Nat. Struct. Mol. Biol.* 2004, **11**:338–345. (7-3)

MacKay, V.L., Mai, B., Waters, L. and Breeden, L.L.: **Early cell cycle box-mediated transcription of *CLN3* and *SWI4* contributes to the proper timing of the G_1-to-S transition in budding yeast.** *Mol. Cell Biol.* 2001, **21**:4140–4148. (10-1)

MacQueen, A.J., Phillips, C.M., Bhalla, N., Weiser, P., Villeneuve, A.M. and Dernburg, A.F.: **Chromosome sites play dual roles to establish homologous synapsis during meiosis in *C. elegans*.** *Cell* 2005, **123**:1037–1050. (9-3)

Maddox, P., Straight, A., Coughlin, P., Mitchison, T.J. and Salmon, E.D.: **Direct observation of microtubule dynamics at kinetochores in *Xenopus* extract spindles: implications for spindle mechanics.** *J. Cell Biol.* 2003, **162**:377–382. (6-11, 6-12)

Maiato, H., DeLuca, J., Salmon, E.D. and Earnshaw, W.C.: **The dynamic kinetochore-microtubule interface.** *J. Cell Sci.* 2004, **117**:5461–5477. (6-5)

Maiato, H., Khodjakov, A. and Rieder, C.L.: ***Drosophila* CLASP is required for the incorporation of microtubule subunits into fluxing kinetochore fibres.** *Nat. Cell Biol.* 2005, **7**:42–47. (6-11)

Maiato, H., Rieder, C.L. and Khodjakov, A.: **Kinetochore-driven formation of kinetochore fibers contributes to spindle assembly during animal mitosis.** *J. Cell Biol.* 2004, **167**:831–840. (6-9)

Mailand, N. and Diffley, J.F.: **CDKs promote DNA replication origin licensing in human cells by protecting Cdc6 from APC/C-dependent proteolysis.** *Cell* 2005, **122**:915–926. (4-4)

Mailand, N., Podtelejnikov, A.V., Groth, A., Mann, M., Bartek, J. and Lukas, J.: **Regulation of G_2/M events by Cdc25A through phosphorylation-dependent modulation of its stability.** *EMBO J.* 2002, **21**:5911–5920. (5-4, 5-5)

Maison, C. and Almouzni, G.: **HP1 and the dynamics of heterochromatin maintenance.** *Nat. Rev. Mol. Cell Biol.* 2004, **5**:296–304. (4-12, 4-13)

Marahrens, Y. and Stillman, B.: **A yeast chromosomal origin of DNA replication defined by multiple functional elements.** *Science* 1992, **255**:817–823. (4-2)

Margolis, R.L.: **Tetraploidy and tumor development.** *Cancer Cell* 2005, **8**:353–354. (12-7)

Marston, A.L. and Amon, A.: **Meiosis: cell-cycle controls shuffle and deal.** *Nat. Rev. Mol. Cell Biol.* 2004, **5**:983–997. (9-1, 9-6, 9-8)

Marston, A.L., Lee, B.H. and Amon, A.: **The Cdc14 phosphatase and the FEAR network control meiotic spindle disassembly and chromosome segregation.** *Dev. Cell* 2003, **4**:711–726. (9-8)

Martin, D.E. and Hall, M.N.: **The expanding TOR signaling network.** *Curr. Opin. Cell Biol.* 2005, **17**:158–166. (10-11)

Martin, D.E., Soulard, A. and Hall, M.N.: **TOR regulates ribosomal protein gene expression via PKA and the Forkhead transcription factor FHL1.** *Cell* 2004, **119**:969–979. (10-11)

Marumoto, T., Zhang, D. and Saya, H.: **Aurora-A—a guardian of poles.** *Nat. Rev. Cancer* 2005, **5**:42–50. (5-7, 6-6)

Marzluff, W.F. and Duronio, R.J.: **Histone mRNA expression: multiple levels of cell cycle regulation and important developmental consequences.** *Curr. Opin. Cell Biol.* 2002, **14**:692–699. (4-10)

Massagué, J.: **G1 cell-cycle control and cancer.** *Nature* 2004, **432**:298–306. (12-3)

Masui, Y. and Markert, C.L.: **Cytoplasmic control of nuclear behavior during meiotic maturation of frog oocytes.** *J. Exp. Zool.* 1971, **177**:129–146. (2-3)

Masumoto, H., Muramatsu, S., Kamimura, Y. and Araki, H.: **S-Cdk-dependent phosphorylation of Sld2 essential for chromosomal DNA replication in budding yeast.** *Nature* 2002, **415**:651–655. (4-8)

Matsumura, F.: **Regulation of myosin II during cytokinesis in higher eukaryotes.** *Trends Cell Biol.* 2005, **15**:371–377. (8-1, 8-2)

Matsumura, F., Ono, S., Yamakita, Y., Totsukawa, G. and Yamashiro, S.: **Specific localization of serine 19 phosphorylated myosin II during cell locomotion and mitosis of cultured cells.** *J. Cell Biol.* 1998, **140**:119–129. (8-2)

Mazia, D.: **Mitosis and the physiology of cell division** in *The Cell Vol. III* Brachet, J. and Mirsky, A.E. eds (Academic Press, New York, 1961), 77–412. (5-0)

Mazumdar, A. and Mazumdar, M.: **How one becomes many: blastoderm cellularization in *Drosophila melanogaster*.** *BioEssays* 2002, **24**:1012–1022. (8-6)

McGill, M., Highfield, D.P., Monahan, T.M. and Brinkley, B.R.: **Effects of nucleic acid specific dyes on centrioles of mammalian cells.** *J. Ultrastruct. Res.* 1976, **57**:43–53. (6-3)

McGowan, C.H. and Russell, P.: **The DNA damage response: sensing and signaling.** *Curr. Opin. Cell Biol.* 2004, **16**:629–633. (11-7)

Melo, J. and Toczyski, D.: **A unified view of the DNA-damage checkpoint.** *Curr. Opin. Cell Biol.* 2002, **14**:237–245. (11-0)

Melo, J.A., Cohen, J. and Toczyski, D.P.: **Two checkpoint complexes are independently recruited to sites of DNA damage *in vivo*.** *Genes Dev.* 2001, **15**:2809–2821. (11-3)

Mendenhall, M.D.: **An inhibitor of $p34^{CDC28}$ protein kinase activity from *Saccharomyces cerevisiae*.** *Science* 1993, **259**:216–219. (10-2)

Meneghini, M.D., Wu, M. and Madhani, H.D.: **Conserved histone variant H2A.Z protects euchromatin from the ectopic spread of silent heterochromatin.** *Cell* 2003, **112**:725–736. (4-13)

Meraldi, P., Honda, R. and Nigg, E.A.: **Aurora-A overexpression reveals tetraploidization as a major route to centrosome amplification in $p53^{-/-}$ cells.** *EMBO J.* 2002, **21**:483–492. (12-7)

Meraldi, P., Honda, R. and Nigg, E.A.: **Aurora kinases link chromosome segregation and cell division to cancer susceptibility.** *Curr. Opin. Genet. Dev.* 2004, **14**:29–36. (5-7)

Meyer, C.A., Kramer, I., Dittrich, R., Marzodko, S., Emmerich, J. and Lehner, C.F.: ***Drosophila* $p27^{Dacapo}$ expression during embryogenesis is controlled by a complex regulatory region independent of cell cycle progression.** *Development* 2002, **129**:319–328. (10-9)

Miller, M.E. and Cross, F.R.: **Cyclin specificity: how many wheels do you need on a unicycle?** *J. Cell Sci.* 2001, **114**:1811–1820. (3-5)

Minshull, J., Blow, J.J. and Hunt, T.: **Translation of cyclin mRNA is necessary for extracts of activated *Xenopus* eggs to enter mitosis.** *Cell* 1989, **56**:947–956. (5-3)

Miranda, J.J., De Wulf, P., Sorger, P.K. and Harrison, S.C.: **The yeast DASH complex forms closed rings on microtubules.** *Nat. Struct. Mol. Biol.* 2005, **12**:138–143. (6-5)

Mishima, M., Pavicic, V., Grüneberg, U., Nigg, E.A. and Glotzer, M.: **Cell cycle regulation of central spindle assembly.** *Nature* 2004, **430**:908–913. (8-5)

Mitchison, J.M.: **Growth during the cell cycle.** *Int. Rev. Cytol.* 2003, **226**:165–258. (10-10, 10-12)

Mitchison, J.M.: *The Biology of the Cell Cycle* (Cambridge University Press, Cambridge, 1971). (1-0, 2-0, 2-1, 2-6)

Mitchison, T.J. and Salmon, E.D.: **Mitosis: a history of division.** *Nat. Cell Biol.* 2001, **3**:E17–E21. (5-0, 6-0)

Miyamoto, D.T., Perlman, Z.E., Burbank, K.S., Groen, A.C. and Mitchison, T.J.: **The kinesin Eg5 drives poleward microtubule flux in *Xenopus laevis* egg extract spindles.** *J. Cell Biol.* 2004, **167**:813–818. (6-11)

Moazed, D.: **Common themes in mechanisms of gene silencing.** *Mol. Cell* 2001, **8**:489–498. (4-12, 4-13)

Morgan, D.O.: **Cyclin-dependent kinases: engines, clocks, and microprocessors.** *Annu. Rev. Cell Dev. Biol.* 1997, **13**:261–291. (1-3, 3-1, 3-3)

Morgan, D.O.: **Regulation of the APC and the exit from mitosis.** *Nat. Cell Biol.* 1999, **1**:E47–E53. (5-1, 7-0, 7-5)

Moritz, M., Braunfeld, M.B., Guenebaut, V., Heuser, J. and Agard, D.A.: **Structure of the gamma-tubulin ring complex: a template for microtubule nucleation.** *Nat. Cell Biol.* 2000, **2**:365–370. (6-2)

Murray, A.W. and Hunt, T.: *The Cell Cycle: An Introduction* (Freeman, New York, 1993). (1-0, 2-0, 2-1, 2-3, 2-6, 5-1)

Murray, A.W. and Kirschner, M.W.: **Cyclin synthesis drives the early embryonic cell cycle.** *Nature* 1989, **339**:275–280. (5-3)

Murray, A.W. and Kirschner, M.W.: **Dominoes and clocks: the union of two views of the cell cycle.** *Science* 1989, **246**:614–621. (1-3)

Murray, A.W., Solomon, M. and Kirschner, M.: **The role of cyclin synthesis and degradation in the control of maturation-promoting factor activity.** *Nature* 1989, **339**:280–286. (7-0)

Musacchio, A. and Hardwick, K.G.: **The spindle checkpoint: structural insights into dynamic signalling.** *Nat. Rev. Mol. Cell Biol.* 2002, **3**:731–741. (7-2, 7-3)

Nakamura, H., Morita, T. and Sato, C.: **Structural organizations of replicon domains during DNA synthetic phase in the mammalian nucleus.** *Exp. Cell Res.* 1986, **165**:291–297. (4-2)

Nash, P., Tang, X., Orlicky, S., Chen, Q., Gertler, F.B., Mendenhall, M.D., Sicheri, F., Pawson, T. and Tyers, M.: **Multisite phosphorylation of a CDK inhibitor sets a threshold for the onset of DNA replication.** *Nature* 2001, **414**:514–521. (10-2)

Nasmyth, K.: **How do so few control so many?** *Cell* 2005, **120**:739–746. (7-3)

Nasmyth, K.: **Segregating sister genomes: the molecular biology of chromosome separation.** *Science* 2002, **297**:559–565. (7-0, 7-4)

Nasmyth, K. and Haering, C.H.: **The structure and function of smc and kleisin complexes.** *Annu. Rev. Biochem.* 2005, **74**:595–648. (5-8, 7-4)

Nguyen, V.Q., Co, C. and Li, J.J.: **Cyclin-dependent kinases prevent DNA re-replication through multiple mechanisms.** *Nature* 2001, **411**:1068–1073. (4-4)

Nigg, E.A.: **Centrosome aberrations: cause or consequence of cancer progression?** *Nat. Rev. Cancer* 2002, **2**:815–825. (6-3, 12-7)

Nilsson, N.-O. and Säll, T.: **A model of chiasma reduction of closely formed crossovers.** *J. Theor. Biol.* 1995, **173**:93–98. (9-6)

Novina, C.D. and Sharp, P.A.: **The RNAi revolution.** *Nature* 2004, **430**:161–164. (2-5)

Nowell, P.C.: **The clonal evolution of tumor cell populations.** *Science* 1976, **194**:23–28. (12-0)

Nurse, P.: **Genetic control of cell size at cell division in yeast.** *Nature* 1975, **256**:547–551. (2-2, 10-12)

Nurse, P.: **Universal control mechanism regulating onset of M-phase.** *Nature* 1990, **344**:503–508. (3-3, 5-1, 5-4)

Nyberg, K.A., Michelson, R.J., Putnam, C.W. and Weinert, T.A.: **Toward maintaining the genome: DNA damage and replication checkpoints.** *Annu. Rev. Genet.* 2002, **36**:617–656. (11-0, 11-7)

Orlicky, S., Tang, X., Willems, A., Tyers, M. and Sicheri, F.: **Structural basis for phosphodependent substrate selection and orientation by the SCF^{Cdc4} ubiquitin ligase.** *Cell* 2003, **112**:243–256. (3-9, 10-2)

Oskarsson, T. and Trumpp, A.: **The Myc trilogy: lord of RNA polymerases.** *Nat. Cell Biol.* 2005, **7**:215–217. (10-11)

Ostergren, G.: **The mechanism of co-orientation in bivalents and multivalents. The theory of pulling.** *Hereditas* 1951, **37**:85–156. (6-12)

Page, S.L. and Hawley, R.S.: **Chromosome choreography: the meiotic ballet.** *Science* 2003, **301**:785–789. (9-0, 9-6)

Papoulas, O., Hays, T.S. and Sisson, J.C.: **The golgin Lava lamp mediates dynein-based Golgi movements during *Drosophila* cellularization.** *Nat. Cell Biol.* 2005, **7**:612–618. (8-6)

Pardee, A.B.: **G1 events and regulation of cell proliferation.** *Science* 1989, **246**:603–608. (10-0)

Parry, D.H. and O'Farrell, P.H.: **The schedule of destruction of three mitotic cyclins can dictate the timing of events during exit from mitosis.** *Curr. Biol.* 2001, **11**:671–683. (7-6)

Parry, D.H., Hickson, G.R. and O'Farrell, P.H.: **Cyclin B destruction triggers changes in kinetochore behavior essential for successful anaphase.** *Curr. Biol.* 2003, **13**:647–653. (7-6)

Parsons, D.W., Wang, T.L., Samuels, Y., Bardelli, A., Cummins, J.M., DeLong, L., Silliman, N., Ptak, J., Szabo, S., Willson, J.K., Markowitz, S., Kinzler, K.W., Vogelstein, B., Lengauer, C. and Velculescu, V.E.: **Colorectal cancer: mutations in a signalling pathway.** *Nature* 2005, **436**:792. (12-8)

Pavletich, N.P.: **Mechanisms of cyclin-dependent kinase regulation: structures of Cdks, their cyclin activators, and CIP and Ink4 inhibitors.** *J. Mol. Biol.* 1999, **287**:821–828. (3-4)

Pellicioli, A. and Foiani, M.: **Signal transduction: how rad53 kinase is activated.** *Curr. Biol.* 2005, **15**:R769–R771. (11-3)

Pereira, G. and Schiebel, E.: **Separase regulates INCENP-Aurora B anaphase spindle function through Cdc14.** *Science* 2003, **302**:2120–2124. (7-6)

Peter, M., Gartner, A., Horecka, J., Ammerer, G. and Herskowitz, I.: **FAR1 links the signal transduction pathway to the cell cycle machinery in yeast.** *Cell* 1993, **73**:747–760. (10-3)

Peters, J.M.: **The anaphase-promoting complex: proteolysis in mitosis and beyond.** *Mol. Cell* 2002, **9**:931–943. (3-10, 7-0, 7-1)

Petronczki, M., Siomos, M.F. and Nasmyth, K.: **Un ménage à quatre: the molecular biology of chromosome segregation in meiosis.** *Cell* 2003, **112**:423–440. (9-6, 9-7)

Pickart, C.M.: **Mechanisms underlying ubiquitination.** *Annu. Rev. Biochem.* 2001, **70**:503–533. (3-9)

Pihan, G. and Doxsey, S.J.: **Mutations and aneuploidy: co-conspirators in cancer?** *Cancer Cell* 2003, **4**:89–94. (12-5, 12-6)

Pihan, G.A., Wallace, J., Zhou, Y. and Doxsey, S.J.: **Centrosome abnormalities and chromosome instability occur together in pre-invasive carcinomas.** *Cancer Res.* 2003, **63**:1398–1404. (12-7)

Pinsky, B.A. and Biggins, S.: **The spindle checkpoint: tension versus attachment.** *Trends Cell Biol.* 2005, **15**:486–493. (7-2)

Polymenis, M. and Schmidt, E.V.: **Coupling of cell division to cell growth by translational control of the G_1 cyclin *CLN3* in yeast.** *Genes Dev.* 1997, **11**:2522–2531. (10-12)

Pomerening, J.R., Kim, S.Y. and Ferrell, J.E. Jr: **Systems-level dissection of the cell-cycle oscillator: bypassing positive feedback produces damped oscillations.** *Cell* 2005, **122**:565–578. (3-11, 5-5)

Pomerening, J.R., Sontag, E.D. and Ferrell, J.E. Jr: **Building a cell cycle oscillator: hysteresis and bistability in the activation of Cdc2.** *Nat. Cell Biol.* 2003, **5**:346–351. (3-8, 3-11)

Poot, R.A., Bozhenok, L., van den Berg, D.L., Steffensen, S., Ferreira, F., Grimaldi, M., Gilbert, N., Ferreira, J. and Varga-Weisz, P.D.: **The Williams syndrome transcription factor interacts with PCNA to target chromatin remodelling by ISWI to replication foci.** *Nat. Cell Biol.* 2004, **6**:1236–1244. (4-11)

Pramila, T. Miles, S., GuhaThakurta, D., Jemiolo, D. and Breeden L.L.: **Conserved homeodomain proteins interact with MADS box protein Mcm1 to restrict ECB-dependent transcription to the M/G1 phase of the cell cycle.** *Genes Dev.* 2002, **16**:3034–3045. (10-1)

Prescott, D.M.: *Reproduction of Eukaryotic Cells* (Academic Press, New York, 1976). (1-0)

Primig, M., Williams, R.M., Winzeler, E.A., Tevzadze, G.G., Conway, A.R., Hwang, S.Y., Davis, R.W. and Esposito, R.E.: **The core meiotic transcriptome in budding yeasts.** *Nat. Genet.* 2000, **26**:415–423. (9-1)

Prokopenko, S.N., Brumby, A., O'Keefe, L., Prior, L., He, Y., Saint, R. and Bellen, H.J.: **A putative exchange factor for Rho1 GTPase is required for initiation of cytokinesis in *Drosophila*.** *Genes Dev.* 1999, **13**:2301–2314. (8-2)

Quintyne, N.J., Reing, J.E., Hoffelder, D.R., Gollin, S.M. and Saunders, W.S.: **Spindle multipolarity is prevented by centrosomal clustering.** *Science* 2005, **307**:127–129. (12-7)

Raghuraman, M.K., Winzeler, E.A., Collingwood, D., Hunt, S., Wodicka, L., Conway, A., Lockhart, D.J., Davis, R.W., Brewer, B.J. and Fangman, W.L.: **Replication dynamics of the yeast genome.** *Science* 2001, **294**:115–121. (4-2)

Rajagopalan, H. and Lengauer, C.: **Aneuploidy and cancer.** *Nature* 2004, **432**:338–341. (12-7)

Rajagopalan, H., Jallepalli, P.V., Rago, C., Velculescu, V.E., Kinzler, K.W., Vogelstein, B. and Lengauer, C.: **Inactivation of hCDC4 can cause chromosomal instability.** *Nature* 2004, **428**:77–81. (12-7)

Rajagopalan, H., Nowak, M.A., Vogelstein, B. and Lengauer, C.: **The significance of unstable chromosomes in colorectal cancer.** *Nat. Rev. Cancer* 2003, **3**:695–701. (12-5, 12-8)

Ramaswamy, S., Ross K.N., Lander, E.S. and Golub, T.R.: **A molecular signature of metastasis in primary solid tumors.** *Nat. Genet.* 2003, **33**:49–54. (12-0)

Ramirez, R.D., Morales, C.P., Herbert, B.S., Rohde, J.M., Passons, C., Shay, J.W. and Wright, W.E.: **Putative telomere-independent mechanisms of replicative aging reflect inadequate growth conditions.** *Genes Dev.* 2001, **15**:398–403. (2-5)

Randell, J.C.W., Bowers J.L., Rodriguez, H.K. and Bell, S.P.: **Sequential ATP hydrolysis by Cdc6 and ORC directs loading of the Mcm2-7 helicase.** *Mol. Cell* 2006, **21**:29–39. (4-3)

Rappaport, R.: *Cytokinesis in Animal Cells* (Cambridge University Press, Cambridge, 1996). (8-0, 8-5)

Rauh, N.R., Schmidt, A., Bormann, J., Nigg, E.A. and Mayer, T.U.: **Calcium triggers exit from meiosis II by targeting the APC/C inhibitor XErp1 for degradation.** *Nature* 2005, **437**:1048–1052. (9-8)

Rayman, J.B., Takahashi, Y., Indjeian, V.B., Dannenberg, J.H., Catchpole, S., Watson, R.J., te Riele, H. and Dynlacht, B.D.: **E2F mediates cell cycle-dependent transcriptional repression in vivo by recruitment of an HDAC1/mSin3B corepressor complex.** *Genes Dev.* 2002, **16**:933–947. (10-5)

Reis, T. and Edgar, B.A.: **Negative regulation of dE2F1 by cyclin-dependent kinases controls cell cycle timing.** *Cell* 2004, **117**:253–264. (10-13)

Revenkova, E., Eijpe, M., Heyting, C., Hodges, C.A., Hunt, P.A., Liebe, B., Scherthan, H. and Jessberger, R.: **Cohesin SMC1 beta is required for meiotic chromosome dynamics, sister chromatid cohesion and DNA recombination.** *Nat. Cell Biol.* 2004, **6**:555–562. (9-7)

Reya, T. and Clevers, H.: **Wnt signalling in stem cells and cancer.** *Nature* 2005, **434**:843–850. (12-8)

Richards, E.J. and Elgin, S.C.: **Epigenetic codes for heterochromatin formation and silencing: rounding up the usual suspects.** *Cell* 2002, **108**:489–500. (4-13)

Richardson, H., Lew, D.J., Henze, M., Sugimoto, K. and Reed S.I.: **Cyclin-B homologs in *Saccharomyces cerevisiae* function in S phase and in G2.** *Genes Dev.* 1992, **6**:2021–2034. (5-2)

Ricke, R.M. and Bielinsky, A.K.: **Mcm10 regulates the stability and chromatin association of DNA polymerase-alpha.** *Mol. Cell* 2004, **16**:173–185. (4-8)

Rieder, C.L. and Alexander, S.P.: **Kinetochores are transported poleward along a single astral microtubule during chromosome attachment to the spindle in newt lung cells.** *J. Cell Biol.* 1990, **110**:81–95. (6-9)

Rieder, C.L. and Salmon, E.D.: **The vertebrate cell kinetochore and its roles during mitosis.** *Trends Cell Biol.* 1998, **8**:310–318. (6-12)

Rieder, C.L., Khodjakov, A., Paliulis, L.V., Fortier, T.M., Cole, R.W. and Sluder, G.: **Mitosis in vertebrate somatic cells with two spindles: implications for the metaphase/anaphase transition checkpoint and cleavage.** *Proc. Natl Acad. Sci. USA* 1997, **94**:5107–5112. (7-3)

Roberts, J.M.: **Evolving ideas about cyclins.** *Cell* 1999, **98**:129–132. (3-5)

Robinson, D.N. and Spudich, J.A.: **Towards a molecular understanding of cytokinesis.** *Trends Cell Biol.* 2000, **10**:228–237. (8-0, 8-1)

Roeder, G.S.: **Meiotic chromosomes: it takes two to tango.** *Genes Dev.* 1997, **11**:2600–2621. (9-0, 9-4)

Roeder, G.S. and Bailis, J.M.: **The pachytene checkpoint.** *Trends Genet.* 2000, **16**:395–403. (9-5)

Rosenblatt, J.: **Spindle assembly: asters part their separate ways.** *Nat. Cell Biol.* 2005, **3**:219–222. (6-6)

Rouse, J. and Jackson, S.P.: **Interfaces between the detection, signaling, and repair of DNA damage.** *Science* 2002, **297**:547–551. (11-0)

Rowley, J.D.: **Chromosome translocations: dangerous liaisons revisited.** *Nat. Rev. Cancer* 2001, **1**:245–250. (12-1)

Royou, A., Field, C., Sisson, J.C., Sullivan, W. and Karess, R.: **Reassessing the role and dynamics of nonmuscle myosin II during furrow formation in early *Drosophila* embryos.** *Mol. Biol. Cell* 2004, **15**:838–850. (8-6)

Rubin, S.M., Gall, A.L., Zheng, N. and Pavletich, N.P.: **Structure of the Rb C-terminal domain bound to E2F1-DP1: a mechanism for phosphorylation-induced E2F release.** *Cell* 2005, **123**:1093–1106. (10-5)

Rudner, A.D. and Murray, A.W.: **Phosphorylation by Cdc28 activates the Cdc20-dependent activity of the anaphase-promoting complex.** *J. Cell Biol.* 2000, **149**:1377–1390. (7-1)

Rupeš, I.: **Checking cell size in yeast.** *Trends Genet.* 2002, **18**:479–485. (10-10, 10-12)

Rusche, L.N., Kirchmaier, A.L. and Rine, J.: **The establishment, inheritance, and function of silenced chromatin in *Saccharomyces cerevisiae*.** *Annu. Rev. Biochem.* 2003, **72**:481–516. (4-12, 4-13)

Russell, P. and Nurse, P.: **Negative regulation of mitosis by *wee1*$^{+}$, a gene encoding a protein kinase homolog.** *Cell* 1987, **49**:559–567. (5-4)

Russo, A.A., Jeffrey, P.D. and Pavletich, N.P.: **Structural basis of cyclin-dependent kinase activation by phosphorylation.** *Nat. Struct. Biol.* 1996, **3**:696–700. (3-4)

Russo, A.A., Jeffrey, P.D., Patten, A.K., Massagué, J. and Pavletich, N.P.: **Crystal structure of the p27^{Kip1} cyclin-dependent-kinase inhibitor bound to the cyclin A–Cdk2 complex.** *Nature* 1996, **382**:325–331. (3-6)

Russo, A.A., Tong, L., Lee, J.O., Jeffrey, P.D. and Pavletich, N.P.: **Structural basis for inhibition of the cyclin-dependent kinase Cdk6 by the tumour suppressor p16^{INK4a}.** *Nature* 1998, **395**:237–243. (3-6)

Salmon, E.D.: **Microtubules: a ring for the depolymerization motor.** *Curr. Biol.* 2005, **15**:R299–R302. (6-11, 6-12)

Sampath, S.C., Ohi, R., Leismann, O., Salic, A., Pozniakovski, A. and Funabiki, H.: **The chromosomal passenger complex is required for chromatin-induced microtubule stabilization and spindle assembly.** *Cell* 2004, **118**:187–202. (6-8)

Samuels, Y. and Ericson, K.: **Oncogenic PI3K and its role in cancer.** *Curr. Opin. Oncol.* 2006, **18**:77–82. (12-4)

Sancar, A., Lindsey-Boltz, L.A., Unsal-Kacmaz, K. and Linn, S.: **Molecular mechanisms of mammalian DNA repair and the DNA damage checkpoints.** *Annu. Rev. Biochem.* 2004, **73**:39–85. (11-1)

Sanders, S.L., Portoso, M., Mata, J., Bahler, J., Allshire, R.C. and Kouzarides, T.: **Methylation of histone H4 lysine 20 controls recruitment of Crb2 to sites of DNA damage.** *Cell* 2004, **119**:603–614. (11-3)

Saucedo, L.J., Gao, X., Chiarelli, D.A., Li, L., Pan, D. and Edgar, B.A.: **Rheb promotes cell growth as a component of the insulin/TOR signalling network.** *Nat. Cell Biol.* 2003, **5**:566–571. (10-13)

Sawyers, C.: **Targeted cancer therapy.** *Nature* 2004, **432**:294–297. (12-9)

Schmekel, K. and Daneholt, B.: **Evidence for close contact between recombination nodules and the central element of the synaptonemal complex.** *Chromosome Res.* 1998, **6**:155–159. (9-4)

Schmekel, K., Meuwissen, R.L.J., Dietrich, A.J.J., Vink, A.C.G., van Marle, J., van Veen, H. and Heyting, C.: **Organization of SCP1 protein molecules within synaptonemal complexes of the rat.** *Exp. Cell Res.* 1996, **226**:20–30. (9-4)

Schroeder, T.E.: **Cytokinesis: filaments in the cleavage furrow.** *Exp. Cell Res.* 1968, **53**:272–276. (8-1)

Schwacha, A. and Kleckner, N.: **Interhomolog bias during meiotic recombination: meiotic functions promote a highly differentiated interhomolog-only pathway.** *Cell* 1997, **90**:1123–1135. (9-2)

Schwob, E., Bohm, T., Mendenhall, M.D. and Nasmyth, K.: **The B-type cyclin kinase inhibitor p40^{SIC1} controls the G1 to S transition in *S. cerevisiae*.** *Cell* 1994, **79**:233–244. (10-2)

Segal, M. and Bloom, K.: **Control of spindle polarity and orientation in *Saccharomyces cerevisiae*.** *Trends Cell Biol.* 2001, **11**:160–166. (6-0, 8-4)

Sha, W., Moore, J., Chen, K., Lassaletta, A.D., Yi, C.S., Tyson, J.J. and Sible, J.C.: **Hysteresis drives cell-cycle transitions in *Xenopus laevis* egg extracts.** *Proc. Natl Acad. Sci. USA* 2003, **100**:975–980. (3-8)

Shah, N. P., Nicoll, J.M., Nagar, B., Gorre, M.E., Paquette, R.L., Kuriyan, J. and Sawyers, C.L.: **Multiple BCR-ABL kinase domain mutations confer polyclonal resistance to the tyrosine kinase inhibitor imatinib (STI571) in chronic phase and blast crisis chronic myeloid leukemia.** *Cancer Cell* 2002, **2**:117–125. (12-9)

Shang, C., Hazbun, T.R., Cheeseman, I.M., Aranda, J., Fields, S., Drubin, D.G. and Barnes, G.: **Kinetochore protein interactions and their regulation by the Aurora kinase Ipl1p.** *Mol. Biol. Cell* 2003, **14**:3342–3355. (6-10)

Sharpless, N.E. and DePinho, R.A.: **Cancer: crime and punishment.** *Nature* 2005, **436**:636–637. (12-4)

Sharpless, N.E. and DePinho, R.A.: **Telomeres, stem cells, senescence, and cancer.** *J. Clin. Invest.* 2004, **113**:160–168. (12-6)

Shaulian, E. and Karin, M.: **AP-1 as a regulator of cell life and death.** *Nat. Cell Biol.* 2002, **4**:E131–E136. (10-6)

Sherr, C.J.: **Principles of tumor suppression.** *Cell* 2004, **116**:235–246. (12-1, 12-3, 12-4)

Sherr, C.J.: **The INK4a/ARF network in tumour suppression.** *Nat. Rev. Mol. Cell Biol.* 2001, **2**:731–737. (11-8, 12-4)

Sherr, C.J. and DePinho, R.A.: **Cellular senescence: mitotic clock or culture shock?** *Cell* 2000, **102**:407–410. (11-8)

Sherr, C.J. and McCormick, F.: **The RB and p53 pathways in cancer.** *Cancer Cell* 2002, **2**:103–112. (10-7, 12-3)

Sherr, C.J. and Roberts, J.M.: **CDK inhibitors: positive and negative regulators of G_1-phase progression.** *Genes Dev.* 1999, **13**:1501–1512. (3-6, 10-5, 10-7, 10-8)

Shi, Q. and King, R.W.: **Chromosome nondisjunction yields tetraploid rather than aneuploid cells in human cell lines.** *Nature* 2005, **437**:1038–1042. (12-7)

Shi, Y. and Massagué, J.: **Mechanisms of TGF-beta signaling from cell membrane to the nucleus.** *Cell* 2003, **113**:685–700. (10-7)

Shirayama, M., Toth, A., Galova, M. and Nasmyth, K.: **APC^{Cdc20} promotes exit from mitosis by destroying the anaphase inhibitor Pds1 and cyclin Clb5.** *Nature* 1999, **402**:203–207. (7-4)

Shonn, M.A., Murray, A.L. and Murray, A.W.: **Spindle checkpoint component Mad2 contributes to biorientation of homologous chromosomes.** *Curr. Biol.* 2003, **13**:1979–1984. (9-7)

Shorter, J. and Warren, G.: **Golgi architecture and inheritance.** *Annu. Rev. Cell Dev. Biol.* 2002, **18**:379–420. (6-7)

Shou, W., Seol, J.H., Shevchenko, A., Baskerville, C., Moazed, D., Chen, Z.W., Jang, J., Charbonneau, H. and Deshaies, R.J.: **Exit from mitosis is triggered by Tem1-dependent release of the protein phosphatase Cdc14 from nucleolar RENT complex.** *Cell* 1999, **97**:233–244. (7-5)

Shuster, C.B. and Burgess, D.R.: **Parameters that specify the timing of cytokinesis.** *J. Cell Biol.* 1999, **146**:981–992. (8-5)

Sisson, J.C., Field, C., Ventura, R., Royou, A. and Sullivan, W.: **Lava lamp, a novel peripheral Golgi protein, is required for *Drosophila melanogaster* cellularization.** *J. Cell Biol.* 2000, **151**:905–918. (8-6)

Sogo, J.M., Stahl, H., Koller, T. and Knippers, R.: **Fork reversal and ssDNA accumulation at stalled replication forks owing to checkpoint defects.** *Science* 2002, **297**:599–602. (11-6)

Sogo, J.M., Stahl, H., Koller, T. and Knippers, R.: **Structure of replicating simian virus 40 minichromosomes. The replication fork, core histone segregation and terminal structures.** *J. Mol. Biol.* 1986, **189**:189–204. (4-11)

Solomon, M., Glotzer, M., Lee, T.H., Phillipe, M. and Kirschner, M.: **Cyclin activation of p34^{cdc2}.** *Cell* 1990, **63**:1013–1024. (5-4)

Spellman, P.T., Sherlock, G., Zhang, M.Q., Iyer, V.R., Anders, K., Eisen, M.B., Brown, P.O., Botstein, D. and Futcher, B.: **Comprehensive identification of cell cycle-regulated genes of the yeast *Saccharomyces cerevisiae* by microarray hybridization.** *Mol. Biol. Cell* 1998, **9**:3273–3297. (3-12)

Stack, S.M. and Anderson, L.K.: **Two-dimensional spreads of synaptonemal complexes from solanaceous plants. II. Synapsis in *Lycopersicon esculentum* (tomato).** *Am. J. Bot.* 1986, **73**:264–281. (9-3)

Stegmeier, F., Visintin, R. and Amon, A.: **Separase, polo kinase, the kinetochore protein Slk19, and Spo12 function in a network that controls Cdc14 localization during early anaphase.** *Cell* 2002, **108**:207–220. (7-5)

Stemmann, O., Zou, H., Gerber, S.A., Gygi, S.P. and Kirschner, M.W.: **Dual inhibition of sister chromatid separation at metaphase.** *Cell* 2001, **107**:715–726. (7-4)

Storchova, Z. and Pellman, D.: **From polyploidy to aneuploidy, genome instability and cancer.** *Nat. Rev. Mol. Cell Biol.* 2004, **5**:45–54. (12-5, 12-7)

Storlazzi, A., Tessé, S., Gargano, S., James, F., Kleckner, N. and Zickler, D.: **Meiotic double-strand breaks at the interface of chromosome movement, chromosome remodeling, and reductional division.** *Genes Dev.* 2003, **17**:2675–2687. (9-3)

Straight, A.F. and Field, C.M.: **Microtubules, membranes and cytokinesis.** *Curr. Biol.* 2000, **10**:R760–R770. (8-3)

Straight, A.F., Field, C.M and Mitchison, T.J.: **Anillin binds nonmuscle myosin II and regulates the contractile ring.** *Mol. Biol. Cell* 2005, **16**:193–201. (8-2)

Ström, L., Lindroos, H.B., Shirahige, K. and Sjogren, C.: **Postreplicative recruitment of cohesin to double-strand breaks is required for DNA repair.** *Mol. Cell* 2004, **16**:1003–1015. (11-3)

Stucki, M., Clapperton, J.A., Mohammad, D., Yaffe, M.B., Smerdon, S.J. and Jackson, S.P.: **MDC1 directly binds phosphorylated histone H2AX to regulate cellular responses to DNA double-strand breaks.** *Cell* 2005, **123**:1213–1226. (11-3)

Stukenberg, P.T.: **Triggering p53 after cytokinesis failure.** *J. Cell Biol.* 2004, **165**:607–608. (12-7)

Sullivan, K.F.: **A solid foundation: functional specialization of centromeric chromatin.** *Curr. Opin. Genet. Dev.* 2001, **11**:182–188. (4-12)

Sumara, I., Vorlaufer, E., Stukenberg, P.T., Kelm, O., Redemann, N., Nigg, E.A. and Peters, J.M.: **The dissociation of cohesin from chromosomes in prophase is regulated by Polo-like kinase.** *Mol. Cell* 2002, **9**:515–525. (5-10)

Sumner, A.T.: **Scanning electron microscopy of mammalian chromosomes from prophase to telophase.** *Chromosoma* 1991, **100**:410–418. (5-9)

Swedlow, J.R. and Hirano, T.: **The making of the mitotic chromosome: modern insights into classical questions.** *Mol. Cell* 2003, **11**:557–569. (5-9, 5-10)

Sweeney, F.D., Yang, F., Chi, A., Shabanowitz, J., Hunt, D.F., and Durocher, D.: ***Saccharomyces cerevisiae* Rad9 acts as a Mec1 adaptor to allow Rad53 activation.** *Curr. Biol.* 2005, **15**:1364–1375. (11-3)

Takahashi, T.S. and Walter, J.C.: **Cdc7–Drf1 is a developmentally regulated protein kinase required for the initiation of vertebrate DNA replication.** *Genes Dev.* 2005, **19**:2295–2300. (4-7)

Takahashi, Y., Rayman, J.B. and Dynlacht, B.D.: **Analysis of promoter binding by the E2F and pRB families in vivo: distinct E2F proteins mediate activation and repression.** *Genes Dev.* 2000, **14**:804–816. (10-5)

Takai, H., Smogorzewska, A. and de Lange, T.: **DNA damage foci at dysfunctional telomeres.** *Curr. Biol.* 2003, **13**:1549–1556. (11-8)

Takayama, Y., Kamimura, Y., Okawa, M., Muramatsu, S., Sugino, A. and Araki, H.: **GINS, a novel multiprotein complex required for chromosomal DNA replication in budding yeast.** *Genes Dev.* 2003, **17**:1153–1165. (4-8)

Takizawa, C.G. and Morgan, D.O.: **Control of mitosis by changes in the subcellular location of cyclin B1–Cdk1 and Cdc25C.** *Curr. Opin. Cell Biol.* 2000, **12**:658–665. (5-6)

Tanaka, S. and Diffley, J.F.: **Interdependent nuclear accumulation of budding yeast Cdt1 and Mcm2–7 during G1 phase.** *Nat. Cell Biol.* 2002, **4**:198–207. (4-4)

Tanaka, T.U., Rachidi, N., Janke, C., Pereira, G., Galova, M., Schiebel, E., Stark, M.J. and Nasmyth, K.: **Evidence that the Ipl1-Sli15 (Aurora kinase-INCENP) complex promotes chromosome bi-orientation by altering kinetochore-spindle pole connections.** *Cell* 2002, **108**:317–329. (6-10)

Tercero, J.A., Longhese, M.P. and Diffley, J.F.: **A central role for DNA replication forks in checkpoint activation and response.** *Mol. Cell* 2003, **11**:1323–1336. (11-6)

Tessé, S., Storlazzi, A., Kleckner, N., Gargano, S. and Zickler, D.: **Localization and roles of Ski8p protein in *Sordaria* meiosis and delineation of three mechanistically distinct steps of meiotic homolog juxtaposition.** *Proc. Natl Acad. Sci. USA* 2003, **100**:12865–12870. (9-3)

Thornton, B.R. and Toczyski, D.P.: **Securin and B-cyclin/CDK are the only essential targets of the APC.** *Nat. Cell Biol.* 2003 **5**:1090–1094. (7-4)

Tlsty, T.D. and Hein, P.W.: **Know thy neighbor: stromal cells can contribute oncogenic signals.** *Curr. Opin. Genet. Dev.* 2001, **11**:54–59. (12-2)

Todaro, G.J. and Green, H.: **Quantitative studies of the growth of mouse embryo cells in culture and their development into established lines.** *J. Cell Biol.* 1963, **17**:299–313. (2-5)

Tolliday, N., Bouquin, N. and Li, R.: **Assembly and regulation of the cytokinetic apparatus in budding yeast.** *Curr. Opin. Microbiol.* 2001, **4**:690–695. (8-4)

Toth, A., Rabitsch, K.P., Galova, M., Schleiffer, A., Buonomo, S.B. and Nasmyth, K.: **Functional genomics identifies monopolin: a kinetochore protein required for segregation of homologs during meiosis I.** *Cell* 2000, **103**:1155–1168. (9-6)

Toyoshima-Morimoto, F., Taniguchi, E. and Nishida, E.: **Plk1 promotes nuclear translocation of human Cdc25C during prophase.** *EMBO Rep.* 2002, **3**:341–348. (5-6)

Trimarchi, J.M. and Lees, J.A.: **Sibling rivalry in the E2F family.** *Nat. Rev. Mol. Cell Biol.* 2002, **3**:11–20. (3-12, 10-4, 10-5)

Tsou, M.-F.B. and Stearns, T.: **Controlling centrosome number: licenses and blocks.** *Curr. Opin. Cell Biol.* 2006, **18**:74–84. (6-4)

Tsou, M.-F.B., Ku, W., Hayashi, A. and Rose, L.S.: **PAR-dependent and geometry-dependent mechanisms of spindle positioning.** *J. Cell Biol.* 2003, **160**:845–855. (8-7)

Tsubouchi, T. and Roeder, G.S.: **A synaptonemal complex protein promotes homology-independent centromere coupling.** *Science* 2005, **308**:870–873. (9-3)

Tulu, U.S., Rusan, N.M. and Wadsworth, P.: **Peripheral, non-centrosome-associated microtubules contribute to spindle formation in centrosome-containing cells.** *Curr. Biol.* 2003, **13**:1894–1899. (6-9)

Tung, K.-S., Hong, E.-J.E. and Roeder, G.S.: **The pachytene checkpoint prevents accumulation and phosphorylation of the meiosis-specific transcription factor Ndt80.** *Proc. Natl Acad. Sci. USA* 2000, **97**:12187–12192. (9-5)

Tunquist, B.J. and Maller, J.L.: **Under arrest: cytostatic factor (CSF)-mediated metaphase arrest in vertebrate eggs.** *Genes Dev.* 2003, **17**:683–710. (9-8)

Tyson, J.J., Chen, K. and Novak, B.: **Network dynamics and cell physiology.** *Nat. Rev. Mol. Cell Biol.* 2001, **2**:908–916. (3-13)

Tyson, J.J., Chen, K.C. and Novak, B.: **Sniffers, buzzers, toggles and blinkers: dynamics of regulatory and signaling pathways in the cell.** *Curr. Opin. Cell Biol.* 2003, **15**:221–231. (3-7, 3-8, 3-11, 3-13)

Ubersax, J.A., Woodbury, E.L., Quang, P.N., Paraz, M., Blethrow, J.D., Shah, K., Shokat, K.M. and Morgan, D.O.: **Targets of the cyclin-dependent kinase Cdk1.** *Nature* 2003, **425**:859–864. (3-1)

Uhlmann, F.: **Chromosome cohesion and separation: from men and molecules.** *Curr. Biol.* 2003, **13**:R104–R114. (7-4)

Uhlmann, F., Wernic, D., Poupart, M.A., Koonin, E.V. and Nasmyth, K.: **Cleavage of cohesin by the CD clan protease separin triggers anaphase in yeast.** *Cell* 2000, **103**:375–386. (7-4)

Unal, E., Arbel-Eden, A., Sattler, U., Shroff, R., Lichten, M., Haber, J.E. and Koshland, D.: **DNA damage response pathway uses histone modification to assemble a double-strand break-specific cohesin domain.** *Mol. Cell* 2004, **16**:991–1002. (11-3)

Uto, K., Inoue, D., Shimuta, K., Nakajo, N. and Sagata, N.: **Chk1, but not Chk2, inhibits Cdc25 phosphatases by a novel common mechanism.** *EMBO J.* 2004, **23**:3386–3396. (11-7)

van Drogen, F., Stucke, V.M., Jorritsma, G. and Peter, M.: **MAP kinase dynamics in response to pheromones in budding yeast.** *Nat. Cell Biol.* 2001, **3**:1051–1059. (10-3)

Vanoosthuyse, V. and Hardwick, K.G.: **Bub1 and the multilayered inhibition of Cdc20–APC/C in mitosis.** *Trends Cell Biol.* 2005, **15**:231–233. (7-3)

Varmus, H. and Weinberg, R. A.: *Genes and the Biology of Cancer* (Scientific American Library, New York, 1993). (12-0)

Verma, R., Annan, R.S., Huddleston, M.J., Carr, S.A., Reynard, G. and Deshaies, R.J.: **Phosphorylation of Sic1p by G1 Cdk required for its degradation and entry into S phase.** *Science* 1997, **278**:455–460. (10-2)

VerPlank, L. and Li, R.: **Cell cycle-regulated trafficking of Chs2 controls actomyosin ring stability during cytokinesis.** *Mol. Biol. Cell* 2005, **16**:2529–2543. (8-3)

Verreault, A.: **De novo nucleosome assembly: new pieces in an old puzzle.** *Genes Dev.* 2000, **14**:1430–1438. (4-11)

Vershon, A.K. and Pierce, M.: **Transcriptional regulation of meiosis in yeast.** *Curr. Opin. Cell Biol.* 2000, **12**:334–339. (9-1, 9-5)

Vidanes, G.M., Bonilla, C.Y. and Toczyski, D.P.: **Complicated tails: histone modifications and the DNA damage response.** *Cell* 2005, **121**:973–976. (11-3)

Vogelstein, B. and Kinzler, K.W.: **Cancer genes and the pathways they control.** *Nat. Med.* 2004, **10**:789–799. (12-1, 12-3, 12-4)

Vogelstein, B. and Kinzler, K.W.: *The Genetic Basis of Human Cancer* 2nd ed. (McGraw-Hill, New York, 2002). (12-0)

Vogelstein, B., Lane, D. and Levine, A.J.: **Surfing the p53 network.** *Nature* 2000, **408**:307–310. (11-4)

Vousden, K.H. and Lu, X.: **Live or let die: the cell's response to p53.** *Nat. Rev. Cancer* 2002, **2**:594–604. (11-4, 11-5)

Wadsworth, P. and Khodjakov, A.: ***E pluribus unum*: towards a universal mechanism for spindle assembly.** *Trends Cell Biol.* 2004, **14**:413–419. (6-0, 6-8, 6-9)

Waga, S. and Stillman, B.: **The DNA replication fork in eukaryotic cells.** *Annu. Rev. Biochem.* 1998, **67**:721–751. (4-0, 4-1)

Wahl, G.M. and Carr, A.M.: **The evolution of diverse biological responses to DNA damage: insights from yeast and p53.** *Nat. Cell Biol.* 2001, **3**:E277–E286. (11-0, 11-4, 11-5)

Walczak, C.E., Vernos, I., Mitchison, T.J., Karsenti, E., and Heald, R.: **A model for the proposed roles of different microtubule-based motor proteins in establishing spindle bipolarity.** *Curr. Biol.* 1998, **8**:903–913. (6-8)

Walter, J. and Newport, J.: **Initiation of eukaryotic DNA replication: origin unwinding and sequential chromatin association of Cdc45, RPA, and DNA polymerase alpha.** *Mol. Cell* 2000, **5**:617–627. (4-8)

Walther, T.C., Askjaer, P., Gentzel, M., Habermann, A., Griffiths, G., Wilm, M., Mattaj, I.W. and Hetzer, M.: **RanGTP mediates nuclear pore complex assembly.** *Nature* 2003, **424**:689–694. (7-7)

Wang, H., Gari, E., Verges, E., Gallego, C. and Aldea, M.: **Recruitment of Cdc28 by Whi3 restricts nuclear accumulation of the G1 cyclin–Cdk complex to late G1.** *EMBO J.* 2004, **23**:180–190. (10-1)

Wäsch, R. and Cross, F.: **APC-dependent proteolysis of the mitotic cyclin Clb2 is essential for mitotic exit.** *Nature* 2002, **418**:556–562. (7-5)

Watanabe, Y.: **Shugoshin: guardian spirit at the centromere.** *Curr. Opin. Cell Biol.* 2005, **17**:590–595. (5-10)

Watanabe, Y.: **Sister chromatid cohesion along arms and at centromeres.** *Trends Genet.* 2005, **21**:405–412. (9-7)

Waters, J.C., Chen, R.-H., Murray, A.W. and Salmon, E.D.: **Localization of Mad2 to kinetochores depends on microtubule attachment, not tension.** *J. Cell Biol.* 1998, **141**:1181–1191. (6-9, 7-2)

Wei, R.R., Sorger, P.K. and Harrison, S.C.: **Molecular organization of the Ndc80 complex, an essential kinetochore component.** *Proc. Natl Acad. Sci. USA* 2005, **102**:5363–5367. (6-5)

Weinberg, R.A.: *The Biology of Cancer* (Garland Science, New York, 2006).

Weinert, T.A. and Hartwell, L.H.: **The RAD9 gene controls the cell cycle response to DNA damage in *Saccharomyces cerevisiae*.** *Science* 1988, **241**:317–322. (11-0)

Weir, B., Zhao, X. and Meyerson, M.: **Somatic alterations in the human cancer genome.** *Cancer Cell* 2004, **6**:433–438. (12-1)

Welcker, M., Singer, J., Loeb, K., Grim, J., Bloecher, A., Gurien-West, M., Clurman, B. and Roberts, J.: **Multisite phosphorylation by Cdk2 and GSK3 controls cyclin E degradation.** *Mol. Cell* 2003, **12**:381–392. (10-8)

Westermann, S., Avila-Sakar, A., Wang, H.W., Niederstrasser, H., Wong, J., Drubin, D.G., Nogales, E. and Barnes, G.: **Formation of a dynamic kinetochore–microtubule interface through assembly of the Dam1 ring complex.** *Mol. Cell* 2005, **17**:277–290. (6-5, 6-11)

Westermann, S., Wang, H.W., Avila-Sakar, A., Drubin, D.G., Nogales, E. and Barnes, G.: **The Dam1 kinetochore ring complex moves processively on depolymerizing microtubule ends.** *Nature* 2006, **440**:565–569. (6-5)

Wheatley, S.P., Hinchcliffe, E.H., Glotzer, M., Hyman, A.A., Sluder, G. and Wang, Y.: **CDK1 inactivation regulates anaphase spindle dynamics and cytokinesis *in vivo*.** *J. Cell Biol.* 1997, **138**:385–393. (7-6)

Whitfield, M.L., Zheng, L.X., Baldwin, A., Ohta, T., Hurt, M.M. and Marzluff, W.F.: **Stem-loop binding protein, the protein that binds the 3′ end of histone mRNA, is cell cycle regulated by both translational and post-translational mechanisms.** *Mol. Cell Biol.* 2000, **20**:4188–4198. (4-10)

Wilson, E.B.: *The Cell in Development and Heredity* 3rd ed. (Macmillan, New York, 1925). (1-0, 5-0)

Winey, M., Morgan, G.P., Straight, P.D., Giddings, T.H. Jr and Mastronarde, D.N.: **Three-dimensional ultrastructure of *Saccharomyces cerevisiae* meiotic spindles.** *Mol. Biol. Cell* 2005, **16**:1178–1188. (9-6)

Winters, M.J., Lamson, R.E., Nakanishi, H., Neiman, A.M., Pryciak, P.M.: **A membrane binding domain in the ste5 scaffold synergizes with Gβγ binding to control localization and signaling in pheromone response.** *Mol. Cell* 2005, **20**:21–32. (10-3)

Wittenberg, C. and Reed, S.I.: **Cell cycle-dependent transcription in yeast: promoters, transcription factors, and transcriptomes.** *Oncogene* 2005, **24**:2746–2755. (3-12, 10-1)

Wittmann, T., Hyman, A. and Desai, A.: **The spindle: a dynamic assembly of microtubules and motors.** *Nat. Cell Biol.* 2001, **3**:E28–E34. (6-0)

Wodarz, A.: **Molecular control of cell polarity and asymmetric cell division in *Drosophila* neuroblasts.** *Curr. Opin. Cell Biol.* 2005, **17**:475–481. (8-7)

Wolfe, B.A. and Gould, K.L.: **Split decisions: coordinating cytokinesis in yeast.** *Trends Cell Biol.* 2005, **15**:10–18. (8-4)

Wolpert, L., Beddington, R., Jessell, T., Lawrence, P., Meyerowitz, E. and Smith, J.: *Principles of Development* 2nd ed. (Oxford University Press, Oxford, 2002). (2-3, 2-4)

Wong, C. and Stearns, T.: **Centrosome number is controlled by a centrosome-intrinsic block to reduplication.** *Nat. Cell Biol.* 2003, **5**:539–544. (6-4)

Woodcock, C.L. and Dimitrov, S.: **Higher-order structure of chromatin and chromosomes.** *Curr. Opin. Genet. Dev.* 2001, **11**:130–135. (4-9)

Wozniak, R. and Clarke, P.R.: **Nuclear pores: sowing the seeds of assembly on the chromatin landscape.** *Curr. Biol.* 2003, **13**:R970–R972. (7-7)

Wu, J.Q., Kuhn, J., Kovar, D. and Pollard, T.: **Spatial and temporal pathway for assembly and constriction of the contractile ring in fission yeast cytokinesis.** *Dev. Cell* 2003, **5**:723–734. (8-4)

Xia, G., Luo, X., Habu, T., Rizo, J., Matsumoto, T. and Yu, H.: **Conformation-specific binding of p31^{comet} antagonizes the function of Mad2 in the spindle checkpoint.** *EMBO J.* 2004, **23**:3133–3143. (7-3)

Xu, L., Ajimura, M., Padmore, R., Klein, C. and Kleckner, N.: **NDT80, a meiosis-specific gene required for exit from pachytene in *Saccharomyces cerevisiae*.** *Mol. Cell Biol.* 1995, **15**:6572–6581. (9-5)

Yamasu, K. and Senshu, T.: **Conservative segregation of tetrameric units of H3 and H4 histones during nucleosome replication.** *J. Biochem.* 1990, **107**:15–20. (4-11)

Yang, J., Song, H., Walsh, S., Bardes, E.S. and Kornbluth, S.: **Combinatorial control of cyclin B1 nuclear trafficking through phosphorylation at multiple sites.** *J. Biol. Chem.* 2001, **276**:3604–3609. (5-6)

Ye, X., Wei, Y., Nalepa, G. and Harper, J.W.: **The cyclin E/Cdk2 substrate p220^{NPAT} is required for S-phase entry, histone gene expression, and Cajal body maintenance in human somatic cells.** *Mol. Cell Biol.* 2003, **23**:8586–8600. (4-10)

Yee, K.S. and Vousden, K.H.: **Complicating the complexity of p53.** *Carcinogenesis* 2005, **26**:1317–1322. (11-4, 11-5)

Yokobayashi, S. and Watanabe, Y.: **The kinetochore protein Moa1 enables cohesion-mediated monopolar attachment at meiosis I.** *Cell* 2005, **123**:803–817. (9-6)

Yu, H.: **Regulation of APC-Cdc20 by the spindle checkpoint.** *Curr. Opin. Cell Biol.* 2002, **14**:706–714. (7-2, 7-3)

Yu, H.G. and Koshland, D.: **Chromosome morphogenesis: condensin-dependent cohesin removal during meiosis.** *Cell* 2005, **123**:397–407. (9-4)

Yüce, O., Piekny, A and Glotzer, M.: **An ECT2–centralspindlin complex regulates the localization and function of RhoA.** *J. Cell Biol.* 2005, **170**:571–582. (8-5)

Zetterberg, A. and Larsson, O.: **Kinetic analysis of regulatory events in G1 leading to proliferation or quiescence of Swiss 3T3 cells.** *Proc. Natl Acad. Sci. USA* 1985, **82**:5365–5369. (10-0)

Zetterberg, A., Engstrom, W. and Dafgard, E.: **The relative effects of different types of growth factors on DNA replication, mitosis, and cellular enlargement.** *Cytometry* 1984, **5**:368–375. (10-13)

Zheng, L., Dominski, Z., Yang, X.C., Elms, P., Raska, C.S., Borchers, C.H. and Marzluff, W.F.: **Phosphorylation of stem-loop binding protein (SLBP) on two threonines triggers degradation of SLBP, the sole cell cycle-regulated factor required for regulation of histone mRNA processing, at the end of S phase.** *Mol. Cell Biol.* 2003, **23**:1590–1601. (4-10)

Zheng, N., Fraenkel, E., Pabo, C.O. and Pavletich, N.P.: **Structural basis of DNA recognition by the heterodimeric cell cycle transcription factor E2F–DP.** *Genes Dev.* 1999, **13**:666–674. (3-12)

Zheng, N., Schulman, B.A., Song, L., Miller, J.J., Jeffrey, P.D., Wang, P., Chu, C., Koepp, D.M., Elledge, S.J., Pagano, M., Conaway, R.C., Conaway, J.W., Harper, J.W. and Pavletich, N.P.: **Structure of the Cul1–Rbx1–Skp1–F box^{Skp2} SCF ubiquitin ligase complex.** *Nature* 2002, **416**:703–709. (3-9)

Zhou, B.B. and Elledge, S.J.: **The DNA damage response: putting checkpoints in perspective.** *Nature* 2000, **408**:433–439. (11-0)

Zickler, D. and Kleckner, N.: **Meiotic chromosomes: integrating structure and function.** *Annu. Rev. Genet.* 1999, **33**:603–754. (9-0, 9-3, 9-4)

Zigmond, S.H.: **Formin-induced nucleation of actin filaments.** *Curr. Opin. Cell Biol.* 2004, **16**:99–105. (8-1)

Zou, L. and Elledge, S.J.: **Sensing DNA damage through ATRIP recognition of RPA-ssDNA complexes.** *Science* 2003, **300**:1542–1548. (11-2)

Index

References to figures are in bold, for example **F4-12** refers to Figure 4-12